Foundations of the Unity of Science

Volume II

Committee of Organization

Rudolf Carnap
Philipp Frank
Joergen Joergensen
Charles Morris
Otto Neurath
Louis Rougier

Advisory Committee

Niels Bohr
Egon Brunswik
J. Clay
John Dewey
Federigo Enriques
Herbert Feigl
Clark L. Hull
Waldemar Kaempffert
Victor F. Lenzen
Jan Lukasiewicz
William M. Malisoff
R. von Mises
G. Mannoury
Ernest Nagel
Arne Naess
Hans Reichenbach
Abel Rey
Bertrand Russell
L. Susan Stebbing
Alfred Tarski
Edward C. Tolman
Joseph H. Woodger

Foundations of the Unity of Science

•

Toward an International Encyclopedia of Unified Science

Volume II, Nos. 1-9

Edited by

Otto Neurath Rudolf Carnap

Charles Morris

The University of Chicago Press
Chicago and London

This edition combines in two clothbound volumes the ten monographs of volume I and nine monographs of volume II of the *Foundations of the Unity of Science: Toward an International Encyclopedia of Unified Science*. This work, combined and in its several parts, was formerly titled *International Encyclopedia of Unified Science*.

International Standard Book Number: 0-226-57588-8
Library of Congress Catalog Card Number: 56-553

The University of Chicago Press, Chicago 60637
The University of Chicago Press, Ltd., London

Cloth edition of volume II published 1970
Printed in the United States of America

CONTENTS

Volume II

Foundations of the Social Sciences

Otto Neurath

Foundations of the Social Sciences

Contents:

Foundations of the Social Sciences

Otto Neurath

I. Terminological Empiricism and the Social Sciences

1. Social Sciences and Analysis of Language

By the use of the term 'social sciences' I join together, rather tentatively, scientific disciplines of different kinds. I do not suggest an "architecture of sociology," nor do I suggest a pyramid of the sciences, with sociology forming a part of it.

The classification of the sciences may be interesting in itself and sometimes also stimulating; but the premature drawing of boundary lines among the sciences does not help us much in research work, not even in arranging its results, as long as we are not able to combine the classification of single items with the presentation of interrelations between them. Successful analysis of interrelations between species made classification in biology important, whereas the grouping of the four elements—fire, air, water, earth—in accordance with the four quality couples—dry-hot, hot-wet, wet-cold, cold-dry—mainly led to hairsplitting scholasticism.

Therefore, I shall speak of sociological statements as members of one big family of statements which may be found in volumes filled with results of research on the behavior of tribes, on the behavior of customs and languages as they spread through mankind, on the behavior of whole nations, on the growing-up of the fine arts in human societies, on the behavior of human groups and representative individuals (e.g., of artists, priests, statesmen, pirates, peasants, workers, and other people in various societies), on the patterns of cities, on the behavior of markets and administration, and on the various ways of personal life within various societies. We should include, of course, studies on the behavior of animal groups. And why should we object to the inclusion of plant sociology?

All the techniques for making well-arranged descriptions, finding correlations, and preparing predictions belong to the field of scientific practice with which I have to do here. At the end of the nineteenth century there was a saying that sociology was mainly composed of statements on programs and on what sociology should look like. That would be a stale joke today, since we have made great progress in the social sciences. Nevertheless, up to now the number of lines in sociological books devoted to what are usually called "methodological questions" and to demarcations between single disciplines is not small. That may perhaps be regarded as a sign of uneasiness, owing to difficulties of language and to the difficulties arising out of the question as to how far better descriptive techniques together with the collection of more material, and the evolution of logical techniques and terminological analysis, may be helpful. Up to now we have met an overwhelming number of expressions dealing with social matters. We do not know whether these expressions fit into the same phraseological framework or whether we have to make a choice before we can properly go on.[1]

One frequently overlooks the fact that the elimination of unempiricist expressions is not sufficient to make our arguing safe. Liberally though we may treat it, there remain phrases and sentences which, though they use only empiricist items, nevertheless do not fit into our comprehensive scheme. We meet them, for example, in jurisprudence, in ethics, in pedagogics, and—this is frequently overlooked—in economics. Even when these expressions avoid any "absolute" or "transcendent" allusions, they frequently do nothing but transfer some customs or sayings from the previous to succeeding forms of society which are connected with some traditional words and phrases, with terms such as 'justice,' 'crime,' 'cost,' 'profit.'

We immediately see the vastness of the field if we analyze, for instance, statements on the migration of social groups. Such a statement may be concerned with the vicissitudes of crops and climates, with alterations in the amount and quality of food produced, and with alterations in certain institutions, such as the correlation in the position and behavior of priests and chieftains. Moreover, some sociologists will take into account what biologists stated on the correlation between the alteration of hormone-equilibrium, etc., in animals and their migration; they may transfer, by analogy, these results to the behavior of human groups.[2] Thus we have to join meteorological, ethnological, biological, and chemical arguments. Social scientists, therefore, are interested in a language which enables them to speak of animals, plants, and crystals in the same way, as far as possible, without anticipatively creating distinctions. Not only the unification of the sociological language is at stake, but a much more comprehensive unification and orchestration, which leads us to a lingua franca of unified science.[3]

2. Terminological Empiricism

The discipline which deals with terms, expressions, and the making of a language may be called "terminology" in harmony with the etymology of this word. We cannot be expected to present once and for all time a Universal Jargon as a complete structure. It will always be in the making.[4]

For the study of some peculiarities of terminology we may form schematic model languages composed only of well-defined items and start with a small and compact nucleus of terms. But we shall not be able to build up our Universal Jargon from scratch, deliberately proceeding step by step. We always use a cluster of vaguely defined assumptions and assertions in our empiricist arguments. Each newly introduced scientific discovery may alter, in some way or other, the use of any of our expressions and sentences and their interpretation. That is just the opposite to any kind of *tabula rasa.* A language scheme used for calculatory problems and a Universal Jargon are different in many respects. Not a few misunderstandings and difficulties seem to arise from mixing up both fields in our arguing.

The application of terminological analysis to the social sciences asks for some training and skill. The usual lack of criticism in acquiring terms and phrases is remarkable; students find an overwhelming heap of *termini technici*, attractive catchwords and certain expletive phrases, which afterward appear loaded with terminological explosives. They learn the terms and are not seldom proud when they can find an opportunity to apply them in some way or another. It is difficult to behave reservedly, because that forces people to wait some time, until they have learned a little more about the technique of selecting and using empiricist terms and empiricist arguments. It is hard for a singer to be silent during the period in which he wants to re-educate his voice by means of a new technique. Perhaps this necessary sacrifice is one of the reasons why it is so difficult to find partners prepared to discuss terminological suggestions in detail. A certain attitude seems to be indispensable; we may call it the attitude of "terminological empiricism."

Terminological empiricism, looking at the above-mentioned hypotheses on the migration of human groups, may discover that research workers speak of climate in terms of temperature and amount of rain, that they correlate complexes of temperature-statements and rain-statements with other complexes of statements in which the terms 'amount of wheat,' 'amount of cattle,' 'behavior of hungry men,' and 'behavior of chieftains' appear. All these terms are spatio-temporal terms and may therefore be called "physicalist" terms. (The terms 'statement' or 'term' are physicalist terms also.)

A sociologist, asked how he had obtained these statements, would perhaps answer that they are based on reports collected by field workers and on reports found in chronicles. That finally implies that eyewitness records are the backbone of the account of the wheat observed, the geological strata observed, the chieftains' behavior observed, and the books observed. As far as chronicles are concerned, the observation-statements of the investigator of such a chronicle deal with observation-statements made by persons centuries ago. We see from this latter remark that the term 'observation-statement' is here used as a spatio-temporal term like the terms 'wheat,' 'geological stratum,' 'chieftains' behavior.'

But what about expressions such as 'spirit of a nation,' 'ethos of a religion,' or 'ethical forces'? Sometimes sentences in which such expressions appear can hardly be connected with observation-statements and have to be dropped as parts of metaphysical speculations, but sometimes we may look at them as "metaphorical" sentences and transform them into physicalist statements. Dropping them would often imply an impoverishment of our argument. Let us imagine that a writer wants to maintain that the migration of tribes may sometimes be correlated with the cyclic behavior of the climate, but sometimes with the behavior of chieftains and priests or with the educational organization of the tribes, which cannot be correlated with rain and sun in any simple way.[5] Sometimes a highly complicated and delicate physicalist phraseology will be needed which has to be evolved for such a purpose. Just the lack of such an empiricist phraseology may explain the tendency of scholars, interested in these problems, to use the vocabulary of an unempiricist language. Negative criticism by empiricists without constructive suggestions of suitable expressions frequently appears rather poor and dull.

3. Observation-Statements and Adopted Expressions

If writers speak of the "spirit of nations," we may ask them how they differentiate between various "spirits." Sometimes the writers may answer that this expression implies only different patterns of life, architecture, poetry, law, or something else. Others may maintain that these patterns are nothing but "signs" of some "entity" "behind" the "appearance" of something "real" which "expresses" itself in this way and that the "essence" of the "mental values" is not apprehensible, etc.; I suggest that it would be better not to discuss these remarks further within logical empiricism, because I see no way of transforming them into physicalist statements.

The remarks on patterns may be treated physicalistically because, as we should say, one can assay sentences derived from them by means of observation-statements. (I use the term 'assaying a statement' as a general one, because the common term 'testing a statement' is already to some extent used in a more specific way.) We may ask how the various patterns, for instance, in architecture, can be characterized, and the writers who are prepared to use the term 'pattern' in a physicalist way may answer: "Where buildings are concerned, you will see certain lines and proportions." That enables us to ask field workers what they saw, and perhaps they may answer: "Yes, we have seen such lines and proportions"; or they may answer, "No, we have not seen such lines and proportions."

In both these cases the sentence on patterns appears to be what we may call "assayable" by observation-statements. I suggest that we call all sentences of which we may say they are in harmony with observation-statements, or not in harmony with them, "statements."

The assayability of sentences depends upon the expressions and the grammar used. We may say that we are prepared to "adopt" expressions and rules for their combination when they lead to "assayable sentences," i.e., to "statements."

I shall speak of an attitude as empiricist if there is the tendency to assay any sentence made, directly or indirectly, by means of observation-statements. I hope the reader will not mind using a hyphen between 'observation' and 'statement.' Experience has taught me that otherwise not a few people involuntarily will behave as if they had to handle two items instead of only one. The hyphen prevents our asking "dangerous" questions in this and similar cases, such as how "observation" and "statement" are connected; or, further, how "sense data" and "mind," the "external world" and the "internal world," are connected. In our physicalist language the expressions just mentioned do not appear.

Observation-statements, if formulated very carefully, may be called "protocol-statements." We shall use the term 'protocolist' as equivalent to 'protocol writer.' We may always ask for the "protocolist-name" and for other details—that will depend on the discussion in question.

Instead of saying, "We compare the hypothesis with the facts," I suggest we say, "We compare the statements brought forward by the hypothesis with observation-statements." I suggest using the term 'compare' only where we may make a comparison-statement dealing with two or more items which are characterized by qualificatory terms of the same kind. We may compare the length of a rod with the length of a tree and we may compare the character of one person with the character of another, but I should not speak in the same way of "comparing a statement with a fact." Rather we make the comparative statement, "The statement 'This table has four legs' has one word more than the table has legs," i.e., comparing two amounts—an "amount of words," and an "amount of legs." But I do not think that people who speak of comparing statements with facts will be satisfied by this possibility of speaking.

4. Indistinct—but Univocal

A widespread opinion identifies this cautious use of our expressions with introducing oversharp definitions and eventually with forming barriers for research workers, who—as the saying goes—have better things to do than to deal with the peculiarities of terms.

First of all: ordinary research workers frequently suffer from multivocal expressions, which disturb sound reasoning, and from terms of speculative metaphysics, which confound plain arguing. Highly educated scholars are sometimes so perplexed that they do not find a way out of intricate entanglements until they alter their phraseology.

But we can avoid the hunting-spears of dangerous terms without mercilessly clarifying our language and without butchering harmless arguments in which some doubtful words occur and, above all, without fighting "indistinctness" wherever we meet it. On the contrary, it is precisely the empiricist who cannot go on without handling "indistinctness" everywhere, since he wants to realize the program of assaying assertions by means of observation-statements. Observation-statements cannot but be indistinct in some way or another. The protocolist-name, the observation-term, and other expressions are not very distinct ones. Empiricists cannot refrain from using faint and blurred expressions with rather vague outlines, but they can more or less avoid multivocalness.

Whereas I see no way of fitting Plato's metaphysical generalities on "the Good" into any empiricist framework, I should call the following Homeric lines physicalist, although their expressions (translated by Pope) are anything but distinct and sharply defined:

Amidst an isle, around whose rocky shore
The forests murmur and the surges roar. . . .[6]

Within the framework of physicalist statements multivocalness is avoidable, but not indistinctness: Alexander von Humboldt's *Cosmos*, for example, provides us with a rich empiricist vocabulary which deals with natural regions as comprehensive patterns. Stimulating descriptive statements may be found together with the scientific attitude of an empiricist throughout this book and in Humboldt's other writings.

All over the world there are languages with extensive clusters of empiricist phrases which may be fairly well translated one into the other and, by becoming a certain pattern, may form the nucleus of a universal language of empiricism. In each of these languages the group in question forms the universal language in, let me say, English, French, etc. These different translatable groups may be called "propositionally equivalent." Therefore, 'universal jargon,' in this paper, has nothing to do with any kind of international language; even of the various international languages, such as Esperanto, Ido, Basic, Interglossa, we may regard one section as the Universal Jargon of the respective international language.

This translatability is the basis of our scientific enterprises. We may use records made by Roman or Chinese historians; statements such as "The old man lived in his stone building," "Many trees have been found there," and "The fruit of this tree was sweet" seem to be translatable into almost all languages. There are, of course, difficulties even in translating such statements, but that is mostly due to the overlapping of terms in various languages, not to their lack of empiricist character. Unfortunately, we have no comparative studies on the stock of equivalents of empiricist phrases in the various languages and the Universal Jargon. The translations of the Bible, the Koran, and other books destined for international use teach us a good lesson. The difficulties start when sentences with terms such as 'law,' 'soul,' 'cause,' have to be translated. Of course, the problem will always be that whole sentences have to be translated into whole sentences; but there are sentences which may be regarded as not translatable because they contain certain expressions which belong to speculative metaphysics.

Sometimes scholars speak of the empiricist attitude and of the idealist or metaphysical one, as of opposites, but overlook the fact that a certain number of empiricist statements are common to both groups of schools. The speculative metaphysicians add some assertions, dealing with "eternal values," "mentality," and other "entities"; but they do not speak in this language about cows and calves or dogs and their puppies. They use, in this field, the language we empiricists are accustomed to use throughout all our discussions and meditations, as it has been used without any break, as far as daily life goes, throughout the ages, by preliterate tribes as well as by modern peoples.

We should not think Western people must always be the givers in the making of the Universal Jargon, at least as far as technical terms are concerned. Sometimes language habits, used for a long time by preliterate people, may be successfully introduced into our scientific language and may come to be used by our Universal Jargon, as, for instance, the term 'taboo,' which characterizes a certain behavior.[7] It is obvious that in this common stock we shall find a great many words of doubtful distinctness; but they are, nevertheless, univocal in usage. That just some of the oldest expressions are suitable for the Universal Jargon may be connected with the kind of selection terms have undergone.

Not a few speculative generalities in the social sciences may be regarded as relics of scholasticism, such as "meaning of an institution," "internal causes," etc. Some arguments dealing with social problems, particularly with history, become equivocal because writers sometimes prefer an involved and crooked language where plain empiricist and univocal expressions might sound unpleasant to readers. Instead of saying bluntly, "One human group killed another and destroyed their buildings and books," certain historians prefer to say, "Forced by its historical mission, the nation started spreading its civilization over the earth." With this kind of evasive phraseology we shall not deal in this paper nor with multivocal speculative phrases which occasionally serve an affected obscurity called "depth" by not a few people.

Here we cannot discuss the social sciences in all their parts; there are many books which inform and orientate one about them and their history.[8] I hope readers will be prepared to apply the technique of terminological analysis suggested here to other sociological problems not mentioned in this paper.

5. Unified Science and Cosmic History

We started by pointing to the social sciences as to a cluster of disciplines, each of which may be named for purposes of orientation but not for classification. I think a similar attitude toward all the sciences together is advisable.

The distinction between "mental" sciences and the rest puts human sociology into the drawer labeled "mental sciences"; but, then, the strange situation arises that human wars, human slavery, and human agriculture ought to be treated within the "mental sciences," whereas ant wars, ant slavery, and ant agriculture ought to be treated under a different heading and, as some scientists assume, tackled by means of different procedures and under application of a different language. Many speculative metaphysical arguments are connected with this classification.

Other subdivisions are not fortunate either. There is a label "physical science" affixed, say, to geology. But later on one has to concede that paleontology also belongs to geology and does not fit under the heading mentioned.[9] Sometimes one creates "mixed sciences," but then geology appears even more mixed, because in statements of scientific research it has been mentioned that the surface of the earth is closely related to human history, and one may look at "man as a geological agent." We may say: "Men are connected with alterations of geological structure like rains and rivers,"[10] and therefore we may get statements which speak of correlations between the alterations of human institutions (connected with the construction of dams, plowing, etc.) and the alteration of the surface of the earth and the climate. That implies that sociological statements enter the geological and perhaps also the astronomical departments in full dress. Let me anticipatively say that difficulties in making predictions on human institutions therefore enter the geological sphere, which does not remain watertightly separated from sociology.

Would it not be preferable to treat all statements and all sciences as coordinated and to abandon for good the traditional hierarchy: physical sciences, biological sciences, social sciences, and similar types of "scientific pyramidism"? We should not even look at mechanics as a nonbiological science but prefer to say more cautiously that the statements of mechanics deal in the same way with falling cats and falling stones.[11]

A formalized theory of physics may use symbols and formulas *ad libitum,* as long as it does not try to apply its results to empiricist hypotheses without particular assumptions. We cannot anticipate future hypotheses and should therefore be cautious in assuming too much from expressions which may be used in assaying hypotheses. Even the store of these expressions may be altered during our empiricist work.

I suggest grouping scientific statements where needed for orientation but not thinking of clear-cut departments. The various bodies of statements may overlap one another. Why should we object to that, since we do not object to the overlapping of scientific papers which may be titled "The Po Basin" and "The Alps." Let us regard the sciences as collections of statements, just as we regard such papers. Why should we not admit in the bigger collections—the sciences—what seems reasonable to us in the smaller collections—the papers?

This leads us to the conclusion that we may look at all sciences as dovetailed to such a degree that we may regard them as parts of one science which deals with stars, Milky Ways, earth, plants, animals, human beings, forests, natural regions, tribes, and nations—in short, a comprehensive cosmic history would be the result of such an agglomeration. But such a cosmic history would not appear only as a collection of various separate groups of statements, because we cannot know anticipatively when it may be useful, and when not, to take into account all the statements together in analyzing certain correlations in a certain field.

Some people may not object to this attempt to expand the sphere of history, but they may still remain doubtful whether chemistry and mechanics also fit into this same scheme. Yet surely they do, because just as we may speak of the behavior of plants and animals in a certain period at a certain place, so we may also speak—in harmony with modern physics and cosmology—of the behavior of sodium and levers in a certain period of the cosmic aggregation. This period may be long, but why should we assign any kind of "universality" to chemical and mechanical statements instead of treating them as historical ones?

Cosmic history would, as far as we are using a Universal Jargon throughout all the branches of research, contain the same statements as our unified science. The language of our *Encyclopedia* may, therefore, be regarded as a typical language of history. There is no conflict between physicalism and this program of cosmic history.

6. Accepted and Rejected Statements

In forming our Universal Jargon, we start from our everyday language and look into the vocabularies used by scientists, novelists, metaphysicians, theologians, and other people in all parts of the earth. We have to decide, as mentioned above, which terms we may take over (i.e.,"adopt"), which terms we may drop (i.e., "decline"), and when we think it advisable to coin—hesitatingly, of course—new terms. Some people fear the loss of expressions, but they do not think of the vast number of expressions successively lost during the ages. They do not think sufficiently of the far-reaching word economy and of what it may perform.[12]

Let us assume we make an agreement to adopt certain terms and decline certain others; then sentences composed of expressions which are not connectable with observation-statements may be called "isolated" ones, e.g., sentences formulated by speculative metaphysicians who use expressions such as "thing in itself." As far as "isolated" sentences form an argument, we may analyze its consistency. We shall not do that in this paper, which deals with statements only. When we think that a writer uses isolated sentences unintentionally, we may try, as mentioned above, to "transform" these isolated "sentences" into empiricist "statements"; in this case we shall not speak of "translation." But we shall speak of "translating one statement into another statement" which then may be called "propositionally equivalent." We avoid the phrase, "Both statements express the same proposition," because we have again to fear that "glossogen" questions arise which lead to dangerous reduplications, as Avenarius ("introjection"), particularly, has shown.[13] We avoid all terms which are not physicalist. I think it dangerous to say that "a sentence expresses a proposition"—sometimes writers add that "a proposition is something immaterial"—by "verbalizing it," whereas "such a proposition expresses perceptual experience by conceptualizing it."[14]

A statement may be called "accepted" or "rejected"; or we may say, "The decision remains suspended for the time being." "New York is a Scottish city" would be called a "rejected statement"; but, nevertheless, it remains a "statement." When we read how Alice met a gryphon, we would not say, "That is no statement at all," because we would look around Hyde Park if scientists told us that they had just brought nice gryphon puppies with red claws from abroad. The statement from *Alice's Adventures in Wonderland* we should call a "fiction-statement."

Within the social sciences we do not use commandments. But no pedantry! Mathematics was formulated in ancient Egypt in that way. All our cookery-books use commandments in spite of their wholly empiricist character. You may simply transform these commandments into assertions such as, "By making a combination of the following type, you will get something liked by the majority of your contemporaries in a certain country." Through this transformation these statements do not become scientific, because our cookery-books do not give a survey of the possible variations and of their respective results, but they could do it perhaps. In a similar way, a writer may regard ethical commandments as descriptions of certain "ways of life." And he or some authority may tell us that they like a particular way of life or that there are certain patterns of life which appear again and again in history. But many writers will tell you that they do not acknowledge this transformation as sufficient or "adequate."

To avoid misunderstandings: after, we have transformed commandments into assertions, we are not sure that they are now empiricist ones; they may be metaphysical ones and use expressions such as 'justice,' 'good,' and 'bad' as "absolute" terms (as we have already stressed); we also mentioned that there are expressions which even without metaphysical flavor only transfer certain word-customs from one generation to the next, e.g., expressions such as 'crime' and 'punishment.' In this way we may only discuss as old folklore many sentences of ethics, jurisprudence, pedagogics, and economics, even if they are of unmetaphysical character.

But often we do not decide whether some sayings are merely old folklore—different in various countries—or insufficiently developed "scientific sayings," i.e., "modern international folklore," acknowledged by us as empiricists. On the one hand, we have to be cautious of dropping expressions too quickly as elements of old folklore, because they may be used later on within our new pattern; on the other hand, we should be cautious of adopting expressions into our modern folklore, called scientific language, before we have examined them carefully.

I think our analysis here deals better with statements that belong obviously to the field of scientific arguing, i.e., to the modern international folklore of empiricists, and I shall only touch fields where we feel uneasy at present. I am far from assuming that the great bulk of such doubtful assertions deals with matters of less importance than other assertions within the social sciences.

A few remarks on the phraseology used: the statement, 'This sentence is a statement within our language,' may be sometimes treated as propositionally equivalent to 'The names in it have designata'; whereas the statement, 'This statement is accepted,' may often be treated as propositionally equivalent to 'The names in it have denotata.'[15] We think this proposed wording may prevent us from some speculative sidestepping to which the designatum-denotatum diction perhaps occasionally seduces one.

In discussing the foundations of the social sciences, there is always a certain danger of looking at "speaker," "speech," and "objects" as three actors, as it were, in a play, who may be separated from another, whereas I treat them as items of one aggregation. As items in an empiricist discussion they belong together; there may be calculi in which three items appear separable and nevertheless in some respect resemble the three empiricist items. But the difference may be essential.

This may be regarded as a proper empiricist statement: 'The name-plate 'elephant' has not been fixed to the cage to which it belongs in accordance with a certain custom.' We may speak in the same way of the name-plate 'gray elephant' or 'white snow' with regard to a certain cage or a certain heap in the corner of a court. We suggest saying: "We accept the statement, 'There is an elephant here,' because it fits into a certain cluster composed of observation-statements and other statements in a particular way." (It may be that this elephant-statement does not fit into another group of statements also accepted by us; encyclopedism does not pretend to present a thoroughly consistent system of statements, though our program tries to extend the area of consistency. Therefore the fitting of statements into a cluster of statements is always of importance.)[16]

We should be doubtful even in admitting a definition of 'true' which implies that the saying, 'There is an elephant here,' may be called true if and only if there is an elephant here. Even this sounds like an "absolute" expression, since 'There is an elephant here' is treated absolutely. This "absolute" way of speaking may be called the "ontological" one (reminding us of the ontological tradition since Aristotle). The second way, as suggested in this paper, may be called the "physicalist" one when we stress its thoroughly spatio-temporal and aggregational character, and may be called the "terminological" one when we stress that we do not start any discussions and meditations with ontological assertions but with the comparison of statements with statements and their terminological analysis.[17]

These and similar remarks become particularly important within the field of the social sciences, where, in discussing historical questions and questions dealing with the measurement of sociological correlations, the tendency appears to speak of "true" and "false" statements, of "error," etc., in an "absolute" way.[18] I suggest that we do not even speak of "the truth" as something "which cannot be reached completely but only to some degree," because that implies that "the truth" may be regarded as a kind of "limit" at least, whereas we do not intend to make any statement on this point. By introducing the geographico-historical expression 'accepted by us at a certain time and at a certain place,' we intentionally avoid the "absolutely" used term 'true' (and its opposite 'false'), which we do not know how to fit into a framework based on observation-statements.

It may be possible that one can use a variation of the true-false phraseology within some calculus. That enables us to distinguish between various levels of analysis. Afterwards the results of such a calculus may be applicable in some way or another to empiricist discussion, but we cannot without particular assumptions expect that the correlations within the calculus in which a true-false distinction (related to a certain language) appears, may be regarded as correlations between items in our empiricist field.[19]

How may scientists (all being historians to a certain degree, as we tried to show) discuss observation-statements made by eyewitnesses? What can they do when the statements seem to be incompatible? You may think, for instance, of judicial proceedings. The eyewitness may say, "The dress was gray." You may be on the bench and order the beadle to present the dress, by saying, "I want to discover whether the statement is true or false by observing the color of the dress." And then you may say, "After looking at the dress, I can declare the eyewitness' statement is not true; the dress is blue." If you and the court have the authority to decide what be legally true, the matter is decided; in the republic of the sciences other customs are valid. There is no such authority, and the eyewitness may send you into the witness box and ask questions from the bench and decide together with the court that your saying is false.

What may we infer from this? That in the republic of the sciences we had better abstain altogether from using the terms 'true' and 'false' and their substitutes or weakened derivates at all levels of discussion, as long as we discuss empiricist problems of cosmic history and unified science. And that is what I suggest we do here as far as the social sciences are concerned.

A social scientist who, after careful analysis, rejects certain reports and hypotheses, reaches a state, finally, in which he has to face comprehensive sets of statements which compete with other comprehensive sets of statements. All these sets may be composed of statements which seem to him plausible and acceptable. There is no place for an empiricist question: Which is the "true" set? but only whether the social scientist has sufficient time and energy to try more than one set or to decide that he, in regard to his lack of time and energy —and this is the important point— should work with one of these comprehensive sets only.

It would be in full accordance with this attitude if a historian wrote more than one history of the Thirty Years' War; more than one biography of Galileo, Wallenstein, or Cromwell, selecting sometimes these, sometimes those, acceptable and plausible statements. I should doubt the scientificalness of another historian who declared, "Such multiplicity cannot remain when a serious research worker looks carefully at the material; of course, there may be at first many possibilities, but one and only one will appear to be the best; we have to find the optimum." I suggest being suspicious wherever a writer tells us of an "optimum," except in cases in which a well-defined calculatory situation enables us to speak of an "optimum." It is just in the social sciences that we meet this tendency to speak often of an "optimum." This tendency is perhaps connected with a certain tendency to comprehensive determinism as Laplace formulated it in a very decisive, "absolute" way.

That leads us to a kind of "pluralism," which appears even within physics, as Duhem and others pointed out. In the social sciences this pluralism is more conspicuous, because relatively simple stories are full of hypotheses. But we should not overlook that even the simplest report is based on hypotheses and, therefore, pluralist. In analyzing historical writings, a social scientist should speak of a "Pluri-Cromwell," also of a "Pluri-Cromwell at a certain date" and of a "Pluri-Cromwell on the following day," of "Pluri-London," and of other pluri-items.[20]

A historian presenting only one biography acts like a scholar who presents only one reconstruction model of a city. Obviously a few wall foundations, a few pictures, and a few chapters in an old author allow more than one reconstruction. History and reconstruction models are not so very different from a writer's historical novel.[21] Most historians fill gaps more reluctantly and do not add lively descriptions to their hypotheses. But why should we, as social scientists, object to any reasonable tentative completion of "torsos"? Why should we even object to speaking of the "internal speech" of a person in action, as long as we do not speak of "motives" and "forces" behind the scenes, but only in expressions such as: 'Perhaps he felt pain at this moment,' 'He remembered old conflicts,' or something like that.

7. The Richness of a Sociological Vocabulary

What can a historian do when A says that he was at a certain place yesterday and that no elephant was there, whereas B says that he was at the place in question yesterday and that there was an elephant in the cage. Perhaps the historian is very interested in these statements. He may have reasons to guess that the A-statement harmonizes with other people's statements. The historian may then say that the A-statement may be regarded as a "factual statement" whereas the B-statement may be regarded as a "hallucinatory statement." The statement, 'B has been in a state of hallucination,' would become an accepted factual statement for the historian and people who agree with him.

The statement, 'There has been an elephant at a certain place' (or its negation), may be assayed by observation-statements in which different sensorial terms appear: one person may tell us that he had felt the elephant; another, that he had seen the elephant; a third that he had smelled the elephant. The term 'elephant' is not correlated to a definite sensorial vocabulary but fits into various sensorial vocabularies; it is "intersensorial," as it were. If people tell us that an electric current of a certain character yesterday appeared somewhere, we have to deal with a term which is obviously intersensorial but not unsensorial, because each of the observation-statements by means of which we assay the statement will contain a definite sensorial term. The one may speak of what he has seen, another of what he has tasted in his mouth, a third of what he has heard. In this way we may tell of the periodicity of electricity or of the periodicity of light by means of intersensorial statements. We know the nucleus of an intersensorial language when a blind person calls a table "round" and a person without touch but with sight also calls the table "round." We have not two words for touch-round and sight-round. If we want a statement for our encyclopedia dealing with the table and its shape, we are not interested in two names such as "touch-round" and "sight-round"; but, if we should want to put statements into our encyclopedia dealing with the state of the persons in question, we need some research to find out whether one person's "round" has been a "touch-round" or a "sight-round." The statement that a certain person dealt with "touch-round" only and not with "sight-round" is an intersensorial statement, because we may discuss this question with persons who personally are, e.g., numb and blind and use other sensorial statements for the discussion. The statement that somebody is "blue-seeing" or "round-seeing," etc., is an "interpersonal" statement—but, again, we shall not speak of an "unpersonal" statement. A person-name is connected in some way or another with any statement at which we look as empiricists. We do not here wish to discuss the problem of how poor the sensorial field of a person could be without destroying the communication of arguments. We may teach a blind person the theory of optics, and, what is more, by means of suitable devices, this blind person may make optical experiments himself. Would smell be sufficient to form a kind of Morse code as a minimum of communication and argument? A modern Condillac could analyze this field.[22]

In the social sciences we need physicalist statements of the type, 'Charles has been in a state of feeling pain,' and this statement is of the same type as a statement such as 'Charles has been in a state of elephant-seeing.' Historians of human social life are highly interested in descriptive terms, such as deal with the feeling-tone of persons, their devotion, their fear and hopes.

A person may be described as horror-struck without adding anything else related to his being struck. People after certain injections feel horror as such. Our terminology is not always adapted to this kind of "object-less" terminology. We have to coin a word if we want to tell the state of a person who hears Beethoven or looks at a certain form of architecture. Usually one speaks of the beauty "of" this composition or "of" this architectural form, instead of "this person's being aestheticalized" or, in poetic language, "drunk with beauty." I suggest not to anticipate that this state of feeling could not be fully "object-less." That is one of the reasons why I should avoid the term 'value' in describing a composition in relation to a person's "interest" or whatever phraseology may be used. The term 'valuation' dealing with the state of a person may be useful, since it leads us to communication and social relations, whereas terms such as 'value' very often slide into unphysicalist language with metaphysical man-traps and then the antagonism "internal world" and "external world," and other dangerous reduplications, frequently appear.[23]

By suggesting the abandoning of a phraseology which leads to these dangerous language situations, I do not suggest reducing the number of feeling-tone terms. On the contrary, I see that, as sociologists, we need more than we find in scientific literature, particularly in books on behavioristics ("psychology"). Writers who try to overcome traditional language barriers, such as James Joyce, provide us with a richer vocabulary than we usually have. We need relatively rich vocabularies for the field of happiness, religious devotion, artistic excitement, and we shall often be glad to find indistinct terms such as 'oceanic feeling' in the writings of psychoanalysts and other scholars who deal with this delicate and entangled realm of language. Later on we may define these terms by means of a small number of terms, but in a new way.

Empiricist sociologists certainly like to discuss correlations between types of social structure and types of warfare. Analogous correlations may be found in animal sociology. But I think that there are also other problems, not less empiricist, if treated physicalistically. An extensive literature dealing with correlations between the growing-up of what is called the Calvinist attitude and what is called the capitalist attitude exists. Apart from accepting or rejecting these hypotheses, let us make a few remarks on the procedures involved in doing so.

There is no particular difficulty in speaking of Calvinism or of capitalism as historically given items in certain places and at certain periods. 'Calvinism' and 'capitalism' would be used like terms such as 'spring in certain countries' and 'vegetation in certain countries,' which may be correlated to one another without having rules as to the use of these terms in general; then we shall ask how we may speak of 'spring' in tropical areas and how we may speak of 'vegetation' where only micro-organisms grow. In this way we may reach a more comprehensive phraseology. Let us try to do the same where we speak of "Calvinism" and "capitalism" and their respective human attitudes. Should we apply the term 'Calvinist attitude' to an attitude connected with "Mohammedanism" or only to an attitude connected with "Christianity"? And if we decide to do the latter, how shall we use the term 'Christianity'? Who may be called a Christian? The answer that we shall apply the name to everybody who calls himself one is a scientific joke and not applicable as a serious procedure. Should we call a man a Christian who professes the Christian faith (it remains an open question how to define that even in a rather indistinct way) and acts mercilessly and cruelly by justifying his actions not only by overt but also—in so far as we can judge—by internal "Christian" arguments? Perhaps somebody would suggest starting from a man's behavior; but then we need a phraseology including a man's behavior and his feeling-tone. I think, then, that terms such as 'Calvinist attitude' and 'capitalist attitude' are not well selected and defined; but we shall try to find feeling-tone and behavior terms which serve our purpose better. Later on we may find that a certain feeling-tone and a certain behavior appear just where we speak historically of Calvinism at a certain place in a certain period, and the same may be applied to "capitalism" as a spatio-temporal item. Instead of "Calvinists" we speak of representatives of certain social groups, one of which may be the Calvinist one; and similarly we may speak of feeling-tones and attitudes of representatives of certain social groups, one of which may be the capitalist one. But just this feeling-tone and behavior phraseology is lacking today, and I think that that is one of the reasons why the wonderful material and the fine analyses dealing with these problems very often do not lead to discussable results.[24]

Certain types of human beings may be characterized by certain groups of qualities. (Psychiatrists have found certain correlations between very different items, including the handwriting of a person.) Why should we not some day analyze the behavior of individuals within certain societies, and the behavior of these societies, after elaborating a classificatory scheme together with some hypothesis on correlations? But it does not help much to anticipate such a future, as long as we do not have the technical devices at hand.

How backward this field of phraseology mostly is, may be learned from a confusion in the headings of books. Sometimes you will find expressly stated that certain feeling-tone items may be discussed in "physiology" and other, very similar ones, in "psychology"—without sufficient reason for such a separation. ("Pain," e.g., sometimes appears in "physiology," because we may speak of "cutaneous sensation," whereas "pleasure," which may be regarded as an opposite to pain in the happiness analysis, often appears in "psychology" under the heading "hedonic tone.")[25] I suggest speaking of "behavioristics" for the comprehensive field which includes more than "psychology" and therefore consistently of "behavioristicians" as representatives of such a discipline. People who do not like these terms may regard them as signals only: "Safety first!"

We see how perplexing anthropological and historical analyses sometimes appear to be, because the phraseology is anything but consistent. It will not bolster up any argument to add that something belongs to the "mental world" or that there are "motives behind an action" instead of using a simple correlation phraseology. Generalities, such as "factors of social change," even when they are connected with empiricist statements, do not seem to further descriptions or predictions; they often serve as a kind of healing balsam.

There are certain expressions which are harmless if used for orientation purposes but dangerous if used for classification and correlation. A sociologist may speak of "religious behavior" in one chapter and of "technical behavior" in another. In both cases these expressions may be sufficiently .univocal, even if indistinct; they may, e.g., overlap one another. But if one asks, before a careful analysis of this phraseology has been made, how religious and technical behavior are related to one another, a dangerous situation may arise. I doubt that we are in a position to speak of "physical factors," "biological factors," "psychological factors," and "cultural factors" as more or less coordinated groups. Even expressions such as 'war,' 'revolution,' 'literary movement,' etc., may be used only as orientating expressions as long as we have no comprehensive classification in this field.

I should suggest dropping sentences such as the following: "When a judgment or an idea is given a name and the name is thought of as an entity, having actual existence, then the concept has been reified," because it is full of dangerous terms, which I suggest be abolished. I am thinking of expressions such as 'existence,' 'entity,' 'reality,' 'thing,' 'fact,' 'concept,' 'mind,' 'mental world,' 'physical world,' 'meaning,' 'progress,' 'the beautiful,' 'the good.'

Let me repeat that I do not suggest as our position at the start (as some people presume I and my friends do) "simple basic assertions," "atomic ideas," "sense data," or their phraseological substitutes, i.e., something elementary, primitive, and poor. On the contrary, I suggest our starting with a full lump of irregularities and indistinctness, as our daily speech offers it. Afterward we may find some regularities in it and relate some items even to a calculus with all its exactness and its formulae. Each item of our start may be called a "clot" (German *Ballung;* French *grégat*).[26] Our observation-statements are rich in clots and not in simple items, not even when 'position of a watch hand' or 'length of a rod' appears within such a statement, because observation-statements are always connected with the indistinct protocolist-term and with indistinct observation-terms.

I hold this position even against scholars who want to start analyzing physics by bringing forward some clear-cut "sense data" or something like that; but in the field of the social sciences a similar position would imply a kind of scientific suicide. Social scientists especially need this richness and indistinctness. Only in this way can we cover the history of arts, the history of tools, the history of happiness, the history of engineering, the history of devotion, the history of church organizations.

It is attractive and stimulating to analyze social correlations by means of mathematical techniques which have been so successful, particularly in modern biology. But this does not involve the application of these techniques to all questions in the social sciences of the moment. I myself would not venture to make a statement on this particular subject. We shall see what a great role may be played by unpredictability in the social sciences and also in cosmic history. I already touched on the area of indistinctness in the social sciences. How are these areas and the areas of mathematization to be dovetailed? All finesses of modern logistic cannot overcome this doubtful start; therefore, we have to think of the limitations simultaneously and create a rich vocabulary which should enable us to attempt any kind of scientific enterprise.[27]

II. Scientific Procedures in Sociology

8. Aggregational Program

Within logical empiricism, terminological analysis helps us a good deal. But we should never forget that no terminological care can blot out lack of criticism in accepting statements which have hardly been assayed by observation-statements. Logical empiricism does not provide us with a wizard's sieve which holds all this uncritical stuff, as logic does not prevent us from making huge buildings of metaphysical speculations, consistent in themselves, but unempiricist. Books on racialism may be written in plain empiricist language but nevertheless lack a critical attitude.

It is important to think how we may corroborate arguments and hypotheses or how we may shake them. We pointed out how we, as social scientists, have to start with collecting hypotheses from climatology, biology, chemistry, etc., if we want to make hypotheses on the migration of tribes. That sounds like joining up separate groups of hypotheses, each of which could be assayed independently. But within the program of cosmic history these hypotheses become much more intimately dovetailed. Assaying one hypothesis implies, as it were, assaying all hypotheses together. And, in starting with assaying some hypothesis and going on successfully then with the argument, we may be hampered by findings in different parts of the cosmic history which alter the whole attitude of our arguing.

Assaying statements on tribes or on levers implies assaying the behavior of our cosmic aggregation. We cannot remove the aggregation and should take it into account, at least on principle; a way of arguing, on the one hand, in harmony with certain arguments of Mach, on the other hand, in harmony with certain arguments of Marx.[28]

This aggregational program may also be applied to all detailed problems. We suggest not starting with the antithesis: living being and the environment (as bio-ecology does), but starting with what may be called a "synusia" composed of men, animals, plants, soil, atmosphere, etc.[29] I am using the term 'synusia' in analogy with the term 'symbiosis,' and I hope that the old theological use of the word will not mar our argument. This looking at the behavior of a "synusia" will help us to avoid such bewildering questions as: How does environment influence a group of men and how does this group of men counterinfluence the environment? A "synusia" composed in the way sketched above may present a certain kind of cohesiveness, i.e., continuance of some relations. As the pluralist program of logical empiricism has its counterpart in some metaphysical speculations, so, too, the aggregational program of logical empiricism has its counterpart in some metaphysical speculations, e.g., in what is called "Holism" ("Ganzheitslehre," etc.)[30]

In the social sciences an empiricist aggregational program only pursues trends already in being. It does not suggest that we should characterize any sociological discipline by describing the human relations in question but the "synusia" in question, i.e., an aggregation with its human beings, animals, houses, streets, swamps, and fields. We suggest using a unified language, e.g., "Millenniums ago, when a swamp and a human group met—the human group vanished, the swamp remained; now, the swamp vanishes and the human group remains." We try to avoid any asymmetry in such a phraseology, particularly the asymmetry connected with the traditional cause-effect phraseology.

9. Growing-out-of Phraseology

Objections have been raised from more than one quarter against the cause-effect phraseology.[31] Most people, in speaking of causality, assume that one has to speak of successions which appear whenever the same conditions are given. This phraseology does not take into account that the expression "whenever the same conditions are given" (*ceteris paribus*) is not in harmony with our speaking of "different cosmic aggregations" which characterize what are called "different times" in accordance with modern physics and cosmological hypotheses. Why should we look at "time" as a kind of pipe, having throughout the same type of wall, which has no chemical relations to the items flowing through the pipe? Let us say more cautiously: "In spite of varying cosmic aggregations, some successions appear again and again, presenting various kinds of continuancies." The term 'different times' should, where doubts arise, be translatable into 'when firmament and watches are different,' or something like that.

The cause-effect phraseology confuses many sociological discussions in which one speaks of one "factor" molding another and of the latter re-influencing the former. Sometimes one speaks of "mutual causation" and often gives a clumsy and perplexing account of what we may call correlation of two items within an aggregation. The whole cause-effect phraseology seems to be rooted in some older assumptions.[32]

I think that the cause-effect phraseology is connected with all kinds of tendencies to prefer what we may call "asymmetry of expressions." I cannot but assume that these tendencies may be related to the asymmetry of our speaking and writing. I suggest treating this problem more carefully from a pedagogical point of view. There are many examples in scientific and every-day discussions which present us with this asymmetry. Let us take an example from a field where one would hardly expect such a tendency. In the "logic of classes" one sometimes tried to demonstrate the equality of the two "logical products": 'ab' and 'ba' (similar attempts in other logical disciplines). Let us imagine that our symbols would not be 'a' and 'b' but '○' and '●' and let us further assume that we should use the monogram of these two symbols as the symbol of the logical product: '◉.' It is obvious that we now do not start with two symbols similar to 'ab' and 'ba.' If we thought we needed two symbols, we might perhaps use some index, e.g., '$◉_1$' and '$◉_2$.' Immediately one would ask: Why only two products and not more? Even in our un-monogrammatic language, we could start by looking at more possibilities, e.g., 'ab,' 'ba,' '${}_{b}$,' '^{b}.' There are scholars who think of these questions, but many of them do not touch this point. Usually there is not even a term and a symbol to tell us that 'ab' and 'ba' are not to be regarded as two operations with equal solutions but as what I suggest calling "symbolically equal." I suggested writing it: $ab \equiv ba$. An example: $\sqrt[3]{5} \equiv 5^{\frac{1}{3}}$.[33]

I think it pedagogically fruitful to link up these and similar problems with the terminological analysis of the cause-effect muddle, which is partly connected with the asymmetry tendency in general. The formulas of physics make the cause-effect expressions, which often appear there as a kind of additional phraseology, relatively harmless, but in the social sciences the cause-effect reasoning often enters into the entire argument.

I suggest avoiding the cause-effect phraseology entirely. Some scholars think that one may abandon it where a "higher" scientific level has been reached but not on the "lower" level. This argument does not seem to be very strong, because there are "savages" who do not use the cause-effect phraseology.[34] Some scientists think it a defect in them. Let us speak, as these "savages" do, of "growing out of" or of "arising from" or of "coming out of." We may say, for instance, that certain educational institutions grow out of a certain social aggregation, together with certain administrative institutions, without, from the start, speaking of education as influencing administration or of administration influencing education.

Sometimes anthropologists use the term 'function' in an asymmetric way, sometimes trying to apply biological phraseology in so far as it contains "teleological" elements. The advantage of what is called a "functionalist program" in maintaining studies about correlations, instead of purely historical ones, could be reached without the asymmetric peculiarities of "functionalism." Sometimes asymmetric correlations may appear, but let us avoid their anticipative preestablishment.[35]

Also the "superstructure-substructure" phraseology belongs to this asymmetric cause-effect language. Apart from the difficulty of how to treat the "priority" of the substructure, it is not very simple to transform the phraseology of this hypothesis into an empiricist one. Let us try to do this: "Knowing the organization of the production apparatus of 1700, one may find out by means of induction what legal and artistic activities are related to it. Knowing the legal and artistic activities of 1700 only, one may hardly be able to find to what organization of production they belong. From the organization apparatus of 1700, one may induce something about future production apparatus and from these predictions successfully predict future artistic and legal activities, but one would not be able to write a history of arts, etc., or make predictions in this field, without knowing the artistic and legal activities together with the production apparatus of the period in question." I do not ask whether we acknowledge this hypothesis or not, but it seems to be a suitable empiricist transformation of the superstructure-substructure phraseology, containing a certain asymmetry of prediction as a hypothesis. Whether representatives of this hypothesis will acknowledge the transformation as a proper one is a different question.

10. Predictions Based on Uniformity

Our ancestors seem to have feared that the spring sun might never rise again. It does not seem unlikely that human beings in almost all regions of a certain climate tried various devices to get the spring sun back. Then a kind of Couéism entered human society, and people repeated: "The sun will come again, the sun will come again—we know the sun did come again formerly."

Scientists are no better off than the man in the street as far as their predictions are concerned. They only collect more material dealing with past uniformities. The famous Kepler's laws may be regarded primarily only as statements which deal with certain star positions given by Tycho de Brahe and other astronomers. And one would call it a tremendous discovery if Kepler had only said: "All known positions of Mars may be regarded as points of a certain ellipse."

This statement does not even assert that other points of this ellipse are also likely to be points of the orbit of Mars.

That would be an extension of this statement and would at first not be backed by observation-statements. Perhaps one would afterwards find other astronomers' protocols with respective data, and they might fit into this ellipse too.

But we may regard it as a further extension if one did expect that Mars's future orbit would be on the same ellipse as that observed. The Dutch "Significi" (Mannoury, Van Dantzig, etc.) maintain this careful distinction between a formula which covers given material and an assertion that this formula may cover material not known before—past, present, or future.

If scientists discuss certain data and add further items to these data in accordance with certain mathematical rules, we may speak of "supplementation" of the given data for forming a series in which these data also appear. I suggest using the term 'supplementation' because I have no reason here to distinguish between extrapolation and interpolation, for which we have no common name.

The statement on the fitting of the known positions of Mars into an ellipse is a purely mathematical one, but the statement that we also expect other positions of Mars to fit into this ellipse is an "induction" proper; and also the expectation that the orbit of Mars will be an ellipse in the future may be called an "induction" proper.

Before we start analyzing any induction, we try to find out whether the sentences of the hypothesis in question are empiricist statements, i.e., made in a language which allows comparison with observation-statements. That we declare that some sentence is "assayable" in this way, and the expressions are "adopted" by us, may not be connected with any statement about the way in which we may support the hypothesis in question and about the way in which we may assay it scientifically in the future. Let us distinguish four cases: a group of sentences may be regarded as a group of statements forming an empiricist hypothesis, even if we have no support by given material or any assumption on future assay by observation-statements. (I.)

Of course, we try to bring forward hypotheses which are supported by given material ("induction") and of which we expect that they may be assayed in the future by somebody. (II.)

There are cases in which we may assay a hypothesis, which we did not support by means of given uniformities. (III.)

There are other hypotheses which may be supported by given uniformities in the past, but of which we do not even assume they would be assayed at any time. (IV.)

Case (IV) sometimes occurs in astronomy. There scientists may predict that after some millions of years the cosmic aggregation will be of a temperature which will not allow any living being to live and to collect observation-statements for assaying this prediction. This case, nevertheless, is connected with our other hypotheses, as a result reached by deduction from accepted hypotheses. In the social sciences such predictions will hardly appear.[36]

In all the sciences we try to present case (II) if possible, but all other statements may also be discussed properly. Whether somebody will look for some comet which is insufficiently predicted is a question of the character of an individual, but the discussion of such a possibility remains within the field of empiricism.

The induction, if scientifically made, is based on some "supplementation"; but there are always many such "supplementations" in harmony with our *Encyclopedia*, and therefore it remains a matter of "decision" what kind of "supplementation" we prefer. There are scholars who try to find certain rules for making inductions; but I can hardly imagine how these could be a substitute for decision.

Let us assume that in some empiricist hypothesis the figures 6, 12, and 18 appear. First of all, these figures have very frequently been obtained by some procedure based on certain conventions, i.e., decisions. We cannot say within empiricism, "Exact measurement teaches us. . . . ." In the social sciences, as in other sciences, we often meet collected results of measurement, and then we try to substitute such collections (called "belonging to the same item") by certain "index figures," often called the "correct" figures, whereas the others often are regarded as "errors." Sometimes these terms are harmless, but sometimes they lead into a "truth" phraseology and support an absolutist viewpoint. In the formulas of calculatory schemes we may speak of "exact measurement" and therefore of "error" also. But we cannot apply the results of the calculatory schemes to the aggregational discussions without making certain assumptions.

If some scholars prefer to work within a calculatory scheme of some "universal laws" of the type "Every A has the behavior *m*," they may say that any "universal law" may be destroyed definitely by any contradicting hypothesis which seems to show that "this single A has not the behavior *m*." They may add that no "universal law" could be "verified" by empiricist statements, because that would need an infinite number of "good" statements. This calculatory scheme, with an asymmetry very attractive to many scholars, is not valid for aggregational arguments which include all sociological sciences.

In the field of aggregational analysis we always have to deal with more than one possibility. There is more than one convention possible in making index figures, more than one in making "supplementations." Let us assume that we decided to accept the figures 6, 12, and 18 and that we wanted to try some "supplementation"; then 6, 9, 12, 15, 18, or 6, 12, 18, 24, 30, or −18, −12, −6, 0, 6, 12, 18, 12, 6, 0, −6, −12, −18, and many others may be in harmony with our three accepted figures. Mathematical procedures may enable us to think of series of which we did not think before, but the decision remains, namely, which of the possible "supplementations" may be used for our "induction."

Supporting an "induction" implies producing uniformities, already given, combined with uniformities in "trend," etc., for producing possible predictions. Of course, in harmony with logical empiricism we ask for the elimination of contradictions where possible and for the collecting of uniformities; but, finally, a margin remains for our decision again and again.

11. Corroborating and Supporting Hypotheses

In cosmic history and particularly in the social sciences we have hardly any use for the refutation-verification asymmetry of certain calculatory schemes. We look at a network of hypotheses only, and we cannot say from which hypotheses certain difficulties arise. We have to start pluralistically all over again. "Supporting" hypotheses by means of material suitable for hypotheses is based on decisions: we are selecting material; "corroborating" hypotheses and "shaking" them by means of "positive instances" and "negative instances"[37] implies decisions; but no "*experimentum crucis*" may invalidate any single hypothesis. Our pluralist attitude makes us suspicious, from the very start, of all such attempts.

We should avoid the usual expression that one selects from possible hypotheses the "best" one, as if it were the normal case to rank them one-dimensionally. The usefulness of any one hypothesis may overlap the usefulness of another, and both hypotheses may present peculiar difficulties which can hardly be compared with each other. The history of the sciences teaches us how often hypotheses grow up and vanish without any possibility of ranking them in such a way.

Let us speak of "positive instances" and "negative instances" not as of single items, which may be enumerated piece by piece,[38] but as of aggregates of arguments. I should therefore say we may "shake" or "corroborate" the assertion of a hypothesis and finally prefer some hypothesis; but I should not suggest saying that we may "confirm a sentence more and more," because even this "weak" statement deals at least with some "limit." This kind of arguing on various "degrees of confirmation" may be useful within calculatory schemes, but we have to make careful analyses before we can apply the results of such calculatory arguing to aggregational studies.

Social scientists sometimes think of physics and astronomy as of an El Dorado of exactness and definiteness, and they assume, frequently, that in this field any kind of contradictions are fatal to hypotheses. Of course, scientists in all sciences try to fit hypotheses into a cluster of other hypotheses, observation-statements, and other accepted statements. But certain defects, e.g., well-described contradictions, do not always induce scientists to discard a hypothesis. They may maintain that this hypothesis is often useful and that there is no other more attractive hypothesis. The "positive instances" play a great role and not only the "negative instances," as in the asymmetric refutation-verification scheme. Newton's gravitation hypothesis has been used in spite of the fact that for about a hundred and fifty years scholars have felt again and again that there were contradictory and ambiguous elements in Newton's hypothesis, but it appeared so successful in analyzing the movements of bodies that only a few scholars really criticized the defects of this hypothesis.[39]

In the social sciences positive results form the backbone of success. Schliemann assumed that Homer's vivid descriptions could hardly be pure invention and Achilles could hardly have been only a river-god; Schliemann dug and found old towns and treasures. Malinowski expected, in accordance with Freud's hypothesis, a Melanesian tale telling of a brother-sister marriage (just because the brother-sister taboo is particularly strong). Negative instances could hardly shake such a hypothesis, but the finding of such a tale corroborated it.

Positive instances "corroborate" hypotheses; therefore, the finding of groups of instances is important. I think this point plays its part in the assumption of many social scientists that the behavior of the masses may be more easily predictable than the behavior of individuals. I think that they sometimes mix up the behavior of the masses which may be regarded as average behavior of individuals, and the behavior of masses which cannot be regarded as such average behavior.

Let us assume that somebody found that the mortality rate of a population has for years been 12 per thousand; then he may perhaps predict (within a network of hypotheses) the continuance of this uniformity. But he would not be able to predict the dying of a certain individual of this population.

But if we wanted to predict the behavior of the various nations after this war, or something concerning the persecution of the Jews in Germany, where can we find material to support any hypothesis dealing with this matter? We hardly know of any persecution of similar dimension; I am not certain whether even the persecution of Jews and Moors in Spain at the end of the Middle Ages and at the beginning of modern times is of a comparable size. But even if we found a few cases, with very different surroundings, can we use these for the making of "supplementations"? Where is the series of cases we need?

We can sometimes predict the behavior of an ant community better, because we have seen many ant communities—but even this allusion to animal communities should be made hesitatingly, since sometimes animal communities alter their behavior in a way not known before. Sometimes we can predict the behavior of single ants better than that of ant communities, because we know many more ants than ant communities. Let us be cautious in stressing the difference between the predictability of groups and individuals.

Since flexibility and adaptability of modern human communities reduce their predictability, many anthropologists seem to prefer the analysis of "preliterate" tribes, which they think are now relatively more stable. Our own "ethnology" (with our jurisprudence, economics, etc.) has usually been treated separately—a strange way of forming scientific branches. All reforms in phraseology will not help much, in this field, as long as one does not try to reduce the tendency to create watertight compartments and the tendency to avoid overlapping disciplines.

We are accustomed to find an unaltered behavior of levers, sodium, the Milky Way, but not of biological items, as far as the growing-up of new types is concerned. How should we predict what would happen if we were to treat the genes of all human beings with X-rays? New unpredictable social orders might appear.[40]

We shall always expect predictable correlations within such new aggregations, new animals, new social patterns; but very often we shall not be able to formulate them before getting new observation-statements. The newly formulated correlations may become obsolete after a short time. Sometimes our own sociological arguing may be connected with the alterations in question.

We see many limitations; nevertheless, it makes a great difference in our decisions dealing with predictions whether we have collected some material or not. The importance of Westermarck, Frazer, and other scholars who tried to present rich comparative material, more or less arranged, may be based on their capacity for destroying fixed opinions on so-called "human nature." Loosening such prejudices is sometimes an activity of social scientists within their own scientific field and within the community as a whole.

Not infrequently certain scientists behave as if they could provide people with many more correlations than is scientifically possible. Sometimes they resemble religious leaders or magicians of the past, pretending to have some superhuman knowledge and cunning.

As social scientists, we have to expect gulfs and gaps everywhere, together with unpredictability, incompleteness, and one-sidedness of our arguing, wherever we may start. Catching one group of hypotheses, we see them connected with all our hypotheses, which cannot be studied as a comprehensive body but only in some way or another, fragmentarily. Even when we look at the cosmic aggregation, we cannot think of a "system" as a limit but have to look at our scientific work from the viewpoint of encyclopedism. We would speak of a "model-encyclopedia" as one formerly spoke of a "model-system." We should not overlook that, in the aggregational field, when we say A is indistinguishable from B and B from C, we cannot declare that, therefore, A is indistinguishable from C, without collecting new observation-statements.

Some scholars think that they are very indulgent if they suggest as a fair compromise the acknowledgment of so-called "working hypotheses" which one cannot expect to be clear cut and proper. That sounds soothing, but our statement is: In all aggregational sciences, and therefore in the social sciences, all hypotheses are, if we adopt this consoling phraseology, nothing but "working hypotheses."

12. Unpredictability within Empiricism

Sometimes the behavior of human groups may be connected with some changes which appear "by chance." Let us imagine that we can predict that a certain unstably situated stone will roll down a slope from a pass but that there is no hypothesis which tells us why we should expect it to roll more to the right than to the left. If the tribes of the right-hand valley are threatened by this stone and an avalanche following the fall of the stone, the history of a continent may become different from the history of the same continent following a migration of tribes connected with the fall of an avalanche into the left-hand valley.

The aggregation formed by stone, snow, valley, tribes, climate, cosmic rays, etc., may be called unstable if even a small variation in the initial state may bring about a tremendous difference in the state of the whole aggregation in question—"tremendous" here from a sociological viewpoint. Our statements, as aggregational ones, deal with such situations and with a plurality of initial states, not with any accurate data of calculatory schemes; therefore, we should not be astonished if such difficulties arise.[41]

It is an open question which social situations may be regarded as unstable. Not a few sociologists seem to try to avoid problems which deal with unstable situations, as if we knew where such were and where not. And not a few, as mentioned above, think they avoid the crux of the matter if they look at the behavior of the masses only.

Sometimes analysis shows us that situations, formerly regarded as unstable, may now be regarded as stable; but sometimes serious difficulties appear where "statements made by individuals or masses" form items within aggregations which one tries to discuss sociologically. "Statements" as social items have to be treated very carefully, because puzzling situations may appear: A scientist may predict that a meteorite will fall and that some people will be killed at a certain place. Should he publish this prediction, people would go away and the prediction could not be in harmony with the future observation-statements made for assaying his prediction. Assuming he only writes down his prediction in his notebook, he may then perhaps be successful in his prediction. If he were to tell people that they were going to hide at the time when the meteorite was due to fall, perhaps his prediction would again be contradicted by observation-statements, because these people, full of mistrust, thinking that the scientist wanted to remove them for his own interest, would remain where they were.

Moreover, if a writer makes a prediction, he has also to take into account the making of the prediction as a social item. We should think of the prophets, who sometimes have been called "men who created the world in harmony with their prophecies, realizing them in this way." Prophets, who may be regarded as "makers," often have been treated as dangerous. Relatively tolerant Roman emperors persecuted magicians and soothsayers, because they sometimes told the people that the emperor would be assassinated by a peasant within two years. In astronomy it makes no difference whether anybody writes a prediction down or not. Predictions made by leading statesmen sometimes may be regarded as information about their intended actions; and those whom we call powerful leaders may imply that they are able to predict successfully their own actions and their results.

But there are problems more puzzling than these. Imagine that somebody wants to make predictions such as the following: "The statement that the formula x may be deduced in a certain way from the formula y cannot be made sooner than in the next century." It is obvious that one cannot accept this statement, because through making the prediction this man also makes the very statement of which he intended to say it will not be made sooner than in the next century.

In the difficult cases we discussed before we started with this puzzle, we might imagine a prophet with tremendous insight who would be able to predict that "this die will fall with six uppermost" or that "this stone will go down into the right valley, and the history of the continent will be shaped in a certain way." But in our last case even a prophet could not make the prediction in question without destroying the success of the prediction by making it. The making of a statement on the building of a house and the making of a statement on the making of a statement act differently, just because we treat a statement physicalistically like a house and not because we start with some metaphysical speculations on the human "mind."

These and similar remarks on unpredictability within empiricist sociology lead to further puzzling results, e.g., that some geological and astronomical connections may also become unpredictable. Usually one thinks of speech as using a small amount of energy, but this amount of energy may be connected with greater energies in a peculiar way: a decision told to other people (as mentioned above), may be connected with geological alterations (building of dams, etc.) which may lastly be connected with an alteration of the orbit of the earth. We learn from this that the unpredictability of certain statements may be connected with the unpredictability of geological or even astronomical changes.

It would be worth studying these and similar problems within the framework of logical empiricism without any slip into speculative metaphysics. How should a nation that could not invent the wheel predict the invention of just this wheel? Nevertheless, we have to maintain that these limitations do not touch our language or our scientific procedures in the social sciences.

13. Unstable Uniformities and Social Engineering

A mechanical engineer may discuss many types of possible aircraft without having any reason for expecting that the realization of any one of his projects has more chance than another. In a similar way, a "social engineer," a "planner," may deal with many possible social patterns without trying to predict which of them will be realized. It may be that, within a society organized in a certain way, one may be able to predict the future of certain alterations better than today—as long as the social order in question is valid—but without being able to predict whether the social order in question will remain for centuries, decades, years, or perhaps only months.

Let us think of an analogy: people may decide by throwing dice whether they will play chess or another game; therefore, they are able to predict something within the game, but not whether chess will be played or another game. It may remain unpredictable what kind of "plan" may be accepted by a parliament, but, after the decision is taken, one may predict many details from the regulations fixed by the accepted plan. But how should we find out when this parliament or another authority will alter the plan? And in what direction?

As long as such uniformity continues—some people may speak of an "artificial uniformity"—one may bring forward very detailed correlations; but by asking when these correlations are valid, we will get the answer: "*rebus sic stantibus*"—and social scientists want to know just these "*res.*" We have very fine studies on market correlations, but we do not know under what conditions these correlations remain valid.[42] Certainly they do not remain valid where no market exists. But to what extent does state interference, for instance, leave such correlations undisturbed?

From this viewpoint "social engineers" are interested in even poorer schemes which present some correlations between human beings and their living conditions within a given aggregation. Then one may ask how an alteration of certain institutions may be connected with the alteration of other items. We know how small the number of items taken into account by such a comparison may be. That is one of the reasons why many scientists prefer the above-mentioned analysis under the very indefinite assumption of "*rebus sic stantibus.*"

There are other points, too, which are of importance. Since new types of society may grow up, the inventors of social patterns or single social institutions may sometimes be better prophets than people who pretend that empiricism is restricted to the analysis of given patterns only—a restriction unknown to architects, technologists, and empiricists in other fields. Sometimes only individual persons may be able to imagine institutions which later are also imagined by the masses. But even the imagination of these single persons is assumed to be correlated with other items of social aggregations and therefore limited. It is superfluous to analyze in detail the coming of such imagined social orders. They may be full of predictability in themselves and may remain unpredictable as such, and this may hold true also for the vanishing of such social orders.

Mechanical engineering as a scientific procedure deals with a selection of certain aggregations, e.g., historically given steam engines and certain planned steam engines, but not with all possible steam engines, whereas scientific mechanics tries to deal with all kinds of possible levers.

This limitation of engineering is perhaps not due to the infinite number of possibilities, as some people assume, but to our limits in finding and handling possible solutions. The number of possible chess games is finite, but we cannot systematically imagine each of all the variations we want; therefore, we have to "invent" new variants. We have not even rules as to how to assay new variants. We watch how good players behave. After some time a successful new variant may be defeated by another master-player's new countermove.

Social inventions are seldom made by means of a well-planned procedure; usually amateurs and novelists bring forward "utopias." The words 'utopia' and 'utopianist' usually include a judgment: a utopia is defined as "an impracticable—ideal—scheme of human perfection and social improvement." People who judge in this way are seldom experts in assaying the practicability of social proposals, and, since the utopias of one period often become the trivialities of the following, we suggest using the term 'utopia' for any kind of invented order, pleasant or unpleasant, plausible or implausible, for maker and reader. "Scientific utopianism" seems to be a fair scientific enterprise, and we may deal with its procedures seriously.

Often it may be useful simply to ask what different types of social orders we may imagine, without starting with the question which social order one thinks likely. As long as one speaks only of a world state, of state federations, of centralized sovereignty, etc., one will not pass beyond the traditional. To get a new type let us imagine, for example, a human community in which overlapping institutions are the rule; one river administration crossing many language and educational regions, which may have nothing to do with police regions, while production organizations may be partly world wide, partly very local.[43] Or let us think of a world community without the use of money as a means of accounting for the whole fabric of production and distribution, without the use of any unit—some people suggest thinking in terms of labor units. In both these cases we may think about historical movements in this direction or not think of them. The analyses of such possibilities are highly scientific as long as they are made carefully and by speaking of the various limitations in imagining social aggregations and predicting them.

It is a widespread misunderstanding that imaginings which do not deal with all the details known today remain defective from the start. Assume somebody inventing a road system without any crossings on the road level, only fly-overs being allowed. Such an organization would obviously avoid car crashes. The man who elaborated such a plan may be without particular knowledge of traffic rules or of the behavior of car-drivers or of accident statistics. On the contrary, sometimes people who know all these things very well cannot formulate simple proposals and dislike solutions for which their expertness is not needed.

From more than one angle, utopianism has been declared unscientific. It is obvious that predictability in the social sciences is limited, but why should "utopianism" not share this fate with "history of the future" ("Philosophy of History," etc.)? Certainly not a few utopianists thought that one could alter society step by step, where others hardly expected that, or they hoped much from single communities of a new type. But these and other hypotheses are not essential for scientific utopianism.[44]

Our phraseology is relatively meager as far as comparative social engineering is concerned; the traditional expressions are usually evolved for transferring customs and habits from one generation to the next. Social engineering, more than anthropology, forces us to create new terms for new types of societies, i.e., for a new folklore just in the making.

14. Social Silhouettes

Mechanical engineers, using old traditions, are accustomed to make statements concerning the effects of machines. They may say of a certain airplane that it transports more weight than another, but that its speed is not so high, the seats less comfortable, the risk of accidents smaller, and so on. Each of these qualifications may be measured by means of its own unit or ranked according to some scale (e.g., comfort of seats). The same seems to be the given procedure in the social sciences. One may compare two social patterns as far as death rate, suicide rate, illiteracy, use of radio sets, and other items may be concerned. Let us speak of various "silhouettes," composed of the items in question,[45] in analogy to individual "profiles."

Some of the sociologists are prepared to handle silhouettes, but not a few try to avoid this type of procedure, based on a multidimensional comparison. They attempt to reach, by some calculations, a single number indicating a "degree of something," and often they cannot even tell us to what this "something" should be correlated and when it remains stable. Very often, where they start by ranking, they treat the "ordinal numbers" of the respective ranks as "cardinal numbers" and perform additions or other more complicated mathematical operations. They try, for instance, to compare revolutions, wars, etc. They rank revolutions according to their extent, violence, etc., by means of conventional scales, and then they introduce some conventions on combining the resulting figures in some way or another, obtaining for each revolution one and only one index number.[46] Similar tendencies appear where "national wealth" is to be compared. In that case, it seems obvious that this unscientific procedure is based on the reckoning in money, which provides us with single figures based on bookkeeping conventions, but which cannot be used as well for any other purpose. That in economics the accountants' conventions, which lead to nice sums, play a great role is more plausible than that this tendency to get "the one sum," i.e., "the index figure" in any case, should also enter other social sciences, where silhouettes frequently seem to be the device best adapted to the arguments in question.

Let us take an old example: the arguments, brought forward by Bortkiewicz against the proposed index figures, remain valid today. Various nations have different mortality rates; one cannot say that where the mortality rate is higher, we may also expect a lower standard of public health. It may be that in one nation the percentage of old people is extraordinarily high and, therefore, the national mortality rate may also be very high, even if in all age groups the mortality rate were smaller than in other nations. The silhouette of mortality rates would tell us what the situation is. But decades ago scholars tried to find a way of ranking nations in one dimension according to their mortality rate and proposed a so-called "standard-population" as basis. They then made the assumption that the age distribution in all nations may be treated as identical with the age distribution of the standard population, and then the respective mortality rates had to be applied to these age groups. From these age-group mortality rates, they formed mortality rates for all the nations and compared these index figures with one another.

The criticism leveled against this proposal has been based on the remark, which is important in all similar cases, that one may make conventions selecting one standard only, but that the order found in this way should remain unchanged if a different standard population has been chosen. But this is not so in our case. Starting with a different "standard population," e.g., with Poland instead of with Sweden, some nations change their relative position in the rank order. Where such criticism can be applied, we should only retain silhouettes.

In brief, the danger arises from the attempt to reach index figures by means of mixing up items, measured in different units or items, obtained by treating marks (results of ranking) as cardinal figures in spite of their being ordinal figures. The enormous advantage we may get from statistical correlations and from any kind of mathematization should not make us careless in this respect. Just as we suggested avoiding asymmetric correlations as long as they are not based on peculiar investigations, I should suggest avoiding index figures and the above-mentioned techniques of transforming ordinal figures of ranking into cardinal figures. I suggest starting with silhouettes, provided that on investigation one did not find that on a different procedure more than that could be done.

15. Ranking Social Items

There are social scientists who carefully try to avoid the seducing attractiveness of index figures and of all such techniques where they are not applicable, but there are not a few scholars who devote much effort to getting marks from experts, which afterwards may be treated by mathematical procedures, as we have indicated above. Sometimes they decline a criticism such as ours as philistine. They stress the point that famous historians speak frankly and freely in vague terms of comparison when they talk to us of various revolutions or wars. Would it not be better, so runs the argument, to make the statement that, for instance, a certain revolution has been violent to a certain degree? Such social scientists then suggest ranking the various qualities of a revolution and computing the conventionally obtained marks in accordance with a convention brought forward by them.

They stress the point that all measurement, being conventional, may also be conventional here.[47] They often overlook the fact that not all conventions are useful. We have seen what criticism could be brought forward against the "standard population"; the same criticism remains valid here. If we altered the figures of the marks and then carried out the addition, we might get a different order of the revolutions.

It remains further open for discussion whether the thoroughly empiricist convention in question is at all suitable for any purpose. Let us assume that somebody has proposed a new conventional measure of human beings, a figure composed of the number of hairs per square inch of head, their weight, age in days, and the length of the right index finger. This well-defined figure will hardly lead us to any correlations, and only correlations could "justify" us in making such a convention.

Therefore, the indistinctness of the above-mentioned historians and other social scientists has some advantage as long as we do not see how we can substitute for these indistinct assertions data based on scientific procedures. As a logical empiricist I would ask for such a procedure, but I do not think it helps us to create some conventional marking of items without particular theories. For a reader who does not adopt the procedure of computation cannot use the results of the computation and would prefer to know the single items of the silhouette, as they would be important in any case. But avoiding this overmathematization should not lead us to oppose mathematization as such, an opposition frequently made from a metaphysical point of view.

How difficult the analysis is may be learned from another old example. In the beginning of psychophysics, scientists tried to discover how people reach the statement that a certain shade of gray is exactly in the middle between white and black. Some scholars would immediately think how much the result would be improved if experts in grays could be asked. (Think of analogous cases, when standard of living or something else is in discussion.) I think we should be doubtful, as long as we have not found out how the estimate is reached. Let us imagine that we ask people to tell us what is the middle figure between 2 and 8. Some people may answer "4," some people "6," some people "between 4 and 6." How would the application of the finest mathematical methods help us? If we were sufficiently clear about the situation, we might perhaps ask the test persons for the geometrical and the arithmetical middle and other middles, separately.

In the social sciences we do not know much about such delicate details; a certain coarseness in our arguing, therefore, sometimes seems to be better, in accordance with our state of scientific analysis, than fine techniques when applied to a perplexing material composed of "clots." Let us look at some problems in ranking connected with planning which present some possibilities of reaching results but also some limitations, and which so often lead social scientists to prefer the index-figure technique at any price. It is remarkable that anthropologists who deal with preliterate tribes without money reckoning are more content with silhouettes than sociologists who deal with Western nations. Scholars who deal with animal sociology hardly think of index figures.

Let us start with the statement that the traditional market society leads to the intentional destruction of milk, coffee, wheat, etc., in periods of hunger and starvation. It is plausible that there are people who imagine some new type of society in which no such destruction could take place. Discussions on rich and poor, on standards of living and similar items, arise. One may ask whether social pattern No. 1 or social pattern No. 2 is the more "efficient."

But "efficient" for what? we shall ask. In accordance with an old tradition, promoted by Bentham and others, somebody may answer, "efficient for human happiness." We see immediately that we cannot measure human happiness, perhaps only rank it. Even if we should adopt a saying, "Somebody is happier when he has a good meal than when he expects starvation," we should not say that he is, for example, $5\frac{3}{4}$ times happier.

Some social scientists maintain that one may speak of the comparison of the happiness possibilities of the same person under different circumstances but not speak of comparing the happiness of various persons. I think that they do not realize that the same person, on various days, may be regarded as two persons who are "genidentical," and, therefore, one may speak of the particular feeling-tones of the same person in different situations in somewhat the same way as of the feeling-tones of two persons.

Scholars who deliberate this difficulty would, I think, be prepared to say that a person who cries and is depressed may be less happy than a person who is obviously enjoying life (though this may depend on the nature of the scientist).

Let us see how far we can get with the assumption that we can compare only the happiness of one person under different conditions. Given a society of two persons, A and B. Let us symbolize the state of happiness of A who enjoys "a" by '(Aa)' and of B who enjoys "b" by '(Bb).' The starting happiness situation of this society may be: (Aa) + (Bb). If we assume that (Aa) = (Ab) and (Bb) $<$ (Ba), then we know that (Aa) + (Bb) $<$ (Ab) + (Ba).

In a society in which barter takes place, when both partners have an advantage from it, the situation (Ab) + (Ba) of increased happiness would not arise. This would arise only if we assume that (Aa) $<$ (Ab); because then A would enjoy the barter too. This implies that another rule, "Exchange takes place when at least one of the partners gains something and neither has a loss," would lead to a higher state of happiness in this small society. We are not starting with an assumption of how human beings behave (how a so-called "*homo economicus*" or his substitute behaves) but in what way the results of various kinds of social behavior are connected with different states of happiness.

Let us assume that (Ab) $<$ (Aa) and (Bb) $<$ (Ba); then we cannot say how the sums behave: (Aa) + (Bb) ? (Ab) + (Ba). Sometimes the assumption that we may compare happiness of various persons may help us, e.g., if (Ab) $<$ (Bb) $<$ (Ba) $<$ (Aa), because then (the above-made comparison remaining unchanged) we may reach (Ab) + (Ba) $<$ (Aa) + (Bb). But we cannot give any answer if (Ab) $<$ (Aa) $<$ (Bb) $<$ (Ba). Therefore, we see that social engineers, planners of social orders, cannot, even if they assume that the happiness of various persons may be compared, always give their answers to the question of how various social orders are connected with the state of happiness of a community.

It is manifest that if social scientists, as they often do, treat the happiness items as measurable and not only as rankable, they get clear-cut sums and can compare the results of various social orders. But I stress the point that we have no scientific aids by which to measure happiness; therefore, social engineers can only describe the distribution of happiness but cannot compare sums of happiness in each case. Sometimes, however, it can be done, as we have seen.

On the other hand, descriptive statements on the happiness of individuals (or of groups of individuals who behave similarly) may be of great importance for people who make decisions with regard to happiness, even when they are not able to add them. Social scientists may characterize "living conditions" by telling of shelter, food, entertainment, friendship, "oceanic feeling," or whatever one thinks of importance. Some of these items may be measured in their own units (e.g., calories, cubic space used in houses, etc.), but in other cases we get only the degrees of certain items which all together constitute what we may call a "silhouette" of living conditions. We may sometimes speak of "higher" and "lower" levels of living conditions. Then we may speak, too, of the "relief" of happiness; for instance, of a "hill" when in one society we meet degree 3 and 6 and of a "plain" if in another society we meet the degree 4 only. Such and similar "felicitological" investigations are properly empiricist, and we may prepare sociological questionnaires which deal with these items. We may start from the same "basis of life" (i.e., an aggregation composed of human beings, animals, swamps, fields, houses, micro-organisms, etc.) and ask what different living conditions will appear by assembling the same items in a different way.

This analysis ("What would grow out of certain behavior?") and the other analysis ("How will people behave?") should not be intermingled; the first is a question of social engineering and the second a question of social history. The happiness analysis is manifestly a very comprehensive one, because we may also speak of the happiness of cats and dogs. We may give descriptive records of happiness as anthropologists even outside the problems of social engineering.

16. Arguing in the Social Sciences

Our main statement is that procedure in all empiricist sciences is the same; yet there are questions of degree: some techniques may be applied more frequently in one science than in another. Therefore, I put certain peculiarities in the foreground which appear particularly in the social sciences—but these also enter all sciences, even cosmic history as such. Here is the field of unpredictability disliked by most empiricists, who want to speak of difficulties only as provisional.[48] We maintain that the social sciences are less "problematic" where human sociology may be treated in the same way as animal sociology or plant sociology. Where language has to be regarded as an important social item and the flexibility and alterability of human societies have to be taken into account, new difficulties appear. Up to now empiricists have not sufficiently realized this situation, thus giving an opportunity to metaphysicians to enter just by this door into the social sciences. We need a literature in the social sciences such as Mach, Duhem, and Poincaré created for physics; the attempts made by Pearson, and others following Mill and his contemporaries, to include the social sciences in the analysis of the sciences are not entirely successful.

Some readers will ask how I can harmonize my statement dealing with the tremendous success in the social sciences and the statements on the limitations described above. How can any efforts help, if we are stayed by barriers which the best prophet could not overcome? How can all these silhouettes help, all these pluridimensional surveys, if we have to make decisions in any case? Should sociology, like other sciences, be restricted to handling problems that are either far removed from us or do not touch our living life? Should it be scientific only where it deals with social fossils?[49]

These questions seem to me serious, and I shall make some remarks on them in the last section of this monograph. Another objection will be made against my restrictive appreciation of the scientific results of what is called "economics." Not a few empiricist friends praise just the scientificality of this branch of the social sciences. I think that judgment partly depends upon the empiricist character of statements made in economics, but that does not imply that these statements present a comprehensive scientific analysis. I shall try to show in the following section that economics is predominantly busy with the systematized transfer of certain traditional institutions. It is a discipline in some respects like jurisprudence, in so far as that discipline does not deal with transcendent speculations, or like ethics, in so far as that discipline deals with the ways of life and not with "eternal values" or other "absolute" items.

This statement, that a discipline deals with the systematized transfer of a verbal tradition, does not imply any criticism. Why should we not transfer traditions? But I think that there are scholars who do not want to acknowledge this descriptive statement as it stands and who want to give the assertions in these disciplines a far wider reach in their scientific capacity. Within all sociological arguing it is important to take into account this difference between a scientific investigation, together with a systematization of the results of such investigations, and a systematized, i.e., scientifically supported, transfer of traditions.

17. Systematized Transfer of Traditions

There appear in the practice of life certain expressions which are empiricist in character but not suitable for use within scientific analysis. They may be regarded as a kind of old folklore, as speech customs which are related to the transfer of certain institutions. One speaks of the "cost" of a production and goes on using this expression not only in our bookkeeping departments but also in social analysis. And here we start with our remarks.

In economics for more than a century people have felt a certain uneasiness as far as phraseology is concerned; the scientific basis of economics is less well founded than the scientific basis of behavioristics, in spite of the latter's shortcomings.[50]

Accountants handling the profit reckoning of firms, ministers of finance handling taxes and customs of whole countries, evolved certain rules for their respective activities which could be written down and taught to the younger clerks and undersecretaries. One of the main points has been the reaching of a fair "difference" between certain "positive" and certain "negative" items recorded in their books. Expressions such as "cost," "profit," and "investment" are plausible results of bookkeeping and reckoning in money. This business is comparable with a chess player's notes of his moves or a railroad manager's notes on transport timetables and similar items of his enterprise. The transfer of chess-playing or railroad administration in a systematized way may be executed, as it were, scientifically, as far as consistency is concerned; but the important point is that the apprentice does not learn the behavior of human beings in general or the making of games or of railroads in general but only the handling of given institutions. It is a long story how economics became a discipline which appeared to be more than an administrative instruction to bookkeepers, ministers of finance, politicians, and other people who wanted to use some historically given institution such as money, taxes, or customs.

Representatives of political economy tried to teach people something about human happiness. And a kind of hybrid, partly dealing with human happiness, partly dealing with bookkeeping, was evolved. But the scientific means they tried to apply have been inferred from bookkeeping and not from happiness analyses.

We should suggest looking at markets and finance and at the whole reckoning in money as an institution like any other, such as funeral rites, golf, rowing, and hunting. To regard money as a historically given institution does not involve any objection to its use—though there may be such objections—but an objection to the application of arguments, valid in the field of higher bookkeeping, to the analysis of social problems and human happiness in general. We should treat this whole field of money reckoning anthropologically as a piece of modern ethnology. And, from the point of view of social engineering, we might also analyze other institutions comparatively, various kinds of money organizations, and—that is important—organizations which are not based on money and nevertheless cover all kinds of modern technology and communication. I think it is obvious that we cannot use money reckoning for assaying the results of the various institutions, some of which are not based on money reckoning.

The topical planning discussions which we mentioned above start mostly, or at least finish, by reckoning in money, or assume that some "reformed" money would form the basis of a future society. The question is not whether the future society will be based on money, in harmony with their predictions, but whether their analysis is scientifically performed in comparing all kinds of societies. Most economists assume expressly or without mentioning it that reckoning in money more or less resembles some other measurement proper. We cannot here discuss this problem; only one remark: the money account sometimes presents decreasing figures, whereas the "basis of life" appears to be improving. The destruction of coffee in Brazil is based on an account in money, not on an account in kind. We may ask how we can compare the results of a society based on money reckoning with a society based on reckoning in kind ("economy in money" compared with "economy in kind"), i.e., how we can look at their efficiency in terms of living conditions. It may be that the result of such an analysis of a world society, based on reckoning in kind and not in money, would be less efficient in terms of living conditions; but the lack of this analysis gives us a hint that economics deals exclusively with a kind of old folklore and not with scientific analysis of social engineering. The money taboo is so general that even in the Soviet Union the study of money-free societies has been abandoned as a kind of "left deviation." Therefore, even there the results of planning have been compared in money figures. Standard prices as a base is seen to provide us with a useful convention. But an alteration in the choice of the standard year alters the order of the result. What in the one case (using a standard year 1) appears to be more successful, looks less successful when we apply standard year 2.

As we may look at economics as a discipline dealing with the transfer of our money institutions, so we may deal with jurisprudence, ethics, pedagogics, and other disciplines in a similar way. Their scientific character is less acknowledged than the scientific character of economics. Jurisprudence, from its start, is full of metaphysical speculations. Successively, some representatives of jurisprudence are becoming sociologists and social engineers who deal with the structure of human societies and their possible alterations. The statements made by some of them about the two worlds of the "ought to be" and of the "is" are sometimes abandoned, though there often remains some distinction between the "external" and the "internal" world, the "internal" world, of course, being reserved for the social sciences.[51]

A thoroughly empiricist reconstruction of jurisprudence would lead not only to the abandonment of all the metaphysical terms used, but also to an alteration of terms such as 'crime,' 'punishment,' etc., which, even in a purely empiricist definition, remain erratic within anthropology and social engineering. I think that in an empiricist sociology the empiricist parts of jurisprudence would spread over various disciplines, and I can hardly believe that something like an "empiricist jurisprudence" as a chapter of sociology will remain.

Ethical studies, so far as they wish to be more than the handing-on of certain customs and commandments to the next generation in a systematized way (and something of this kind seems to be indispensable in any human society), could continue as an analysis of possible paths of life.[52] One can analyze the paths of life of various communities, or utopias prepared by single thinkers, or utopias only elaborated for the purpose of this analysis. One may analyze the connections between the various paths of life and happiness, between the various paths of life and the other human attitudes.

Pedagogics usually deals with the transfer of certain customs and habits, with the transfer of folklore, from one generation to the next. This is an important business, like others, and may be executed to a certain extent in a systematized way. The metaphysical speculation, frequently connected with it, could be abolished; but this discipline would remain a systematized transfer of customs as long as we cannot, as anthropologists or social engineers, analyze education. The wide field of significant educational possibilities has hardly been discussed. There are important parts of education not yet touched by the analysis. Only by chance have educational peculiarities been investigated as ethnological peculiarities. A research worker, for instance, has asked whether the whole examination procedure could be regarded as an initiation rite. This would explain why a certain torture is connected with it, why examinations are not repeated from time to time, and why they are connected with festivals.[53] I do not know whether the results of this investigation are convincing, but it does not start with the assumption that our educational system is in principle a kind of well-built structure, like a railroad system, with relatively few taboos and little old folklore. I repeat that a frequent misunderstanding found in books by empiricists is that acknowledging something as old folklore or taboo implies fighting it. But I cannot see any reason why I should not acknowledge something as folklore or taboo and transfer it to the next generation (without calling it an "eternal value") as I myself adapt my actions to it.

III. Sociology and the Practice of Life

18. Sociology of Sociology

Sociologists deal, among other things, with tools and tales, with the language of magic, theology, jurisprudence, economics, and pedagogics; but also with the language used by the sociologists themselves, with their statements and their habits, i.e., with the behavioristics of sociologists. We are just starting a sociology of the sciences and have not up to now a sufficiently elaborated classification of the sociological hypotheses as such.

We have difficulties even in classifying hypotheses in physics. It has not been very useful, for instance, to classify the optical hypotheses by starting from the dichotomy "undulation theory"—"emission theories." Like all asymmetric beginnings, dichotomies are doubtful tools. It seems to be much more fruitful to start with the main elements of the various hypotheses and to distinguish among Huygens' ray, Newton's ray, and Euler's ray and then to look at the possible combinations of various items. In this way we get not two groups of hypotheses, with the weak "eclectic" hypotheses between, but a whole matrix of hypotheses, characterized by the items in question, to which the scientists say, "Yes" or "No" (e.g., if we look at two qualities only, we get: I: a, b; II: non-a, b; III: a, non-b; IV: non-a, non-b.)[54]

In sociology we have, up to now, no clear trend of hypotheses, as we had no such trend in the hypotheses on electricity in a period in which undulation and emission theories of light already existed. Perhaps at the moment our sociological studies are in their youth.

I have sometimes tried to classify hypotheses on modern social life in the market society by starting from the main items discussed by social scientists. One sees immediately that there are social scientists who put the boom-slump cycles in the center and tried to correlate them with the structure of our market society, whereas other scholars who had perhaps very carefully analyzed some problems of social engineering, did not even devote much time to this item, but spoke vaguely, in general terms, of ups and downs in social life, as one speaks of rhythms in other branches of the sciences.[55] Another item of importance seems to be formed by the statements on predictability. But, as I maintained, I think we shall not be able to start with classification with any success in the near future.

As long as we cannot classify the sociological hypotheses, we can hardly expect great success in finding correlations among sociological hypotheses and other social items.[56] Nevertheless, of course, we try to find something, even before we are able to start from a classification, hoping that just this dissecting of historical material will help us in creating the classification. From an aggregational point of view, we look on these scientific activities as dovetailed.

One difficulty is that we usually try to compare the published opinions of scholars, not their "internal speech," which is sometimes different. Scholars, like other human beings, fear their neighbors, the police, and other people. Even free-lances, who often pretend to be very free, sometimes hesitate to say what they think.[57]

Further, certain statements are used as flags or as tools; in both cases we are uncertain what kind of correlation we may introduce. Voltaire said about fossil shells found on mountains that they may have been relics of pilgrims who lost them on their trip to Jerusalem (shells formed a part of their dress). One usually assumes that these statements are related to his criticism of the Bible. He did not want to support the story of the big flood told by the theologians who used the shells as witnesses. A century later mere theologians, on the other hand, fought geological hypotheses which wanted to recognize more than one flood and volcanic eruptions in addition.

Have I to remind my readers of the Voltaires and theologians of our times in the social sciences, when, e.g., the Soviet Union, the Bankers, Planned Economy or Free Trade are the subjects of discussion?

"Wishful" thinking is often accused of distorting argument. I think that this criticism of "wishful" thinking may be reasonable, but one should not overlook that there is no "neutral" position from which we may judge. And further, love and hate are often good teachers; people may select certain problems and pile up scientifically sound arguments—let us speak of "thinkful" wishing in such cases. People full of "thinkful" wishing may present the same scientific material as people of other habits but may combine it differently. In this way a sociologist may support scientifically some decision without becoming unscientific.

It is not simple to analyze the correlations between people's sayings and the social aggregation to which they belong, and it is particularly difficult to do this where sociologists are concerned. We maintained how difficult it is to speak of the correlations between "Calvinist" and "capitalist" attitudes. We have fine questionnaires as far as preliterate tribes are concerned but hardly any when we try to ask sociologists how they themselves behave in arguing and writing. Since sociological predictions are directly connected with actions which create what has been predicted, the difficulty is very understandable. More than in other sciences, taboos and old folklore come into the picture when human actions are under discussion.

19. Argument, Decision, Action

These statements on difficulties in the social sciences will be challenged by different kinds of people. On the one hand, people who like straight formulas will be disappointed, but also people who "want to know what to do"; in both cases clear-cut material, one solution, is wanted. But social scientists cannot provide either for mankind. Men of action, who try to make statements on their doings, speak of "actions made in accordance with one's own decision (internal speech)" or of "conscious actions," and want to co-ordinate one resolute decision with one resolute action. But how can they get one resolute decision from a pluralist argument?

Scientific social engineering and planning may bring forward more plans, and historians of the future will perhaps depict more possibilities than up to now. And then people will ask how scientists can help when they provide us with so many possibilities.

Let us consider an analogy. Imagine a gambler who wants to win a lot of money and has no knowledge of gambling banks and their chances. An expert may tell him about ten such banks and their traditional risks. What more can the expert do for him? I know that not a few people will shudder when they realize that I compare essential parts of their lives with gambling. In discussions on death rates, in discussions on other social problems, they are accustomed to that. What would you think of a gambler who declares that he will start with some fantastic theory which teaches him definitely how to win in any case? Or of a gambler who asks crystal-gazers to tell him how to proceed?

But in our own society, perhaps, a majority of people want to have something which enables them to make decisions in accordance with one and only one guiding rule, with one statement on the future, or something like that. One gets the impression that through the ages people who have to make sudden decisions frequently, such as generals, statesmen, industrialists, heroes, adventurers, pirates, are more devoted to the sayings of astrologers, soothsayers, and other guides of this type than are the rest of their contemporaries. Perhaps the penetrant doctrines dealing with the "iron laws of history" and of the "evidence which dictates that some attempts are hopeless from the start" belong to the family of augury, omens, and palmistry.

But there are other people who do not want to have a definite goal. They may like risks or like avoiding risks, but bearing uncertainty with calmness, they are prepared to act resolutely; they may behave loyally and try to love one another, without any other "justification" than that it is just their adopted path of life. They may act mercifully and kindly without asking for any "eternal values" or for "commandments given by a deity." They may be prepared even to make great sacrifices for the possibility of living within a tolerant environment, living in an atmosphere of friendship, security, scientific freedom, and "oceanic feeling," which enables people to enrich the pattern of their lives.

Of course, some people who do not speak of "eternal values" and their substitutes may be intolerant, may persecute one another, and may be interested only in good food and nice clothes. But persecution, intolerance, and appreciation of food and clothes have also been found in societies which promoted "eternal values"; indeed, even in societies which promoted love as an important item within their commandments. I cannot see much correlation between arguing and acting, as far as love and hate are concerned. Anthropologists tell us of small tribes in which aggression is less habitual than in great societies in which love has been preached for centuries. Scientists may find out more about this situation, by means of formulae or without them.

Not much scientific research has dealt with these problems of social and individual behavior. There are not a few topical questions which have not found a thorough analysis. We should not decline a hypothesis as unscientific which asserts that in a certain period in certain countries "church ties," "religious instruction," etc., are connected with a way of life characterized by a democratic attitude, toleration, preparedness to take promises seriously, being merciful, etc., whereas the breaking of church ties and the weakening of religious instruction may shake the habits of the above-mentioned ways of life. Then people who decide to promote a scientific attitude would be faced with a dilemma, such as that supporting a scientific attitude may imply weakening of faith in certain statements made by religious bodies, and, in accordance with the hypothesis mentioned above, the weakening of toleration and thereby the weakening of the scientific attitude. Whereas the supporting of certain statements made by religious instructors may be connected with the weakening of the scientific attitude and thus also of toleration which is basically connected with a scientific attitude. Scientifically, we should perhaps say that a certain mixture of scientific attitude and unscientific attitude always remains and that in a certain situation it may be difficult to estimate the items of such a mixture.

The hypothesis under consideration, when regarded as a social item, may be correlated with certain behavior. It might happen that people who are convinced of the validity of this hypothesis would cease to be tolerant, merciful, loyal, etc., when they felt their belief in certain religious statements vanishing, because they were accustomed to this connection by training. Whereas other people who did not know of this hypothesis might remain tolerant, merciful, and loyal in a period in which their belief in certain religious statements was vanishing. Could we then object to a person's guess that perhaps promoting the close connection between the belief in certain religious statements and a certain kind of behavior (being tolerant, democratic, merciful, loyal, etc.) is endangering this way of life in a period in which the church ties are loosened and the belief in certain religious statements too? And would it seem unreasonable if persons who look at this situation in this way suggest promoting in all people toleration, democratic life, mercy, loyalty, etc., independently of church ties and of the scientific attitude?

As social scientists, we have today to say that we lack scientific material dealing with this matter seriously. But just this information gives free scope to many kinds of decisions. People may promote the increase of religious instruction or the increase of scientific instruction. They can do it with what may be called "a good scientific conscience," but they become unscientific when they maintain that their decisions, their propaganda, are based on far-reaching experiential statements.

The pluralist attitude of arguing enables conservative people, who like traditions, to bring forward the lack of predictability. They may object to an abandonment of the open fireplace, assuming that it may be connected with an abandonment of some personal kindness. (The fireplace familiarity perhaps excludes some controversial talk and is connected with toleration.) People who want to alter social and private life may, however, go on declaring that nobody is able to predict that the happier future they pursue will remain unreachable.

In a tolerant world community empiricists and nonempiricists may live together peacefully, whereas in an intolerant atmosphere when somebody wants to fight, the flag is not far away, and any kind of strange alliance may appear, as sociologists show us. Then even small differences in verbal formulations may be used as banners in fighting one another.

Even where sociologists cannot make predictions, they may provide men of action or meditation with empiricist material. One could not predict reasonably that the gipsies would be left as social "fossils"; but if somebody tells us that some adaptations of social groups will be unavoidable, we may tell him about the gipsies. We argue differently and act differently, when we know the material provided by the social sciences. One sees immediately how important it is to have proper descriptive statements. Not only are predictions for long periods important but also statements dealing carefully with topical situations; one may find uniformities within the old material; one may find new uniformities in the newest statements too. And even our restricted scientific work may be dovetailed into our social and private life in various ways. Just what we are doing in this paper may have some importance, because altering our scientific language is cohesive with altering our social and private life. There is no exterritoriality for sociologists, or for other scientists, and this is not always sufficiently acknowledged. Sociologists are not only outside their scientific field in arguing, deciding, and acting like other human beings; they also argue, decide, and act like other human beings within their scientific field.

Imagine sailors who, far out at sea, transform the shape of their clumsy vessel from a more circular to a more fishlike one. They make use of some drifting timber, besides the timber of the old structure, to modify the skeleton and the hull of their vessel. But they cannot put the ship in dock in order to start from scratch. During their work they stay on the old structure and deal with heavy gales and thundering waves. In transforming their ship they take care that dangerous leakages do not occur. A new ship grows out of the old one, step by step —and while they are still building, the sailors may already be thinking of a new structure, and they will not always agree with one another. The whole business will go on in a way we cannot even anticipate today.

That is our fate.

NOTES

1. George A. Lundberg, *Social Research* (1942), p. 85 (statistical remarks on sociological terms).

2. E.g., Walter Heape, *Emigration, Migration, and Nomadism* (Cambridge, 1931), pp. 6, 101.

3. Alexander Lesser sponsors co-operation and collaborative effort dealing with "unified social science" in "Functionalism in Social Anthropology," *American Anthropologist*, XXXVII (1935), 387.

4. Otto Neurath, "Universal Jargon and Terminology," *Aristotelian Society Proceedings*, 1940–41, p. 127.

5. E.g., compare Ellsworth Huntington, *The Pulse of Asia* (Boston, 1907), with Arnold J. Toynbee, *A Study of History* (London, 1934 ff.), III, 432, 447 ff.

6. Homer, *Odyssey*, trans. Alexander Pope, Book I, verse 50.

7. A. R. Radcliffe-Brown, *Taboo* (Cambridge, 1939), p. 8.

8. Pitirim Sorokin, *Contemporary Sociological Theories* (London, 1937); R. H. Lowie, *The History of Ethnological Theory* (London, 1937). The *Encyclopaedia of the Social Sciences* is a rich mine of information. For stimulating reading see the series, *The Social Sciences: Their Relations in Theory and Teaching* (London, 1936, 1937, 1938) (reports of conferences held under the auspices of the Institute of Sociology), with papers by Ernest Barker, J. A. Hobson, T. H. Marshall, M. Postan, M. Oakeshott, G. N. Clark, J. L. Stocks, D. W. Brogan, H. J. Laski, J. R. Hicks, P. Sargant Florence, G. F. Shove, Karl Mannheim, Morris Ginsberg, A. M. Carr-Saunders, Joseph Needam, S. Zuckermann, E. E. Evans-Pritchard, Raymond Firth, John Layard, Edward Glover, Godfrey H. Thomson, B. A. Wortley, H. Campion, R. G. D. Allen, John Hilton, Brinley Thomas, Lionel Robbins, Maurice Dobb, C. A. Mace, H. A. Mess, John Macmurray, G. C. Field, and L. S. Stebbing.

9. Charles S. Peirce, *Collected Papers* (Cambridge: Harvard University Press, I, 120: "Geognosy applies to physics as well as biology."

10. R. L. Sherlock, *Man as Geological Agent* (London, 1922), p. 333.

11. Otto Neurath, "The Departmentalization of Unified Science," *Erkenntnis* (*Journal of Unified Science*), VII, 270 ff.

12. Discussed by C. K. Ogden in his writings on "Basic English" and by Lancelot Hogben in his writings on "Interglossa." Both, stressing the importance of word economy, think that about a thousand words are sufficient for defining all other terms.

13. Otto Neurath, *Einheitswissenschaft und Psychologie* (Wien, 1933), p. 27 ("Verdoppelungen"). In "Heft 1, Einheitswissenschaft," edited by Otto Neurath, in collaboration with Rudolf Carnap, Philipp Frank, and Hans Hahn.

14. Andrew Paul Ushenko, *The Problems of Logic* (London, 1941), p. 175.

15. Charles Morris, *Foundations of the Theory of Signs* ("International Encyclopedia of Unified Science," Vol. I, No. 2 [Chicago, 1938]), p. 5.

16. Critics sometimes called this approach the "coherence theory of truth." This is misleading, because this name is traditionally applied to certain

metaphysical speculations: C. R. Morris, *Idealistic Logic* (London, 1933), pp. 182 ff.

17. I think this is more or less in harmony with Carnap's suggestions in *The Logical Syntax of Language* (London, 1937), pp. 286, 239. His "material" mode of speech more or less covers our "ontological," and his "formal," our "physicalist" or "terminological" mode of speech.

18. E.g., Catharine Morris Cox, *The Early Mental Traits of Three Hundred Geniuses* (Stanford: Stanford University Press, 1926), pp. 81 ("the *true* IQ"), 26, 28.

19. I think that "semantics," as evolved by Carnap and Tarski, will support many kinds of calculus analysis, but I feel uneasy when thinking of its application to empiricist arguments and the danger of slipping into "ontological" ways of arguing.

20. Pluralist tendencies appear frequently. William James in his *A Pluralist Universe* (1909) tried to connect Bergson and Peirce. He and others did not avoid metaphysical speculations (e.g., James, *Memories and Studies* [1911], p. 369, "A Pluralist Mystic"). A useful survey is Jean Wahl, *The Pluralist Philosophies of England and America*, trans. Fred Rothwell (London, 1925).

21. Gaetano Salvemini, *Historian and Scientist* (London, 1939), p. 18, on "literary biographer." Fine remarks in this book on "moral training," p. 190.

22. Condillac's *Treatise on the Sensations* and Diderot's *Letter to the Blind.*

23. John Dewey, *Theory of Valuation* ("International Encyclopedia of Unified Science," Vol. II, No. 4 [Chicago, 1939]).

24. The writings of Max Weber; interesting criticism, e.g., by Roman Catholic writers: J. B. Kraus, *Scholastik, Puritanismus und Kapitalismus* (Munich, 1930); Amintore Fanfani, *Catholicism, Protestantism and Capitalism*, trans. T. Parsons (London, 1935). The defects of terminology and arguments, e.g., in remarks by Tawney (Foreword to Weber's *The Protestant Ethics . . . .* [London, 1930]) on the "confusion of causes and occasions" as if we had a theory which supports this kind of expression. Defects of terminology are even manifest in the fine analysis made by Svend Ranulf in *The Jealousy of the Gods and Criminal Law at Athens* (London and Copenhagen, 1933) and *Moral Indignation and Middle Class Psychology* (Copenhagen, 1938). Ranulf tries to be particularly cautious and reaches many useful results.

25. The literature on this subject becomes gradually more serious and systematized, e.g., Robert S. Woodworth, *Experimental Psychology* (London, 1938), pp. 234 ff.; F. Paul Han, *The Laws of Feeling*, trans. Ogden (London, 1930); Charles Hartshorne, *The Philosophy and Psychology of Sensation* (Chicago, 1934).

26. R. Bouvier suggested the term '*grégat*' to Otto Neurath, "L'Encyclopédie comme 'modèle,' " *Revue des synthèse*, October, 1936, p. 190.

27. The question of how far we may improve our coarse and insufficiently classified material by means of refined techniques seems to be stimulatingly discussed by Ethel Shanas in "A Critique of Dodd's *Dimensions of Society*," *American Journal of Sociology*. Vol. XLVIII (September, 1942), and by Henry Ozanne in "A Critique of a Critique: Further Comment on S. C. Dodd's *Dimensions of Society*," *American Journal of Sociology*, Vol. XLVIII (March, 1943).

28. I should like to call such a sociology a Marxist one, because Marx and Engels were more pluralist than others and started with a scientific physicalist approach; but an increasing number of scientists regard Marxism as a doctrine, based on "The Unity of Opposites; Qualities which negate themselves and lead through Negation to Synthesis; Quantity passing into Quality" and similar sentences which do not seem to be transformable into empiricist expressions.

29. Animal ecology leads in this direction (see Charles Elton, *Animal Ecology* [1927], with Huxley's introduction, pp. 35 ff.).

30. E.g., M. J. W. Bews, *Human Ecology* (London, 1935), with an introduction by General Smuts, pp. x, xi, where he speaks of reality as a pattern of patterns and of the vision of this harmony, what the gods fed on and what mortals strive for, according to Plato.

31. Philipp Frank, *Das Kausalgesetz und seine Grenzen* (Wien, 1931); William H. George, *The Scientist in Action: A Scientific Study of his Methods* (London, 1938), contrasting the description or patterning theory of science (Mach, Kirchhoff, Karl Pearson, and Hobson) with the cause-effect explanation (pp. 33, 342). He objects to expressions such as "real," "existence," "exists," "one theory nearer to the truth than another," "crucial experiment"; but he says "in precisely the same circumstances very similar things can be observed," an expression we should drop from our carefully selected language.

32. Hans Kelsen, "Die Entstehung des Kausalgesetzes aus dem Vergeltungsprinzip," *Journal of Unified Science*, 1939, p. 69.

33. Otto Neurath, "Definitionsgleichheit und symbolische Gleichheit," *Archiv für systematische Philosophie*, 1910, p. 142, and L. Susan Stebbing's remarks on Russell in *A Modern Introduction to Logic* (2d ed.; London, 1933), p. 506, n. 2.

34. O. Demetrucopulon Lee, "A Primitive System of Value," *Philosophy of Science*, 1940, p. 355.

35. One should recognize asymmetry more when it appears. See Carl Britton, *Communication* (London, 1939), p. 62, on asymmetry of solipsism.

36. Carnap on the discussion between Schlick and Lewis, "Testability and Meaning," *Philosophy of Science*, October, 1936, p. 423.

37. G. Ch. Lichtenberg, as so frequently he does, makes a good remark on this subject in his notes: "Es ist ein grosser Unterschied zwischen etwas glauben und das Gegenteil nicht glauben koennen. Ich kann sehr oft etwas glauben, ohne es beweisen zu koennen, so wie ich etwas nicht glaube, ohne es wiederlegen zu koennen. Die Seite, die ich nehme wird nicht durch stricten Beweis, sondern durch das Uebergewicht bestimmt."

38. Hans Reichenbach tried to test separate items, an attempt which has been criticized, e.g., by Hilde Geiringer, *Journal of Unified Science*, V, 277 ff., and VIII, 151 ff.

39. Finlay Freundlich explained this point in his lecture at Utrecht in 1938: "Ueber den gegenwaertigen Stand der empirischen Begruendung der Relativitaetstheorie."

40. Lancelot Hogben in his *Nature and Nurture* (London, 1933) maintains that one should not think that biological statements lead to sociological theories without any difficulties (pp. 9, 116 ff., 121).

41. James Clerk Maxwell in an essay of 1873 discussed these questions and mentioned sociology in a very stimulating way (Lewis Campell and William Garnett, *The Life of James Clerk Maxwell* [London, 1882], p. 438).

42. Astonishing results were reached, e.g., by Henry Schultz in *The Theory and Measurement of Demand* (Chicago, 1938), where he presented far-reaching correlations between various items.

43. Otto Neurath, *International Planning for Freedom*, chapter on "International Planning in the Making," *New Commonwealth Quarterly* (now *London Quarterly of World Affairs*), July, 1942, p. 24.

44. Marx and Engels tried to discredit what they called "unscientific utopianism." This may be connected with the lack of interest in planning which characterizes Continental socialists. Ballod-Atlanticus' in *Der Zukunftstaat* (Popper-Lynkeus and Bellamy published their suggestions before Ballod) is an exception. On discussion of possibilities see: John Langdon-Davies, *A Short History of the Future* (London, 1936).

45. Otto Neurath, *Modern Man in the Making* (London, New York, 1939), pp. 61, 145, continuing an old tradition.

46. Sorokin started in this way. This tendency to use ordinal figures as cardinal ones and to compute them can be found in many writings; Wood's *Historiometry* or J. L. Moreno's sociometric studies, which provide us with much stimulating and important material, are examples of it.

47. Lionel Robbins, like many others, thinks his confession that certain measurements are purely conventional (national income, national capital, etc.) forms a kind of definite justification, *An Essay on the Nature and Significance of Economic Science* (London, 1940), pp. x, 57.

48. E.g., Lundberg, *Social Research*, 1942, p. 21.

49. The remarkable book by Mill's friend, George Cornewall Lewis, *A Treatise on the Methods of Observation and Reasoning in Politics* (London, 1852), is full of defects. Nevertheless I think it is of importance. Some remarks sound very modern: "A good logical method . . . . abridges mental labour and renders an equal amount of exertion more productive" (p. 4); "It prevents waste of labour, in striving after an exactness which is not attainable . . . ." (p. 7). Chapter iv, "On the Technical Language of Politics," follows Whewell. Chapter xxiv, "Upon Prediction in Politics," should be mentioned in any history of scientific procedures, just because the author tries to analyze actual life. T. H. Pear wrote about avoiding topical problems, e.g., "Psychologists and Culture," *Bulletin of John Ryland's Library*, Vol. XXIII, No. 2 (October, 1939).

50. Very characteristic are the defects in the whole terminological approach made by Malthus, *The Principles of Political Economy* (2d ed., 1836), pp. 21, 47; in Richard Whately's *Elements of Logic* a great many of the "ambiguous terms" in his appendix are economic ones, a topic on which the Archbishop also lectured. L. M. Fraser, in *Economic Thought and Language* (London, 1937), depicts the situation without suggesting a way out. Others criticize the terminological situation too (e.g., P. Sargant Florence, *Uplift in Economics* [London, 1929], and in other writings). Marshall and others try to justify what they call "elasticity" in the use of terms (*Money Credit and Commerce* [London, 1929], (p. 13.

There are some attempts, partly supported by associations, to unify sociological terminology, but they do not start from a comprehensive analysis.

The remarks on "folklore" may be complemented by statements in Thurman W. Arnold's *The Folklore of Capitalism* (New York, 1937).

51. For the former viewpoint, Hans Kelsen; for the latter, Felix Kaufmann—both connected with logical empiricism without accepting it.

52. E.g., Charles Morris, *Paths of Life* (New York, 1942), I think some classificatory remarks are oversimplified, but there are approaches in harmony with the above-given outlines.

53. J. C. Flugel, "The Examination as Initiation Rite and Anxiety Situation," *International Journal of Psychoanalysis*, XX (1939), 275.

54. Otto Neurath, "Principielles zur Geschichte der Optik," *Archiv fuer Geschichte der Naturwissenschaften*, 1915.

55. The crisis problem appears to be focused in the nineteenth century, e.g., by Sismondi, Karl Marx, Henry George, and Wilhelm Neurath, whereas authors like Ricardo do not take it seriously; neither do writers like Pareto, in his various books, repeated in his *The Mind and Society*, trans. A. Livingston (1935), IV, 2338.

56. I mentioned how difficult it is to characterize 'Calvinism,' 'Christianity,' and similar terms. How should we correlate Plato's assertions with other items, as long as people describe Plato in very different ways? The majority of descriptions hardly take note of his totalitarian warlike attitude, of his merciless and cruel social education. Wieland's *Aristipp* did not appeal very much to German scientists; Crossman's *Plato To-day* and Fite's *The Platonic Legend* seem not to be much influential in scientific circles, nor Kelsen's analysis of Plato.

57. Mark Twain in the Introduction to his *What Is Man?* (1905) says: "Why did they not speak out? Because they dreaded (and could not bear) the disapproval of the people around them. . . . . The same reason has restrained me."

INDEX OF TERMS

The Structure of Scientific Revolutions

Second Edition, Enlarged

Thomas S. Kuhn

The Structure of Scientific Revolutions

Contents:

The Structure of Scientific Revolutions

Thomas S. Kuhn

Preface

The essay that follows is the first full published report on a project originally conceived almost fifteen years ago. At that time I was a graduate student in theoretical physics already within sight of the end of my dissertation. A fortunate involvement with an experimental college course treating physical science for the non-scientist provided my first exposure to the history of science. To my complete surprise, that exposure to out-of-date scientific theory and practice radically undermined some of my basic conceptions about the nature of science and the reasons for its special success.

Those conceptions were ones I had previously drawn partly from scientific training itself and partly from a long-standing avocational interest in the philosophy of science. Somehow, whatever their pedagogic utility and their abstract plausibility, those notions did not at all fit the enterprise that historical study displayed. Yet they were and are fundamental to many discussions of science, and their failures of verisimilitude therefore seemed thoroughly worth pursuing. The result was a drastic shift in my career plans, a shift from physics to history of science and then, gradually, from relatively straightforward historical problems back to the more philosophical concerns that had initially led me to history. Except for a few articles, this essay is the first of my published works in which these early concerns are dominant. In some part it is an attempt to explain to myself and to friends how I happened to be drawn from science to its history in the first place.

My first opportunity to pursue in depth some of the ideas set forth below was provided by three years as a Junior Fellow of the Society of Fellows of Harvard University. Without that period of freedom the transition to a new field of study would have been far more difficult and might not have been achieved. Part of my time in those years was devoted to history of science proper. In particular I continued to study the writings of Alexandre Koyré and first encountered those of Emile Meyerson, Hélène Metzger, and Anneliese Maier.[1] More clearly than most other recent scholars, this group has shown what it was like to think scientifically in a period when the canons of scientific thought were very different from those current today. Though I increasingly question a few of their particular historical interpretations, their works, together with A. O. Lovejoy's *Great Chain of Being*, have been second only to primary source materials in shaping my conception of what the history of scientific ideas can be.

Much of my time in those years, however, was spent exploring fields without apparent relation to history of science but in which research now discloses problems like the ones history was bringing to my attention. A footnote encountered by chance led me to the experiments by which Jean Piaget has illuminated both the various worlds of the growing child and the process of transition from one to the next.[2] One of my colleagues set me to reading papers in the psychology of perception, particularly the Gestalt psychologists; another introduced me to B. L. Whorf's speculations about the effect of language on world view; and W. V. O. Quine opened for me the philosophical puzzles of the analytic-synthetic distinction.[3] That is the sort of random exploration that the Society of Fellows permits, and only through it could I have encountered Ludwik Fleck's almost unknown monograph, *Entstehung und Entwicklung einer wis-*

1 Particularly influential were Alexandre Koyré, *Etudes Galiléennes* (3 vols.; Paris, 1939); Emile Meyerson, *Identity and Reality*, trans. Kate Loewenberg (New York, 1930); Hélène Metzger, *Les doctrines chimiques en France du début du XVIIe à la fin du XVIIIe siècle* (Paris, 1923), and *Newton, Stahl, Boerhaave et la doctrine chimique* (Paris, 1930); and Anneliese Maier, *Die Vorläufer Galileis im 14. Jahrhundert* ("Studien zur Naturphilosophie der Spätscholastik"; Rome, 1949).

2 Because they displayed concepts and processes that also emerge directly from the history of science, two sets of Piaget's investigations proved particularly important: *The Child's Conception of Causality*, trans. Marjorie Gabain (London, 1930), and *Les notions de mouvement et de vitesse chez l'enfant* (Paris, 1946).

3 Whorf's papers have since been collected by John B. Carroll, *Language, Thought, and Reality—Selected Writings of Benjamin Lee Whorf* (New York, 1956). Quine has presented his views in "Two Dogmas of Empiricism," reprinted in his *From a Logical Point of View* (Cambridge, Mass., 1953), pp. 20–46.

senschaftlichen Tatsache (Basel, 1935), an essay that anticipates many of my own ideas. Together with a remark from another Junior Fellow, Francis X. Sutton, Fleck's work made me realize that those ideas might require to be set in the sociology of the scientific community. Though readers will find few references to either these works or conversations below, I am indebted to them in more ways than I can now reconstruct or evaluate.

During my last year as a Junior Fellow, an invitation to lecture for the Lowell Institute in Boston provided a first chance to try out my still developing notion of science. The result was a series of eight public lectures, delivered during March, 1951, on "The Quest for Physical Theory." In the next year I began to teach history of science proper, and for almost a decade the problems of instructing in a field I had never systematically studied left little time for explicit articulation of the ideas that had first brought me to it. Fortunately, however, those ideas proved a source of implicit orientation and of some problem-structure for much of my more advanced teaching. I therefore have my students to thank for invaluable lessons both about the viability of my views and about the techniques appropriate to their effective communication. The same problems and orientation give unity to most of the dominantly historical, and apparently diverse, studies I have published since the end of my fellowship. Several of them deal with the integral part played by one or another metaphysic in creative scientific research. Others examine the way in which the experimental bases of a new theory are accumulated and assimilated by men committed to an incompatible older theory. In the process they describe the type of development that I have below called the "emergence" of a new theory or discovery. There are other such ties besides.

The final stage in the development of this essay began with an invitation to spend the year 1958–59 at the Center for Advanced Studies in the Behavioral Sciences. Once again I was able to give undivided attention to the problems discussed below. Even more important, spending the year in a community composed predominantly of social scientists confronted me with unanticipated problems about the differences between such communities and those of the natural scientists among whom I had been trained. Particularly, I was struck by the number and extent of the overt disagreements between social scientists about the nature of legitimate scientific problems and methods. Both history and acquaintance made me doubt that practitioners of the natural sciences possess firmer or more permanent answers to such questions than their colleagues in social science. Yet, somehow, the practice of astronomy, physics, chemistry, or biology normally fails to evoke the controversies over fundamentals that today often seem endemic among, say, psychologists or sociologists. Attempting to discover the source of that difference led me to recognize the role in scientific research of what I have since called "paradigms." These I take to be universally recognized scientific achievements that for a time provide model problems and solutions to a community of practitioners. Once that piece of my puzzle fell into place, a draft of this essay emerged rapidly.

The subsequent history of that draft need not be recounted here, but a few words must be said about the form that it has preserved through revisions. Until a first version had been completed and largely revised, I anticipated that the manuscript would appear exclusively as a volume in the *Encyclopedia of Unified Science.* The editors of that pioneering work had first solicited it, then held me firmly to a commitment, and finally waited with extraordinary tact and patience for a result. I am much indebted to them, particularly to Charles Morris, for wielding the essential goad and for advising me about the manuscript that resulted. Space limits of the *Encyclopedia* made it necessary, however, to present my views in an extremely condensed and schematic form. Though subsequent events have somewhat relaxed those restrictions and have made possible simultaneous independent publication, this work remains an essay rather than the full-scale book my subject will ultimately demand.

Since my most fundamental objective is to urge a change in

the perception and evaluation of familiar data, the schematic character of this first presentation need be no drawback. On the contrary, readers whose own research has prepared them for the sort of reorientation here advocated may find the essay form both more suggestive and easier to assimilate. But it has disadvantages as well, and these may justify my illustrating at the very start the sorts of extension in both scope and depth that I hope ultimately to include in a longer version. Far more historical evidence is available than I have had space to exploit below. Furthermore, that evidence comes from the history of biological as well as of physical science. My decision to deal here exclusively with the latter was made partly to increase this essay's coherence and partly on grounds of present competence. In addition, the view of science to be developed here suggests the potential fruitfulness of a number of new sorts of research, both historical and sociological. For example, the manner in which anomalies, or violations of expectation, attract the increasing attention of a scientific community needs detailed study, as does the emergence of the crises that may be induced by repeated failure to make an anomaly conform. Or again, if I am right that each scientific revolution alters the historical perspective of the community that experiences it, then that change of perspective should affect the structure of postrevolutionary textbooks and research publications. One such effect—a shift in the distribution of the technical literature cited in the footnotes to research reports—ought to be studied as a possible index to the occurrence of revolutions.

The need for drastic condensation has also forced me to forego discussion of a number of major problems. My distinction between the pre- and the post-paradigm periods in the development of a science is, for example, much too schematic. Each of the schools whose competition characterizes the earlier period is guided by something much like a paradigm; there are circumstances, though I think them rare, under which two paradigms can coexist peacefully in the later period. Mere possession of a paradigm is not quite a sufficient criterion for the developmental transition discussed in Section II. More important, except in occasional brief asides, I have said nothing about the role of technological advance or of external social, economic, and intellectual conditions in the development of the sciences. One need, however, look no further than Copernicus and the calendar to discover that external conditions may help to transform a mere anomaly into a source of acute crisis. The same example would illustrate the way in which conditions outside the sciences may influence the range of alternatives available to the man who seeks to end a crisis by proposing one or another revolutionary reform.[4] Explicit consideration of effects like these would not, I think, modify the main theses developed in this essay, but it would surely add an analytic dimension of first-rate importance for the understanding of scientific advance.

Finally, and perhaps most important of all, limitations of space have drastically affected my treatment of the philosophical implications of this essay's historically oriented view of science. Clearly, there are such implications, and I have tried both to point out and to document the main ones. But in doing so I have usually refrained from detailed discussion of the various positions taken by contemporary philosophers on the corresponding issues. Where I have indicated skepticism, it has more often been directed to a philosophical attitude than to any one of its fully articulated expressions. As a result, some of those who know and work within one of those articulated positions may feel that I have missed their point. I think they will be wrong, but this essay is not calculated to convince them. To attempt that would have required a far longer and very different sort of book.

The autobiographical fragments with which this preface

[4] These factors are discussed in T. S. Kuhn, *The Copernican Revolution: Planetary Astronomy in the Development of Western Thought* (Cambridge, Mass., 1957), pp. 122–32, 270–71. Other effects of external intellectual and economic conditions upon substantive scientific development are illustrated in my papers, "Conservation of Energy as an Example of Simultaneous Discovery," *Critical Problems in the History of Science*, ed. Marshall Clagett (Madison, Wis., 1959), pp. 321–56; "Engineering Precedent for the Work of Sadi Carnot," *Archives internationales d'histoire des sciences*, XIII (1960), 247–51; and "Sadi Carnot and the Cagnard Engine," *Isis*, LII (1961), 567–74. It is, therefore, only with respect to the problems discussed in this essay that I take the role of external factors to be minor.

opens will serve to acknowledge what I can recognize of my main debt both to the works of scholarship and to the institutions that have helped give form to my thought. The remainder of that debt I shall try to discharge by citation in the pages that follow. Nothing said above or below, however, will more than hint at the number and nature of my personal obligations to the many individuals whose suggestions and criticisms have at one time or another sustained and directed my intellectual development. Too much time has elapsed since the ideas in this essay began to take shape; a list of all those who may properly find some signs of their influence in its pages would be almost coextensive with a list of my friends and acquaintances. Under the circumstances, I must restrict myself to the few most significant influences that even a faulty memory will never entirely suppress.

It was James B. Conant, then president of Harvard University, who first introduced me to the history of science and thus initiated the transformation in my conception of the nature of scientific advance. Ever since that process began, he has been generous of his ideas, criticisms, and time—including the time required to read and suggest important changes in the draft of my manuscript. Leonard K. Nash, with whom for five years I taught the historically oriented course that Dr. Conant had started, was an even more active collaborator during the years when my ideas first began to take shape, and he has been much missed during the later stages of their development. Fortunately, however, after my departure from Cambridge, his place as creative sounding board and more was assumed by my Berkeley colleague, Stanley Cavell. That Cavell, a philosopher mainly concerned with ethics and aesthetics, should have reached conclusions quite so congruent to my own has been a constant source of stimulation and encouragement to me. He is, furthermore, the only person with whom I have ever been able to explore my ideas in incomplete sentences. That mode of communication attests an understanding that has enabled him to point me the way through or around several major barriers encountered while preparing my first manuscript.

Since that version was drafted, many other friends have helped with its reformulation. They will, I think, forgive me if I name only the four whose contributions proved most far-reaching and decisive: Paul K. Feyerabend of Berkeley, Ernest Nagel of Columbia, H. Pierre Noyes of the Lawrence Radiation Laboratory, and my student, John L. Heilbron, who has often worked closely with me in preparing a final version for the press. I have found all their reservations and suggestions extremely helpful, but I have no reason to believe (and some reason to doubt) that either they or the others mentioned above approve in its entirety the manuscript that results.

My final acknowledgments, to my parents, wife, and children, must be of a rather different sort. In ways which I shall probably be the last to recognize, each of them, too, has contributed intellectual ingredients to my work. But they have also, in varying degrees, done something more important. They have, that is, let it go on and even encouraged my devotion to it. Anyone who has wrestled with a project like mine will recognize what it has occasionally cost them. I do not know how to give them thanks.

T. S. K.

BERKELEY, CALIFORNIA
February 1962

I. Introduction: A Role for History

History, if viewed as a repository for more than anecdote or chronology, could produce a decisive transformation in the image of science by which we are now possessed. That image has previously been drawn, even by scientists themselves, mainly from the study of finished scientific achievements as these are recorded in the classics and, more recently, in the textbooks from which each new scientific generation learns to practice its trade. Inevitably, however, the aim of such books is persuasive and pedagogic; a concept of science drawn from them is no more likely to fit the enterprise that produced them than an image of a national culture drawn from a tourist brochure or a language text. This essay attempts to show that we have been misled by them in fundamental ways. Its aim is a sketch of the quite different concept of science that can emerge from the historical record of the research activity itself.

Even from history, however, that new concept will not be forthcoming if historical data continue to be sought and scrutinized mainly to answer questions posed by the unhistorical stereotype drawn from science texts. Those texts have, for example, often seemed to imply that the content of science is uniquely exemplified by the observations, laws, and theories described in their pages. Almost as regularly, the same books have been read as saying that scientific methods are simply the ones illustrated by the manipulative techniques used in gathering textbook data, together with the logical operations employed when relating those data to the textbook's theoretical generalizations. The result has been a concept of science with profound implications about its nature and development.

If science is the constellation of facts, theories, and methods collected in current texts, then scientists are the men who, successfully or not, have striven to contribute one or another element to that particular constellation. Scientific development becomes the piecemeal process by which these items have been added, singly and in combination, to the ever growing stockpile that constitutes scientific technique and knowledge. And history of science becomes the discipline that chronicles both these successive increments and the obstacles that have inhibited their accumulation. Concerned with scientific development, the historian then appears to have two main tasks. On the one hand, he must determine by what man and at what point in time each contemporary scientific fact, law, and theory was discovered or invented. On the other, he must describe and explain the congeries of error, myth, and superstition that have inhibited the more rapid accumulation of the constituents of the modern science text. Much research has been directed to these ends, and some still is.

In recent years, however, a few historians of science have been finding it more and more difficult to fulfil the functions that the concept of development-by-accumulation assigns to them. As chroniclers of an incremental process, they discover that additional research makes it harder, not easier, to answer questions like: When was oxygen discovered? Who first conceived of energy conservation? Increasingly, a few of them suspect that these are simply the wrong sorts of questions to ask. Perhaps science does not develop by the accumulation of individual discoveries and inventions. Simultaneously, these same historians confront growing difficulties in distinguishing the "scientific" component of past observation and belief from what their predecessors had readily labeled "error" and "superstition." The more carefully they study, say, Aristotelian dynamics, phlogistic chemistry, or caloric thermodynamics, the more certain they feel that those once current views of nature were, as a whole, neither less scientific nor more the product of human idiosyncrasy than those current today. If these out-of-date beliefs are to be called myths, then myths can be produced by the same sorts of methods and held for the same sorts of reasons that now lead to scientific knowledge. If, on the other hand, they are to be called science, then science has included bodies of belief quite incompatible with the ones we hold today. Given these alternatives, the historian must choose the latter. Out-of-date theories are not in principle unscientific because they have been discarded. That choice, however, makes it difficult to see scientific development as a process of accretion. The same historical research that displays the difficulties in isolating individual inventions and discoveries gives ground for profound doubts about the cumulative process through which these individual contributions to science were thought to have been compounded.

The result of all these doubts and difficulties is a historiographic revolution in the study of science, though one that is still in its early stages. Gradually, and often without entirely realizing they are doing so, historians of science have begun to ask new sorts of questions and to trace different, and often less than cumulative, developmental lines for the sciences. Rather than seeking the permanent contributions of an older science to our present vantage, they attempt to display the historical integrity of that science in its own time. They ask, for example, not about the relation of Galileo's views to those of modern science, but rather about the relationship between his views and those of his group, i.e., his teachers, contemporaries, and immediate successors in the sciences. Furthermore, they insist upon studying the opinions of that group and other similar ones from the viewpoint—usually very different from that of modern science—that gives those opinions the maximum internal coherence and the closest possible fit to nature. Seen through the works that result, works perhaps best exemplified in the writings of Alexandre Koyré, science does not seem altogether the same enterprise as the one discussed by writers in the older historiographic tradition. By implication, at least, these historical studies suggest the possibility of a new image of science. This essay aims to delineate that image by making explicit some of the new historiography's implications.

What aspects of science will emerge to prominence in the course of this effort? First, at least in order of presentation, is the insufficiency of methodological directives, by themselves, to dictate a unique substantive conclusion to many sorts of scientific questions. Instructed to examine electrical or chemical phenomena, the man who is ignorant of these fields but who knows what it is to be scientific may legitimately reach any one of a number of incompatible conclusions. Among those legitimate possibilities, the particular conclusions he does arrive at are probably determined by his prior experience in other fields, by the accidents of his investigation, and by his own individual makeup. What beliefs about the stars, for example, does he bring to the study of chemistry or electricity? Which of the many conceivable experiments relevant to the new field does he elect to perform first? And what aspects of the complex phenomenon that then results strike him as particularly relevant to an elucidation of the nature of chemical change or of electrical affinity? For the individual, at least, and sometimes for the scientific community as well, answers to questions like these are often essential determinants of scientific development. We shall note, for example, in Section II that the early developmental stages of most sciences have been characterized by continual competition between a number of distinct views of nature, each partially derived from, and all roughly compatible with, the dictates of scientific observation and method. What differentiated these various schools was not one or another failure of method—they were all "scientific"—but what we shall come to call their incommensurable ways of seeing the world and of practicing science in it. Observation and experience can and must drastically restrict the range of admissible scientific belief, else there would be no science. But they cannot alone determine a particular body of such belief. An apparently arbitrary element, compounded of personal and historical accident, is always a formative ingredient of the beliefs espoused by a given scientific community at a given time.

That element of arbitrariness does not, however, indicate that any scientific group could practice its trade without some set of received beliefs. Nor does it make less consequential the particular constellation to which the group, at a given time, is in fact committed. Effective research scarcely begins before a scientific community thinks it has acquired firm answers to questions like the following: What are the fundamental entities of which the universe is composed? How do these interact with each other and with the senses? What questions may legitimately be asked about such entities and what techniques employed in seeking solutions? At least in the mature sciences, answers (or full substitutes for answers) to questions like these are firmly embedded in the educational initiation that prepares and licenses the student for professional practice. Because that education is both rigorous and rigid, these answers come to exert a deep hold on the scientific mind. That they can do so does much to account both for the peculiar efficiency of the normal research activity and for the direction in which it proceeds at any given time. When examining normal science in Sections III, IV, and V, we shall want finally to describe that research as a strenuous and devoted attempt to force nature into the conceptual boxes supplied by professional education. Simultaneously, we shall wonder whether research could proceed without such boxes, whatever the element of arbitrariness in their historic origins and, occasionally, in their subsequent development.

Yet that element of arbitrariness is present, and it too has an important effect on scientific development, one which will be examined in detail in Sections VI, VII, and VIII. Normal science, the activity in which most scientists inevitably spend almost all their time, is predicated on the assumption that the scientific community knows what the world is like. Much of the success of the enterprise derives from the community's willingness to defend that assumption, if necessary at considerable cost. Normal science, for example, often suppresses fundamental novelties because they are necessarily subversive of its basic commitments. Nevertheless, so long as those commitments retain an element of the arbitrary, the very nature of normal research ensures that novelty shall not be suppressed for very long. Sometimes a normal problem, one that ought to be solvable by known rules and procedures, resists the reiterated onslaught of the ablest members of the group within whose competence it falls. On other occasions a piece of equipment designed and constructed for the purpose of normal research fails to perform in the anticipated manner, revealing an anomaly that cannot, despite repeated effort, be aligned with professional expectation. In these and other ways besides, normal science repeatedly goes astray. And when it does—when, that is, the profession can no longer evade anomalies that subvert the existing tradition of scientific practice—then begin the extraordinary investigations that lead the profession at last to a new set of commitments, a new basis for the practice of science. The extraordinary episodes in which that shift of professional commitments occurs are the ones known in this essay as scientific revolutions. They are the tradition-shattering complements to the tradition-bound activity of normal science.

The most obvious examples of scientific revolutions are those famous episodes in scientific development that have often been labeled revolutions before. Therefore, in Sections IX and X, where the nature of scientific revolutions is first directly scrutinized, we shall deal repeatedly with the major turning points in scientific development associated with the names of Copernicus, Newton, Lavoisier, and Einstein. More clearly than most other episodes in the history of at least the physical sciences, these display what all scientific revolutions are about. Each of them necessitated the community's rejection of one time-honored scientific theory in favor of another incompatible with it. Each produced a consequent shift in the problems available for scientific scrutiny and in the standards by which the profession determined what should count as an admissible problem or as a legitimate problem-solution. And each transformed the scientific imagination in ways that we shall ultimately need to describe as a transformation of the world within which scientific work was done. Such changes, together with the controversies that almost always accompany them, are the defining characteristics of scientific revolutions.

These characteristics emerge with particular clarity from a study of, say, the Newtonian or the chemical revolution. It is, however, a fundamental thesis of this essay that they can also be retrieved from the study of many other episodes that were not so obviously revolutionary. For the far smaller professional group affected by them, Maxwell's equations were as revolutionary as Einstein's, and they were resisted accordingly. The invention of other new theories regularly, and appropriately, evokes the same response from some of the specialists on whose area of special competence they impinge. For these men the new theory implies a change in the rules governing the prior practice of normal science. Inevitably, therefore, it reflects upon much scientific work they have already successfully completed. That is why a new theory, however special its range of application, is seldom or never just an increment to what is already known. Its assimilation requires the reconstruction of prior theory and the re-evaluation of prior fact, an intrinsically revolutionary process that is seldom completed by a single man and never overnight. No wonder historians have had difficulty in dating precisely this extended process that their vocabulary impels them to view as an isolated event.

Nor are new inventions of theory the only scientific events that have revolutionary impact upon the specialists in whose domain they occur. The commitments that govern normal science specify not only what sorts of entities the universe does contain, but also, by implication, those that it does not. It follows, though the point will require extended discussion, that a discovery like that of oxygen or X-rays does not simply add one more item to the population of the scientist's world. Ultimately it has that effect, but not until the professional community has re-evaluated traditional experimental procedures, altered its conception of entities with which it has long been familiar, and, in the process, shifted the network of theory through which it deals with the world. Scientific fact and theory are not categorically separable, except perhaps within a single tradition of normal-scientific practice. That is why the unexpected discovery is not simply factual in its import and why the scientist's world is qualitatively transformed as well as quantitatively enriched by fundamental novelties of either fact or theory.

This extended conception of the nature of scientific revolutions is the one delineated in the pages that follow. Admittedly the extension strains customary usage. Nevertheless, I shall continue to speak even of discoveries as revolutionary, because it is just the possibility of relating their structure to that of, say, the Copernican revolution that makes the extended conception seem to me so important. The preceding discussion indicates how the complementary notions of normal science and of scientific revolutions will be developed in the nine sections immediately to follow. The rest of the essay attempts to dispose of three remaining central questions. Section XI, by discussing the textbook tradition, considers why scientific revolutions have previously been so difficult to see. Section XII describes the revolutionary competition between the proponents of the old normal-scientific tradition and the adherents of the new one. It thus considers the process that should somehow, in a theory of scientific inquiry, replace the confirmation or falsification procedures made familiar by our usual image of science. Competition between segments of the scientific community is the only historical process that ever actually results in the rejection of one previously accepted theory or in the adoption of another. Finally, Section XIII will ask how development through revolutions can be compatible with the apparently unique character of scientific progress. For that question, however, this essay will provide no more than the main outlines of an answer, one which depends upon characteristics of the scientific community that require much additional exploration and study.

Undoubtedly, some readers will already have wondered whether historical study can possibly effect the sort of conceptual transformation aimed at here. An entire arsenal of dichotomies is available to suggest that it cannot properly do so. History, we too often say, is a purely descriptive discipline. The theses suggested above are, however, often interpretive and sometimes normative. Again, many of my generalizations are about the sociology or social psychology of scientists; yet at least a few of my conclusions belong traditionally to logic or epistemology. In the preceding paragraph I may even seem to have violated the very influential contemporary distinction between "the context of discovery" and "the context of justification." Can anything more than profound confusion be indicated by this admixture of diverse fields and concerns?

Having been weaned intellectually on these distinctions and others like them, I could scarcely be more aware of their import and force. For many years I took them to be about the nature of knowledge, and I still suppose that, appropriately recast, they have something important to tell us. Yet my attempts to apply them, even *grosso modo,* to the actual situations in which knowledge is gained, accepted, and assimilated have made them seem extraordinarily problematic. Rather than being elementary logical or methodological distinctions, which would thus be prior to the analysis of scientific knowledge, they now seem integral parts of a traditional set of substantive answers to the very questions upon which they have been deployed. That circularity does not at all invalidate them. But it does make them parts of a theory and, by doing so, subjects them to the same scrutiny regularly applied to theories in other fields. If they are to have more than pure abstraction as their content, then that content must be discovered by observing them in application to the data they are meant to elucidate. How could history of science fail to be a source of phenomena to which theories about knowledge may legitimately be asked to apply?

II. The Route to Normal Science

In this essay, 'normal science' means research firmly based upon one or more past scientific achievements, achievements that some particular scientific community acknowledges for a time as supplying the foundation for its further practice. Today such achievements are recounted, though seldom in their original form, by science textbooks, elementary and advanced. These textbooks expound the body of accepted theory, illustrate many or all of its successful applications, and compare these applications with exemplary observations and experiments. Before such books became popular early in the nineteenth century (and until even more recently in the newly matured sciences), many of the famous classics of science fulfilled a similar function. Aristotle's *Physica,* Ptolemy's *Almagest,* Newton's *Principia* and *Opticks,* Franklin's *Electricity,* Lavoisier's *Chemistry,* and Lyell's *Geology*—these and many other works served for a time implicitly to define the legitimate problems and methods of a research field for succeeding generations of practitioners. They were able to do so because they shared two essential characteristics. Their achievement was sufficiently unprecedented to attract an enduring group of adherents away from competing modes of scientific activity. Simultaneously, it was sufficiently open-ended to leave all sorts of problems for the redefined group of practitioners to resolve.

Achievements that share these two characteristics I shall henceforth refer to as 'paradigms,' a term that relates closely to 'normal science.' By choosing it, I mean to suggest that some accepted examples of actual scientific practice—examples which include law, theory, application, and instrumentation together—provide models from which spring particular coherent traditions of scientific research. These are the traditions which the historian describes under such rubrics as 'Ptolemaic astronomy' (or 'Copernican'), 'Aristotelian dynamics' (or 'Newtonian'), 'corpuscular optics' (or 'wave optics'), and so on. The study of paradigms, including many that are far more specialized than those named illustratively above, is what mainly prepares the student for membership in the particular scientific community with which he will later practice. Because he there joins men who learned the bases of their field from the same concrete models, his subsequent practice will seldom evoke overt disagreement over fundamentals. Men whose research is based on shared paradigms are committed to the same rules and standards for scientific practice. That commitment and the apparent consensus it produces are prerequisites for normal science, i.e., for the genesis and continuation of a particular research tradition.

Because in this essay the concept of a paradigm will often substitute for a variety of familiar notions, more will need to be said about the reasons for its introduction. Why is the concrete scientific achievement, as a locus of professional commitment, prior to the various concepts, laws, theories, and points of view that may be abstracted from it? In what sense is the shared paradigm a fundamental unit for the student of scientific development, a unit that cannot be fully reduced to logically atomic components which might function in its stead? When we encounter them in Section V, answers to these questions and to others like them will prove basic to an understanding both of normal science and of the associated concept of paradigms. That more abstract discussion will depend, however, upon a previous exposure to examples of normal science or of paradigms in operation. In particular, both these related concepts will be clarified by noting that there can be a sort of scientific research without paradigms, or at least without any so unequivocal and so binding as the ones named above. Acquisition of a paradigm and of the more esoteric type of research it permits is a sign of maturity in the development of any given scientific field.

If the historian traces the scientific knowledge of any selected group of related phenomena backward in time, he is likely to encounter some minor variant of a pattern here illustrated from the history of physical optics. Today's physics textbooks tell the student that light is photons, i.e., quantum-mechanical entities that exhibit some characteristics of waves and some of particles. Research proceeds accordingly, or rather according to the more elaborate and mathematical characterization from which this usual verbalization is derived. That characterization of light is, however, scarcely half a century old. Before it was developed by Planck, Einstein, and others early in this century, physics texts taught that light was transverse wave motion, a conception rooted in a paradigm that derived ultimately from the optical writings of Young and Fresnel in the early nineteenth century. Nor was the wave theory the first to be embraced by almost all practitioners of optical science. During the eighteenth century the paradigm for this field was provided by Newton's *Opticks,* which taught that light was material corpuscles. At that time physicists sought evidence, as the early wave theorists had not, of the pressure exerted by light particles impinging on solid bodies.[1]

These transformations of the paradigms of physical optics are scientific revolutions, and the successive transition from one paradigm to another via revolution is the usual developmental pattern of mature science. It is not, however, the pattern characteristic of the period before Newton's work, and that is the contrast that concerns us here. No period between remote antiquity and the end of the seventeenth century exhibited a single generally accepted view about the nature of light. Instead there were a number of competing schools and subschools, most of them espousing one variant or another of Epicurean, Aristotelian, or Platonic theory. One group took light to be particles emanating from material bodies; for another it was a modification of the medium that intervened between the body and the eye; still another explained light in terms of an interaction of the medium with an emanation from the eye; and there were other combinations and modifications besides. Each of the corresponding schools derived strength from its relation to some particular metaphysic, and each emphasized, as para-

[1] Joseph Priestley, *The History and Present State of Discoveries Relating to Vision, Light, and Colours* (London, 1772), pp. 385–90.

digmatic observations, the particular cluster of optical phenomena that its own theory could do most to explain. Other observations were dealt with by *ad hoc* elaborations, or they remained as outstanding problems for further research.[2]

At various times all these schools made significant contributions to the body of concepts, phenomena, and techniques from which Newton drew the first nearly uniformly accepted paradigm for physical optics. Any definition of the scientist that excludes at least the more creative members of these various schools will exclude their modern successors as well. Those men were scientists. Yet anyone examining a survey of physical optics before Newton may well conclude that, though the field's practitioners were scientists, the net result of their activity was something less than science. Being able to take no common body of belief for granted, each writer on physical optics felt forced to build his field anew from its foundations. In doing so, his choice of supporting observation and experiment was relatively free, for there was no standard set of methods or of phenomena that every optical writer felt forced to employ and explain. Under these circumstances, the dialogue of the resulting books was often directed as much to the members of other schools as it was to nature. That pattern is not unfamiliar in a number of creative fields today, nor is it incompatible with significant discovery and invention. It is not, however, the pattern of development that physical optics acquired after Newton and that other natural sciences make familiar today.

The history of electrical research in the first half of the eighteenth century provides a more concrete and better known example of the way a science develops before it acquires its first universally received paradigm. During that period there were almost as many views about the nature of electricity as there were important electrical experimenters, men like Hauksbee, Gray, Desaguliers, Du Fay, Nollett, Watson, Franklin, and others. All their numerous concepts of electricity had something in common—they were partially derived from one or an-

[2] Vasco Ronchi, *Histoire de la lumière*, trans. Jean Taton (Paris, 1956), chaps. i–iv.

other version of the mechanico-corpuscular philosophy that guided all scientific research of the day. In addition, all were components of real scientific theories, of theories that had been drawn in part from experiment and observation and that partially determined the choice and interpretation of additional problems undertaken in research. Yet though all the experiments were electrical and though most of the experimenters read each other's works, their theories had no more than a family resemblance.[3]

One early group of theories, following seventeenth-century practice, regarded attraction and frictional generation as the fundamental electrical phenomena. This group tended to treat repulsion as a secondary effect due to some sort of mechanical rebounding and also to postpone for as long as possible both discussion and systematic research on Gray's newly discovered effect, electrical conduction. Other "electricians" (the term is their own) took attraction and repulsion to be equally elementary manifestations of electricity and modified their theories and research accordingly. (Actually, this group is remarkably small—even Franklin's theory never quite accounted for the mutual repulsion of two negatively charged bodies.) But they had as much difficulty as the first group in accounting simultaneously for any but the simplest conduction effects. Those effects, however, provided the starting point for still a third group, one which tended to speak of electricity as a "fluid" that could run through conductors rather than as an "effluvium" that emanated from non-conductors. This group, in its turn, had difficulty reconciling its theory with a number of attractive and

[3] Duane Roller and Duane H. D. Roller, *The Development of the Concept of Electric Charge: Electricity from the Greeks to Coulomb* ("Harvard Case Histories in Experimental Science," Case 8; Cambridge, Mass., 1954); and I. B. Cohen, *Franklin and Newton: An Inquiry into Speculative Newtonian Experimental Science and Franklin's Work in Electricity as an Example Thereof* (Philadelphia, 1956), chaps. vii–xii. For some of the analytic detail in the paragraph that follows in the text, I am indebted to a still unpublished paper by my student John L. Heilbron. Pending its publication, a somewhat more extended and more precise account of the emergence of Franklin's paradigm is included in T. S. Kuhn, "The Function of Dogma in Scientific Research," in A. C. Crombie (ed.), "Symposium on the History of Science, University of Oxford, July 9–15, 1961," to be published by Heinemann Educational Books, Ltd.

repulsive effects. Only through the work of Franklin and his immediate successors did a theory arise that could account with something like equal facility for very nearly all these effects and that therefore could and did provide a subsequent generation of "electricians" with a common paradigm for its research.

Excluding those fields, like mathematics and astronomy, in which the first firm paradigms date from prehistory and also those, like biochemistry, that arose by division and recombination of specialties already matured, the situations outlined above are historically typical. Though it involves my continuing to employ the unfortunate simplification that tags an extended historical episode with a single and somewhat arbitrarily chosen name (e.g., Newton or Franklin), I suggest that similar fundamental disagreements characterized, for example, the study of motion before Aristotle and of statics before Archimedes, the study of heat before Black, of chemistry before Boyle and Boerhaave, and of historical geology before Hutton. In parts of biology—the study of heredity, for example—the first universally received paradigms are still more recent; and it remains an open question what parts of social science have yet acquired such paradigms at all. History suggests that the road to a firm research consensus is extraordinarily arduous.

History also suggests, however, some reasons for the difficulties encountered on that road. In the absence of a paradigm or some candidate for paradigm, all of the facts that could possibly pertain to the development of a given science are likely to seem equally relevant. As a result, early fact-gathering is a far more nearly random activity than the one that subsequent scientific development makes familiar. Furthermore, in the absence of a reason for seeking some particular form of more recondite information, early fact-gathering is usually restricted to the wealth of data that lie ready to hand. The resulting pool of facts contains those accessible to casual observation and experiment together with some of the more esoteric data retrievable from established crafts like medicine, calendar making, and metallurgy. Because the crafts are one readily accessible source of facts that could not have been casually discovered, technology has often played a vital role in the emergence of new sciences.

But though this sort of fact-collecting has been essential to the origin of many significant sciences, anyone who examines, for example, Pliny's encyclopedic writings or the Baconian natural histories of the seventeenth century will discover that it produces a morass. One somehow hesitates to call the literature that results scientific. The Baconian "histories" of heat, color, wind, mining, and so on, are filled with information, some of it recondite. But they juxtapose facts that will later prove revealing (e.g., heating by mixture) with others (e.g., the warmth of dung heaps) that will for some time remain too complex to be integrated with theory at all.[4] In addition, since any description must be partial, the typical natural history often omits from its immensely circumstantial accounts just those details that later scientists will find sources of important illumination. Almost none of the early "histories" of electricity, for example, mention that chaff, attracted to a rubbed glass rod, bounces off again. That effect seemed mechanical, not electrical.[5] Moreover, since the casual fact-gatherer seldom possesses the time or the tools to be critical, the natural histories often juxtapose descriptions like the above with others, say, heating by antiperistasis (or by cooling), that we are now quite unable to confirm.[6] Only very occasionally, as in the cases of ancient statics, dynamics, and geometrical optics, do facts collected with so little guidance from pre-established theory speak with sufficient clarity to permit the emergence of a first paradigm.

This is the situation that creates the schools characteristic of the early stages of a science's development. No natural history can be interpreted in the absence of at least some implicit body

[4] Compare the sketch for a natural history of heat in Bacon's *Novum Organum*, Vol. VIII of *The Works of Francis Bacon*, ed. J. Spedding, R. L. Ellis, and D. D. Heath (New York, 1869), pp. 179–203.

[5] Roller and Roller, *op. cit.*, pp. 14, 22, 28, 43. Only after the work recorded in the last of these citations do repulsive effects gain general recognition as unequivocally electrical.

[6] Bacon, *op. cit.*, pp. 235, 337, says, "Water slightly warm is more easily frozen than quite cold." For a partial account of the earlier history of this strange observation, see Marshall Clagett, *Giovanni Marliani and Late Medieval Physics* (New York, 1941), chap. iv.

of intertwined theoretical and methodological belief that permits selection, evaluation, and criticism. If that body of belief is not already implicit in the collection of facts—in which case more than "mere facts" are at hand—it must be externally supplied, perhaps by a current metaphysic, by another science, or by personal and historical accident. No wonder, then, that in the early stages of the development of any science different men confronting the same range of phenomena, but not usually all the same particular phenomena, describe and interpret them in different ways. What is surprising, and perhaps also unique in its degree to the fields we call science, is that such initial divergences should ever largely disappear.

For they do disappear to a very considerable extent and then apparently once and for all. Furthermore, their disappearance is usually caused by the triumph of one of the pre-paradigm schools, which, because of its own characteristic beliefs and preconceptions, emphasized only some special part of the too sizable and inchoate pool of information. Those electricians who thought electricity a fluid and therefore gave particular emphasis to conduction provide an excellent case in point. Led by this belief, which could scarcely cope with the known multiplicity of attractive and repulsive effects, several of them conceived the idea of bottling the electrical fluid. The immediate fruit of their efforts was the Leyden jar, a device which might never have been discovered by a man exploring nature casually or at random, but which was in fact independently developed by at least two investigators in the early 1740's.[7] Almost from the start of his electrical researches, Franklin was particularly concerned to explain that strange and, in the event, particularly revealing piece of special apparatus. His success in doing so provided the most effective of the arguments that made his theory a paradigm, though one that was still unable to account for quite all the known cases of electrical repulsion.[8] To be accepted as a paradigm, a theory must seem better than its competitors, but

[7] Roller and Roller, *op. cit.*, pp. 51–54.

[8] The troublesome case was the mutual repulsion of negatively charged bodies, for which see Cohen, *op. cit.*, pp. 491–94, 531–43.

it need not, and in fact never does, explain all the facts with which it can be confronted.

What the fluid theory of electricity did for the subgroup that held it, the Franklinian paradigm later did for the entire group of electricians. It suggested which experiments would be worth performing and which, because directed to secondary or to overly complex manifestations of electricity, would not. Only the paradigm did the job far more effectively, partly because the end of interschool debate ended the constant reiteration of fundamentals and partly because the confidence that they were on the right track encouraged scientists to undertake more precise, esoteric, and consuming sorts of work.[9] Freed from the concern with any and all electrical phenomena, the united group of electricians could pursue selected phenomena in far more detail, designing much special equipment for the task and employing it more stubbornly and systematically than electricians had ever done before. Both fact collection and theory articulation became highly directed activities. The effectiveness and efficiency of electrical research increased accordingly, providing evidence for a societal version of Francis Bacon's acute methodological dictum: "Truth emerges more readily from error than from confusion."[10]

We shall be examining the nature of this highly directed or paradigm-based research in the next section, but must first note briefly how the emergence of a paradigm affects the structure of the group that practices the field. When, in the development of a natural science, an individual or group first produces a synthesis able to attract most of the next generation's practitioners, the older schools gradually disappear. In part their disappear-

[9] It should be noted that the acceptance of Franklin's theory did not end quite all debate. In 1759 Robert Symmer proposed a two-fluid version of that theory, and for many years thereafter electricians were divided about whether electricity was a single fluid or two. But the debates on this subject only confirm what has been said above about the manner in which a universally recognized achievement unites the profession. Electricians, though they continued divided on this point, rapidly concluded that no experimental tests could distinguish the two versions of the theory and that they were therefore equivalent. After that, both schools could and did exploit all the benefits that the Franklinian theory provided (*ibid.*, pp. 543–46, 548–54).

[10] Bacon, *op. cit.*, p. 210.

ance is caused by their members' conversion to the new paradigm. But there are always some men who cling to one or another of the older views, and they are simply read out of the profession, which thereafter ignores their work. The new paradigm implies a new and more rigid definition of the field. Those unwilling or unable to accommodate their work to it must proceed in isolation or attach themselves to some other group.[11] Historically, they have often simply stayed in the departments of philosophy from which so many of the special sciences have been spawned. As these indications hint, it is sometimes just its reception of a paradigm that transforms a group previously interested merely in the study of nature into a profession or, at least, a discipline. In the sciences (though not in fields like medicine, technology, and law, of which the principal *raison d'être* is an external social need), the formation of specialized journals, the foundation of specialists' societies, and the claim for a special place in the curriculum have usually been associated with a group's first reception of a single paradigm. At least this was the case between the time, a century and a half ago, when the institutional pattern of scientific specialization first developed and the very recent time when the paraphernalia of specialization acquired a prestige of their own.

The more rigid definition of the scientific group has other consequences. When the individual scientist can take a paradigm for granted, he need no longer, in his major works, attempt to build his field anew, starting from first principles and justify-

[11] The history of electricity provides an excellent example which could be duplicated from the careers of Priestley, Kelvin, and others. Franklin reports that Nollet, who at mid-century was the most influential of the Continental electricians, "lived to see himself the last of his Sect, except Mr. B.—his Eleve and immediate Disciple" (Max Farrand [ed.], *Benjamin Franklin's Memoirs* [Berkeley, Calif., 1949], pp. 384–86). More interesting, however, is the endurance of whole schools in increasing isolation from professional science. Consider, for example, the case of astrology, which was once an integral part of astronomy. Or consider the continuation in the late eighteenth and early nineteenth centuries of a previously respected tradition of "romantic" chemistry. This is the tradition discussed by Charles C. Gillispie in "The *Encyclopédie* and the Jacobin Philosophy of Science: A Study in Ideas and Consequences," *Critical Problems in the History of Science*, ed. Marshall Clagett (Madison, Wis., 1959), pp. 255–89; and "The Formation of Lamarck's Evolutionary Theory," *Archives internationales d'histoire des sciences*, XXXVII (1956), 323–38.

ing the use of each concept introduced. That can be left to the writer of textbooks. Given a textbook, however, the creative scientist can begin his research where it leaves off and thus concentrate exclusively upon the subtlest and most esoteric aspects of the natural phenomena that concern his group. And as he does this, his research communiqués will begin to change in ways whose evolution has been too little studied but whose modern end products are obvious to all and oppressive to many. No longer will his researches usually be embodied in books addressed, like Franklin's *Experiments . . . on Electricity* or Darwin's *Origin of Species,* to anyone who might be interested in the subject matter of the field. Instead they will usually appear as brief articles addressed only to professional colleagues, the men whose knowledge of a shared paradigm can be assumed and who prove to be the only ones able to read the papers addressed to them.

Today in the sciences, books are usually either texts or retrospective reflections upon one aspect or another of the scientific life. The scientist who writes one is more likely to find his professional reputation impaired than enhanced. Only in the earlier, pre-paradigm, stages of the development of the various sciences did the book ordinarily possess the same relation to professional achievement that it still retains in other creative fields. And only in those fields that still retain the book, with or without the article, as a vehicle for research communication are the lines of professionalization still so loosely drawn that the layman may hope to follow progress by reading the practitioners' original reports. Both in mathematics and astronomy, research reports had ceased already in antiquity to be intelligible to a generally educated audience. In dynamics, research became similarly esoteric in the later Middle Ages, and it recaptured general intelligibility only briefly during the early seventeenth century when a new paradigm replaced the one that had guided medieval research. Electrical research began to require translation for the layman before the end of the eighteenth century, and most other fields of physical science ceased to be generally accessible in the nineteenth. During the same two centuries similar transitions can be isolated in the various parts of the biological sciences. In parts of the social sciences they may well be occurring today. Although it has become customary, and is surely proper, to deplore the widening gulf that separates the professional scientist from his colleagues in other fields, too little attention is paid to the essential relationship between that gulf and the mechanisms intrinsic to scientific advance.

Ever since prehistoric antiquity one field of study after another has crossed the divide between what the historian might call its prehistory as a science and its history proper. These transitions to maturity have seldom been so sudden or so unequivocal as my necessarily schematic discussion may have implied. But neither have they been historically gradual, coextensive, that is to say, with the entire development of the fields within which they occurred. Writers on electricity during the first four decades of the eighteenth century possessed far more information about electrical phenomena than had their sixteenth-century predecessors. During the half-century after 1740, few new sorts of electrical phenomena were added to their lists. Nevertheless, in important respects, the electrical writings of Cavendish, Coulomb, and Volta in the last third of the eighteenth century seem further removed from those of Gray, Du Fay, and even Franklin than are the writings of these early eighteenth-century electrical discoverers from those of the sixteenth century.[12] Sometime between 1740 and 1780, electricians were for the first time enabled to take the foundations of their field for granted. From that point they pushed on to more concrete and recondite problems, and increasingly they then reported their results in articles addressed to other electricians rather than in books addressed to the learned world at large. As a group they achieved what had been gained by astronomers in antiquity

[12] The post-Franklinian developments include an immense increase in the sensitivity of charge detectors, the first reliable and generally diffused techniques for measuring charge, the evolution of the concept of capacity and its relation to a newly refined notion of electric tension, and the quantification of electrostatic force. On all of these see Roller and Roller, *op. cit.*, pp. 66–81; W. C. Walker, "The Detection and Estimation of Electric Charges in the Eighteenth Century," *Annals of Science*, I (1936), 66–100; and Edmund Hoppe, *Geschichte der Elektrizität* (Leipzig, 1884), Part I, chaps. iii–iv.

and by students of motion in the Middle Ages, of physical optics in the late seventeenth century, and of historical geology in the early nineteenth. They had, that is, achieved a paradigm that proved able to guide the whole group's research. Except with the advantage of hindsight, it is hard to find another criterion that so clearly proclaims a field a science.

III. The Nature of Normal Science

What then is the nature of the more professional and esoteric research that a group's reception of a single paradigm permits? If the paradigm represents work that has been done once and for all, what further problems does it leave the united group to resolve? Those questions will seem even more urgent if we now note one respect in which the terms used so far may be misleading. In its established usage, a paradigm is an accepted model or pattern, and that aspect of its meaning has enabled me, lacking a better word, to appropriate 'paradigm' here. But it will shortly be clear that the sense of 'model' and 'pattern' that permits the appropriation is not quite the one usual in defining 'paradigm.' In grammar, for example, '*amo, amas, amat*' is a paradigm because it displays the pattern to be used in conjugating a large number of other Latin verbs, e.g., in producing '*laudo, laudas, laudat.*' In this standard application, the paradigm functions by permitting the replication of examples any one of which could in principle serve to replace it. In a science, on the other hand, a paradigm is rarely an object for replication. Instead, like an accepted judicial decision in the common law, it is an object for further articulation and specification under new or more stringent conditions.

To see how this can be so, we must recognize how very limited in both scope and precision a paradigm can be at the time of its first appearance. Paradigms gain their status because they are more successful than their competitors in solving a few problems that the group of practitioners has come to recognize as acute. To be more successful is not, however, to be either completely successful with a single problem or notably successful with any large number. The success of a paradigm—whether Aristotle's analysis of motion, Ptolemy's computations of planetary position, Lavoisier's application of the balance, or Maxwell's mathematization of the electromagnetic field—is at the start largely a promise of success discoverable in selected and still incomplete examples. Normal science consists in the actualization of that promise, an actualization achieved by extending the knowledge of those facts that the paradigm displays as particularly revealing, by increasing the extent of the match between those facts and the paradigm's predictions, and by further articulation of the paradigm itself.

Few people who are not actually practitioners of a mature science realize how much mop-up work of this sort a paradigm leaves to be done or quite how fascinating such work can prove in the execution. And these points need to be understood. Mopping-up operations are what engage most scientists throughout their careers. They constitute what I am here calling normal science. Closely examined, whether historically or in the contemporary laboratory, that enterprise seems an attempt to force nature into the preformed and relatively inflexible box that the paradigm supplies. No part of the aim of normal science is to call forth new sorts of phenomena; indeed those that will not fit the box are often not seen at all. Nor do scientists normally aim to invent new theories, and they are often intolerant of those invented by others.[1] Instead, normal-scientific research is directed to the articulation of those phenomena and theories that the paradigm already supplies.

Perhaps these are defects. The areas investigated by normal science are, of course, minuscule; the enterprise now under discussion has drastically restricted vision. But those restrictions, born from confidence in a paradigm, turn out to be essential to the development of science. By focusing attention upon a small range of relatively esoteric problems, the paradigm forces scientists to investigate some part of nature in a detail and depth that would otherwise be unimaginable. And normal science possesses a built-in mechanism that ensures the relaxation of the restrictions that bound research whenever the paradigm from which they derive ceases to function effectively. At that point scientists begin to behave differently, and the nature of their research problems changes. In the interim, however, during the

[1] Bernard Barber, "Resistance by Scientists to Scientific Discovery," *Science*, CXXXIV (1961), 596–602.

period when the paradigm is successful, the profession will have solved problems that its members could scarcely have imagined and would never have undertaken without commitment to the paradigm. And at least part of that achievement always proves to be permanent.

To display more clearly what is meant by normal or paradigm-based research, let me now attempt to classify and illustrate the problems of which normal science principally consists. For convenience I postpone theoretical activity and begin with fact-gathering, that is, with the experiments and observations described in the technical journals through which scientists inform their professional colleagues of the results of their continuing research. On what aspects of nature do scientists ordinarily report? What determines their choice? And, since most scientific observation consumes much time, equipment, and money, what motivates the scientist to pursue that choice to a conclusion?

There are, I think, only three normal foci for factual scientific investigation, and they are neither always nor permanently distinct. First is that class of facts that the paradigm has shown to be particularly revealing of the nature of things. By employing them in solving problems, the paradigm has made them worth determining both with more precision and in a larger variety of situations. At one time or another, these significant factual determinations have included: in astronomy—stellar position and magnitude, the periods of eclipsing binaries and of planets; in physics—the specific gravities and compressibilities of materials, wave lengths and spectral intensities, electrical conductivities and contact potentials; and in chemistry—composition and combining weights, boiling points and acidity of solutions, structural formulas and optical activities. Attempts to increase the accuracy and scope with which facts like these are known occupy a significant fraction of the literature of experimental and observational science. Again and again complex special apparatus has been designed for such purposes, and the invention, construction, and deployment of that apparatus have demanded first-rate talent, much time, and considerable financial backing. Synchrotrons and radiotelescopes are only the most recent examples of the lengths to which research workers will go if a paradigm assures them that the facts they seek are important. From Tycho Brahe to E. O. Lawrence, some scientists have acquired great reputations, not from any novelty of their discoveries, but from the precision, reliability, and scope of the methods they developed for the redetermination of a previously known sort of fact.

A second usual but smaller class of factual determinations is directed to those facts that, though often without much intrinsic interest, can be compared directly with predictions from the paradigm theory. As we shall see shortly, when I turn from the experimental to the theoretical problems of normal science, there are seldom many areas in which a scientific theory, particularly if it is cast in a predominantly mathematical form, can be directly compared with nature. No more than three such areas are even yet accessible to Einstein's general theory of relativity.[2] Furthermore, even in those areas where application is possible, it often demands theoretical and instrumental approximations that severely limit the agreement to be expected. Improving that agreement or finding new areas in which agreement can be demonstrated at all presents a constant challenge to the skill and imagination of the experimentalist and observer. Special telescopes to demonstrate the Copernican prediction of annual parallax; Atwood's machine, first invented almost a century after the *Principia*, to give the first unequivocal demonstration of Newton's second law; Foucault's apparatus to show that the speed of light is greater in air than in water; or the gigantic scintillation counter designed to demonstrate the existence of

[2] The only long-standing check point still generally recognized is the precession of Mercury's perihelion. The red shift in the spectrum of light from distant stars can be derived from considerations more elementary than general relativity, and the same may be possible for the bending of light around the sun, a point now in some dispute. In any case, measurements of the latter phenomenon remain equivocal. One additional check point may have been established very recently: the gravitational shift of Mossbauer radiation. Perhaps there will soon be others in this now active but long dormant field. For an up-to-date capsule account of the problem, see L. I. Schiff, "A Report on the NASA Conference on Experimental Tests of Theories of Relativity," *Physics Today,* XIV (1961), 42–48.

the neutrino—these pieces of special apparatus and many others like them illustrate the immense effort and ingenuity that have been required to bring nature and theory into closer and closer agreement.[3] That attempt to demonstrate agreement is a second type of normal experimental work, and it is even more obviously dependent than the first upon a paradigm. The existence of the paradigm sets the problem to be solved; often the paradigm theory is implicated directly in the design of apparatus able to solve the problem. Without the *Principia,* for example, measurements made with the Atwood machine would have meant nothing at all.

A third class of experiments and observations exhausts, I think, the fact-gathering activities of normal science. It consists of empirical work undertaken to articulate the paradigm theory, resolving some of its residual ambiguities and permitting the solution of problems to which it had previously only drawn attention. This class proves to be the most important of all, and its description demands its subdivision. In the more mathematical sciences, some of the experiments aimed at articulation are directed to the determination of physical constants. Newton's work, for example, indicated that the force between two unit masses at unit distance would be the same for all types of matter at all positions in the universe. But his own problems could be solved without even estimating the size of this attraction, the universal gravitational constant; and no one else devised apparatus able to determine it for a century after the *Principia* appeared. Nor was Cavendish's famous determination in the 1790's the last. Because of its central position in physical theory, improved values of the gravitational constant have been the object of repeated efforts ever since by a number of outstanding

[3] For two of the parallax telescopes, see Abraham Wolf, *A History of Science, Technology, and Philosophy in the Eighteenth Century* (2d ed.; London, 1952), pp. 103–5. For the Atwood machine, see N. R. Hanson, *Patterns of Discovery* (Cambridge, 1958), pp. 100–102, 207–8. For the last two pieces of special apparatus, see M. L. Foucault, "Méthode générale pour mesurer la vitesse de la lumière dans l'air et les milieux transparants. Vitesses relatives de la lumière dans l'air et dans l'eau . . . ," *Comptes rendus . . . de l'Académie des sciences,* XXX (1850), 551–60; and C. L. Cowan, Jr., *et al.,* "Detection of the Free Neutrino: A Confirmation," *Science,* CXXIV (1956), 103–4.

experimentalists.[4] Other examples of the same sort of continuing work would include determinations of the astronomical unit, Avogadro's number, Joule's coefficient, the electronic charge, and so on. Few of these elaborate efforts would have been conceived and none would have been carried out without a paradigm theory to define the problem and to guarantee the existence of a stable solution.

Efforts to articulate a paradigm are not, however, restricted to the determination of universal constants. They may, for example, also aim at quantitative laws: Boyle's Law relating gas pressure to volume, Coulomb's Law of electrical attraction, and Joule's formula relating heat generated to electrical resistance and current are all in this category. Perhaps it is not apparent that a paradigm is prerequisite to the discovery of laws like these. We often hear that they are found by examining measurements undertaken for their own sake and without theoretical commitment. But history offers no support for so excessively Baconian a method. Boyle's experiments were not conceivable (and if conceived would have received another interpretation or none at all) until air was recognized as an elastic fluid to which all the elaborate concepts of hydrostatics could be applied.[5] Coulomb's success depended upon his constructing special apparatus to measure the force between point charges. (Those who had previously measured electrical forces using ordinary pan balances, etc., had found no consistent or simple regularity at all.) But that design, in turn, depended upon the previous recognition that every particle of electric fluid acts upon every other at a distance. It was for the force between such particles—the only force which might safely be assumed

[4] J. H. P[oynting] reviews some two dozen measurements of the gravitational constant between 1741 and 1901 in "Gravitation Constant and Mean Density of the Earth," *Encyclopaedia Britannica* (11th ed.; Cambridge, 1910–11), XII, 385–89.

[5] For the full transplantation of hydrostatic concepts into pneumatics, see *The Physical Treatises of Pascal*, trans. I. H. B. Spiers and A. G. H. Spiers, with an introduction and notes by F. Barry (New York, 1937). Torricelli's original introduction of the parallelism ("We live submerged at the bottom of an ocean of the element air") occurs on p. 164. Its rapid development is displayed by the two main treatises.

a simple function of distance—that Coulomb was looking.[6] Joule's experiments could also be used to illustrate how quantitative laws emerge through paradigm articulation. In fact, so general and close is the relation between qualitative paradigm and quantitative law that, since Galileo, such laws have often been correctly guessed with the aid of a paradigm years before apparatus could be designed for their experimental determination.[7]

Finally, there is a third sort of experiment which aims to articulate a paradigm. More than the others this one can resemble exploration, and it is particularly prevalent in those periods and sciences that deal more with the qualitative than with the quantitative aspects of nature's regularity. Often a paradigm developed for one set of phenomena is ambiguous in its application to other closely related ones. Then experiments are necessary to choose among the alternative ways of applying the paradigm to the new area of interest. For example, the paradigm applications of the caloric theory were to heating and cooling by mixtures and by change of state. But heat could be released or absorbed in many other ways—e.g., by chemical combination, by friction, and by compression or absorption of a gas—and to each of these other phenomena the theory could be applied in several ways. If the vacuum had a heat capacity, for example, heating by compression could be explained as the result of mixing gas with void. Or it might be due to a change in the specific heat of gases with changing pressure. And there were several other explanations besides. Many experiments were undertaken to elaborate these various possibilities and to distinguish between them; all these experiments arose from the caloric theory as paradigm, and all exploited it in the design of experiments and in the interpretation of results.[8] Once the phe-

[6] Duane Roller and Duane H. D. Roller, *The Development of the Concept of Electric Charge: Electricity from the Greeks to Coulomb* ("Harvard Case Histories in Experimental Science," Case 8; Cambridge, Mass., 1954), pp. 66–80.

[7] For examples, see T. S. Kuhn, "The Function of Measurement in Modern Physical Science," *Isis,* LII (1961), 161–93.

[8] T. S. Kuhn, "The Caloric Theory of Adiabatic Compression," *Isis,* XLIX (1958), 132–40.

nomenon of heating by compression had been established, all further experiments in the area were paradigm-dependent in this way. Given the phenomenon, how else could an experiment to elucidate it have been chosen?

Turn now to the theoretical problems of normal science, which fall into very nearly the same classes as the experimental and observational. A part of normal theoretical work, though only a small part, consists simply in the use of existing theory to predict factual information of intrinsic value. The manufacture of astronomical ephemerides, the computation of lens characteristics, and the production of radio propagation curves are examples of problems of this sort. Scientists, however, generally regard them as hack work to be relegated to engineers or technicians. At no time do very many of them appear in significant scientific journals. But these journals do contain a great many theoretical discussions of problems that, to the non-scientist, must seem almost identical. These are the manipulations of theory undertaken, not because the predictions in which they result are intrinsically valuable, but because they can be confronted directly with experiment. Their purpose is to display a new application of the paradigm or to increase the precision of an application that has already been made.

The need for work of this sort arises from the immense difficulties often encountered in developing points of contact between a theory and nature. These difficulties can be briefly illustrated by an examination of the history of dynamics after Newton. By the early eighteenth century those scientists who found a paradigm in the *Principia* took the generality of its conclusions for granted, and they had every reason to do so. No other work known to the history of science has simultaneously permitted so large an increase in both the scope and precision of research. For the heavens Newton had derived Kepler's Laws of planetary motion and also explained certain of the observed respects in which the moon failed to obey them. For the earth he had derived the results of some scattered observations on pendulums and the tides. With the aid of additional but *ad hoc* assumptions, he had also been able to derive Boyle's Law and an important formula for the speed of sound in air. Given the state of science at the time, the success of the demonstrations was extremely impressive. Yet given the presumptive generality of Newton's Laws, the number of these applications was not great, and Newton developed almost no others. Furthermore, compared with what any graduate student of physics can achieve with those same laws today, Newton's few applications were not even developed with precision. Finally, the *Principia* had been designed for application chiefly to problems of celestial mechanics. How to adapt it for terrestrial applications, particularly for those of motion under constraint, was by no means clear. Terrestrial problems were, in any case, already being attacked with great success by a quite different set of techniques developed originally by Galileo and Huyghens and extended on the Continent during the eighteenth century by the Bernoullis, d'Alembert, and many others. Presumably their techniques and those of the *Principia* could be shown to be special cases of a more general formulation, but for some time no one saw quite how.[9]

Restrict attention for the moment to the problem of precision. We have already illustrated its empirical aspect. Special equipment—like Cavendish's apparatus, the Atwood machine, or improved telescopes—was required in order to provide the special data that the concrete applications of Newton's paradigm demanded. Similar difficulties in obtaining agreement existed on the side of theory. In applying his laws to pendulums, for example, Newton was forced to treat the bob as a mass point in order to provide a unique definition of pendulum length. Most of his theorems, the few exceptions being hypothetical and preliminary, also ignored the effect of air resistance. These were sound physical approximations. Nevertheless, as approximations they restricted the agreement to be expected

[9] C. Truesdell, "A Program toward Rediscovering the Rational Mechanics of the Age of Reason," *Archive for History of the Exact Sciences,* I (1960), 3–36, and "Reactions of Late Baroque Mechanics to Success, Conjecture, Error, and Failure in Newton's *Principia,*" *Texas Quarterly,* X (1967), 281–97. T. L. Hankins, "The Reception of Newton's Second Law of Motion in the Eighteenth Century," *Archives internationales d'histoire des sciences,* XX (1967), 42–65.

between Newton's predictions and actual experiments. The same difficulties appear even more clearly in the application of Newton's theory to the heavens. Simple quantitative telescopic observations indicate that the planets do not quite obey Kepler's Laws, and Newton's theory indicates that they should not. To derive those laws, Newton had been forced to neglect all gravitational attraction except that between individual planets and the sun. Since the planets also attract each other, only approximate agreement between the applied theory and telescopic observation could be expected.[10]

The agreement obtained was, of course, more than satisfactory to those who obtained it. Excepting for some terrestrial problems, no other theory could do nearly so well. None of those who questioned the validity of Newton's work did so because of its limited agreement with experiment and observation. Nevertheless, these limitations of agreement left many fascinating theoretical problems for Newton's successors. Theoretical techniques were, for example, required for treating the motions of more than two simultaneously attracting bodies and for investigating the stability of perturbed orbits. Problems like these occupied many of Europe's best mathematicians during the eighteenth and early nineteenth century. Euler, Lagrange, Laplace, and Gauss all did some of their most brilliant work on problems aimed to improve the match between Newton's paradigm and observation of the heavens. Many of these figures worked simultaneously to develop the mathematics required for applications that neither Newton nor the contemporary Continental school of mechanics had even attempted. They produced, for example, an immense literature and some very powerful mathematical techniques for hydrodynamics and for the problem of vibrating strings. These problems of application account for what is probably the most brilliant and consuming scientific work of the eighteenth century. Other examples could be discovered by an examination of the post-paradigm period in the development of thermodynamics, the wave theory of light, electromagnetic theory, or any other branch of science whose fundamental laws are

[10] Wolf, *op. cit.*, pp. 75–81, 96–101; and William Whewell, *History of the*

fully quantitative. At least in the more mathematical sciences, most theoretical work is of this sort.

But it is not all of this sort. Even in the mathematical sciences there are also theoretical problems of paradigm articulation; and during periods when scientific development is predominantly qualitative, these problems dominate. Some of the problems, in both the more quantitative and more qualitative sciences, aim simply at clarification by reformulation. The *Principia*, for example, did not always prove an easy work to apply, partly because it retained some of the clumsiness inevitable in a first venture and partly because so much of its meaning was only implicit in its applications. For many terrestrial applications, in any case, an apparently unrelated set of Continental techniques seemed vastly more powerful. Therefore, from Euler and Lagrange in the eighteenth century to Hamilton, Jacobi, and Hertz in the nineteenth, many of Europe's most brilliant mathematical physicists repeatedly endeavored to reformulate mechanical theory in an equivalent but logically and aesthetically more satisfying form. They wished, that is, to exhibit the explicit and implicit lessons of the *Principia* and of Continental mechanics in a logically more coherent version, one that would be at once more uniform and less equivocal in its application to the newly elaborated problems of mechanics.[11]

Similar reformulations of a paradigm have occurred repeatedly in all of the sciences, but most of them have produced more substantial changes in the paradigm than the reformulations of the *Principia* cited above. Such changes result from the empirical work previously described as aimed at paradigm articulation. Indeed, to classify that sort of work as empirical was arbitrary. More than any other sort of normal research, the problems of paradigm articulation are simultaneously theoretical and experimental; the examples given previously will serve equally well here. Before he could construct his equipment and make measurements with it, Coulomb had to employ electrical theory to determine how his equipment should be built. The consequence of his measurements was a refinement in that

[11] René Dugas, *Histoire de la mécanique* (Neuchatel, 1950), Books IV–V.

theory. Or again, the men who designed the experiments that were to distinguish between the various theories of heating by compression were generally the same men who had made up the versions being compared. They were working both with fact and with theory, and their work produced not simply new information but a more precise paradigm, obtained by the elimination of ambiguities that the original from which they worked had retained. In many sciences, most normal work is of this sort.

These three classes of problems—determination of significant fact, matching of facts with theory, and articulation of theory—exhaust, I think, the literature of normal science, both empirical and theoretical. They do not, of course, quite exhaust the entire literature of science. There are also extraordinary problems, and it may well be their resolution that makes the scientific enterprise as a whole so particularly worthwhile. But extraordinary problems are not to be had for the asking. They emerge only on special occasions prepared by the advance of normal research. Inevitably, therefore, the overwhelming majority of the problems undertaken by even the very best scientists usually fall into one of the three categories outlined above. Work under the paradigm can be conducted in no other way, and to desert the paradigm is to cease practicing the science it defines. We shall shortly discover that such desertions do occur. They are the pivots about which scientific revolutions turn. But before beginning the study of such revolutions, we require a more panoramic view of the normal-scientific pursuits that prepare the way.

IV. Normal Science as Puzzle-solving

Perhaps the most striking feature of the normal research problems we have just encountered is how little they aim to produce major novelties, conceptual or phenomenal. Sometimes, as in a wave-length measurement, everything but the most esoteric detail of the result is known in advance, and the typical latitude of expectation is only somewhat wider. Coulomb's measurements need not, perhaps, have fitted an inverse square law; the men who worked on heating by compression were often prepared for any one of several results. Yet even in cases like these the range of anticipated, and thus of assimilable, results is always small compared with the range that imagination can conceive. And the project whose outcome does not fall in that narrower range is usually just a research failure, one which reflects not on nature but on the scientist.

In the eighteenth century, for example, little attention was paid to the experiments that measured electrical attraction with devices like the pan balance. Because they yielded neither consistent nor simple results, they could not be used to articulate the paradigm from which they derived. Therefore, they remained *mere* facts, unrelated and unrelatable to the continuing progress of electrical research. Only in retrospect, possessed of a subsequent paradigm, can we see what characteristics of electrical phenomena they display. Coulomb and his contemporaries, of course, also possessed this later paradigm or one that, when applied to the problem of attraction, yielded the same expectations. That is why Coulomb was able to design apparatus that gave a result assimilable by paradigm articulation. But it is also why that result surprised no one and why several of Coulomb's contemporaries had been able to predict it in advance. Even the project whose goal is paradigm articulation does not aim at the *unexpected* novelty.

But if the aim of normal science is not major substantive novelties—if failure to come near the anticipated result is usually failure as a scientist—then why are these problems undertaken at all? Part of the answer has already been developed. To scientists, at least, the results gained in normal research are significant because they add to the scope and precision with which the paradigm can be applied. That answer, however, cannot account for the enthusiasm and devotion that scientists display for the problems of normal research. No one devotes years to, say, the development of a better spectrometer or the production of an improved solution to the problem of vibrating strings simply because of the importance of the information that will be obtained. The data to be gained by computing ephemerides or by further measurements with an existing instrument are often just as significant, but those activities are regularly spurned by scientists because they are so largely repetitions of procedures that have been carried through before. That rejection provides a clue to the fascination of the normal research problem. Though its outcome can be anticipated, often in detail so great that what remains to be known is itself uninteresting, the way to achieve that outcome remains very much in doubt. Bringing a normal research problem to a conclusion is achieving the anticipated in a new way, and it requires the solution of all sorts of complex instrumental, conceptual, and mathematical puzzles. The man who succeeds proves himself an expert puzzle-solver, and the challenge of the puzzle is an important part of what usually drives him on.

The terms 'puzzle' and 'puzzle-solver' highlight several of the themes that have become increasingly prominent in the preceding pages. Puzzles are, in the entirely standard meaning here employed, that special category of problems that can serve to test ingenuity or skill in solution. Dictionary illustrations are 'jigsaw puzzle' and 'crossword puzzle,' and it is the characteristics that these share with the problems of normal science that we now need to isolate. One of them has just been mentioned. It is no criterion of goodness in a puzzle that its outcome be intrinsically interesting or important. On the contrary, the really pressing problems, e.g., a cure for cancer or the design of a lasting peace, are often not puzzles at all, largely because they may not have any solution. Consider the jigsaw puzzle whose pieces are selected at random from each of two different puzzle boxes. Since that problem is likely to defy (though it might not) even the most ingenious of men, it cannot serve as a test of skill in solution. In any usual sense it is not a puzzle at all. Though intrinsic value is no criterion for a puzzle, the assured existence of a solution is.

We have already seen, however, that one of the things a scientific community acquires with a paradigm is a criterion for choosing problems that, while the paradigm is taken for granted, can be assumed to have solutions. To a great extent these are the only problems that the community will admit as scientific or encourage its members to undertake. Other problems, including many that had previously been standard, are rejected as metaphysical, as the concern of another discipline, or sometimes as just too problematic to be worth the time. A paradigm can, for that matter, even insulate the community from those socially important problems that are not reducible to the puzzle form, because they cannot be stated in terms of the conceptual and instrumental tools the paradigm supplies. Such problems can be a distraction, a lesson brilliantly illustrated by several facets of seventeenth-century Baconianism and by some of the contemporary social sciences. One of the reasons why normal science seems to progress so rapidly is that its practitioners concentrate on problems that only their own lack of ingenuity should keep them from solving.

If, however, the problems of normal science are puzzles in this sense, we need no longer ask why scientists attack them with such passion and devotion. A man may be attracted to science for all sorts of reasons. Among them are the desire to be useful, the excitement of exploring new territory, the hope of finding order, and the drive to test established knowledge. These motives and others besides also help to determine the particular problems that will later engage him. Furthermore, though the result is occasional frustration, there is good reason why motives like these should first attract him and then lead him on.[1] The scientific enterprise as a whole does from time to time prove useful, open up new territory, display order, and test long-accepted belief. Nevertheless, *the individual* engaged on a normal research problem *is almost never doing any one of these things.* Once engaged, his motivation is of a rather different sort. What then challenges him is the conviction that, if only he is skilful enough, he will succeed in solving a puzzle that no one before has solved or solved so well. Many of the greatest scientific minds have devoted all of their professional attention to demanding puzzles of this sort. On most occasions any particular field of specialization offers nothing else to do, a fact that makes it no less fascinating to the proper sort of addict.

Turn now to another, more difficult, and more revealing aspect of the parallelism between puzzles and the problems of normal science. If it is to classify as a puzzle, a problem must be characterized by more than an assured solution. There must also be rules that limit both the nature of acceptable solutions and the steps by which they are to be obtained. To solve a jigsaw puzzle is not, for example, merely "to make a picture." Either a child or a contemporary artist could do that by scattering selected pieces, as abstract shapes, upon some neutral ground. The picture thus produced might be far better, and would certainly be more original, than the one from which the puzzle had been made. Nevertheless, such a picture would not be a solution. To achieve that all the pieces must be used, their plain sides must be turned down, and they must be interlocked without forcing until no holes remain. Those are among the rules that govern jigsaw-puzzle solutions. Similar restrictions upon the admissible solutions of crossword puzzles, riddles, chess problems, and so on, are readily discovered.

If we can accept a considerably broadened use of the term

[1] The frustrations induced by the conflict between the individual's role and the over-all pattern of scientific development can, however, occasionally be quite serious. On this subject, see Lawrence S. Kubie, "Some Unsolved Problems of the Scientific Career," *American Scientist,* XLI (1953), 596–613; and XLII (1954), 104–12.

'rule'—one that will occasionally equate it with 'established viewpoint' or with 'preconception'—then the problems accessible within a given research tradition display something much like this set of puzzle characteristics. The man who builds an instrument to determine optical wave lengths must not be satisfied with a piece of equipment that merely attributes particular numbers to particular spectral lines. He is not just an explorer or measurer. On the contrary, he must show, by analyzing his apparatus in terms of the established body of optical theory, that the numbers his instrument produces are the ones that enter theory as wave lengths. If some residual vagueness in the theory or some unanalyzed component of his apparatus prevents his completing that demonstration, his colleagues may well conclude that he has measured nothing at all. For example, the electron-scattering maxima that were later diagnosed as indices of electron wave length had no apparent significance when first observed and recorded. Before they became measures of anything, they had to be related to a theory that predicted the wave-like behavior of matter in motion. And even after that relation was pointed out, the apparatus had to be redesigned so that the experimental results might be correlated unequivocally with theory.[2] Until those conditions had been satisfied, no problem had been solved.

Similar sorts of restrictions bound the admissible solutions to theoretical problems. Throughout the eighteenth century those scientists who tried to derive the observed motion of the moon from Newton's laws of motion and gravitation consistently failed to do so. As a result, some of them suggested replacing the inverse square law with a law that deviated from it at small distances. To do that, however, would have been to change the paradigm, to define a new puzzle, and not to solve the old one. In the event, scientists preserved the rules until, in 1750, one of them discovered how they could successfully be applied.[3]

[2] For a brief account of the evolution of these experiments, see page 4 of C. J. Davisson's lecture in *Les prix Nobel en 1937* (Stockholm, 1938).

[3] W. Whewell, *History of the Inductive Sciences* (rev. ed.; London, 1847), II, 101–5, 220–22.

Only a change in the rules of the game could have provided an alternative.

The study of normal-scientific traditions discloses many additional rules, and these provide much information about the commitments that scientists derive from their paradigms. What can we say are the main categories into which these rules fall?[4] The most obvious and probably the most binding is exemplified by the sorts of generalizations we have just noted. These are explicit statements of scientific law and about scientific concepts and theories. While they continue to be honored, such statements help to set puzzles and to limit acceptable solutions. Newton's Laws, for example, performed those functions during the eighteenth and nineteenth centuries. As long as they did so, quantity-of-matter was a fundamental ontological category for physical scientists, and the forces that act between bits of matter were a dominant topic for research.[5] In chemistry the laws of fixed and definite proportions had, for a long time, an exactly similar force—setting the problem of atomic weights, bounding the admissible results of chemical analyses, and informing chemists what atoms and molecules, compounds and mixtures were.[6] Maxwell's equations and the laws of statistical thermodynamics have the same hold and function today.

Rules like these are, however, neither the only nor even the most interesting variety displayed by historical study. At a level lower or more concrete than that of laws and theories, there is, for example, a multitude of commitments to preferred types of instrumentation and to the ways in which accepted instruments may legitimately be employed. Changing attitudes toward the role of fire in chemical analyses played a vital part in the de-

[4] I owe this question to W. O. Hagstrom, whose work in the sociology of science sometimes overlaps my own.

[5] For these aspects of Newtonianism, see I. B. Cohen, *Franklin and Newton: An Inquiry into Speculative Newtonian Experimental Science and Franklin's Work in Electricity as an Example Thereof* (Philadelphia, 1956), chap. vii, esp. pp. 255–57, 275–77.

[6] This example is discussed at length near the end of Section X.

velopment of chemistry in the seventeenth century.[7] Helmholtz, in the nineteenth, encountered strong resistance from physiologists to the notion that physical experimentation could illuminate their field.[8] And in this century the curious history of chemical chromatography again illustrates the endurance of instrumental commitments that, as much as laws and theory, provide scientists with rules of the game.[9] When we analyze the discovery of X-rays, we shall find reasons for commitments of this sort.

Less local and temporary, though still not unchanging characteristics of science, are the higher level, quasi-metaphysical commitments that historical study so regularly displays. After about 1630, for example, and particularly after the appearance of Descartes's immensely influential scientific writings, most physical scientists assumed that the universe was composed of microscopic corpuscles and that all natural phenomena could be explained in terms of corpuscular shape, size, motion, and interaction. That nest of commitments proved to be both metaphysical and methodological. As metaphysical, it told scientists what sorts of entities the universe did and did not contain: there was only shaped matter in motion. As methodological, it told them what ultimate laws and fundamental explanations must be like: laws must specify corpuscular motion and interaction, and explanation must reduce any given natural phenomenon to corpuscular action under these laws. More important still, the corpuscular conception of the universe told scientists what many of their research problems should be. For example, a chemist who, like Boyle, embraced the new philosophy gave particular attention to reactions that could be viewed as transmutations. More clearly than any others these displayed the process of corpuscular rearrangement that must underlie all

[7] H. Metzger, *Les doctrines chimiques en France du début du XVIIe siècle à la fin du XVIIIe siècle* (Paris, 1923), pp. 359–61; Marie Boas, *Robert Boyle and Seventeenth-Century Chemistry* (Cambridge, 1958), pp. 112–15.

[8] Leo Königsberger, *Hermann von Helmholtz*, trans. Francis A. Welby (Oxford, 1906), pp. 65–66.

[9] James E. Meinhard, "Chromatography: A Perspective," *Science*, CX (1949), 387–92.

chemical change.[10] Similar effects of corpuscularism can be observed in the study of mechanics, optics, and heat.

Finally, at a still higher level, there is another set of commitments without which no man is a scientist. The scientist must, for example, be concerned to understand the world and to extend the precision and scope with which it has been ordered. That commitment must, in turn, lead him to scrutinize, either for himself or through colleagues, some aspect of nature in great empirical detail. And, if that scrutiny displays pockets of apparent disorder, then these must challenge him to a new refinement of his observational techniques or to a further articulation of his theories. Undoubtedly there are still other rules like these, ones which have held for scientists at all times.

The existence of this strong network of commitments—conceptual, theoretical, instrumental, and methodological—is a principal source of the metaphor that relates normal science to puzzle-solving. Because it provides rules that tell the practitioner of a mature specialty what both the world and his science are like, he can concentrate with assurance upon the esoteric problems that these rules and existing knowledge define for him. What then personally challenges him is how to bring the residual puzzle to a solution. In these and other respects a discussion of puzzles and of rules illuminates the nature of normal scientific practice. Yet, in another way, that illumination may be significantly misleading. Though there obviously are rules to which all the practitioners of a scientific specialty adhere at a given time, those rules may not by themselves specify all that the practice of those specialists has in common. Normal science is a highly determined activity, but it need not be entirely determined by rules. That is why, at the start of this essay, I introduced shared paradigms rather than shared rules, assumptions, and points of view as the source of coherence for normal research traditions. Rules, I suggest, derive from paradigms, but paradigms can guide research even in the absence of rules.

[10] For corpuscularism in general, see Marie Boas, "The Establishment of the Mechanical Philosophy," *Osiris*, X (1952), 412–541. For its effects on Boyle's chemistry, see T. S. Kuhn, "Robert Boyle and Structural Chemistry in the Seventeenth Century," *Isis*, XLIII (1952), 12–36.

V. The Priority of Paradigms

To discover the relation between rules, paradigms, and normal science, consider first how the historian isolates the particular loci of commitment that have just been described as accepted rules. Close historical investigation of a given specialty at a given time discloses a set of recurrent and quasi-standard illustrations of various theories in their conceptual, observational, and instrumental applications. These are the community's paradigms, revealed in its textbooks, lectures, and laboratory exercises. By studying them and by practicing with them, the members of the corresponding community learn their trade. The historian, of course, will discover in addition a penumbral area occupied by achievements whose status is still in doubt, but the core of solved problems and techniques will usually be clear. Despite occasional ambiguities, the paradigms of a mature scientific community can be determined with relative ease.

The determination of shared paradigms is not, however, the determination of shared rules. That demands a second step and one of a somewhat different kind. When undertaking it, the historian must compare the community's paradigms with each other and with its current research reports. In doing so, his object is to discover what isolable elements, explicit or implicit, the members of that community may have *abstracted* from their more global paradigms and deployed as rules in their research. Anyone who has attempted to describe or analyze the evolution of a particular scientific tradition will necessarily have sought accepted principles and rules of this sort. Almost certainly, as the preceding section indicates, he will have met with at least partial success. But, if his experience has been at all like my own, he will have found the search for rules both more difficult and less satisfying than the search for paradigms. Some of the generalizations he employs to describe the community's shared beliefs will present no problems. Others, however, including some of those used as illustrations above, will seem a shade too strong. Phrased in just that way, or in any other way he can imagine, they would almost certainly have been rejected by some members of the group he studies. Nevertheless, if the coherence of the research tradition is to be understood in terms of rules, some specification of common ground in the corresponding area is needed. As a result, the search for a body of rules competent to constitute a given normal research tradition becomes a source of continual and deep frustration.

Recognizing that frustration, however, makes it possible to diagnose its source. Scientists can agree that a Newton, Lavoisier, Maxwell, or Einstein has produced an apparently permanent solution to a group of outstanding problems and still disagree, sometimes without being aware of it, about the particular abstract characteristics that make those solutions permanent. They can, that is, agree in their *identification* of a paradigm without agreeing on, or even attempting to produce, a full *interpretation* or *rationalization* of it. Lack of a standard interpretation or of an agreed reduction to rules will not prevent a paradigm from guiding research. Normal science can be determined in part by the direct inspection of paradigms, a process that is often aided by but does not depend upon the formulation of rules and assumptions. Indeed, the existence of a paradigm need not even imply that any full set of rules exists.[1]

Inevitably, the first effect of those statements is to raise problems. In the absence of a competent body of rules, what restricts the scientist to a particular normal-scientific tradition? What can the phrase 'direct inspection of paradigms' mean? Partial answers to questions like these were developed by the the late Ludwig Wittgenstein, though in a very different context. Because that context is both more elementary and more familiar, it will help to consider his form of the argument first. What need we know, Wittgenstein asked, in order that we

[1] Michael Polanyi has brilliantly developed a very similar theme, arguing that much of the scientist's success depends upon "tacit knowledge," i.e., upon knowledge that is acquired through practice and that cannot be articulated explicitly. See his *Personal Knowledge* (Chicago, 1958), particularly chaps. v and vi.

apply terms like 'chair,' or 'leaf,' or 'game' unequivocally and without provoking argument?[2]

That question is very old and has generally been answered by saying that we must know, consciously or intuitively, what a chair, or leaf, or game *is*. We must, that is, grasp some set of attributes that all games and that only games have in common. Wittgenstein, however, concluded that, given the way we use language and the sort of world to which we apply it, there need be no such set of characteristics. Though a discussion of *some* of the attributes shared by a *number* of games or chairs or leaves often helps us learn how to employ the corresponding term, there is no set of characteristics that is simultaneously applicable to all members of the class and to them alone. Instead, confronted with a previously unobserved activity, we apply the term 'game' because what we are seeing bears a close "family resemblance" to a number of the activities that we have previously learned to call by that name. For Wittgenstein, in short, games, and chairs, and leaves are natural families, each constituted by a network of overlapping and crisscross resemblances. The existence of such a network sufficiently accounts for our success in identifying the corresponding object or activity. Only if the families we named overlapped and merged gradually into one another—only, that is, if there were no *natural* families—would our success in identifying and naming provide evidence for a set of common characteristics corresponding to each of the class names we employ.

Something of the same sort may very well hold for the various research problems and techniques that arise within a single normal-scientific tradition. What these have in common is not that they satisfy some explicit or even some fully discoverable set of rules and assumptions that gives the tradition its character and its hold upon the scientific mind. Instead, they may relate by resemblance and by modeling to one or another part of the scientific corpus which the community in question al-

[2] Ludwig Wittgenstein, *Philosophical Investigations*, trans. G. E. M. Anscombe (New York, 1953), pp. 31–36. Wittgenstein, however, says almost nothing about the sort of world necessary to support the naming procedure he outlines. Part of the point that follows cannot therefore be attributed to him.

ready recognizes as among its established achievements. Scientists work from models acquired through education and through subsequent exposure to the literature often without quite knowing or needing to know what characteristics have given these models the status of community paradigms. And because they do so, they need no full set of rules. The coherence displayed by the research tradition in which they participate may not imply even the existence of an underlying body of rules and assumptions that additional historical or philosophical investigation might uncover. That scientists do not usually ask or debate what makes a particular problem or solution legitimate tempts us to suppose that, at least intuitively, they know the answer. But it may only indicate that neither the question nor the answer is felt to be relevant to their research. Paradigms may be prior to, more binding, and more complete than any set of rules for research that could be unequivocally abstracted from them.

So far this point has been entirely theoretical: paradigms *could* determine normal science without the intervention of discoverable rules. Let me now try to increase both its clarity and urgency by indicating some of the reasons for believing that paradigms actually do operate in this manner. The first, which has already been discussed quite fully, is the severe difficulty of discovering the rules that have guided particular normal-scientific traditions. That difficulty is very nearly the same as the one the philosopher encounters when he tries to say what all games have in common. The second, to which the first is really a corollary, is rooted in the nature of scientific education. Scientists, it should already be clear, never learn concepts, laws, and theories in the abstract and by themselves. Instead, these intellectual tools are from the start encountered in a historically and pedagogically prior unit that displays them with and through their applications. A new theory is always announced together with applications to some concrete range of natural phenomena; without them it would not be even a candidate for acceptance. After it has been accepted, those same applications or others accompany the theory into the textbooks from which the future practitioner will learn his trade. They are not there merely as embroidery or even as documentation. On the contrary, the process of learning a theory depends upon the study of applications, including practice problem-solving both with a pencil and paper and with instruments in the laboratory. If, for example, the student of Newtonian dynamics ever discovers the meaning of terms like 'force,' 'mass,' 'space,' and 'time,' he does so less from the incomplete though sometimes helpful definitions in his text than by observing and participating in the application of these concepts to problem-solution.

That process of learning by finger exercise or by doing continues throughout the process of professional initiation. As the student proceeds from his freshman course to and through his doctoral dissertation, the problems assigned to him become more complex and less completely precedented. But they continue to be closely modeled on previous achievements as are the problems that normally occupy him during his subsequent independent scientific career. One is at liberty to suppose that somewhere along the way the scientist has intuitively abstracted rules of the game for himself, but there is little reason to believe it. Though many scientists talk easily and well about the particular individual hypotheses that underlie a concrete piece of current research, they are little better than laymen at characterizing the established bases of their field, its legitimate problems and methods. If they have learned such abstractions at all, they show it mainly through their ability to do successful research. That ability can, however, be understood without recourse to hypothetical rules of the game.

These consequences of scientific education have a converse that provides a third reason to suppose that paradigms guide research by direct modeling as well as through abstracted rules. Normal science can proceed without rules only so long as the relevant scientific community accepts without question the particular problem-solutions already achieved. Rules should therefore become important and the characteristic unconcern about them should vanish whenever paradigms or models are felt to be insecure. That is, moreover, exactly what does occur. The preparadigm period, in particular, is regularly marked by frequent and deep debates over legitimate methods, problems, and standards of solution, though these serve rather to define schools than to produce agreement. We have already noted a few of these debates in optics and electricity, and they played an even larger role in the development of seventeenth-century chemistry and of early nineteenth-century geology.[3] Furthermore, debates like these do not vanish once and for all with the appearance of a paradigm. Though almost non-existent during periods of normal science, they recur regularly just before and during scientific revolutions, the periods when paradigms are first under attack and then subject to change. The transition from Newtonian to quantum mechanics evoked many debates about both the nature and the standards of physics, some of which still continue.[4] There are people alive today who can remember the similar arguments engendered by Maxwell's electromagnetic theory and by statistical mechanics.[5] And earlier still, the assimilation of Galileo's and Newton's mechanics gave rise to a particularly famous series of debates with Aristotelians, Cartesians, and Leibnizians about the standards legitimate to science.[6] When scientists disagree about whether the fundamental problems of their field have been solved, the search for rules gains a function that it does not ordinarily possess. While

[3] For chemistry, see H. Metzger, *Les doctrines chimiques en France du début du XVIIe à la fin du XVIIIe siècle* (Paris, 1923), pp. 24–27, 146–49; and Marie Boas, *Robert Boyle and Seventeenth-Century Chemistry* (Cambridge, 1958), chap. ii. For geology, see Walter F. Cannon, "The Uniformitarian-Catastrophist Debate," *Isis,* LI (1960), 38–55; and C. C. Gillispie, *Genesis and Geology* (Cambridge, Mass., 1951), chaps. iv–v.

[4] For controversies over quantum mechanics, see Jean Ullmo, *La crise de la physique quantique* (Paris, 1950), chap. ii.

[5] For statistical mechanics, see René Dugas, *La théorie physique au sens de Boltzmann et ses prolongements modernes* (Neuchatel, 1959), pp. 158–84, 206–19. For the reception of Maxwell's work, see Max Planck, "Maxwell's Influence in Germany," in *James Clerk Maxwell: A Commemoration Volume, 1831–1931* (Cambridge, 1931), pp. 45–65, esp. pp. 58–63; and Silvanus P. Thompson, *The Life of William Thomson Baron Kelvin of Largs* (London, 1910), II, 1021–27.

[6] For a sample of the battle with the Aristotelians, see A. Koyré, "A Documentary History of the Problem of Fall from Kepler to Newton," *Transactions of the American Philosophical Society,* XLV (1955), 329–95. For the debates with the Cartesians and Leibnizians, see Pierre Brunet, *L'introduction des théories de Newton en France au XVIIIe siècle* (Paris, 1931); and A. Koyré, *From the Closed World to the Infinite Universe* (Baltimore, 1957), chap. xi.

paradigms remain secure, however, they can function without agreement over rationalization or without any attempted rationalization at all.

A fourth reason for granting paradigms a status prior to that of shared rules and assumptions can conclude this section. The introduction to this essay suggested that there can be small revolutions as well as large ones, that some revolutions affect only the members of a professional subspecialty, and that for such groups even the discovery of a new and unexpected phenomenon may be revolutionary. The next section will introduce selected revolutions of that sort, and it is still far from clear how they can exist. If normal science is so rigid and if scientific communities are so close-knit as the preceding discussion has implied, how can a change of paradigm ever affect only a small subgroup? What has been said so far may have seemed to imply that normal science is a single monolithic and unified enterprise that must stand or fall with any one of its paradigms as well as with all of them together. But science is obviously seldom or never like that. Often, viewing all fields together, it seems instead a rather ramshackle structure with little coherence among its various parts. Nothing said to this point should, however, conflict with that very familiar observation. On the contrary, substituting paradigms for rules should make the diversity of scientific fields and specialties easier to understand. Explicit rules, when they exist, are usually common to a very broad scientific group, but paradigms need not be. The practitioners of widely separated fields, say astronomy and taxonomic botany, are educated by exposure to quite different achievements described in very different books. And even men who, being in the same or in closely related fields, begin by studying many of the same books and achievements may acquire rather different paradigms in the course of professional specialization.

Consider, for a single example, the quite large and diverse community constituted by all physical scientists. Each member of that group today is taught the laws of, say, quantum mechanics, and most of them employ these laws at some point in their research or teaching. But they do not all learn the same applications of these laws, and they are not therefore all affected in the same ways by changes in quantum-mechanical practice. On the road to professional specialization, a few physical scientists encounter only the basic principles of quantum mechanics. Others study in detail the paradigm applications of these principles to chemistry, still others to the physics of the solid state, and so on. What quantum mechanics means to each of them depends upon what courses he has had, what texts he has read, and which journals he studies. It follows that, though a change in quantum-mechanical law will be revolutionary for all of these groups, a change that reflects only on one or another of the paradigm applications of quantum mechanics need be revolutionary only for the members of a particular professional subspecialty. For the rest of the profession and for those who practice other physical sciences, that change need not be revolutionary at all. In short, though quantum mechanics (or Newtonian dynamics, or electromagnetic theory) is a paradigm for many scientific groups, it is not the same paradigm for them all. Therefore, it can simultaneously determine several traditions of normal science that overlap without being coextensive. A revolution produced within one of these traditions will not necessarily extend to the others as well.

One brief illustration of specialization's effect may give this whole series of points additional force. An investigator who hoped to learn something about what scientists took the atomic theory to be asked a distinguished physicist and an eminent chemist whether a single atom of helium was or was not a molecule. Both answered without hesitation, but their answers were not the same. For the chemist the atom of helium was a molecule because it behaved like one with respect to the kinetic theory of gases. For the physicist, on the other hand, the helium atom was not a molecule because it displayed no molecular spectrum.[7] Presumably both men were talking of the same par-

[7] The investigator was James K. Senior, to whom I am indebted for a verbal report. Some related issues are treated in his paper, "The Vernacular of the Laboratory," *Philosophy of Science*, XXV (1958), 163–68.

ticle, but they were viewing it through their own research training and practice. Their experience in problem-solving told them what a molecule must be. Undoubtedly their experiences had had much in common, but they did not, in this case, tell the two specialists the same thing. As we proceed we shall discover how consequential paradigm differences of this sort can occasionally be.

VI. Anomaly and the Emergence of Scientific Discoveries

Normal science, the puzzle-solving activity we have just examined, is a highly cumulative enterprise, eminently successful in its aim, the steady extension of the scope and precision of scientific knowledge. In all these respects it fits with great precision the most usual image of scientific work. Yet one standard product of the scientific enterprise is missing. Normal science does not aim at novelties of fact or theory and, when successful, finds none. New and unsuspected phenomena are, however, repeatedly uncovered by scientific research, and radical new theories have again and again been invented by scientists. History even suggests that the scientific enterprise has developed a uniquely powerful technique for producing surprises of this sort. If this characteristic of science is to be reconciled with what has already been said, then research under a paradigm must be a particularly effective way of inducing paradigm change. That is what fundamental novelties of fact and theory do. Produced inadvertently by a game played under one set of rules, their assimilation requires the elaboration of another set. After they have become parts of science, the enterprise, at least of those specialists in whose particular field the novelties lie, is never quite the same again.

We must now ask how changes of this sort can come about, considering first discoveries, or novelties of fact, and then inventions, or novelties of theory. That distinction between discovery and invention or between fact and theory will, however, immediately prove to be exceedingly artificial. Its artificiality is an important clue to several of this essay's main theses. Examining selected discoveries in the rest of this section, we shall quickly find that they are not isolated events but extended episodes with a regularly recurrent structure. Discovery commences with the awareness of anomaly, i.e., with the recognition that nature has somehow violated the paradigm-induced expectations that govern normal science. It then continues with a more or less extended exploration of the area of anomaly. And it closes only when the paradigm theory has been adjusted so that the anomalous has become the expected. Assimilating a new sort of fact demands a more than additive adjustment of theory, and until that adjustment is completed—until the scientist has learned to see nature in a different way—the new fact is not quite a scientific fact at all.

To see how closely factual and theoretical novelty are intertwined in scientific discovery examine a particularly famous example, the discovery of oxygen. At least three different men have a legitimate claim to it, and several other chemists must, in the early 1770's, have had enriched air in a laboratory vessel without knowing it.[1] The progress of normal science, in this case of pneumatic chemistry, prepared the way to a breakthrough quite thoroughly. The earliest of the claimants to prepare a relatively pure sample of the gas was the Swedish apothecary, C. W. Scheele. We may, however, ignore his work since it was not published until oxygen's discovery had repeatedly been announced elsewhere and thus had no effect upon the historical pattern that most concerns us here.[2] The second in time to establish a claim was the British scientist and divine, Joseph Priestley, who collected the gas released by heated red oxide of mercury as one item in a prolonged normal investigation of the "airs" evolved by a large number of solid substances. In 1774 he identified the gas thus produced as nitrous oxide and in 1775, led by further tests, as common air with less than its usual quantity of phlogiston. The third claimant, Lavoisier, started the work that led him to oxygen after Priestley's experiments of 1774 and possibly as the result of a hint from Priestley. Early in

[1] For the still classic discussion of oxygen's discovery, see A. N. Meldrum, *The Eighteenth-Century Revolution in Science—the First Phase* (Calcutta, 1930), chap. v. An indispensable recent review, including an account of the priority controversy, is Maurice Daumas, *Lavoisier, théoricien et expérimentateur* (Paris, 1955), chaps. ii–iii. For a fuller account and bibliography, see also T. S. Kuhn, "The Historical Structure of Scientific Discovery," *Science*, CXXXVI (June 1, 1962), 760–64.

[2] See, however, Uno Bocklund, "A Lost Letter from Scheele to Lavoisier," *Lychnos*, 1957–58, pp. 39 62, for a different evaluation of Scheele's role.

1775 Lavoisier reported that the gas obtained by heating the red oxide of mercury was "air itself entire without alteration [except that] . . . it comes out more pure, more respirable."[3] By 1777, probably with the assistance of a second hint from Priestley, Lavoisier had concluded that the gas was a distinct species, one of the two main constituents of the atmosphere, a conclusion that Priestley was never able to accept.

This pattern of discovery raises a question that can be asked about every novel phenomenon that has ever entered the consciousness of scientists. Was it Priestley or Lavoisier, if either, who first discovered oxygen? In any case, when was oxygen discovered? In that form the question could be asked even if only one claimant had existed. As a ruling about priority and date, an answer does not at all concern us. Nevertheless, an attempt to produce one will illuminate the nature of discovery, because there is no answer of the kind that is sought. Discovery is not the sort of process about which the question is appropriately asked. The fact that it is asked—the priority for oxygen has repeatedly been contested since the 1780's—is a symptom of something askew in the image of science that gives discovery so fundamental a role. Look once more at our example. Priestley's claim to the discovery of oxygen is based upon his priority in isolating a gas that was later recognized as a distinct species. But Priestley's sample was not pure, and, if holding impure oxygen in one's hands is to discover it, that had been done by everyone who ever bottled atmospheric air. Besides, if Priestley was the discoverer, when was the discovery made? In 1774 he thought he had obtained nitrous oxide, a species he already knew; in 1775 he saw the gas as dephlogisticated air, which is still not oxygen or even, for phlogistic chemists, a quite unexpected sort of gas. Lavoisier's claim may be stronger, but it presents the same problems. If we refuse the palm to Priestley, we cannot award it to Lavoisier for the work of 1775 which led

[3] J. B. Conant, *The Overthrow of the Phlogiston Theory: The Chemical Revolution of 1775–1789* ("Harvard Case Histories in Experimental Science," Case 2; Cambridge, Mass., 1950), p. 23. This very useful pamphlet reprints many of the relevant documents.

him to identify the gas as the "air itself entire." Presumably we wait for the work of 1776 and 1777 which led Lavoisier to see not merely the gas but what the gas was. Yet even this award could be questioned, for in 1777 and to the end of his life Lavoisier insisted that oxygen was an atomic "principle of acidity" and that oxygen gas was formed only when that "principle" united with caloric, the matter of heat.[4] Shall we therefore say that oxygen had not yet been discovered in 1777? Some may be tempted to do so. But the principle of acidity was not banished from chemistry until after 1810, and caloric lingered until the 1860's. Oxygen had become a standard chemical substance before either of those dates.

Clearly we need a new vocabulary and concepts for analyzing events like the discovery of oxygen. Though undoubtedly correct, the sentence, "Oxygen was discovered," misleads by suggesting that discovering something is a single simple act assimilable to our usual (and also questionable) concept of seeing. That is why we so readily assume that discovering, like seeing or touching, should be unequivocally attributable to an individual and to a moment in time. But the latter attribution is always impossible, and the former often is as well. Ignoring Scheele, we can safely say that oxygen had not been discovered before 1774, and we would probably also say that it had been discovered by 1777 or shortly thereafter. But within those limits or others like them, any attempt to date the discovery must inevitably be arbitrary because discovering a new sort of phenomenon is necessarily a complex event, one which involves recognizing both *that* something is and *what* it is. Note, for example, that if oxygen were dephlogisticated air for us, we should insist without hesitation that Priestley had discovered it, though we would still not know quite when. But if both observation and conceptualization, fact and assimilation to theory, are inseparably linked in discovery, then discovery is a process and must take time. Only when all the relevant conceptual categories are prepared in advance, in which case the phenomenon would not

[4] H. Metzger, *La philosophie de la matière chez Lavoisier* (Paris, 1935); and Daumas, *op. cit.*, chap. vii.

be of a new sort, can discovering *that* and discovering *what* occur effortlessly, together, and in an instant.

Grant now that discovery involves an extended, though not necessarily long, process of conceptual assimilation. Can we also say that it involves a change in paradigm? To that question, no general answer can yet be given, but in this case at least, the answer must be yes. What Lavoisier announced in his papers from 1777 on was not so much the discovery of oxygen as the oxygen theory of combustion. That theory was the keystone for a reformulation of chemistry so vast that it is usually called the chemical revolution. Indeed, if the discovery of oxygen had not been an intimate part of the emergence of a new paradigm for chemistry, the question of priority from which we began would never have seemed so important. In this case as in others, the value placed upon a new phenomenon and thus upon its discoverer varies with our estimate of the extent to which the phenomenon violated paradigm-induced anticipations. Notice, however, since it will be important later, that the discovery of oxygen was not by itself the cause of the change in chemical theory. Long before he played any part in the discovery of the new gas, Lavoisier was convinced both that something was wrong with the phlogiston theory and that burning bodies absorbed some part of the atmosphere. That much he had recorded in a sealed note deposited with the Secretary of the French Academy in 1772.[5] What the work on oxygen did was to give much additional form and structure to Lavoisier's earlier sense that something was amiss. It told him a thing he was already prepared to discover—the nature of the substance that combustion removes from the atmosphere. That advance awareness of difficulties must be a significant part of what enabled Lavoisier to see in experiments like Priestley's a gas that Priestley had been unable to see there himself. Conversely, the fact that a major paradigm revision was needed to see what Lavoisier saw must be the principal reason why Priestley was, to the end of his long life, unable to see it.

[5] The most authoritative account of the origin of Lavoisier's discontent is Henry Guerlac, *Lavoisier—the Crucial Year: The Background and Origin of His First Experiments on Combustion in 1772* (Ithaca, N.Y., 1961).

Two other and far briefer examples will reinforce much that has just been said and simultaneously carry us from an elucidation of the nature of discoveries toward an understanding of the circumstances under which they emerge in science. In an effort to represent the main ways in which discoveries can come about, these examples are chosen to be different both from each other and from the discovery of oxygen. The first, X-rays, is a classic case of discovery through accident, a type that occurs more frequently than the impersonal standards of scientific reporting allow us easily to realize. Its story opens on the day that the physicist Roentgen interrupted a normal investigation of cathode rays because he had noticed that a barium platino-cyanide screen at some distance from his shielded apparatus glowed when the discharge was in process. Further investigations—they required seven hectic weeks during which Roentgen rarely left the laboratory—indicated that the cause of the glow came in straight lines from the cathode ray tube, that the radiation cast shadows, could not be deflected by a magnet, and much else besides. Before announcing his discovery, Roentgen had convinced himself that his effect was not due to cathode rays but to an agent with at least some similarity to light.[6]

Even so brief an epitome reveals striking resemblances to the discovery of oxygen: before experimenting with red oxide of mercury, Lavoisier had performed experiments that did not produce the results anticipated under the phlogiston paradigm; Roentgen's discovery commenced with the recognition that his screen glowed when it should not. In both cases the perception of anomaly—of a phenomenon, that is, for which his paradigm had not readied the investigator—played an essential role in preparing the way for perception of novelty. But, again in both cases, the perception that something had gone wrong was only the prelude to discovery. Neither oxygen nor X-rays emerged without a further process of experimentation and assimilation. At what point in Roentgen's investigation, for example, ought we say that X-rays had actually been discovered? Not, in any

[6] L. W. Taylor, *Physics, the Pioneer Science* (Boston, 1941), pp. 790–94; and T. W. Chalmers, *Historic Researches* (London, 1949), pp. 218–19.

case, at the first instant, when all that had been noted was a glowing screen. At least one other investigator had seen that glow and, to his subsequent chagrin, discovered nothing at all.[7] Nor, it is almost as clear, can the moment of discovery be pushed forward to a point during the last week of investigation, by which time Roentgen was exploring the properties of the new radiation he had *already* discovered. We can only say that X-rays emerged in Würzburg between November 8 and December 28, 1895.

In a third area, however, the existence of significant parallels between the discoveries of oxygen and of X-rays is far less apparent. Unlike the discovery of oxygen, that of X-rays was not, at least for a decade after the event, implicated in any obvious upheaval in scientific theory. In what sense, then, can the assimilation of that discovery be said to have necessitated paradigm change? The case for denying such a change is very strong. To be sure, the paradigms subscribed to by Roentgen and his contemporaries could not have been used to predict X-rays. (Maxwell's electromagnetic theory had not yet been accepted everywhere, and the particulate theory of cathode rays was only one of several current speculations.) But neither did those paradigms, at least in any obvious sense, prohibit the existence of X-rays as the phlogiston theory had prohibited Lavoisier's interpretation of Priestley's gas. On the contrary, in 1895 accepted scientific theory and practice admitted a number of forms of radiation—visible, infrared, and ultraviolet. Why could not X-rays have been accepted as just one more form of a well-known class of natural phenomena? Why were they not, for example, received in the same way as the discovery of an additional chemical element? New elements to fill empty places in the periodic table were still being sought and found in Roentgen's day. Their pursuit was a standard project for normal science, and success was an occasion only for congratulations, not for surprise.

[7] E. T. Whittaker, *A History of the Theories of Aether and Electricity,* I (2d ed.; London, 1951), 358, n. 1. Sir George Thomson has informed me of a second near miss. Alerted by unaccountably fogged photographic plates, Sir William Crookes was also on the track of the discovery.

X-rays, however, were greeted not only with surprise but with shock. Lord Kelvin at first pronounced them an elaborate hoax.[8] Others, though they could not doubt the evidence, were clearly staggered by it. Though X-rays were not prohibited by established theory, they violated deeply entrenched expectations. Those expectations, I suggest, were implicit in the design and interpretation of established laboratory procedures. By the 1890's cathode ray equipment was widely deployed in numerous European laboratories. If Roentgen's apparatus had produced X-rays, then a number of other experimentalists must for some time have been producing those rays without knowing it. Perhaps those rays, which might well have other unacknowledged sources too, were implicated in behavior previously explained without reference to them. At the very least, several sorts of long familiar apparatus would in the future have to be shielded with lead. Previously completed work on normal projects would now have to be done again because earlier scientists had failed to recognize and control a relevant variable. X-rays, to be sure, opened up a new field and thus added to the potential domain of normal science. But they also, and this is now the more important point, changed fields that had already existed. In the process they denied previously paradigmatic types of instrumentation their right to that title.

In short, consciously or not, the decision to employ a particular piece of apparatus and to use it in a particular way carries an assumption that only certain sorts of circumstances will arise. There are instrumental as well as theoretical expectations, and they have often played a decisive role in scientific development. One such expectation is, for example, part of the story of oxygen's belated discovery. Using a standard test for "the goodness of air," both Priestley and Lavoisier mixed two volumes of their gas with one volume of nitric oxide, shook the mixture over water, and measured the volume of the gaseous residue. The previous experience from which this standard procedure had evolved assured them that with atmospheric air the residue

[8] Silvanus P. Thompson, *The Life of Sir William Thomson Baron Kelvin of Largs* (London, 1910), II, 1125.

would be one volume and that for any other gas (or for polluted air) it would be greater. In the oxygen experiments both found a residue close to one volume and identified the gas accordingly. Only much later and in part through an accident did Priestley renounce the standard procedure and try mixing nitric oxide with his gas in other proportions. He then found that with quadruple the volume of nitric oxide there was almost no residue at all. His commitment to the original test procedure—a procedure sanctioned by much previous experience—had been simultaneously a commitment to the non-existence of gases that could behave as oxygen did.[9]

Illustrations of this sort could be multiplied by reference, for example, to the belated identification of uranium fission. One reason why that nuclear reaction proved especially difficult to recognize was that men who knew what to expect when bombarding uranium chose chemical tests aimed mainly at elements from the upper end of the periodic table.[10] Ought we conclude from the frequency with which such instrumental commitments prove misleading that science should abandon standard tests and standard instruments? That would result in an inconceivable method of research. Paradigm procedures and applications are as necessary to science as paradigm laws and theories, and they have the same effects. Inevitably they restrict the phenomenological field accessible for scientific investigation at any

[9] Conant, *op. cit.*, pp. 18–20.

[10] K. K. Darrow, "Nuclear Fission," *Bell System Technical Journal*, XIX (1940), 267–89. Krypton, one of the two main fission products, seems not to have been identified by chemical means until after the reaction was well understood. Barium, the other product, was almost identified chemically at a late stage of the investigation because, as it happened, that element had to be added to the radioactive solution to precipitate the heavy element for which nuclear chemists were looking. Failure to separate that added barium from the radioactive product finally led, after the reaction had been repeatedly investigated for almost five years, to the following report: "As chemists we should be led by this research . . . to change all the names in the preceding [reaction] schema and thus write Ba, La, Ce instead of Ra, Ac, Th. But as 'nuclear chemists,' with close affiliations to physics, we cannot bring ourselves to this leap which would contradict all previous experience of nuclear physics. It may be that a series of strange accidents renders our results deceptive" (Otto Hahn and Fritz Strassman, "Uber den Nachweis und das Verhalten der bei der Bestrahlung des Urans mittels Neutronen entstehended Erdalkalimetalle," *Die Naturwissenschaften*, XXVII [1939], 15).

given time. Recognizing that much, we may simultaneously see an essential sense in which a discovery like X-rays necessitates paradigm change—and therefore change in both procedures and expectations—for a special segment of the scientific community. As a result, we may also understand how the discovery of X-rays could seem to open a strange new world to many scientists and could thus participate so effectively in the crisis that led to twentieth-century physics.

Our final example of scientific discovery, that of the Leyden jar, belongs to a class that may be described as theory-induced. Initially, the term may seem paradoxical. Much that has been said so far suggests that discoveries predicted by theory in advance are parts of normal science and result in no *new sort* of fact. I have, for example, previously referred to the discoveries of new chemical elements during the second half of the nineteenth century as proceeding from normal science in that way. But not all theories are paradigm theories. Both during pre-paradigm periods and during the crises that lead to large-scale changes of paradigm, scientists usually develop many speculative and unarticulated theories that can themselves point the way to discovery. Often, however, that discovery is not quite the one anticipated by the speculative and tentative hypothesis. Only as experiment and tentative theory are together articulated to a match does the discovery emerge and the theory become a paradigm.

The discovery of the Leyden jar displays all these features as well as the others we have observed before. When it began, there was no single paradigm for electrical research. Instead, a number of theories, all derived from relatively accessible phenomena, were in competition. None of them succeeded in ordering the whole variety of electrical phenomena very well. That failure is the source of several of the anomalies that provide background for the discovery of the Leyden jar. One of the competing schools of electricians took electricity to be a fluid, and that conception led a number of men to attempt bottling the fluid by holding a water-filled glass vial in their hands and touching the water to a conductor suspended from an active electrostatic generator. On removing the jar from the machine and touching the water (or a conductor connected to it) with his free hand, each of these investigators experienced a severe shock. Those first experiments did not, however, provide electricians with the Leyden jar. That device emerged more slowly, and it is again impossible to say just when its discovery was completed. The initial attempts to store electrical fluid worked only because investigators held the vial in their hands while standing upon the ground. Electricians had still to learn that the jar required an outer as well as an inner conducting coating and that the fluid is not really stored in the jar at all. Somewhere in the course of the investigations that showed them this, and which introduced them to several other anomalous effects, the device that we call the Leyden jar emerged. Furthermore, the experiments that led to its emergence, many of them performed by Franklin, were also the ones that necessitated the drastic revision of the fluid theory and thus provided the first full paradigm for electricity.[11]

To a greater or lesser extent (corresponding to the continuum from the shocking to the anticipated result), the characteristics common to the three examples above are characteristic of all discoveries from which new sorts of phenomena emerge. Those characteristics include: the previous awareness of anomaly, the gradual and simultaneous emergence of both observational and conceptual recognition, and the consequent change of paradigm categories and procedures often accompanied by resistance. There is even evidence that these same characteristics are built into the nature of the perceptual process itself. In a psychological experiment that deserves to be far better known outside the trade, Bruner and Postman asked experimental subjects to identify on short and controlled exposure a series of playing cards. Many of the cards were normal, but some were made anoma-

[11] For various stages in the Leyden jar's evolution, see I. B. Cohen, *Franklin and Newton: An Inquiry into Speculative Newtonian Experimental Science and Franklin's Work in Electricity as an Example Thereof* (Philadelphia, 1956), pp. 385-86, 400-406, 452-67, 506-7. The last stage is described by Whittaker, *op. cit.*, pp. 50-52.

lous, e.g., a red six of spades and a black four of hearts. Each experimental run was constituted by the display of a single card to a single subject in a series of gradually increased exposures. After each exposure the subject was asked what he had seen, and the run was terminated by two successive correct identifications.[12]

Even on the shortest exposures many subjects identified most of the cards, and after a small increase all the subjects identified them all. For the normal cards these identifications were usually correct, but the anomalous cards were almost always identified, without apparent hesitation or puzzlement, as normal. The black four of hearts might, for example, be identified as the four of either spades or hearts. Without any awareness of trouble, it was immediately fitted to one of the conceptual categories prepared by prior experience. One would not even like to say that the subjects had seen something different from what they identified. With a further increase of exposure to the anomalous cards, subjects did begin to hesitate and to display awareness of anomaly. Exposed, for example, to the red six of spades, some would say: That's the six of spades, but there's something wrong with it—the black has a red border. Further increase of exposure resulted in still more hesitation and confusion until finally, and sometimes quite suddenly, most subjects would produce the correct identification without hesitation. Moreover, after doing this with two or three of the anomalous cards, they would have little further difficulty with the others. A few subjects, however, were never able to make the requisite adjustment of their categories. Even at forty times the average exposure required to recognize normal cards for what they were, more than 10 per cent of the anomalous cards were not correctly identified. And the subjects who then failed often experienced acute personal distress. One of them exclaimed: "I can't make the suit out, whatever it is. It didn't even look like a card that time. I don't know what color it is now or whether it's a spade or a heart. I'm

[12] J. S. Bruner and Leo Postman, "On the Perception of Incongruity: A Paradigm," *Journal of Personality*, XVIII (1949), 206–23.

not even sure now what a spade looks like. My God!"[13] In the next section we shall occasionally see scientists behaving this way too.

Either as a metaphor or because it reflects the nature of the mind, that psychological experiment provides a wonderfully simple and cogent schema for the process of scientific discovery. In science, as in the playing card experiment, novelty emerges only with difficulty, manifested by resistance, against a background provided by expectation. Initially, only the anticipated and usual are experienced even under circumstances where anomaly is later to be observed. Further acquaintance, however, does result in awareness of something wrong or does relate the effect to something that has gone wrong before. That awareness of anomaly opens a period in which conceptual categories are adjusted until the initially anomalous has become the anticipated. At this point the discovery has been completed. I have already urged that that process or one very much like it is involved in the emergence of all fundamental scientific novelties. Let me now point out that, recognizing the process, we can at last begin to see why normal science, a pursuit not directed to novelties and tending at first to suppress them, should nevertheless be so effective in causing them to arise.

In the development of any science, the first received paradigm is usually felt to account quite successfully for most of the observations and experiments easily accessible to that science's practitioners. Further development, therefore, ordinarily calls for the construction of elaborate equipment, the development of an esoteric vocabulary and skills, and a refinement of concepts that increasingly lessens their resemblance to their usual common-sense prototypes. That professionalization leads, on the one hand, to an immense restriction of the scientist's vision and to a considerable resistance to paradigm change. The science has become increasingly rigid. On the other hand, within those areas to which the paradigm directs the attention of the

[13] *Ibid.*, p. 218. My colleague Postman tells me that, though knowing all about the apparatus and display in advance, he nevertheless found looking at the incongruous cards acutely uncomfortable.

group, normal science leads to a detail of information and to a precision of the observation-theory match that could be achieved in no other way. Furthermore, that detail and precision-of-match have a value that transcends their not always very high intrinsic interest. Without the special apparatus that is constructed mainly for anticipated functions, the results that lead ultimately to novelty could not occur. And even when the apparatus exists, novelty ordinarily emerges only for the man who, knowing *with precision* what he should expect, is able to recognize that something has gone wrong. Anomaly appears only against the background provided by the paradigm. The more precise and far-reaching that paradigm is, the more sensitive an indicator it provides of anomaly and hence of an occasion for paradigm change. In the normal mode of discovery, even resistance to change has a use that will be explored more fully in the next section. By ensuring that the paradigm will not be too easily surrendered, resistance guarantees that scientists will not be lightly distracted and that the anomalies that lead to paradigm change will penetrate existing knowledge to the core. The very fact that a significant scientific novelty so often emerges simultaneously from several laboratories is an index both to the strongly traditional nature of normal science and to the completeness with which that traditional pursuit prepares the way for its own change.

VII. Crisis and the Emergence of Scientific Theories

All the discoveries considered in Section VI were causes of or contributors to paradigm change. Furthermore, the changes in which these discoveries were implicated were all destructive as well as constructive. After the discovery had been assimilated, scientists were able to account for a wider range of natural phenomena or to account with greater precision for some of those previously known. But that gain was achieved only by discarding some previously standard beliefs or procedures and, simultaneously, by replacing those components of the previous paradigm with others. Shifts of this sort are, I have argued, associated with all discoveries achieved through normal science, excepting only the unsurprising ones that had been anticipated in all but their details. Discoveries are not, however, the only sources of these destructive-constructive paradigm changes. In this section we shall begin to consider the similar, but usually far larger, shifts that result from the invention of new theories.

Having argued already that in the sciences fact and theory, discovery and invention, are not categorically and permanently distinct, we can anticipate overlap between this section and the last. (The impossible suggestion that Priestley first discovered oxygen and Lavoisier then invented it has its attractions. Oxygen has already been encountered as discovery; we shall shortly meet it again as invention.) In taking up the emergence of new theories we shall inevitably extend our understanding of discovery as well. Still, overlap is not identity. The sorts of discoveries considered in the last section were not, at least singly, responsible for such paradigm shifts as the Copernican, Newtonian, chemical, and Einsteinian revolutions. Nor were they responsible for the somewhat smaller, because more exclusively professional, changes in paradigm produced by the wave theory of light, the dynamical theory of heat, or Maxwell's electromagnetic theory. How can theories like these arise from normal science, an activity even less directed to their pursuit than to that of discoveries?

If awareness of anomaly plays a role in the emergence of new sorts of phenomena, it should surprise no one that a similar but more profound awareness is prerequisite to all acceptable changes of theory. On this point historical evidence is, I think, entirely unequivocal. The state of Ptolemaic astronomy was a scandal before Copernicus' announcement.[1] Galileo's contributions to the study of motion depended closely upon difficulties discovered in Aristotle's theory by scholastic critics.[2] Newton's new theory of light and color originated in the discovery that none of the existing pre-paradigm theories would account for the length of the spectrum, and the wave theory that replaced Newton's was announced in the midst of growing concern about anomalies in the relation of diffraction and polarization effects to Newton's theory.[3] Thermodynamics was born from the collision of two existing nineteenth-century physical theories, and quantum mechanics from a variety of difficulties surrounding black-body radiation, specific heats, and the photoelectric effect.[4] Furthermore, in all these cases except that of Newton the awareness of anomaly had lasted so long and penetrated so deep that one can appropriately describe the fields affected by it as in a state of growing crisis. Because it demands large-scale paradigm destruction and major shifts in the problems and techniques of normal science, the emergence of new theories is generally preceded by a period of pronounced professional in-

1 A. R. Hall, *The Scientific Revolution, 1500–1800* (London, 1954), p. 16.

2 Marshall Clagett, *The Science of Mechanics in the Middle Ages* (Madison, Wis., 1959), Parts II–III. A. Koyré displays a number of medieval elements in Galileo's thought in his *Etudes Galiléennes* (Paris, 1939), particularly Vol. I.

3 For Newton, see T. S. Kuhn, "Newton's Optical Papers," in *Isaac Newton's Papers and Letters in Natural Philosophy*, ed. I. B. Cohen (Cambridge, Mass., 1958), pp. 27–45. For the prelude to the wave theory, see E. T. Whittaker, *A History of the Theories of Aether and Electricity*, I (2d ed.; London, 1951), 94–109; and W. Whewell, *History of the Inductive Sciences* (rev. ed.; London, 1847), II, 396–466.

4 For thermodynamics, see Silvanus P. Thompson, *Life of William Thomson Baron Kelvin of Largs* (London, 1910), I, 266–81. For the quantum theory, see Fritz Reiche, *The Quantum Theory*, trans. H. S. Hatfield and H. L. Brose (London, 1922), chaps. i–ii.

security. As one might expect, that insecurity is generated by the persistent failure of the puzzles of normal science to come out as they should. Failure of existing rules is the prelude to a search for new ones.

Look first at a particularly famous case of paradigm change, the emergence of Copernican astronomy. When its predecessor, the Ptolemaic system, was first developed during the last two centuries before Christ and the first two after, it was admirably successful in predicting the changing positions of both stars and planets. No other ancient system had performed so well; for the stars, Ptolemaic astronomy is still widely used today as an engineering approximation; for the planets, Ptolemy's predictions were as good as Copernicus'. But to be admirably successful is never, for a scientific theory, to be completely successful. With respect both to planetary position and to precession of the equinoxes, predictions made with Ptolemy's system never quite conformed with the best available observations. Further reduction of those minor discrepancies constituted many of the principal problems of normal astronomical research for many of Ptolemy's successors, just as a similar attempt to bring celestial observation and Newtonian theory together provided normal research problems for Newton's eighteenth-century successors. For some time astronomers had every reason to suppose that these attempts would be as successful as those that had led to Ptolemy's system. Given a particular discrepancy, astronomers were invariably able to eliminate it by making some particular adjustment in Ptolemy's system of compounded circles. But as time went on, a man looking at the net result of the normal research effort of many astronomers could observe that astronomy's complexity was increasing far more rapidly than its accuracy and that a discrepancy corrected in one place was likely to show up in another.[5]

Because the astronomical tradition was repeatedly interrupted from outside and because, in the absence of printing, communication between astronomers was restricted, these dif-

[5] J. L. E. Dreyer, *A History of Astronomy from Thales to Kepler* (2d ed.; New York, 1953), chaps. xi–xii.

ficulties were only slowly recognized. But awareness did come. By the thirteenth century Alfonso X could proclaim that if God had consulted him when creating the universe, he would have received good advice. In the sixteenth century, Copernicus' co-worker, Domenico da Novara, held that no system so cumbersome and inaccurate as the Ptolemaic had become could possibly be true of nature. And Copernicus himself wrote in the Preface to the *De Revolutionibus* that the astronomical tradition he inherited had finally created only a monster. By the early sixteenth century an increasing number of Europe's best astronomers were recognizing that the astronomical paradigm was failing in application to its own traditional problems. That recognition was prerequisite to Copernicus' rejection of the Ptolemaic paradigm and his search for a new one. His famous preface still provides one of the classic descriptions of a crisis state.[6]

Breakdown of the normal technical puzzle-solving activity is not, of course, the only ingredient of the astronomical crisis that faced Copernicus. An extended treatment would also discuss the social pressure for calendar reform, a pressure that made the puzzle of precession particularly urgent. In addition, a fuller account would consider medieval criticism of Aristotle, the rise of Renaissance Neoplatonism, and other significant historical elements besides. But technical breakdown would still remain the core of the crisis. In a mature science—and astronomy had become that in antiquity—external factors like those cited above are principally significant in determining the timing of breakdown, the ease with which it can be recognized, and the area in which, because it is given particular attention, the breakdown first occurs. Though immensely important, issues of that sort are out of bounds for this essay.

If that much is clear in the case of the Copernican revolution, let us turn from it to a second and rather different example, the crisis that preceded the emergence of Lavoisier's oxygen theory of combustion. In the 1770's many factors combined to generate

[6] T. S. Kuhn, *The Copernican Revolution* (Cambridge, Mass., 1957), pp. 135–43.

a crisis in chemistry, and historians are not altogether agreed about either their nature or their relative importance. But two of them are generally accepted as of first-rate significance: the rise of pneumatic chemistry and the question of weight relations. The history of the first begins in the seventeenth century with development of the air pump and its deployment in chemical experimentation. During the following century, using that pump and a number of other pneumatic devices, chemists came increasingly to realize that air must be an active ingredient in chemical reactions. But with a few exceptions—so equivocal that they may not be exceptions at all—chemists continued to believe that air was the only sort of gas. Until 1756, when Joseph Black showed that fixed air (CO_2) was consistently distinguishable from normal air, two samples of gas were thought to be distinct only in their impurities.[7]

After Black's work the investigation of gases proceeded rapidly, most notably in the hands of Cavendish, Priestley, and Scheele, who together developed a number of new techniques capable of distinguishing one sample of gas from another. All these men, from Black through Scheele, believed in the phlogiston theory and often employed it in their design and interpretation of experiments. Scheele actually first produced oxygen by an elaborate chain of experiments designed to dephlogisticate heat. Yet the net result of their experiments was a variety of gas samples and gas properties so elaborate that the phlogiston theory proved increasingly little able to cope with laboratory experience. Though none of these chemists suggested that the theory should be replaced, they were unable to apply it consistently. By the time Lavoisier began his experiments on airs in the early 1770's, there were almost as many versions of the phlogiston theory as there were pneumatic chemists.[8] That

[7] J. R. Partington, *A Short History of Chemistry* (2d ed.; London, 1951), pp. 48–51, 73–85, 90–120.

[8] Though their main concern is with a slightly later period, much relevant material is scattered throughout J. R. Partington and Douglas McKie's "Historical Studies on the Phlogiston Theory," *Annals of Science*, II (1937), 361–404; III (1938), 1–58, 337–71; and IV (1939), 337–71.

proliferation of versions of a theory is a very usual symptom of crisis. In his preface, Copernicus complained of it as well.

The increasing vagueness and decreasing utility of the phlogiston theory for pneumatic chemistry were not, however, the only source of the crisis that confronted Lavoisier. He was also much concerned to explain the gain in weight that most bodies experience when burned or roasted, and that again is a problem with a long prehistory. At least a few Islamic chemists had known that some metals gain weight when roasted. In the seventeenth century several investigators had concluded from this same fact that a roasted metal takes up some ingredient from the atmosphere. But in the seventeenth century that conclusion seemed unnecessary to most chemists. If chemical reactions could alter the volume, color, and texture of the ingredients, why should they not alter weight as well? Weight was not always taken to be the measure of quantity of matter. Besides, weight-gain on roasting remained an isolated phenomenon. Most natural bodies (e.g., wood) lose weight on roasting as the phlogiston theory was later to say they should.

During the eighteenth century, however, these initially adequate responses to the problem of weight-gain became increasingly difficult to maintain. Partly because the balance was increasingly used as a standard chemical tool and partly because the development of pneumatic chemistry made it possible and desirable to retain the gaseous products of reactions, chemists discovered more and more cases in which weight-gain accompanied roasting. Simultaneously, the gradual assimilation of Newton's gravitational theory led chemists to insist that gain in weight must mean gain in quantity of matter. Those conclusions did not result in rejection of the phlogiston theory, for that theory could be adjusted in many ways. Perhaps phlogiston had negative weight, or perhaps fire particles or something else entered the roasted body as phlogiston left it. There were other explanations besides. But if the problem of weight-gain did not lead to rejection, it did lead to an increasing number of special studies in which this problem bulked large. One of them, "On phlogiston considered as a substance with weight and [analyzed] in terms of the weight changes it produces in bodies with which it unites," was read to the French Academy early in 1772, the year which closed with Lavoisier's delivery of his famous sealed note to the Academy's Secretary. Before that note was written a problem that had been at the edge of the chemist's consciousness for many years had become an outstanding unsolved puzzle.[9] Many different versions of the phlogiston theory were being elaborated to meet it. Like the problems of pneumatic chemistry, those of weight-gain were making it harder and harder to know what the phlogiston theory was. Though still believed and trusted as a working tool, a paradigm of eighteenth-century chemistry was gradually losing its unique status. Increasingly, the research it guided resembled that conducted under the competing schools of the pre-paradigm period, another typical effect of crisis.

Consider now, as a third and final example, the late nineteenth century crisis in physics that prepared the way for the emergence of relativity theory. One root of that crisis can be traced to the late seventeenth century when a number of natural philosophers, most notably Leibniz, criticized Newton's retention of an updated version of the classic conception of absolute space.[10] They were very nearly, though never quite, able to show that absolute positions and absolute motions were without any function at all in Newton's system; and they did succeed in hinting at the considerable aesthetic appeal a fully relativistic conception of space and motion would later come to display. But their critique was purely logical. Like the early Copernicans who criticized Aristotle's proofs of the earth's stability, they did not dream that transition to a relativistic system could have observational consequences. At no point did they relate their views to any problems that arose when applying Newtonian theory to nature. As a result, their views died with

[9] H. Guerlac, *Lavoisier—the Crucial Year* (Ithaca, N.Y., 1961). The entire book documents the evolution and first recognition of a crisis. For a clear statement of the situation with respect to Lavoisier, see p. 35.

[10] Max Jammer, *Concepts of Space: The History of Theories of Space in Physics* (Cambridge, Mass., 1954), pp. 114–24.

them during the early decades of the eighteenth century to be resurrected only in the last decades of the nineteenth when they had a very different relation to the practice of physics.

The technical problems to which a relativistic philosophy of space was ultimately to be related began to enter normal science with the acceptance of the wave theory of light after about 1815, though they evoked no crisis until the 1890's. If light is wave motion propagated in a mechanical ether governed by Newton's Laws, then both celestial observation and terrestrial experiment become potentially capable of detecting drift through the ether. Of the celestial observations, only those of aberration promised sufficient accuracy to provide relevant information, and the detection of ether-drift by aberration measurements therefore became a recognized problem for normal research. Much special equipment was built to resolve it. That equipment, however, detected no observable drift, and the problem was therefore transferred from the experimentalists and observers to the theoreticians. During the central decades of the century Fresnel, Stokes, and others devised numerous articulations of the ether theory designed to explain the failure to observe drift. Each of these articulations assumed that a moving body drags some fraction of the ether with it. And each was sufficiently successful to explain the negative results not only of celestial observation but also of terrestrial experimentation, including the famous experiment of Michelson and Morley.[11] There was still no conflict excepting that between the various articulations. In the absence of relevant experimental techniques, that conflict never became acute.

The situation changed again only with the gradual acceptance of Maxwell's electromagnetic theory in the last two decades of the nineteenth century. Maxwell himself was a Newtonian who believed that light and electromagnetism in general were due to variable displacements of the particles of a mechanical ether. His earliest versions of a theory for electricity and

[11] Joseph Larmor, *Aether and Matter . . . Including a Discussion of the Influence of the Earth's Motion on Optical Phenomena* (Cambridge, 1900), pp. 6-20, 320–22.

magnetism made direct use of hypothetical properties with which he endowed this medium. These were dropped from his final version, but he still believed his electromagnetic theory compatible with some articulation of the Newtonian mechanical view.[12] Developing a suitable articulation was a challenge for him and his successors. In practice, however, as has happened again and again in scientific development, the required articulation proved immensely difficult to produce. Just as Copernicus' astronomical proposal, despite the optimism of its author, created an increasing crisis for existing theories of motion, so Maxwell's theory, despite its Newtonian origin, ultimately produced a crisis for the paradigm from which it had sprung.[13] Furthermore, the locus at which that crisis became most acute was provided by the problems we have just been considering, those of motion with respect to the ether.

Maxwell's discussion of the electromagnetic behavior of bodies in motion had made no reference to ether drag, and it proved very difficult to introduce such drag into his theory. As a result, a whole series of earlier observations designed to detect drift through the ether became anomalous. The years after 1890 therefore witnessed a long series of attempts, both experimental and theoretical, to detect motion with respect to the ether and to work ether drag into Maxwell's theory. The former were uniformly unsuccessful, though some analysts thought their results equivocal. The latter produced a number of promising starts, particularly those of Lorentz and Fitzgerald, but they also disclosed still other puzzles and finally resulted in just that proliferation of competing theories that we have previously found to be the concomitant of crisis.[14] It is against that historical setting that Einstein's special theory of relativity emerged in 1905.

These three examples are almost entirely typical. In each case a novel theory emerged only after a pronounced failure in the

[12] R. T. Glazebrook, *James Clerk Maxwell and Modern Physics* (London, 1896), chap. ix. For Maxwell's final attitude, see his own book, *A Treatise on Electricity and Magnetism* (3d ed.; Oxford, 1892), p. 470.

[13] For astronomy's role in the development of mechanics, see Kuhn, *op. cit.*, chap. vii.

[14] Whittaker, *op. cit.*, I, 386–410; and II (London, 1953), 27–40.

normal problem-solving activity. Furthermore, except for the case of Copernicus in which factors external to science played a particularly large role, that breakdown and the proliferation of theories that is its sign occurred no more than a decade or two before the new theory's enunciation. The novel theory seems a direct response to crisis. Note also, though this may not be quite so typical, that the problems with respect to which breakdown occurred were all of a type that had long been recognized. Previous practice of normal science had given every reason to consider them solved or all but solved, which helps to explain why the sense of failure, when it came, could be so acute. Failure with a new sort of problem is often disappointing but never surprising. Neither problems nor puzzles yield often to the first attack. Finally, these examples share another characteristic that may help to make the case for the role of crisis impressive: the solution to each of them had been at least partially anticipated during a period when there was no crisis in the corresponding science; and in the absence of crisis those anticipations had been ignored.

The only complete anticipation is also the most famous, that of Copernicus by Aristarchus in the third century B.C. It is often said that if Greek science had been less deductive and less ridden by dogma, heliocentric astronomy might have begun its development eighteen centuries earlier than it did.[15] But that is to ignore all historical context. When Aristarchus' suggestion was made, the vastly more reasonable geocentric system had no needs that a heliocentric system might even conceivably have fulfilled. The whole development of Ptolemaic astronomy, both its triumphs and its breakdown, falls in the centuries after Aristarchus' proposal. Besides, there were no obvious reasons for taking Aristarchus seriously. Even Copernicus' more elaborate proposal was neither simpler nor more accurate than Ptolemy's system. Available observational tests, as we shall see more clear-

[15] For Aristarchus' work, see T. L. Heath, *Aristarchus of Samos: The Ancient Copernicus* (Oxford, 1913), Part II. For an extreme statement of the traditional position about the neglect of Aristarchus' achievement, see Arthur Koestler, *The Sleepwalkers: A History of Man's Changing Vision of the Universe* (London, 1959), p. 50.

ly below, provided no basis for a choice between them. Under those circumstances, one of the factors that led astronomers to Copernicus (and one that could not have led them to Aristarchus) was the recognized crisis that had been responsible for innovation in the first place. Ptolemaic astronomy had failed to solve its problems; the time had come to give a competitor a chance. Our other two examples provide no similarly full anticipations. But surely one reason why the theories of combustion by absorption from the atmosphere—theories developed in the seventeenth century by Rey, Hooke, and Mayow—failed to get a sufficient hearing was that they made no contact with a recognized trouble spot in normal scientific practice.[16] And the long neglect by eighteenth- and nineteenth-century scientists of Newton's relativistic critics must largely have been due to a similar failure in confrontation.

Philosophers of science have repeatedly demonstrated that more than one theoretical construction can always be placed upon a given collection of data. History of science indicates that, particularly in the early developmental stages of a new paradigm, it is not even very difficult to invent such alternates. But that invention of alternates is just what scientists seldom undertake except during the pre-paradigm stage of their science's development and at very special occasions during its subsequent evolution. So long as the tools a paradigm supplies continue to prove capable of solving the problems it defines, science moves fastest and penetrates most deeply through confident employment of those tools. The reason is clear. As in manufacture so in science—retooling is an extravagance to be reserved for the occasion that demands it. The significance of crises is the indication they provide that an occasion for retooling has arrived.

[16] Partington, *op. cit.*, pp. 78–85.

VIII. The Response to Crisis

Let us then assume that crises are a necessary precondition for the emergence of novel theories and ask next how scientists respond to their existence. Part of the answer, as obvious as it is important, can be discovered by noting first what scientists never do when confronted by even severe and prolonged anomalies. Though they may begin to lose faith and then to consider alternatives, they do not renounce the paradigm that has led them into crisis. They do not, that is, treat anomalies as counterinstances, though in the vocabulary of philosophy of science that is what they are. In part this generalization is simply a statement from historic fact, based upon examples like those given above and, more extensively, below. These hint what our later examination of paradigm rejection will disclose more fully: once it has achieved the status of paradigm, a scientific theory is declared invalid only if an alternate candidate is available to take its place. No process yet disclosed by the historical study of scientific development at all resembles the methodological stereotype of falsification by direct comparison with nature. That remark does not mean that scientists do not reject scientific theories, or that experience and experiment are not essential to the process in which they do so. But it does mean—what will ultimately be a central point—that the act of judgment that leads scientists to reject a previously accepted theory is always based upon more than a comparison of that theory with the world. The decision to reject one paradigm is always simultaneously the decision to accept another, and the judgment leading to that decision involves the comparison of both paradigms with nature *and* with each other.

There is, in addition, a second reason for doubting that scientists reject paradigms because confronted with anomalies or counterinstances. In developing it my argument will itself foreshadow another of this essay's main theses. The reasons for doubt sketched above were purely factual; they were, that is, themselves counterinstances to a prevalent epistemological theory. As such, if my present point is correct, they can at best help to create a crisis or, more accurately, to reinforce one that is already very much in existence. By themselves they cannot and will not falsify that philosophical theory, for its defenders will do what we have already seen scientists doing when confronted by anomaly. They will devise numerous articulations and *ad hoc* modifications of their theory in order to eliminate any apparent conflict. Many of the relevant modifications and qualifications are, in fact, already in the literature. If, therefore, these epistemological counterinstances are to constitute more than a minor irritant, that will be because they help to permit the emergence of a new and different analysis of science within which they are no longer a source of trouble. Furthermore, if a typical pattern, which we shall later observe in scientific revolutions, is applicable here, these anomalies will then no longer seem to be simply facts. From within a new theory of scientific knowledge, they may instead seem very much like tautologies, statements of situations that could not conceivably have been otherwise.

It has often been observed, for example, that Newton's second law of motion, though it took centuries of difficult factual and theoretical research to achieve, behaves for those committed to Newton's theory very much like a purely logical statement that no amount of observation could refute.[1] In Section X we shall see that the chemical law of fixed proportion, which before Dalton was an occasional experimental finding of very dubious generality, became after Dalton's work an ingredient of a definition of chemical compound that no experimental work could by itself have upset. Something much like that will also happen to the generalization that scientists fail to reject paradigms when faced with anomalies or counterinstances. They could not do so and still remain scientists.

Though history is unlikely to record their names, some men have undoubtedly been driven to desert science because of

[1] See particularly the discussion in N. R. Hanson, *Patterns of Discovery* (Cambridge, 1958), pp. 99–105.

their inability to tolerate crisis. Like artists, creative scientists must occasionally be able to live in a world out of joint—elsewhere I have described that necessity as "the essential tension" implicit in scientific research.[2] But that rejection of science in favor of another occupation is, I think, the only sort of paradigm rejection to which counterinstances by themselves can lead. Once a first paradigm through which to view nature has been found, there is no such thing as research in the absence of any paradigm. To reject one paradigm without simultaneously substituting another is to reject science itself. That act reflects not on the paradigm but on the man. Inevitably he will be seen by his colleagues as "the carpenter who blames his tools."

The same point can be made at least equally effectively in reverse: there is no such thing as research without counterinstances. For what is it that differentiates normal science from science in a crisis state? Not, surely, that the former confronts no counterinstances. On the contrary, what we previously called the puzzles that constitute normal science exist only because no paradigm that provides a basis for scientific research ever completely resolves all its problems. The very few that have ever seemed to do so (e.g., geometric optics) have shortly ceased to yield research problems at all and have instead become tools for engineering. Excepting those that are exclusively instrumental, every problem that normal science sees as a puzzle can be seen, from another viewpoint, as a counterinstance and thus as a source of crisis. Copernicus saw as counterinstances what most of Ptolemy's other successors had seen as puzzles in the match between observation and theory. Lavoisier saw as a counterinstance what Priestley had seen as a successfully solved puzzle in the articulation of the phlogiston theory. And Einstein saw as counterinstances what Lorentz, Fitzgerald, and others had seen as puzzles in the articulation of Newton's and Max-

[2] T. S. Kuhn, "The Essential Tension: Tradition and Innovation in Scientific Research," in *The Third (1959) University of Utah Research Conference on the Identification of Creative Scientific Talent,* ed. Calvin W. Taylor (Salt Lake City, 1959), pp. 162–77. For the comparable phenomenon among artists, see Frank Barron, "The Psychology of Imagination," *Scientific American,* CXCIX (September, 1958), 151–66, esp. 160.

well's theories. Furthermore, even the existence of crisis does not by itself transform a puzzle into a counterinstance. There is no such sharp dividing line. Instead, by proliferating versions of the paradigm, crisis loosens the rules of normal puzzle-solving in ways that ultimately permit a new paradigm to emerge. There are, I think, only two alternatives: either no scientific theory ever confronts a counterinstance, or all such theories confront counterinstances at all times.

How can the situation have seemed otherwise? That question necessarily leads to the historical and critical elucidation of philosophy, and those topics are here barred. But we can at least note two reasons why science has seemed to provide so apt an illustration of the generalization that truth and falsity are uniquely and unequivocally determined by the confrontation of statement with fact. Normal science does and must continually strive to bring theory and fact into closer agreement, and that activity can easily be seen as testing or as a search for confirmation or falsification. Instead, its object is to solve a puzzle for whose very existence the validity of the paradigm must be assumed. Failure to achieve a solution discredits only the scientist and not the theory. Here, even more than above, the proverb applies: "It is a poor carpenter who blames his tools." In addition, the manner in which science pedagogy entangles discussion of a theory with remarks on its exemplary applications has helped to reinforce a confirmation-theory drawn predominantly from other sources. Given the slightest reason for doing so, the man who reads a science text can easily take the applications to be the evidence for the theory, the reasons why it ought to be believed. But science students accept theories on the authority of teacher and text, not because of evidence. What alternatives have they, or what competence? The applications given in texts are not there as evidence but because learning them is part of learning the paradigm at the base of current practice. If applications were set forth as evidence, then the very failure of texts to suggest alternative interpretations or to discuss problems for which scientists have failed to produce paradigm solutions would convict their authors of extreme bias. There is not the slightest reason for such an indictment.

How, then, to return to the initial question, do scientists respond to the awareness of an anomaly in the fit between theory and nature? What has just been said indicates that even a discrepancy unaccountably larger than that experienced in other applications of the theory need not draw any very profound response. There are always some discrepancies. Even the most stubborn ones usually respond at last to normal practice. Very often scientists are willing to wait, particularly if there are many problems available in other parts of the field. We have already noted, for example, that during the sixty years after Newton's original computation, the predicted motion of the moon's perigee remained only half of that observed. As Europe's best mathematical physicists continued to wrestle unsuccessfully with the well-known discrepancy, there were occasional proposals for a modification of Newton's inverse square law. But no one took these proposals very seriously, and in practice this patience with a major anomaly proved justified. Clairaut in 1750 was able to show that only the mathematics of the application had been wrong and that Newtonian theory could stand as before.[3] Even in cases where no mere mistake seems quite possible (perhaps because the mathematics involved is simpler or of a familiar and elsewhere successful sort), persistent and recognized anomaly does not always induce crisis. No one seriously questioned Newtonian theory because of the long-recognized discrepancies between predictions from that theory and both the speed of sound and the motion of Mercury. The first discrepancy was ultimately and quite unexpectedly resolved by experiments on heat undertaken for a very different purpose; the second vanished with the general theory of relativity after a crisis that it had had no role in creating.[4] Apparent-

[3] W. Whewell, *History of the Inductive Sciences* (rev. ed.; London, 1847), II, 220–21.

[4] For the speed of sound, see T. S. Kuhn, "The Caloric Theory of Adiabatic Compression," *Isis*, XLIV (1958), 136–37. For the secular shift in Mercury's perihelion, see E. T. Whittaker, *A History of the Theories of Aether and Electricity*, II (London, 1953), 151, 179.

ly neither had seemed sufficiently fundamental to evoke the malaise that goes with crisis. They could be recognized as counterinstances and still be set aside for later work.

It follows that if an anomaly is to evoke crisis, it must usually be more than just an anomaly. There are always difficulties somewhere in the paradigm-nature fit; most of them are set right sooner or later, often by processes that could not have been foreseen. The scientist who pauses to examine every anomaly he notes will seldom get significant work done. We therefore have to ask what it is that makes an anomaly seem worth concerted scrutiny, and to that question there is probably no fully general answer. The cases we have already examined are characteristic but scarcely prescriptive. Sometimes an anomaly will clearly call into question explicit and fundamental generalizations of the paradigm, as the problem of ether drag did for those who accepted Maxwell's theory. Or, as in the Copernican revolution, an anomaly without apparent fundamental import may evoke crisis if the applications that it inhibits have a particular practical importance, in this case for calendar design and astrology. Or, as in eighteenth-century chemistry, the development of normal science may transform an anomaly that had previously been only a vexation into a source of crisis: the problem of weight relations had a very different status after the evolution of pneumatic-chemical techniques. Presumably there are still other circumstances that can make an anomaly particularly pressing, and ordinarily several of these will combine. We have already noted, for example, that one source of the crisis that confronted Copernicus was the mere length of time during which astronomers had wrestled unsuccessfully with the reduction of the residual discrepancies in Ptolemy's system.

When, for these reasons or others like them, an anomaly comes to seem more than just another puzzle of normal science, the transition to crisis and to extraordinary science has begun. The anomaly itself now comes to be more generally recognized as such by the profession. More and more attention is devoted to it by more and more of the field's most eminent men. If it still continues to resist, as it usually does not, many of them may come to view its resolution as *the* subject matter of their discipline. For them the field will no longer look quite the same as it had earlier. Part of its different appearance results simply from the new fixation point of scientific scrutiny. An even more important source of change is the divergent nature of the numerous partial solutions that concerted attention to the problem has made available. The early attacks upon the resistant problem will have followed the paradigm rules quite closely. But with continuing resistance, more and more of the attacks upon it will have involved some minor or not so minor articulation of the paradigm, no two of them quite alike, each partially successful, but none sufficiently so to be accepted as paradigm by the group. Through this proliferation of divergent articulations (more and more frequently they will come to be described as *ad hoc* adjustments), the rules of normal science become increasingly blurred. Though there still is a paradigm, few practitioners prove to be entirely agreed about what it is. Even formerly standard solutions of solved problems are called in question.

When acute, this situation is sometimes recognized by the scientists involved. Copernicus complained that in his day astronomers were so "inconsistent in these [astronomical] investigations . . . that they cannot even explain or observe the constant length of the seasonal year." "With them," he continued, "it is as though an artist were to gather the hands, feet, head and other members for his images from diverse models, each part excellently drawn, but not related to a single body, and since they in no way match each other, the result would be monster rather than man."[5] Einstein, restricted by current usage to less florid language, wrote only, "It was as if the ground had been pulled out from under one, with no firm foundation to be seen anywhere, upon which one could have built."[6] And Wolfgang Pauli, in the months before Heisenberg's paper on matrix

[5] Quoted in T. S. Kuhn, *The Copernican Revolution* (Cambridge, Mass., 1957), p. 138.

[6] Albert Einstein, "Autobiographical Note," in *Albert Einstein: Philosopher-Scientist,* ed. P. A. Schilpp (Evanston, Ill., 1949), p. 45.

mechanics pointed the way to a new quantum theory, wrote to a friend, "At the moment physics is again terribly confused. In any case, it is too difficult for me, and I wish I had been a movie comedian or something of the sort and had never heard of physics." That testimony is particularly impressive if contrasted with Pauli's words less than five months later: "Heisenberg's type of mechanics has again given me hope and joy in life. To be sure it does not supply the solution to the riddle, but I believe it is again possible to march forward."[7]

Such explicit recognitions of breakdown are extremely rare, but the effects of crisis do not entirely depend upon its conscious recognition. What can we say these effects are? Only two of them seem to be universal. All crises begin with the blurring of a paradigm and the consequent loosening of the rules for normal research. In this respect research during crisis very much resembles research during the pre-paradigm period, except that in the former the locus of difference is both smaller and more clearly defined. And all crises close in one of three ways. Sometimes normal science ultimately proves able to handle the crisis-provoking problem despite the despair of those who have seen it as the end of an existing paradigm. On other occasions the problem resists even apparently radical new approaches. Then scientists may conclude that no solution will be forthcoming in the present state of their field. The problem is labelled and set aside for a future generation with more developed tools. Or, finally, the case that will most concern us here, a crisis may end with the emergence of a new candidate for paradigm and with the ensuing battle over its acceptance. This last mode of closure will be considered at length in later sections, but we must anticipate a bit of what will be said there in order to complete these remarks about the evolution and anatomy of the crisis state.

The transition from a paradigm in crisis to a new one from which a new tradition of normal science can emerge is far from a cumulative process, one achieved by an articulation or exten-

[7] Ralph Kronig, "The Turning Point," in *Theoretical Physics in the Twentieth Century: A Memorial Volume to Wolfgang Pauli*, ed. M. Fierz and V. F. Weisskopf (New York, 1960), pp. 22, 25–26. Much of this article describes the crisis in quantum mechanics in the years immediately before 1925.

sion of the old paradigm. Rather it is a reconstruction of the field from new fundamentals, a reconstruction that changes some of the field's most elementary theoretical generalizations as well as many of its paradigm methods and applications. During the transition period there will be a large but never complete overlap between the problems that can be solved by the old and by the new paradigm. But there will also be a decisive difference in the modes of solution. When the transition is complete, the profession will have changed its view of the field, its methods, and its goals. One perceptive historian, viewing a classic case of a science's reorientation by paradigm change, recently described it as "picking up the other end of the stick," a process that involves "handling the same bundle of data as before, but placing them in a new system of relations with one another by giving them a different framework."[8] Others who have noted this aspect of scientific advance have emphasized its similarity to a change in visual gestalt: the marks on paper that were first seen as a bird are now seen as an antelope, or vice versa.[9] That parallel can be misleading. Scientists do not see something *as* something else; instead, they simply see it. We have already examined some of the problems created by saying that Priestley saw oxygen as dephlogisticated air. In addition, the scientist does not preserve the gestalt subject's freedom to switch back and forth between ways of seeing. Nevertheless, the switch of gestalt, particularly because it is today so familiar, is a useful elementary prototype for what occurs in full-scale paradigm shift.

The preceding anticipation may help us recognize crisis as an appropriate prelude to the emergence of new theories, particularly since we have already examined a small-scale version of the same process in discussing the emergence of discoveries. Just because the emergence of a new theory breaks with one tradition of scientific practice and introduces a new one conducted under different rules and within a different universe of

[8] Herbert Butterfield, *The Origins of Modern Science, 1300–1800* (London, 1949), pp. 1–7.

[9] Hanson, *op. cit.*, chap. i.

discourse, it is likely to occur only when the first tradition is felt to have gone badly astray. That remark is, however, no more than a prelude to the investigation of the crisis-state, and, unfortunately, the questions to which it leads demand the competence of the psychologist even more than that of the historian. What is extraordinary research like? How is anomaly made lawlike? How do scientists proceed when aware only that something has gone fundamentally wrong at a level with which their training has not equipped them to deal? Those questions need far more investigation, and it ought not all be historical. What follows will necessarily be more tentative and less complete than what has gone before.

Often a new paradigm emerges, at least in embryo, before a crisis has developed far or been explicitly recognized. Lavoisier's work provides a case in point. His sealed note was deposited with the French Academy less than a year after the first thorough study of weight relations in the phlogiston theory and before Priestley's publications had revealed the full extent of the crisis in pneumatic chemistry. Or again, Thomas Young's first accounts of the wave theory of light appeared at a very early stage of a developing crisis in optics, one that would be almost unnoticeable except that, with no assistance from Young, it had grown to an international scientific scandal within a decade of the time he first wrote. In cases like these one can say only that a minor breakdown of the paradigm and the very first blurring of its rules for normal science were sufficient to induce in someone a new way of looking at the field. What intervened between the first sense of trouble and the recognition of an available alternate must have been largely unconscious.

In other cases, however—those of Copernicus, Einstein, and contemporary nuclear theory, for example—considerable time elapses between the first consciousness of breakdown and the emergence of a new paradigm. When that occurs, the historian may capture at least a few hints of what extraordinary science is like. Faced with an admittedly fundamental anomaly in theory, the scientist's first effort will often be to isolate it more precisely and to give it structure. Though now aware that they cannot be quite right, he will push the rules of normal science harder than ever to see, in the area of difficulty, just where and how far they can be made to work. Simultaneously he will seek for ways of magnifying the breakdown, of making it more striking and perhaps also more suggestive than it had been when displayed in experiments the outcome of which was thought to be known in advance. And in the latter effort, more than in any other part of the post-paradigm development of science, he will look almost like our most prevalent image of the scientist. He will, in the first place, often seem a man searching at random, trying experiments just to see what will happen, looking for an effect whose nature he cannot quite guess. Simultaneously, since no experiment can be conceived without some sort of theory, the scientist in crisis will constantly try to generate speculative theories that, if successful, may disclose the road to a new paradigm and, if unsuccessful, can be surrendered with relative ease.

Kepler's account of his prolonged struggle with the motion of Mars and Priestley's description of his response to the proliferation of new gases provide classic examples of the more random sort of research produced by the awareness of anomaly.[10] But probably the best illustrations of all come from contemporary research in field theory and on fundamental particles. In the absence of a crisis that made it necessary to see just how far the rules of normal science could stretch, would the immense effort required to detect the neutrino have seemed justified? Or, if the rules had not obviously broken down at some undisclosed point, would the radical hypothesis of parity non-conservation have been either suggested or tested? Like much other research in physics during the past decade, these experiments were in part attempts to localize and define the source of a still diffuse set of anomalies.

This sort of extraordinary research is often, though by no

[10] For an account of Kepler's work on Mars, see J. L. E. Dreyer, *A History of Astronomy from Thales to Kepler* (2d ed.; New York, 1953), pp. 380–93. Occasional inaccuracies do not prevent Dreyer's précis from providing the material needed here. For Priestley, see his own work, esp. *Experiments and Observations on Different Kinds of Air* (London, 1774–75).

means generally, accompanied by another. It is, I think, particularly in periods of acknowledged crisis that scientists have turned to philosophical analysis as a device for unlocking the riddles of their field. Scientists have not generally needed or wanted to be philosophers. Indeed, normal science usually holds creative philosophy at arm's length, and probably for good reasons. To the extent that normal research work can be conducted by using the paradigm as a model, rules and assumptions need not be made explicit. In Section V we noted that the full set of rules sought by philosophical analysis need not even exist. But that is not to say that the search for assumptions (even for non-existent ones) cannot be an effective way to weaken the grip of a tradition upon the mind and to suggest the basis for a new one. It is no accident that the emergence of Newtonian physics in the seventeenth century and of relativity and quantum mechanics in the twentieth should have been both preceded and accompanied by fundamental philosophical analyses of the contemporary research tradition.[11] Nor is it an accident that in both these periods the so-called thought experiment should have played so critical a role in the progress of research. As I have shown elsewhere, the analytical thought experimentation that bulks so large in the writings of Galileo, Einstein, Bohr, and others is perfectly calculated to expose the old paradigm to existing knowledge in ways that isolate the root of crisis with a clarity unattainable in the laboratory.[12]

With the deployment, singly or together, of these extraordinary procedures, one other thing may occur. By concentrating scientific attention upon a narrow area of trouble and by preparing the scientific mind to recognize experimental anomalies for what they are, crisis often proliferates new discoveries. We have already noted how the awareness of crisis distinguishes

[11] For the philosophical counterpoint that accompanied seventeenth-century mechanics, see René Dugas, *La mécanique au XVIIe siècle* (Neuchatel, 1954), particularly chap. xi. For the similar nineteenth-century episode, see the same author's earlier book, *Histoire de la mécanique* (Neuchatel, 1950), pp. 419–43.

[12] T. S. Kuhn, "A Function for Thought Experiments," in *Mélanges Alexandre Koyré,* ed. R. Taton and I. B. Cohen, to be published by Hermann (Paris) in 1963.

Lavoisier's work on oxygen from Priestley's; and oxygen was not the only new gas that the chemists aware of anomaly were able to discover in Priestley's work. Or again, new optical discoveries accumulated rapidly just before and during the emergence of the wave theory of light. Some, like polarization by reflection, were a result of the accidents that concentrated work in an area of trouble makes likely. (Malus, who made the discovery, was just starting work for the Academy's prize essay on double refraction, a subject widely known to be in an unsatisfactory state.) Others, like the light spot at the center of the shadow of a circular disk, were predictions from the new hypothesis, ones whose success helped to transform it to a paradigm for later work. And still others, like the colors of scratches and of thick plates, were effects that had often been seen and occasionally remarked before, but that, like Priestley's oxygen, had been assimilated to well-known effects in ways that prevented their being seen for what they were.[13] A similar account could be given of the multiple discoveries that, from about 1895, were a constant concomitant of the emergence of quantum mechanics.

Extraordinary research must have still other manifestations and effects, but in this area we have scarcely begun to discover the questions that need to be asked. Perhaps, however, no more are needed at this point. The preceding remarks should suffice to show how crisis simultaneously loosens the stereotypes and provides the incremental data necessary for a fundamental paradigm shift. Sometimes the shape of the new paradigm is foreshadowed in the structure that extraordinary research has given to the anomaly. Einstein wrote that before he had any substitute for classical mechanics, he could see the interrelation between the known anomalies of black-body radiation, the photoelectric effect, and specific heats.[14] More often no such structure is consciously seen in advance. Instead, the new paradigm, or a sufficient hint to permit later articulation, emerges

[13] For the new optical discoveries in general, see V. Ronchi, *Histoire de la lumière* (Paris, 1956), chap. vii. For the earlier explanation of one of these effects, see J. Priestley, *The History and Present State of Discoveries Relating to Vision, Light and Colours* (London, 1772), pp. 498–520.

[14] Einstein, *loc. cit.*

all at once, sometimes in the middle of the night, in the mind of a man deeply immersed in crisis. What the nature of that final stage is—how an individual invents (or finds he has invented) a new way of giving order to data now all assembled—must here remain inscrutable and may be permanently so. Let us here note only one thing about it. Almost always the men who achieve these fundamental inventions of a new paradigm have been either very young or very new to the field whose paradigm they change.[15] And perhaps that point need not have been made explicit, for obviously these are the men who, being little committed by prior practice to the traditional rules of normal science, are particularly likely to see that those rules no longer define a playable game and to conceive another set that can replace them.

The resulting transition to a new paradigm is scientific revolution, a subject that we are at long last prepared to approach directly. Note first, however, one last and apparently elusive respect in which the material of the last three sections has prepared the way. Until Section VI, where the concept of anomaly was first introduced, the terms 'revolution' and 'extraordinary science' may have seemed equivalent. More important, neither term may have seemed to mean more than 'non-normal science,' a circularity that will have bothered at least a few readers. In practice, it need not have done so. We are about to discover that a similar circularity is characteristic of scientific theories. Bothersome or not, however, that circularity is no longer unqualified. This section of the essay and the two preceding have educed numerous criteria of a breakdown in normal scientific activity, criteria that do not at all depend upon whether breakdown is succeeded by revolution. Confronted with anomaly or

[15] This generalization about the role of youth in fundamental scientific research is so common as to be a cliché. Furthermore, a glance at almost any list of fundamental contributions to scientific theory will provide impressionistic confirmation. Nevertheless, the generalization badly needs systematic investigation. Harvey C. Lehman (*Age and Achievement* [Princeton, 1953]) provides many useful data; but his studies make no attempt to single out contributions that involve fundamental reconceptualization. Nor do they inquire about the special circumstances, if any, that may accompany relatively late productivity in the sciences.

with crisis, scientists take a different attitude toward existing paradigms, and the nature of their research changes accordingly. The proliferation of competing articulations, the willingness to try anything, the expression of explicit discontent, the recourse to philosophy and to debate over fundamentals, all these are symptoms of a transition from normal to extraordinary research. It is upon their existence more than upon that of revolutions that the notion of normal science depends.

IX. The Nature and Necessity of Scientific Revolutions

These remarks permit us at last to consider the problems that provide this essay with its title. What are scientific revolutions, and what is their function in scientific development? Much of the answer to these questions has been anticipated in earlier sections. In particular, the preceding discussion has indicated that scientific revolutions are here taken to be those non-cumulative developmental episodes in which an older paradigm is replaced in whole or in part by an incompatible new one. There is more to be said, however, and an essential part of it can be introduced by asking one further question. Why should a change of paradigm be called a revolution? In the face of the vast and essential differences between political and scientific development, what parallelism can justify the metaphor that finds revolutions in both?

One aspect of the parallelism must already be apparent. Political revolutions are inaugurated by a growing sense, often restricted to a segment of the political community, that existing institutions have ceased adequately to meet the problems posed by an environment that they have in part created. In much the same way, scientific revolutions are inaugurated by a growing sense, again often restricted to a narrow subdivision of the scientific community, that an existing paradigm has ceased to function adequately in the exploration of an aspect of nature to which that paradigm itself had previously led the way. In both political and scientific development the sense of malfunction that can lead to crisis is prerequisite to revolution. Furthermore, though it admittedly strains the metaphor, that parallelism holds not only for the major paradigm changes, like those attributable to Copernicus and Lavoisier, but also for the far smaller ones associated with the assimilation of a new sort of phenomenon, like oxygen or X-rays. Scientific revolutions, as we noted at the end of Section V, need seem revolutionary only to those whose paradigms are affected by them. To outsiders they may, like the Balkan revolutions of the early twentieth century, seem normal parts of the developmental process. Astronomers, for example, could accept X-rays as a mere addition to knowledge, for their paradigms were unaffected by the existence of the new radiation. But for men like Kelvin, Crookes, and Roentgen, whose research dealt with radiation theory or with cathode ray tubes, the emergence of X-rays necessarily violated one paradigm as it created another. That is why these rays could be discovered only through something's first going wrong with normal research.

This genetic aspect of the parallel between political and scientific development should no longer be open to doubt. The parallel has, however, a second and more profound aspect upon which the significance of the first depends. Political revolutions aim to change political institutions in ways that those institutions themselves prohibit. Their success therefore necessitates the partial relinquishment of one set of institutions in favor of another, and in the interim, society is not fully governed by institutions at all. Initially it is crisis alone that attenuates the role of political institutions as we have already seen it attenuate the role of paradigms. In increasing numbers individuals become increasingly estranged from political life and behave more and more eccentrically within it. Then, as the crisis deepens, many of these individuals commit themselves to some concrete proposal for the reconstruction of society in a new institutional framework. At that point the society is divided into competing camps or parties, one seeking to defend the old institutional constellation, the others seeking to institute some new one. And, once that polarization has occurred, *political recourse fails.* Because they differ about the institutional matrix within which political change is to be achieved and evaluated, because they acknowledge no supra-institutional framework for the adjudication of revolutionary difference, the parties to a revolutionary conflict must finally resort to the techniques of mass persuasion, often including force. Though revolutions have had a vital role in the evolution of political institutions, that role depends upon their being partially extrapolitical or extrainstitutional events.

The remainder of this essay aims to demonstrate that the historical study of paradigm change reveals very similar characteristics in the evolution of the sciences. Like the choice between competing political institutions, that between competing paradigms proves to be a choice between incompatible modes of community life. Because it has that character, the choice is not and cannot be determined merely by the evaluative procedures characteristic of normal science, for these depend in part upon a particular paradigm, and that paradigm is at issue. When paradigms enter, as they must, into a debate about paradigm choice, their role is necessarily circular. Each group uses its own paradigm to argue in that paradigm's defense.

The resulting circularity does not, of course, make the arguments wrong or even ineffectual. The man who premises a paradigm when arguing in its defense can nonetheless provide a clear exhibit of what scientific practice will be like for those who adopt the new view of nature. That exhibit can be immensely persuasive, often compellingly so. Yet, whatever its force, the status of the circular argument is only that of persuasion. It cannot be made logically or even probabilistically compelling for those who refuse to step into the circle. The premises and values shared by the two parties to a debate over paradigms are not sufficiently extensive for that. As in political revolutions, so in paradigm choice—there is no standard higher than the assent of the relevant community. To discover how scientific revolutions are effected, we shall therefore have to examine not only the impact of nature and of logic, but also the techniques of persuasive argumentation effective within the quite special groups that constitute the community of scientists.

To discover why this issue of paradigm choice can never be unequivocally settled by logic and experiment alone, we must shortly examine the nature of the differences that separate the proponents of a traditional paradigm from their revolutionary successors. That examination is the principal object of this section and the next. We have, however, already noted numerous examples of such differences, and no one will doubt that history can supply many others. What is more likely to be doubted than their existence—and what must therefore be considered first—is that such examples provide essential information about the nature of science. Granting that paradigm rejection has been a historic fact, does it illuminate more than human credulity and confusion? Are there intrinsic reasons why the assimilation of either a new sort of phenomenon or a new scientific theory must demand the rejection of an older paradigm?

First notice that if there are such reasons, they do not derive from the logical structure of scientific knowledge. In principle, a new phenomenon might emerge without reflecting destructively upon any part of past scientific practice. Though discovering life on the moon would today be destructive of existing paradigms (these tell us things about the moon that seem incompatible with life's existence there), discovering life in some less well-known part of the galaxy would not. By the same token, a new theory does not have to conflict with any of its predecessors. It might deal exclusively with phenomena not previously known, as the quantum theory deals (but, significantly, not exclusively) with subatomic phenomena unknown before the twentieth century. Or again, the new theory might be simply a higher level theory than those known before, one that linked together a whole group of lower level theories without substantially changing any. Today, the theory of energy conservation provides just such links between dynamics, chemistry, electricity, optics, thermal theory, and so on. Still other compatible relationships between old and new theories can be conceived. Any and all of them might be exemplified by the historical process through which science has developed. If they were, scientific development would be genuinely cumulative. New sorts of phenomena would simply disclose order in an aspect of nature where none had been seen before. In the evolution of science new knowledge would replace ignorance rather than replace knowledge of another and incompatible sort.

Of course, science (or some other enterprise, perhaps less effective) might have developed in that fully cumulative manner. Many people have believed that it did so, and most still seem to suppose that cumulation is at least the ideal that historical development would display if only it had not so often been distorted by human idiosyncrasy. There are important reasons for that belief. In Section X we shall discover how closely the view of science-as-cumulation is entangled with a dominant epistemology that takes knowledge to be a construction placed directly upon raw sense data by the mind. And in Section XI we shall examine the strong support provided to the same historiographic schema by the techniques of effective science pedagogy. Nevertheless, despite the immense plausibility of that ideal image, there is increasing reason to wonder whether it can possibly be an image of *science*. After the pre-paradigm period the assimilation of all new theories and of almost all new sorts of phenomena has in fact demanded the destruction of a prior paradigm and a consequent conflict between competing schools of scientific thought. Cumulative acquisition of unanticipated novelties proves to be an almost non-existent exception to the rule of scientific development. The man who takes historic fact seriously must suspect that science does not tend toward the ideal that our image of its cumulativeness has suggested. Perhaps it is another sort of enterprise.

If, however, resistant facts can carry us that far, then a second look at the ground we have already covered may suggest that cumulative acquisition of novelty is not only rare in fact but improbable in principle. Normal research, which *is* cumulative, owes its success to the ability of scientists regularly to select problems that can be solved with conceptual and instrumental techniques close to those already in existence. (That is why an excessive concern with useful problems, regardless of their relation to existing knowledge and technique, can so easily inhibit scientific development.) The man who is striving to solve a problem defined by existing knowledge and technique is not, however, just looking around. He knows what he wants to achieve, and he designs his instruments and directs his thoughts accordingly. Unanticipated novelty, the new discovery, can emerge only to the extent that his anticipations about nature and his instruments prove wrong. Often the importance of the resulting discovery will itself be proportional to the extent and stubbornness of the anomaly that foreshadowed it. Obviously, then, there must be a conflict between the paradigm that discloses anomaly and the one that later renders the anomaly lawlike. The examples of discovery through paradigm destruction examined in Section VI did not confront us with mere historical accident. There is no other effective way in which discoveries might be generated.

The same argument applies even more clearly to the invention of new theories. There are, in principle, only three types of phenomena about which a new theory might be developed. The first consists of phenomena already well explained by existing paradigms, and these seldom provide either motive or point of departure for theory construction. When they do, as with the three famous anticipations discussed at the end of Section VII, the theories that result are seldom accepted, because nature provides no ground for discrimination. A second class of phenomena consists of those whose nature is indicated by existing paradigms but whose details can be understood only through further theory articulation. These are the phenomena to which scientists direct their research much of the time, but that research aims at the articulation of existing paradigms rather than at the invention of new ones. Only when these attempts at articulation fail do scientists encounter the third type of phenomena, the recognized anomalies whose characteristic feature is their stubborn refusal to be assimilated to existing paradigms. This type alone gives rise to new theories. Paradigms provide all phenomena except anomalies with a theory-determined place in the scientist's field of vision.

But if new theories are called forth to resolve anomalies in the relation of an existing theory to nature, then the successful new theory must somewhere permit predictions that are different from those derived from its predecessor. That difference could not occur if the two were logically compatible. In the process of being assimilated, the second must displace the first. Even a theory like energy conservation, which today seems a logical superstructure that relates to nature only through independently established theories, did not develop historically without paradigm destruction. Instead, it emerged from a crisis in which an essential ingredient was the incompatibility between Newtonian dynamics and some recently formulated consequences of the caloric theory of heat. Only after the caloric theory had been rejected could energy conservation become part of science.[1] And only after it had been part of science for some time could it come to seem a theory of a logically higher type, one not in conflict with its predecessors. It is hard to see how new theories could arise without these destructive changes in beliefs about nature. Though logical inclusiveness remains a permissible view of the relation between successive scientific theories, it is a historical implausibility.

A century ago it would, I think, have been possible to let the case for the necessity of revolutions rest at this point. But today, unfortunately, that cannot be done because the view of the subject developed above cannot be maintained if the most prevalent contemporary interpretation of the nature and function of scientific theory is accepted. That interpretation, closely associated with early logical positivism and not categorically rejected by its successors, would restrict the range and meaning of an accepted theory so that it could not possibly conflict with any later theory that made predictions about some of the same natural phenomena. The best-known and the strongest case for this restricted conception of a scientific theory emerges in discussions of the relation between contemporary Einsteinian dynamics and the older dynamical equations that descend from Newton's *Principia*. From the viewpoint of this essay these two theories are fundamentally incompatible in the sense illustrated by the relation of Copernican to Ptolemaic astronomy: Einstein's theory can be accepted only with the recognition that Newton's was wrong. Today this remains a minority view.[2] We must therefore examine the most prevalent objections to it.

[1] Silvanus P. Thompson, *Life of William Thomson Baron Kelvin of Largs* (London, 1910), I, 266–81.

[2] See, for example, the remarks by P. P. Wiener in *Philosophy of Science*, XXV (1958), 298.

The gist of these objections can be developed as follows. Relativistic dynamics cannot have shown Newtonian dynamics to be wrong, for Newtonian dynamics is still used with great success by most engineers and, in selected applications, by many physicists. Furthermore, the propriety of this use of the older theory can be proved from the very theory that has, in other applications, replaced it. Einstein's theory can be used to show that predictions from Newton's equations will be as good as our measuring instruments in all applications that satisfy a small number of restrictive conditions. For example, if Newtonian theory is to provide a good approximate solution, the relative velocities of the bodies considered must be small compared with the velocity of light. Subject to this condition and a few others, Newtonian theory seems to be derivable from Einsteinian, of which it is therefore a special case.

But, the objection continues, no theory can possibly conflict with one of its special cases. If Einsteinian science seems to make Newtonian dynamics wrong, that is only because some Newtonians were so incautious as to claim that Newtonian theory yielded entirely precise results or that it was valid at very high relative velocities. Since they could not have had any evidence for such claims, they betrayed the standards of science when they made them. In so far as Newtonian theory was ever a truly scientific theory supported by valid evidence, it still is. Only extravagant claims for the theory—claims that were never properly parts of science—can have been shown by Einstein to be wrong. Purged of these merely human extravagances, Newtonian theory has never been challenged and cannot be.

Some variant of this argument is quite sufficient to make any theory ever used by a significant group of competent scientists immune to attack. The much-maligned phlogiston theory, for example, gave order to a large number of physical and chemical phenomena. It explained why bodies burned—they were rich in phlogiston—and why metals had so many more properties in common than did their ores. The metals were all compounded from different elementary earths combined with phlogiston, and the latter, common to all metals, produced common properties. In addition, the phlogiston theory accounted for a number of reactions in which acids were formed by the combustion of substances like carbon and sulphur. Also, it explained the decrease of volume when combustion occurs in a confined volume of air—the phlogiston released by combustion "spoils" the elasticity of the air that absorbed it, just as fire "spoils" the elasticity of a steel spring.[3] If these were the only phenomena that the phlogiston theorists had claimed for their theory, that theory could never have been challenged. A similar argument will suffice for any theory that has ever been successfully applied to any range of phenomena at all.

But to save theories in this way, their range of application must be restricted to those phenomena and to that precision of observation with which the experimental evidence in hand already deals.[4] Carried just a step further (and the step can scarcely be avoided once the first is taken), such a limitation prohibits the scientist from claiming to speak "scientifically" about any phenomenon not already observed. Even in its present form the restriction forbids the scientist to rely upon a theory in his own research whenever that research enters an area or seeks a degree of precision for which past practice with the theory offers no precedent. These prohibitions are logically unexceptionable. But the result of accepting them would be the end of the research through which science may develop further.

By now that point too is virtually a tautology. Without commitment to a paradigm there could be no normal science. Furthermore, that commitment must extend to areas and to degrees of precision for which there is no full precedent. If it did not, the paradigm could provide no puzzles that had not already been solved. Besides, it is not only normal science that depends upon commitment to a paradigm. If existing theory binds the

[3] James B. Conant, *Overthrow of the Phlogiston Theory* (Cambridge, 1950), pp. 13–16; and J. R. Partington, *A Short History of Chemistry* (2d ed.; London, 1951), pp. 85–88. The fullest and most sympathetic account of the phlogiston theory's achievements is by H. Metzger, *Newton, Stahl, Boerhaave et la doctrine chimique* (Paris, 1930), Part II.

[4] Compare the conclusions reached through a very different sort of analysis by R. B. Braithwaite, *Scientific Explanation* (Cambridge, 1953), pp. 50–87, esp. p. 76.

scientist only with respect to existing applications, then there can be no surprises, anomalies, or crises. But these are just the signposts that point the way to extraordinary science. If positivistic restrictions on the range of a theory's legitimate applicability are taken literally, the mechanism that tells the scientific community what problems may lead to fundamental change must cease to function. And when that occurs, the community will inevitably return to something much like its pre-paradigm state, a condition in which all members practice science but in which their gross product scarcely resembles science at all. Is it really any wonder that the price of significant scientific advance is a commitment that runs the risk of being wrong?

More important, there is a revealing logical lacuna in the positivist's argument, one that will reintroduce us immediately to the nature of revolutionary change. Can Newtonian dynamics really be *derived* from relativistic dynamics? What would such a derivation look like? Imagine a set of statements, E_1, E_2, . . . , E_n, which together embody the laws of relativity theory. These statements contain variables and parameters representing spatial position, time, rest mass, etc. From them, together with the apparatus of logic and mathematics, is deducible a whole set of further statements including some that can be checked by observation. To prove the adequacy of Newtonian dynamics as a special case, we must add to the E_i's additional statements, like $(v/c)^2 << 1$, restricting the range of the parameters and variables. This enlarged set of statements is then manipulated to yield a new set, N_1, N_2, . . . , N_m, which is identical in form with Newton's laws of motion, the law of gravity, and so on. Apparently Newtonian dynamics has been derived from Einsteinian, subject to a few limiting conditions.

Yet the derivation is spurious, at least to this point. Though the N_i's are a special case of the laws of relativistic mechanics, they are not Newton's Laws. Or at least they are not unless those laws are reinterpreted in a way that would have been impossible until after Einstein's work. The variables and parameters that in the Einsteinian E_i's represented spatial position, time, mass, etc., still occur in the N_i's; and they there still represent Einsteinian space, time, and mass. But the physical referents of these Einsteinian concepts are by no means identical with those of the Newtonian concepts that bear the same name. (Newtonian mass is conserved; Einsteinian is convertible with energy. Only at low relative velocities may the two be measured in the same way, and even then they must not be conceived to be the same.) Unless we change the definitions of the variables in the N_i's, the statements we have derived are not Newtonian. If we do change them, we cannot properly be said to have *derived* Newton's Laws, at least not in any sense of "derive" now generally recognized. Our argument has, of course, explained why Newton's Laws ever seemed to work. In doing so it has justified, say, an automobile driver in acting as though he lived in a Newtonian universe. An argument of the same type is used to justify teaching earth-centered astronomy to surveyors. But the argument has still not done what it purported to do. It has not, that is, shown Newton's Laws to be a limiting case of Einstein's. For in the passage to the limit it is not only the forms of the laws that have changed. Simultaneously we have had to alter the fundamental structural elements of which the universe to which they apply is composed.

This need to change the meaning of established and familiar concepts is central to the revolutionary impact of Einstein's theory. Though subtler than the changes from geocentrism to heliocentrism, from phlogiston to oxygen, or from corpuscles to waves, the resulting conceptual transformation is no less decisively destructive of a previously established paradigm. We may even come to see it as a prototype for revolutionary reorientations in the sciences. Just because it did not involve the introduction of additional objects or concepts, the transition from Newtonian to Einsteinian mechanics illustrates with particular clarity the scientific revolution as a displacement of the conceptual network through which scientists view the world.

These remarks should suffice to show what might, in another philosophical climate, have been taken for granted. At least for scientists, most of the apparent differences between a discarded scientific theory and its successor are real. Though an out-of-date theory can always be viewed as a special case of its up-to-date successor, it must be transformed for the purpose. And the transformation is one that can be undertaken only with the advantages of hindsight, the explicit guidance of the more recent theory. Furthermore, even if that transformation were a legitimate device to employ in interpreting the older theory, the result of its application would be a theory so restricted that it could only restate what was already known. Because of its economy, that restatement would have utility, but it could not suffice for the guidance of research.

Let us, therefore, now take it for granted that the differences between successive paradigms are both necessary and irreconcilable. Can we then say more explicitly what sorts of differences these are? The most apparent type has already been illustrated repeatedly. Successive paradigms tell us different things about the population of the universe and about that population's behavior. They differ, that is, about such questions as the existence of subatomic particles, the materiality of light, and the conservation of heat or of energy. These are the substantive differences between successive paradigms, and they require no further illustration. But paradigms differ in more than substance, for they are directed not only to nature but also back upon the science that produced them. They are the source of the methods, problem-field, and standards of solution accepted by any mature scientific community at any given time. As a result, the reception of a new paradigm often necessitates a redefinition of the corresponding science. Some old problems may be relegated to another science or declared entirely "unscientific." Others that were previously non-existent or trivial may, with a new paradigm, become the very archetypes of significant scientific achievement. And as the problems change, so, often, does the standard that distinguishes a real scientific solution from a mere metaphysical speculation, word game, or mathematical play. The normal-scientific tradition that emerges from a scientific revolution is not only incompatible but often actually incommensurable with that which has gone before.

The impact of Newton's work upon the normal seventeenth-century tradition of scientific practice provides a striking example of these subtler effects of paradigm shift. Before Newton was born the "new science" of the century had at last succeeded in rejecting Aristotelian and scholastic explanations expressed in terms of the essences of material bodies. To say that a stone fell because its "nature" drove it toward the center of the universe had been made to look a mere tautological word-play, something it had not previously been. Henceforth the entire flux of sensory appearances, including color, taste, and even weight, was to be explained in terms of the size, shape, position, and motion of the elementary corpuscles of base matter. The attribution of other qualities to the elementary atoms was a resort to the occult and therefore out of bounds for science. Molière caught the new spirit precisely when he ridiculed the doctor who explained opium's efficacy as a soporific by attributing to it a dormitive potency. During the last half of the seventeenth century many scientists preferred to say that the round shape of the opium particles enabled them to sooth the nerves about which they moved.[5]

In an earlier period explanations in terms of occult qualities had been an integral part of productive scientific work. Nevertheless, the seventeenth century's new commitment to mechanico-corpuscular explanation proved immensely fruitful for a number of sciences, ridding them of problems that had defied generally accepted solution and suggesting others to replace them. In dynamics, for example, Newton's three laws of motion are less a product of novel experiments than of the attempt to reinterpret well-known observations in terms of the motions and interactions of primary neutral corpuscles. Consider just one concrete illustration. Since neutral corpuscles could act on each other only by contact, the mechanico-corpuscular view of nature directed scientific attention to a brand-new subject of study, the alteration of particulate motions by collisions. Descartes announced the problem and provided its first putative

[5] For corpuscularism in general, see Marie Boas, "The Establishment of the Mechanical Philosophy," *Osiris*, X (1952), 412–541. For the effect of particle-shape on taste, see *ibid.*, p. 483.

solution. Huyghens, Wren, and Wallis carried it still further, partly by experimenting with colliding pendulum bobs, but mostly by applying previously well-known characteristics of motion to the new problem. And Newton embedded their results in his laws of motion. The equal "action" and "reaction" of the third law are the changes in quantity of motion experienced by the two parties to a collision. The same change of motion supplies the definition of dynamical force implicit in the second law. In this case, as in many others during the seventeenth century, the corpuscular paradigm bred both a new problem and a large part of that problem's solution.[6]

Yet, though much of Newton's work was directed to problems and embodied standards derived from the mechanico-corpuscular world view, the effect of the paradigm that resulted from his work was a further and partially destructive change in the problems and standards legitimate for science. Gravity, interpreted as an innate attraction between every pair of particles of matter, was an occult quality in the same sense as the scholastics' "tendency to fall" had been. Therefore, while the standards of corpuscularism remained in effect, the search for a mechanical explanation of gravity was one of the most challenging problems for those who accepted the *Principia* as paradigm. Newton devoted much attention to it and so did many of his eighteenth-century successors. The only apparent option was to reject Newton's theory for its failure to explain gravity, and that alternative, too, was widely adopted. Yet neither of these views ultimately triumphed. Unable either to practice science without the *Principia* or to make that work conform to the corpuscular standards of the seventeenth century, scientists gradually accepted the view that gravity was indeed innate. By the mid-eighteenth century that interpretation had been almost universally accepted, and the result was a genuine reversion (which is not the same as a retrogression) to a scholastic standard. Innate attractions and repulsions joined size, shape, posi-

[6] R. Dugas, *La mécanique au XVIIe siècle* (Neuchatel, 1954), pp. 177–85, 284–98, 345–56.

tion, and motion as physically irreducible primary properties of matter.[7]

The resulting change in the standards and problem-field of physical science was once again consequential. By the 1740's, for example, electricians could speak of the attractive "virtue" of the electric fluid without thereby inviting the ridicule that had greeted Molière's doctor a century before. As they did so, electrical phenomena increasingly displayed an order different from the one they had shown when viewed as the effects of a mechanical effluvium that could act only by contact. In particular, when electrical action-at-a-distance became a subject for study in its own right, the phenomenon we now call charging by induction could be recognized as one of its effects. Previously, when seen at all, it had been attributed to the direct action of electrical "atmospheres" or to the leakages inevitable in any electrical laboratory. The new view of inductive effects was, in turn, the key to Franklin's analysis of the Leyden jar and thus to the emergence of a new and Newtonian paradigm for electricity. Nor were dynamics and electricity the only scientific fields affected by the legitimization of the search for forces innate to matter. The large body of eighteenth-century literature on chemical affinities and replacement series also derives from this supramechanical aspect of Newtonianism. Chemists who believed in these differential attractions between the various chemical species set up previously unimagined experiments and searched for new sorts of reactions. Without the data and the chemical concepts developed in that process, the later work of Lavoisier and, more particularly, of Dalton would be incomprehensible.[8] Changes in the standards governing permissible problems, concepts, and explanations can transform a science. In the next section I shall even suggest a sense in which they transform the world.

[7] I. B. Cohen, *Franklin and Newton: An Inquiry into Speculative Newtonian Experimental Science and Franklin's Work in Electricity as an Example Thereof* (Philadelphia, 1956), chaps. vi–vii.

[8] For electricity, see *ibid*, chaps. viii–ix. For chemistry, see Metzger, *op. cit.*, Part I.

Other examples of these nonsubstantive differences between successive paradigms can be retrieved from the history of any science in almost any period of its development. For the moment let us be content with just two other and far briefer illustrations. Before the chemical revolution, one of the acknowledged tasks of chemistry was to account for the qualities of chemical substances and for the changes these qualities underwent during chemical reactions. With the aid of a small number of elementary "principles"—of which phlogiston was one—the chemist was to explain why some substances are acidic, others metalline, combustible, and so forth. Some success in this direction had been achieved. We have already noted that phlogiston explained why the metals were so much alike, and we could have developed a similar argument for the acids. Lavoisier's reform, however, ultimately did away with chemical "principles," and thus ended by depriving chemistry of some actual and much potential explanatory power. To compensate for this loss, a change in standards was required. During much of the nineteenth century failure to explain the qualities of compounds was no indictment of a chemical theory.[9]

Or again, Clerk Maxwell shared with other nineteenth-century proponents of the wave theory of light the conviction that light waves must be propagated through a material ether. Designing a mechanical medium to support such waves was a standard problem for many of his ablest contemporaries. His own theory, however, the electromagnetic theory of light, gave no account at all of a medium able to support light waves, and it clearly made such an account harder to provide than it had seemed before. Initially, Maxwell's theory was widely rejected for those reasons. But, like Newton's theory, Maxwell's proved difficult to dispense with, and as it achieved the status of a paradigm, the community's attitude toward it changed. In the early decades of the twentieth century Maxwell's insistence upon the existence of a mechanical ether looked more and more like lip service, which it emphatically had not been, and the attempts to design such an ethereal medium were abandoned. Scientists no

[9] E. Meyerson, *Identity and Reality* (New York, 1930), chap. x.

longer thought it unscientific to speak of an electrical "displacement" without specifying what was being displaced. The result, again, was a new set of problems and standards, one which, in the event, had much to do with the emergence of relativity theory.[10]

These characteristic shifts in the scientific community's conception of its legitimate problems and standards would have less significance to this essay's thesis if one could suppose that they always occurred from some methodologically lower to some higher type. In that case their effects, too, would seem cumulative. No wonder that some historians have argued that the history of science records a continuing increase in the maturity and refinement of man's conception of the nature of science.[11] Yet the case for cumulative development of science's problems and standards is even harder to make than the case for cumulation of theories. The attempt to explain gravity, though fruitfully abandoned by most eighteenth-century scientists, was not directed to an intrinsically illegitimate problem; the objections to innate forces were neither inherently unscientific nor metaphysical in some pejorative sense. There are no external standards to permit a judgment of that sort. What occurred was neither a decline nor a raising of standards, but simply a change demanded by the adoption of a new paradigm. Furthermore, that change has since been reversed and could be again. In the twentieth century Einstein succeeded in explaining gravitational attractions, and that explanation has returned science to a set of canons and problems that are, in this particular respect, more like those of Newton's predecessors than of his successors. Or again, the development of quantum mechanics has reversed the methodological prohibition that originated in the chemical revolution. Chemists now attempt, and with great success, to explain the color, state of aggregation, and other qualities of the substances used and produced in their laboratories. A similar rever-

[10] E. T. Whittaker, *A History of the Theories of Aether and Electricity*, II (London, 1953), 28–30.

[11] For a brilliant and entirely up-to-date attempt to fit scientific development into this Procrustean bed, see C. C. Gillispie, *The Edge of Objectivity: An Essay in the History of Scientific Ideas* (Princeton, 1960).

sal may even be underway in electromagnetic theory. Space, in contemporary physics, is not the inert and homogenous substratum employed in both Newton's and Maxwell's theories; some of its new properties are not unlike those once attributed to the ether; we may someday come to know what an electric displacement is.

By shifting emphasis from the cognitive to the normative functions of paradigms, the preceding examples enlarge our understanding of the ways in which paradigms give form to the scientific life. Previously, we had principally examined the paradigm's role as a vehicle for scientific theory. In that role it functions by telling the scientist about the entities that nature does and does not contain and about the ways in which those entities behave. That information provides a map whose details are elucidated by mature scientific research. And since nature is too complex and varied to be explored at random, that map is as essential as observation and experiment to science's continuing development. Through the theories they embody, paradigms prove to be constitutive of the research activity. They are also, however, constitutive of science in other respects, and that is now the point. In particular, our most recent examples show that paradigms provide scientists not only with a map but also with some of the directions essential for map-making. In learning a paradigm the scientist acquires theory, methods, and standards together, usually in an inextricable mixture. Therefore, when paradigms change, there are usually significant shifts in the criteria determining the legitimacy both of problems and of proposed solutions.

That observation returns us to the point from which this section began, for it provides our first explicit indication of why the choice between competing paradigms regularly raises questions that cannot be resolved by the criteria of normal science. To the extent, as significant as it is incomplete, that two scientific schools disagree about what is a problem and what a solution, they will inevitably talk through each other when debating the relative merits of their respective paradigms. In the partially circular arguments that regularly result, each paradigm will be shown to satisfy more or less the criteria that it dictates for itself and to fall short of a few of those dictated by its opponent. There are other reasons, too, for the incompleteness of logical contact that consistently characterizes paradigm debates. For example, since no paradigm ever solves all the problems it defines and since no two paradigms leave all the same problems unsolved, paradigm debates always involve the question: Which problems is it more significant to have solved? Like the issue of competing standards, that question of values can be answered only in terms of criteria that lie outside of normal science altogether, and it is that recourse to external criteria that most obviously makes paradigm debates revolutionary. Something even more fundamental than standards and values is, however, also at stake. I have so far argued only that paradigms are constitutive of science. Now I wish to display a sense in which they are constitutive of nature as well.

X. Revolutions as Changes of World View

Examining the record of past research from the vantage of contemporary historiography, the historian of science may be tempted to exclaim that when paradigms change, the world itself changes with them. Led by a new paradigm, scientists adopt new instruments and look in new places. Even more important, during revolutions scientists see new and different things when looking with familiar instruments in places they have looked before. It is rather as if the professional community had been suddenly transported to another planet where familiar objects are seen in a different light and are joined by unfamiliar ones as well. Of course, nothing of quite that sort does occur: there is no geographical transplantation; outside the laboratory everyday affairs usually continue as before. Nevertheless, paradigm changes do cause scientists to see the world of their research-engagement differently. In so far as their only recourse to that world is through what they see and do, we may want to say that after a revolution scientists are responding to a different world.

It is as elementary prototypes for these transformations of the scientist's world that the familiar demonstrations of a switch in visual gestalt prove so suggestive. What were ducks in the scientist's world before the revolution are rabbits afterwards. The man who first saw the exterior of the box from above later sees its interior from below. Transformations like these, though usually more gradual and almost always irreversible, are common concomitants of scientific training. Looking at a contour map, the student sees lines on paper, the cartographer a picture of a terrain. Looking at a bubble-chamber photograph, the student sees confused and broken lines, the physicist a record of familiar subnuclear events. Only after a number of such transformations of vision does the student become an inhabitant of the scientist's world, seeing what the scientist sees and responding as the scientist does. The world that the student then enters is not, however, fixed once and for all by the nature of the environment, on the one hand, and of science, on the other. Rather, it is determined jointly by the environment and the particular normal-scientific tradition that the student has been trained to pursue. Therefore, at times of revolution, when the normal-scientific tradition changes, the scientist's perception of his environment must be re-educated—in some familiar situations he must learn to see a new gestalt. After he has done so the world of his research will seem, here and there, incommensurable with the one he had inhabited before. That is another reason why schools guided by different paradigms are always slightly at cross-purposes.

In their most usual form, of course, gestalt experiments illustrate only the nature of perceptual transformations. They tell us nothing about the role of paradigms or of previously assimilated experience in the process of perception. But on that point there is a rich body of psychological literature, much of it stemming from the pioneering work of the Hanover Institute. An experimental subject who puts on goggles fitted with inverting lenses initially sees the entire world upside down. At the start his perceptual apparatus functions as it had been trained to function in the absence of the goggles, and the result is extreme disorientation, an acute personal crisis. But after the subject has begun to learn to deal with his new world, his entire visual field flips over, usually after an intervening period in which vision is simply confused. Thereafter, objects are again seen as they had been before the goggles were put on. The assimilation of a previously anomalous visual field has reacted upon and changed the field itself.[1] Literally as well as metaphorically, the man accustomed to inverting lenses has undergone a revolutionary transformation of vision.

The subjects of the anomalous playing-card experiment discussed in Section VI experienced a quite similar transformation. Until taught by prolonged exposure that the universe contained

[1] The original experiments were by George M. Stratton, "Vision without Inversion of the Retinal Image," *Psychological Review,* IV (1897), 341–60, 463–81. A more up-to-date review is provided by Harvey A. Carr, *An Introduction to Space Perception* (New York, 1935), pp. 18–57.

anomalous cards, they saw only the types of cards for which previous experience had equipped them. Yet once experience had provided the requisite additional categories, they were able to see all anomalous cards on the first inspection long enough to permit any identification at all. Still other experiments demonstrate that the perceived size, color, and so on, of experimentally displayed objects also varies with the subject's previous training and experience.[2] Surveying the rich experimental literature from which these examples are drawn makes one suspect that something like a paradigm is prerequisite to perception itself. What a man sees depends both upon what he looks at and also upon what his previous visual-conceptual experience has taught him to see. In the absence of such training there can only be, in William James's phrase, "a bloomin' buzzin' confusion."

In recent years several of those concerned with the history of science have found the sorts of experiments described above immensely suggestive. N. R. Hanson, in particular, has used gestalt demonstrations to elaborate some of the same consequences of scientific belief that concern me here.[3] Other colleagues have repeatedly noted that history of science would make better and more coherent sense if one could suppose that scientists occasionally experienced shifts of perception like those described above. Yet, though psychological experiments are suggestive, they cannot, in the nature of the case, be more than that. They do display characteristics of perception that *could* be central to scientific development, but they do not demonstrate that the careful and controlled observation exercised by the research scientist at all partakes of those characteristics. Furthermore, the very nature of these experiments makes any direct demonstration of that point impossible. If historical example is to make these psychological experiments seem rele-

[2] For examples, see Albert H. Hastorf, "The Influence of Suggestion on the Relationship between Stimulus Size and Perceived Distance," *Journal of Psychology,* XXIX (1950), 195–217; and Jerome S. Bruner, Leo Postman, and John Rodrigues, "Expectations and the Perception of Color," *American Journal of Psychology,* LXIV (1951), 216–27.

[3] N. R. Hanson, *Patterns of Discovery* (Cambridge, 1958), chap. i.

vant, we must first notice the sorts of evidence that we may and may not expect history to provide.

The subject of a gestalt demonstration knows that his perception has shifted because he can make it shift back and forth repeatedly while he holds the same book or piece of paper in his hands. Aware that nothing in his environment has changed, he directs his attention increasingly not to the figure (duck or rabbit) but to the lines on the paper he is looking at. Ultimately he may even learn to see those lines without seeing either of the figures, and he may then say (what he could not legitimately have said earlier) that it is these lines that he really sees but that he sees them alternately *as* a duck and *as* a rabbit. By the same token, the subject of the anomalous card experiment knows (or, more accurately, can be persuaded) that his perception must have shifted because an external authority, the experimenter, assures him that regardless of what he *saw,* he was *looking at* a black five of hearts all the time. In both these cases, as in all similar psychological experiments, the effectiveness of the demonstration depends upon its being analyzable in this way. Unless there were an external standard with respect to which a switch of vision could be demonstrated, no conclusion about alternate perceptual possibilities could be drawn.

With scientific observation, however, the situation is exactly reversed. The scientist can have no recourse above or beyond what he sees with his eyes and instruments. If there were some higher authority by recourse to which his vision might be shown to have shifted, then that authority would itself become the source of his data, and the behavior of his vision would become a source of problems (as that of the experimental subject is for the psychologist). The same sorts of problems would arise if the scientist could switch back and forth like the subject of the gestalt experiments. The period during which light was "sometimes a wave and sometimes a particle" was a period of crisis—a period when something was wrong—and it ended only with the development of wave mechanics and the realization that light was a self-consistent entity different from both waves and particles. In the sciences, therefore, if perceptual switches accompany paradigm changes, we may not expect scientists to attest to these changes directly. Looking at the moon, the convert to Copernicanism does not say, "I used to see a planet, but now I see a satellite." That locution would imply a sense in which the Ptolemaic system had once been correct. Instead, a convert to the new astronomy says, "I once took the moon to be (or saw the moon as) a planet, but I was mistaken." That sort of statement does recur in the aftermath of scientific revolutions. If it ordinarily disguises a shift of scientific vision or some other mental transformation with the same effect, we may not expect direct testimony about that shift. Rather we must look for indirect and behavioral evidence that the scientist with a new paradigm sees differently from the way he had seen before.

Let us then return to the data and ask what sorts of transformations in the scientist's world the historian who believes in such changes can discover. Sir William Herschel's discovery of Uranus provides a first example and one that closely parallels the anomalous card experiment. On at least seventeen different occasions between 1690 and 1781, a number of astronomers, including several of Europe's most eminent observers, had seen a star in positions that we now suppose must have been occupied at the time by Uranus. One of the best observers in this group had actually seen the star on four successive nights in 1769 without noting the motion that could have suggested another identification. Herschel, when he first observed the same object twelve years later, did so with a much improved telescope of his own manufacture. As a result, he was able to notice an apparent disk-size that was at least unusual for stars. Something was awry, and he therefore postponed identification pending further scrutiny. That scrutiny disclosed Uranus' motion among the stars, and Herschel therefore announced that he had seen a new comet! Only several months later, after fruitless attempts to fit the observed motion to a cometary orbit, did Lexell suggest that the orbit was probably planetary.[4] When that suggestion was accepted, there were several fewer stars and one more planet in the world of the professional astronomer. A celestial body that

[4] Peter Doig, *A Concise History of Astronomy* (London, 1950), pp. 115–16.

had been observed off and on for almost a century was seen differently after 1781 because, like an anomalous playing card, it could no longer be fitted to the perceptual categories (star or comet) provided by the paradigm that had previously prevailed.

The shift of vision that enabled astronomers to see Uranus, the planet, does not, however, seem to have affected only the perception of that previously observed object. Its consequences were more far-reaching. Probably, though the evidence is equivocal, the minor paradigm change forced by Herschel helped to prepare astronomers for the rapid discovery, after 1801, of the numerous minor planets or asteroids. Because of their small size, these did not display the anomalous magnification that had alerted Herschel. Nevertheless, astronomers prepared to find additional planets were able, with standard instruments, to identify twenty of them in the first fifty years of the nineteenth century.[5] The history of astronomy provides many other examples of paradigm-induced changes in scientific perception, some of them even less equivocal. Can it conceivably be an accident, for example, that Western astronomers first saw change in the previously immutable heavens during the half-century after Copernicus' new paradigm was first proposed? The Chinese, whose cosmological beliefs did not preclude celestial change, had recorded the appearance of many new stars in the heavens at a much earlier date. Also, even without the aid of a telescope, the Chinese had systematically recorded the appearance of sunspots centuries before these were seen by Galileo and his contemporaries.[6] Nor were sunspots and a new star the only examples of celestial change to emerge in the heavens of Western astronomy immediately after Copernicus. Using traditional instruments, some as simple as a piece of thread, late sixteenth-century astronomers repeatedly discovered that comets wandered at will through the space previously reserved for the

[5] Rudolph Wolf, *Geschichte der Astronomie* (Munich, 1877), pp. 513–15, 683–93. Notice particularly how difficult Wolf's account makes it to explain these discoveries as a consequence of Bode's Law.

[6] Joseph Needham, *Science and Civilization in China*, III (Cambridge, 1959), 423–29, 434–36.

immutable planets and stars.[7] The very ease and rapidity with which astronomers saw new things when looking at old objects with old instruments may make us wish to say that, after Copernicus, astronomers lived in a different world. In any case, their research responded as though that were the case.

The preceding examples are selected from astronomy because reports of celestial observation are frequently delivered in a vocabulary consisting of relatively pure observation terms. Only in such reports can we hope to find anything like a full parallelism between the observations of scientists and those of the psychologist's experimental subjects. But we need not insist on so full a parallelism, and we have much to gain by relaxing our standard. If we can be content with the everyday use of the verb 'to see,' we may quickly recognize that we have already encountered many other examples of the shifts in scientific perception that accompany paradigm change. The extended use of 'perception' and of 'seeing' will shortly require explicit defense, but let me first illustrate its application in practice.

Look again for a moment at two of our previous examples from the history of electricity. During the seventeenth century, when their research was guided by one or another effluvium theory, electricians repeatedly saw chaff particles rebound from, or fall off, the electrified bodies that had attracted them. At least that is what seventeenth-century observers said they saw, and we have no more reason to doubt their reports of perception than our own. Placed before the same apparatus, a modern observer would see electrostatic repulsion (rather than mechanical or gravitational rebounding), but historically, with one universally ignored exception, electrostatic repulsion was not seen as such until Hauksbee's large-scale apparatus had greatly magnified its effects. Repulsion after contact electrification was, however, only one of many new repulsive effects that Hauksbee saw. Through his researches, rather as in a gestalt switch, repulsion suddenly became *the* fundamental manifestation of electrification, and it was then attraction that needed to be ex-

[7] T. S. Kuhn, *The Copernican Revolution* (Cambridge, Mass., 1957), pp. 206–9.

plained.[8] The electrical phenomena visible in the early eighteenth century were both subtler and more varied than those seen by observers in the seventeenth century. Or again, after the assimilation of Franklin's paradigm, the electrician looking at a Leyden jar saw something different from what he had seen before. The device had become a condenser, for which neither the jar shape nor glass was required. Instead, the two conducting coatings—one of which had been no part of the original device—emerged to prominence. As both written discussions and pictorial representations gradually attest, two metal plates with a non-conductor between them had become the prototype for the class.[9] Simultaneously, other inductive effects received new descriptions, and still others were noted for the first time.

Shifts of this sort are not restricted to astronomy and electricity. We have already remarked some of the similar transformations of vision that can be drawn from the history of chemistry. Lavoisier, we said, saw oxygen where Priestley had seen dephlogisticated air and where others had seen nothing at all. In learning to see oxygen, however, Lavoisier also had to change his view of many other more familiar substances. He had, for example, to see a compound ore where Priestley and his contemporaries had seen an elementary earth, and there were other such changes besides. At the very least, as a result of discovering oxygen, Lavoisier saw nature differently. And in the absence of some recourse to that hypothetical fixed nature that he "saw differently," the principle of economy will urge us to say that after discovering oxygen Lavoisier worked in a different world.

I shall inquire in a moment about the possibility of avoiding this strange locution, but first we require an additional example of its use, this one deriving from one of the best known parts of the work of Galileo. Since remote antiquity most people have seen one or another heavy body swinging back and forth on a string or chain until it finally comes to rest. To the Aristotelians,

[8] Duane Roller and Duane H. D. Roller, *The Development of the Concept of Electric Charge* (Cambridge, Mass., 1954), pp. 21–29.

[9] See the discussion in Section VII and the literature to which the reference there cited in note 9 will lead.

who believed that a heavy body is moved by its own nature from a higher position to a state of natural rest at a lower one, the swinging body was simply falling with difficulty. Constrained by the chain, it could achieve rest at its low point only after a tortuous motion and a considerable time. Galileo, on the other hand, looking at the swinging body, saw a pendulum, a body that almost succeeded in repeating the same motion over and over again ad infinitum. And having seen that much, Galileo observed other properties of the pendulum as well and constructed many of the most significant and original parts of his new dynamics around them. From the properties of the pendulum, for example, Galileo derived his only full and sound arguments for the independence of weight and rate of fall, as well as for the relationship between vertical height and terminal velocity of motions down inclined planes.[10] All these natural phenomena he saw differently from the way they had been seen before.

Why did that shift of vision occur? Through Galileo's individual genius, of course. But note that genius does not here manifest itself in more accurate or objective observation of the swinging body. Descriptively, the Aristotelian perception is just as accurate. When Galileo reported that the pendulum's period was independent of amplitude for amplitudes as great as 90°, his view of the pendulum led him to see far more regularity than we can now discover there.[11] Rather, what seems to have been involved was the exploitation by genius of perceptual possibilities made available by a medieval paradigm shift. Galileo was not raised completely as an Aristotelian. On the contrary, he was trained to analyze motions in terms of the impetus theory, a late medieval paradigm which held that the continuing motion of a heavy body is due to an internal power implanted in it by the projector that initiated its motion. Jean Buridan and Nicole Oresme, the fourteenth-century scholastics who brought the impetus theory to its most perfect formulations, are the first men

[10] Galileo Galilei, *Dialogues concerning Two New Sciences,* trans. H. Crew and A. de Salvio (Evanston, Ill., 1946), pp. 80–81, 162–66.

[11] *Ibid.*, pp. 91–94, 244.

known to have seen in oscillatory motions any part of what Galileo saw there. Buridan describes the motion of a vibrating string as one in which impetus is first implanted when the string is struck; the impetus is next consumed in displacing the string against the resistance of its tension; tension then carries the string back, implanting increasing impetus until the mid-point of motion is reached; after that the impetus displaces the string in the opposite direction, again against the string's tension, and so on in a symmetric process that may continue indefinitely. Later in the century Oresme sketched a similar analysis of the swinging stone in what now appears as the first discussion of a pendulum.[12] His view is clearly very close to the one with which Galileo first approached the pendulum. At least in Oresme's case, and almost certainly in Galileo's as well, it was a view made possible by the transition from the original Aristotelian to the scholastic impetus paradigm for motion. Until that scholastic paradigm was invented, there were no pendulums, but only swinging stones, for the scientist to see. Pendulums were brought into existence by something very like a paradigm-induced gestalt switch.

Do we, however, really need to describe what separates Galileo from Aristotle, or Lavoisier from Priestley, as a transformation of vision? Did these men really *see* different things when *looking at* the same sorts of objects? Is there any legitimate sense in which we can say that they pursued their research in different worlds? Those questions can no longer be postponed, for there is obviously another and far more usual way to describe all of the historical examples outlined above. Many readers will surely want to say that what changes with a paradigm is only the scientist's interpretation of observations that themselves are fixed once and for all by the nature of the environment and of the perceptual apparatus. On this view, Priestley and Lavoisier both saw oxygen, but they interpreted their observations differently; Aristotle and Galileo both saw pendu-

[12] M. Clagett, *The Science of Mechanics in the Middle Ages* (Madison, Wis., 1959), pp. 537–38, 570.

lums, but they differed in their interpretations of what they both had seen.

Let me say at once that this very usual view of what occurs when scientists change their minds about fundamental matters can be neither all wrong nor a mere mistake. Rather it is an essential part of a philosophical paradigm initiated by Descartes and developed at the same time as Newtonian dynamics. That paradigm has served both science and philosophy well. Its exploitation, like that of dynamics itself, has been fruitful of a fundamental understanding that perhaps could not have been achieved in another way. But as the example of Newtonian dynamics also indicates, even the most striking past success provides no guarantee that crisis can be indefinitely postponed. Today research in parts of philosophy, psychology, linguistics, and even art history, all converge to suggest that the traditional paradigm is somehow askew. That failure to fit is also made increasingly apparent by the historical study of science to which most of our attention is necessarily directed here.

None of these crisis-promoting subjects has yet produced a viable alternate to the traditional epistemological paradigm, but they do begin to suggest what some of that paradigm's characteristics will be. I am, for example, acutely aware of the difficulties created by saying that when Aristotle and Galileo looked at swinging stones, the first saw constrained fall, the second a pendulum. The same difficulties are presented in an even more fundamental form by the opening sentences of this section: though the world does not change with a change of paradigm, the scientist afterward works in a different world. Nevertheless, I am convinced that we must learn to make sense of statements that at least resemble these. What occurs during a scientific revolution is not fully reducible to a reinterpretation of individual and stable data. In the first place, the data are not unequivocally stable. A pendulum is not a falling stone, nor is oxygen dephlogisticated air. Consequently, the data that scientists collect from these diverse objects are, as we shall shortly see, themselves different. More important, the process by which either the individual or the community makes the transition from constrained fall to the pendulum or from dephlogisticated air to oxygen is not one that resembles interpretation. How could it do so in the absence of fixed data for the scientist to interpret? Rather than being an interpreter, the scientist who embraces a new paradigm is like the man wearing inverting lenses. Confronting the same constellation of objects as before and knowing that he does so, he nevertheless finds them transformed through and through in many of their details.

None of these remarks is intended to indicate that scientists do not characteristically interpret observations and data. On the contrary, Galileo interpreted observations on the pendulum, Aristotle observations on falling stones, Musschenbroek observations on a charge-filled bottle, and Franklin observations on a condenser. But each of these interpretations presupposed a paradigm. They were parts of normal science, an enterprise that, as we have already seen, aims to refine, extend, and articulate a paradigm that is already in existence. Section III provided many examples in which interpretation played a central role. Those examples typify the overwhelming majority of research. In each of them the scientist, by virtue of an accepted paradigm, knew what a datum was, what instruments might be used to retrieve it, and what concepts were relevant to its interpretation. Given a paradigm, interpretation of data is central to the enterprise that explores it.

But that interpretive enterprise—and this was the burden of the paragraph before last—can only articulate a paradigm, not correct it. Paradigms are not corrigible by normal science at all. Instead, as we have already seen, normal science ultimately leads only to the recognition of anomalies and to crises. And these are terminated, not by deliberation and interpretation, but by a relatively sudden and unstructured event like the gesalt switch. Scientists then often speak of the "scales falling from the eyes" or of the "lightning flash" that "inundates" a previously obscure puzzle, enabling its components to be seen in a new way that for the first time permits its solution. On other occasions the relevant illumination comes in sleep.[13] No ordinary sense of the term 'interpretation' fits these flashes of intuition through which a new paradigm is born. Though such intuitions depend upon the experience, both anomalous and congruent, gained with the old paradigm, they are not logically or piecemeal linked to particular items of that experience as an interpretation would be. Instead, they gather up large portions of that experience and transform them to the rather different bundle of experience that will thereafter be linked piecemeal to the new paradigm but not to the old.

To learn more about what these differences in experience can be, return for a moment to Aristotle, Galileo, and the pendulum. What data did the interaction of their different paradigms and their common environment make accessible to each of them? Seeing constrained fall, the Aristotelian would measure (or at least discuss—the Aristotelian seldom measured) the weight of the stone, the vertical height to which it had been raised, and the time required for it to achieve rest. Together with the resistance of the medium, these were the conceptual categories deployed by Aristotelian science when dealing with a falling body.[14] Normal research guided by them could not have produced the laws that Galileo discovered. It could only—and by another route it did—lead to the series of crises from which Galileo's view of the swinging stone emerged. As a result of those crises and of other intellectual changes besides, Galileo saw the swinging stone quite differently. Archimedes' work on floating bodies made the medium non-essential; the impetus theory rendered the motion symmetrical and enduring; and Neoplatonism directed Galileo's attention to the motion's circu-

13 [Jacques] Hadamard, *Subconscient intuition, et logique dans la recherche scientifique* (*Conférence faite au Palais de la Découverte le 8 Décembre 1945* [Alençon, n.d.]), pp. 7–8. A much fuller account, though one exclusively restricted to mathematical innovations, is the same author's *The Psychology of Invention in the Mathematical Field* (Princeton, 1949).

14 T. S. Kuhn, "A Function for Thought Experiments," in *Mélanges Alexandre Koyré*, ed. R. Taton and I. B. Cohen, to be published by Hermann (Paris) in 1963.

lar form.[15] He therefore measured only weight, radius, angular displacement, and time per swing, which were precisely the data that could be interpreted to yield Galileo's laws for the pendulum. In the event, interpretation proved almost unnecessary. Given Galileo's paradigms, pendulum-like regularities were very nearly accessible to inspection. How else are we to account for Galileo's discovery that the bob's period is entirely independent of amplitude, a discovery that the normal science stemming from Galileo had to eradicate and that we are quite unable to document today. Regularities that could not have existed for an Aristotelian (and that are, in fact, nowhere precisely exemplified by nature) were consequences of immediate experience for the man who saw the swinging stone as Galileo did.

Perhaps that example is too fanciful since the Aristotelians recorded no discussions of swinging stones. On their paradigm it was an extraordinarily complex phenomenon. But the Aristotelians did discuss the simpler case, stones falling without uncommon constraints, and the same differences of vision are apparent there. Contemplating a falling stone, Aristotle saw a change of state rather than a process. For him the relevant measures of a motion were therefore total distance covered and total time elapsed, parameters which yield what we should now call not speed but average speed.[16] Similarly, because the stone was impelled by its nature to reach its final resting point, Aristotle saw the relevant distance parameter at any instant during the motion as the distance *to* the final end point rather than as that *from* the origin of motion.[17] Those conceptual parameters underlie and give sense to most of his well-known "laws of motion." Partly through the impetus paradigm, however, and partly through a doctrine known as the latitude of forms, scholastic criticism changed this way of viewing motion. A stone moved by impetus gained more and more of it while receding from its

[15] A. Koyré, *Etudes Galiléennes* (Paris, 1939), I, 46–51; and "Galileo and Plato," *Journal of the History of Ideas*, IV (1943), 400–428.

[16] Kuhn, "A Function for Thought Experiments," in *Mélanges Alexandre Koyré* (see n. 14 for full citation).

[17] Koyré, *Etudes* . . . , II, 7–11.

starting point; distance from rather than distance to therefore became the revelant parameter. In addition, Aristotle's notion of speed was bifurcated by the scholastics into concepts that soon after Galileo became our average speed and instantaneous speed. But when seen through the paradigm of which these conceptions were a part, the falling stone, like the pendulum, exhibited its governing laws almost on inspection. Galileo was not one of the first men to suggest that stones fall with a uniformly accelerated motion.[18] Furthermore, he had developed his theorem on this subject together with many of its consequences before he experimented with an inclined plane. That theorem was another one of the network of new regularities accessible to genius in the world determined jointly by nature and by the paradigms upon which Galileo and his contemporaries had been raised. Living in that world, Galileo could still, when he chose, explain why Aristotle had seen what he did. Nevertheless, the immediate content of Galileo's experience with falling stones was not what Aristotle's had been.

It is, of course, by no means clear that we need be so concerned with "immediate experience"—that is, with the perceptual features that a paradigm so highlights that they surrender their regularities almost upon inspection. Those features must obviously change with the scientist's commitments to paradigms, but they are far from what we ordinarily have in mind when we speak of the raw data or the brute experience from which scientific research is reputed to proceed. Perhaps immediate experience should be set aside as fluid, and we should discuss instead the concrete operations and measurements that the scientist performs in his laboratory. Or perhaps the analysis should be carried further still from the immediately given. It might, for example, be conducted in terms of some neutral observation-language, perhaps one designed to conform to the retinal imprints that mediate what the scientist sees. Only in one of these ways can we hope to retrieve a realm in which experience is again stable once and for all—in which the pendulum and constrained fall are not different perceptions but rather

[18] Clagett, *op. cit.*, chaps. iv, vi, and ix.

different interpretations of the unequivocal data provided by observation of a swinging stone.

But is sensory experience fixed and neutral? Are theories simply man-made interpretations of given data? The epistemological viewpoint that has most often guided Western philosophy for three centuries dictates an immediate and unequivocal, Yes! In the absence of a developed alternative, I find it impossible to relinquish entirely that viewpoint. Yet it no longer functions effectively, and the attempts to make it do so through the introduction of a neutral language of observations now seem to me hopeless.

The operations and measurements that a scientist undertakes in the laboratory are not "the given" of experience but rather "the collected with difficulty." They are not what the scientist sees—at least not before his research is well advanced and his attention focused. Rather, they are concrete indices to the content of more elementary perceptions, and as such they are selected for the close scrutiny of normal research only because they promise opportunity for the fruitful elaboration of an accepted paradigm. Far more clearly than the immediate experience from which they in part derive, operations and measurements are paradigm-determined. Science does not deal in all possible laboratory manipulations. Instead, it selects those relevant to the juxtaposition of a paradigm with the immediate experience that that paradigm has partially determined. As a result, scientists with different paradigms engage in different concrete laboratory manipulations. The measurements to be performed on a pendulum are not the ones relevant to a case of constrained fall. Nor are the operations relevant for the elucidation of oxygen's properties uniformly the same as those required when investigating the characteristics of dephlogisticated air.

As for a pure observation-language, perhaps one will yet be devised. But three centuries after Descartes our hope for such an eventuality still depends exclusively upon a theory of perception and of the mind. And modern psychological experimentation is rapidly proliferating phenomena with which that theory can scarcely deal. The duck-rabbit shows that two men with the same retinal impressions can see different things; the inverting lenses show that two men with different retinal impressions can see the same thing. Psychology supplies a great deal of other evidence to the same effect, and the doubts that derive from it are readily reinforced by the history of attempts to exhibit an actual language of observation. No current attempt to achieve that end has yet come close to a generally applicable language of pure percepts. And those attempts that come closest share one characteristic that strongly reinforces several of this essay's main theses. From the start they presuppose a paradigm, taken either from a current scientific theory or from some fraction of everyday discourse, and they then try to eliminate from it all non-logical and non-perceptual terms. In a few realms of discourse this effort has been carried very far and with fascinating results. There can be no question that efforts of this sort are worth pursuing. But their result is a language that—like those employed in the sciences—embodies a host of expectations about nature and fails to function the moment these expectations are violated. Nelson Goodman makes exactly this point in describing the aims of his *Structure of Appearance:* "It is fortunate that nothing more [than phenomena known to exist] is in question; for the notion of 'possible' cases, of cases that do not exist but might have existed, is far from clear."[19] No language thus restricted to reporting a world fully known in advance can produce mere neutral and objective reports on "the given." Philosophical investigation has not yet provided even a hint of what a language able to do that would be like.

Under these circumstances we may at least suspect that scientists are right in principle as well as in practice when they treat

[19] N. Goodman, *The Structure of Appearance* (Cambridge, Mass., 1951), pp. 4–5. The passage is worth quoting more extensively: "If all and only those residents of Wilmington in 1947 that weigh between 175 and 180 pounds have red hair, then 'red-haired 1947 resident of Wilmington' and '1947 resident of Wilmington weighing between 175 and 180 pounds' may be joined in a constructional definition. . . . The question whether there 'might have been' someone to whom one but not the other of these predicates would apply has no bearing . . . once we have determined that there is no such person. . . . It is fortunate that nothing more is in question; for the notion of 'possible' cases, of cases that do not exist but might have existed, is far from clear."

oxygen and pendulums (and perhaps also atoms and electrons) as the fundamental ingredients of their immediate experience. As a result of the paradigm-embodied experience of the race, the culture, and, finally, the profession, the world of the scientist has come to be populated with planets and pendulums, condensers and compound ores, and other such bodies besides. Compared with these objects of perception, both meter stick readings and retinal imprints are elaborate constructs to which experience has direct access only when the scientist, for the special purposes of his research, arranges that one or the other should do so. This is not to suggest that pendulums, for example, are the only things a scientist could possibly see when looking at a swinging stone. (We have already noted that members of another scientific community could see constrained fall.) But it is to suggest that the scientist who looks at a swinging stone can have no experience that is in principle more elementary than seeing a pendulum. The alternative is not some hypothetical "fixed" vision, but vision through another paradigm, one which makes the swinging stone something else.

All of this may seem more reasonable if we again remember that neither scientists nor laymen learn to see the world piecemeal or item by item. Except when all the conceptual and manipulative categories are prepared in advance—e.g., for the discovery of an additional transuranic element or for catching sight of a new house—both scientists and laymen sort out whole areas together from the flux of experience. The child who transfers the word 'mama' from all humans to all females and then to his mother is not just learning what 'mama' means or who his mother is. Simultaneously he is learning some of the differences between males and females as well as something about the ways in which all but one female will behave toward him. His reactions, expectations, and beliefs—indeed, much of his perceived world—change accordingly. By the same token, the Copernicans who denied its traditional title 'planet' to the sun were not only learning what 'planet' meant or what the sun was. Instead, they were changing the meaning of 'planet' so that it could continue to make useful distinctions in a world where all celestial bodies, not just the sun, were seen differently from the way they had been seen before. The same point could be made about any of our earlier examples. To see oxygen instead of dephlogisticated air, the condenser instead of the Leyden jar, or the pendulum instead of constrained fall, was only one part of an integrated shift in the scientist's vision of a great many related chemical, electrical, or dynamical phenomena. Paradigms determine large areas of experience at the same time.

It is, however, only after experience has been thus determined that the search for an operational definition or a pure observation-language can begin. The scientist or philosopher who asks what measurements or retinal imprints make the pendulum what it is must already be able to recognize a pendulum when he sees one. If he saw constrained fall instead, his question could not even be asked. And if he saw a pendulum, but saw it in the same way he saw a tuning fork or an oscillating balance, his question could not be answered. At least it could not be answered in the same way, because it would not be the same question. Therefore, though they are always legitimate and are occasionally extraordinarily fruitful, questions about retinal imprints or about the consequences of particular laboratory manipulations presuppose a world already perceptually and conceptually subdivided in a certain way. In a sense such questions are parts of normal science, for they depend upon the existence of a paradigm and they receive different answers as a result of paradigm change.

To conclude this section, let us henceforth neglect retinal impressions and again restrict attention to the laboratory operations that provide the scientist with concrete though fragmentary indices to what he has already seen. One way in which such laboratory operations change with paradigms has already been observed repeatedly. After a scientific revolution many old measurements and manipulations become irrelevant and are replaced by others instead. One does not apply all the same tests to oxygen as to dephlogisticated air. But changes of this sort are never total. Whatever he may then see, the scientist after a revolution is still looking at the same world. Furthermore, though he may previously have employed them differently, much of his language and most of his laboratory instruments are still the same as they were before. As a result, postrevolutionary science invariably includes many of the same manipulations, performed with the same instruments and described in the same terms, as its prerevolutionary predecessor. If these enduring manipulations have been changed at all, the change must lie either in their relation to the paradigm or in their concrete results. I now suggest, by the introduction of one last new example, that both these sorts of changes occur. Examining the work of Dalton and his contemporaries, we shall discover that one and the same operation, when it attaches to nature through a different paradigm, can become an index to a quite different aspect of nature's regularity. In addition, we shall see that occasionally the old manipulation in its new role will yield different concrete results.

Throughout much of the eighteenth century and into the nineteenth, European chemists almost universally believed that the elementary atoms of which all chemical species consisted were held together by forces of mutual affinity. Thus a lump of silver cohered because of the forces of affinity between silver corpuscles (until after Lavoisier these corpuscles were themselves thought of as compounded from still more elementary particles). On the same theory silver dissolved in acid (or salt in water) because the particles of acid attracted those of silver (or the particles of water attracted those of salt) more strongly than particles of these solutes attracted each other. Or again, copper would dissolve in the silver solution and precipitate silver, because the copper-acid affinity was greater than the affinity of acid for silver. A great many other phenomena were explained in the same way. In the eighteenth century the theory of elective affinity was an admirable chemical paradigm, widely and sometimes fruitfully deployed in the design and analysis of chemical experimentation.[20]

Affinity theory, however, drew the line separating physical

[20] H. Metzger, *Newton, Stahl, Boerhaave et la doctrine chimique* (Paris, 1930), pp. 34–68.

mixtures from chemical compounds in a way that has become unfamiliar since the assimilation of Dalton's work. Eighteenth-century chemists did recognize two sorts of processes. When mixing produced heat, light, effervescence or something else of the sort, chemical union was seen to have taken place. If, on the other hand, the particles in the mixture could be distinguished by eye or mechanically separated, there was only physical mixture. But in the very large number of intermediate cases—salt in water, alloys, glass, oxygen in the atmosphere, and so on—these crude criteria were of little use. Guided by their paradigm, most chemists viewed this entire intermediate range as chemical, because the processes of which it consisted were all governed by forces of the same sort. Salt in water or oxygen in nitrogen was just as much an example of chemical combination as was the combination produced by oxidizing copper. The arguments for viewing solutions as compounds were very strong. Affinity theory itself was well attested. Besides, the formation of a compound accounted for a solution's observed homogeneity. If, for example, oxygen and nitrogen were only mixed and not combined in the atmosphere, then the heavier gas, oxygen, should settle to the bottom. Dalton, who took the atmosphere to be a mixture, was never satisfactorily able to explain oxygen's failure to do so. The assimilation of his atomic theory ultimately created an anomaly where there had been none before.[21]

One is tempted to say that the chemists who viewed solutions as compounds differed from their successors only over a matter of definition. In one sense that may have been the case. But that sense is not the one that makes definitions mere conventional conveniences. In the eighteenth century mixtures were not fully distinguished from compounds by operational tests, and perhaps they could not have been. Even if chemists had looked for such tests, they would have sought criteria that made the solution a compound. The mixture-compound distinction was part of their paradigm—part of the way they viewed their whole

[21] *Ibid.*, pp. 124–29, 139–48. For Dalton, see Leonard K. Nash, *The Atomic-Molecular Theory* ("Harvard Case Histories in Experimental Science," Case 4; Cambridge, Mass., 1950), pp. 14–21.

field of research—and as such it was prior to any particular laboratory test, though not to the accumulated experience of chemistry as a whole.

But while chemistry was viewed in this way, chemical phenomena exemplified laws different from those that emerged with the assimilation of Dalton's new paradigm. In particular, while solutions remained compounds, no amount of chemical experimentation could by itself have produced the law of fixed proportions. At the end of the eighteenth century it was widely known that *some* compounds ordinarily contained fixed proportions by weight of their constituents. For some categories of reactions the German chemist Richter had even noted the further regularities now embraced by the law of chemical equivalents.[22] But no chemist made use of these regularities except in recipes, and no one until almost the end of the century thought of generalizing them. Given the obvious counterinstances, like glass or like salt in water, no generalization was possible without an abandonment of affinity theory and a reconceptualization of the boundaries of the chemist's domain. That consequence became explicit at the very end of the century in a famous debate between the French chemists Proust and Berthollet. The first claimed that all chemical reactions occurred in fixed proportion, the latter that they did not. Each collected impressive experimental evidence for his view. Nevertheless, the two men necessarily talked through each other, and their debate was entirely inconclusive. Where Berthollet saw a compound that could vary in proportion, Proust saw only a physical mixture.[23] To that issue neither experiment nor a change of definitional convention could be relevant. The two men were as fundamentally at cross-purposes as Galileo and Aristotle had been.

This was the situation during the years when John Dalton undertook the investigations that led finally to his famous chemical atomic theory. But until the very last stages of those investiga-

[22] J. R. Partington, *A Short History of Chemistry* (2d ed.; London, 1951), pp. 161–63.

[23] A. N. Meldrum, "The Development of the Atomic Theory: (1) Berthollet's Doctrine of Variable Proportions," *Manchester Memoirs*, LIV (1910), 1–16.

tions, Dalton was neither a chemist nor interested in chemistry. Instead, he was a meteorologist investigating the, for him, physical problems of the absorption of gases by water and of water by the atmosphere. Partly because his training was in a different specialty and partly because of his own work in that specialty, he approached these problems with a paradigm different from that of contemporary chemists. In particular, he viewed the mixture of gases or the absorption of a gas in water as a physical process, one in which forces of affinity played no part. To him, therefore, the observed homogeneity of solutions was a problem, but one which he thought he could solve if he could determine the relative sizes and weights of the various atomic particles in his experimental mixtures. It was to determine these sizes and weights that Dalton finally turned to chemistry, supposing from the start that, in the restricted range of reactions that he took to be chemical, atoms could only combine one-to-one or in some other simple whole-number ratio.[24] That natural assumption did enable him to determine the sizes and weights of elementary particles, but it also made the law of constant proportion a tautology. For Dalton, any reaction in which the ingredients did not enter in fixed proportion was *ipso facto* not a purely chemical process. A law that experiment could not have established before Dalton's work, became, once that work was accepted, a constitutive principle that no single set of chemical measurements could have upset. As a result of what is perhaps our fullest example of a scientific revolution, the same chemical manipulations assumed a relationship to chemical generalization very different from the one they had had before.

Needless to say, Dalton's conclusions were widely attacked when first announced. Berthollet, in particular, was never convinced. Considering the nature of the issue, he need not have been. But to most chemists Dalton's new paradigm proved convincing where Proust's had not been, for it had implications far wider and more important than a new criterion for distinguish-

[24] L. K. Nash, "The Origin of Dalton's Chemical Atomic Theory," *Isis*, XLVII (1956), 101–16.

ing a mixture from a compound. If, for example, atoms could combine chemically only in simple whole-number ratios, then a re-examination of existing chemical data should disclose examples of multiple as well as of fixed proportions. Chemists stopped writing that the two oxides of, say, carbon contained 56 per cent and 72 per cent of oxygen by weight; instead they wrote that one weight of carbon would combine either with 1.3 or with 2.6 weights of oxygen. When the results of old manipulations were recorded in this way, a 2:1 ratio leaped to the eye; and this occurred in the analysis of many well-known reactions and of new ones besides. In addition, Dalton's paradigm made it possible to assimilate Richter's work and to see its full generality. Also, it suggested new experiments, particularly those of Gay-Lussac on combining volumes, and these yielded still other regularities, ones that chemists had not previously dreamed of. What chemists took from Dalton was not new experimental laws but a new way of practicing chemistry (he himself called it the "new system of chemical philosophy"), and this proved so rapidly fruitful that only a few of the older chemists in France and Britain were able to resist it.[25] As a result, chemists came to live in a world where reactions behaved quite differently from the way they had before.

As all this went on, one other typical and very important change occurred. Here and there the very numerical data of chemistry began to shift. When Dalton first searched the chemical literature for data to support his physical theory, he found some records of reactions that fitted, but he can scarcely have avoided finding others that did not. Proust's own measurements on the two oxides of copper yielded, for example, an oxygen weight-ratio of 1.47:1 rather than the 2:1 demanded by the atomic theory; and Proust is just the man who might have been expected to achieve the Daltonian ratio.[26] He was, that is, a fine

[25] A. N. Meldrum, "The Development of the Atomic Theory: (6) The Reception Accorded to the Theory Advocated by Dalton," *Manchester Memoirs,* LV (1911), 1–10.

[26] For Proust, see Meldrum, "Berthollet's Doctrine of Variable Proportions," *Manchester Memoirs,* LIV (1910), 8. The detailed history of the gradual changes in measurements of chemical composition and of atomic weights has yet to be written, but Partington, *op. cit.*, provides many useful leads to it.

experimentalist, and his view of the relation between mixtures and compounds was very close to Dalton's. But it is hard to make nature fit a paradigm. That is why the puzzles of normal science are so challenging and also why measurements undertaken without a paradigm so seldom lead to any conclusions at all. Chemists could not, therefore, simply accept Dalton's theory on the evidence, for much of that was still negative. Instead, even after accepting the theory, they had still to beat nature into line, a process which, in the event, took almost another generation. When it was done, even the percentage composition of well-known compounds was different. The data themselves had changed. That is the last of the senses in which we may want to say that after a revolution scientists work in a different world.

XI. The Invisibility of Revolutions

We must still ask how scientific revolutions close. Before doing so, however, a last attempt to reinforce conviction about their existence and nature seems called for. I have so far tried to display revolutions by illustration, and the examples could be multiplied *ad nauseam*. But clearly, most of them, which were deliberately selected for their familiarity, have customarily been viewed not as revolutions but as additions to scientific knowledge. That same view could equally well be taken of any additional illustrations, and these would probably be ineffective. I suggest that there are excellent reasons why revolutions have proved to be so nearly invisible. Both scientists and laymen take much of their image of creative scientific activity from an authoritative source that systematically disguises—partly for important functional reasons—the existence and significance of scientific revolutions. Only when the nature of that authority is recognized and analyzed can one hope to make historical example fully effective. Furthermore, though the point can be fully developed only in my concluding section, the analysis now required will begin to indicate one of the aspects of scientific work that most clearly distinguishes it from every other creative pursuit except perhaps theology.

As the source of authority, I have in mind principally textbooks of science together with both the popularizations and the philosophical works modeled on them. All three of these categories—until recently no other significant sources of information about science have been available except through the practice of research—have one thing in common. They address themselves to an already articulated body of problems, data, and theory, most often to the particular set of paradigms to which the scientific community is committed at the time they are written. Textbooks themselves aim to communicate the vocabulary and syntax of a contemporary scientific language. Popularizations attempt to describe these same applications in a language closer to that of everyday life. And philosophy of science, particularly that of the English-speaking world, analyzes the logical structure of the same completed body of scientific knowledge. Though a fuller treatment would necessarily deal with the very real distinctions between these three genres, it is their similarities that most concern us here. All three record the stable *outcome* of past revolutions and thus display the bases of the current normal-scientific tradition. To fulfill their function they need not provide authentic information about the way in which those bases were first recognized and then embraced by the profession. In the case of textbooks, at least, there are even good reasons why, in these matters, they should be systematically misleading.

We noted in Section II that an increasing reliance on textbooks or their equivalent was an invariable concomitant of the emergence of a first paradigm in any field of science. The concluding section of this essay will argue that the domination of a mature science by such texts significantly differentiates its developmental pattern from that of other fields. For the moment let us simply take it for granted that, to an extent unprecedented in other fields, both the layman's and the practitioner's knowledge of science is based on textbooks and a few other types of literature derived from them. Textbooks, however, being pedagogic vehicles for the perpetuation of normal science, have to be rewritten in whole or in part whenever the language, problem-structure, or standards of normal science change. In short, they have to be rewritten in the aftermath of each scientific revolution, and, once rewritten, they inevitably disguise not only the role but the very existence of the revolutions that produced them. Unless he has personally experienced a revolution in his own lifetime, the historical sense either of the working scientist or of the lay reader of textbook literature extends only to the outcome of the most recent revolutions in the field.

Textbooks thus begin by truncating the scientist's sense of his discipline's history and then proceed to supply a substitute for what they have eliminated. Characteristically, textbooks of science contain just a bit of history, either in an introductory chapter or, more often, in scattered references to the great heroes of an earlier age. From such references both students and professionals come to feel like participants in a long-standing historical tradition. Yet the textbook-derived tradition in which scientists come to sense their participation is one that, in fact, never existed. For reasons that are both obvious and highly functional, science textbooks (and too many of the older histories of science) refer only to that part of the work of past scientists that can easily be viewed as contributions to the statement and solution of the texts' paradigm problems. Partly by selection and partly by distortion, the scientists of earlier ages are implicitly represented as having worked upon the same set of fixed problems and in accordance with the same set of fixed canons that the most recent revolution in scientific theory and method has made seem scientific. No wonder that textbooks and the historical tradition they imply have to be rewritten after each scientific revolution. And no wonder that, as they are rewritten, science once again comes to seem largely cumulative.

Scientists are not, of course, the only group that tends to see its discipline's past developing linearly toward its present vantage. The temptation to write history backward is both omnipresent and perennial. But scientists are more affected by the temptation to rewrite history, partly because the results of scientific research show no obvious dependence upon the historical context of the inquiry, and partly because, except during crisis and revolution, the scientist's contemporary position seems so secure. More historical detail, whether of science's present or of its past, or more responsibility to the historical details that are presented, could only give artificial status to human idiosyncrasy, error, and confusion. Why dignify what science's best and most persistent efforts have made it possible to discard? The depreciation of historical fact is deeply, and probably functionally, ingrained in the ideology of the scientific profession, the same profession that places the highest of all values upon factual details of other sorts. Whitehead caught the unhistorical spirit of the scientific community when he wrote, "A science that hesitates to forget its founders is lost." Yet he was not quite right, for the sciences, like other professional enterprises, do need their heroes and do preserve their names. Fortunately, instead of forgetting these heroes, scientists have been able to forget or revise their works.

The result is a persistent tendency to make the history of science look linear or cumulative, a tendency that even affects scientists looking back at their own research. For example, all three of Dalton's incompatible accounts of the development of his chemical atomism make it appear that he was interested from an early date in just those chemical problems of combining proportions that he was later famous for having solved. Actually those problems seem only to have occurred to him with their solutions, and then not until his own creative work was very nearly complete.[1] What all of Dalton's accounts omit are the revolutionary effects of applying to chemistry a set of questions and concepts previously restricted to physics and meteorology. That is what Dalton did, and the result was a reorientation toward the field, a reorientation that taught chemists to ask new questions about and to draw new conclusions from old data.

Or again, Newton wrote that Galileo had discovered that the constant force of gravity produces a motion proportional to the square of the time. In fact, Galileo's kinematic theorem does take that form when embedded in the matrix of Newton's own dynamical concepts. But Galileo said nothing of the sort. His discussion of falling bodies rarely alludes to forces, much less to a uniform gravitational force that causes bodies to fall.[2] By crediting to Galileo the answer to a question that Galileo's paradigms did not permit to be asked, Newton's account hides the effect of a small but revolutionary reformulation in the questions that scientists asked about motion as well as in the

1 L. K. Nash, "The Origins of Dalton's Chemical Atomic Theory," *Isis,* XLVII (1956), 101–16.

2 For Newton's remark, see Florian Cajori (ed.), *Sir Isaac Newton's Mathematical Principles of Natural Philosophy and His System of the World* (Berkeley, Calif., 1946), p. 21. The passage should be compared with Galileo's own discussion in his *Dialogues concerning Two New Sciences,* trans. H. Crew and A. de Salvio (Evanston, Ill., 1946), pp. 154–76.

answers they felt able to accept. But it is just this sort of change in the formulation of questions and answers that accounts, far more than novel empirical discoveries, for the transition from Aristotelian to Galilean and from Galilean to Newtonian dynamics. By disguising such changes, the textbook tendency to make the development of science linear hides a process that lies at the heart of the most significant episodes of scientific development.

The preceding examples display, each within the context of a single revolution, the beginnings of a reconstruction of history that is regularly completed by postrevolutionary science texts. But in that completion more is involved than a multiplication of the historical misconstructions illustrated above. Those misconstructions render revolutions invisible; the arrangement of the still visible material in science texts implies a process that, if it existed, would deny revolutions a function. Because they aim quickly to acquaint the student with what the contemporary scientific community thinks it knows, textbooks treat the various experiments, concepts, laws, and theories of the current normal science as separately and as nearly seriatim as possible. As pedagogy this technique of presentation is unexceptionable. But when combined with the generally unhistorical air of science writing and with the occasional systematic misconstructions discussed above, one strong impression is overwhelmingly likely to follow: science has reached its present state by a series of individual discoveries and inventions that, when gathered together, constitute the modern body of technical knowledge. From the beginning of the scientific enterprise, a textbook presentation implies, scientists have striven for the particular objectives that are embodied in today's paradigms. One by one, in a process often compared to the addition of bricks to a building, scientists have added another fact, concept, law, or theory to the body of information supplied in the contemporary science text.

But that is not the way a science develops. Many of the puzzles of contemporary normal science did not exist until after the most recent scientific revolution. Very few of them can be traced back to the historic beginning of the science within which they now occur. Earlier generations pursued their own problems with their own instruments and their own canons of solution. Nor is it just the problems that have changed. Rather the whole network of fact and theory that the textbook paradigm fits to nature has shifted. Is the constancy of chemical composition, for example, a mere fact of experience that chemists could have discovered by experiment within any one of the worlds within which chemists have practiced? Or is it rather one element—and an indubitable one, at that—in a new fabric of associated fact and theory that Dalton fitted to the earlier chemical experience as a whole, changing that experience in the process? Or by the same token, is the constant acceleration produced by a constant force a mere fact that students of dynamics have always sought, or is it rather the answer to a question that first arose only within Newtonian theory and that that theory could answer from the body of information available before the question was asked?

These questions are here asked about what appear as the piecemeal-discovered facts of a textbook presentation. But obviously, they have implications as well for what the text presents as theories. Those theories, of course, do "fit the facts," but only by transforming previously accessible information into facts that, for the preceding paradigm, had not existed at all. And that means that theories too do not evolve piecemeal to fit facts that were there all the time. Rather, they emerge together with the facts they fit from a revolutionary reformulation of the preceding scientific tradition, a tradition within which the knowledge-mediated relationship between the scientist and nature was not quite the same.

One last example may clarify this account of the impact of textbook presentation upon our image of scientific development. Every elementary chemistry text must discuss the concept of a chemical element. Almost always, when that notion is introduced, its origin is attributed to the seventeenth-century chemist, Robert Boyle, in whose *Sceptical Chymist* the attentive reader will find a definition of 'element' quite close to that in use today. Reference to Boyle's contribution helps to make the neophyte aware that chemistry did not begin with the sulfa drugs; in addition, it tells him that one of the scientist's traditional tasks is to invent concepts of this sort. As a part of the pedagogic arsenal that makes a man a scientist, the attribution is immensely successful. Nevertheless, it illustrates once more the pattern of historical mistakes that misleads both students and laymen about the nature of the scientific enterprise.

According to Boyle, who was quite right, his "definition" of an element was no more than a paraphrase of a traditional chemical concept; Boyle offered it only in order to argue that no such thing as a chemical element exists; as history, the textbook version of Boyle's contribution is quite mistaken.[3] That mistake, of course, is trivial, though no more so than any other misrepresentation of data. What is not trivial, however, is the impression of science fostered when this sort of mistake is first compounded and then built into the technical structure of the text. Like 'time,' 'energy,' 'force,' or 'particle,' the concept of an element is the sort of textbook ingredient that is often not invented or discovered at all. Boyle's definition, in particular, can be traced back at least to Aristotle and forward through Lavoisier into modern texts. Yet that is not to say that science has possessed the modern concept of an element since antiquity. Verbal definitions like Boyle's have little scientific content when considered by themselves. They are not full logical specifications of meaning (if there are such), but more nearly pedagogic aids. The scientific concepts to which they point gain full significance only when related, within a text or other systematic presentation, to other scientific concepts, to manipulative procedures, and to paradigm applications. It follows that concepts like that of an element can scarcely be invented independent of context. Furthermore, given the context, they rarely require invention because they are already at hand. Both Boyle and Lavoisier changed the chemical significance of 'element' in important ways. But they did not invent the notion

[3] T. S. Kuhn, "Robert Boyle and Structural Chemistry in the Seventeenth Century," *Isis*, XLIII (1952), 26–29.

or even change the verbal formula that serves as its definition. Nor, as we have seen, did Einstein have to invent or even explicitly redefine 'space' and 'time' in order to give them new meaning within the context of his work.

What then was Boyle's historical function in that part of his work that includes the famous "definition"? He was a leader of a scientific revolution that, by changing the relation of 'element' to chemical manipulation and chemical theory, transformed the notion into a tool quite different from what it had been before and transformed both chemistry and the chemist's world in the process.[4] Other revolutions, including the one that centers around Lavoisier, were required to give the concept its modern form and function. But Boyle provides a typical example both of the process involved at each of these stages and of what happens to that process when existing knowledge is embodied in a textbook. More than any other single aspect of science, that pedagogic form has determined our image of the nature of science and of the role of discovery and invention in its advance.

[4] Marie Boas, in her *Robert Boyle and Seventeenth-Century Chemistry* (Cambridge, 1958), deals in many places with Boyle's positive contributions to the evolution of the concept of a chemical element.

XII. The Resolution of Revolutions

The textbooks we have just been discussing are produced only in the aftermath of a scientific revolution. They are the bases for a new tradition of normal science. In taking up the question of their structure we have clearly missed a step. What is the process by which a new candidate for paradigm replaces its predecessor? Any new interpretation of nature, whether a discovery or a theory, emerges first in the mind of one or a few individuals. It is they who first learn to see science and the world differently, and their ability to make the transition is facilitated by two circumstances that are not common to most other members of their profession. Invariably their attention has been intensely concentrated upon the crisis-provoking problems; usually, in addition, they are men so young or so new to the crisis-ridden field that practice has committed them less deeply than most of their contemporaries to the world view and rules determined by the old paradigm. How are they able, what must they do, to convert the entire profession or the relevant professional subgroup to their way of seeing science and the world? What causes the group to abandon one tradition of normal research in favor of another?

To see the urgency of those questions, remember that they are the only reconstructions the historian can supply for the philosopher's inquiry about the testing, verification, or falsification of established scientific theories. In so far as he is engaged in normal science, the research worker is a solver of puzzles, not a tester of paradigms. Though he may, during the search for a particular puzzle's solution, try out a number of alternative approaches, rejecting those that fail to yield the desired result, he is not testing the *paradigm* when he does so. Instead he is like the chess player who, with a problem stated and the board physically or mentally before him, tries out various alternative moves in the search for a solution. These trial attempts, whether by the chess player or by the scientist, are trials only of themselves, not of the rules of the game. They are possible only so long as the paradigm itself is taken for granted. Therefore, paradigm-testing occurs only after persistent failure to solve a noteworthy puzzle has given rise to crisis. And even then it occurs only after the sense of crisis has evoked an alternate candidate for paradigm. In the sciences the testing situation never consists, as puzzle-solving does, simply in the comparison of a single paradigm with nature. Instead, testing occurs as part of the competition between two rival paradigms for the allegiance of the scientific community.

Closely examined, this formulation displays unexpected and probably significant parallels to two of the most popular contemporary philosophical theories about verification. Few philosophers of science still seek absolute criteria for the verification of scientific theories. Noting that no theory can ever be exposed to all possible relevant tests, they ask not whether a theory has been verified but rather about its probability in the light of the evidence that actually exists. And to answer that question one important school is driven to compare the ability of different theories to explain the evidence at hand. That insistence on comparing theories also characterizes the historical situation in which a new theory is accepted. Very probably it points one of the directions in which future discussions of verification should go.

In their most usual forms, however, probabilistic verification theories all have recourse to one or another of the pure or neutral observation-languages discussed in Section X. One probabilistic theory asks that we compare the given scientific theory with all others that might be imagined to fit the same collection of observed data. Another demands the construction in imagination of all the tests that the given scientific theory might conceivably be asked to pass.[1] Apparently some such construction is necessary for the computation of specific probabilities, absolute or relative, and it is hard to see how such a construction can

[1] For a brief sketch of the main routes to probabilistic verification theories, see Ernest Nagel, *Principles of the Theory of Probability*, Vol. I, No. 6, of *International Encyclopedia of Unified Science*, pp. 60–75.

possibly be achieved. If, as I have already urged, there can be no scientifically or empirically neutral system of language or concepts, then the proposed construction of alternate tests and theories must proceed from within one or another paradigm-based tradition. Thus restricted it would have no access to all possible experiences or to all possible theories. As a result, probabilistic theories disguise the verification situation as much as they illuminate it. Though that situation does, as they insist, depend upon the comparison of theories and of much widespread evidence, the theories and observations at issue are always closely related to ones already in existence. Verification is like natural selection: it picks out the most viable among the actual alternatives in a particular historical situation. Whether that choice is the best that could have been made if still other alternatives had been available or if the data had been of another sort is not a question that can usefully be asked. There are no tools to employ in seeking answers to it.

A very different approach to this whole network of problems has been developed by Karl R. Popper who denies the existence of any verification procedures at all.[2] Instead, he emphasizes the importance of falsification, i.e., of the test that, because its outcome is negative, necessitates the rejection of an established theory. Clearly, the role thus attributed to falsification is much like the one this essay assigns to anomalous experiences, i.e., to experiences that, by evoking crisis, prepare the way for a new theory. Nevertheless, anomalous experiences may not be identified with falsifying ones. Indeed, I doubt that the latter exist. As has repeatedly been emphasized before, no theory ever solves all the puzzles with which it is confronted at a given time; nor are the solutions already achieved often perfect. On the contrary, it is just the incompleteness and imperfection of the existing data-theory fit that, at any time, define many of the puzzles that characterize normal science. If any and every failure to fit were ground for theory rejection, all theories ought to be rejected at all times. On the other hand, if only severe failure

[2] K. R. Popper, *The Logic of Scientific Discovery* (New York, 1959), esp. chaps. i–iv.

to fit justifies theory rejection, then the Popperians will require some criterion of "improbability" or of "degree of falsification." In developing one they will almost certainly encounter the same network of difficulties that has haunted the advocates of the various probabilistic verification theories.

Many of the preceding difficulties can be avoided by recognizing that both of these prevalent and opposed views about the underlying logic of scientific inquiry have tried to compress two largely separate processes into one. Popper's anomalous experience is important to science because it evokes competitors for an existing paradigm. But falsification, though it surely occurs, does not happen with, or simply because of, the emergence of an anomaly or falsifying instance. Instead, it is a subsequent and separate process that might equally well be called verification since it consists in the triumph of a new paradigm over the old one. Furthermore, it is in that joint verification-falsification process that the probabilist's comparison of theories plays a central role. Such a two-stage formulation has, I think, the virtue of great verisimilitude, and it may also enable us to begin explicating the role of agreement (or disagreement) between fact and theory in the verification process. To the historian, at least, it makes little sense to suggest that verification is establishing the agreement of fact with theory. All historically significant theories have agreed with the facts, but only more or less. There is no more precise answer to the question whether or how well an individual theory fits the facts. But questions much like that can be asked when theories are taken collectively or even in pairs. It makes a great deal of sense to ask which of two actual and competing theories fits the facts *better*. Though neither Priestley's nor Lavoisier's theory, for example, agreed precisely with existing observations, few contemporaries hesitated more than a decade in concluding that Lavoisier's theory provided the better fit of the two.

This formulation, however, makes the task of choosing between paradigms look both easier and more familiar than it is. If there were but one set of scientific problems, one world within which to work on them, and one set of standards for their solution, paradigm competition might be settled more or less routinely by some process like counting the number of problems solved by each. But, in fact, these conditions are never met completely. The proponents of competing paradigms are always at least slightly at cross-purposes. Neither side will grant all the non-empirical assumptions that the other needs in order to make its case. Like Proust and Berthollet arguing about the composition of chemical compounds, they are bound partly to talk through each other. Though each may hope to convert the other to his way of seeing his science and its problems, neither may hope to prove his case. The competition between paradigms is not the sort of battle that can be resolved by proofs.

We have already seen several reasons why the proponents of competing paradigms must fail to make complete contact with each other's viewpoints. Collectively these reasons have been described as the incommensurability of the pre- and postrevolutionary normal-scientific traditions, and we need only recapitulate them briefly here. In the first place, the proponents of competing paradigms will often disagree about the list of problems that any candidate for paradigm must resolve. Their standards or their definitions of science are not the same. Must a theory of motion explain the cause of the attractive forces between particles of matter or may it simply note the existence of such forces? Newton's dynamics was widely rejected because, unlike both Aristotle's and Descartes's theories, it implied the latter answer to the question. When Newton's theory had been accepted, a question was therefore banished from science. That question, however, was one that general relativity may proudly claim to have solved. Or again, as disseminated in the nineteenth century, Lavoisier's chemical theory inhibited chemists from asking why the metals were so much alike, a question that phlogistic chemistry had both asked and answered. The transition to Lavoisier's paradigm had, like the transition to Newton's, meant a loss not only of a permissible question but of an achieved solution. That loss was not, however, permanent either. In the twentieth century questions about the qualities of chemical substances have entered science again, together with some answers to them.

More is involved, however, than the incommensurability of standards. Since new paradigms are born from old ones, they ordinarily incorporate much of the vocabulary and apparatus, both conceptual and manipulative, that the traditional paradigm had previously employed. But they seldom employ these borrowed elements in quite the traditional way. Within the new paradigm, old terms, concepts, and experiments fall into new relationships one with the other. The inevitable result is what we must call, though the term is not quite right, a misunderstanding between the two competing schools. The laymen who scoffed at Einstein's general theory of relativity because space could not be "curved"—it was not that sort of thing—were not simply wrong or mistaken. Nor were the mathematicians, physicists, and philosophers who tried to develop a Euclidean version of Einstein's theory.[3] What had previously been meant by space was necessarily flat, homogeneous, isotropic, and unaffected by the presence of matter. If it had not been, Newtonian physics would not have worked. To make the transition to Einstein's universe, the whole conceptual web whose strands are space, time, matter, force, and so on, had to be shifted and laid down again on nature whole. Only men who had together undergone or failed to undergo that transformation would be able to discover precisely what they agreed or disagreed about. Communication across the revolutionary divide is inevitably partial. Consider, for another example, the men who called Copernicus mad because he proclaimed that the earth moved. They were not either just wrong or quite wrong. Part of what they meant by 'earth' was fixed position. Their earth, at least, could not be moved. Correspondingly, Copernicus' innovation was not simply to move the earth. Rather, it was a whole new way of regarding the problems of physics and astronomy,

[3] For lay reactions to the concept of curved space, see Philipp Frank, *Einstein, His Life and Times,* trans. and ed. G. Rosen and S. Kusaka (New York, 1947), pp. 142–46. For a few of the attempts to preserve the gains of general relativity within a Euclidean space, see C. Nordmann, *Einstein and the Universe,* trans. J. McCabe (New York, 1922), chap. ix.

one that necessarily changed the meaning of both 'earth' and 'motion.'[4] Without those changes the concept of a moving earth was mad. On the other hand, once they had been made and understood, both Descartes and Huyghens could realize that the earth's motion was a question with no content for science.[5]

These examples point to the third and most fundamental aspect of the incommensurability of competing paradigms. In a sense that I am unable to explicate further, the proponents of competing paradigms practice their trades in different worlds. One contains constrained bodies that fall slowly, the other pendulums that repeat their motions again and again. In one, solutions are compounds, in the other mixtures. One is embedded in a flat, the other in a curved, matrix of space. Practicing in different worlds, the two groups of scientists see different things when they look from the same point in the same direction. Again, that is not to say that they can see anything they please. Both are looking at the world, and what they look at has not changed. But in some areas they see different things, and they see them in different relations one to the other. That is why a law that cannot even be demonstrated to one group of scientists may occasionally seem intuitively obvious to another. Equally, it is why, before they can hope to communicate fully, one group or the other must experience the conversion that we have been calling a paradigm shift. Just because it is a transition between incommensurables, the transition between competing paradigms cannot be made a step at a time, forced by logic and neutral experience. Like the gestalt switch, it must occur all at once (though not necessarily in an instant) or not at all.

How, then, are scientists brought to make this transposition? Part of the answer is that they are very often not. Copernicanism made few converts for almost a century after Copernicus' death. Newton's work was not generally accepted, particularly on the Continent, for more than half a century after the *Prin-*

[4] T. S. Kuhn, *The Copernican Revolution* (Cambridge, Mass., 1957), chaps. iii, iv, and vii. The extent to which heliocentrism was more than a strictly astronomical issue is a major theme of the entire book.

[5] Max Jammer, *Concepts of Space* (Cambridge, Mass., 1954), pp. 118–24.

cipia appeared.[6] Priestley never accepted the oxygen theory, nor Lord Kelvin the electromagnetic theory, and so on. The difficulties of conversion have often been noted by scientists themselves. Darwin, in a particularly perceptive passage at the end of his *Origin of Species,* wrote: "Although I am fully convinced of the truth of the views given in this volume . . . , I by no means expect to convince experienced naturalists whose minds are stocked with a multitude of facts all viewed, during a long course of years, from a point of view directly opposite to mine. . . . [B]ut I look with confidence to the future,–to young and rising naturalists, who will be able to view both sides of the question with impartiality."[7] And Max Planck, surveying his own career in his *Scientific Autobiography,* sadly remarked that "a new scientific truth does not triumph by convincing its opponents and making them see the light, but rather because its opponents eventually die, and a new generation grows up that is familiar with it."[8]

These facts and others like them are too commonly known to need further emphasis. But they do need re-evaluation. In the past they have most often been taken to indicate that scientists, being only human, cannot always admit their errors, even when confronted with strict proof. I would argue, rather, that in these matters neither proof nor error is at issue. The transfer of allegiance fom paradigm to paradigm is a conversion experience that cannot be forced. Lifelong resistance, particularly from those whose productive careers have committed them to an older tradition of normal science, is not a violation of scientific standards but an index to the nature of scientific research itself. The source of resistance is the assurance that the older paradigm will ultimately solve all its problems, that nature can be shoved

[6] I. B. Cohen, *Franklin and Newton: An Inquiry into Speculative Newtonian Experimental Science and Franklin's Work in Electricity as an Example Thereof* (Philadelphia, 1956), pp. 93–94.

[7] Charles Darwin, *On the Origin of Species . . .* (authorized edition from 6th English ed.; New York, 1889), II, 295–96.

[8] Max Planck, *Scientific Autobiography and Other Papers,* trans. F. Gaynor (New York, 1949), pp. 33–34.

into the box the paradigm provides. Inevitably, at times of revolution, that assurance seems stubborn and pigheaded as indeed it sometimes becomes. But it is also something more. That same assurance is what makes normal or puzzle-solving science possible. And it is only through normal science that the professional community of scientists succeeds, first, in exploiting the potential scope and precision of the older paradigm and, then, in isolating the difficulty through the study of which a new paradigm may emerge.

Still, to say that resistance is inevitable and legitimate, that paradigm change cannot be justified by proof, is not to say that no arguments are relevant or that scientists cannot be persuaded to change their minds. Though a generation is sometimes required to effect the change, scientific communities have again and again been converted to new paradigms. Furthermore, these conversions occur not despite the fact that scientists are human but because they are. Though some scientists, particularly the older and more experienced ones, may resist indefinitely, most of them can be reached in one way or another. Conversions will occur a few at a time until, after the last holdouts have died, the whole profession will again be practicing under a single, but now a different, paradigm. We must therefore ask how conversion is induced and how resisted.

What sort of answer to that question may we expect? Just because it is asked about techniques of persuasion, or about argument and counterargument in a situation in which there can be no proof, our question is a new one, demanding a sort of study that has not previously been undertaken. We shall have to settle for a very partial and impressionistic survey. In addition, what has already been said combines with the result of that survey to suggest that, when asked about persuasion rather than proof, the question of the nature of scientific argument has no single or uniform answer. Individual scientists embrace a new paradigm for all sorts of reasons and usually for several at once. Some of these reasons—for example, the sun worship that helped make Kepler a Copernican—lie outside the apparent sphere of science entirely.[9] Others must depend upon idiosyncrasies of autobiography and personality. Even the nationality or the prior reputation of the innovator and his teachers can sometimes play a significant role.[10] Ultimately, therefore, we must learn to ask this question differently. Our concern will not then be with the arguments that in fact convert one or another individual, but rather with the sort of community that always sooner or later re-forms as a single group. That problem, however, I postpone to the final section, examining meanwhile some of the sorts of argument that prove particularly effective in the battles over paradigm change.

Probably the single most prevalent claim advanced by the proponents of a new paradigm is that they can solve the problems that have led the old one to a crisis. When it can legitimately be made, this claim is often the most effective one possible. In the area for which it is advanced the paradigm is known to be in trouble. That trouble has repeatedly been explored, and attempts to remove it have again and again proved vain. "Crucial experiments"—those able to discriminate particularly sharply between the two paradigms—have been recognized and attested before the new paradigm was even invented. Copernicus thus claimed that he had solved the long-vexing problem of the length of the calendar year, Newton that he had reconciled terrestrial and celestial mechanics, Lavoisier that he had solved the problems of gas-identity and of weight relations, and Einstein that he had made electrodynamics compatible with a revised science of motion.

Claims of this sort are particularly likely to succeed if the new paradigm displays a quantitative precision strikingly better than

[9] For the role of sun worship in Kepler's thought, see E. A. Burtt, *The Metaphysical Foundations of Modern Physical Science* (rev. ed.; New York, 1932), pp. 44–49.

[10] For the role of reputation, consider the following: Lord Rayleigh, at a time when his reputation was established, submitted to the British Association a paper on some paradoxes of electrodynamics. His name was inadvertently omitted when the paper was first sent, and the paper itself was at first rejected as the work of some "paradoxer." Shortly afterwards, with the author's name in place, the paper was accepted with profuse apologies (R. J. Strutt, 4th Baron Rayleigh, *John William Strutt, Third Baron Rayleigh* [New York, 1924], p. 228).

its older competitor. The quantitative superiority of Kepler's Rudolphine tables to all those computed from the Ptolemaic theory was a major factor in the conversion of astronomers to Copernicanism. Newton's success in predicting quantitative astronomical observations was probably the single most important reason for his theory's triumph over its more reasonable but uniformly qualitative competitors. And in this century the striking quantitative success of both Planck's radiation law and the Bohr atom quickly persuaded many physicists to adopt them even though, viewing physical science as a whole, both these contributions created many more problems than they solved.[11]

The claim to have solved the crisis-provoking problems is, however, rarely sufficient by itself. Nor can it always legitimately be made. In fact, Copernicus' theory was not more accurate than Ptolemy's and did not lead directly to any improvement in the calendar. Or again, the wave theory of light was not, for some years after it was first announced, even as successful as its corpuscular rival in resolving the polarization effects that were a principal cause of the optical crisis. Sometimes the looser practice that characterizes extraordinary research will produce a candidate for paradigm that initially helps not at all with the problems that have evoked crisis. When that occurs, evidence must be drawn from other parts of the field as it often is anyway. In those other areas particularly persuasive arguments can be developed if the new paradigm permits the prediction of phenomena that had been entirely unsuspected while the old one prevailed.

Copernicus' theory, for example, suggested that planets should be like the earth, that Venus should show phases, and that the universe must be vastly larger than had previously been supposed. As a result, when sixty years after his death the telescope suddenly displayed mountains on the moon, the phases of Venus, and an immense number of previously unsuspected stars,

[11] For the problems created by the quantum theory, see F. Reiche, *The Quantum Theory* (London, 1922), chaps. ii, vi–ix. For the other examples in this paragraph, see the earlier references in this section.

those observations brought the new theory a great many converts, particularly among non-astronomers.[12] In the case of the wave theory, one main source of professional conversions was even more dramatic. French resistance collapsed suddenly and relatively completely when Fresnel was able to demonstrate the existence of a white spot at the center of the shadow of a circular disk. That was an effect that not even he had anticipated but that Poisson, initially one of his opponents, had shown to be a necessary if absurd consequence of Fresnel's theory.[13] Because of their shock value and because they have so obviously not been "built into" the new theory from the start, arguments like these prove especially persuasive. And sometimes that extra strength can be exploited even though the phenomenon in question had been observed long before the theory that accounts for it was first introduced. Einstein, for example, seems not to have anticipated that general relativity would account with precision for the well-known anomaly in the motion of Mercury's perihelion, and he experienced a corresponding triumph when it did so.[14]

All the arguments for a new paradigm discussed so far have been based upon the competitors' comparative ability to solve problems. To scientists those arguments are ordinarily the most significant and persuasive. The preceding examples should leave no doubt about the source of their immense appeal. But, for reasons to which we shall shortly revert, they are neither individually nor collectively compelling. Fortunately, there is also another sort of consideration that can lead scientists to reject an old paradigm in favor of a new. These are the arguments, rarely made entirely explicit, that appeal to the individual's sense of the appropriate or the aesthetic—the new theory is said to be "neater," "more suitable," or "simpler" than the old. Probably

12 Kuhn, *op. cit.*, pp. 219–25.

13 E. T. Whittaker, *A History of the Theories of Aether and Electricity*, I (2d ed.; London, 1951), 108.

14 See *ibid.*, II (1953), 151–80, for the development of general relativity. For Einstein's reaction to the precise agreement of the theory with the observed motion of Mercury's perihelion, see the letter quoted in P. A. Schilpp (ed.), *Albert Einstein, Philosopher-Scientist* (Evanston, Ill., 1949), p. 101.

such arguments are less effective in the sciences than in mathematics. The early versions of most new paradigms are crude. By the time their full aesthetic appeal can be developed, most of the community has been persuaded by other means. Nevertheless, the importance of aesthetic considerations can sometimes be decisive. Though they often attract only a few scientists to a new theory, it is upon those few that its ultimate triumph may depend. If they had not quickly taken it up for highly individual reasons, the new candidate for paradigm might never have been sufficiently developed to attract the allegiance of the scientific community as a whole.

To see the reason for the importance of these more subjective and aesthetic considerations, remember what a paradigm debate is about. When a new candidate for paradigm is first proposed, it has seldom solved more than a few of the problems that confront it, and most of those solutions are still far from perfect. Until Kepler, the Copernican theory scarcely improved upon the predictions of planetary position made by Ptolemy. When Lavoisier saw oxygen as "the air itself entire," his new theory could cope not at all with the problems presented by the proliferation of new gases, a point that Priestley made with great success in his counterattack. Cases like Fresnel's white spot are extremely rare. Ordinarily, it is only much later, after the new paradigm has been developed, accepted, and exploited that apparently decisive arguments—the Foucault pendulum to demonstrate the rotation of the earth or the Fizeau experiment to show that light moves faster in air than in water—are developed. Producing them is part of normal science, and their role is not in paradigm debate but in postrevolutionary texts.

Before those texts are written, while the debate goes on, the situation is very different. Usually the opponents of a new paradigm can legitimately claim that even in the area of crisis it is little superior to its traditional rival. Of course, it handles some problems better, has disclosed some new regularities. But the older paradigm can presumably be articulated to meet these challenges as it has met others before. Both Tycho Brahe's earth-centered astronomical system and the later versions of the phlogiston theory were responses to challenges posed by a new candidate for paradigm, and both were quite successful.[15] In addition, the defenders of traditional theory and procedure can almost always point to problems that its new rival has not solved but that for their view are no problems at all. Until the discovery of the composition of water, the combustion of hydrogen was a strong argument for the phlogiston theory and against Lavoisier's. And after the oxygen theory had triumphed, it could still not explain the preparation of a combustible gas from carbon, a phenomenon to which the phlogistonists had pointed as strong support for their view.[16] Even in the area of crisis, the balance of argument and counterargument can sometimes be very close indeed. And outside that area the balance will often decisively favor the tradition. Copernicus destroyed a time-honored explanation of terrestrial motion without replacing it; Newton did the same for an older explanation of gravity, Lavoisier for the common properties of metals, and so on. In short, if a new candidate for paradigm had to be judged from the start by hard-headed people who examined only relative problem-solving ability, the sciences would experience very few major revolutions. Add the counterarguments generated by what we previously called the incommensurability of paradigms, and the sciences might experience no revolutions at all.

But paradigm debates are not really about relative problem-solving ability, though for good reasons they are usually couched in those terms. Instead, the issue is which paradigm should in the future guide research on problems many of which neither competitor can yet claim to resolve completely. A decision between alternate ways of practicing science is called for, and in the circumstances that decision must be based less on

[15] For Brahe's system, which was geometrically entirely equivalent to Copernicus', see J. L. E. Dreyer, *A History of Astronomy from Thales to Kepler* (2d ed.; New York, 1953), pp. 359–71. For the last versions of the phlogiston theory and their success, see J. R. Partington and D. McKie, "Historical Studies of the Phlogiston Theory," *Annals of Science,* IV (1939), 113–49.

[16] For the problem presented by hydrogen, see J. R. Partington, *A Short History of Chemistry* (2d ed.; London, 1951), p. 134. For carbon monoxide, see H. Kopp, *Geschichte der Chemie,* III (Braunschweig, 1845), 294–96.

past achievement than on future promise. The man who embraces a new paradigm at an early stage must often do so in defiance of the evidence provided by problem-solving. He must, that is, have faith that the new paradigm will succeed with the many large problems that confront it, knowing only that the older paradigm has failed with a few. A decision of that kind can only be made on faith.

That is one of the reasons why prior crisis proves so important. Scientists who have not experienced it will seldom renounce the hard evidence of problem-solving to follow what may easily prove and will be widely regarded as a will-o'-the-wisp. But crisis alone is not enough. There must also be a basis, though it need be neither rational nor ultimately correct, for faith in the particular candidate chosen. Something must make at least a few scientists feel that the new proposal is on the right track, and sometimes it is only personal and inarticulate aesthetic considerations that can do that. Men have been converted by them at times when most of the articulable technical arguments pointed the other way. When first introduced, neither Copernicus' astronomical theory nor De Broglie's theory of matter had many other significant grounds of appeal. Even today Einstein's general theory attracts men principally on aesthetic grounds, an appeal that few people outside of mathematics have been able to feel.

This is not to suggest that new paradigms triumph ultimately through some mystical aesthetic. On the contrary, very few men desert a tradition for these reasons alone. Often those who do turn out to have been misled. But if a paradigm is ever to triumph it must gain some first supporters, men who will develop it to the point where hardheaded arguments can be produced and multiplied. And even those arguments, when they come, are not individually decisive. Because scientists are reasonable men, one or another argument will ultimately persuade many of them. But there is no single argument that can or should persuade them all. Rather than a single group conversion, what occurs is an increasing shift in the distribution of professional allegiances.

At the start a new candidate for paradigm may have few supporters, and on occasions the supporters' motives may be suspect. Nevertheless, if they are competent, they will improve it, explore its possibilities, and show what it would be like to belong to the community guided by it. And as that goes on, if the paradigm is one destined to win its fight, the number and strength of the persuasive arguments in its favor will increase. More scientists will then be converted, and the exploration of the new paradigm will go on. Gradually the number of experiments, instruments, articles, and books based upon the paradigm will multiply. Still more men, convinced of the new view's fruitfulness, will adopt the new mode of practicing normal science, until at last only a few elderly hold-outs remain. And even they, we cannot say, are wrong. Though the historian can always find men—Priestley, for instance—who were unreasonable to resist for as long as they did, he will not find a point at which resistance becomes illogical or unscientific. At most he may wish to say that the man who continues to resist after his whole profession has been converted has *ipso facto* ceased to be a scientist.

XIII. Progress through Revolutions

The preceding pages have carried my schematic description of scientific development as far as it can go in this essay. Nevertheless, they cannot quite provide a conclusion. If this description has at all caught the essential structure of a science's continuing evolution, it will simultaneously have posed a special problem: Why should the enterprise sketched above move steadily ahead in ways that, say, art, political theory, or philosophy does not? Why is progress a perquisite reserved almost exclusively for the activities we call science? The most usual answers to that question have been denied in the body of this essay. We must conclude it by asking whether substitutes can be found.

Notice immediately that part of the question is entirely semantic. To a very great extent the term 'science' is reserved for fields that do progress in obvious ways. Nowhere does this show more clearly than in the recurrent debates about whether one or another of the contemporary social sciences is really a science. These debates have parallels in the pre-paradigm periods of fields that are today unhesitatingly labeled science. Their ostensible issue throughout is a definition of that vexing term. Men argue that psychology, for example, is a science because it possesses such and such characteristics. Others counter that those characteristics are either unnecessary or not sufficient to make a field a science. Often great energy is invested, great passion aroused, and the outsider is at a loss to know why. Can very much depend upon a *definition* of 'science'? Can a definition tell a man whether he is a scientist or not? If so, why do not natural scientists or artists worry about the definition of the term? Inevitably one suspects that the issue is more fundamental. Probably questions like the following are really being asked: Why does my field fail to move ahead in the way that, say, physics does? What changes in technique or method or ideology would enable it to do so? These are not, however, questions that could respond to an agreement on definition. Furthermore, if precedent from the natural sciences serves, they will cease to be a source of concern not when a definition is found, but when the groups that now doubt their own status achieve consensus about their past and present accomplishments. It may, for example, be significant that economists argue less about whether their field is a science than do practitioners of some other fields of social science. Is that because economists know what science is? Or is it rather economics about which they agree?

That point has a converse that, though no longer simply semantic, may help to display the inextricable connections between our notions of science and of progress. For many centuries, both in antiquity and again in early modern Europe, painting was regarded as *the* cumulative discipline. During those years the artist's goal was assumed to be representation. Critics and historians, like Pliny and Vasari, then recorded with veneration the series of inventions from foreshortening through chiaroscuro that had made possible successively more perfect representations of nature.[1] But those are also the years, particularly during the Renaissance, when little cleavage was felt between the sciences and the arts. Leonardo was only one of many men who passed freely back and forth between fields that only later became categorically distinct.[2] Furthermore, even after that steady exchange had ceased, the term 'art' continued to apply as much to technology and the crafts, which were also seen as progressive, as to painting and sculpture. Only when the latter unequivocally renounced representation as their goal and began to learn again from primitive models did the cleavage we now take for granted assume anything like its present depth. And even today, to switch fields once more, part of our difficulty in seeing the profound differences between science and technology must relate to the fact that progress is an obvious attribute of both fields.

[1] E. H. Gombrich, *Art and Illusion: A Study in the Psychology of Pictorial Representation* (New York, 1960), pp. 11–12.

[2] *Ibid.*, p. 97; and Giorgio de Santillana, "The Role of Art in the Scientific Renaissance," in *Critical Problems in the History of Science*, ed. M. Clagett (Madison, Wis., 1959), pp. 33–65.

It can, however, only clarify, not solve, our present difficulty to recognize that we tend to see as science any field in which progress is marked. There remains the problem of understanding why progress should be so noteworthy a characteristic of an enterprise conducted with the techniques and goals this essay has described. That question proves to be several in one, and we shall have to consider each of them separately. In all cases but the last, however, their resolution will depend in part upon an inversion of our normal view of the relation between scientific activity and the community that practices it. We must learn to recognize as causes what have ordinarily been taken to be effects. If we can do that, the phrases 'scientific progress' and even 'scientific objectivity' may come to seem in part redundant. In fact, one aspect of the redundancy has just been illustrated. Does a field make progress because it is a science, or is it a science because it makes progress?

Ask now why an enterprise like normal science should progress, and begin by recalling a few of its most salient characteristics. Normally, the members of a mature scientific community work from a single paradigm or from a closely related set. Very rarely do different scientific communities investigate the same problems. In those exceptional cases the groups hold several major paradigms in common. Viewed from within any single community, however, whether of scientists or of non-scientists, the result of successful creative work *is* progress. How could it possibly be anything else? We have, for example, just noted that while artists aimed at representation as their goal, both critics and historians chronicled the progress of the apparently united group. Other creative fields display progress of the same sort. The theologian who articulates dogma or the philosopher who refines the Kantian imperatives contributes to progress, if only to that of the group that shares his premises. No creative school recognizes a category of work that is, on the one hand, a creative success, but is not, on the other, an addition to the collective achievement of the group. If we doubt, as many do, that non-scientific fields make progress, that cannot be because individual schools make none. Rather, it must be because there are always competing schools, each of which constantly questions the very foundations of the others. The man who argues that philosophy, for example, has made no progress emphasizes that there are still Aristotelians, not that Aristotelianism has failed to progress.

These doubts about progress arise, however, in the sciences too. Throughout the pre-paradigm period when there is a multiplicity of competing schools, evidence of progress, except within schools, is very hard to find. This is the period described in Section II as one during which individuals practice science, but in which the results of their enterprise do not add up to science as we know it. And again, during periods of revolution when the fundamental tenets of a field are once more at issue, doubts are repeatedly expressed about the very possibility of continued progress if one or another of the opposed paradigms is adopted. Those who rejected Newtonianism proclaimed that its reliance upon innate forces would return science to the Dark Ages. Those who opposed Lavoisier's chemistry held that the rejection of chemical "principles" in favor of laboratory elements was the rejection of achieved chemical explanation by those who would take refuge in a mere name. A similar, though more moderately expressed, feeling seems to underlie the opposition of Einstein, Bohm, and others, to the dominant probabilistic interpretation of quantum mechanics. In short, it is only during periods of normal science that progress seems both obvious and assured. During those periods, however, the scientific community could view the fruits of its work in no other way.

With respect to normal science, then, part of the answer to the problem of progress lies simply in the eye of the beholder. Scientific progress is not different in kind from progress in other fields, but the absence at most times of competing schools that question each other's aims and standards makes the progress of a normal-scientific community far easier to see. That, however, is only part of the answer and by no means the most important part. We have, for example, already noted that once the reception of a common paradigm has freed the scientific community from the need constantly to re-examine its first principles, the members of that community can concentrate exclusively upon the subtlest and most esoteric of the phenomena that concern it. Inevitably, that does increase both the effectiveness and the efficiency with which the group as a whole solves new problems. Other aspects of professional life in the sciences enhance this very special efficiency still further.

Some of these are consequences of the unparalleled insulation of mature scientific communities from the demands of the laity and of everyday life. That insulation has never been complete—we are now discussing matters of degree. Nevertheless, there are no other professional communities in which individual creative work is so exclusively addressed to and evaluated by other members of the profession. The most esoteric of poets or the most abstract of theologians is far more concerned than the scientist with lay approbation of his creative work, though he may be even less concerned with approbation in general. That difference proves consequential. Just because he is working only for an audience of colleagues, an audience that shares his own values and beliefs, the scientist can take a single set of standards for granted. He need not worry about what some other group or school will think and can therefore dispose of one problem and get on to the next more quickly than those who work for a more heterodox group. Even more important, the insulation of the scientific community from society permits the individual scientist to concentrate his attention upon problems that he has good reason to believe he will be able to solve. Unlike the engineer, and many doctors, and most theologians, the scientist need not choose problems because they urgently need solution and without regard for the tools available to solve them. In this respect, also, the contrast between natural scientists and many social scientists proves instructive. The latter often tend, as the former almost never do, to defend their choice of a research problem—e.g., the effects of racial discrimination or the causes of the business cycle—chiefly in terms of the social importance of achieving a solution. Which group would one then expect to solve problems at a more rapid rate?

The effects of insulation from the larger society are greatly intensified by another characteristic of the professional scientific community, the nature of its educational initiation. In music, the graphic arts, and literature, the practitioner gains his education by exposure to the works of other artists, principally earlier artists. Textbooks, except compendia of or handbooks to original creations, have only a secondary role. In history, philosophy, and the social sciences, textbook literature has a greater significance. But even in these fields the elementary college course employs parallel readings in original sources, some of them the "classics" of the field, others the contemporary research reports that practitioners write for each other. As a result, the student in any one of these disciplines is constantly made aware of the immense variety of problems that the members of his future group have, in the course of time, attempted to solve. Even more important, he has constantly before him a number of competing and incommensurable solutions to these problems, solutions that he must ultimately evaluate for himself.

Contrast this situation with that in at least the contemporary natural sciences. In these fields the student relies mainly on textbooks until, in his third or fourth year of graduate work, he begins his own research. Many science curricula do not ask even graduate students to read in works not written specially for students. The few that do assign supplementary reading in research papers and monographs restrict such assignments to the most advanced courses and to materials that take up more or less where the available texts leave off. Until the very last stages in the education of a scientist, textbooks are systematically substituted for the creative scientific literature that made them possible. Given the confidence in their paradigms, which makes this educational technique possible, few scientists would wish to change it. Why, after all, should the student of physics, for example, read the works of Newton, Faraday, Einstein, or Schrödinger, when everything he needs to know about these works is recapitulated in a far briefer, more precise, and more systematic form in a number of up-to-date textbooks?

Without wishing to defend the excessive lengths to which this type of education has occasionally been carried, one cannot help but notice that in general it has been immensely effective.

Of course, it is a narrow and rigid education, probably more so than any other except perhaps in orthodox theology. But for normal-scientific work, for puzzle-solving within the tradition that the textbooks define, the scientist is almost perfectly equipped. Furthermore, he is well equipped for another task as well—the generation through normal science of significant crises. When they arise, the scientist is not, of course, equally well prepared. Even though prolonged crises are probably reflected in less rigid educational practice, scientific training is not well designed to produce the man who will easily discover a fresh approach. But so long as somebody appears with a new candidate for paradigm—usually a young man or one new to the field—the loss due to rigidity accrues only to the individual. Given a generation in which to effect the change, individual rigidity is compatible with a community that can switch from paradigm to paradigm when the occasion demands. Particularly, it is compatible when that very rigidity provides the community with a sensitive indicator that something has gone wrong.

In its normal state, then, a scientific community is an immensely efficient instrument for solving the problems or puzzles that its paradigms define. Furthermore, the result of solving those problems must inevitably be progress. There is no problem here. Seeing that much, however, only highlights the second main part of the problem of progress in the sciences. Let us therefore turn to it and ask about progress through extraordinary science. Why should progress also be the apparently universal concomitant of scientific revolutions? Once again, there is much to be learned by asking what else the result of a revolution could be. Revolutions close with a total victory for one of the two opposing camps. Will that group ever say that the result of its victory has been something less than progress? That would be rather like admitting that they had been wrong and their opponents right. To them, at least, the outcome of revolution must be progress, and they are in an excellent position to make certain that future members of their community will see past history in the same way. Section XI described in detail the techniques by which this is accomplished, and we have just recurred to a closely related aspect of professional scientific life. When it repudiates a past paradigm, a scientific community simultaneously renounces, as a fit subject for professional scrutiny, most of the books and articles in which that paradigm had been embodied. Scientific education makes use of no equivalent for the art museum or the library of classics, and the result is a sometimes drastic distortion in the scientist's perception of his discipline's past. More than the practitioners of other creative fields, he comes to see it as leading in a straight line to the discipline's present vantage. In short, he comes to see it as progress. No alternative is available to him while he remains in the field.

Inevitably those remarks will suggest that the member of a mature scientific community is, like the typical character of Orwell's *1984*, the victim of a history rewritten by the powers that be. Furthermore, that suggestion is not altogether inappropriate. There are losses as well as gains in scientific revolutions, and scientists tend to be peculiarly blind to the former.[3] On the other hand, no explanation of progress through revolutions may stop at this point. To do so would be to imply that in the sciences might makes right, a formulation which would again not be entirely wrong if it did not suppress the nature of the process and of the authority by which the choice between paradigms is made. If authority alone, and particularly if nonprofessional authority, were the arbiter of paradigm debates, the outcome of those debates might still be revolution, but it would not be *scientific* revolution. The very existence of science depends upon vesting the power to choose between paradigms in the members of a special kind of community. Just how special that community must be if science is to survive and grow may be indicated by the very tenuousness of humanity's hold on the scientific enterprise. Every civilization of which we have records

3 Historians of science often encounter this blindness in a particularly striking form. The group of students who come to them from the sciences is very often the most rewarding group they teach. But it is also usually the most frustrating at the start. Because science students "know the right answers," it is particularly difficult to make them analyze an older science in its own terms.

has possessed a technology, an art, a religion, a political system, laws, and so on. In many cases those facets of civilization have been as developed as our own. But only the civilizations that descend from Hellenic Greece have possessed more than the most rudimentary science. The bulk of scientific knowledge is a product of Europe in the last four centuries. No other place and time has supported the very special communities from which scientific productivity comes.

What are the essential characteristics of these communities? Obviously, they need vastly more study. In this area only the most tentative generalizations are possible. Nevertheless, a number of requisites for membership in a professional scientific group must already be strikingly clear. The scientist must, for example, be concerned to solve problems about the behavior of nature. In addition, though his concern with nature may be global in its extent, the problems on which he works must be problems of detail. More important, the solutions that satisfy him may not be merely personal but must instead be accepted as solutions by many. The group that shares them may not, however, be drawn at random from society as a whole, but is rather the well-defined community of the scientist's professional compeers. One of the strongest, if still unwritten, rules of scientific life is the prohibition of appeals to heads of state or to the populace at large in matters scientific. Recognition of the existence of a uniquely competent professional group and acceptance of its role as the exclusive arbiter of professional achievement has further implications. The group's members, as individuals and by virtue of their shared training and experience, must be seen as the sole possessors of the rules of the game or of some equivalent basis for unequivocal judgments. To doubt that they shared some such basis for evaluations would be to admit the existence of incompatible standards of scientific achievement. That admission would inevitably raise the question whether truth in the sciences can be one.

This small list of characteristics common to scientific communities has been drawn entirely from the practice of normal science, and it should have been. That is the activity for which the scientist is ordinarily trained. Note, however, that despite its small size the list is already sufficient to set such communities apart from all other professional groups. And note, in addition, that despite its source in normal science the list accounts for many special features of the group's response during revolutions and particularly during paradigm debates. We have already observed that a group of this sort must see a paradigm change as progress. Now we may recognize that the perception is, in important respects, self-fulfilling. The scientific community is a supremely efficient instrument for maximizing the number and precision of the problem solved through paradigm change.

Because the unit of scientific achievement is the solved problem and because the group knows well which problems have already been solved, few scientists will easily be persuaded to adopt a viewpoint that again opens to question many problems that had previously been solved. Nature itself must first undermine professional security by making prior achievements seem problematic. Furthermore, even when that has occurred and a new candidate for paradigm has been evoked, scientists will be reluctant to embrace it unless convinced that two all-important conditions are being met. First, the new candidate must seem to resolve some outstanding and generally recognized problem that can be met in no other way. Second, the new paradigm must promise to preserve a relatively large part of the concrete problem-solving ability that has accrued to science through its predecessors. Novelty for its own sake is not a desideratum in the sciences as it is in so many other creative fields. As a result, though new paradigms seldom or never possess all the capabilities of their predecessors, they usually preserve a great deal of the most concrete parts of past achievement and they always permit additional concrete problem-solutions besides.

To say this much is not to suggest that the ability to solve problems is either the unique or an unequivocal basis for paradigm choice. We have already noted many reasons why there can be no criterion of that sort. But it does suggest that a community of scientific specialists will do all that it can to ensure the continuing growth of the assembled data that it can treat with precision and detail. In the process the community will sustain losses. Often some old problems must be banished. Frequently, in addition, revolution narrows the scope of the community's professional concerns, increases the extent of its specialization, and attenuates its communication with other groups, both scientific and lay. Though science surely grows in depth, it may not grow in breadth as well. If it does so, that breadth is manifest mainly in the proliferation of scientific specialties, not in the scope of any single specialty alone. Yet despite these and other losses to the individual communities, the nature of such communities provides a virtual guarantee that both the list of problems solved by science and the precision of individual problem-solutions will grow and grow. At least, the nature of the community provides such a guarantee if there is any way at all in which it can be provided. What better criterion than the decision of the scientific group could there be?

These last paragraphs point the directions in which I believe a more refined solution of the problem of progress in the sciences must be sought. Perhaps they indicate that scientific progress is not quite what we had taken it to be. But they simultaneously show that a sort of progress will inevitably characterize the scientific enterprise so long as such an enterprise survives. In the sciences there need not be progress of another sort. We may, to be more precise, have to relinquish the notion, explicit or implicit, that changes of paradigm carry scientists and those who learn from them closer and closer to the truth.

It is now time to notice that until the last very few pages the term 'truth' had entered this essay only in a quotation from Francis Bacon. And even in those pages it entered only as a source for the scientist's conviction that incompatible rules for doing science cannot coexist except during revolutions when the profession's main task is to eliminate all sets but one. The developmental process described in this essay has been a process of evolution *from* primitive beginnings—a process whose successive stages are characterized by an increasingly detailed and refined understanding of nature. But nothing that has been or will be said makes it a process of evolution *toward* anything. Inevitably that lacuna will have disturbed many readers. We are all deeply accustomed to seeing science as the one enterprise that draws constantly nearer to some goal set by nature in advance.

But need there be any such goal? Can we not account for both science's existence and its success in terms of evolution from the community's state of knowledge at any given time? Does it really help to imagine that there is some one full, objective, true account of nature and that the proper measure of scientific achievement is the extent to which it brings us closer to that ultimate goal? If we can learn to substitute evolution-from-what-we-do-know for evolution-toward-what-we-wish-to-know, a number of vexing problems may vanish in the process. Somewhere in this maze, for example, must lie the problem of induction.

I cannot yet specify in any detail the consequences of this alternate view of scientific advance. But it helps to recognize that the conceptual transposition here recommended is very close to one that the West undertook just a century ago. It is particularly helpful because in both cases the main obstacle to transposition is the same. When Darwin first published his theory of evolution by natural selection in 1859, what most bothered many professionals was neither the notion of species change nor the possible descent of man from apes. The evidence pointing to evolution, including the evolution of man, had been accumulating for decades, and the idea of evolution had been suggested and widely disseminated before. Though evolution, as such, did encounter resistance, particularly from some religious groups, it was by no means the greatest of the difficulties the Darwinians faced. That difficulty stemmed from an idea that was more nearly Darwin's own. All the well-known pre-Darwinian evolutionary theories—those of Lamarck, Chambers, Spencer, and the German *Naturphilosophen*—had taken evolution to be a goal-directed process. The "idea" of man and of the contemporary flora and fauna was thought to have been present from the first creation of life, perhaps in the mind of God. That idea or plan had provided the direction and the guiding force to the entire evolutionary process. Each new stage of evolutionary development was a more perfect realization of a plan that had been present from the start.[4]

For many men the abolition of that teleological kind of evolution was the most significant and least palatable of Darwin's suggestions.[5] The *Origin of Species* recognized no goal set either by God or nature. Instead, natural selection, operating in the given environment and with the actual organisms presently at hand, was responsible for the gradual but steady emergence of more elaborate, further articulated, and vastly more specialized organisms. Even such marvelously adapted organs as the eye and hand of man—organs whose design had previously provided powerful arguments for the existence of a supreme artificer and an advance plan—were products of a process that moved steadily *from* primitive beginnings but *toward* no goal. The belief that natural selection, resulting from mere competition between organisms for survival, could have produced man together with the higher animals and plants was the most difficult and disturbing aspect of Darwin's theory. What could 'evolution,' 'development,' and 'progress' mean in the absence of a specified goal? To many people, such terms suddenly seemed self-contradictory.

The analogy that relates the evolution of organisms to the evolution of scientific ideas can easily be pushed too far. But with respect to the issues of this closing section it is very nearly perfect. The process described in Section XII as the resolution of revolutions is the selection by conflict within the scientific community of the fittest way to practice future science. The net result of a sequence of such revolutionary selections, separated by periods of normal research, is the wonderfully adapted set of instruments we call modern scientific knowledge. Successive stages in that developmental process are marked by an increase in articulation and specialization. And the entire process may have occurred, as we now suppose biological evolution did,

[4] Loren Eiseley, *Darwin's Century: Evolution and the Men Who Discovered It* (New York, 1958), chaps. ii, iv–v.

[5] For a particularly acute account of one prominent Darwinian's struggle with this problem, see A. Hunter Dupree, *Asa Gray, 1810–1888* (Cambridge, Mass., 1959), pp. 295–306, 355–83.

without benefit of a set goal, a permanent fixed scientific truth, of which each stage in the development of scientific knowledge is a better exemplar.

Anyone who has followed the argument this far will nevertheless feel the need to ask why the evolutionary process should work. What must nature, including man, be like in order that science be possible at all? Why should scientific communities be able to reach a firm consensus unattainable in other fields? Why should consensus endure across one paradigm change after another? And why should paradigm change invariably produce an instrument more perfect in any sense than those known before? From one point of view those questions, excepting the first, have already been answered. But from another they are as open as they were when this essay began. It is not only the scientific community that must be special. The world of which that community is a part must also possess quite special characteristics, and we are no closer than we were at the start to knowing what these must be. That problem—What must the world be like in order that man may know it?—was not, however, created by this essay. On the contrary, it is as old as science itself, and it remains unanswered. But it need not be answered in this place. Any conception of nature compatible with the growth of science by proof is compatible with the evolutionary view of science developed here. Since this view is also compatible with close observation of scientific life, there are strong arguments for employing it in attempts to solve the host of problems that still remain.

Postscript—1969

It has now been almost seven years since this book was first published.[1] In the interim both the response of critics and my own further work have increased my understanding of a number of the issues it raises. On fundamentals my viewpoint is very nearly unchanged, but I now recognize aspects of its initial formulation that create gratuitous difficulties and misunderstandings. Since some of those misunderstandings have been my own, their elimination enables me to gain ground that should ultimately provide the basis for a new version of the book.[2] Meanwhile, I welcome the chance to sketch needed revisions, to comment on some reiterated criticisms, and to suggest directions in which my own thought is presently developing.[3]

Several of the key difficulties of my original text cluster about the concept of a paradigm, and my discussion begins with them.[4] In the subsection that follows at once, I suggest the desirability of disentangling that concept from the notion of a scientific community, indicate how this may be done, and discuss some signifi-

[1] This postscript was first prepared at the suggestion of my onetime student and longtime friend, Dr. Shigeru Nakayama of the University of Tokyo, for inclusion in his Japanese translation of this book. I am grateful to him for the idea, for his patience in awaiting its fruition, and for permission to include the result in the English language edition.

[2] For this edition I have attempted no systematic rewriting, restricting alterations to a few typographical errors plus two passages which contained isolable errors. One of these is the description of the role of Newton's *Principia* in the development of eighteenth-century mechanics on pp. 92-95, above. The other concerns the response to crises on p. 146.

[3] Other indications will be found in two recent essays of mine: "Reflection on My Critics," in Imre Lakatos and Alan Musgrave (eds.), *Criticism and the Growth of Knowledge* (Cambridge, 1970); and "Second Thoughts on Paradigms," in Frederick Suppe (ed.), *The Structure of Scientific Theories* (Urbana, Ill., 1970 or 1971), both currently in press. I shall cite the first of these essays below as "Reflections" and the volume in which it appears as *Growth of Knowledge;* the second essay will be referred to as "Second Thoughts."

[4] For particularly cogent criticism of my initial presentation of paradigms see: Margaret Masterman, "The Nature of a Paradigm," in *Growth of Knowledge;* and Dudley Shapere, "The Structure of Scientific Revolutions," *Philosophical Review,* LXXIII (1964), 383–94.

cant consequences of the resulting analytic separation. Next I consider what occurs when paradigms are sought by examining the behavior of the members of a *previously determined* scientific community. That procedure quickly discloses that in much of the book the term 'paradigm' is used in two different senses. On the one hand, it stands for the entire constellation of beliefs, values, techniques, and so on shared by the members of a given community. On the other, it denotes one sort of element in that constellation, the concrete puzzle-solutions which, employed as models or examples, can replace explicit rules as a basis for the solution of the remaining puzzles of normal science. The first sense of the term, call it the sociological, is the subject of Subsection 2, below; Subsection 3 is devoted to paradigms as exemplary past achievements.

Philosophically, at least, this second sense of 'paradigm' is the deeper of the two, and the claims I have made in its name are the main sources for the controversies and misunderstandings that the book has evoked, particularly for the charge that I make of science a subjective and irrational enterprise. These issues are considered in Subsections 4 and 5. The first argues that terms like 'subjective' and 'intuitive' cannot appropriately be applied to the components of knowledge that I have described as tacitly embedded in shared examples. Though such knowledge is not, without essential change, subject to paraphrase in terms of rules and criteria, it is nevertheless systematic, time tested, and in some sense corrigible. Subsection 5 applies that argument to the problem of choice between two incompatible theories, urging in brief conclusion that men who hold incommensurable viewpoints be thought of as members of different language communities and that their communication problems be analyzed as problems of translation. Three residual issues are discussed in the concluding Subsections, 6 and 7. The first considers the charge that the view of science developed in this book is through-and-through relativistic. The second begins by inquiring whether my argument really suffers, as has been said, from a confusion between the descriptive and the normative modes; it concludes with brief remarks on a topic deserving a separate essay: the extent to which the book's main theses may legitimately be applied to fields other than science.

1. *Paradigms and Community Structure*

The term 'paradigm' enters the preceding pages early, and its manner of entry is intrinsically circular. A paradigm is what the members of a scientific community share, *and,* conversely, a scientific community consists of men who share a paradigm. Not all circularities are vicious (I shall defend an argument of similar structure late in this postscript), but this one is a source of real difficulties. Scientific communities can and should be isolated without prior recourse to paradigms; the latter can then be discovered by scrutinizing the behavior of a given community's members. If this book were being rewritten, it would therefore open with a discussion of the community structure of science, a topic that has recently become a significant subject of sociological research and that historians of science are also beginning to take seriously. Preliminary results, many of them still unpublished, suggest that the empirical techniques required for its exploration are non-trivial, but some are in hand and others are sure to be developed.[5] Most practicing scientists respond at once to questions about their community affiliations, taking for granted that responsibility for the various current specialties is distributed among groups of at least roughly determinate membership. I shall therefore here assume that more systematic means for their identification will be found. Instead of presenting preliminary research results, let me briefly articulate the intuitive notion of community that underlies much in the earlier chapters of this book. It is a notion now widely shared by scientists, sociologists, and a number of historians of science.

[5] W. O. Hagstrom, *The Scientific Community* (New York, 1965), chaps. iv and v; D. J. Price and D. de B. Beaver, "Collaboration in an Invisible College," *American Psychologist,* XXI (1966), 1011–18; Diana Crane, "Social Structure in a Group of Scientists: A Test of the 'Invisible College' Hypothesis," *American Sociological Review,* XXXIV (1969), 335–52; N. C. Mullins, *Social Networks among Biological Scientists,* (Ph.D. diss., Harvard University, 1966), and "The Micro-Structure of an Invisible College: The Phage Group" (paper delivered at an annual meeting of the American Sociological Association, Boston, 1968).

A scientific community consists, on this view, of the practitioners of a scientific specialty. To an extent unparalleled in most other fields, they have undergone similar educations and professional initiations; in the process they have absorbed the same technical literature and drawn many of the same lessons from it. Usually the boundaries of that standard literature mark the limits of a scientific subject matter, and each community ordinarily has a subject matter of its own. There are schools in the sciences, communities, that is, which approach the same subject from incompatible viewpoints. But they are far rarer there than in other fields; they are always in competition; and their competition is usually quickly ended. As a result, the members of a scientific community see themselves and are seen by others as the men uniquely responsible for the pursuit of a set of shared goals, including the training of their successors. Within such groups communication is relatively full and professional judgment relatively unanimous. Because the attention of different scientific communities is, on the other hand, focused on different matters, professional communication across group lines is sometimes arduous, often results in misunderstanding, and may, if pursued, evoke significant and previously unsuspected disagreement.

Communities in this sense exist, of course, at numerous levels. The most global is the community of all natural scientists. At an only slightly lower level the main scientific professional groups are communities: physicists, chemists, astronomers, zoologists, and the like. For these major groupings, community membership is readily established except at the fringes. Subject of highest degree, membership in professional societies, and journals read are ordinarily more than sufficient. Similar techniques will also isolate major subgroups: organic chemists, and perhaps protein chemists among them, solid-state and high-energy physicists, radio astronomers, and so on. It is only at the next lower level that empirical problems emerge. How, to take a contemporary example, would one have isolated the phage group prior to its public acclaim? For this purpose one must have recourse to attendance at special conferences, to the distribution of draft manuscripts or galley proofs prior to publication, and above all to formal and informal communication networks including those discovered in correspondence and in the linkages among citations.[6] I take it that the job can and will be done, at least for the contemporary scene and the more recent parts of the historical. Typically it may yield communities of perhaps one hundred members, occasionally significantly fewer. Usually individual scientists, particularly the ablest, will belong to several such groups either simultaneously or in succession.

Communities of this sort are the units that this book has presented as the producers and validators of scientific knowledge. Paradigms are something shared by the members of such groups. Without reference to the nature of these shared elements, many aspects of science described in the preceding pages can scarcely be understood. But other aspects can, though they are not independently presented in my original text. It is therefore worth noting, before turning to paradigms directly, a series of issues that require reference to community structure alone.

Probably the most striking of these is what I have previously called the transition from the pre- to the post-paradigm period in the development of a scientific field. That transition is the one sketched above in Section II. Before it occurs, a number of schools compete for the domination of a given field. Afterward, in the wake of some notable scientific achievement, the number of schools is greatly reduced, ordinarily to one, and a more efficient mode of scientific practice begins. The latter is generally esoteric and oriented to puzzle-solving, as the work of a group can be only when its members take the foundations of their field for granted.

The nature of that transition to maturity deserves fuller discussion than it has received in this book, particularly from those concerned with the development of the contemporary social

[6] Eugene Garfield, *The Use of Citation Data in Writing the History of Science* (Philadelphia: Institute of Scientific Information, 1964); M. M. Kessler, "Comparison of the Results of Bibliographic Coupling and Analytic Subject Indexing," *American Documentation,* XVI (1965), 223–33; D. J. Price, "Networks of Scientific Papers," *Science,* CIL (1965), 510–15.

sciences. To that end it may help to point out that the transition need not (I now think should not) be associated with the first acquisition of a paradigm. The members of all scientific communities, including the schools of the "pre-paradigm" period, share the sorts of elements which I have collectively labelled 'a paradigm.' What changes with the transition to maturity is not the presence of a paradigm but rather its nature. Only after the change is normal puzzle-solving research possible. Many of the attributes of a developed science which I have above associated with the acquisition of a paradigm I would therefore now discuss as consequences of the acquisition of the sort of paradigm that identifies challenging puzzles, supplies clues to their solution, and guarantees that the truly clever practitioner will succeed. Only those who have taken courage from observing that their own field (or school) has paradigms are likely to feel that something important is sacrificed by the change.

A second issue, more important at least to historians, concerns this book's implicit one-to-one identification of scientific communities with scientific subject matters. I have, that is, repeatedly acted as though, say, 'physical optics,' 'electricity,' and 'heat' must name scientific communities because they do name subject matters for research. The only alternative my text has seemed to allow is that all these subjects have belonged to the physics community. Identifications of that sort will not, however, usually withstand examination, as my colleagues in history have repeatedly pointed out. There was, for example, no physics community before the mid-nineteenth century, and it was then formed by the merger of parts of two previously separate communities, mathematics and natural philosophy *(physique expérimentale).* What is today the subject matter for a single broad community has been variously distributed among diverse communities in the past. Other narrower subjects, for example heat and the theory of matter, have existed for long periods without becoming the special province of any single scientific community. Both normal science and revolutions are, however, community-based activities. To discover and analyze them, one must first unravel the changing community structure of the sciences over time. A paradigm governs, in the first instance, not a subject matter but rather a group of practitioners. Any study of paradigm-directed or of paradigm-shattering research must begin by locating the responsible group or groups.

When the analysis of scientific development is approached in that way, several difficulties which have been foci for critical attention are likely to vanish. A number of commentators have, for example, used the theory of matter to suggest that I drastically overstate the unanimity of scientists in their allegiance to a paradigm. Until comparatively recently, they point out, those theories have been topics for continuing disagreement and debate. I agree with the description but think it no counter-example. Theories of matter were not, at least until about 1920, the special province or the subject matter for any scientific community. Instead, they were tools for a large number of specialists' groups. Members of different communities sometimes chose different tools and criticized the choice made by others. Even more important, a theory of matter is not the sort of topic on which the members of even a single community must necessarily agree. The need for agreement depends on what it is the community does. Chemistry in the first half of the nineteenth century provides a case in point. Though several of the community's fundamental tools—constant proportion, multiple proportion, and combining weights—had become common property as a result of Dalton's atomic theory, it was quite possible for chemists, after the event, to base their work on these tools and to disagree, sometimes vehemently, about the existence of atoms.

Some other difficulties and misunderstandings will, I believe, be dissolved in the same way. Partly because of the examples I have chosen and partly because of my vagueness about the nature and size of the relevant communities, a few readers of this book have concluded that my concern is primarily or exclusively with major revolutions such as those associated with Copernicus, Newton, Darwin, or Einstein. A clearer delineation of community structure should, however, help to enforce the rather different impression I have tried to create. A revolution is for me a special sort of change involving a certain sort of reconstruction of group commitments. But it need not be a large change, nor need it seem revolutionary to those outside a single community, consisting perhaps of fewer than twenty-five people. It is just because this type of change, little recognized or discussed in the literature of the philosophy of science, occurs so regularly on this smaller scale that revolutionary, as against cumulative, change so badly needs to be understood.

One last alteration, closely related to the preceding, may help to facilitate that understanding. A number of critics have doubted whether crisis, the common awareness that something has gone wrong, precedes revolutions so invariably as I have implied in my original text. Nothing important to my argument depends, however, on crises' being an absolute prerequisite to revolutions; they need only be the usual prelude, supplying, that is, a self-correcting mechanism which ensures that the rigidity of normal science will not forever go unchallenged. Revolutions may also be induced in other ways, though I think they seldom are. In addition, I would now point out what the absence of an adequate discussion of community structure has obscured above: crises need not be generated by the work of the community that experiences them and that sometimes undergoes revolution as a result. New instruments like the electron microscope or new laws like Maxwell's may develop in one specialty and their assimilation create crisis in another.

2. *Paradigms as the Constellation of Group Commitments*

Turn now to paradigms and ask what they can possibly be. My original text leaves no more obscure or important question. One sympathetic reader, who shares my conviction that 'paradigm' names the central philosophical elements of the book, prepared a partial analytic index and concluded that the term is used in at least twenty-two different ways.[7] Most of those differences are, I now think, due to stylistic inconsistencies (e.g., Newton's Laws are sometimes a paradigm, sometimes parts of a paradigm, and

[7] Masterman, *op. cit.*

sometimes paradigmatic), and they can be eliminated with relative ease. But, with that editorial work done, two very different usages of the term would remain, and they require separation. The more global use is the subject of this subsection; the other will be considered in the next.

Having isolated a particular community of specialists by techniques like those just discussed, one may usefully ask: What do its members share that accounts for the relative fulness of their professional communication and the relative unanimity of their professional judgments? To that question my original text licenses the answer, a paradigm or set of paradigms. But for this use, unlike the one to be discussed below, the term is inappropriate. Scientists themselves would say they share a theory or set of theories, and I shall be glad if the term can ultimately be recaptured for this use. As currently used in philosophy of science, however, 'theory' connotes a structure far more limited in nature and scope than the one required here. Until the term can be freed from its current implications, it will avoid confusion to adopt another. For present purposes I suggest 'disciplinary matrix': 'disciplinary' because it refers to the common possession of the practitioners of a particular discipline; 'matrix' because it is composed of ordered elements of various sorts, each requiring further specification. All or most of the objects of group commitment that my original text makes paradigms, parts of paradigms, or paradigmatic are constituents of the disciplinary matrix, and as such they form a whole and function together. They are, however, no longer to be discussed as though they were all of a piece. I shall not here attempt an exhaustive list, but noting the main sorts of components of a disciplinary matrix will both clarify the nature of my present approach and simultaneously prepare for my next main point.

One important sort of component I shall label 'symbolic generalizations,' having in mind those expressions, deployed without question or dissent by group members, which can readily be cast in a logical form like $(x)(y)(z)\phi(x, y, z)$. They are the formal or the readily formalizable components of the disciplinary matrix. Sometimes they are found already in symbolic form: $f = ma$ or $I = V/R$. Others are ordinarily expressed in words: "elements combine in constant proportion by weight," or "action equals reaction." If it were not for the general acceptance of expressions like these, there would be no points at which group members could attach the powerful techniques of logical and mathematical manipulation in their puzzle-solving enterprise. Though the example of taxonomy suggests that normal science can proceed with few such expressions, the power of a science seems quite generally to increase with the number of symbolic generalizations its practioners have at their disposal.

These generalizations look like laws of nature, but their function for group members is not often that alone. Sometimes it is: for example the Joule-Lenz Law, $H = RI^2$. When that law was discovered, community members already knew what H, R, and I stood for, and these generalizations simply told them something about the behavior of heat, current, and resistance that they had not known before. But more often, as discussion earlier in the book indicates, symbolic generalizations simultaneously serve a second function, one that is ordinarily sharply separated in analyses by philosophers of science. Like $f = ma$ or $I = V/R$, they function in part as laws but also in part as definitions of some of the symbols they deploy. Furthermore, the balance between their inseparable legislative and definitional force shifts over time. In another context these points would repay detailed analysis, for the nature of the commitment to a law is very different from that of commitment to a definition. Laws are often corrigible piecemeal, but definitions, being tautologies, are not. For example, part of what the acceptance of Ohm's Law demanded was a redefinition of both 'current' and 'resistance'; if those terms had continued to mean what they had meant before, Ohm's Law could not have been right; that is why it was so strenuously opposed as, say, the Joule-Lenz Law was not.[8] Probably that situation is typical. I currently suspect that

[8] For significant parts of this episode see: T. M. Brown, "The Electric Current in Early Nineteenth-Century French Physics," *Historical Studies in the Physical Sciences,* I (1969), 61–103, and Morton Schagrin, "Resistance to Ohm's Law," *American Journal of Physics,* XXI (1963), 536–47.

all revolutions involve, among other things, the abandonment of generalizations the force of which had previously been in some part that of tautologies. Did Einstein show that simultaneity was relative or did he alter the notion of simultaneity itself? Were those who heard paradox in the phrase 'relativity of simultaneity' simply wrong?

Consider next a second type of component of the disciplinary matrix, one about which a good deal has been said in my original text under such rubrics as 'metaphysical paradigms' or 'the metaphysical parts of paradigms.' I have in mind shared commitments to such beliefs as: heat is the kinetic energy of the constituent parts of bodies; all perceptible phenomena are due to the interaction of qualitatively neutral atoms in the void, or, alternatively, to matter and force, or to fields. Rewriting the book now I would describe such commitments as beliefs in particular models, and I would expand the category models to include also the relatively heuristic variety: the electric circuit may be regarded as a steady-state hydrodynamic system; the molecules of a gas behave like tiny elastic billiard balls in random motion. Though the strength of group commitment varies, with nontrivial consequences, along the spectrum from heuristic to ontological models, all models have similar functions. Among other things they supply the group with preferred or permissible analogies and metaphors. By doing so they help to determine what will be accepted as an explanation and as a puzzle-solution; conversely, they assist in the determination of the roster of unsolved puzzles and in the evaluation of the importance of each. Note, however, that the members of scientific communities may not have to share even heuristic models, though they usually do so. I have already pointed out that membership in the community of chemists during the first half of the nineteenth century did not demand a belief in atoms.

A third sort of element in the disciplinary matrix I shall here describe as values. Usually they are more widely shared among different communities than either symbolic generalizations or models, and they do much to provide a sense of community to natural scientists as a whole. Though they function at all times, their particular importance emerges when the members of a particular community must identify crisis or, later, choose between incompatible ways of practicing their discipline. Probably the most deeply held values concern predictions: they should be accurate; quantitative predictions are preferable to qualitative ones; whatever the margin of permissible error, it should be consistently satisfied in a given field; and so on. There are also, however, values to be used in judging whole theories: they must, first and foremost, permit puzzle-formulation and solution; where possible they should be simple, self-consistent, and plausible, compatible, that is, with other theories currently deployed. (I now think it a weakness of my original text that so little attention is given to such values as internal and external consistency in considering sources of crisis and factors in theory choice.) Other sorts of values exist as well—for example, science should (or need not) be socially useful—but the preceding should indicate what I have in mind.

One aspect of shared values does, however, require particular mention. To a greater extent than other sorts of components of the disciplinary matrix, values may be shared by men who differ in their application. Judgments of accuracy are relatively, though not entirely, stable from one time to another and from one member to another in a particular group. But judgments of simplicity, consistency, plausibility, and so on often vary greatly from individual to individual. What was for Einstein an insupportable inconsistency in the old quantum theory, one that rendered the pursuit of normal science impossible, was for Bohr and others a difficulty that could be expected to work itself out by normal means. Even more important, in those situations where values must be applied, different values, taken alone, would often dictate different choices. One theory may be more accurate but less consistent or plausible than another; again the old quantum theory provides an example. In short, though values are widely shared by scientists and though commitment to them is both deep and constitutive of science, the application of values is sometimes considerably affected by the features of individual personality and biography that differentiate the members of the group.

To many readers of the preceding chapters, this characteristic of the operation of shared values has seemed a major weakness of my position. Because I insist that what scientists share is not sufficient to command uniform assent about such matters as the choice between competing theories or the distinction between an ordinary anomaly and a crisis-provoking one, I am occasionally accused of glorifying subjectivity and even irrationality.[9] But that reaction ignores two characteristics displayed by value judgments in any field. First, shared values can be important determinants of group behavior even though the members of the group do not all apply them in the same way. (If that were not the case, there would be no *special* philosophic problems about value theory or aesthetics.) Men did not all paint alike during the periods when representation was a primary value, but the developmental pattern of the plastic arts changed drastically when that value was abandoned.[10] Imagine what would happen in the sciences if consistency ceased to be a primary value. Second, individual variability in the application of shared values may serve functions essential to science. The points at which values must be applied are invariably also those at which risks must be taken. Most anomalies are resolved by normal means; most proposals for new theories do prove to be wrong. If all members of a community responded to each anomaly as a source of crisis or embraced each new theory advanced by a colleague, science would cease. If, on the other hand, no one reacted to anomalies or to brand-new theories in high-risk ways, there would be few or no revolutions. In matters like these the resort to shared values rather than to shared rules governing individual choice may be the community's way of distributing risk and assuring the long-term success of its enterprise.

Turn now to a fourth sort of element in the disciplinary matrix, not the only other kind but the last I shall discuss here. For it the term 'paradigm' would be entirely appropriate, both philologi-

[9] See particularly: Dudley Shapere, "Meaning and Scientific Change," in *Mind and Cosmos: Essays in Contemporary Science and Philosophy,* The University of Pittsburgh Series in the Philosophy of Science, III (Pittsburgh, 1966), 41–85; Israel Scheffler, *Science and Subjectivity* (New York, 1967); and the essays of Sir Karl Popper and Imre Lakatos in *Growth of Knowledge.*

[10] See the discussion at the beginning of Section XIII, above.

cally and autobiographically; this is the component of a group's shared commitments which first led me to the choice of that word. Because the term has assumed a life of its own, however, I shall here substitute 'exemplars.' By it I mean, initially, the concrete problem-solutions that students encounter from the start of their scientific education, whether in laboratories, on examinations, or at the ends of chapters in science texts. To these shared examples should, however, be added at least some of the technical problem-solutions found in the periodical literature that scientists encounter during their post-educational research careers and that also show them by example how their job is to be done. More than other sorts of components of the disciplinary matrix, differences between sets of exemplars provide the community fine-structure of science. All physicists, for example, begin by learning the same exemplars: problems such as the inclined plane, the conical pendulum, and Keplerian orbits; instruments such as the vernier, the calorimeter, and the Wheatstone bridge. As their training develops, however, the symbolic generalizations they share are increasingly illustrated by different exemplars. Though both solid-state and field-theoretic physicists share the Schrödinger equation, only its more elementary applications are common to both groups.

3. *Paradigms as Shared Examples*

The paradigm as shared example is the central element of what I now take to be the most novel and least understood aspect of this book. Exemplars will therefore require more attention than the other sorts of components of the disciplinary matrix. Philosophers of science have not ordinarily discussed the problems encountered by a student in laboratories or in science texts, for these are thought to supply only practice in the application of what the student already knows. He cannot, it is said, solve problems at all unless he has first learned the theory and some rules for applying it. Scientific knowledge is embedded in theory and rules; problems are supplied to gain facility in their application. I have tried to argue, however, that this localization of the cognitive content of science is wrong. After the student has done many problems, he may gain only added facility by solving more. But at the start and for some time after, doing problems is learning consequential things about nature. In the absence of such exemplars, the laws and theories he has previously learned would have little empirical content.

To indicate what I have in mind I revert briefly to symbolic generalizations. One widely shared example is Newton's Second Law of Motion, generally written as $f = ma$. The sociologist, say, or the linguist who discovers that the corresponding expression is unproblematically uttered and received by the members of a given community will not, without much additional investigation, have learned a great deal about what either the expression or the terms in it mean, about how the scientists of the community attach the expression to nature. Indeed, the fact that they accept it without question and use it as a point at which to introduce logical and mathematical manipulation does not of itself imply that they agree at all about such matters as meaning and application. Of course they do agree to a considerable extent, or the fact would rapidly emerge from their subsequent conversation. But one may well ask at what point and by what means they have come to do so. How have they learned, faced with a given experimental situation, to pick out the relevant forces, masses, and accelerations?

In practice, though this aspect of the situation is seldom or never noted, what students have to learn is even more complex than that. It is not quite the case that logical and mathematical manipulation are applied directly to $f = ma$. That expression proves on examination to be a law-sketch or a law-schema. As the student or the practicing scientist moves from one problem situation to the next, the symbolic generalization to which such manipulations apply changes. For the case of free fall, $f = ma$ becomes $mg = m\frac{d^2s}{dt^2}$; for the simple pendulum it is transformed to $mg \sin\theta = -ml\frac{d^2\theta}{dt^2}$; for a pair of interacting harmonic oscillators it becomes two equations, the first of which may be written

$m_1\frac{d^2s_1}{dt^2} + k_1s_1 = k_2(s_2 - s_1 + d)$; and for more complex situations, such as the gyroscope, it takes still other forms, the family resemblance of which to $f = ma$ is still harder to discover. Yet, while learning to identify forces, masses, and accelerations in a variety of physical situations not previously encountered, the student has also learned to design the appropriate version of $f = ma$ through which to interrelate them, often a version for which he has encountered no literal equivalent before. How has he learned to do this?

A phenomenon familiar to both students of science and historians of science provides a clue. The former regularly report that they have read through a chapter of their text, understood it perfectly, but nonetheless had difficulty solving a number of the problems at the chapter's end. Ordinarily, also, those difficulties dissolve in the same way. The student discovers, with or without the assistance of his instructor, a way to see his problem as *like* a problem he has already encountered. Having seen the resemblance, grasped the analogy between two or more distinct problems, he can interrelate symbols and attach them to nature in the ways that have proved effective before. The law-sketch, say $f = ma$, has functioned as a tool, informing the student what similarities to look for, signaling the gestalt in which the situation is to be seen. The resultant ability to see a variety of situations as like each other, as subject for $f = ma$ or some other symbolic generalization, is, I think, the main thing a student acquires by doing exemplary problems, whether with a pencil and paper or in a well-designed laboratory. After he has completed a certain number, which may vary widely from one individual to the next, he views the situations that confront him as a scientist in the same gestalt as other members of his specialists' group. For him they are no longer the same situations he had encountered when his training began. He has meanwhile assimilated a time-tested and group-licensed way of seeing.

The role of acquired similarity relations also shows clearly in the history of science. Scientists solve puzzles by modeling them on previous puzzle-solutions, often with only minimal recourse to symbolic generalizations. Galileo found that a ball rolling down an incline acquires just enough velocity to return it to the same vertical height on a second incline of any slope, and he learned to see that experimental situation as like the pendulum with a point-mass for a bob. Huyghens then solved the problem of the center of oscillation of a physical pendulum by imagining that the extended body of the latter was composed of Galilean point-pendula, the bonds between which could be instantaneously released at any point in the swing. After the bonds were released, the individual point-pendula would swing freely, but their collective center of gravity when each attained its highest point would, like that of Galileo's pendulum, rise only to the height from which the center of gravity of the extended pendulum had begun to fall. Finally, Daniel Bernoulli discovered how to make the flow of water from an orifice resemble Huyghens' pendulum. Determine the descent of the center of gravity of the water in tank and jet during an infinitesimal interval of time. Next imagine that each particle of water afterward moves separately upward to the maximum height attainable with the velocity acquired during that interval. The ascent of the center of gravity of the individual particles must then equal the descent of the center of gravity of the water in tank and jet. From that view of the problem the long-sought speed of efflux followed at once.[11]

That example should begin to make clear what I mean by learning from problems to see situations as like each other, as subjects for the application of the same scientific law or law-sketch. Simultaneously it should show why I refer to the consequential knowledge of nature acquired while learning the similarity relationship and thereafter embodied in a way of viewing

[11] For the example, see: René Dugas, *A History of Mechanics,* trans. J. R. Maddox (Neuchatel, 1955), pp. 135–36, 186–93, and Daniel Bernoulli, *Hydrodynamica, sive de viribus et motibus fluidorum, commentarii opus academicum* (Strasbourg, 1738), Sec. iii. For the extent to which mechanics progressed during the first half of the eighteenth century by modelling one problem-solution on another, see Clifford Truesdell, "Reactions of Late Baroque Mechanics to Success, Conjecture, Error, and Failure in Newton's *Principia,*" *Texas Quarterly,* X (1967), 238–58.

physical situations rather than in rules or laws. The three problems in the example, all of them exemplars for eighteenth-century mechanicians, deploy only one law of nature. Known as the Principle of *vis viva,* it was usually stated as: "Actual descent equals potential ascent." Bernoulli's application of the law should suggest how consequential it was. Yet the verbal statement of the law, taken by itself, is virtually impotent. Present it to a contemporary student of physics, who knows the words and can do all these problems but now employs different means. Then imagine what the words, though all well known, can have said to a man who did not know even the problems. For him the generalization could begin to function only when he learned to recognize "actual descents" and "potential ascents" as ingredients of nature, and that is to learn something, prior to the law, about the situations that nature does and does not present. That sort of learning is not acquired by exclusively verbal means. Rather it comes as one is given words together with concrete examples of how they function in use; nature and words are learned together. To borrow once more Michael Polanyi's useful phrase, what results from this process is "tacit knowledge" which is learned by doing science rather than by acquiring rules for doing it.

4. *Tacit Knowledge and Intuition*

That reference to tacit knowledge and the concurrent rejection of rules isolates another problem that has bothered many of my critics and seemed to provide a basis for charges of subjectivity and irrationality. Some readers have felt that I was trying to make science rest on unanalyzable individual intuitions rather than on logic and law. But that interpretation goes astray in two essential respects. First, if I am talking at all about intuitions, they are not individual. Rather they are the tested and shared possessions of the members of a successful group, and the novice acquires them through training as a part of his preparation for group-membership. Second, they are not in principle unanalyzable. On the contrary, I am currently experimenting with a computer program designed to investigate their properties at an elementary level.

About that program I shall have nothing to say here,[12] but even mention of it should make my most essential point. When I speak of knowledge embedded in shared exemplars, I am not referring to a mode of knowing that is less systematic or less analyzable than knowledge embedded in rules, laws, or criteria of identification. Instead I have in mind a manner of knowing which is miscontrued if reconstructed in terms of rules that are first abstracted from exemplars and thereafter function in their stead. Or, to put the same point differently, when I speak of acquiring from exemplars the ability to recognize a given situation as like some and unlike others that one has seen before, I am not suggesting a process that is not potentially fully explicable in terms of neuro-cerebral mechanism. Instead I am claiming that the explication will not, by its nature, answer the question, "Similar with respect to what?" That question is a request for a rule, in this case for the criteria by which particular situations are grouped into similarity sets, and I am arguing that the temptation to seek criteria (or at least a full set) should be resisted in this case. It is not, however, system but a particular sort of system that I am opposing.

To give that point substance, I must briefly digress. What follows seems obvious to me now, but the constant recourse in my original text to phrases like "the world changes" suggests that it has not always been so. If two people stand at the same place and gaze in the same direction, we must, under pain of solipsism, conclude that they receive closely similar stimuli. (If both could put their eyes at the same place, the stimuli would be identical.) But people do not see stimuli; our knowledge of them is highly theoretical and abstract. Instead they have sensations, and we are under no compulsion to suppose that the sensations of our two viewers are the same. (Sceptics might remember that color blindness was nowhere noticed until John Dalton's description of it in 1794.) On the contrary, much

[12] Some information on this subject can be found in "Second Thoughts."

neural processing takes place between the receipt of a stimulus and the awareness of a sensation. Among the few things that we know about it with assurance are: that very different stimuli can produce the same sensations; that the same stimulus can produce very different sensations; and, finally, that the route from stimulus to sensation is in part conditioned by education. Individuals raised in different societies behave on some occasions as though they saw different things. If we were not tempted to identify stimuli one-to-one with sensations, we might recognize that they actually do so.

Notice now that two groups, the members of which have systematically different sensations on receipt of the same stimuli, do *in some sense* live in different worlds. We posit the existence of stimuli to explain our perceptions of the world, and we posit their immutability to avoid both individual and social solipsism. About neither posit have I the slightest reservation. But our world is populated in the first instance not by stimuli but by the objects of our sensations, and these need not be the same individual to individual or group to group. To the extent, of course, that individuals belong to the same group and thus share education, language, experience, and culture, we have good reason to suppose that their sensations are the same. How else are we to understand the fulness of their communication and the communality of their behavioral responses to their environment? They must see things, process stimuli, in much the same ways. But where the differentiation and specialization of groups begins, we have no similar evidence for the immutability of sensation. Mere parochialism, I suspect, makes us suppose that the route from stimuli to sensation is the same for the members of all groups.

Returning now to exemplars and rules, what I have been trying to suggest, in however preliminary a fashion, is this. One of the fundamental techniques by which the members of a group, whether an entire culture or a specialists' sub-community within it, learn to see the same things when confronted with the same stimuli is by being shown examples of situations that their predecessors in the group have already learned to see as like each other and as different from other sorts of situations. These similar situations may be successive sensory presentations of the same individual—say of mother, who is ultimately recognized on sight as what she is and as different from father or sister. They may be presentations of the members of natural families, say of swans on the one hand and of geese on the other. Or they may, for the members of more specialized groups, be examples of the Newtonian situation, of situations, that is, that are alike in being subject to a version of the symbolic form $f = ma$ and that are different from those situations to which, for example, the law-sketches of optics apply.

Grant for the moment that something of this sort does occur. Ought we say that what has been acquired from exemplars is rules and the ability to apply them? That description is tempting because our seeing a situation as like ones we have encountered before must be the result of neural processing, fully governed by physical and chemical laws. In this sense, once we have learned to do it, recognition of similarity must be as fully systematic as the beating of our hearts. But that very parallel suggests that recognition may also be involuntary, a process over which we have no control. If it is, then we may not properly conceive it as something we manage by applying rules and criteria. To speak of it in those terms implies that we have access to alternatives, that we might, for example, have disobeyed a rule, or misapplied a criterion, or experimented with some other way of seeing.[13] Those, I take it, are just the sorts of things we cannot do.

Or, more precisely, those are things we cannot do until after we have had a sensation, perceived something. Then we do often seek criteria and put them to use. Then we may engage in interpretation, a deliberative process by which we choose among alternatives as we do not in perception itself. Perhaps, for example, something is odd about what we have seen (remember the anomalous playing cards). Turning a corner we see mother

[13] This point might never have needed making if all laws were like Newton's and all rules like the Ten Commandments. In that case the phrase 'breaking a law' would be nonsense, and a rejection of rules would not seem to imply a process not governed by law. Unfortunately, traffic laws and similar products of legislation can be broken, which makes the confusion easy.

entering a downtown store at a time we had thought she was home. Contemplating what we have seen we suddenly exclaim, "That wasn't mother, for she has red hair!" Entering the store we see the woman again and cannot understand how she could have been taken for mother. Or, perhaps we see the tail feathers of a waterfowl feeding from the bottom of a shallow pool. Is it a swan or a goose? We contemplate what we have seen, mentally comparing the tail feathers with those of swans and geese we have seen before. Or, perhaps, being proto-scientists, we simply want to know some general characteristic (the whiteness of swans, for example) of the members of a natural family we can already recognize with ease. Again, we contemplate what we have previously perceived, searching for what the members of the given family have in common.

These are all deliberative processes, and in them we do seek and deploy criteria and rules. We try, that is, to interpret sensations already at hand, to analyze what is for us the given. However we do that, the processes involved must ultimately be neural, and they are therefore governed by the same *physicochemical* laws that govern perception on the one hand and the beating of our hearts on the other. But the fact that the system obeys the same laws in all three cases provides no reason to suppose that our neural apparatus is programmed to operate the same way in interpretation as in perception or in either as in the beating of our hearts. What I have been opposing in this book is therefore the attempt, traditional since Descartes but not before, to analyze perception as an interpretive process, as an unconscious version of what we do after we have perceived.

What makes the integrity of perception worth emphasizing is, of course, that so much past experience is embodied in the neural apparatus that transforms stimuli to sensations. An appropriately programmed perceptual mechanism has survival value. To say that the members of different groups may have different perceptions when confronted with the same stimuli is not to imply that they may have just any perceptions at all. In many environments a group that could not tell wolves from dogs could not endure. Nor would a group of nuclear physicists today survive as scientists if unable to recognize the tracks of alpha particles and electrons. It is just because so very few ways of seeing will do that the ones that have withstood the tests of group use are worth transmitting from generation to generation. Equally, it is because they have been selected for their success over historic time that we must speak of the experience and knowledge of nature embedded in the stimulus-to-sensation route.

Perhaps 'knowledge' is the wrong word, but there are reasons for employing it. What is built into the neural process that transforms stimuli to sensations has the following characteristics: it has been transmitted through education; it has, by trial, been found more effective than its historical competitors in a group's current environment; and, finally, it is subject to change both through further education and through the discovery of misfits with the environment. Those are characteristics of knowledge, and they explain why I use the term. But it is strange usage, for one other characteristic is missing. We have no direct access to what it is we know, no rules or generalizations with which to express this knowledge. Rules which could supply that access would refer to stimuli not sensations, and stimuli we can know only through elaborate theory. In its absence, the knowledge embedded in the stimulus-to-sensation route remains tacit.

Though it is obviously preliminary and need not be correct in all details, what has just been said about sensation is meant literally. At the very least it is a hypothesis about vision which should be subject to experimental investigation though probably not to direct check. But talk like this of seeing and sensation here also serves metaphorical functions as it does in the body of the book. We do not *see* electrons, but rather their tracks or else bubbles of vapor in a cloud chamber. We do not *see* electric currents at all, but rather the needle of an ammeter or galvanometer. Yet in the preceding pages, particularly in Section X, I have repeatedly acted as though we did perceive theoretical entities like currents, electrons, and fields, as though we learned to do so from examination of exemplars, and as though in these cases too it would be wrong to replace talk of seeing with talk of criteria and interpretation. The metaphor that transfers 'seeing' to contexts like these is scarcely a sufficient basis for such claims. In the long run it will need to be eliminated in favor of a more literal mode of discourse.

The computer program referred to above begins to suggest ways in which that may be done, but neither available space nor the extent of my present understanding permits my eliminating the metaphor here.[14] Instead I shall try briefly to bulwark it. Seeing water droplets or a needle against a numerical scale is a primitive perceptual experience for the man unacquainted with cloud chambers and ammeters. It thus requires contemplation, analysis, and interpretation (or else the intervention of external authority) before conclusions can be reached about electrons or currents. But the position of the man who has learned about these instruments and had much exemplary experience with them is very different, and there are corresponding differences in the way he processes the stimuli that reach him from them. Regarding the vapor in his breath on a cold winter afternoon, his sensation may be the same as that of a layman, but viewing a cloud chamber he sees (here literally) not droplets but the tracks of electrons, alpha particles, and so on. Those tracks are, if you will, criteria that he interprets as indices of the presence of the corresponding particles, but that route is both shorter and different from the one taken by the man who interprets droplets.

Or consider the scientist inspecting an ammeter to determine the number against which the needle has settled. His sensation probably is the same as the layman's, particularly if the latter has

[14] For readers of "Second Thoughts" the following cryptic remarks may be leading. The possibility of immediate recognition of the members of natural families depends upon the existence, after neural processing, of empty perceptual space between the families to be discriminated. If, for example, there were a perceived continuum of waterfowl ranging from geese to swans, we should be compelled to introduce a specific criterion for distinguishing them. A similar point can be made for unobservable entities. If a physical theory admits the existence of nothing else like an electric current, then a small number of criteria, which may vary considerably from case to case, will suffice to identify currents even though there is no set of rules that specifies the necessary and sufficient conditions for the identification. That point suggests a plausible corollary which may be more important. Given a set of necessary and sufficient conditions for identifying a theoretical entity, that entity can be eliminated from the ontology of a theory by substitution. In the absence of such rules, however, these entities are not eliminable; the theory then demands their existence.

read other sorts of meters before. But he has seen the meter (again often literally) in the context of the entire circuit, and he knows something about its internal structure. For him the needle's position is a criterion, but only of *the value* of the current. To interpret it he need determine only on which scale the meter is to be read. For the layman, on the other hand, the needle's position is not a criterion of anything except itself. To interpret it, he must examine the whole layout of wires, internal and external, experiment with batteries and magnets, and so on. In the metaphorical no less than in the literal use of 'seeing,' interpretation begins where perception ends. The two processes are not the same, and what perception leaves for interpretation to complete depends drastically on the nature and amount of prior experience and training.

5. *Exemplars, Incommensurability, and Revolutions*

What has just been said provides a basis for clarifying one more aspect of the book: my remarks on incommensurability and its consequences for scientists debating the choice between successive theories.[15] In Sections X and XII I have argued that the parties to such debates inevitably see differently certain of the experimental or observational situations to which both have recourse. Since the vocabularies in which they discuss such situations consist, however, predominantly of the same terms, they must be attaching some of those terms to nature differently, and their communication is inevitably only partial. As a result, the superiority of one theory to another is something that cannot be proved in the debate. Instead, I have insisted, each party must try, by persuasion, to convert the other. Only philosophers have seriously misconstrued the intent of these parts of my argument. A number of them, however, have reported that I believe the following:[16] the proponents of incommensurable theories

[15] The points that follow are dealt with in more detail in Secs. v and vi of "Reflections."

[16] See the works cited in note 9, above, and also the essay by Stephen Toulmin in *Growth of Knowledge*.

cannot communicate with each other at all; as a result, in a debate over theory-choice there can be no recourse to *good* reasons; instead theory must be chosen for reasons that are ultimately personal and subjective; some sort of mystical apperception is responsible for the decision actually reached. More than any other parts of the book, the passages on which these misconstructions rest have been responsible for charges of irrationality.

Consider first my remarks on proof. The point I have been trying to make is a simple one, long familiar in philosophy of science. Debates over theory-choice cannot be cast in a form that fully resembles logical or mathematical proof. In the latter, premises and rules of inference are stipulated from the start. If there is disagreement about conclusions, the parties to the ensuing debate can retrace their steps one by one, checking each against prior stipulation. At the end of that process one or the other must concede that he has made a mistake, violated a previously accepted rule. After that concession he has no recourse, and his opponent's proof is then compelling. Only if the two discover instead that they differ about the meaning or application of stipulated rules, that their prior agreement provides no sufficient basis for proof, does the debate continue in the form it inevitably takes during scientific revolutions. That debate is about premises, and its recourse is to persuasion as a prelude to the possibility of proof.

Nothing about that relatively familiar thesis implies either that there are no good reasons for being persuaded or that those reasons are not ultimately decisive for the group. Nor does it even imply that the reasons for choice are different from those usually listed by philosophers of science: accuracy, simplicity, fruitfulness, and the like. What it should suggest, however, is that such reasons function as values and that they can thus be differently applied, individually and collectively, by men who concur in honoring them. If two men disagree, for example, about the relative fruitfulness of their theories, or if they agree about that but disagree about the relative importance of fruitfulness and, say, scope in reaching a choice, neither can be convicted of a mistake. Nor is either being unscientific. There is no neutral algorithm for theory-choice, no systematic decision procedure which, properly applied, must lead each individual in the group to the same decision. In this sense it is the community of specialists rather than its individual members that makes the effective decision. To understand why science develops as it does, one need not unravel the details of biography and personality that lead each individual to a particular choice, though that topic has vast fascination. What one must understand, however, is the manner in which a particular set of shared values interacts with the particular experiences shared by a community of specialists to ensure that most members of the group will ultimately find one set of arguments rather than another decisive.

That process is persuasion, but it presents a deeper problem. Two men who perceive the same situation differently but nevertheless employ the same vocabulary in its discussion must be using words differently. They speak, that is, from what I have called incommensurable viewpoints. How can they even hope to talk together much less to be persuasive. Even a preliminary answer to that question demands further specification of the nature of the difficulty. I suppose that, at least in part, it takes the following form.

The practice of normal science depends on the ability, acquired from exemplars, to group objects and situations into similarity sets which are primitive in the sense that the grouping is done without an answer to the question, "Similar with respect to what?" One central aspect of any revolution is, then, that some of the similarity relations change. Objects that were grouped in the same set before are grouped in different ones afterward and vice versa. Think of the sun, moon, Mars, and earth before and after Copernicus; of free fall, pendular, and planetary motion before and after Galileo; or of salts, alloys, and a sulpuhur–iron filing mix before and after Dalton. Since most objects within even the altered sets continue to be grouped together, the names of the sets are usually preserved. Nevertheless, the transfer of a subset is ordinarily part of a critical change in the network of interrelations among them. Transferring the metals from the set of compounds to the set of elements played an essential role in the emergence of a new theory of combustion, of acidity, and of physical and chemical combination. In short order those changes had spread through all of chemistry. Not surprisingly, therefore, when such redistributions occur, two men whose discourse had previously proceeded with apparently full understanding may suddenly find themselves responding to the same stimulus with incompatible descriptions and generalizations. Those difficulties will not be felt in all areas of even their scientific discourse, but they will arise and will then cluster most densely about the phenomena upon which the choice of theory most centrally depends.

Such problems, though they first become evident in communication, are not merely linguistic, and they cannot be resolved simply by stipulating the definitions of troublesome terms. Because the words about which difficulties cluster have been learned in part from direct application to exemplars, the participants in a communication breakdown cannot say, "I use the word 'element' (or 'mixture,' or 'planet,' or 'unconstrained motion') in ways determined by the following criteria." They cannot, that is, resort to a neutral language which both use in the same way and which is adequate to the statement of both their theories or even of both those theories' empirical consequences. Part of the difference is prior to the application of the languages in which it is nevertheless reflected.

The men who experience such communication breakdowns must, however, have some recourse. The stimuli that impinge upon them are the same. So is their general neural apparatus, however differently programmed. Furthermore, except in a small, if all-important, area of experience even their neural programming must be very nearly the same, for they share a history, except the immediate past. As a result, both their everyday and most of their scientific world and language are shared. Given that much in common, they should be able to find out a great deal about how they differ. The techniques required are not, however, either straightforward, or comfortable, or parts of the scientist's normal arsenal. Scientists rarely recognize them for quite what they are, and they seldom use them for longer than is required to induce conversion or convince themselves that it will not be obtained.

Briefly put, what the participants in a communication breakdown can do is recognize each other as members of different language communities and then become translators.[17] Taking the differences between their own intra- and inter-group discourse as itself a subject for study, they can first attempt to discover the terms and locutions that, used unproblematically within each community, are nevertheless foci of trouble for inter-group discussions. (Locutions that present no such difficulties may be homophonically translated.) Having isolated such areas of difficulty in scientific communication, they can next resort to their shared everyday vocabularies in an effort further to elucidate their troubles. Each may, that is, try to discover what the other would see and say when presented with a stimulus to which his own verbal response would be different. If they can sufficiently refrain from explaining anomalous behavior as the consequence of mere error or madness, they may in time become very good predictors of each other's behavior. Each will have learned to translate the other's theory and its consequences into his own language and simultaneously to describe in his language the world to which that theory applies. That is what the historian of science regularly does (or should) when dealing with out-of-date scientific theories.

Since translation, if pursued, allows the participants in a communication breakdown to experience vicariously something of the merits and defects of each other's points of view, it is a potent tool both for persuasion and for conversion. But even persuasion need not succeed, and, if it does, it need not be

[17] The already classic source for most of the relevant aspects of translation is W. V. O. Quine, *Word and Object* (Cambridge, Mass., and New York, 1960), chaps. i and ii. But Quine seems to assume that two men receiving the same stimulus must have the same sensation and therefore has little to say about the extent to which a translator must be able to *describe* the world to which the language being translated applies. For the latter point see, E. A. Nida, "Linguistics and Ethnology in Translation Problems," in Del Hymes (ed.), *Language and Culture in Society* (New York, 1964), pp. 90–97.

accompanied or followed by conversion. The two experiences are not the same, an important distinction that I have only recently fully recognized.

To persuade someone is, I take it, to convince him that one's own view is superior and ought therefore supplant his own. That much is occasionally achieved without recourse to anything like translation. In its absence many of the explanations and problem-statements endorsed by the members of one scientific group will be opaque to the other. But each language community can usually produce from the start a few concrete research results that, though describable in sentences understood in the same way by both groups, cannot yet be accounted for by the other community in its own terms. If the new viewpoint endures for a time and continues to be fruitful, the research results verbalizable in this way are likely to grow in number. For some men such results alone will be decisive. They can say: I don't know how the proponents of the new view succeed, but I must learn; whatever they are doing, it is clearly right. That reaction comes particularly easily to men just entering the profession, for they have not yet acquired the special vocabularies and commitments of either group.

Arguments statable in the vocabulary that both groups use in the same way are not, however, usually decisive, at least not until a very late stage in the evolution of the opposing views. Among those already admitted to the profession, few will be persuaded without some recourse to the more extended comparisons permitted by translation. Though the price is often sentences of great length and complexity (think of the Proust-Berthollet controversy conducted without recourse to the term 'element'), many additional research results can be *translated* from one community's language into the other's. As translation proceeds, furthermore, some members of each community may also begin vicariously to understand how a statement previously opaque could seem an explanation to members of the opposing group. The availability of techniques like these does not, of course, guarantee persuasion. For most people translation is a threatening process, and it is entirely foreign to normal science.

Counter-arguments are, in any case, always available, and no rules prescribe how the balance must be struck. Nevertheless, as argument piles on argument and as challenge after challenge is successfully met, only blind stubbornness can at the end account for continued resistance.

That being the case, a second aspect of translation, long familiar to both historians and linguists, becomes crucially important. To translate a theory or worldview into one's own language is not to make it one's own. For that one must go native, discover that one is thinking and working in, not simply translating out of, a language that was previously foreign. That transition is not, however, one that an individual may make or refrain from making by deliberation and choice, however good his reasons for wishing to do so. Instead, at some point in the process of learning to translate, he finds that the transition has occurred, that he has slipped into the new language without a decision having been made. Or else, like many of those who first encountered, say, relativity or quantum mechanics in their middle years, he finds himself fully persuaded of the new view but nevertheless unable to internalize it and be at home in the world it helps to shape. Intellectually such a man has made his choice, but the conversion required if it is to be effective eludes him. He may use the new theory nonetheless, but he will do so as a foreigner in a foreign environment, an alternative available to him only because there are natives already there. His work is parasitic on theirs, for he lacks the constellation of mental sets which future members of the community will acquire through education.

The conversion experience that I have likened to a gestalt switch remains, therefore, at the heart of the revolutionary process. Good reasons for choice provide motives for conversion and a climate in which it is more likely to occur. Translation may, in addition, provide points of entry for the neural reprogramming that, however inscrutable at this time, must underlie conversion. But neither good reasons nor translation constitute conversion, and it is that process we must explicate in order to understand an essential sort of scientific change.

6. *Revolutions and Relativism*

One consequence of the position just outlined has particularly bothered a number of my critics.[18] They find my viewpoint relativistic, particularly as it is developed in the last section of this book. My remarks about translation highlight the reasons for the charge. The proponents of different theories are like the members of different language-culture communities. Recognizing the parallelism suggests that in some sense both groups may be right. Applied to culture and its development that position is relativistic.

But applied to science it may not be, and it is in any case far from *mere* relativism in a respect that its critics have failed to see. Taken as a group or in groups, practitioners of the developed sciences are, I have argued, fundamentally puzzle-solvers. Though the values that they deploy at times of theory-choice derive from other aspects of their work as well, the demonstrated ability to set up and to solve puzzles presented by nature is, in case of value conflict, the dominant criterion for most members of a scientific group. Like any other value, puzzle-solving ability proves equivocal in application. Two men who share it may nevertheless differ in the judgments they draw from its use. But the behavior of a community which makes it preeminent will be very different from that of one which does not. In the sciences, I believe, the high value accorded to puzzle-solving ability has the following consequences.

Imagine an evolutionary tree representing the development of the modern scientific specialties from their common origins in, say, primitive natural philosophy and the crafts. A line drawn up that tree, never doubling back, from the trunk to the tip of some branch would trace a succession of theories related by descent. Considering any two such theories, chosen from points not too near their origin, it should be easy to design a list of criteria that would enable an uncommitted observer to distinguish the earlier from the more recent theory time after time. Among

[18] Shapere, "Structure of Scientific Revolutions," and Popper in *Growth of Knowledge*.

the most useful would be: accuracy of prediction, particularly of quantitative prediction; the balance between esoteric and everyday subject matter; and the number of different problems solved. Less useful for this purpose, though also important determinants of scientific life, would be such values as simplicity, scope, and compatibility with other specialties. Those lists are not yet the ones required, but I have no doubt that they can be completed. If they can, then scientific development is, like biological, a unidirectional and irreversible process. Later scientific theories are better than earlier ones for solving puzzles in the often quite different environments to which they are applied. That is not a relativist's position, and it displays the sense in which I am a convinced believer in scientific progress.

Compared with the notion of progress most prevalent among both philosophers of science and laymen, however, this position lacks an essential element. A scientific theory is usually felt to be better than its predecessors not only in the sense that it is a better instrument for discovering and solving puzzles but also because it is somehow a better representation of what nature is really like. One often hears that successive theories grow ever closer to, or approximate more and more closely to, the truth. Apparently generalizations like that refer not to the puzzle-solutions and the concrete predictions derived from a theory but rather to its ontology, to the match, that is, between the entities with which the theory populates nature and what is "really there."

Perhaps there is some other way of salvaging the notion of 'truth' for application to whole theories, but this one will not do. There is, I think, no theory-independent way to reconstruct phrases like 'really there'; the notion of a match between the ontology of a theory and its "real" counterpart in nature now seems to me illusive in principle. Besides, as a historian, I am impressed with the implausability of the view. I do not doubt, for example, that Newton's mechanics improves on Aristotle's and that Einstein's improves on Newton's as instruments for puzzle-solving. But I can see in their succession no coherent direction of ontological development. On the contrary, in some important respects, though by no means in all, Einstein's general theory of relativity is closer to Aristotle's than either of them is to Newton's. Though the temptation to describe that position as relativistic is understandable, the description seems to me wrong. Conversely, if the position be relativism, I cannot see that the relativist loses anything needed to account for the nature and development of the sciences.

7. *The Nature of Science*

I conclude with a brief discussion of two recurrent reactions to my original text, the first critical, the second favorable, and neither, I think, quite right. Though the two relate neither to what has been said so far nor to each other, both have been sufficiently prevalent to demand at least some response.

A few readers of my original text have noticed that I repeatedly pass back and forth between the descriptive and the normative modes, a transition particularly marked in occasional passages that open with, "But that is not what scientists do," and close by claiming that scientists ought not do so. Some critics claim that I am confusing description with prescription, violating the time-honored philosophical theorem: 'Is' cannot imply 'ought.'[19]

That theorem has, in practice, become a tag, and it is no longer everywhere honored. A number of contemporary philosophers have discovered important contexts in which the normative and the descriptive are inextricably mixed.[20] 'Is' and 'ought' are by no means always so separate as they have seemed. But no recourse to the subtleties of contemporary linguistic philosophy is needed to unravel what has seemed confused about this aspect of my position. The preceding pages present a viewpoint or theory about the nature of science, and, like other philosophies of science, the theory has consequences for the way in which scientists should behave if their enterprise is to succeed. Though

[19] For one of many examples, see P. K. Feyerabend's essay in *Growth of Knowledge*.

[20] Stanley Cavell, *Must We Mean What We Say?* (New York, 1969), chap. i.

it need not be right, any more than any other theory, it provides a legitimate basis for reiterated 'oughts' and 'shoulds.' Conversely, one set of reasons for taking the theory seriously is that scientists, whose methods have been developed and selected for their success, do in fact behave as the theory says they should. My descriptive generalizations are evidence for the theory precisely because they can also be derived from it, whereas on other views of the nature of science they constitute anomalous behavior.

The circularity of that argument is not, I think, vicious. The consequences of the viewpoint being discussed are not exhausted by the observations upon which it rested at the start. Even before this book was first published, I had found parts of the theory it presents a useful tool for the exploration of scientific behavior and development. Comparison of this postscript with the pages of the original may suggest that it has continued to play that role. No merely circular point of view can provide such guidance.

To one last reaction to this book, my answer must be of a different sort. A number of those who have taken pleasure from it have done so less because it illuminates science than because they read its main theses as applicable to many other fields as well. I see what they mean and would not like to discourage their attempts to extend the position, but their reaction has nevertheless puzzled me. To the extent that the book portrays scientific development as a succession of tradition-bound periods punctuated by non-cumulative breaks, its theses are undoubtedly of wide applicability. But they should be, for they are borrowed from other fields. Historians of literature, of music, of the arts, of political development, and of many other human activities have long described their subjects in the same way. Periodization in terms of revolutionary breaks in style, taste, and institutional structure have been among their standard tools. If I have been original with respect to concepts like these, it has mainly been by applying them to the sciences, fields which had been widely thought to develop in a different way. Conceivably the notion of a paradigm as a concrete achievement, an exemplar, is a second contribution. I suspect, for example, that some of the notorious difficulties surrounding the notion of style in the arts may vanish if paintings can be seen to be modeled on one another rather than produced in conformity to some abstracted canons of style.[21]

This book, however, was intended also to make another sort of point, one that has been less clearly visible to many of its readers. Though scientific development may resemble that in other fields more closely than has often been supposed, it is also strikingly different. To say, for example, that the sciences, at least after a certain point in their development, progress in a way that other fields do not, cannot have been all wrong, whatever progress itself may be. One of the objects of the book was to examine such differences and begin accounting for them.

Consider, for example, the reiterated emphasis, above, on the absence or, as I should now say, on the relative scarcity of competing schools in the developed sciences. Or remember my remarks about the extent to which the members of a given scientific community provide the only audience and the only judges of that community's work. Or think again about the special nature of scientific education, about puzzle-solving as a goal, and about the value system which the scientific group deploys in periods of crisis and decision. The book isolates other features of the same sort, none necessarily unique to science but in conjunction setting the activity apart.

About all these features of science there is a great deal more to be learned. Having opened this postscript by emphasizing the need to study the community structure of science, I shall close by underscoring the need for similar and, above all, for comparative study of the corresponding communities in other fields. How does one elect and how is one elected to membership in a particular community, scientific or not? What is the process and what are the stages of socialization to the group? What does the group collectively see as its goals; what deviations, individual or collective, will it tolerate; and how does it control the impermissible aberration? A fuller understanding of science will de-

[21] For this point as well as a more extended discussion of what is special about the sciences, see T. S. Kuhn, "Comment [on the Relations of Science and Art]," *Comparative Studies in Philosophy and History*, XI (1969), 403–12.

pend on answers to other sorts of questions as well, but there is no area in which more work is so badly needed. Scientific knowledge, like language, is intrinsically the common property of a group or else nothing at all. To understand it we shall need to know the special characteristics of the groups that create and use it.

Science and the Structure of Ethics

Abraham Edel

Science and the Structure of Ethics

Contents:

Science and the Structure of Ethics

Abraham Edel

I. The Nature and Complexity of the Problem

1. Issues in the "Relation of Science and Ethics"

Traditional views about the aloofness of ethics from science embody traditional conceptions of man and of science. Such slogans as "Science deals with the quantitative, not the qualitative," "Science deals with nature, not spirit," and "Science is theoretical, ethics is practical" give way before logical and mathematical analyses of order, the progress of psychology, the established importance of abstract theory in applied science. Contemporary reassessments in the philosophy of science as well as the tremendous advance of twentieth-century science call for recasting the problem of the relation of science and ethics in a fresh perspective.[1]

The complexity of the problem is evident from the several ways in which it can be formulated. A familiar way is purely *logical:* Can ethical propositions be deduced from scientific propositions? This leads to a theoretical impasse or, at best, a long detour. A more promising way is *logico-scientific:* for given meanings of 'science' and of 'ethics,' what patterns of relations (logical, psychological or sociological, pragmatic-instrumental, historical) can be envisaged and which can actually be found? How have these relation-patterns changed with shifting conceptions of the nature of science and ethics? A third formulation is necessitated by the discovery of changeable components in the patterns: How far is the relation of science and ethics an *evaluative problem,* requiring policy determination?

In such restructuring of the problem we must distinguish *the place of scientific results in ethical theory, the role of scientific method in ethical theory,* and *the impact of the scientific temper in ethical theory.*

2. The Place of Scientific Results in Ethical Theory

It is generally agreed that scientific results furnish the best way of improving means-judgments in ethics; but there is still a strong tendency to concentrate solely on this function. Accordingly, little has been done toward analyzing other points at which scientific results enter into ethical theories. One consequence is that there is no ready way to study the shifts that may be required in an ethical theory because of fresh advances in psychology or the social sciences or of assessing the extent of such relations.

A major source of difficulty in determining the relations of scientific results to ethics stems from the fringe-boundaries of the concept. In the physical sciences we can specify quite readily what are 'scientific results.' In the psychological and social fields, some would maintain that we have few results, but only areas of promise—models or schemata rather than specific established theories. It is not our purpose here to appraise the state of development of the several areas. Many questions for which there are not yet answers have moved sufficiently far into the domain of science to be regarded as scientific questions. For example, the psychology of perception has results; the psychology of thinking is more disputable; the relation of willing and thinking shares in the difficulties of the latter. We may distinguish scientific questions with established results as answers, scientific questions with competing theories as proposed answers, scientific questions with even the direction of answers unclear. In some cases it may not even be clear to which of the existent sciences the enterprise of answering is to be assigned.

On the other hand, the concept of 'scientific results' should not be limited to general theories. It covers also 'scientific findings,' where what is discovered is the existence of fresh specimens—a hitherto unknown metal, or species of insects, or kinship pattern, or a different moral code. Some theorists have attempted to draw sharp distinctions between fact-finding history and generalizing science. For some purposes sharp lines may be useful if achievable; but for considering the place of scientific results in ethics, particular findings of history or anthropology that throw any light or suggest any lessons come well within the scope of 'scientific results.'

What kinds of scientific results can we expect to have some relevance for ethics? In a broad sense, any that add to our picture of the world in which man lives and acts, that throw light upon the nature of man and his capacities, social relations, and experiences. Biology furnishes a picture of the constitution of the human animal, his place in evolutionary development, his instinctual equipment and basic drives, his interaction with the environment. Psychology explores many human powers basic to ethics. There are studies of pleasure and pain, of feelings and emotions, volition and inner conflict, basic needs, mechanisms of defense and conditions of insight, development of character and personality, varieties and conditions of moral feelings and their phenomenological description, psychiatric materials on amorality and on moral rigidity, patterns and qualities of interpersonal relationships, and so on. The social sciences, whether general as anthropology and sociology, or oriented to specific phases of institutional life as economics and political science, or specifically problem-centered as education or penology, offer a constantly growing fund of information and suggested generalization. They depict the basic and recurrent problems of societies, the variety of patterns in which men have tried to solve them, the kinds of group organizations men have elaborated, the extent of their success or failure, conditions and problems of stability and change, the human costs of social organizations of different types, and so forth. Included within their scope is the social history of morality itself and of ethical theorizing. Similarly, history insofar as it goes beyond the reconstruction of particular events furnishes some accounts or suggestions about the growth of basic aspirations and strivings, about the emergence and career of various ideals, about major instrumentalities and the impact of different kinds of problems and events on human hopes, and it provides some materials at least for speculation on what under what conditions is likely to be unavoidable, and what room there is for men's reconstructive efforts.[2] These are the kinds of materials with whose impact in ethical theory we are concerned.

The extent of such scientific results—the reach of science in dealing with human phenomena—depends on whether the state of the field is such as to set limits to applicability of scientific method. The field must be sufficiently determinate to admit of isolable and recurrent elements, and this means that the phenomena must not be unmanageably complex and unmanageably unstable. For general reference, let us speak of the *indeterminacy of a field* as constituted by its *field instability* and *field complexity*. Some of the ancients thought that the physical world was in such constant flux that no stable knowledge of it was attainable. (Aristotle tells us that this view of Heraclitus led Plato to turn to the purely mathematical domain as the prototype of knowledge.) Many of the moderns believe that the human field is characterized by a major indeterminacy of one sort or another. Theories about these are equivalent to theories about the presumed unavoidable limits that the sciences of man will reach in their development. Part of the task of tracing the place of scientific results within ethical theory is to render explicit such field assumptions operating as the basis of judgments about the nature of ethics, so that they can be assessed in terms of the scientific evidence.

One small part of the relevant area of scientific results is the scientific study of directly moral phenomena. It is obvious, though insufficiently stressed,[3] that there are no restrictions on the scientific study of these phenomena. Whatever be the interpretation within ethical theory of moral utterances such as 'I ought to do *X*' or '*Y* is good,' there is for each the corresponding descriptive statement that a person is having a certain kind of experience. Whether the phenomenon it describes can be *successfully* studied scientifically depends in turn on the state of the field.

3. The Role of Scientific Method in Ethical Theory

How far can the methods of defining, locating phenomena, isolating data, discriminating observations, carrying out experiments or carefully controlled observations, dealing with alternative hypotheses, utilizing logical techniques of classifying and systematizing, verifying and establishing a body of reliable knowledge in general form, and so forth, be applied in ethics? How far is ethics in fact in a prescientific stage, but capable of being developed? Or is this a misguided illusion? What are the necessary conditions for such a development and how far does or can ethics satisfy them? On many of these issues contemporary moral philosophers are seriously divided.

There are general attempts to argue that if scientific method proved applicable *within ethics,* that is, if utterances of the form '*X* is good' and 'I ought to do *Y*' could be established scientifically, the body of such assertions would constitute a science and not ethics. A theoretical science, it is argued, acquires a body of well-demarcated phenomena and looks for laws on the basis of which there can be prediction, and theories that will systematize the laws. A practical science aims not to explain but to secure certain results in production or conduct. Ethics is prescriptive; it tells a man what he should do, not what he will do; it is therefore practical. But, of course, engineering, medicine, psychoanalytic therapy, education, are all practical disciplines, and yet they are markedly different in the extent to which they are scientific. Engineering is predominantly scientific, medicine in many parts. Education is scarcely so, and psychoanalytic practitioners sometimes stoutly defend the view that theirs is an art rather than a science. But educational theory as part of the study of cultural transmission and psychoanalytic psychology as part of the psychology of personality are clearly scientific in intent if not in accomplishment. Even if they were so in accomplishment, it would still be a distinct question how far the practical discipline could or could not apply their results and utilize scientific methods.

In a similar vein it is sometimes argued that if 'theory' means the same in 'an ethical theory' and 'a physical theory,' then an ethical theory is a theory *about* morality rather than within the ethical field; it is a scientific theory rather than a formulation yielding moral guidance. Certainly, a theory about fishing is not fishing, and a theory about advertising is not advertising, any more than a theory about motion is itself a motion. But a theory about fishing may be the most practical way to help organize successful fishing, and a theory about advertising may help restructure advertising activities. Similarly, a theory about morality may be the only way to *understand* morality as well as the most effective guide.

What is really involved in these arguments is whether ethics requires a "logic of the will" distinctive in type from the procedures of scientific inquiry. This is a legitimate problem, but the answer does not follow from the mere acceptance of practical aims in the discipline. It is more helpful to pinpoint the differences that are presumed to exist in ways of treating concepts, in expectations of the possibility or impossibility of working out verification procedures, and so forth.

The problem of the role of scientific method within ethical theory is not coincident with the place of scientific results within ethical theory. Suppose it were a discovery of psychology that man is so irrational that on basic questions of human purpose men were incapable of thinking logically or rationally. Then obviously, since using scientific method involves reflecting rationally, men could not use scientific method in moral processes. It would thus be a scientific result that scientific method had no role in ethics! (In fact, irrationalist claims usually embody some assumptions about the nature of the human will which yield this type of result.) However, this example further suggests that, while the two questions are different, their answers are interrelated. If the human field is sufficiently determinate and scientific results have a constitutive place in ethics, then the wider use of scientific method within ethical theory may be possible. Whether to take advantage of this possibility would be a decision of methodological policy. To see what considerations enter into such a decision requires a detailed examination of the structure of an ethical theory and the nature of its conceptual and methodological apparatus.

4. The Impact of the Scientific Temper in Ethical Theory

The contemporary scene has witnessed increased impact of a reflective analytical-empirical outlook on ethical theorizing. Calls for clarity, for the separation of analytic from empirical issues, and of decisional from both, are frequent. Concern among philosophers with the logic of ethics and with the language of ethics, and the rise among anthropologists of empirical value studies, further reflect the scientific trend.

In the case of the scientific temper, the evaluative character of the relation of science and ethics emerges quite clearly, for there are alternatives. There is also the mystical temper, or the activist temper. Which to employ is a policy decision. Even more, it is clearly an ethical decision; for a choice of temper is in effect a selection of a virtue-set. Compare this with the question of the place of scientific results in ethics. Here there may be little choice. If there really are gaps in any ethical theory which have to be filled in somehow by a conception of the self or a conception of the relation of intellect and will, then the only choice may be whether to use outworn scientific results or contemporary scientific results. (Of course, there is always the alternative of using myth or ideology or of embodying partial results geared to produce a particular ethical outcome at all costs.)

Policy decisions in ethical theory share with all basic policy decisions the problem of vindication—a complex but not insuperable problem. The attempt to go as far as possible toward the use of science in ethics is part of a unified philosophy of man and his world, which presents a promising present policy. The bases for policy decision in dealing with scientific results and scientific method in ethical theory will be considered in the body of this work. The question of evaluating the virtues which constitute the scientific temper is a particular ethical problem which lies beyond the scope of this general study. It will be touched on briefly at the end.

5. Moralities

The history of writings on ethics shows that ethical theories are reflective attempts to understand and help guide morality. Understanding often involves providing principles for systematizing and interpreting; guidance may take the form of adding confidence and furnishing basic direction through justifications or offering organized programs of modification. Ethics is a reflective enterprise; moralities (frequently called 'moral codes') have a more overtly regulative character. In addition, actual moral codes and propounded ethical theories have had varying relations and sometimes relatively independent historical careers. On the other hand, the intimacy of their relation is witnessed by the fact that the answer to the question "What is a morality?" is a fundamental part of an ethical theory.

In the array of propounded ethical theories there are many different answers to this latter question. Some locate their phenomena as men's moral judgments which they construe as individual acts and examine psychologically or phenomenologically; here are to be found theories of moral requiredness, indefinable moral qualities, morality as some dimension of interest or appetition, and so on. Others, starting equally with individual moral judgments, give them a linguistic cast; moral statements are identified by the presence of moral terms ('ought,' 'good,' 'duty,' etc.), and ethical theory takes the form of logical analysis. Others, still focusing on individual discourse and looking for the "normative force" of ethical utterances, find their basic phenomena in specific functions of the utterance —expressive, hortatory, commendatory, etc. A rather different initial approach is found in those who approach morality as a sociocultural phenomenon and locate their initial data in group processes and institutional forms.

A full delineation of a morality has to include all the features which different views invoke for varying theoretical purposes. A morality embraces many qualities and processes: rules enjoining or forbidding selected types of action, selected character-traits cultivated or avoided, selected patterns of goals and means. These in turn are organized in the behavior and consciousness of people in a variety of ways—a conception of the community concerned and a responsible person, a set of ethical terms and some scheme of systematization of discourse involving them, some pattern of justification processes, some types of sanctions, some selection from the range of human feelings tied in with these regulative procedures. Thus typical Western morality has its code of ten commandments; its virtues and vices; its goods of happiness or salvation or success; its universalistic moral community in which "the dignity of man" is taken to insure that everyone counts; its notion of individual responsibility; its language of 'ought,' 'right,' 'wrong,' 'good,' etc.; its religious or utilitarian justifications; its heaven-and-hell or internal-conscience sanctions; its concentration on guilt and shame. The ideal of scientific description of moralities would be a moral map of the globe over the history of man, just as one could have a linguistic map or a map of religions. This presupposes coordinates of description, principles of classification, and some determination of configuration-types.[4]

6. Methodological Approaches in Ethics

If the role of science in ethics is to be fully explored, it is important to include all the methodological approaches operative in ethical theory today, dispersed though they may be. Four standpoints are suggested as pertinent.[5]

(1) The analytic approach provides reflective examination of the conceptual apparatus of ethics—the meanings, uses, and relations of ethical terms; formation rules and modes of reasoning; ways of justification; and so forth. It helps distinguish stipulative, factual, and purposive components. Its outcome consists in conceptual schemes and methodological guides.

(2) Descriptive approaches focus on the experiences and processes that are identified either as moral or as relevant to the solution of moral questions. There are various descriptive modes—the familiar behavioral and introspective modes, phenomenological description reporting qualities and relations in the field of awareness, sociocultural description adapted to group phenomena, historical description concentrating on succession of phenomena over large time-spans.

(3) Causal-explanatory approaches look for the conditions and contexts of the occurrence of moral phenomena, experiences, processes. In whatever terms these phenomena be described, there is always the question of relating them, as they occur, to other phases of the life and internal economy of the individual and to the other phases of operation of the society at a given point of development of its history and culture.

(4) Evaluative approaches involve the employment of standards to assess whatever is the focus of attention. They are operative in decision, in the justification or criticism of decision, in the establishment of principles and policies, in the formation and reiteration of commitments, in the development, stabilization, and alteration of standards themselves.

Each of these standpoints brings to bear methods increasingly refined in contemporary logic, science, and human experience. The lessons of modern logic, mathematics, and linguistic inquiry give force to analytic sifting in ethics; for example, the traditional overready resort to moral "axioms" rested on appeal to arithmetic and Euclid, and must reckon afresh with contemporary interpretations of mathematics. Modes of description and observation have been refined in the controversies of psychological schools and in both social science approaches and phenomenological philosophies. Causal investigation opens the door to what can be learned about the career of morality and ethical ideas in the light of various psychological, cultural, social, and historical schools of thought. Evaluative processes in other fields—in law and education, as well as economics and politics, even procedures of standard-development in pure and applied physical science—help throw light on comparable moral processes.

In any particular inquiry in ethical theory, care must be taken to recognize which enterprise is being carried on at what point and in what respect. Take, for example, the ethical problem of *obligation*. It is an analytic problem to determine the possible meanings and various uses of 'ought,' how far its use differs in such expressions as 'I ought' and 'He ought,' whether it is meaningful to say that something 'ought to be' or only that it 'ought to be done,' whether 'ought' implies 'can' and in what senses, and so on. It is a descriptive problem to give a clear representation of the feeling of obligation, or the pangs of conscience, or the occurrence of requiredness in the field of awareness, or the type of interpersonal relation that is designated by such phrases as 'recognizing a claim by another upon oneself'—all this on the assumption that we have in these various cases some definite experience we are referring to. It is a causal-explanatory problem to discover the conditions (whether they be psychological or sociocultural) under which we find a harsh conscience or peremptory obligation-feelings, or under which the obligation is felt as imposed from without or issuing from one's self. It is an evaluative problem to determine whether promise-keeping is a moderately weighty or an absolute commitment or whether, if possible, men should cultivate humanistic rather than authoritarian consciences.[6] Confusions clearly arise if one simply asks "What is obligation?" without distinguishing the type of inquiry intended.

7. The Structure of an Ethical Theory

What are we to understand by (i) *an ethical theory,* (ii) *the structure of a given ethical theory,* and (iii) *the structure of ethics?*

(i) An ethical theory provides, broadly speaking, an analysis of the basic concepts and methods of a morality, a descriptive account of the types of phenomena involved, an explanation of the relations of the morality to the fuller context of human life, and procedures for using the morality in deciding and evaluating. There have obviously been many ethical theories (e.g., Platonic, Stoic, Kantian, Utilitarian). In the textual and historical materials from which each of these is extracted, we can distinguish between the morality as a pattern of virtues, rules, and social policies, sanctions and feelings, and so on, and the ethical theory of the human makeup and its strivings, the basic order and its necessities, the conceptual analyses of 'good' and 'obligation' which are offered to expound and support the morality.

(ii) To speak of 'the structure of a given ethical theory' is to call attention to the kinds of problems to which the particular ethical theory is addressing itself in its attempt to understand and guide the morality with which it is associated. On the whole, the structure of traditional and contemporary ethical theories becomes clear if we see their assumptions about the human field in its world background, the concepts and methods they employ, their account of the way in which standards are established and decisions made, their specific interpretations of freedom and responsibility.

(iii) To speak of 'the structure of ethics' is to go beyond the particular notion of 'the structure of a given ethical theory.' It assumes a certain community of enterprise in different ethical theories—that they are addressing themselves to the same central problems however different their solutions. This need not be an unexamined assumption; it is possible to render the basic problems explicit and to study the variety of their formulation and the limits beyond which we would cease to be dealing with a common field.

If the problems are comparable, it becomes possible to suggest co-ordinates for the mapping of ethical theories, to look for possibly invariant features, and to analyze the role that different parts of theories play in the functioning of the theories themselves. To explore the relation of science and the structure of ethics is to go along a major avenue of such comparative judgments.

II. The Theory of Existential Perspectives

8. The Concept of an Existential Perspective (EP)

A striking fact in the comparative study of ethical theories is their diversity. There are individual and social types, otherworldly and this-worldly, inner-oriented and outer-oriented, intuitive and empirical, rigorous and genial, conformity-directed and welfare-oriented, and so on.[7] The issues among them include historical factors and valuational conflicts. Our concern here is to discover how far the differences are due to different answers to scientific questions, operative within the ethical theories. Our heuristic principle is that this factor plays a major role in the inner structure of ethical theories, that it is sound policy to render it systematically explicit, and that a sizable part of the evaluation of ethical theories (and therefore of the task of reconstruction in ethical theory) centers about it.

From the point of view of this inquiry, ethical theories divide into three kinds: those that specify a basis in scientific results, those that specify a trans-scientific basis (usually theological or metaphysical), and those that ultimately deny the relevance of any existential basis at all to value theory. Our procedure of inquiry will be as follows:

a) To construct a general or abstract concept—the *existential perspective of an ethical theory*, hereafter abbreviated to EP—to refer to assumptions about the world and human nature, images of man, etc., operative in an ethical theory. Such a general concept leaves open the types of specialization it will take in different ethical theories—e.g., metaphysical, psychological, historical. As here used, the term 'existential' refers to a way of viewing *existence*. It is to be distinguished from the term 'existentialist,' which has come to designate one special theory about existence. An existentialist EP would thus be one type of EP.[8]

b) To explore the *overtly scientific* types of EP and suggest the extent to which differences in the ethical theories that have such EP's stem from their reliance on different sciences or selective emphases in the psychological and social sciences or different stages in the development of these sciences.

c) To explore *theological and metaphysical* types of EP to see how far they embody different answers to scientific questions, and what role these play in the ethical theories that have such EP's.

d) To criticize the claims of "purist" ethical theories that disparage the role of existential assumptions in the inner workings of ethical theory and to suggest that they have a hidden or "displaced" EP behind the façade of autonomy. This will be called a *transcendence* EP, and the same kinds of questions can be asked about it as about the theological and metaphysical types.

e) To suggest criteria for the evaluation of EP's as part of the general task of securing a more adequate ethical theory in the modern world.

The EP of a given ethical theory is its view of the world and its properties, man's nature and condition, insofar as these enter into its understanding of moral processes and moral judgments. In looking for the EP of a given ethical theory, we could ask how far it employs:

A. a particular view of the world and its constituents
   a particular theory or model of its processes and mechanisms
B. a particular view of the nature of man and his dominant aims
   a particular reference to unavoidables in life and action (birth, maturation, reproduction, aging, death)
   a particular theory of men's faculties—intellectual, emotional, practical—and their relations and an image of the self
C. a particular image of community—its nature, bonds, extent
D. a particular consequent view of the degree of knowability of the world, human life, community, and their processes
   a particular view of determinateness and indeterminateness in their patterning
   a particular expectation of dominant dangers (e.g., illness, human aggression, etc.)
   a particular assessment of control-possibilities, including (where it exists) any estimate of morality itself as a control-instrument.

In recent sociological and anthropological theory there has been some movement toward a configurational concept in describing a group's values. For example, Kluckhohn adapts the concept of a 'value-orientation' to this purpose and defines it as "a generalized and organized conception, influencing behavior, of nature, of man's place in it, of man's relation to man, and of the desirable and nondesirable as they relate to man-environment and interhuman relations."[9] Redfield explores "world views" as "the way a people characteristically look outward upon the universe," "the structure of things as man is aware of them."[10] Others explore "ideologies" and their components, or varieties of philosophical assumptions.[11]

Such materials are very suggestive for EP dimensions likely to be overlooked if we concentrate on the Western ethical tradition. The EP concept, however, is a narrower one in two respects. It is concerned with the stage-setting for morality, which need not coincide with *total* world-outlook, and it aims at winnowing out analytically the *existential* aspects, no matter how fused they may be in the people's outlook with value-components.

A useful device for working out EP dimensions is to employ a theatrical metaphor: the EP is the *stage-setting* for the performance embodied in the ethical processes.[12] Thus we may think of the scenic *space* of an ethical theory: moral problems are set for a man as an issue of what is going on inside himself, between himself and other individuals, between himself and a group, between himself and the cosmos. Or, again, it may set them as primarily group problems or whole-world problems. Similarly, the *time* may be here-and-now (focusing on the quality of present experience), the past (e.g., moral duties construed as debt to one's ancestors), the future (governed by distant goals), temporally indeterminate or "aorist" (concerned with the eternal).

What of the *dramatis personae* for the performance? Some EP's include gods as well as men; some have men alone. Some have whole groups, cast as nations, species, or classes. Others use the central dramatic figure of the individual man. Where the stage is further limited, the actor may be some part of man—a soul, a rational element, a superego, a host of desires, or instincts, drives, and needs.

As for *typical lines of action,* the script often assigns well-demarcated roles. For example, in religious stage-settings, the Greek gods act like exalted humans with all their passions and conflicts, the God of the Decalogue like an exalted father; in West Africa the stage contains both Ancestral Spirits concerned with the behavior of their lineal descendants and individualized Fates.[13] Nations or peoples unfold destinies or accumulate power, species struggle for survival, classes for domination. Souls seek salvation or beatitude in relation to a divine figure; sometimes they are torn in the struggle of divine and satanic beings. Rational elements do everything, from maintaining Kantian consistency or engaging in Benthamite calculation to restraining strong drives in Pueblo Indian conformity. Superegos specialize in inflicting guilt-feelings or parade in judicial robes. Drives press on, needs seek satisfaction; hosts of desires usually stumble over one another until conditioned into harmonious equilibria. Some plots require richly decked stage-settings; some, very bare ones.

Similar suggestions can come from the typical environment of the action: the non-human world as means to be manipulated, or as possessed of compelling power, as forcing certain unavoidables or presenting dangers. The stage may be brightly lit (intelligible) or obscure and murky (inscrutable). What is its *steadiness*—absolute fixity or near-chaos (in which a good Stoic is ready to face anything)? What is the degree of *determinacy in plot structure*—a fixed plan with no alternatives, or plenty of room for choice? What is the *central focus* of the characters and their action? Two stage-settings may both include the individual and the group. But in one the group is external to the individual, an instrumental help or hindrance to his goals; in the other, the individual is essentially a group being. Both differ from an EP that leaves the individual out and speaks only of group survival and expansion. (Psychologically, the question of central focus may be seen here as one of self-involvement.)

A concluding warning: we should never underestimate the ingenuity of stage-designers (or the complexities of human life): Hegelian ethics uses a simultaneous two-scene setting—the world process in one, the individual consciousness with its pressing passions in the other—and central focus is on the cunning of reason which builds the action of the former on the materials of the latter.

9. Role of Existential Perspective within an Ethical Theory

How far EP answers are answers to scientific questions, and how influential they are in ethical theories, is a basic question in the rest of this work. By way of anticipation, let us illustrate the range of influence in one type of theory with an overtly scientific EP—the Utilitarianism of Bentham and Mill.[14] The scenic space of its EP is this world of men; its time is the period of worldly life. Its dramatis personae are individuals treated as units, not communities or nations. Its typical lines of action are pursuit of pleasure and avoidance of pain; so strongly is this stated that we are likely to overlook the concomitant activity of forming associations according to psychological laws. Causality is taken for granted in the operations of man and nature; men age and die; there is always a strongly operative background of other people; individuals are very different in their sensibilities and the circumstances that determine them; religious forces enter the picture only through beliefs in them which have an influence on conduct. The stage is fairly luminous: men can reason and calculate; they need not simply hold on to blind custom. There is sufficient steadiness and determinacy so that men can successfully plan and generalize and apply the lessons of their experience; what is more, one can count on their following certain lines once they fully realize that it means increased happiness and once they have formed the appropriate associations. The central focus in the theory is social, in the sense that there is a primary interest in the shape of social forms and institutions; at times there is a double focus, with shuttling between individual and society.

How does this particular EP tie in with the remaining constituents of the structure of Utilitarian ethics? The reference is not to causal determination but to relations varying from logical implication through instrumental and functional gearings to different degrees of parallelism or similarity. A number of tie-ups stand out quite clearly. The picture of men's basic strivings is directly incorporated into the definitions of ethical terms: 'good' is analyzed as pleasure or what produces a pleasure surplus, and 'ought' in terms of the line of action conforming to the greatest happiness principle, and comparably for other ethical terms. The syntax of ethical expressions follows consistently; for example, the subject term in the expression '*X* is good' in its primary use is an experience term; comparative terms (e.g., 'better') are analyzed in terms of degree or quantity of pleasure, as in the familiar felicific calculus; and so on. The assumed rational abilities of men and the luminosity and determinateness of the stage make it possible to ascribe an inductive methodology for ethics, with lessons of experience about what is productive of the greatest happiness. Operative indices for applying moral rules are linked to assumptions in the stage-setting; for example, since sensibilities differ, every man is his own best judge of what pleases him. Techniques for social stabilization of the morality similarly are linked to assumptions about the way in which men are influenced—not merely the detailed theory of sanctions as specific forms of pleasure and pain inducement but the concentration on such fields as economic relations, law, education, and formation of social habit rather than merely on moral discourse, as some ethical theories are prone to do. These areas of application also express the central focus, with its aim to provide standards for social judgment and social reform. We need not enter into the content-values of the theory; some, such as the high appreciation of rational activity as a pleasure, certainly parallel the place they occupy in the EP. But we may complete this brief survey by noting that the very nature and tasks of ethical theorizing are worked out in a way congruent with the EP in its several dimensions—ethical theorizing is itself a sharpening of the tools for calculation or reckoning so that the pursuit of happiness will be more rather than less effective, far-reaching in plan rather than blind or tradition-bound.

Just as in Utilitarian ethics views of men's world, nature, motivations, interactions, guide and help shape formal and methodological elements, so too in other ethical theories we can trace a similar influence, though the precise pattern of influence varies with the content.

10. Overtly Scientific Existential Perspectives: Physical and Biological

Overtly scientific EP's may be extracted from ethical theories that have actually been propounded and also by looking to the specific sciences to see what their offerings are for ethical theory.

Physical stage-settings have been offered in principle, but scarcely more. A Democritus or a Hobbes may construe man as a bundle of particles or physical motions, but he is soon given a physiological and even a psychological costume before he begins to interact with his fellows. Sometimes the physical description provides a model for the transfer of some properties to human action (e.g., self-maintenance, self-preservation), or some of the "laws" of human behavior are cast in a mold analogous to the laws of physics. Physical EP's therefore appear usually as physical-biological or physical-psychological.

Yet it would be a mistake to ignore the power exercised even by the promissory notes of this EP. Its ethical potential is sometimes far-reaching. It eliminates interpretations of duty or obligation in terms of gods or souls. It points a direction for the reduction of ethical concepts—for example, of good to feeling pleased or goal-seeking, and of conscience to a special form of fear, and these in turn to an internal movement or a drop in tension and to a disorganization state or special tensions, and so on into chemical-electrical processes. The causal emphasis in the physical stage-setting usually means an interpretation of will-acts as resultant phenomena occurring in a lawlike manner, as in Hobbes's description of will as "the last appetite, or aversion, immediately adhering to the action, or to the omission thereof."[15] Nor are we dealing merely with philosophical history. Physical stage-settings have increased significance today in relation to the live interest in the mechanism of "thinking-machines" and the attempts to work out concepts of teleological mechanisms. Similarly, construing the body as a stabilized machine with determinate modes of action becomes readily translatable into a system of inherent or unalterable tendencies seeking fulfilment, and so into an egoistic individualism.

Biological stage-settings derive their concepts from several sources. One is the study of the individual organism as it develops and functions. The focus of such an EP is on intraorganic units or on individuals. Many an ethical stage is set with 'impulses' or more determinate 'instincts' made into stable unquestioned starting-points to which ethical activity is held responsible.

Other biological EP's focus on groups—small, or whole populations, or even humanity as a single global population—and devote special attention to properties characterizing the evolution of the species or the biological history of populations. Sometimes we find a single strand selected for a central role, such as the struggle for existence in nineteenth-century survival-of-the-fittest ethics, or the phenomenon of mutual aid, as in Kropotkin's ethics.[16] Sometimes we find an organization property, such as adaptive harmony or adjustment, or maintenance of a special kind of equilibrium providing key slogans for the ethical processes enacted on their stage. Social-organism theories of ethics exploit the analogy and call on biology to provide the picture of how men became so integrated in their human social reactions;[17] or the homeostatic model is transferred to social organization and to ethical regulation. Sometimes attention is fastened on a developmental thread or a persistent trend, such as increase of population, spread of the area of human control, development of larger co-operative aggregates, and so forth; these tend to be cast as criteria of "progress."

A third type of biological EP sets a full-scale evolutionary stage. While its concepts are evolutionary, its materials usually go far beyond biology. Herbert Spencer's picture of evolution in all fields from chaotic homogeneity to organized heterogeneity is a classic illustration. Such stage-settings often deal consciously with the concepts of ethics and the structure of ethical theory, as Spencer sought to do, and attempt to incorporate changing elements within ethics into their account of evolution itself. The strongly ideological character of Spencer's system has obscured its theoretical scope.[18]

11. Overtly Scientific Existential Perspectives: Psychological

Psychological EP's have been perhaps most common in modern ethical theories. Their philosophical articulation has tended to follow the lines of the development of psychological theory itself.

The classical *introspective* type is most clearly seen in the theoretical formulations of the British empirical tradition from Hume to J. S. Mill. The building-blocks of ethics are feelings or sentiments, introspectively discernible states of consciousness —whether pleasures and pains, acts of sympathy, driving passions, or more deliberative calculations expressive of self-love and fear.

To locate the elementary units and exhibit the mode of combination—for example, to pin down sympathy and show how it is built up into feelings of duty, as Adam Smith does,[19] or to pinpoint pleasures and pains and show modes of measurement of large-scale "lots" in terms of their components, as Bentham does[20]—constitutes a great part of the task of such ethical inquiry. A sizable part of contemporary empirical ethics retains this tradition. Bertrand Russell's stage-setting is the matrix of desires with a special focus on the conflict of impulse and intelligence;[21] Schlick's focus is on the pleasantness or unpleasantness of ideas in the motivation of willing;[22] and a great part of general value theory is cast in affective terms of desire and feeling (described as states of consciousness) as the essence of the value phenomenon.

With the rise of *behaviorist* tendencies and the revolt against introspection—especially in ethics because of its dualistic associations—an emphasis on observable behavior replaced the compounding of states of consciousness. The psychological stage-setting was conceived to be an extension of the biological picture of the organism and its adaptive responses. Directional concepts could not be wholly discarded but could be seen as selective tendencies, propensities to action, and the like, in the human organism.

Ralph Barton Perry's *General Theory of Value* (1926) exhibits clearly this type of EP. The basic concept of *interest* in terms of which Perry identifies value is given a biological basis and a psychological articulation: "Interested or purposive action is action adopted because the anticipatory responses which it arouses coincide with the unfulfilled or implicit phase of a governing propensity."[23] Modes of interest, the role of cognition, types of integration, are studied, but this does not carry us on to a social setting of the stage; on the contrary, Perry says, "we are concerned only with society in so far as it is a composition of subjects who interact *interestedly*, or are integrated in and through their interests."[24] Two influences in the growth of behaviorist psychology enrich this type of stage-setting. The first is the liberalization of behaviorism in such work as E. C. Tolman's,[25] in which 'goal-seeking behavior' comes into its own, with rigorously controlled criteria for identifying the phenomenon. The corresponding ethical stage-setting is still individual-psychological, with distinctively human aspects found in the proliferation and organization of the scheme of purposes in group environments and in the role of cognition mediating the rise and fulfilment of goal-seeking. The second influence is the impact of Freudian depth and basic need emphasis, which led the behavioral stage-settings to give a more explicit place to need-concepts, without however surrendering the insistence on observable verifiability. The difference that these two influences make in the refashioning of the psychological stage-setting is apparent if we compare Perry's early work with the mapping of the structure of appetition and aversion that we find in Stephen Pepper's recent *Sources of Value*.[26]

Freudian psychology and its derivative schools have provided materials for a whole class of stage-settings that have been explicitly applied to ethics. The analyses of conscience, character-development, internal conflict, have an obvious bearing on questions of obligation, virtue, and freedom in ethical theory. More sharply, the Freudian stage is specified by the career of the instincts in the development of the individual, the differentiation of the id in the rise of the ego and the superego, their modes of interaction and conflict, the types of sublimation or neurotic distortion or breakdown that ensues—in general, a theory of the maturation of personality. In the maturation process a pleasure principle is modified by a reality principle which admits of deferment in gratification; conscience is depicted as a specialized development of anxiety in intrafamilial reactions of the oedipal phase; and morality is generally construed as a repressive mechanism against aggressive tendencies. The Freudian stage reaches back into a biological stratum, at least in Freud's later biological derivation of the Life instinct and the Death instinct. It reaches out toward a social structuring by seeking to find the "cement" in interpersonal relations that make wider group association possible, and it assumes historical contours by picturing the growth of civilization as the increase of group affiliative extent, resting on a constantly greater repression.[27]

Variant theories of personality within the broad psychoanalytic outlook produce many readjustments in the ethical stage. Conceptions of an independent conflict-free source of ego development rooted in processes of perception and motion[28] yield a less repressive picture of the tasks of morality and a greater role for a positive picture of the self. Theories that tie aggressiveness to frustration rather than an instinctive base yield an EP with more room for action.[29] There are variants in the specific steps of the developmental process, the outlines of character-formation, the role assigned to cultural influences, and so on, all of which have different ethical potentials and some of which have been applied to ethical theory.[30] Some psychoanalytical writers, such as Sullivan, shift the whole focus of the inquiry to an emphasis on interpersonal relations.[31] As an EP this ceases to be individualistic-psychological and becomes at least two-person psychological; its ethical import, though not as yet analyzed, would seem to be considerable and not unlike the broadening in those psychological schools that stress a constitutive role for cultural elements. Finally, whatever the variety of psychoanalytical approaches, there are some findings with considerable ethical impact—a conception of basic human needs as scientifically discoverable and as demanding a place in any EP, or a recognition that there are techniques of systematic self-deception, which challenges the ultimacy of individual introspective value reports and renders them in principle partially corrigible.

Phenomenological psychological approaches are the modern heirs of the older scrutiny of consciousness and the moral sentiments. Their scientific development in contemporary Gestalt theory embodies a revolt against the behaviorist neglect of the "meaningful" aspects of experience and yet an unreadiness to equate these with the older introspective picture of states of consciousness. In the writings of some psychologists and some philosophers[32] the stage is set with men as beings who have a level of experience with qualities of its own as moral, and who find in the field of this experience such contents as "meanings" and "values," and who are capable of inspecting the field and enunciating relations. There is the confidence that general and stable results can be achieved by phenomenological inspection of moral experience just as laws have been found for the visual field and other areas of perception phenomena. There is no attempt to restrict causal research into men as bearers of such fields, but this is construed as an external enterprise, involving the correlation of physical or physiological variables with field variables.[33]

There are, of course, serious attempts today to develop a unified psychological model of man in which phenomenological and behavioral methods will be employed together with conceptions of underlying needs and developmental processes. The success of such attempts would have coresponding effects on the available EP's of the psychological types.

12. Overtly Scientific Existential Perspectives: Sociocultural and Sociohistorical

Many ethical theories that present no systematic picture of their existential assumptions are prone to rely on an interpersonal or general sociocultural account of the context in which ethical problems arise. For example, emotive theory focuses on disagreement in attitude, which involves at least a two-person group, in setting up its fundamental paradigms for interpreting ethical language.[34] The same is true of prescriptive types, for which the distinctive function of ethics—commending, advising, persuading, etc.—is clearly interpersonal. Where a sociocultural EP is explicitly advocated, it represents a theoretical conclusion that morality is in essence a group or social phenomenon, like language, religion, or law. Such a view criticizes psychological EP's for casting the individual's relations to others as a balance of internal forces within him; an attempt is made instead to account for the very rise and properties of a self in interpersonal and social terms. Historical emphasis on change similarly is said to be excessive in ignoring those perennial and unavoidable features in all periods that make it possible to treat morality as a persisting structure in human life.

Perhaps the best illustration among twentieth-century ethical theories of a consciously sociocultural type is John Dewey's. It regards man as a natural biological organism possessing basic biological needs; but these needs are held socially patterned. Dewey's ethics is also constantly concerned with psychological questions; but his psychology is itself of a behavioral-social type, rejecting the introspective approach as a residue of dualism. His fundamental concept of *habit* is thus a concept of social, not individual, psychology. It is more akin to custom, as patterned culture in action. In their interaction, habits constitute the self, the will, and character; and all features of the mental life introspectively ascertained are to be explained by them, rather than the reverse. The moral situation itself is defined in terms of "the mutual modification of habits by one another,"[35] and reflective morality arises out of conflicts in habits or customary morality. Purely biological needs can function in the ethical process only as generalized impulse coming to the fore in the conflict of habits. The spotlight in Dewey's ethics comes to rest on the sociocultural.[36] The framework of moral conceptions expresses the permanent contours of associated living.[37] The content of problems, and so of morality, is recognized to be undergoing constant transformation historically, but, instead of moving into a specifically sociohistorical stage-setting, Dewey focuses on the method of successful solution of problems as the perennial clue in ethical understanding.

A considerable part of current sociological treatment of value questions seems to be moving toward a sociocultural EP. The social group is often pictured as engaged in such tasks as maintaining its cultural pattern, adjusting to changing forces, harmonizing its conflicts; sometimes there is reference to more specific but nevertheless perennial problems—continuing its population, transmitting its skills, satisfying its material needs, organizing social relations among age and sex groups, and so on. Ethical processes emerge as control processes and refinement-adjustment mechanisms or cohesion-forces, embodying some degree of deliberation. Such analyses, while carried out in terms of providing social understanding of value structures, are equally available in constructing ethical theories.

Sociohistorical EP's are differentiated from sociocultural not by their recognition of change but by a different estimate of its theoretical significance. They find historical variables or epoch-parameters strategically situated within ethical theory, so that their neglect would constitute a misreading of ethical processes. In a specifically sociohistorical EP, every present social state is seen as a more or less temporary equilibrium of forces in a process of continual change. On this view ethical processes have a directional significance; the particular plot depends on how the historical process is itself read, on what degree of unity it is found to possess, and also on the kind of biological, psychological, and sociocultural subsets that may be associated with the historical setting.

For specifically sociohistorical EP's, Marxian theory in its historical-materialist aspects provides a clear illustration. Here major stress falls on the sociohistorical specificity in the materials of ethics, both content and structure. Given the familiar Marxian picture of historical development in terms of the growth of the forces of production in the history of mankind, the changing systems of relations of production, the conflict of social classes, consciousness in all its forms is taken to reflect the needs, pressures, and tensions of this matrix and the defense, critique, and projection of solutions for the interests involved and emerging. Thus the moralities at any period will reflect—both in the sense of causal origin out of previously existent materials and in the sense of being geared to—the basic needs of the dominant productive modes, the particular form of the relations of production, the particular stage of the struggle of classes (at least during the historical period in which classes constitute crucial social phenomena). While many elements of moral content change with relative rapidity, what appears as more lasting or more perennial or more structural or even as abstract is not free of historical dependence but represents historical recurrences and similarities (though often with changed content) or what is grounded in somewhat longer-lasting features of economic structure or social situations or expresses directional trends in social evolution. To attempt to understand moral phenomena and ethical structures without the full sociohistorical stage-setting is taken to risk distortion. Not only does "Thou shalt not steal" acquire its meaning from the existent property-forms but the idea of a divine command corresponds to a particular stage of historical development, the ideal of justice even in its universal aspects represents the cry against the exploitative atmosphere of class domination, and the hopes and aims underlying it reflect the effort of men for wider productive freedom.[38]

The special force of sociohistorical EP's comes from combination of wide historical sweep with insistence on epoch or period or even particular-moment specificity. Wide sweeps are especially to be found in the integration of biological evolutionary with historical vistas. Thus Herbert Spencer delineates the changes that take place in ethical conceptions as mankind moves through successive stages; or Kropotkin moves from mutual aid among the animal species to a picture of the historical struggles of the co-operative phases of mutual living against the coercive.[39] Julian Huxley's "Evolutionary Ethics"[40] shows many points at which biological, psychological, and historical components can be fused in a broad-vista stage-setting. Thus there is the wider background of the emergence of man with a particular biological equipment; the specific psychological mechanisms by which conduct is charged with feeling and an order maintained through the period of growth; and the historical shift in the very functions of ethics corresponding to the central focus of human problems in the light of the degree of social development—from an ethics of solidarity required for survival, to one of group-domination, to a contemporary task of providing opportunities and safeguarding the possibilities of mankind's future development.

13. Science in Theological and Metaphysical Existential Perspectives

Theological EP's, found in the ethical theories of the supernaturalistic religions, are characterized by the central role given to the concept of the supernatural or the divine. Western religions exhibit considerable differences in their picture of God's properties, man's derivation from God, God's intentions for man. This variety is considerably increased if we go beyond the Western tradition. The ethical potential of God's properties is often considerable. For example, usually God is taken to be more or less on man's side and to have a dependable character. One may conjecture how the ethical process might be conceived if God were construed as omnipotent enemy or as utterly arbitrary in mode of action—the latter a suggestion approached in medieval views of the primacy of will over reason in God. In some primitive religions supernatural action is in fact sometimes seen as arbitrary and even malevolent.

Despite their concern with the supernatural, scientific issues arise in many ways in such EP's, especially in assumptions about the character of human feelings, human responses to sanctions, human cognitive faculties, the causality of human predicaments, the influence of the non-human environment, and so on. The patterning of psychological or social or historical components within theological EP's is often sufficiently marked to differentiate them as theological-psychological, theological-historical, etc. And within each there are marked selective emphases. The score for the orchestration of the human feelings goes from the stress on calculated benefits of eternal happiness as reward for obedience, to filial emotions, and beyond that to refined feelings of respect, dependence, sense of finitude, guilt, anguish. Some recent religious stage-settings are almost wholly psychological in their content. On the other hand, traditional theological ethics has often used a wide historical stage, with God as author of the plot. The story varies from the Augustinian development from creation to resurrection to the belief of a Persian sect that the world is waiting for seven great goblets to be filled with human tears. Variety is also found in the sanctions imbedded in the modes of divine action in relation to man—from emphasis on reward and punishment here or hereafter to theories of grace and the ways in which it may occur. Problems of the extent of control are focused in the range of beliefs about free will and predestination. Questions of the possibility of man's knowing the divine take shape in rational, mystical, authoritarian-revealed and individual-intuitive religious epistemologies, with different imbedded conceptions of the character of the human condition. Theories of environmental influence also play their parts. Just as Augustine found it necessary to reckon with the belief that men's fate is written in the stars, so subsequent religious ethics had to shed the Augustinian picture of demons and come to deal with psychological forces in human guilt-phenomena. Similarly today, primitive religions moving into the modern world have to reckon with the impact of the knowledge of germs on their beliefs that illness had a moral punitive character. One of the chief analytic tasks in examining theological EP's is to sift out what are the scientific questions; for example, how much of the concept of sin as an experience of the sinner is covered by psychoanalytic accounts of guilt-feeling and its configurations. There are no doubt differences of orientation here between religious and secularist approaches to such questions, but the inquiry of the extent to which science penetrates can be common to both. It is worth noting that even questions of argumentation in theological accounts can be considered from this point of view; for example, how far in the variety of arguments for the existence of God is there reliance on specific accounts of the nature of the human intellect, of memory, of the character of knowledge (for example, of the absolute truth of mathematics), of the moral feelings, and so on.

What is the influence within a theological ethical theory of the specific answers given to all these scientific questions in its EP? Research has not been organized along this line of inquiry to any considerable extent. But the historical comparative picture of religious differences on ethical questions prompts the suggestion that the influence is a great one. The reach of science, however, may go even beyond this to deal in an indirect way with other parts of the theological EP. Thus a question like the existence and properties of God may be considered in terms of the scientific study of the ways in which these ideas function in the ethical process in the lives and thoughts of men. For example, a comparative study of theological EP's suggests that it is not the mere assumption of the existence of the divine but the specific properties of the divine that carry the burden of ethical functioning. The Kantian thesis that "it was the moral ideas that gave rise to that concept of the Divine Being which we now hold to be correct,"[41] taken out of its religious argumentation context and viewed as a historical hypothesis, points to the further thesis that the properties assigned to the divine being at any period have in fact been guided by concern with the type of ethical process which will be the outcome. This is a scientific-historical thesis of major importance in dealing with the relations of religion and ethics, but in any case its logical status is that of a purely scientific generalization.[42]

In similar fashion, it can be seen that *metaphysical* EP's do not go wholly beyond the reach of science because they are metaphysical—and irrespective of the nature of the metaphysics asserted. Let us take as an illustration the teleological types. Here the metaphysical elaboration predominates, although a religious element remains at the periphery. The world is set as exhibiting purpose in its structure, and man is given a place in the purposive scheme. He is pictured as fundamentally trying to go in a certain direction, whether conscious of it or not. The Aristotelian stage-setting puts man at the top in a plurality of natures seeking expression. It has biological, psychological, and social components, but is definitely non-historical. The Hegelian stage, by contrast, provides an organized historical progression according to a fixed logic. What the individual is, how he is composed, and what he strives for are referred to the wider pattern. Ethical processes are given a meaning and reference in terms of this framework. While the teleological outlook may be readily incorporated in a religious perspective, with deity prescribing the basic purposes—and was so incorporated in the fusion of the Hebraic-Christian and the Greek philosophical traditions—it must also be remembered that, in ancient philosophy, teleology was also the dominant form that science took, so that it constituted a stage in specific scientific endeavors.

Teleological EP's operate primarily with the notion of a *plan in things*. The plan has a locus either in every individual who has his own nature, or more often in the species every member of which is endowed with the same nature, or in the group or historically continuous people conceived as a single entity, or in the whole of the world as a unitary being. A mode of operation is also specified for the plan; theological teleologies often use sheer fiat or act of will, Aristotelian teleology has nature work like the artist or craftsman with the plan as a kind of immanent blueprint, post-Cartesian teleologies model the plan operation on their theories of the relation of mind to matter, and in idealist (Hegelian-type) teleologies the plan becomes the "logic" or spirit working its way out in the appearances of the world or its historical unfolding. The relation of the individual to the plan takes a different form in each of these types.

The core of the teleological EP lies in the way it employs the human-nature concept. Its concept of nature combines three strands: how an entity acts universally or for the most part (regular or lawlike behavior), what is native or primitive in its constitution (its unlearned behavior or the path of maturation and action in the absence of "distorting forces"), and what it is in some basic sense striving to attain and so (with an implicit definition of 'good' as goal in basic striving) what is the good for it. As long as we assume, say, in the case of man, a stable system operating according to plan, the accounts of regular mode of expression, inherent drives, and basically satisfying activity will yield correlated results. Temporary divergences or irregularities will be seen as accidents rather than mutations, as temporary lags rather than basic changes of direction.

Even the pursuit of evil will be seen as but a distorted way of pursuing the good. Men follow the apparent good, says Aristotle, which may not coincide with their real good; even the devil, says Aquinas, was not naturally wicked, and the will can tend to something only as a good to its nature; the next stage coming into historical being, according to Hegel, must be higher, because it is a next step in the movement of the World-Spirit toward the fulfilment of freedom. In all these forms, clearly, there is assumption of stability in the underlying plan. It is the feeling that this stability needs some grounding which no doubt accounts for the reaching-out of a formalized teleological stage-setting at its borders, either to a religious framework or to a scientific account of basic forces guaranteeing the system as a whole.

Now, whatever metaphysical form the teleological account takes, its picture of stability, of dependable human motivations, of the frustrations consequent on distorting forces, embodies numerous questions requiring scientific answers. (In this respect there is a clear isomorphism between, for example, the ancient teleologies and the contemporary concern with teleological mechanisms, self-stabilizing or homeostatic systems.) And with respect to the elements of a non-scientific type, the sociology of knowledge as a science can study their valuational functioning in the same way as was indicated above for theological stage-settings.

14. Science in Transcendence Existential Perspectives

Our concern here is with theories that disparage the role of existential assumptions in the internal workings of ethical theory and, in varying degrees, propound the autonomy of ethics. For example, G. E. Moore, in his early review of Brentano's *The Origin of the Knowledge of Right and Wrong*, says: "The great merit of this view over all except Sidgwick's is its recognition that all truths of the form 'This is good in itself' are logically independent of any truth about what exists. No ethical proposition of this form is such that, if a certain thing exists, it is true, whereas, if that thing does not exist, it is false. All such ethical truths are true, whatever the nature of the world may be."[43] In his *Principia ethica* Moore fashions the now-familiar concept of the naturalistic fallacy to brand any interpretation of 'good' in terms that are natural, theological, metaphysical, or in any way descriptive of existential entities.[44]

Such a purist approach is not, however, the property of analytic schools alone. In a phenomenological vein, N. Hartmann argues for a realm of value over and above the sensory scientific and theological-metaphysical domains and constantly stresses the purity of this transcending domain, accessible to an intuitive sensibility. In some forms of contemporary existentialism the focus is not on an object of transcendence so much as an act of transcendence, in which the self finds its absolute freedom in the process of deciding. By grouping such varied views (and others) together as a class of transcendence EP's, we imply that there is somehow here a common view of existence and that it furnishes a special type of stage-setting for ethics. But we have the task of showing that in such views the stage-setting is inarticulate or incomplete or even displaced and that, when uncovered, it is found to pose scientific questions.

In Kant, from whom stem most modern claims for the autonomy of morals—that morality is not a function of any existential situation but is unique or *sui generis;* that ethical processes somehow transcend existence and tell it what it ought to be—there is little difficulty in discovering an incomplete stage-setting. The autonomy stress is directed primarily against hedonistic views or any that make obligation a function of desires, passions, sentiments. But there is one feeling Kant wishes to maintain as ethically relevant; this is awe, or respect, which he denies to be a natural sentiment. He is perfectly ready to set broad existence conditions for morality, both in the portrayal of man as in tension between two worlds and in construing his account of the categorical imperative as an exhibition of man's rational nature. (In fact, the account is said to hold for all rational beings other than man if there be such in the universe.) The outcome is simply that Kant is setting a stage but is precluded by his own theory of knowledge from investigating it scientifically. Hence it remains incompletely presented, and there remain large gaps and obscurities in his ethics.[45]

As long as the classical ontological schemes were dominant, transcendence was teleologically tinged, and emphasis was on the object of transcendence to which the individual was rising, whether God or Truth or Reality or the Good. When Kant appointed himself receiver in the alleged bankruptcy of metaphysics, the object moved into the background to be veiled in obscurity, leaving the act of transcendence alone on the stage. It took many forms, beginning with the transcendental ego of Kant's theory. In some philosophers this was the opening wedge for a modern idealism. Sometimes the central focus on transcendence is quite explicit and deliberate. For example, T. H. Green, who stood in the full shadow of the evolutionary theory and its spreading applications in ethics, looked to the epistemological transcendence of man explaining his world, and the moral transcendence in the consciousness of a moral ideal, to stem the tide of naturalism.[46]

The pure-axiological ethics, whether cast in the phenomenological vein or in the British analytic vein, restores an object-orientation freed from traditional teleologies by relying on a cognitive transcendence act. The clue for finding the existential perspective in such ethical theories is to suspect a displacement to the mode of cognition.[47] Neither Moore nor Hartmann rejects the reference to the structure of existence in the *application* of ethics. For Moore, what is right is what will *produce* the greatest good, implying a causal ordering of existence. And, for Hartmann, the positive, as contrasted with the ideal ought-to-be, and so the ought-to-do, will depend on the tension between the structure of existence and the ideal domain. But, with respect to the cognition of the good or the ideal ought-to-be, Moore relies on a kind of self-evidence, and Hartmann on a kind of valuational sensitivity in grasping the ideal.[48] There is a spurious simplicity in their claims: the quality of goodness is there before you, it is self-evident, you either see it or do not, and there is nothing to argue about; the value is grasped by those who have the appropriate sensitivity, not by those who are morally blind or pass by on the other side of the street. It is this feature which prompts the search for existential presuppositions not in the further exploration of the object pointed to or the language of realistic Platonesque ideal entities describing the object but in the mode of cognition. The simple clarity of the insistence on non-natural qualities or ideal objects is unavoidably coupled with vagueness or obscurity in describing the mode of cognition. Hence arises the hypothesis—which I believe a full-length analysis of their ethical philosophies would substantiate—that the EP is to be found in some conception of the self and its cognitive-affective activities.

Contemporary existentialist and existentially tinged philosophies have focused most explicitly and most sharply on the act of transcendence. Reinhold Niebuhr's whole theological and ethical outlook rests on the capacity of the human spirit "of standing continually outside itself in terms of infinite regression,"[49] which indicates an essential homelessness of the human spirit; Niebuhr sees this as the basis of human freedom, creativity, and uniqueness. Although Sartre criticizes both the tendency in the later work of Husserl to turn toward a transcendental ego and the idealist-religious strain in other existentialists, he is himself building a concept of absolute freedom and responsibility on the ability of the individual in choice somehow to transcend all determinants and established guides.[50] Whatever our estimate of these pictures is to be, and whatever the outcome sought by their advocates, there can be little doubt that we have here, in the sense analyzed originally, an EP, but narrowed down to a particular view of the self in action—a striving of man beyond himself, a kind of self-extricating process.

Now the study of the self and its features is in principle a scientific question. It is, of course, possible that there are limits to its scope, but these could be discovered only in its study, not as antecedent postulates. Thus, no matter what form the transcendence stage-settings for ethics may take, such questions as the cognitive beholding or the affective sensibility of man, or, in turn, the aloneness of the human spirit, its activity of rising above or its endlessly regressing movement, its creation of a psychic distance or gap from its object of beholding, and no doubt many more types of subtle phenomena are all serious materials for scientific scrutiny both as phenomena and as elements that enter into ethical processes.

15. Evaluation of Existential Perspectives

The purpose of a comparative study of EP's is ultimately to stimulate the construction of a more adequate one for contemporary ethical theory. This involves evaluation, and so some marks of adequacy. Let us specify a set of terms for such a discussion. We shall call a feature that is appealed to in an evaluative inquiry a *reference point.* A class of reference points is a *standpoint.* So, for example, we can ask of an existential perspective whether it is clearly formulated and whether it is consistent; these are reference points, shaping up into or expressing a logical standpoint. A reference point becomes an *evaluative criterion* when it is assigned a positive or negative value in an evaluative reckoning; consistency almost always, and clarity usually, in a reflective enterprise has a positive value. A unified system of criteria we shall call a *standard.* Sometimes the standard is first on the scene and generates criteria; sometimes the criteria are first, fragmentary and scattered, and are unified by a theory of their relation or by some underlying purpose. We shall meet these concepts later in considering evaluative processes. Here we are concerned with suggesting some standards and some standpoints for evaluating EP's.

Logical standards.—These include criteria of conceptual clarity, consistency, and a variety of points of methodological refinement. These on the whole are sufficiently accepted in most contexts to be presented as a standard.

Many concepts playing a central role in the EP's sketched above lack conceptual clarity. For example, biological-psychological EP's that see human life as a struggle for power are notoriously vague in explicating the notion of power itself. The history of hedonism shows the ambiguities of 'pleasure'—whether a feeling, an act of preference, or a psychological surrogate for some pattern of social activity. Similarly, 'self-preservation' often enters the stage-setting as designating a biological tendency to keep alive, and then shifts in the ethical process to a psychological tendency to achieve one's ideals; only if the connections are established for the transformation by showing how the self grows and changes qualitatively in its effort to survive is confusion avoided, but this means enlarging the biological EP to a biopsychological type.

Inconsistency arises from conflicting properties of the entities involved in EP's. Hedonism has often been criticized for setting the human quest as maximizing pleasures; for, it is said, feelings cannot be summated. The problem of evil in traditional theological EP's in effect charges inconsistency in the description of God as omnipotent, omniscient, and good while allowing evil to exist.

Methodological refinement refers to the way in which a theory is constructed for use—how precise are its operative tests, how clear the relation of its parts, or the distinctions between what are data and what is interpretation, and so on. For example, the concept of human nature in the teleological EP's is often criticized on the ground that it has no dependable tests for the natural when its combined indices fall apart. It was realized after Darwin that the human-nature concept in its account of regularities is systematic-descriptive, not causal-explanatory; that the idea of what is native covers a number of different concepts—now represented by such ideas as instincts and drives, and still requiring careful analysis; that there is no a priori guaranty that any of these "natures" will remain fixed or that they may not contain incongruent tendencies or forces working against one another; that the idea of good may be linked in different ways to man's nature so that whatever is natural is not necessarily *ipso facto* good—in fact, it may be a residue of previous evolutionary developments no longer serving a constructive role.

In similar fashion, EP's that severely limit the span of human sentiments—as Hobbes's view of self-love as covering the generous feelings—often lack methodological refinement, failing to analyze adequately the logic by which other alleged sentiments are to be "reduced" to those admitted on the stage. Or theological EP's that speak in terms of God's will may fail to provide a means of determinately applying the concept.

Failure to distinguish data from interpretation characterizes especially many transcendence EP's that go from exhibiting the transcendence act on the part of the self to some ontological assertion about man and his world. But comparative study shows that the same phenomenon may be variously interpreted: it leads to an "environment of eternity" in Niebuhr's theology; is imbedded by Santayana in a materialist view of man as "a portion of the natural flux" with a moving center and maintaining equilibrium by striving for the ideal; or issues simply in a common-sense explication of "the systematic elusiveness of 'I,' " as Ryle carries it out.[51]

The logical standpoint is probably never wholly denied a place among criteria; and, of its elements, consistency is most likely to be admitted, for its absence is frustrating. The degree of clarity achievable and of methodological precision is sometimes taken to rest on the extent of human powers and the orderliness of the subject matter; if so, their admission as criteria becomes a partial consequence of the application of truth-criteria. This indirectly admits them in any case, where the discovery of truth is taken to be a scientific matter, since logical criteria are internal to scientific method. But if the stage is deliberately set to be obscure because the theme and action require candlelight and looming shadows, then it is likely to be associated with a metaphysics of irrationalism, which often, in spite of this, pays its tribute to logic by seeking to exhibit its own reasonableness!

Truth standards.—Since an EP purports to give a picture of ethically relevant aspects of the world, man's nature and condition, the accuracy of any part of the picture may be called to account, and, insofar as science has penetrated any field, to scientific account. Truth is thus imbedded in the aims of an EP; it is not merely seeking to give a pleasing account. (Of course, if the truth of life prove too horrible, some will always be found to shift the valuation and ask for a soothing delusion.)

Differences of degree and shading in truth-criticism are not without significance. An EP may be charged with outright falsity on the claim that its entities do not exist, as the atheist denies the existence of divine beings central to a religious ethics. Or the charge may be partial falsity or misintepretation of data, as critics of Freud's death-instinct would recognize the importance of self-hatred phenomena but deny their underived instinctive basis. Or the charge may be that of taking literally what is only to be seen as mythical or figurative representation—as stage-settings in terms of a "spirit of a nation" determining the focal duties of a people may be said to turn a symbolic expression for cultural tendencies or patterning into a collective ghost. Sometimes the charge combines truth-criteria with other standpoints. For example, the charge may be partial truth, misleading because of what is neglected; this includes also a failure in comprehensiveness. Or the indictment of "ideology" charges some falsity combined with the subserving of narrow or personal interests, sometimes with the element of self-deceit. (Here we sometimes find the language of criticism which speaks of a "false consciousness" interposed between the observer and the real phenomena, or the psychological language of a "screen," or the sociological analysis of a "lag'" in intellectual habits.)

Of special interest is a particular truth-criticism which we may label as "lacking independence." This characterizes an EP whose sole evidence lies in the phenomena of the moral domain. Thus Kant sets the stage with man as having free will; but, according to Kant, the only evidence for it is the demand for it implicit in the moral consciousness. Similarly, if an EP included the picture of man as inherently altruistic in a world of completely selfish action, and offered as evidence that the morality embodied the injunction to love thy neighbor and was meaningless without a deep-seated human-nature base for it, it would lack independence. This does not imply that moral phenomena do not constitute evidence, and one could conjecture about what a world would be like in which they would be the sole evidence. But, clearly, an EP that is independently established is in a much stronger position. Psychological and theological EP's have both been charged at various times with lacking independence, but their response has tended to be different. When it is claimed that the pictures of man as selfish or aggressive or cooperative and naturally sympathetic have been geared to the advocacy of particular social outlooks in the history of political theory, psychology as a science attempts to shed these elements as value-intrusions and insists on clearer empirical marks of the traits involved and, where possible, experiments to determine properties of man's nature. In religious theory, on the other hand, the growth of self-consciousness about the gearing of religious EP's to ethical demands (the hypothesis indicated above) has often been taken to point to the center of the religious outlook.

The standpoint of comprehensiveness or completeness.—One EP may be more complete or comprehensive than another. This is not so simple a standpoint as it seems, for there is always the question: comprehensive enough to accomplish what? We could answer by referring to "the tasks of ethics" and so speak of a standard of comprehensiveness; but, since the tasks have not been systematized in contemporary theory, it is more prudent to limit ourselves to speaking of a standpoint.

In general, the situation is analogous to that in a play: there is no virtue in sheer crowding of the stage, but the stage is not comprehensively set if the action requires props which are not there. There is a clear sense in which—granted they are both true—a stage-setting which describes cultural properties in addition to psychological properties is more comprehensively set for application to moral processes. On the other hand, an EP that is less comprehensive—for example, a physical or biological one —may maintain itself by a promissory note, that the qualities of human life and personal relations which are the common currency of ethics can be furnished by future correlation to physical and biological processes. This would seem to show that comprehensiveness is a function not of the number of entities involved in the stage-setting but of their properties and theoretical explanatory power.

Since comprehensiveness as a criterion is relative to some assumption of the jobs that ethics has to perform, a circle may arise which it is well to note explicitly. The conception of the tasks of ethics is itself part of a fully expanded ethical theory and so is probably to some extent a function of an implicit EP. Hence the point may be reached where a narrow EP justifies its own incompleteness by narrowing the tasks of ethics. A good example is to be found in G. E. Moore's ethics. In his denial of the relevance of the character of existence to ultimate ethical judgment he has in fact made the stage so bare that the only kind of action possible is ethical star-gazing, not ethical navigation. The tasks of ethics are narrowed to knowledge, not practice; few procedures are developed for transforming acts of vision into methods of guiding human action; and it is then found on the basis of the constructions offered that ethics is practically impotent to challenge any moral rule that happens to exist.[52] Another example on a large scale is the modern narrowing of ethics to personal rather than social issues. Its individualistic stages—usually of a psychological type—so narrow the conceived tasks of ethics that little relevant advice can be given to social policy.

Fortunately, from a theoretical point of view, the circle is rarely completely closed. The pressures that generate moralities are insistent enough to keep a wider notion of the tasks of ethics open, so that comprehensive adequacy remains a workable criterion. The comparative conclusion that EP's constructed in narrow terms tend to broaden out to cover a wider field—whether by shifting the meaning of their central concepts or by spreading wider their theoretical nets (as, for instance, we saw, Freudian theory moved out over social relations and even historical development)—and the historical lesson that theories offered first in a limited way tend to pick up supplementation (as Utilitarian ethics in the 19th century gathered a biological base and a historical scope in the evolutionary utilitarians) both tend to suggest that comprehensiveness can be regarded as a criterion implicitly admitted even where not explicitly acknowledged. But even in the extreme case where the circle described above has been closed, the issue is simply shifted directly to the varying conceptions of the tasks of ethics—a partly historical, partly psychological, partly policy-decision question.

Orientational and functional standpoints.—There are many interesting and potentially valuable approaches to EP's stemming from psychological, social, and historical study of their impact and modes of functioning. These are perhaps unified enough to be regarded as constituting an orientational and a functional standpoint.

The orientational standpoint asks questions that are perhaps less causal than phenomenological. How, if men set the stage in a given way, would they tend to feel and act in making ethical judgments? Would the EP carry with it an active outlook, or passivity and resignation, or a constant feeling of being shoved around? Would it prompt a rational process or a mustering of feeling and pressure? Do religious EP's prompt to resignation in the sense of consolation or of acquiescence? Do they have an authoritarian potential or a liberating quality? On a teleological EP a man may have the exalted sense of being drawn to the light. On a mechanistic one he may have a sense of being propelled, whether the forces be the impact of molecules, neuromotor reactions, conditioned reflexes, repressed desires, or propaganda. Karen Horney criticizes Freud's theory as leading analyzed people "not to take a stand toward anything without making the reservation that probably their judgment is merely an expression of unconscious preferences or dislikes" and as jeopardizing the spontaneity and depth of emotional experience.[53] The more comprehensive naturalist EP's convey the sense of a man in continual dynamic interaction with his surroundings, physical and social, and regarding his choices as a responsible fashioning of himself. In general, then, the orientational standpoint involves generalizations about attitude-tendency. It thus specializes in what we may call the *virtue-potential* of a stage-setting.

Such judgments are by no means simple. There are rarely one-to-one relations between the stage-setting and the attitude associated with it; fatalism may bring passive acceptance or a destiny-activism. The actual virtues that emerge in the ethical theory are a function of the whole picture and not of the stage-setting alone. For example, an EP whose general orientation is to realistic appraisal of existing conditions may yield stoical acquiescence in a world that is realistically discoverable as hard but quite different virtues in a world that is realistically found full of opportunity.

The virtue-potential of an EP may be explored with respect to each of the EP dimensions mapped above, such as the space-time features or the steadiness and intelligibility of the world, or it may be related to configurational properties, such as the qualities of being well equipped or at a loss, or the degree of room for action. William James treats a free-will doctrine in such a fashion when he identifies it as a general cosmological theory of *promise*.[54] Similarly, his attacks on the notion of the Absolute and his advocacy of a pluralistic approach were directed primarily against an attitude of a closed world to which one could only become resigned, in favor of an open world with room for human initiative.

There is no sharp break between the orientational and the functional standpoint of inquiry, but there is a marked difference in emphasis and direction. While the orientational standpoint is largely phenomenological, the functional is concerned with psychological, social, and historical roles and services and is more intimately related to causal-explanatory inquiries. Thus, for example, the question raised whether a specific theological EP involved acquiescence or consolation was orientational; if we probe into psychological relations and ask whether it releases guilt or intensifies it, whether it functions as a projective system or as a realistic facing of the totality of things, or into the history of religions and ask whether they serve as specific opiates or vigorous social organizers, we are following the functional path. Similarly, it is an orientational question whether a physical EP carries an atmosphere of promise or of coercion; but it is a functional inquiry, resting on causal analysis, to differentiate the optimistic eighteenth-century mechanism, with its promise of a clean slate and the remaking of man, from the twentieth-century pessimistic counterpart which identifies being a machine with being manipulated.

Inquiry along such functional standpoints is clearly scientific in type and method. All the relations between belief and feeling that a developed psychology may discover and all the lessons of the theory of personality and its mechanisms may have application in studying the psychological functioning of EP's. Comparably in sociohistorical functional inquiry we have to ask how the stage-setting fits into group aims, whose aims, and distinguish ascertainable real aims from apparent aims. Thus a hierarchical picture of the cosmos as a basis for ethics may in a particular age and society serve the social function of a fixed and stratified feudal system. Similarly, an equalitarian hedonistic stage-setting, with every man pictured as knowing best what will give him pleasure, may function socially to support and justify a laissez faire economy. Only a scientific functional analysis could explain why the teleological idea of the natural carries a conservative acquiescence pattern in one age and the idea of the native or inherent a revolutionary pattern at another time. Similarly, analysis would have to distinguish where the social conditions operated as cause of the EP's adoption, where a pre-existent EP was put to a fresh social use, and where the social function was itself a constitutive part of the EP.[55]

Whether such orientational and functional standpoints can generate evaluative criteria and standards is a distinct issue. In certain contexts they could quite readily; for example, where a given end was widely accepted, they could provide criteria for the *successful functioning* of the individual and the social field with respect to that end. In general, they can furnish standards only to the extent that there develop evaluative concepts of *a healthy personality* and at least minimum agreed-on concepts of *social well-being*. But these in turn presuppose well-established scientific theory, as well as some shared human purposes.

If the evaluation of EP's takes the indicated shape, the place of scientific inquiry and scientific results in the fashioning of an EP for contemporary ethics is a large one. The EP raises precisely the kind of questions for which scientific answers are required, even if the roster of the sciences of the day cannot yet answer them. How far an EP will also require some components of a non-scientific type (metaphysical, theological, etc.) is likely to remain the subject of controversy. The more sanguine may anticipate a situation in ethics comparable to that in biology, where non-scientific formulations were forced back outside the field into the status of philosophical speculations rather than active participants in internal decisions. The less sanguine may expect that limits to the reach of science in ethics will be discovered by science in its own forward movement. At a minimum, it is likely that the area of scientific answers will become increasingly central within the operations of ethical theory. For example, it will not be a metaphysical or theological interpretation of the human will, of emotions, or of self, that will bear the theoretical burden in ethics but the properties of such processes discovered in psychology and the social sciences, forcing their recognition on all alternative interpretations.

As to particular sciences, it looks as though most of the purely physical and biological EP's that have been offered as self-sufficient for ethics are inadequate in terms of the criterion of comprehensiveness. Even contemporary biologists have stopped talking of biological ethics in nineteenth-century fashion and think more in terms of biological bases for ethics. The psychological components in an EP are of central importance, but, whatever form they take, it does not look as though they can be cast purely in introspective terms or purely in behavioral terms or purely in phenomenological terms. The psychological concepts entering into EP's will probably have to represent configurations of various of these elements (including depth elements) resting on empirical correlations and theoretical unifications. The issue between a psychological EP and a social EP (whether the latter be sociocultural or sociohistorical) is not as yet conclusively resolved. If it is individualistic-psychological, it must embody a clear theory of interpersonal and social relations; and, if it is social, it must make unambiguous room for individual mutation and variation. Favoring the former is the importance of individual choice and decision as a central phenomenon in the field of ethics; favoring the latter is the fact that morality, like law or religion or science, is a social form.[56] But the underlying scientific issue is whether an individualistic psychology is really possible, or whether the inroads of social and cultural materials in the explanations of individual behavior and development do not call eventually for a more integrated model of man overcoming the present dichotomies. From this point of view, the formulation of a contemporary EP cast in scientific terms need not bind itself to the present mold of the sciences of man; it can follow in outline the direction in which it sees the sciences to be moving, although it may thus leave gaps in its detail and so perhaps be compelled to leave some theoretical issues in ethics with sets of alternative answers.

Again, it looks as though any form of EP will have to reckon with the fact of change in human life and, in dealing with every aspect—whether discourse and usage, feeling and motivation, institutions and cultural forms—be attentive to the possible impact in both the content and the structure of ethics of directional transformations. This means that eternities, universals, absolutes, structures, if they are to be offered in the EP, will have to take the form of invariants, grounded or evidenced constancies, methodological necessities, or even stipulated values.

Whether, if a social EP is decided on, it will have to be specifically sociohistorical or can remain generally sociocultural is also an undecided question. Probably it depends on the rate of change to be found in the human field. If there are structures relevant for ethics that have remained constant over the whole of human history and are likely to remain so, and, if they are central enough, then a sociocultural EP may be soundest policy. If there have been fundamental modifications in basic structure, then a sociohistorical EP is better policy. In any case, a sociohistorical EP would have to be formulated so as to allow for possible generalization across societies and epochs; and a sociocultural EP would have to allow for variations in specific detail and fashion subsidiary categories for determining relevance in time and place. Such questions of the extent of invariance and its grounds, although they point to different methodological policies in the present structuring of the field, do not constitute issues of principle. They are determinable questions of science and history, even though the answers may be long in coming into sight.

Since the concept of an EP as a systematic tool of analysis in ethical theory is only being fashioned, theses about such outcomes of evaluation cannot here be carried further. But, in the light of even the present state of knowledge, we cannot be satisfied with anything less than a unified EP which integrates the biological and psychological with the sociocultural and historical. The chief obstacle is the present lack of integration in the sciences themselves. This is, however, a scientific problem of which they are clearly conscious.

III. The Role of Science in Conceptual and Methodological Analysis

16. Conceptual-Methodological Frameworks

In contemporary ethical theory we often find questions of the following sort:

What are the distinctive terms to be employed in moral discourse (e.g., 'good,' 'right,' 'wrong,' 'ought,' 'virtue,' etc.)? How are they to be related and moral sentences formed out of them? (For example: "Is 'right' to be defined in terms of 'good' or are these two independent terms? If 'good' is used as a predicate term, is the subject-form unrestricted, or can only an experience properly be described as good? Are 'ought'-sentences really disguised imperatives?)

Can we properly define moral terms by non-moral terms? What kinds of relations are to be permitted between moral terms and descriptive terms referring to experiences, feelings, phenomenal qualities, contexts of human processes, and so forth? Are we to allow equivalence-definitions or other types of semantical rules? Or is the relation to take the form of some specified contextual function (e.g., to express feeling or to commend)?

What relations of moral sentences are to be permitted? Are there logical relations such as consistency between moral utterances or more generalized material relations of coherence? What form does organization or systematization take within a morality? Are there laws and systems of laws? Or, in some other sense, hierarchies of norms? Or rough collections of discrete decisions with at most family resemblances? Or other forms of patterning?

What modes of certification do moral sentences allow? Is it some type of cognition or empirical verification, or some type of feeling-sensitivity, or some type of willing or commitment-acceptance?

What modes of reasoning and justification are appropriate to the moral field? Is ethical reasoning deductive in form, or inductive? Or has it its own type of logic, with its own criteria of 'good reasons'? Is there a logic of choice as distinct from a logic of thought? What is the nature of the process of application and decision? Can decision be rational or do decisions ultimately "just happen"?

Answers to these and hosts of kindred questions about how the concepts and methods of morality are to be analyzed, construed, employed, and applied give us a fairly clear indication of the *conceptual-methodological framework* of an ethical theory. They furnish a kind of *logical profile* of that theory. Comparative study of logical profiles raises questions of their evaluation, and attempts to reconstruct the methodological framework of ethics call for framework policy decisions.

Our inquiry here is twofold: (*a*) How far are specific results of the sciences presupposed in the occurrence or adoption of a specific logical profile in an ethical theory? (*b*) How far is it possible to employ scientific method in ethics, as against regarding it as an overextension of a misleading model?

a) Scientific results were seen to play a large part in the constitution and evaluation of EP's; such influence carries over into frameworks if logical profiles take different shape according to the specific EP imbedded in the ethical theory. In the preceding chapter we started with EP's and suggested the scope of their influence; in the present chapter we start with frameworks, utilizing illustrations of framework problems, and look for points of indispensable EP reference. For example, one traditional kind of theory links the meaning of 'good' to *self-realization*, another to an act of *commitment*, another to *pleasure*. These may all be regarded as *EP variables*, since their meaning is furnished by the specific account of the self, of the will, of pleasure. (The picture of the self in self-realization theories of modern idealism differs markedly from that in contemporary psychoanalytically grounded ethics; even 'pleasure' in the hedonistic tradition has different interpretations.) The heuristic principle in the inquiries of this chapter is that whatever framework problem is analyzed we will find its answer to be in some significant measure grounded in the values assigned to the EP variables in the ethical theory. There is no one way in which scientific results in an EP enter into all frameworks; it is the task of comparative research to sketch the actual variety with which scientific approaches and judgments embodied in different EP's enter different logical frameworks. Nor is it being maintained that scientific considerations are the sole determinants of the logical profile. Comparable study would have to be directed to language-habits and to pragmatic or purposive elements.[57] But it is clear that they enter unavoidably with considerable determinative force and that they have accordingly an important place as grounds of policy decision in framework reconstruction.

b) Under what conditions would it be possible to structure ethics as a scientific enterprise, rather than as an art-enterprise, or a practical enterprise of some sort? Under what conditions would this not be possible?

For scientific method to be applicable in ethics, the human field must be sufficiently determinate to provide the following: (i) A concept of *moral phenomena* (including *moral experience*) sufficient to mark off an area of inquiry (observed qualities, feelings, determinate will-acts or behavioral processes or human interrelations in definite types of contexts) either as distinctively moral or at least as an area in which interpretation of moral terms is to be sought or pointer-readings for verification of moral statements to be discriminated. (ii) A set of moral terms and definite ways for linking them to the established area of moral phenomena. (iii) Some meaning for generalization or systematization in the reiteration of experience or phases of experience or some more complex invariance or descriptive patterning. (This is the condition of possible regularity or "law-likeness" resting on some isolability among the phenomena.) (iv) Some mode of verification or certification for the generalizations and some procedures of validation or justification for working principles of a higher order. (v) Some modes of application and decision so that the systems of generalization will have relevance to the practical tasks of morality.

Even if moral utterances were wholly expressive or "blind-volitional," scientific method in ethics would be possible if there is determinateness and lawfulness in the occurrence of the expressive utterances or blind-volitional acts. If a high degree of such regularity were found and a systematic explanatory theory developed, a concomitant or correlated descriptive use of moral terms could arise, just as constructs together with operations replace initial quality terms in any of the customary sciences. On the current philosophical scene, ethical theories of an emotive and prescriptivist type have tended to grow conceptions of validation or of "good reasons" even while insisting on the practical nature of ethics and the practical interpretation of ethical concepts.[58] This amounts to recognizing that there is a degree of determinacy in the field; where it is overlooked in one part of a theory it will come up in another part. In general, if a conception of the task of ethics as practical is offered, the question whether the practical task can be most effectively carried out by a descriptive or theoretical or practical interpretation of moral terms is not itself a *practical* but a *scientific* question guiding framework policy.[59]

The denial of the possible utility of scientific method has to establish the kind of conditions in the field—for example, an intrinsic arbitrariness in the will—which will rule out any way of satisfying the conditions stated above. Dostoevski thus points to "an interest which introduces general confusion into everything"[60] and speaks of the independence of the will at all costs, which may even mean a man acting against his own interests! To determine whether such an interest exists and probes deeply into the nature of man or whether it is a clinical symptom would appear to necessitate a scientific scrutiny of its bases of operation.

The five sections that follow deal with each in turn of the conditions for the possible use of scientific method in ethics. At the same time, however, the discussion of framework questions is oriented to discovering the pivotal role of EP variables in framework decisions.

17. Are There Workable Concepts of Moral Phenomena and Moral Experience?

Consider, for example, the following familiar utterances:

It was my responsibility. My conscience won't let me do it. It is obviously the thing to do. It was a strong temptation, but I resisted it. It was a strong temptation, so I succumbed to it. I felt that was no excuse. I recognize he has a claim on me but. . . . Here I take my stand; I cannot do otherwise. That's outrageous. What else could a man do and still live with himself? That's unfair. I sympathize with his predicament. Have you no scruples about doing it? Have you no compunctions? That would be giving up what I've worked for all my life. It was a courageous thing to do. You'll never regret it. That would be a wonderfully satisfying way to live. I was so ashamed of myself. I felt as if I had been dragged through the mud. Surely we are deeply committed to this. Would you want your child to be like that? It's not worthwhile doing. That was a fine experience. I can see that it is his ideal, but it does not attract me. It's a matter of simple loyalty. I did promise, so I shall do it. After all, I am a member of the group, so I shall bear my share of the costs.

Such a list could be continued indefinitely, moving off in different directions. There are hosts of simple valuings—enjoyments, delights, and satisfactions and their opposites. There are diverse sets of feelings—varieties of guilt, shame, awe, respect, indignation, gratitude, sympathy, care. There are classes of interpersonal reactions—admirations and recriminations, with hosts of finely shaded adjectives. There are apprehended qualities of experience, such as finding something congruent or fitting, or frustrating, or ominous and overshadowing. There are moral-model relationships, such as the experience one has in regarding someone as an authority or as an ideal-figure. There are reflective experiences, such as what one would have chosen if he had had a clearer view or been less excited or what he would have recommended if he had been more disinterested. In all these cases we could ask what kind of experience is taking place—what kind of tasting or perceiving or feeling or willing or reflecting. Or we could ask what kind of phenomenon is taking place—what kind of qualities are appearing, what object-relations existing, what personal relations being manifested, and so on.

How sharp is the demarcation and the articulation of this realm? There have been many attempts to specify a single mark of the moral, as if moral experiences when isolated would be as simple as the sound of a bell or as unique as the taste of a persimmon. Underlying EP's sometimes have a limiting effect on the kind of data explored. Thus the various individualistic psychological EP's in dealing with obligation experiences tend to look inside the individual, and come out with different types of guilt or shame or remorse feelings. Interpersonal and social EP's identify obligation experiences as directly transindividual and so concentrate on claims and counterclaims, rights and corresponding duty relationships, and recognition of whole-institutional demands. A comparative perspective is required to insure extracting the full range of phenomena. Moral experience might turn out to be a complex orchestral experience with many instruments playing and even with cultural variations in the score.[61]

The field of relevant phenomena falls into fairly distinct groups: (*a*) *the desire-aspiration-satisfaction* group, including acts of desire, striving, goal-seeking, aspiration, pleasures and satisfactions, and so forth (and their opposites), as well as the recognition of the objects of these acts; (*b*) *binding-authority* phenomena, including consciousness of demands and claims and ties, as well as feelings of remorse and promptings of conscience; (*c*) acts of *appreciation and depreciation,* including reflective reaction to persons and traits, acts and situations.

These groups provide sufficient basis for interpreting moral terms and for furnishing pointer-readings in verification. This does not have to wait for a fully developed account of precisely what falls within the moral domain. A carefully identified phenomenon—an act of approval, a feeling of guilt, a feeling of satisfaction under controlled conditions—can serve as a verifying instance for a particular moral statement, whether or not it ranks as a moral experience, just as a pointer-displacement verifies the presence of an electrical property without being itself an electrical event. However unsettled the precise marks of a moral experience, gathering a wide pool of possibly relevant experiences is a firm starting-point for extending scientific method in ethics.

18. Ethical Concept-Families and Their Existential Linkage

It has long been recognized that there are three major families of concepts in the ethical linguistic community. One, including 'good,' 'bad,' 'desirable,' and the like, may be called the *good-family,* although nowadays it is perhaps more common to speak of value-terms. The second, including 'right,' 'wrong,' 'ought,' 'duty,' etc., may be called the *obligation-family.* The third is the *virtue-family,* with its broods of specific virtues and vices.[62]

Some existential linkage for such ethical concepts is a necessary condition for the applicability of scientific method to the ethical domain. This is too often discussed as if it required equating each ethical term such as 'good' with some lower-order descriptive predicate such as 'pleasant' or 'is desired' and as if the rejection of such an equation ended the possibility of scientific method in ethics. This is an undue restriction on inquiry.[63] It is also unduly entwined with controversies over descriptivism and non-cognitivism, that is, whether an ethical statement is a descriptive report or serves some other function. The whole inquiry of the existential linkage of ethical concepts acquires wider scope by attentiveness to variations within the ethical tradition and to comparable problems in the philosophy of science generally.

Historically, different ethical concepts have always been closely associated with the various groups of moral experiences indicated in the previous section: the good-family with the desire-aspiration-satisfaction group, the obligation-family with the binding-authority phenomena, and the virtue-family with at least a large part of the appreciation or reflective-approval group. Hereafter, let us use the term 'domain' to cover a group of phenomena as associated with a family of concepts. We have thus the good-domain, the obligation-domain, and the virtue-domain.

We can also note historical shifts in the dominance of the concept-families. In ancient times ethical theory confidently assumed that, if we knew the human good, everything else would fall into place. Medieval ethics seems to have thrust contractual ties and mutual obligations into a more prominent theoretical position. Kantian and post-Kantian ethics have intrenched the concept of obligation as almost definitory of ethics. Virtue concepts made inroads in ancient times by such devices as the Stoic construction of virtue as the primary content of the good and in some modern periods by the central place given to the moral ideal of character and personality. The historical careers of the concepts appears to reflect the relevance of different moral experience groupings to the institutional and historical problems of the day.

Comparative inventory of analyses offered in various theories for each of the central concepts shows that there is always a reference to some portion of the field of moral phenomena. For example, we find obligation analyzed as

a voice of veto or command (Socrates' demon, or typical accounts of conscience as a still small voice)
a sense of "office" or a job to be done (Stoic)
awe or respect for law or rationality (Kant)
a sense of overwhelming pressure (one of Bergson's two senses, assimilating it to habit)
a sense of aspiration or attraction (the other of Bergson's senses, assimilating it to the ideal; Plato's analysis of the tug of the Good)
a sense of a governing whole, or a choice by the whole self (Bosanquet, modern idealistic philosophy)
a type of debt (Nietzsche)
a contractual-type of commitment (Socrates in Plato's *Crito*)
a type of reasoning directed to maximization, or to harmonizing conflicting aims (e.g., Bentham, Santayana)
a type of sentiment, such as a pyramiding of sympathetic responses (e.g., Adam Smith)
a sense of loyalty or commitment (Royce)
an anxiety embodying developmental derivatives (Freud)
a vectorial quality of requiredness (Köhler, Mandelbaum)[64]

and, of course, many other ways. The formal features of each conception seem to reflect the material properties of specific moral experiences, in some cases with rich content, in others with only abstract outline.

Analytic controversies, historical considerations, and comparative lessons combine to suggest that it is worth differentiating more systematically the various ways in which ethical terms may be linked to existential entities, qualities, and procedures. Well-developed accounts of such problems in the philosophy of science prove helpful. We find at least five different types of linkage which throw considerable light on possibilities in ethics.

(i) The equation of an ethical term with a descriptive predicate so that we have the necessary and sufficient conditions for applying the term. This is currently rejected as not feasible, but it must be allowed to remain as a possibility. For example, on a descriptivist approach there might turn out to be a unique phenomenological-field property, such as requiredness, equated precisely with a basic use of 'good' or 'right' as a fundamental term. On a non-cognitivist position—say, an expressive or emotive theory—there might turn out to be a unique feeling or emotion for each different basic ethical term (e.g., horror for 'wrong,' approval for 'good'). Although not formulated logically as a definition, it might be offered as a precise model for analyzing the situation in which the term is properly employed. Lewis Feuer proposes that, "corresponding to different social structures with their different personality-forms, there will likewise be diverse ethical languages each with its specific psychoanalytical characterization."[65] For example, for 'This is good' he suggests: for a liberal society, "I like this, and, since we are so much alike, you probably would like it too"; for a Calvinist society: "I dislike this, but was compelled by my father to accept it; now, having identified myself with him, whatever resentment I harbor against him will be deflected toward my own children, who will suffer as I did." Whatever the adequacy of such an account, the attempt itself shows that the search for definition surrogates for ethical terms is not to be barred by an expressive non-cognitivism in ethical theory.

(ii) Non-ethical terms may be offered as an interpretation or model for a system of relations in which ethical terms have already been elaborated, as a physical model may interpret a set of geometric postulates. Though ethical theories have not explicitly employed such procedures, analysis of the relations of ethical terms is sometimes of this sort. Let us construct a simple hypothetical example, with three fairly familiar stipulations about the mutual relations of selected ethical terms:

1. If a person has a *duty* to perform some act, there is some person or persons who have a *right* to its performance.
2. If a person has a *right* to some act, then there is some *good* which he will derive from its performance.
3. Every *duty* has a *ground* in some character of the particular situation or previous situations.

One model which readily suggests itself is that of *debt*. To have a duty may be interpreted as to owe a debt; to have a right is to be a creditor; a good is some proprietary object or service which is the subject matter of the transaction. The ground is the "value" conferred in the loan. This model appears to satisfy the three "postulates." Whether it proves adequate in the long run depends, of course, on the relation of these three to the rest of the ethical theory—additional stipulations, factual assumptions, etc. And, since these are generally rarely worked out explicitly, there is a large area of arbitrary employment of models.

That an ethics in terms of debt can be worked out in detail is clear from its actual occurrence, for example, in Japan, as described by Ruth Benedict in *The Chrysanthemum and the Sword.*[66] There are infinite debts and finite debts, debts repayable only in kind and those repayable in money, etc. Note, in general, how a debt ethics would have to reconstruct familiar obligations in our society. The first postulate demands that for every obligation we find a creditor; the third, that it involve a present or past value received. Thus social obligations would have to be construed as debts to God, to ancestors, even to society collectively for its role in fashioning the individual. Obligations to the future are ruled out, because future generations have not given us a value received. Presumably they could be construed as loans; actually, in Japanese ethics obligations to one's children are seen as repayment of debts to one's parents.[67]

It is, of course, possible that an alternative model may be offered for the same postulate set. For example, *contract* would also be possible, since every debt itself could be regarded as a contract, with the parties as debtor and creditor, the object intended the good, and motives of the parties the ground. A contract interpretation would probably considerably liberalize the obligation system; historically, in the Western world, it has involved a wider individualism.

(iii) A third linkage of ethical terms to existential materials may be compared to the role of operational definitions. An ethical theory may co-ordinate '*X* is wrong' with a first-person introspective operation such as 'When I contemplate myself as having done *X*, I feel remorse.' This does not give the full meaning of 'wrong' but provides a means of partial identification after which fuller exploration of the material identified can then be carried out.

Some traditional ethical analyses may be construed along these lines. In the good-family the test of what is desired for its own sake is perhaps the most prominent. If this is thought of as a kind of operational definition, then we can see the specification of normal conditions—that one must not be in a disturbed condition, or ignorant of what one is doing, or have built up contrary habits, etc.—as comparable to making sure that measurements of length are reckoned at standard temperature and pressure. In the obligation-family the repeated attempts to refine feelings of conscience and remorse, to differentiate them from fear of consequences or pain of loss or hurt to self-love or self-esteem, may be seen as sharpening "pointer-readings." In the virtue-family there was developed the notion of the "ideal spectator" who has his sympathetic and other reactions in a cool hour or from the vantage-point of a "disinterested" observer.

(iv) A fourth mode of linkage lies in the discovery of *empirical* indices correlated with the application of previously established linkages for ethical terms. Suppose 'wrong' is linked to the (operational) remorse procedure. It might then be found that, in some domains of publicly performed actions, social disapproval was a fairly regular accompaniment of acts which produce remorse. Without confusing remorse with fear of social disapproval, it might still be possible to use social disapproval as an empirical index for 'wrong' in that domain.

An operational specification may be outworn and take its place as an empirical index for limited domains, as when a child acquiring a wider understanding of obligation than simply 'what his parents emphatically demand of him' may continue to use this test as at least a first index of duty. Or an empirical index may prove more reliable once a general theory has been developed and so take the place of the original operation with which it was empirically correlated, as with most reflective persons the feeling of remorse itself becomes more of an empirical index than an identifying operation, and the fact that we have hurt others may move more into the role of operative test.

(v) A fifth type of linkage may be extracted from the work of the linguistic analysts, in spite of the fact that they usually regard it as exhibiting the practical rather than the scientific character of ethics. This type joins ethical terms with functional contexts. Ethical terms are treated as having jobs to do; different ethical expressions may be doing similar jobs or the same ethical terms doing different jobs in different contexts.[68] For example, 'ought' may have the function of criticizing, advising, deciding; it may have different significance where a person is deciding what he is to do ('I ought') and where he is specifying a role ('A man in such-and-such a position ought to . . .'). Ethical terms may vary in contexts of general social legislation and individual operative applications, peer-age groups and cross-generation groups, action-problem contexts and educational policy contexts, as well as spectator contexts and participant contexts; and so on. A realistic job-mapping involves a thorough understanding of what is going on in a field in all its institutional and cultural background. In a sense, this becomes the application of an anthropological orientation to the analysis of specific moral patterns.

The five types of linkage by no means exhaust the possible existential connections of concepts. All sorts of other types may play important parts. For example, in frequent explications of 'duty' as an action that is productive of the greatest good, material concepts of consequences or effects enter into the very relationship of the two ethical terms. And in such a familiar slogan as "'Ought' implies 'can'" there is the suggestion of necessary material conditions for the application of an ethical term, already canvassed in the conception of a stage-setting. But, whatever further developments are possible, the five types indicated open a wide path for a treatment in ethics quite close in spirit to that of scientific inquiry.

19. Organization, Generalization, Systematization

Because the familiar concept of *moral law* has occupied a central position in modern ethics, an effort is required to look at the domains of moral experience and their conceptualizations and ask whether different patterns of organization are possible and appropriate and where lawlike generalization fits or does not fit. Let us examine the possibilities in each of the three major ethical domains.

(i) The good-domain has experimented with several organizations, reflecting the particular stage of scientific study of its phenomena—will, desire, feeling—in short, the complex history of psychology. The *means-ends* category emerges as the general way of systematizing goal-striving. Under classical teleological psychologies it is given a *hierarchical* specialization with the supreme end as the highest good. Obligation phenomena fall into place as means to achieving the good, and virtues represent character-traits similarly oriented. Comparative value is identified with degrees of completeness in achieving the good. Generalizations are possible within this scheme; they are universal or for-the-most-part statements about men's striving for goals or the frequency of success of means in leading to ends. As teleological underpinnings (in fixed human-nature metaphysical EP's or in theological EP's) give way in modern times, this scheme of organization is somewhat transformed. With the removal of fixed ends, the hierarchical character suffers. A number of different types emerge. Some experiment with *part-whole* relations, as in the idealist organic philosophies in which the completeness of the whole replaces the final end. Some develop the concept of *ideals* as organizing and guiding foci. Some turn to the biological aspects of the scientific picture of desire phenomena and fashion concepts of *underlying drives* or *guiding propensities.* A great many model their mode of organization on satisfaction phenomena; hence the great variety of affective theories of value. Of these, pleasure theory is the most prominent, developing its ethical concepts on what it takes to be a universal theory of motivation. Comparative value becomes a scale of measurement of greatest happiness, or maximum preferences. Obligation phenomena are attuned to maximum-generating instrumentalities, and virtues become the safest educational self-investments yielding the steadiest happiness-return. Finally, on the contemporary scene, as new conceptions of the relations of phenomena arise, altered schemes are proposed. Scientific conceptions of homeostatic or self-stabilizing systems have provided a model for combining both goal-seeking and regulative aspects in a more coherent way, although unable as yet to integrate the phenomenal and the affective or to conceptualize shifts in basic goal-direction that are found in human life in historical change. In any case, it is becoming increasingly clear that an adequate organization of this area can follow only on an adequate unified theory of man.[69]

(ii) In the obligation-domain the dominant conception of

moral law traditionally fused authoritarian command, rational order, and universal generality. (In Kant, for example, the content of command turns out to be to universalize, a rational being is identified as one who governs himself by law, and the form in which morality issues is that of laws.) This particular amalgam may represent a modern Western specialization in the pattern of conscience. However, an authoritative command need not be universal in form.[70] Rationality permits of a variety of forms and even involves some appreciation of the conditions of uniqueness. And the very idea of moral laws in the sense of obligation-universals is only one of a variety of forms that obligation phenomena and their analysis may bring to light.

If laws be studied as types of rules, then at least the following may be distinguished (using in explication a prohibitory predicate):[71] (*a*) *Must-rules*—the act is never to be done, could not conceivably be justified (cf. religious notion of utterly damning, moral notion of infinite obligation as contrasted with finite, ordinary notion of utterly unforgiveable act). (*b*) *Always-rules*—this act should never be done in any case that will actually occur; although conceivable conditions might justify it, they will not take place. (These are acts of finite but very high obligation.) (*c*) *Phase rules* (or *break-only-with-regret rules*)—the act is to be regarded as diminishing the result in any value-reckoning; it is a negative weight but may be outweighed; even when outweighed the drag of its weight remains.[72] (*d*) *For-the-most-part rules*—rough frequency with which the act turns out to be wrong. There are no doubt other types which a careful survey of different moralities would discover; e.g., Linton calls attention to "Do not do this, but if you do it, go about it in this fashion."[73]

The many notions fused in the moral law concept and the diversity of forms point up the scientific issues underlying the occurrence of given organization patterns. We can then investigate what degrees of determinateness in the field of moral phenomena are required to support what kind of rule-type and, in an evaluative enterprise such as working out standards for a given field, pose the question which should be strict standards, which flexible, which partial, and so on.

(iii) Since theories of virtue and vice are centrally anchored to the phenomena of character and character-formation, organizing schemes in this area reflect directly the underlying EP's. Plato interprets virtues as the qualities of different parts of the soul, fitting the whole for successful pursuit of life's quest. The medieval religious outlook sees them as qualities of spirit unified by acceptance or turning away from divine will. A Hobbesian materialism sees them as traits reasonably directed toward maintenance of peace and order. A Benthamite utilitarianism sees them as unified in a prudential pursuit of the general happiness. A Marxian materialism sees them as the character-types fashioned by the productive processes and relations of man's material life and selectively reinforced in the light of dominant class needs. A Nietzschean voluntarism sees them as the will asserting or thwarting itself in a heroic or Hebraic-Christian configuration, respectively. A Millian liberalism sees them as forms of reasonable individual initiative in the human pursuit of general happiness. While an objective idealism finds the unity of virtue in the historical growth of the moral ideal, a Deweyan emphasis on the role of ethical reflection in mediating change sees virtues as methodological qualities in the successful pursuit of human interests. In contemporary scientific studies of virtues the influence of underlying scientific theory on organizing ideas is explicit. Behaviorist trends see virtues as habits established in manifold areas, endowed with no central unity, bound to specific contexts of application and sanctions. Phenomenologically inclined psychologies look for phenomenal invariants, general essences, to differentiate, for example, sympathy from pity and mutual respect from mutual advantage. Psychoanalytical psychologies look for character-types reflecting the growth of personality and the typical deviations and distortions at different stages. Culturally oriented psychologies look for personality patterns (each with its specific virtue-set) corresponding to dominant culture patterns transmitted in socialization (educational) processes. In all these cases the same lesson is reached for the virtue field as for the previous ones: just as 'means,' 'ends,' 'moral law,' and the rest will be found to take their shape from the theories of human goal-seeking, appetition, regulative volition, and so forth, and the conditions of their exercise, so categories for understanding virtues express the answers given in the psychology of personality and its development.

(iv) With the development of twentieth-century linguistic-logical analysis, many of the formal framework questions have been brought into sharp focus. Two tendencies have been found in this analysis. One is to carry out as formal a logical reconstruction as possible: to formulate explicitly syntactic and semantic rules and, where possible, to provide a deductive system adequate to the field. The other analyzes actual linguistic usages, not to develop powerful logical instruments, but rather to render linguistic habits explicit and remove paradoxical cases.[74] In the carrying-out of both types of analysis, a point is reached where alternatives have to be faced. The formalist may develop different whole systems of deontic logic, and the ethical theorist will have to decide which to use or which is most adequate for ethical inquiries. Similarly, the informalist may be faced with conflicting patterns of linguistic usage and so have to decide which is correct or which is more serviceable in the light of specific aims.[75] At this point, questions of existing language-habits or conceptual patterns, questions of aims or purposes in the specific field, and factual or scientific questions about the materials in the field arise. How relevant is the third consideration for decision in formal framework questions? The brief illustrations that follow are from issues that have at various times agitated ethical theory.

a) One problem concerns the subject-type to be permitted in sentences with ethical predicates, e.g., of the form '*X* is right (wrong),' '*X* is good (bad).' If we line up some alternative answers we find:

Only a will-act or decision can be right or wrong (Kant, Schlick).

'Right' should be used only of behavioral acts, not act plus motive (W. D. Ross).

Moral judgments are not really passed on intentions but on persons (Westermarck).

Any kind of thing can be good.

Only a conscious experience can be good; everything else can be only a source or cause of good.[76]

How can we decide, whether in determining the syntax of ethical sentences or in determining correct use, which path to follow? The kinds of arguments used throw some light on the issues. Schlick thinks only decisions deserve the name of conduct because they are in fact the direction-molders of human life. Presumably, if this assumption turned out false—if decisions simply registered trends that had already stabilized themselves—this change in the psychological theory of the role of the will would remove the ground for this selection of subject-type. Westermarck is explicitly appealing to a psychology of the emotions in moral judgment. Those who insist on experiences as sole possible subject for 'good' may already have in mind a specific interpretation like 'pleasure' or else be operating in general with an EP in which value is 'subjective.' Ross offers his view as a proposal—that 'right' be confined to behavioral *acts* and that 'morally good' be used for *actions* (differentiated as an act done for a certain motive)[77]—for two chief reasons. First, good consequences and good motives do not always coincide; presumably, the refinement would spare us the paradox of saying "He did a wrong act" when a man helped another from a selfish motive such as reward. (But this introduces the possibility of other paradoxical-sounding expressions, such as "It is morally good to do such-and-such a wrong act.") Second, Ross finds it difficult to allow that it is my duty to have a given motive or do an act from a given motive, since it cannot be my duty to do what is not in my power. Hence the psychological assumption that motives are not within our power is one of the grounds in his stipulation of subject-type. Without entering into the psychological question of such control, and the parallel problem whether men can be told to control their feelings—which carries us into the whole character of repression and its effects in human life—it is clear that the decision on appropriate subject-type here becomes to a great extent a function of the state of psychological knowledge and control.

b) Another type of problem concerns the mutual relations of ethical terms. Take the perennial issue of the relation of the right and the good. Here again, no matter how we approach it, we end up with alternative accounts to which we may affix labels:

The *goodist* structure: 'right' is to be defined as 'productive of the greatest good.'

The *rightist* structure: 'good' is to be defined in terms of 'right'; for example, to call anything good is to say that it is what a good man would choose, and a good man is one who does what is right.

The *co-ordinate* structure: 'right' and 'good' are separate and co-ordinate terms; propositions relating them are synthetic; any interpretations for them are distinct.[78]

The formulation "What is the relation of the right and the good" telescopes several different types of inquiry, which can be untangled by looking at the sort of considerations deemed relevant to decision. It cannot be a purely analytic problem if the three structures are all found in use, and if—as seems likely—each can be extended to cover the whole field of preanalytic phenomena.

Descriptive issues enter either directly or as dominating considerations in an evaluative proposal. For example, the great strength of the co-ordinate structure probably comes from men's recognition that this corresponds to the picture of their phenomenal field: they experience consciously a conflict of duty and interest rather than a conflict of two interests. Those who seek to relate the right and the good may question this phenomenological picture either by presenting a different one or by questioning its ultimacy in terms of an explanatory account. G. E. Moore would say that what he means by 'right' is what is productive of the greatest good; or the same goodist formulation is cast in a linguistic mold by Nowell-Smith when he takes 'good' to be expressive of a pro-attitude and argues that "pro-words are logically prior to deontological words" in the sense that "they form part of the contextual background in which alone deontological words can be understood, while the reverse is not the case."[79] If there really are two different phenomenological pictures, then the issue between them has to shift to an *explanatory* account. Each side will offer its own scientific (probably psychological) theory. For example, Westermarck argues that the concept of duty can never be reduced to that of goodness, because the former springs from the emotion of moral disapproval while the latter springs from that of moral approval, and—this is the crucial point—he finds these to be psychologically quite different in their sources.[80] On the other hand, Utilitarianism seems to have a complex logic supporting its goodist structure. The full theory of human psychology and social and historical development, it is believed, will show obligation experiences in all their detail to be a function of the pursuit of aims by men, under specific conditions of development, with more important paths of striving becoming associated with more demanding phenomenal qualities and more stringent feelings. Where the obligation experiences go off on their own in contrast to value experiences, it may be construed as a distortion, or psychological lag, or lack of clarity about aims, as well as (perhaps most importantly) the conflict of aims in different groups on the historical scene. Therefore, the goodist structure is taken to represent the most reasonable long-range interpretation of the dynamics of moral phenomena.

Clearly, these issues express the conflict of different EP's, and it is the answer to the scientific questions which is prerequisite to the theoretical decision. When the issue is eventually posed as an *evaluative* one, questions of the purposes that ethical theory and structures of this sort serve will also, of course, be relevant.

c) As a third illustration, take the question of the correct logical form for ethical sentences. In the last few decades, we find such views as:

"Killing is wrong" is really a way of saying "Thou shalt not kill."

"Stealing is wrong" is to be interpreted as "Stealing!!!" where the exclamation marks indicate an expression of horror.

"Friendship is good" is to be interpreted as "Would that everyone desired friendship."[81]

These were only a beginning; they were followed by the full force of emotive theory and a variety of increasingly sophisticated formulations.[82] In contemporary theory the range of these alternatives tends to be worked on in two ways. The formalists construct logical systems, with each of these notions providing the central undefined concept. Thus we can have a logic of imperatives, of optatives, of permissives, and so forth—not excluding a logic using the indicative form with such a concept as 'better' in a central position.[83] The other path is to map functional contexts corresponding to each particular alternative—indicatives instruct or inform, imperatives affect the will, optatives influence aspiration, and so on. Whether the work is done in the formalist or the informalist mode, the outcome is a set of alternatives, and the issue of decision once again has to be located as either analytic, descriptive, causal-explanatory, or evaluative.

Once again, the analytic approach discovers and works out alternatives but cannot decide among them without reference to some aims or some factual assumptions. To say that indicative formulations are really an unclear way of expressing commands is to claim that contexts of indicative ethical use will in fact be found to be command contexts. Or else the analysis may be making a methodological proposal that ethical utterances be construed as commands because this relieves us of having to regard ethical utterances as true or false. (Defenders of the ethical indicative would regard this as begging the question, and some logicians have even tried to construe imperatives so that truth values could be assigned to them.)[84]

Inquiry is compelled therefore to go into descriptive and explanatory questions, either to find existential interpretation for the central terms of formalist systems or to explore the human activities in the contexts selected as distinctive by the informalists. In either case we are carried into psychology—the nature of commanding and the authoritarian situation and the type of personality involved; the nature of wishes, idle and efficacious types; the place of intelligence and knowledge in motivation; the ways of influencing people and the respective force of subtle urging and providing insight; and so on. Similarly, we may be led into the social sciences—the actual moral codes in different peoples and cultures or subcultures and social strata; the extent to which they take the form of, so to speak, an indicative morality, an imperative morality, an optative morality, and so on; and the social, cultural, and historical conditions under which they develop.[85] Such scientific study would give us a wider basis of knowledge for assessing the different forms. We would know how far a given form of discourse—say, again, the imperative type—arises in or is sustained by authoritarian institutions and repressive psychological techniques, how far it is part of a special personality-structure, how far it successfully maintains that structure or touches off inner rebellion. We would also have educational generalizations about the comparative effects of commanding, or showing and predicting consequences, or cultivating responsibility through common modes of decision, or a host of comparable alternatives. As a consequence, our deferred methodological policy decision of the appropriate logical form for ethical sentences would represent the result of an evaluation in the light of assumed aims or purposes, on the basis of the results of scientific inquiry. Hypothetical alternatives may be envisaged. For example, if we found that always and everywhere morality embodied a dominant concern with repressing of otherwise irrepressible aggression, that the most successful technique of repression was peremptory will or command, that the imperative form of discourse touched off this technique—in short, if we took one of the traditional pessimistic views of human nature—then the imperative structure of ethical sentences would be both natural and represent sound methodological policy. If we took the more optimistic liberal assumptions about man, his development and prospects, the imperativist form might very well represent an authoritarian vestige, and some aspirative form—optative or indicative, depending on the role we saw knowledge taking—a sounder policy. And so on. And decision among these views is ultimately a scientific question.

The three illustrations of formal framework questions and

their character show how—though not in any one or uniform way—scientific considerations are relevant to decision no matter in what terms the decision problem is initially cast. The discussion of justification modes that follows continues this lesson in a kindred area.

20. Validation, Verification, Reasoning, Justification

These terms tend (with considerable shifting and uneasiness) to be used as follows: 'validation' in general for an exhibition of adequacy; 'verification' where the crucial appeal is to some form of experience; 'reasoning' for some legitimate form of inference; 'justification' in a blanket way for any acceptable direction of appeal under criticism. Let us use 'justification' as the generic term. The salient justification-candidates in ethics have been (i) intuition of a universal, (ii) perceptual disclosure of a particular, (iii) deductive derivation, (iv) inductive establishment, (v) normative (persuasive) success, and (vi) furnishing good-reasons.

(i) *Intuition of a universal.*—Intuition is generally mistrusted in contemporary philosophy as a mode of grasping truth. But intuition-claims have covered a variety of possibilities requiring separation. In ethics we find:

Stipulative definitions, such as "Courage is virtue with respect to fear and confidence." (If this is regarded as lexical definition, reporting usage, it is not, of course, intuitive.)

Abstract schema with blanks to be filled in. "Justice is giving every man his due," if not stipulative definition of 'justice,' is almost a sentential function, with 'due' a blank to be filled in by some apportionment-criterion.

Analytically true or false statements. "It is wrong to misuse social institutions," which Baier offers as part of absolute morality,[86] seems analytic, since the idea of 'misuse' already includes that of using wrongly or badly. (It may be an empirical invariant, if what it means is that every actual morality will be found to contain such a conception, to which it will invariantly assign a negative value.)

Theorems in insufficiently explicit systems. "To prefer a lesser good for one's self rather than a greater good for another is wrong," often urged as axiomatic, is usually derivative from certain definitions of 'good' and 'wrong' whose effect is to rule out the expression 'my good' and make all size estimates of good impersonal.[87]

Basic or abstract value-affirmations. Even so apparently methodological a proposed intuition as "Hereafter *as such* is to be regarded neither less nor more than Now"[88] in effect is recommending a rational attitude oriented to the totality of life, for which there are conceivable value alternatives. More specific basic intuitions turn out to be vehicles for special forms of life or social organization: e.g., a man is entitled to the full produce of his labor, or a man in distress ought to be helped. In the long run, basic value-axioms are assessed not by one's intuitive response but by the system of life they organize, which is a more complicated evaluative process.

Phenomenological reports. Some contemporary claims for intuitive moral universals of a lower order—e.g., "Cruelty is wrong"—insist that self-evidence characterizes the object of beholding; there is no appeal to a mode of intuiting as a process that of itself guarantees reliability.[89] In that case, self-evidence is best interpreted as a phenomenological property of the object in the field, like a man having an honest face or a sinister look. In this sense, 'intuitive' does not carry the connotation of true or self-justifying. Thus "Cruelty is wrong" as self-evident either would be a single-person datum concerning his phenomenological field or would involve the further claim that all men would give the same report for a similarly structured field. (Compare Karl Duncker's argument that the same valuation will be found for all phenomenal fields in which the meaning grasped is the same.)[90] This is no longer a simple intuitive inquiry but a transcultural one in which phenomenological reports play only the role of verifying observations—in short, a scientific inquiry.

Whatever interpretation is taken of a given intuition-claim, it therefore requires further justification along one or another of the lines indicated.

(ii) *Perceptual disclosure of a particular.*—A long tradition from Aristotle's day takes a singular moral judgment to be certified in a kind of perceptual act, enabling it thereafter to serve as a hard datum in verifying general propositions. Here again we find several types:

Immediate affective *acts* or *states* as moral phenomena: for example, being pleased or feeling obliged. As acts or states, they require conceptualization and interpretation. Hence they are not hard data but can in practice be refined into serviceable pointer-readings.

Immediate *reports* about the occurrence of affective acts or states. These present conceptualization and so are in principle corrigible. For example, a report that a person is pleased opens the way to considering whether it is pure pleasure or relief from anxiety or any of a variety of qualities that a developed psychology might find. Hence whether it is a simple pleasure-report or a more complex obligation report or commitment report or any other phenomenological report, it becomes a reliable cognitive disclosure only to the extent that its significance is determinable by reference to scientific theory.

A synthesis in the particular situation of complex factors apparently not capable of prior analytic separation and compounding: a man grasps directly in a complex situation where his duty lies. This is perhaps the most typical and also most useful of the claims for intuitive particular disclosure. Here the procedure is less like that of scientific verification by hard data than that of application involving selective sense of relevance and synthesis of claims arising from different features, together with an ability not to overlook what may be pertinent. This operation, like that of the engineer or the judge, is necessary; but justification is another matter, lying in unraveling the strands and assessing their weight.

(iii) *Deductive derivation.*—How far ethical statements are justified by reasoning from an established theoretical system or some fragment thereof depends in the long run on how far such systems come to be established. Some deduction there will always be; on the other hand, the rationalist's dream of a complete code deductively formulated, is no doubt just a dream. How far ethics can move in such a direction depends on the objective character of the human field and the order it actually proves to have. If it supports generalizations, and stable definitions with well-demarcated modes of application, if the indeterminacy points in choosing among rules, in interpreting them, are not too great, then some measure of useful deductive form will be achievable.

(iv) *Inductive establishment.*—This again depends on a certain degree of stability and determinacy in the field. That there be existential links for ethical terms is a necessary condition; but that the material to which the link goes have some order and that it be discoverable are further conditions. Thus the use of induction in a utilitarian morality depends on there being reliable regularities in what brings people pleasure. Similarly, a morality that speaks of resolving problems or satisfying needs can be inductive only to the extent to which there is dependable achievable knowledge of needs and problems and modes of satisfaction.

(v) *Normative success.*—At a certain point, one segment of contemporary ethical theory lost faith in the possibilities of any of the above methods of ethical justification. The emotive theory embodied this shift. It gave ethical terms the specific existential linkage of reference to a context of mutual persuasive effort in cases of disagreement. Ethical reasoning was construed not as a logical (deductive or inductive) relation between statements (premises and conclusion) but as a causal relation between beliefs and attitudes as events. Inductive evidence could be used to establish beliefs, but the relation between the belief and the attitude as causal was subject to individual differences. Therefore the concept of validity in ethics is in fact abandoned.[91] The realities of the situation are pressure, influence, persuasion. A method of justification is itself an object of advocacy, specifying a pattern of transition from factual premises to ethical conclusions.

In assuming no dependable regularity in cognitive-emotional or cognitive-volitional relations, this view anticipates a particular outcome of the study of the human field.

(vi) *Furnishing good-reasons.*—A widespread current approach suggests that even in the absence of fixed *general* criteria for going from factual premises to ethical conclusions, ordinary language supplies a validation concept for ethics. There are in different contexts acknowledged patterns of good reasons for deciding or acting in given ways. There are informal or "unscheduled" logics or implicit rules of what is relevant and what is not.[92] Ethical terms are interpreted practically by reference to the contexts of choosing, deciding, advising, etc. Validity consists in conforming to the contextual rules. A good reason for giving someone a book is that you borrowed it or promised to return it. It is a good reason for obeying a moral rule that it is part of the code, for changing an item in the code that it will reduce human suffering, and so on.[93]

In this mode of analysis some regard reasons as reasons for doing or deciding or being under obligation rather than as reasons for believing or statements logically supporting (deductively or inductively) other (ethical) statements. Even in analyzing statements, there is sometimes a considerable extension of the concept of reasons, so that assertions relating properties are translated so as to specify reasons. Instead of saying "Promises ought to be kept," it will be said: "The fact that I promised is a good reason for my doing the act (or for my being under obligation to do it)." For example, Baier translates "What shall I do?" and "What ought I to do?" into "What is the best thing to do?" and this in turn means 'the course supported by the best reasons.'[94] This would be helpful if the analysis of 'the best reasons' constituted an improvement on that of 'good' or 'ought.' But we are told that we act "in accordance with what we take to be the best reasons . . . because we *want* to follow the best reasons";[95] "We mean by the word 'reason' something that can make us do things."[96] But in the end "The criteria of 'best course of action' are linked with what we mean by 'the good life.' " And, "Our very purpose in 'playing the reasoning game' is to maximize satisfactions and minimize frustrations."[97]

There is a kind of arbitrariness in translating everything into the language of good reasons, because no new mode of justification is introduced thereby. The mode of justification is either that the language habits have that pattern (or contextual logic) or else simply that it is intuitively clear that making a promise is a good reason for keeping it. In some respects the situation is parallel to translating a material general statement in science into a material principle of inference. And perhaps the reason is the same—it stresses the conventional role rather than the empirical character. But it must not be forgotten that this should be done only to a well-established statement.

If, however, the immediacy or intuitive character of the apprehension of good reasons is stressed, then the whole approach can be most clearly seen as a variant of a phenomenological analysis. The good-reasons relation between a fact and an action is a kind of vector in the field when both are contemplated or "beheld" side by side. But if so, then we have argued above, this is not a mode of justification but a datum or an observation-statement which may serve to confirm or establish some statement or theory relating phenomenological to non-phenomenological data.[98]

Such an interpretation of the good-reasons relation, especially as applied to the relation of beliefs and acts, rather than statements, helps bring some order into the uses of 'reasons.' We can then see that the concept of reasons varies with the method of inquiry. If it is analytic method, the relations are logical, and reasons are premises. If it is descriptive, the relations between reasons and what they are reasons for is phenomenological. If it is causal, then reasons are provided when we have given a causal explanation. If it is evaluative method, reasons are furnished for given acts when we have indicated the ends or purposes that the acts serve. In short, the very concept of reasons has a contextual variability depending on the method of inquiry.

In general, the different processes considered in this section rest more profoundly than is commonly realized on the extent of the determinacy of the field. Intuition of a universal is practically applicable where there is considerable stability in phenomenological field characteristics. Perceptual disclosure of particulars will be appealed to where complexity is great and the disorder high, so that judgment seems practical art rather than calculated conclusion. Deductive derivation, as we saw, has extended applicability only in the conditions of greatest stability and relative steadiness of classification. Inductive establishment presupposes a moderate degree of determinacy sufficient to support some verification. Normative success as typical analysis of justification betokens that field condition in which the context of ethical discourse is clear but few dependable features characterize the relations in the field. The good-reasons approach sees fragmentary contexts of relative stability. It is not our purpose to decide here what is the state of the field but to note that *the various justification processes and theories about them reflect such estimates.* This is the sense in which we suggested at the outset that methodological policy decisions embodied some anticipation of scientific results.

21. Application and Evaluative Processes

As we move into the theory of application and evaluative processes, the specificity of situations becomes greater, conditions of special fields are added to the general contours of the human field, decision is directed from generalized principles to particular concrete situations culminating with the here-and-now. Here is where different bodies of factual knowledge converge, skill and insight combine with "know-how," informal procedural principles of relevance and selection begin to loom large. Here the significance of factual materials is unavoidable. But here, too, questions of complexity and subtle differences in evaluation have greater force. Weights and measures so often fail to be at hand that scientific aid seems out of sight. Although we cannot enter into a systematic examination of application theory, a few extended suggestions may be offered by way of comment.

(i) There is a pervasive evaluative process going on at the grass roots in human life, constantly raising reference-points, building them into evaluative criteria, fashioning these into partial standards, and greeting a more systematized morality as a guide in an ongoing process. Certainly, there is no dearth of reference-points, and we often become conscious of them when they are already functioning as criteria. You stretch your hand out for an apple and are already looking for one that is firm and red-cheeked; the apple industry has to go further and articulate a whole theory of grading.[99] You want a drink of water; do you want it enough to interrupt what you are doing? Even in so simple a desire you are faced with the standpoint of its strength and the collateral effects of satisfying it; a social enterprise like education has to work out careful standards of discipline and permissiveness for all stages in the growth of the young. Large-scale social processes—sorting and quality measurement in industry, legal regulation in realms as diverse as health conditions in ocupations, or the marks of substandard housing, or the fair value of a public utility as a basis for rate adjustment, or the meaning of reasonable caution or negligence in automobile driving, and so on—show clearly the interplay of empirical, technical, theoretical, and ideal or purposive elements that enter into standard-formation.

Ethical theory (with the exception of general value theory)[100] has not sufficiently explored the richness and variety of possible criteria in concrete evaluative processes within the moral domain. Consider, for example, criteria that arise when a man or a group not merely contemplates a proposed means or end but deliberates whether *to adopt* it. For means, one might suggest: Will the means if acted on produce the end (its *effectuality*)? What is the quality of its performance (*efficiency*)? How does the use of the means affect collateral ends (*constructiveness* or *destructiveness*)? If there is a considerable investment of energy and resources in providing the means, for what other ends can they be used (their *multivalence*)? What is the *liberating-power* of alternative means to the same end? To what human needs or drives may the means-activity itself give expression (its *expressiveness*)? How satisfying or enjoyable is the means-activity itself (its *luster*)? How far will a particular means if utilized tend to become an end in its own right (its "*telicity*")? What is the resultant *cost* of employing a particular means, in terms of the disvalue of the means-activity, the end-concomitants, and the consequences? For ends: How *attractive* is the end envisaged by itself? How far capable of occurring without means-components (its *purity*)? How long-lasting (its *permanence*)? In relation to other goals, what support does the occurrence of one end give to others (*constructiveness* or *destructiveness*)? What is its *area* in the field of endeavor of the given person or group? In relation to means, what is its *attainability* and its *cost*? In relation to the personal and social economy, what is the strength of the underlying drives and problems to which it is addressed (its *depth*)? With respect to these, what is its *role*? In the light of its role, does it prove to be spurious (e.g., rationalization) or authentic (insightful and realistic in grounding), that is, how *genuine* is it? What is the degree of *satisfaction* that it brings?

Criteria for evaluating ideals will overlap with those for ends but may have novel elements corresponding to the nature and role of the ideal as a human phenomenon.[101] In the obligation-family a fresh set of reference-points may prove relevant; for example, *stringency* plays the part here that attractiveness plays in ends. Similarly, virtues can be estimated from the point of view of *utility* as well as *attractiveness,* and, once they are set in a psychological and social context, there will be criteria of psychological *maturity* of a given virtue, standpoints provided by the social forms for which the virtues act as *girders*. And so on. It is also possible that reference-points may rise initially from scientific exploration. For example, Erikson, with an eye on the development from babyhood to maturity, suggests that criteria may be charted for decisive ego-victories at given points.[102] This suggests standpoints reflecting an acceptable theory of the development of the self. These might in turn be reflected in a revision of criteria of obligation (for example, a clearer picture of sincerity as against rigidity in the sense of duty) or of virtue (for example, in giving body to the idea of mature virtues as against reaction-formation types). And the development of a reliable social science has comparable effects. The route of such standard-formation is from the science through the EP to the development of criteria.

(ii) At the same time that there is this creative process of building up evaluative criteria from below, and consolidating them into standards, there is also a parallel process going on higher up. This consists in gearing the morality as a more or less stabilized system, to the tasks of application in a determinate setting. Elements of the morality—whether significant content, methods or procedures, special features manifested in special persons, and so on—move into a guiding role for evaluative decision because they are able to exercise the office of standards in the existing situation. They are selected for this in a way not unlike that in which natural entities in virtue of their regularity of process come to function as clocks, or special commodities like grain or gold come to function as media of exchange and modes of reckoning value. A comparative examination of the kinds of standards that emerge as a morality prepares for action adds to our understanding of evaluative processes.

Most prominent perhaps is the *paramount ideal* type. Ends of generality and abstractness that organize a great area of effort are prone to grow into ideals and serve as standards in appraisal. Important feelings, fundamental drives, and obligation-relations of broad scope may also function in this way. Happiness, harmony, justice, and the satisfaction of one's whole nature has each at times stood out as a key standard. Virtues too have sometimes grown into the role of central ideal, so that the very notion of "the moral ideal" came to mean a standard of character-development. These are being considered as standards rather than as ends or as the content of "the good life" when they are elevated to a supreme position for guiding conduct. In some cases, even a fairly concrete end may gain in scope and influence and come to furnish a standard of this sort for particular lives or particular groups.

Method standards also occur, reflecting the methodological properties of different EP's. Bentham will want to know if a decision has carried out a rational calculation; Kant will look to its conscientiousness; Dewey will look to its reflective character; and so on.

Individual models furnish functioning standards more widely than we are likely to think. The man of practical wisdom, the exalted leader, the impartial spectator, the saint, are familiar examples. To sense the spirit of his action, although not able to describe it adequately or reduce it to rule, provides a kind of dynamic sensitivity. It is not only persons who constitute models. As Parker points out, every striking event—a deep satisfaction or great deed, or even a first experience—may rise in memory to provide a touchstone for anything resembling it.[103] It resembles the clear case in definitional theory where a formal definition has not been achieved.

Jurisdictional standards tend to be overlooked in spite of their great importance. One issue is declared a matter of duty; another, a matter of taste. This is an academic matter; that one calls for action. Here we have a compromise situation; there, only a question of expediency; there, strictly a matter of principle. It is not always easy to see what criteria are being employed. Sometimes they are rough judgments of urgency, of importance, of complexity calling for deliberation, and so on. And there are hosts of relevance criteria in different areas of content that may predetermine the kind of result appraisal will bring. In some critical cases the pivotal point may lie in calling these relevance criteria into account for their own appraisal on some other standard.

One of the least recognized standard-types, although in practice it may be one of the most important, is the *indispensable-means* type. A means of wide scope and critical importance because many ends require it comes into central attention and functions as a widespread standard. The need for peace or the need for industrialization in a great part of the world becomes such a standard. In the extreme case its logical form may be simply as follows: Let every party concerned specify his own ends without even revealing them. Then the general proposition is asserted that, for all or almost all such ends actually held, it will be found that they cannot in fact be achieved without the common means. Hence the means becomes the immediately applicable standard. (Hobbes on peace and security is the obvious historical instance.) Braithwaite makes a similar point in another context: "In this Kingdom are many mansions. It is more reasonable to seek to enter this Kingdom by the only known modes of entry than to postpone the attempt until assured as to which, if any, of the mansions is the ultimate end of the quest."[104]

Finally, we refer to a whole host of *totality* standards. These are characterized by reference to a whole which is to provide the basic standard. It may be a whole life of the individual[105] or the idea of the good of the whole community, as in many ethical theories, both individualistic and organic. Or it may be a whole-history ideal, such as the growth of freedom of mankind collectively.

(iii) The outcome of criteria development from below and standard-formation from above may be a greater stability in moral judgment than the complexity of the processes might suggest. How stable it is depends—as in the theoretical problems of morality discussed earlier—on the determinateness of the field. And this, which also is basic to the policy decision how far to employ scientific method in application problems, cannot be settled a priori. Here, again, the results of a scientific examination of the field are prior to the policy decision about method. How necessary this independent examination is can be suggested by showing—paradoxical as it may sound—a few types of cases in which the standards in application may prove *more dependable* than the morality that is being applied.

The clearest illustration is the indispensable-means standard just considered among the inventory of types. If its achievement is a prolonged affair and affects a great part of life in a given epoch, then it can operate as a dependable standard in spite of the disagreement about ends.

Sometimes a moral standard may be stronger than the general moral principle on which it rests. Take, for example, the standard of honesty in the scientific profession—that is, of a scientist in his scientific work. The general moral principle of honesty as a virtue has a long history. It merges sometimes with wider virtues of integrity in a human being and authenticity in interpersonal relations. It has hard going in some institutional fields—for example, in the business world, where "Caveat emptor" could long reign supreme, or in political fields, where international morality countenances espionage and counterespionage. There has been some progress in the justification of honesty as a basic moral virtue, especially in the light of recent psychological advances in the understanding of the scope of self-deceit and the importance of insight in human well-being; and it is likely that major agreement could be achieved on its acceptance as a phase-rule. There is still, however, a large gap between honesty as a moral virtue and honesty as part of the functioning standard in many areas of application. But, as a standard in scientific work in the modern world, little discussion is needed, in the light of the role, goals, and co-operative procedure of inquiry in the field. The standard does not answer whether men should become scientists, but it is scarcely shakable that if you become one you are obligated to follow the standard of honesty in your work.

Another context is the convergence of different underlying moralities on a given conclusion within a limited range: Christian, Utilitarian, Marxian moralities all include the idea that racial discrimination is wrong. Of course, the moralities converging here might yield varying conclusions beyond the ordinary range, just as two mathematical formulas may yield the same finite set over a given range in the series and diverge beyond, or a physical law maintain a given form only within certain limiting conditions. But if the standard is set for the given range, and it is probable that the problems to be faced will arise within that range, then the standard may turn out to be supported by all conflicting moralities. It may therefore be more dependable in its domain than any one morality in general.

Another possibility is that a standard may rest on the applications of those parts of moralities that are invariant—commonly recognized goods, or isolated obligations—or it may rest in some cases on compromise portions of a morality, as where the standard calls for the act while morality argues about the motive, for example, in what often appears as standards of "common decency" in neighborly relations.

Fuller logical exploration is needed of the types of specific relations between elements of the morality and elements of the standards that apply it. But it may be possible to construct a consolidated standard for a given age cutting across major cultural and ethical theory differences. Such an integrated standard would have a role in evaluative processes similar to that played in factual determinations by the body of existent scientific knowledge of a given period. This attempt has been outlined elsewhere by the present writer in a concept of the *valuational base,* as an interlocking structure of human knowledge and human striving which embodies the best available knowledge of man's aspirations and conditions.[106] Its constituents would be fundamental human needs, perennial aspirations and major goals in their specific sociohistorical forms, central necessary conditions, and critical contingent factors. And a study of the way in which items would be included or excluded from this base showed clearly the penetrating role that findings of the psychological and social sciences might play in formulating the standard.

IV. Decision, Freedom, and Responsibility

22. Evaluative Processes in Unstructured Situations

Underlying the treatment of the place of science in ethics has been the hypothesis that the human field has a greater degree of *determinacy* than we are prone to assign to it and that problems in human life which generate and support standards and in which standards are utilized present themselves as more or less *structured* because the field has sufficient order in terms of continuing conditions, continuing aims, and modes of interaction. Let us differentiate the terms, thinking of the field as the scientist studies it as *determinate,* and the field as it appears phenomenologically in the view of the person making decisions as *structured.* Now if the determinacy of a field either because of too great a complexity or too great an instability falls below a certain point, the very problems seem to lose their structure. Established standards no longer seem to apply, and new standards seem incapable of establishment. There is a problem, but no definitive analysis even of what it is. What is the character of evaluative processes at such points?

There are probably many different types of ill-structured fields, and evaluative processes will take different shape depending on the factors that give rise to complexity and instability. Let us briefly survey sample types along a range from the highly structured to the practically unstructured.

In the "normal" highly structured types there are well-established standards and clear criteria of jurisdiction. Problems are treated by bringing to bear existent aims or existent knowledge to broaden or sharpen the standard, or by otherwise "harnessing" the indeterminacy.

A complex area with considerable variability may be given the jurisdictional mark of "to be settled according to taste." This is often misunderstood as the absence of standards, whereas it may be an extension of a libertarian ideal to broaden the area of individual choice; this is obvious in the area of food preferences, less obvious in that of reading preferences, and becomes a controversial issue in the modern world in the advocacy, for example, of planned parenthood as expressive of parental choice. Again, in dealing with fringe-areas, the problem is felt less as one of lack of standards than as a specific question of determining the jurisdiction of existent standards. For example, there may be doubt whether an issue is a professional matter and so governed by the stricter rules of that profession, or a personal matter and so more open to subjective considerations provided they are "sincere," or a legal matter subject to established law. In general, however complex the task, the evaluative process is felt as one of *applying* existent standards; although recognizing the role of interstitial "legislation," such a view never feels it as wholly unbound or arbitrary.

Where the indeterminacy arises from complexity of conditions, or from changing conditions, but not from changing aims, the field of decision becomes less specifically structured, but by no means wholly unstructured. The shift may be from content-standards to model-standards or method-standards. These may vary from assignment of fixed or authoritative locus of decision —papal infallibility, the decision of a controlling party, or the procedurally determined expression of popular will—to a general philosophy of method (e.g., Dewey's reorientation of ethics).[107] Up to a point such a substitution of decision-method for content may hold. But if the indeterminacy involves large components of conflicting or changing aims, the very methods as standards may become a battleground.

Where the indeterminacy is clearly grounded in conflict or change of aims, the decision situation becomes structured in the broadest value terms. Virtue-standards may come to the fore, and, as in late Roman Stoicism with its stress on the precariousness of life and the folly of committing one's self to specific aims and hopes, the fixed point of decision becomes the maintenance of personal integrity. Or else most generalized totality-standards are invoked; one is not told what to decide but simply to remember that one is deciding for one's whole life or for all mankind.

The end of the range is found in the extreme situation where the very aims of men are regarded as indeterminate and the conditions of life as utterly precarious. Here the situation of decision looms as central and almost solitary. Because of the great influence of this structure in contemporary ethics, it is important to understand its EP. When it speaks of decision, it is thinking not of the slow change in standards—as the American people changed their standard of responsibility for unemployment over the period of the Great Depression—or of cumulative changes in an individual's aims over decades, but of the lone individual in the here-and-now; it extends the category of *decision* to cover practically every turn that a man can make. (And especially since responsibility is tied up with decision, it has momentous ethical consequences.) But in the very extension, any support for decision is itself removed. If you follow established standards, it is an implicit decision to follow them; if you apply a principle, it is a fresh decision—a resubscription or a readoption of the principle in the very act. Hence decision itself, devouring any possible basis on which it might rest, becomes seen as pure act. It becomes the absolutely cut-off (free) evaluative process, which we may call *the Sartrean leap.*[108] In more analytic fashion the same point is embodied in current prescriptivist emphases in ethics that insist on reformulating the meaning of ethical terms by reference to the first-person or agent's present decisional situation.

Putting aside metaphysical or analytic dogmas, we can extract from such a perspective a number of insights about evaluative processes in unstructured situations.

(i) There is indeed a decisional element implicit in all deliberative use of standards, even in what seems to be automatic subsumption of a situation under a rule, for there is a jurisdictional assumption in turning it over to this rule. Probing for this decisional element has been everywhere useful in laying bare value assumptions—particularly in the philosophy of law, where the deductive model of interpreting and applying pre-existent law often had marked ideological uses. The complete mapping of the components of standard-application must always include the value assumptions, even though in the areas under investigation they may be practically negligible *because they are constant and deeply rooted.*

(ii) It is, however, by no means clear that such an unearthing of value assumptions means that they have to be reviewed. We have to distinguish the cases where bringing to consciousness means *bringing into question* from those in which it means *bringing into recognition.* The thoroughly legitimate inference that the possible scope of evaluative processes is endless and that nothing is exempt should not be confused with the assumption that consciousness always has a subverting role and that where it does not it is because there has been some fresh perpetuating volition. This is the issue we may epitomize as whether the Socratic quest for knowing oneself is really *finding* or *creating*. The present point is that there is no *general* answer. Sometimes it is recognizing what we are like and what commitments we have. Sometimes it is altering in the act of discovering. Exploration of these processes is best done by using 'finding' and 'creating' as empirical categories for psychological research rather than giving a valuational cast to the phenomena under investigation by some blanket appelation. This means probing for further knowledge about the relation of consciousness and volition, the extent of permanent unavoidable needs that play a constitutive role in the self, the influence of different conditions in motivational processes, perennial sources of internal conflict, and so on.

(iii) In the extreme case there might seem to remain at least one structured feature in any decision situation—the initiating question reduced simply to "What shall I do?" still means that *something is to be done.* If there is no remaining sense of aims, conditions, determinate dangers, definite frustrations, then its structural residue is at least a call for help. Below such minima, its phenomenological structure as a decision situation vanishes but its status as a phenomenon-to-be-understood still remains, with all its qualities of suffering and anguish. It is philosophical—especially existentialist—attempts to interpret such phenomena as in some sense furnishing a profound truth about the human situation and the nature of ethics that makes it important to show clearly the opposing avenues of theoretical solution. And these, it becomes very evident, bring us back to issues of underlying EP's.

The essence of the scientific analysis here is to insist that phenomenological lack of structure need not mean actual indeterminacy in the situation. The field conditions for the lack may be quite determinate. Psychological accounts of frustration in unstructured situations show that there are definite aims that are being blocked. Sociological accounts of anomie and its consequences do not deny that there are definite social and individual needs that have become entangled and channeled in such a way as to produce confusion in the consciousness of standards. In revolutionary situations in history, new social patterns have already formed even while consciousness is cast in terms of the older breakdown. So too in such personal experiences as dramatic conversion or pervasive object-less anxiety or dread. There is no need to take the one as a leap by a bare self behind phenomena or the other as a special metaphysical indicator. There are scientific problems enough in their study, but they are specific problems of discerning accumulating forces, the conditions of qualitative change, and so forth.

The essence of the opposite view is that the phenomenological feature of lack of structure in decision situations is central because it provides points at which we can break through the local blinders of ordinary situations. Wisdom in ethics comes from concentrating on the marginal indeterminacy precisely because it reveals to us the human predicament. Anguish and dread therefore have ontological significance and are not to be taken as generalized anxiety detached from its object or in some sense accumulated in human processes. In actual philosophical contexts, of course, such theses are tied up with specific EP conceptions, whether of ultimate subjectivity or the central reliability of phenomenological method, and so forth; but basic focus is on the phenomena and the issue of their interpretation.

Perhaps the scope of these issues and the place of science in their resolution can best be suggested by a central illustration—attitudes to death. At one extreme we find the Spinozistic maxim that the free man is not concerned with death; at the other we find the view that to see a situation in the light of death is the only way to be sure of authenticity in its understanding. Such opposing views might charge each other, respectively, with morbidity and perversity, or with escapism, bland optimism, and naïveté. The battleground extends over many a field. The existence or non-existence of basic anxiety, whether if it exists it represents the "rock bottom" of man or a frustration reaction, the nature and role of insight, the genuineness of man's desire for immortality or whether it is a reflex of an unsatisfied life, the possibilities of a happy life without sacrifice of depth, the fear of death and whether it is genuine or a realistic feeling of "lack of time" or a cover for unconscious ideas—all these and many more issues are involved in resolving the major theses. Obviously, in spite of the ideological use of every item in this inventory, the basic problem is part of the growing area of the psychological and cultural and historical study of human reactions. It may yet be possible to distinguish whose optimistic attitudes, having what qualities and arising under what conditions, are escape attitudes, and whose despair attitudes or "outsider" attitudes, with what qualities under what conditions, represent basically clinical symptoms, and to see how far propositions about death and non-being are really propositions about ways of facing life, or how far the reverse is basically true. If the broad outlines of a scientific philosophy today incline against the general thesis of the primacy of death, it is not an unphilosophical optimism but a refusal to accept a limited phenomenological picture of a given moment in the historical development of mankind under specific conditions as in itself authentic, without relating it to the broader accumulation of knowledge—biological, psychological, historical—that man has acquired of himself.

Our survey of the range of situations from the highly structured to the barely structured suggests that scientific results have a place throughout—that the controversies are analyzable into questions that are essentially scientific in type. Scientific method would seem to be applicable in actual evaluative processes according to the extent to which the situations are structured and in different degrees and with different types of standards according to the types of structuring. But beyond the limits of its applicability at any time, there arises the factual question whether the growth of scientific knowledge of man will increase the range of applicability of the method and also the methodological policy problem whether the conceptual apparatus of ethical theory should be increasingly geared toward such an extension. On the analysis carried out, such decision involves a reckoning of the dependable projection of the curve of increasing knowledge. It also raises questions, however, of limiting conditions, and in the tradition of ethical theory this focuses on the problem of freedom of the will.

23. Toward a Scientific Study of Freedom and Responsibility

In popular consciousness the problem of free will stands as a perennial road-block to the relation of science and ethics.[109] Science is tied to determinism, ethics demands responsibility, responsibility requires free will, free will is antithetical to determinism. How can we project a scientific study of this domain?

(i) *Location of phenomena and descriptive concepts.*—Two bodies of conceptualized phenomena can be found. One is the *voluntaristic group:* the sense of freedom, various shades of willingness from jumping at the opportunity, acting eagerly, to intending, trying; again, the sense of acting unwillingly or being coerced, intimidated, provoked, compelled, of slipping into an act, of acting against one's better judgment, or half-heartedly, of succumbing to temptation. The second is the *responsibility group:* praising, blaming, finding fault, feeling at fault; offering excuses, holding liable; holding responsible, feeling responsible, finding meritorious.

These groups of phenomena and concepts extend far beyond these samples. They need study in many terms: descriptive and analytic; first-personal contexts, third-personal, interpersonal; border-lines with other moral phenomena (e.g., feeling responsible and feeling guilty about, or holding liable and making claims on). The search for phenomena can fruitfully go into experiential and conceptual elements in religion and law as well as morality: expiation, grace, forgiveness, retribution, vengeance, punishment, crime, liability of animals and things, obsession, addiction, scapegoating. Linguistic analysis has to distinguish different uses so as to untangle the phenomena: e.g., 'fault' in the legal maxim "No liability without fault" refers to some criterion of voluntary behavior, whereas in 'finding at fault' it is often equivalent to 'holding responsible.'

The central aim is to identify clearly the specific phenomena involved and provide descriptive concepts, thus separating the phenomena from interpretations.

(ii) *Causal-explanatory study.*—Such inquiry finds many aids in the psychological and sociocultural sciences. Depth psychology furnishes explanatory concepts concerning origins and functioning of guilt-feeling, mechanisms of identification and internalization of standards, functioning of the superego and emergence of the ego-ideal, phenomena of self-accusation and mechanisms of lifting the load of guilt, and so on.[110] Social and historical explanations are directed to functions and changes in responsibility patterns and in voluntaristic patterns.

For example, what explains the abandonment of holding animals responsible, or perhaps why it lasted so long in some quarters? Under what conditions do societies establish distinctions between voluntary and involuntary commission of an offense and narrow down punishment to the former? Under what conditions do patterns of kin responsibility yield to individual responsibility? Why does Oedipus feel so profoundly guilty while Adam Smith can go so far as to speak of the "fallacious sense of guilt" of Oedipus and Jocasta?[111] Can we explain the growth of a concept of "liability without fault" in modern times in terms of the rise of large-scale industrial enterprises with statistically regular mishaps, so that bearing an economic burden becomes distinct from accepting guilt? And so on.

(iii) *Inadequacy of present theoretical categories.*—Even partial undertaking of such studies shows the difficulty of conceptualizing so vast a field under the two general concepts of freedom and responsibility. What credence then is to be given to general formulas relating the two groups—whether it be a philosophical slogan such as "Responsibility is meaningless without freedom" or a legal maxim such as "No liability without fault" (in the voluntaristic sense of 'fault')?

If the generalizations are synthetic statements, counter-instances can readily be found. In law fault becomes stretched from intentional act to negligent act to implicit undertaking to obligation implicit in the situation whether undertaken or not and even when unaware.[112] Similarly, the philosophical slogan runs counter to: religious views of predestination combined with responsibility, legal concepts of vicarious liability (e.g., of master for servant), the sense of unavoidable guilt in some ethical theories,[113] of original sin in Western religions, and so on. Now it may be possible to discount such an array of counter-instances by subtle analytic differentiation of some, rejection of others as incorrect, and so on. But this will involve working out a theory of the two groups of phenomena on the basis of careful study. Any attempt to short-cut this study by relying on an analytically or intuitively necessary connection runs the risk of exalting a possibly provincial consciousness into a universal law.

There are also evaluative questions in the relation of phenomena in the two groups. For example: If a man is provoked, should his punishment be less? If he enters a conspiracy half-heartedly, is his responsibility greater or less? How far should a man be held responsible for acts of others, and under what conditions? Is coercion always bad, and forgiveness always good (at least as phase rules)? It is therefore possible that the general rule that responsibility is meaningless without freedom, as well as its legal counterpart, may be operating evaluatively—for example, as the expression of the growth of modern individualism in the past few centuries, demanding that the individual be freed from obligations to which he has not given explicit assent.[114]

In fact, a critical inquiry should go further and seek the very basis of unity embodied in the theoretical categories rather than taking them intuitively. The general concept of responsibility has seemed to have a firmness of outline which the concept of free will has almost wholly lacked. It is oversimple to say that this is because the former is a practical concept and the latter theoretical. Both have practical and theoretical components. Perhaps the apparent clarity of the responsibility concept comes from the fact that it is anchored to a standardized family of rewards and punishments; the concept of voluntary action lacks this visible reminder. But we have no really adequate theory of why we punish or of how effectively punishment deters and blame hinders or what blaming does to the spirit of a human being (e.g., whether it arouses a constructive sense of shame or subtly threatens with a loss of love).[115] And while it is clear that voluntaristic phenomena depend for their understanding on a theory of the will, we do not yet have a full-fledged psychology which can give us an account of the role of thinking and feeling in the phenomena of will, and all these in relation to a conception of the self and its development and its ways of dealing with internal conflict. In spite of rapid contemporary psychological progress, in this theoretical area there is still more program than accomplishment.

Nor is it useful to go to the metaphysical tradition in the free-will problem. Historical study shows that the very meaning of the problem undergoes change with the underlying EP in which it is cast. Given supernaturalistic and dualistic stage-settings, free will is assigned as a central property to soul or mind, as against the causality or determinism operative in and among bodies. Given a deterministic materialism or naturalism, the very locus of the problem is found at other points; for example, the achievement under favorable causal conditions of a certain type of character, or the development of a certain type of consciousness. There is continuity in the sense that all are concerned with furnishing some theory of voluntaristic phenomena and relating them to responsibility phenomena. But the free-will problem is seen to be a derivative problem, in the sense of requiring for its understanding and formulation location within a given world view. This does not, of course, mean that it does not require solution.

24. Toward a Strategy for Solution of the Free-Will Problem in Ethics

A strategy for solution of the problem of free will and responsibility in the light of the kind of study mapped above is primarily one of separating the strands and relating them either to the area of ethical theory to which they are found connected or else to the specific EP from which they stem and the fate of whose evaluation they share. Many strands furnish what turn out to be answers to some part of the whole network of problems.

(i) In some pragmatic accounts, to find a man responsible is not to attribute to him some antecedent voluntaristic phenomenon but to affirm that he ought to be punished or rewarded. The reasons may be various, dependent on field, purposes of the enterprise, general values.[116] Whether this is an adequate general account or not, it does call attention to one strand: that some large part of responsibility theory can be seen as a vehicle of standard-formation for the application of sanctions. It even verges in some cases—for example, in assigning liability for accidents which are "nobody's fault"—on a simple working-out of principles of distribution of gains and losses as one works out principles of taxation.

(ii) We can look in traditional accounts of freedom for the portrait of the free man—whether the Stoic with complete self-possession, the Spinozistic free man in a determinist world with clear knowledge of his aims and conditions and viewing his actions in terms of determination from the essential nature of the whole, the Kantian conscientious universalistic self-regulator, the Marxian free man as fully conscious of necessity as possible in his time and turning his knowledge into social action for the further achievement of human values, the Russellian free man defying matter and worshiping at the shrine of his own values, the Freudian free man with insight into himself strengthening his ego to extend its sphere, or a host of others.[117] We can separate them from the issues of belief in determinism or metaphysical free will and see how far they function as proposals for virtue configurations, to be assessed as ethical standards.

(iii) Some judgments of coercion, intimidation, provocation, etc., can also be seen as ways of determining permissible processes in institutional and interpersonal relations. Deciding what is or is not coercion in contractual relations, or in labor relations, or in the relation of superior and inferior in the military organization may in effect be valuational decision on how far the state may interfere to provide social security, what permissible weapons there may be in business-labor tension, what limits there are to obedience to commands of superiors (cf. the problems of responsibility in the Nuremberg trials).

(viii) Some stress the phenomenological datum of the sense of freedom in the first-person act of decision, as wholly removed from scientific scrutiny and explanation. This can, however, be seen as a problem in the psychological theory of the self. It is by no means solved at the present stage of inquiry and may itself prove to contain a fusion of strands or cover a multiple rather than a single issue. Certainly one would have to distinguish the feeling of the reality of the act of will (the sense that a self or an I is involved in the process), the consciousness of the I as the source or starting-point of the act, the sense of efficacy of the will (that something different or novel may come from the fact that I willed this rather than that), the sense of possible alternatives (that I could have willed otherwise), and perhaps many other elements. In spite of the numerous attempts to extract a transcendental ego from these phenomenological data, there appears to be nothing here that a growing conception of the self as a special kind of dynamical system in the career of the individual organism may not take in its stride. Naturalistic theories have tried to exhibit the sense of freedom sometimes as a sign of the congruity of the path chosen with the basic desires of the agent, sometimes as the sense of absence of coercion by others, sometimes as the cognitive recognition that the act was in some quite specific sense "avoidable." Another possibility, as yet insufficiently explored, is that the situation of choice is itself structured in such a way as (phenomenologically) to create a gap (cognitive "distance") between the self and the field, so that determinism cannot be taken to apply to the self without putting the self into the field.[120] Hence, *in the act of choosing*, the sense of freedom would be phenomenologically meaningful; but in the subsequent study of his choice even by the agent himself, determinism would be likewise meaningful. At the present stage of the growth of the psychological study of the self, what is needed is a theory that will serve to relate the organic, the behavioral, the phenomenological, the developmental, and even the cultural and the social approaches.

How far the strategy here envisaged may serve to solve the complex problems in the freedom and responsibility area de-

(iv) Other judgments of coercion or compulsion center on the voluntaristic phenomena and attempt to establish or invoke empirical criteria of assent or extent of internal harmony in decision. Schlick went so far as to assert that the issue of free will versus determinism was a pseudo-issue and that voluntary decision versus compulsion from others was the only issue that made sense.[118] In any case, it is one of the issues, and its scope is broadened by psychological knowledge of compulsions, of unconscious influence of authority-figures, etc. Similarly, increased social knowledge of the strength of social forces can help stabilize the indices of coercion.

(v) Part of the traditional free-will controversy concerned primarily whether man is to be studied as part of nature or is in some sense outside the order of nature. Contemporary scientific philosophies regard this issue as settled by this time, and the full incorporation of man within the natural world as established.

(vi) Part of the argument for free will has centered on the possibility of novelty in the world, which seemed to be ruled out by a complete or hard determinism. In one form or another, the advance of the philosophy of science has furnished analyses in which complete predictability to the last fine shade is not an essential mark of determinism and in which the inability to predict in *advance* does not entail the inability to find causes *after* the novel phenomenon has come into existence.[119] (In the theory of mind, this has meant that a scientific approach is not inherently bound to epiphenomenalism.)

(vii) A step beyond issues of novelty have come attempts to work out a principle of creativity as somehow anti-deterministic. Here, again, once the possibility of novelty is granted, the phenomena of creativity become detached from specific free-willist doctrine. In empirical terms it is possible to map the points of creativity in biological evolution, in individual biography, in history. That this is not incapable of association with a causal approach can be seen in Marxian materialism, in which freedom is identified as the growth of effective consciousness in human life with the evolution of the material-social world.

pends of course on the results of the researches envisaged. The very least we may expect are a broader presentation of the phenomena and a breakthrough in the stalemate of the traditional issue.

25. The Creative Temper and the Scientific Temper

At the opening of this work we posed the problems of the place of scientific results, the role of scientific method, and the impact of the scientific temper in ethical theory. Scientific results have been found to have an unavoidable place in the understanding of ethical theories and an indispensable position in their evaluation. The utilization of scientific method in ethics has been shown to be genuinely possible, with the extent of its practicability dependent on the extent of determinateness found in the human field. It has not been the task of this work to carry out an inventory of the scientific materials that establish the requisite extent of this determinateness. The policy recommendation that scientific method be utilized in ethics—as distinct from the logical and methodological anlysis of its possibility—involves a recognition of the understanding the growth of the sciences of man has already brought to ethics. There is also the promise of further understanding with the further growth of these inquiries, especially when directed to the phenomena of the moral field itself. And there is the recognition that alternative policies have not proved fruitful in the development of ethical theory and the suspicion that too often they themselves rest on some *ersatz* science.

Decision about the scientific temper is a further question. In part, of course, it would be necessitated in a policy decision to utilize scientific method in ethics. But that this advance-guard of the scientific outlook should be allowed to permeate the field of ethical inquiry, no matter how far and in what detail scientific method be utilizable, is itself an ethical question. For the scientific temper would function in ethics as a set of character-traits and that is as a virtue-constellation. To assess its impact is therefore to estimate both what has been and what ought to be the place of these virtues in ethics. It is, in short, part of the theory of virtues in a morality.

Since we have not attempted to go into the field of evaluative morality in this work, the full answer lies beyond our present scope. There is, however, a basic consideration which may be offered by way of recommendation for these virtues. Man's whole development is tied up with the progress of knowledge and its application; the virtues of reflective assessment of experience, wide range of investigation, sober weighing of evidence, and the rest of the familiar list have certainly earned their keep. What is more, ethical theory itself has always been a reflective enterprise, with rare and impassioned exceptions. It might be well to try out the explicit development of the whole field in a scientific spirit.

Perhaps the chief criticisms of the scientific temper have centered on misunderstandings of its nature—the contrast between the scientific and the creative. This reflects what is sometimes the parochial scientism of a limited period, the blind emphasis on the mechanical and the measurable. A less stereotyped view of the scientific temper is to be found in its greatest geniuses and in the history of science at critical periods. There it is the imaginative, the grasp of new possibilities, the ability to see things never seen before and to pose questions in a new way, to strip off blinders rather than to impose a new brand of them, which strike the investigator. Then the scientific temper and the creative temper become two faces of a single coin. It would indeed be a strange retribution if mankind, so prone to seek its salvation in the act, to conjure up romanticisms of the heart and the will, were to find the stoutest ally for both heart and will in the quest for knowledge.

Notes

CHAPTER I

1. I should like to express my indebtedness to the National Science Foundation for a grant, during the academic year 1959–60, to continue my work on the relation of science and ethics. The present monograph is a part of this work.

2. Cf. Abraham Edel, *Ethical Judgment: The Use of Science in Ethics* (Glencoe, Ill.: Free Press, 1955), chaps. v–viii.

3. Exceptions will be found in Charles Morris, *Signs, Language and Behavior* (New York: Prentice-Hall, Inc., 1946), pp. 230–33; DeWitt H. Parker, *The Philosophy of Value* (Ann Arbor: University of Michigan Press, 1957), p. 87. For general discussion of this problem cf. Abraham Edel, *Method in Ethical Theory* (Indianapolis: Bobbs-Merrill Co., 1963), chap. vii.

4. Cf. May Edel and Abraham Edel, *Anthropology and Ethics* (Springfield, Ill.: Charles C Thomas, 1959).

5. My *Method in Ethical Theory* is devoted to an exposition of these four standpoints.

6. Cf. Erich Fromm, *Man for Himself* (New York: Rinehart & Co., 1947), pp. 141–72.

CHAPTER II

7. Cf. Edel and Edel, *op. cit.*, chap. xiv.

8. Reference to some types of existentialist perspectives will be found in section 14 below.

9. Clyde Kluckhohn and others, "Values and Value-Orientation in the Theory of Action," in Talcott Parsons and Edward A. Shils (eds.), *Toward a General Theory of Action* (Cambridge, Mass.: Harvard University Press, 1951), p. 411.

10. Robert Redfield, *The Primitive World and Its Transformations* (Ithaca, N.Y.: Cornell University Press, 1953), pp. 85, 86. See the whole of chap. iv.

11. Cf. Daryll Forde (ed.), *African Worlds: Studies in the Cosmological Ideas and Social Values of African Peoples* (New York: Oxford University Press [for the International African Institute], 1954); Ethel Albert, "The Classification of Values: A Method and Illustration," *American Anthropologist*, LVIII (April, 1956), 221–48; Florence Kluckhohn, "Dominant and Substitute Profiles of Cultural Orientations: Their Significance for the Analysis of Social Stratification," *Social Forces*, XXVIII (1950), 376–93; A. I. Hallowell, *Culture and Experience* (Philadelphia: University of Pennsylvania Press, 1955), esp. chap. iv, and chaps. viii, ix, and xi; John J. Honigmann, *The World of Man* (New York: Harper & Bros., 1959), chaps. xxxiv–xxxix.

12. The stage-setting concept was worked out for ethics in my "Coordinates of Criticism in Ethical Theory," *Philosophy and Phenomenological Research*, VII (June, 1947), esp. pp. 550–54 (reprinted in Edel, *Method in Ethical Theory*). It was suggested by A. F. Bentley's procedure in schematizing psychological approaches in his *Behavior, Knowledge, Fact* (Bloomington. Ind.: Principia Press, 1935), esp. Part I. Cf. also the mode of schematic representation in Egon Brunswik, *The Conceptual Framework of Psychology*, in this *Encyclopedia*, Vol. I, No. 10.

13. Cf. Meyer Fortes, *Oedipus and Job in West African Religion* (Cambridge: Cambridge University Press, 1959).

14. Jeremy Bentham, *Principles of Morals and Legislation* (new edition corrected by author, 1823; republished, Oxford: Clarendon Press, 1907); J. S. Mill, *Utilitarianism* (1863) (reprinted, New York: Liberal Arts Press, 1949).
15. Thomas Hobbes, *Leviathan* (London, 1651).
16. Cf. Herbert Spencer, *Social Statics* (1850) (New York: D. Appleton & Co., 1865); Peter Kropotkin, *Mutual Aid* (1902) (Penguin Books, Ltd., 1939).
17. Cf. Leslie Stephen, *The Science of Ethics* (New York: G. P. Putnam's Sons, 1882), esp. chap. iii.
18. Herbert Spencer, *The Principles of Ethics* (New York: D. Appleton & Co., 1896–97), Vols. I–II.
19. Adam Smith, *The Theory of the Moral Sentiments* (London, 1759; 6th ed., 1790), Part III (reprinted in part in *Adam Smith's Moral and Political Philosophy* [New York: Hafner Publishing Co., 1948]).
20. Bentham, *op. cit.*, chap. iv.
21. Bertrand Russell, *Human Society in Ethics and Politics* (New York: Simon & Schuster, 1955).
22. Moritz Schlick, *Problems of Ethics* (New York: Prentice-Hall, Inc., 1939), chap. ii.
23. Ralph Barton Perry, *General Theory of Value* (New York: Longmans, Green & Co., 1926), p. 209.
24. *Ibid.*, p. 471.
25. E. C. Tolman, *Purposive Behavior in Animals and Men* (New York: Century Co., 1932); see also the collected papers in *Behavior and Psychological Man* (Berkeley: University of California Press, 1958).
26. Stephen C. Pepper, *The Sources of Value* (Berkeley: University of California Press, 1958).
27. For a general picture see Sigmund Freud, *New Introductory Lectures on Psycho-analysis* (New York: W. W. Norton & Co., 1933). The biological, social, and historical theses are found, respectively, in *Beyond the Pleasure Principle* (New York: Boni & Liveright), *Group Psychology and the Analysis of the Ego* (New York: Boni & Liveright), and *Civilization and Its Discontents* (New York: Jonathan Cape & Harrison Smith, 1930).
28. E.g., Heinz Hartmann, "Ego Psychology and the Problem of Adaptation," in *Organization and Pathology of Thought*, ed. David Rapaport (New York: Columbia University Press, 1951).
29. Such theories are found in Freudian, neo-Freudian, and general psychological schools (cf. Otto Fenichel, *The Psychoanalytic Theory of Neurosis* [New York: W. W. Norton & Co., 1945], pp. 59–61; Fromm, *op. cit.*, pp. 210–26; and John Dollard *et al.*, *Frustration and Aggression* [New Haven, Conn.: Yale University Press, 1939]).
30. Cf. Fromm, *op. cit.*; Karen Horney, *The Neurotic Personality of Our Time* (New York: W. W. Norton & Co., 1937); Erik H. Erikson, *Childhood and Society* (New York: W. W. Norton & Co., 1950); J. C. Flugel, *Man, Morals and Society* (New York: International Universities Press, 1945); Abram Kardiner, *The Psychological Frontiers of Society* (New York: Columbia University Press, 1945).
31. Harry Stack Sullivan, *Conceptions of Modern Psychiatry* (2d ed.; New York: W. W. Norton & Co., 1953).
32. Among psychologists cf. Wolfgang Köhler, *The Place of Value in a World of Facts* (New York: Liveright Publishing Corp., 1938), chap. iii; Karl Duncker, "Ethical Relativity? (An Enquiry into the Psychology of Ethics)," *Mind*, XLVIII (January, 1939), 39–57; Solomon E. Asch, *Social Psychology* (New York: Prentice-Hall, Inc., 1952), chaps. ii, xii, and xiii. Among philosophers, Maurice

Mandelbaum, *The Phenomenology of Moral Experience* (Glencoe, Ill.: Free Press, 1955); Nicolai Hartmann, *Ethics* (New York: Macmillan Co., 1932), Vols. I–III; and Max Scheler, *The Nature of Sympathy* (London: Routledge & Kegan Paul, Ltd., 1954). Hartmann and Scheler go beyond descriptive phenomenology, with which alone we are here concerned.

33. Cf. sections on phenomenological description in Edel, *Method in Ethical Theory*, chaps. viii and xi.

34. Cf. Charles L. Stevenson, *Ethics and Language* (New Haven, Conn.: Yale University Press, 1944), chaps. i, ii.

35. John Dewey, *Human Nature and Conduct* (New York: Modern Library, 1930), p. 39.

36. In an article, "Some Questions about Value?" *Journal of Philosophy*, LI (August 17, 1944), 455, Dewey asks: "Are values and valuations such that they can be treated on a psychological basis of an allegedly 'individual' kind? Or are they so definitely and completely socio-cultural that they can be effectively dealt with only in that context?"

37. John Dewey, in Part II of Dewey and Tufts, *Ethics* (rev. ed.; New York: Henry Holt & Co., 1932). Cf. his concluding statement (pp. 342–44).

38. The outlines of the Marxian theory of ethics are clear in Frederick Engels, *Anti-Duehring* (New York: International Publishers), chaps. ix–xi and Part III.

39. Spencer's mode of analysis is clear in the detailed discussion of his *The Principles of Ethics*. For Kropotkin see *op. cit.*, chaps. iii–viii; also his *Ethics, Origins and Development* (New York: Dial Press, 1924).

40. Julian Huxley, "Evolutionary Ethics," in T. H. Huxley and Julian Huxley, *Touchstone for Ethics* (New York: Harper & Bros., 1947).

41. *Immanuel Kant's Critique of Pure Reason*, trans. Norman Kemp Smith (London: Macmillan & Co., 1929), p. 643. For illustration of Kant's detailed elaboration of his approach see, e.g., *Immanuel Kant's Religion within the Limits of Reason Alone*, trans. T. M. Greene and H. H. Hudson (Chicago: Open Court Publishing Co., 1934), p. 131.

42. The kinds of arguments used in traditional doctrinal conflicts constitute illuminating materials for such an inquiry (cf. *Documents of the Christian Church*, ed. Henry Bettenson [New York: Oxford University Press, 1947]).

43. *International Journal of Ethics*, October, 1903, p. 116.

44. G. E. Moore, *Principia ethica* (Cambridge: Cambridge University Press, 1903), chap. i.

45. Compare John Laird's comment in his *An Enquiry into Moral Notions* (London: George Allen & Unwin Ltd., 1935), p. 106: "What Kant called 'respect' for the moral law is a ghost from Sinai, a crepuscular thing that sins against the natural light. Therefore consistent Kantians must either bring divinity into their ethics, not as a consequence but as part of the analytic of their fundamental conceptions, or else retire to purely terrene ramparts and abandon their view that the moral fact is unique of its kind."

46. T. H. Green, *Prolegomena to Ethics* (Oxford: Clarendon Press, 1883). The issue is clearly posed in the Introduction.

47. Other clues occasionally useful in detecting an implicit EP are: conceptions of the tasks of ethics, contexts of application, guiding models.

48. Moore, *op. cit.*, pp. 143 ff.; Hartmann, *op. cit.*, I, 218 ff.

49. Reinhold Niebuhr, *The Nature and Destiny of Man* (New York: Charles Scribner's Sons, 1943), I, 13–14.

50. Jean-Paul Sartre, *The Transcendence of the Ego* (New York: Noonday Press, 1957); *Existentialism and Humanism* (London: Methuen & Co., 1948).

51. Niebuhr, *op. cit.*, I, 122–25; George Santayana, *Reason in Society* (New York: Charles Scribner's Sons, 1930), pp. 3–4; Gilbert Ryle, *The Concept of Mind* (London: Hutchinson's University Library, 1949), pp. 195–98.

52. For a detailed criticism along these lines see Abraham Edel, "The Logical Structure of G. E. Moore's Ethical Theory," in *The Philosophy of G. E. Moore*, ed. P. A. Schilpp ("Library of Living Philosophers," Vol. IV [Evanston: Northwestern University, 1942]), esp. pp. 169–76.

53. Karen Horney, *New Ways in Psychoanalysis* (New York: W. W. Norton & Co., 1939), pp. 187–88.

54. William James, *Pragmatism* (1907) (New York: Longmans, Green & Co., 1947), p. 119.

55. For a general discussion of different relations in which ideas can stand to social context see Abraham Edel, "Context and Content in the Theory of Ideas," in *Philosophy for the Future*, ed. R. W. Sellars, V. J. McGill, and M. Farber (New York: Macmillan Co., 1949).

56. For an attempt to work out sociocultural categories for structuring ethical theory see Edel, *Method in Ethical Theory*, chap. ix.

CHAPTER III

57. Cf. Edel, "Coordinates of Criticism in Ethical Theory," *op. cit.*

58. Cf. R. B. Brandt, "The Status of Empirical Assertion Theories in Ethics," *Mind*, LXI (October, 1952), 458–79.

59. Cf. Edel, *Method in Ethical Theory*, chap. vi, sec. 6.

60. Fyodor Dostoevski, *Letters from the Underworld* ("Everyman" ed.; New York: E. P. Dutton & Co., 1945), p. 27.

61. For an examination of the problem of the mark of the moral, and for consideration of different candidates, see Edel and Edel, *op. cit.*, chap. ii; also *Method in Ethical Theory*, chap. vii, section on "What Are the Data of Ethics?"

62. For a study of the three families and some problems in their relation see Laird, *op. cit.*

63. For a study of the variety that may be embraced in the notion of defining ethical terms cf. Edel, *Method in Ethical Theory*, chap. v.

64. For Socratic and Platonic views, Plato's *Apology, Crito, Symposium*. For Stoics, Epictetus' *Enchiridion* and Marcus Aurelius' *Meditations*. For Kant, his *Critique of Practical Reason* and his *Fundamental Principles of the Metaphysic of Morals;* his *Lectures on Ethics* (London: Methuen & Co., 1930) (compiled out of students' lecture-notes) contains illuminating materials on the psychology of morals. For Nietzsche, his *Genealogy of Morals*. For Bentham, Köhler, Mandelbaum, works cited in previous notes. For Freud, works cited in n. 27, above. Also Bernard Bosanquet, *The Principle of Individuality and Value* (London: Macmillan & Co., 1912), Lecture IV; Josiah Royce, *The Philosophy of Loyalty* (New York: Macmillan Co., 1911); Henri Bergson, *The Two Sources of Morality and Religion* (New York: Henry Holt & Co., 1935), chap. i.

65. Lewis Samuel Feuer, *Psychoanalysis and Ethics* (Springfield, Ill.: Charles C Thomas, 1955), p. 21.

66. Ruth Benedict, *The Chrysanthemum and the Sword* (New York: Houghton Mifflin Co., 1946), chaps. v–vii; see especially the schematic table of obligations, p. 116.

67. *Ibid.*, p. 102.

68. P. H. Nowell-Smith, *Ethics* (London: Penguin Books, 1954), chap. vii.

69. See works, cited in previous notes, by Bentham, Bosanquet, Dewey, Parker, Pepper, Perry. For homeostatic models cf. Anatol Rapoport, "Homeostasis Re-

considered," in Roy R. Grinker (ed.), *Towards a Unified Theory of Human Behavior* (New York: Basic Books, 1956).

70. It is sometimes held, however, that an authoritative particular command requires justification by reference to universals. This would seem to depend on the meaning of 'authoritative'; also, it may be that the demand for a universal element creeps in through the notion of justification. For a defense of the universalist condition see R. B. Brandt, *Ethical Theory* (Englewood Cliffs, N.J.: Prentice-Hall, Inc., 1959), chap. ii.

71. These are presented in greater detail in Edel, *Ethical Judgment,* chap. ii.

72. The concept of *phase rule* is intended to meet the kind of problems for which W. D. Ross employs the notion of *prima facie duty* (*The Right and the Good* [Oxford: Clarendon Press, 1930], pp. 24–29); or John Dewey distinguishes *principle* from *rule* (Dewey and Tufts, *op. cit.*, pp. 304–5); or, in terms of differentiating effect within the individual's consciousness, Hartmann develops the notion of *unavoidable guilt* (*op. cit.*, I, 299–302; cf. II, 281–85).

73. Ralph Linton, *The Study of Man* (New York: D. Appleton–Century Co., 1936), p. 433. The context is a discussion of patterns of misconduct.

74. Cf. P. F. Strawson, "Construction and Analysis," in A. J. Ayer *et al.*, *The Revolution in Philosophy* (London: Macmillan & Co., 1956).

75. Cf. Edel, *Method in Ethical Theory,* chap. iv.

76. For Kant see works cited above, n. 64; Schlick, *op. cit.*, p. 32; Edward Westermarck, *Ethical Relativity* (New York: Harcourt Brace & Co., 1932), pp. 152–53; Ross, *op. cit.*, pp. 6–7. The view that there is no restriction on the subject in the case of 'good' appears to be G. E. Moore's in *Principia ethica,* although later (in "Is Goodness a Quality?" *Aristotelian Society,* Supplementary Volume XI [1932], 122–24) he suggests translating 'good' into 'worth having for its own sake,' which limits the subject to experiences. The hedonist would, of course, share this limitation for different reasons.

77. Ross, *op. cit.,* pp. 6–7; *Foundations of Ethics* (Oxford: Clarendon Press, 1939), p. 115.

78. In modern ethics the traditional contrast of Utilitarian and Kantian ethics illustrates the contrast of goodist and rightist frameworks. In twentieth-century ethics G. E. Moore's *Principia ethica* is a clear illustration of the goodist type; see also H. W. B. Joseph, *Some Problems of Ethics* (Oxford: Clarendon Press, 1931). The insistence on primacy for 'right' or 'ought' is found in H. A. Prichard, *Moral Obligation* (Oxford: Clarendon Press, 1949), especially Essay 5. Hartmann's translation of value into an Ought-to-Be carries a rightist connotation (*op. cit.*). Recent works sometimes make a formal translation along these lines; e.g., Everett W. Hall, in *What Is Value?* (London: Routledge & Kegan Paul Ltd., 1952) considers '*a* is good' as meaning 'There is a property, X, such that *a* ought to exemplify X and *a* does exemplify X' (p. 178). The coordinate structure is a basic thesis in Ross (works cited in nn. 72 and 77). Insistence on fundamental separation of value and obligation is found in C. I. Lewis, *An Analysis of Knowledge and Valuation* (LaSalle, Ill.: Open Court Publishing Co., 1946), chaps. xiii and xvi. It takes the form of a separation of evaluation and obligation as two functions of language, in Alexander Sesonske's *Value and Obligation* (Berkeley: University of California Press, 1957). A different mode of separation (between theory of duties and theory of ideals) is found in Leonard Nelson's *System of Ethics* (New Haven, Conn.: Yale University Press, 1956). Dewey makes an interesting attempt at reconciling the conflict of the right and the good (Dewey and Tufts, *op. cit.*, pp. 249 ff.).

79. Nowell-Smith, *op. cit.*, p. 133.

80. Westermarck, *op. cit.*, p. 122.

81. Cf., respectively, Rudolf Carnap, *Philosophy and Logical Syntax* (London: Kegan Paul, 1935), pp. 22–26; A. J. Ayer, *Language, Truth and Logic* (London: Victor Gollancz, 1936), chap. vi; Bertrand Russell, *Religion and Science* (New York: Henry Holt & Co., 1935), p. 235.

82. E.g., Hall, *op. cit.*, esp. chap. 6; R. M. Hare, *The Language of Morals* (Oxford: Clarendon Press, 1952), chap. xii. See also n. 83.

83. For a useful brief outline of this area see A. R. Anderson and O. K. Moore, "The Formal Analysis of Normative Concepts," *American Sociological Review*, XXII (February, 1957), 9–17. For a system embodying 'better' as fundamental concept see Soren Halldén, *On the Logic of 'Better'* ("Library of Theoria," No. 2 [Lund: C. W. K. Gleerup, 1957]).

84. E.g., Anderson and Moore say: "Certain writers have evinced an extreme reluctance, for philosophical and grammatical reasons, to admit any very close relation between propositions and commands. In our opinion this is an open question, to be decided by constructing logical systems whose utility can be tested in scientific practice" (*op. cit.*, p. 15). They take as their point of departure the suggestion by H. G. Bohnert ("The Semiotic Status of Commands," *Philosophy of Science*, XII [July, 1945], 302–15) that "Do A" is to be taken as elliptical for "Either you will do A, or else S," where S refers to a sanction. For a criticism of Bohnert's view see Hall, *op. cit.*, pp. 131–32, and Hare, *op. cit.*, pp. 7–8. There is an interesting parallel between this kind of logical attempt and Justice Holmes's attempt to construe a contract as assumption of risk, that is, a prediction that something will come to pass or else certain "damages" will be assessed (see Oliver Wendell Holmes, Jr., *The Common Law* [1881] [Boston: Little, Brown & Co., 1938], pp. 298 ff.).

85. Edel and Edel, *op. cit.*, pp. 127–29 and chap. xiv.

86. Kurt Baier, *The Moral Point of View* (Ithaca, N.Y.: Cornell University Press, 1958), p. 183.

87. Moore, *Principia ethica*, pp. 97–99.

88. Henry Sidgwick, *The Methods of Ethics* (6th ed.; London: Macmillan & Co. Ltd., 1901), pp. 381 ff. Cf. C. I. Lewis, *op. cit.*, pp. 483 ff., 492 ff.

89. Moore, *Principia ethica*, pp. 143–44, quite clearly differentiates his view from much of traditional intuitionism.

90. Duncker, *op. cit.*

91. Stevenson, *op. cit.*, chap. vii.

92. This term is Gilbert Ryle's. For his conception of informal logic see his *Dilemmas* (Cambridge: Cambridge University Press, 1956), chap. viii.

93. Stephen E. Toulmin, *The Place of Reason in Ethics* (Cambridge: Cambridge University Press, 1953).

94. Baier, *op. cit.*, pp. 86–87.

95. *Ibid.*, pp. 142–43.

96. *Ibid.*, p. 260.

97. *Ibid.*, p. 301.

98. A parallel analysis would hold if we give an activist interpretation to furnishing good reasons. In that case the linguistic formulation would be serving the purpose of "announcing a policy of action" (Stuart Hampshire, *Thought and Action* [London: Chatto & Windus, 1959], p. 129). Hampshire adds: "I do not helplessly encounter reasons for action; I acknowledge certain things as reasons for action." The relation between phenomenological and activist components is a separate question.

99. Cf. J. O. Urmson's attempt to interpret 'good' as a grading label, in his "On Grading," *Mind*, LIX (1950), 145–69.

100. Bentham's felicific calculus is a classic source in the search for dimensions

of value measurement (*op. cit.*, chap. iv). For attention to these problems in value theory cf. Perry, *op. cit.*, chap. xxi, and his *Realms of Value* (Cambridge, Mass.: Harvard University Press, 1954), chap. iv; John Laird, *The Idea of Value* (Cambridge: Cambridge University Press, 1929), chap. x; Parker, *op. cit.*, chaps. v–vii; Charles Morris, *Varieties of Human Value* (Chicago: University of Chicago Press, 1956). While generic value criteria have been explored, criteria in the specific moral field tend to be overlooked, often on the assumption that they are special applications of the generic.

101. Edel, *Method in Ethical Theory*, chap. xv.

102. Erikson, *op. cit.*, p. 218 and chap. vii; cf. also pp. 230–34.

103. Parker, *op. cit.*, pp. 152–53. He models this on the touchstone method of criticism in literature, expounded by Matthew Arnold.

104. R. B. Braithwaite, *Moral Principles and Inductive Policies* (Annual Philosophical Lecture, Henriette Herz Trust, British Academy, 1950 [*Proceedings of the British Academy* (London), XXXVI, 68]).

105. Compare the way in which C. I. Lewis casts his basic axiom for obligation (*op. cit.*, pp. 503–10).

106. Edel, *Ethical Judgment*, chap. ix.

CHAPTER IV

107. Cf. the works of Dewey cited in nn. 35 and 37, above. See also his *Theory of Valuation*, in this *Encyclopedia*, Vol. II, No. 4.

108. Sartre, *Existentialism and Humanism.*

109. For a recent presentation of views cf. *Determinism and Freedom in the Age of Modern Science: A Philosophical Symposium*, ed. Sidney Hook (New York University Press, 1958); cf. also Wilfrid Sellars and John Hospers (eds.), *Readings in Ethical Theory* (New York: Appleton-Century-Crofts, Inc., 1952), Part VII; Herbert Fingarette, "Psychoanalytic Perspectives on Moral Guilt and Responsibility: A Re-evaluation," *Philosophy and Phenomenological Research*, XVI (September 1955), 18–36.

110. Flugel (*op. cit.*) discusses such materials with special reference to morals.

111. Smith, *The Theory of the Moral Sentiments*, p. 135.

112. For a brief sketch see Roscoe Pound, *An Introduction to the Philosophy of Law* (New Haven, Conn.: Yale University Press, 1922), chap. iv.

113. Hartmann, *op. cit.*, I, 299–302.

114. For an interesting treatment of the growth of this individualism in relation to the legal category of contract cf. Morris R. Cohen, *Law and the Social Order* (New York: Harcourt, Brace & Co., 1933), pp. 74–88.

115. For some of the complexities in the analysis of shame see Gerhart Piers and Milton B. Singer, *Shame and Guilt* (Springfield, Ill.: Charles C Thomas, 1953). For the impact of shame in the growth of an individual see Helen M. Lynd, *On Shame and the Search for Identity* (New York: Harcourt, Brace & Co., 1958). For blame see Herbert Fingarette, "Blame: Its Motive and Meaning in Everyday Life," *Psychoanalytic Review*, XLIV (April, 1957), 193–211.

116. J. S. Mill's *Utilitarianism*, chap. v, inclines at several points to such a mode of analysis. For an explicit application to the theory of praise and blame see Dewey, *Human Nature and Conduct*, e.g., pp. 314 ff. For a parallel mode of thought in legal theory cf. Oliver Wendell Holmes, "The Path of the Law," in *Collected Legal Papers* (New York: Harcourt, Brace & Co., 1920).

117. Benedict de Spinoza, *Ethics*, Part V; Bertrand Russell, "A Free Man's Worship," in *Mysticism and Logic* (New York: W. W. Norton & Co., 1929). For Stoic view see n. 64 above; Kantian, n. 64; Marxian, n. 38; Freudian, n. 27.

118. Schlick, *op. cit.*, chap. vii.

119. For consideration of the problems of novelty and emergence cf. A. O. Lovejoy, "The Meanings of 'Emergence' and Its Modes," *Proceedings of the Sixth International Congress of Philosophy* (1926), pp. 20–33; Ernest Nagel, "The Meaning of Reduction in the Natural Sciences," in Robert C. Stauffer (ed.), *Science and Civilization* (Madison: University of Wisconsin Press, 1949); Abraham Edel, *The Theory and Practice of Philosophy* (New York: Harcourt, Brace & Co., 1946), pp. 48–64; Gustav Bergmann, "Holism, Historicism and Emergence," *Philosophy of Science*, XI (October, 1944), 209–21; P. E. Meehl and Wilfrid Sellars, "The Concept of Emergence," in H. Feigl and M. Scriven (eds.), *The Foundations of Science and the Concepts of Psychology and Psychoanalysis* (Minneapolis: University of Minnesota Press, 1956), pp. 239–52; Paul Oppenheim and Hilary Putnam, "Unity of Science as a Working Hypothesis," in H. Feigl, M. Scriven, and G. Maxwell (eds.), *Concepts, Theories and the Mind-Body Problem* (Minneapolis: University of Minnesota Press, 1958), pp. 3–36; R. W. Sellars, V. J. McGill, and M. Farber (eds.), *Philosophy for the Future* (New York: Macmillan Co., 1949), essays by J. B. S. Haldane, T. C. Schneirla, and B. J. Stern.

120. Gilbert Ryle, in the reference cited above, n. 51, seems to me to be developing such a conception, though in different terms.

Theory of Valuation

John Dewey

Theory of Valuation

Contents:

Theory of Valuation

John Dewey

I. Its Problems

A skeptically inclined person viewing the present state of the discussion of valuing and values might find reason for concluding that a great ado is being made about very little, possibly about nothing at all. For the existing state of discussion shows not only that there is a great difference of opinion about the proper theoretical interpretation to be put upon facts, which might be a healthy sign of progress, but also that there is great disagreement as to what the facts are to which theory applies, and indeed whether there are any facts to which a theory of value can apply. For a survey of the current literature of the subject discloses that views on the subject range from the belief, at one extreme, that so-called "values" are but emotional epithets or mere ejaculations, to the belief, at the other extreme, that a priori necessary standardized, rational values are the principles upon which art, science, and morals depend for their validity. And between these two conceptions lies a number of intermediate views. The same survey will also disclose that discussion of the subject of "values" is profoundly affected by epistemological theories about idealism and realism and by metaphysical theories regarding the "subjective" and the "objective."

Given a situation of this sort, it is not easy to find a starting-point which is not compromised in advance. For what seems on the surface to be a proper starting-point may in fact be simply the conclusion of some prior epistemological or metaphysical theory. Perhaps it is safest to begin by asking how it is that the problem of valuation-theory has come to bulk so largely in recent discussions. Have there been any factors in intellectual history which have produced such marked changes in scientific attitudes and conceptions as to throw the problem into relief?

When one looks at the problem of valuation in this context, one is at once struck by the fact that the sciences of astronomy, physics, chemistry, etc., do not contain expressions that by any stretch of the imagination can be regarded as standing for value-facts or conceptions. But, on the other hand, all deliberate, all planned human conduct, personal and collective, seems to be influenced, if not controlled, by estimates of value or worth of ends to be attained. Good sense in practical affairs is generally identified with a sense of relative values. This contrast between natural science and human affairs apparently results in a bifurcation, amounting to a radical split. There seems to be no ground common to the conceptions and methods that are taken for granted in all physical matters and those that appear to be most important in respect to human activities. Since the propositions of the natural sciences concern matters-of-fact and the relations between them, and since such propositions constitute the subject matter acknowledged to possess preeminent scientific standing, the question inevitably arises whether scientific propositions about the direction of human conduct, about any situation into which the idea of *should* enters, are possible; and, if so, of what sort they are and the grounds upon which they rest.

The elimination of value-conceptions from the science of nonhuman phenomena is, from a historical point of view, comparatively recent. For centuries, until, say, the sixteenth and seventeenth centuries, nature was supposed to be what it is because of the presence within it of *ends*. In their very capacity as ends they represented complete or *perfect* Being. All natural changes were believed to be striving to actualize these ends as the goals toward which they moved by their own nature. Classic philosophy identified *ens*, *verum*, and *bonum*, and the identification was taken to be an expression of the constitution of nature as the object of natural science. In such a context there was no call and no place for any *separate* problem of valuation and values, since what are now termed values were taken to be integrally incorporated in the very structure of the world. But when teleological considerations were eliminated from one natural science after another, and finally from the sciences of physiology and biology, the problem of value arose as a separate problem.

If it is asked why it happened that, with the exclusion from nature of conceptions of ends and of striving to attain them, the conception of values did not entirely drop out—as did, for example, that of phlogiston—the answer is suggested by what has been said about the place of conceptions and estimates of value in distinctively human affairs. Human behavior *seems* to be influenced, if not controlled, by considerations such as are expressed in the words 'good-bad,' 'right-wrong,' 'admirable-hideous,' etc. All conduct that is not simply either blindly impulsive or mechanically routine seems to involve valuations. The problem of valuation is thus closely associated with the problem of the structure of the sciences of *human* activities and *human* relations. When the problem of valuation is placed in this context, it begins to be clear that the problem is one of moment. The various and conflicting theories that are entertained about valuation also take on significance. For those who hold that the field of scientifically warranted propositions is exhausted in the field of propositions of physics and chemistry will be led to hold that there are no genuine value-propositions or judgments, no propositions that state (affirm or deny) anything about values capable of support and test by experimental evidence. Others, who accept the distinction between the nonpersonal field and the personal or human field as one of two separate fields of existence, the physical and the mental or psychical, will hold that the elimination of value-categories from the physical field makes it clear that they are located in the mental. A third school employs the fact that value-expressions are not found in the physical sciences as proof that the subject matter of the physical sciences is only partial (sometimes called merely "phenomenal") and hence needs to be supplemented by a "higher" type of subject matter and knowledge in which value-categories are supreme over those of factual existence.

The views just listed are typical but not exhaustive. They are listed not so much to indicate the theme of discussion as to help delimit the central problem about which discussions turn, often, apparently, without being aware of their source; namely, the problem of the possibility of genuine propositions about the direction of human affairs. Were it possible, it would probably be desirable to discuss this problem with a minimum of explicit reference to value-expressions. For much ambiguity has been imported into discussion of the latter from outside epistemological and psychological sources. Since this mode of approach is not possible under existing circumstances, this introductory section will conclude with some remarks about certain linguistic expressions purporting to designate distinctive value-facts.

1. The expression 'value' is used as a verb and a noun, and there is a basic dispute as to which sense is primary. If there are things that are values or that have the property of value apart from connection with any activity, then the verb 'to value' is derivative. For in this case an act of apprehension is called valuation simply because of the object it grasps. If, however, the active sense, designated by a verb, is primary, then the noun 'value' designates what common speech calls a *valuable*—something that is the object of a certain kind of activity. For example, things which exist independently of being valued, like diamonds or mines and forests, are valuable when they are the objects of certain human activities. There are many nouns designating things not in their primary existence but as the material or objectives (as when something is called a target) of activities. The question whether this holds in the case of a thing (or the property) called value is one of the matters involved in controversy. Take, for example, the following quotations. Value is said to be "best defined as the qualitative content of an apprehending process. . . . . It is a given qualitative content present to attention or intuition." This statement would seem to take 'value' as primarily a noun, or at least an adjective, designating an object or its intrinsic quality. But when the same author goes on to speak of the process of intuiting and apprehending, he says: "What seems to distinguish

the act of valuing from the bare act of intuiting is that the former is qualified, to a noticeable degree, by feeling. . . . . It consciously discriminates some specific content. But the act of valuing is also emotional; it is the conscious expression of an interest, a motor-affective attitude." This passage gives the opposite impression of the one previously cited. Nor is the matter made clearer when it is further said that "the value-quality or content of the experience has been distinguished from the value-act or psychological attitude of which this content is the immediate object"—a position that seems like an attempt to solve a problem by riding two horses going in opposite directions.

Furthermore, when attention is confined to the usage of the verb 'to value,' we find that common speech exhibits a double usage. For a glance at the dictionary will show that in ordinary speech the words 'valuing' and 'valuation' are verbally employed to designate both *prizing*, in the sense of holding precious, dear (and various other nearly equivalent activities, like honoring, regarding highly), and *appraising* in the sense of *putting* a value upon, *assigning* value to. This is an activity of rating, an act that involves comparison, as is explicit, for example, in appraisals in money terms of goods and services. The double meaning is significant because there is implicit in it one of the basic issues regarding valuation. For in *prizing*, emphasis falls upon something having definite *personal* reference, which, like all activities of distinctively personal reference, has an aspectual quality called emotional. Valuation as *appraisal*, however, is primarily concerned with a relational property of objects so that an intellectual aspect is uppermost of the same general sort that is found in '*estimate*' as distinguished from the personal-emotional word '*esteem*.' That the same verb is employed in both senses suggests the problem upon which schools are divided at the present time. Which of the two references is basic in its implications? Are the two activities separate or are they complementary? In connection with etymological history, it is suggestive (though, of course, in no way conclusive) that 'praise,' 'prize,' and 'price' are all derived from the same Latin word; that 'appreciate' and 'appraise' were once used interchangeably; and that 'dear' is still used as equivalent both to 'precious' and to 'costly' in monetary price. While the dual significance of the word as used in ordinary speech raises a problem, the question of linguistic usage is further extended—not to say confused—by the fact that current theories often identify the verb 'to value' with 'to enjoy' in the sense of receiving pleasure or gratification from something, finding it agreeable; and also with 'to enjoy' in the active sense of *concurring* in an activity and its outcome.

2. If we take certain words commonly regarded as value-expressions, we find no agreement in theoretical discussions as to their proper status. There are, for example, those who hold that 'good' means *good for*, useful, serviceable, helpful; while 'bad' means harmful, detrimental—a conception which contains implicitly a complete theory of valuation. Others hold that a sharp difference exists between good in the sense of 'good for' and that which is 'good in itself.' Again, as just noted, there are those who hold that 'pleasant' and 'gratifying' are value-expressions of the first rank, while others would not give them standing as primary value-expressions. There is also dispute as to the respective status of 'good' and 'right' as value-words.

The conclusion is that verbal usage gives us little help. Indeed, when it is used to give direction to the discussion, it proves confusing. The most that reference to linguistic expressions can do at the outset is to point out certain problems. These problems may be used to delimit the topic under discussion. As far, then, as the terminology of the present discussion is concerned, the word 'valuation' will be used, both verbally and as a noun, as the most neutral in its theoretical implications, leaving it to further discussion to determine its connection with *prizing*, *appraising*, *enjoying*, etc.

II. Value-Expression as Ejaculatory

Discussion will begin with consideration of the most extreme of the views which have been advanced. This view affirms that value-expressions cannot be constituents of propositions, that is, of sentences which affirm or deny, because they are purely ejaculatory. Such expressions as 'good,' 'bad,' 'right,' 'wrong,' 'lovely,' 'hideous,' etc., are regarded as of the same nature as interjections; or as phenomena like blushing, smiling, weeping; or/and as stimuli to move others to act in certain ways—much as one says "Gee" to oxen or "Whoa" to a horse. They do not say or state anything, not even about feelings; they merely evince or manifest the latter.

The following quotations represent this view: "If I say to some one, 'You acted wrongly in stealing that money,' I am not *stating* anything more than if I had simply said 'You stole that money.' . . . . It is as if I had said 'You stole that money' in a peculiar tone of horror, or written it with the addition of some special exclamation marks. The tone . . . . merely serves to show that the expression is attended by certain feelings in the speaker." And again: "Ethical terms do not serve only to express feelings. They are calculated also to arouse feeling and so to stimulate action. . . . . Thus the sentence 'It is your duty to tell the truth' may be regarded both as the expression of a certain sort of ethical feeling about truthfulness and as the expression of the command 'Tell the truth.' . . . . In the sentence 'It is good to tell the truth' the command has become little more than a suggestion." On what grounds the writer calls the terms and the "feelings" of which he speaks "ethical" does not appear. Nevertheless, applying this adjective to the feelings seems to involve some objective ground for discriminating and identifying them as of a certain kind, a conclusion inconsistent with the position taken. But, ignoring this fact, we pass on to a further illustration: "In saying 'tolerance is a virtue' I should not be making a statement about my own feelings or about anything else. I should simply be evincing my own feelings, which is not at all the same thing as saying that I have them." Hence "it is impossible to dispute about questions of value," for sentences that do not say or state anything whatever cannot, a fortiori, be incompatible with one another. Cases of apparent dispute or of opposed statements are, if they have any meaning at all, reducible to differences regarding the facts of the case—as there might be a dispute whether a man performed the particular action called stealing or lying. Our hope or expectation is that if "we can get an opponent to agree with us about the empirical facts of the case he will adopt the same moral attitude toward them as we do"—though once more it is not evident why the attitude is called "moral" rather than "magical," "belligerent," or any one of thousands of adjectives that might be selected at random.

Discussion will proceed, as has previously been intimated, by analyzing the facts that are appealed to and not by discussing the merits of the theory in the abstract. Let us begin with phenomena that admittedly say nothing, like the first cries of a baby, his first smiles, or his early cooings, gurglings, and squeals. When it is said that they "express feelings," there is a dangerous ambiguity in the words 'feelings' and 'express.' What is clear in the case of tears or smiles ought to be clear in the case of sounds involuntarily uttered. They are not in themselves expressive. They are constituents of a larger organic condition. They are facts of organic behavior and are *not* in any sense whatever value-expressions. They may, however, be taken by other persons as *signs* of an organic state, and, so taken, *qua* signs or treated as *symptoms*, they evoke certain responsive forms of behavior in these other persons. A baby cries. The mother takes the cry as a sign the baby is hungry or that a pin is pricking it, and so acts to change the organic condition inferred to exist by using the cry as an evidential sign.

Then, as the baby matures, it becomes aware of the connection that exists between a certain cry, the activity evoked, and the consequences produced in response to it. The cry (gesture, posture) is now made *in order* to evoke the activity and in order to experience the consequences of that activity. Just as with respect to the original response there is a difference between the activity that is merely *caused* by the cry as a stimulus (as the cry of a child may awaken a sleeping mother before she is even aware there is a cry) and an activity that is evoked by the cry interpreted as a *sign* or evidence of something, so there is a difference between the original cry—which may properly be called purely ejaculatory—and the cry made on purpose, that is, with the intent to evoke a response that will have certain consequences. The latter cry exists in the medium of language; it is a linguistic sign that not only says something but is intended to say, to convey, to tell.

What is it which is then told or stated? In connection with this question, a fatal ambiguity in the word 'feelings' requires notice. For perhaps the view will be propounded that at most all that is communicated is the existence of certain feelings along perhaps with a desire to obtain other feelings in consequence of the activity evoked in another person. But any such view (*a*) goes contrary to the obvious facts with which the account began and (*b*) introduces a totally superfluous not to say empirically unverifiable matter. (*a*) For what we started with was not a feeling but an organic condition of which a cry, or tears, or a smile, or a blush, is a constituent part. (*b*) The word 'feelings' is accordingly either a strictly behavioral term, a name for the total organic state of which the cry or gesture is a part, or it is a word which is introduced entirely gratuitously. The phenomena in question are events in the course of the life of an organic being, not differing from taking food or gaining weight. But just as a gain in weight may be taken as a sign or evidence of proper feeding, so the cry may be taken as a sign or evidence of some special occurrence in organic life.

The phrase 'evincing feeling,' whether or not 'evincing' is taken as a synonym of 'expressing,' has, then, no business in the report of what takes place. The original activity—crying, smiling, weeping, squealing—is, as we have seen, a part of a larger organic state, so the phrase does not apply to it. When the cry or bodily attitude is purposely made, it is not a feeling that is evinced or expressed. Overt linguistic behavior is undertaken so as to obtain a change in organic conditions—a change to occur as the result of some behavior undertaken by some other person. Take another simple example: A smacking of the lips is or may be part of the original behavioral action called taking food. In one social group the noise made in smacking the lips is treated as a sign of boorishness or of "bad manners." Hence as the young grow in power of muscular control, they are taught to inhibit this activity. In another social group smacking the lips and the accompanying noise are taken as a sign that a guest is properly aware of what the host has provided. Both cases are completely describable in terms of observable modes of behavior and their respective observable consequences.

The serious problem in this connection is why the word 'feelings' is introduced in the theoretical account, since it is unnecessary in report of what actually happens. There is but one reasonable answer. The word is brought in from an alleged psychological theory which is couched in mentalistic terms, or in terms of alleged states of an inner consciousness or something of that sort. Now it is irrelevant and unnecessary to ask in connection with events before us whether there are in fact such inner states. For, even if there be such states, they are by description wholly private, accessible only to private inspection. Consequently, even if there were a legitimate introspectionist theory of states of consciousness or of feelings as purely mentalistic, there is no justification for borrowing from this theory in giving an account of the occurrences under examination. The reference to "feelings" is superfluous and gratuitous, moreover, because the important part of the account given is the use of "value-expressions" to influence the conduct of others by evoking certain responses from them. From the standpoint of an empirical report it is meaningless, since the interpretation is couched in terms of something not open to public inspection and verification. If there are "feelings" of the kind mentioned, there cannot be any assurance that any given word when used by two different persons even refers to the same thing, since the thing is not open to common observation and description.

Confining further consideration, then, to the part of the account that has an empirical meaning, namely, the existence of organic activities which evoke certain responses from others and which are capable of being employed with a view to evoking them, the following statements are warranted: (1) The phenomena in question are *social* phenomena where 'social' means simply that there is a form of behavior of the nature of an interaction or transaction between two or more persons. Such an interpersonal activity exists whenever one person—as a mother or nurse—treats a sound made by another person incidentally to a more extensive organic behavior *as a sign*, and responds to it in that capacity instead of reacting to it in its primary existence. The interpersonal activity is even more evident when the item of organic personal behavior in question takes place *for the sake of* evoking a certain kind of response from other persons. If, then, we follow the writer in locating value-expressions where he located them, we are led, after carrying out the required elimination of the ambiguity of 'expression' and the irrelevance of 'feeling,' to the conclusions that value-expressions have to do with or are involved in the behavioral relations of persons to one another. (2) Taken as signs (and, a fortiori, when used as signs) gestures, postures, and words are linguistic symbols. They say something and are of the nature of propositions. Take, for example, the case of a person who assumes the posture appropriate to an ailing person and who utters sounds such as the latter person would ordinarily make. It is then a legitimate subject of inquiry whether the person is genuinely ailing and incapacitated for work or is malingering. The conclusions obtained as a result of the inquiries undertaken will certainly "evoke" from other persons very different kinds of responsive behavior. The investigation is carried on to determine what is the actual case of things that are empirically observable; it is not about inner "feelings." Physicians have worked out experimental tests that have a high degree of reliability. Every parent and schoolteacher learns to be on guard against the assuming by a child of certain facial "expressions" and bodily attitudes for the purpose of causing inferences to be drawn which are the source of favor on the part of the adult. In such cases (they could easily be extended to include more complex matters) the propositions that embody the inference are likely to be in error when only a short segment of behavior is observed and are likely to be warranted when they rest upon a prolonged segment or upon a variety of carefully scrutinized data—traits that the propositions in question have in common with all genuine physical propositions. (3) So far the question has not been raised as to whether the propositions that occur in the course of interpersonal behavioral situations are or are not of the nature of valuation-propositions. The conclusions reached are hypothetical. *If* the expressions involved are valuation-expressions, as this particular school takes them to be, *then* it follows (i) that valuation-phenomena are social or interpersonal phenomena and (ii) that they are such as to provide material for propositions about observable events—propositions subject to empirical test and verification or refutation. But so far the hypothesis remains a hypothesis. It raises the question whether the statements which occur with a view to influencing the activity of others, so as to call out from them certain modes of activity having certain consequences, are phenomena falling under the head of valuation.

Take, for example, the case of a person calling "Fire!" or "Help!" There can be no doubt of the intent to influence the conduct of others in order to bring about certain consequences capable of observation and of statement in propositions. The expressions, taken in their observable context, say something of a complex character. When analyzed, what is said is (i) that there exists a situation that will have obnoxious consequences; (ii) that the person uttering the expressions is unable to cope with the situation; and (iii) that an improved situation is anticipated in case the assistance of others is obtained. All three of these matters are capable of being tested by empirical evidence, since they all refer to things that are observable. The proposition in which the content of the last point (the anticipation) is stated is capable, for example, of being tested by observation of what happens in a particular case. Previous observations may substantiate the conclusion that in any case objectionable consequences are much less likely to happen if the linguistic sign is employed in order to obtain the assistance it is designed to evoke.

Examination shows certain resemblances between these cases and those previously examined which, according to the passage quoted, contain valuation-expressions. The propositions refer directly to an *existing* situation and indirectly to a *future* situation which it is intended and desired to produce. The expressions noted are employed as intermediaries to bring about the desired change from present to future conditions. In the set of illustrative cases that was first examined, certain valuation-words, like 'good' and 'right,' explicitly appear; in the second set there are no *explicit* value-expressions. The cry for aid, however, when taken in connection with its existential context, affirms in effect, although not in so many words, that the situation with reference to which the cry is made is "bad." It is "bad" in the sense that it is objected to, while a future situation which is *better* is anticipated, provided the cry evokes a certain response. The analysis may seem to be unnecessarily detailed. But, unless in each set of examples the existential context is made clear, the verbal expressions that are employed can be made to mean anything or nothing. When the contexts are taken into account, what emerges are propositions assigning a relatively negative value to existing conditions; a comparatively positive value to a prospective set of conditions; and intermediate propositions (which may or may not contain a valuation-expression) intended to evoke activities that will bring about a transformation from one state of affairs to another. There are thus involved (i) aversion to an existing situation and attraction toward a prospective possible situation and (ii) a *specifiable and testable relation between the latter as an end and certain activities as means for accomplishing it.* Two problems for further discussion are thus set. One of them is the relation of active or behavioral attitudes to what may be called (for the purpose of identification) *liking* and *disliking*, while the other is the relation of valuation to things as means-end.

III. Valuation as Liking and Disliking

That liking and disliking in their connection with valuation are to be considered in terms of observable and identifiable modes of behavior follows from what is stated in the previous section. As behavioral the adjective 'affective-motor' is applicable, although care must be taken not to permit the "affective" quality to be interpreted in terms of private "feelings"—an interpretation that nullifies the active and observable element expressed in 'motor.' For the "motor" takes place in the public and observable world, and, like anything else taking place there, has observable conditions and consequences. When, then, the word 'liking' is used as a name for a mode of behavior (not as a name for a private and inaccessible feeling), what sort of activities does it stand for? What is its designatum? This inquiry is forwarded by noting that the words 'caring' and 'caring for' are, as modes of behavior, closely connected with 'liking,' and that other substantially equivalent words are 'looking out for or after,' 'cherishing,' 'being devoted to,' 'attending to,' in the sense of 'tending', 'ministering to,' 'fostering'—words that all seem to be variants of what is referred to by 'prizing,' which, as we saw earlier, is one of the two main significations recognized by the dictionary. When these words are taken in the behavioral sense, or as naming activities that take place so as to maintain or procure certain conditions, it is possible to demarcate what is designated by them from things designated by such an ambiguous word as 'enjoy.' For the latter word may point to a condition of *receiving* gratification *from* something already in existence, apart from any affective-motor action exerted as a condition of its production or continued existence. Or it may refer to precisely the latter activity, in which case 'to enjoy' is a synonym for the activity of taking delight in an effort, having a certain overtone of relishing, which "takes pains," as we say, to perpetuate *the existence of conditions* from which gratification is received. Enjoying in this active sense is marked by energy expended to secure the conditions that are the source of the gratification.

The foregoing remarks serve the purpose of getting theory away from a futile task of trying to assign signification to words in isolation from objects as designata. We are led instead to evocation of specifiable existential situations and to observation of what takes place in them. We are directed to observe whether energy is put forth to call into existence or to maintain in existence certain conditions; in ordinary language, to note whether effort is evoked, whether pains are taken to bring about the existence of certain conditions rather than others, the need for expenditure of energy showing that there exist conditions adverse to what is wanted. The mother who professes to prize her child and to enjoy (in the active sense of the word) the child's companionship but who systematically neglects the child and does not seek out occasions for being with the child is deceiving herself; if she makes, in addition, demonstrative signs of affection—like fondling—only when others are present, she is presumably trying to deceive them also. It is by observations of behavior—which observations (as the last illustration suggests) may need to be extended over a considerable space-time —that the existence and description of valuations have to be determined. Observation of the amount of energy expended and the length of time over which it persists enables qualifying adjectives like 'slight' and 'great' to be warrantably prefixed to a given valuation. The direction the energy is observed to take, as toward and away from, enables grounded discrimination to be made between "positive" and "negative" valuations. If there are "feelings" existing in addition, their existence has nothing to do with any verifiable proposition that can be made about a valuation.

Because valuations in the sense of prizing and caring for occur only when it is necessary to bring something into existence which is lacking, or to conserve in existence something which is menaced by outside conditions, valuation *involves* desiring. The latter is to be distinguished from mere wishing in the sense in which wishes occur in the absence of effort. "If wishes were horses, beggars would ride." There is something lacking, and it would be gratifying if it were present, but there is either no energy expended to bring what is absent into existence or else, under the given conditions, no expenditure of effort would bring it into existence—as when the baby is said to cry for the moon, and when infantile adults indulge in dreams about how nice everything would be if things were only different. The *designata* in the cases to which the names 'desiring' and 'wishing' are respectively applied are basically different. When, accordingly, 'valuation' is defined in terms of desiring, the prerequisite is a treatment of desire in terms of the existential context in which it arises and functions. If 'valuation' is defined in terms of desire as something initial and complete in itself, there is nothing by which to discriminate one desire from another and hence no way in which to measure the worth of different valuations in comparison with one another. Desires are desires, and that is all that can be said. Furthermore, desire is then conceived of as *merely* personal and hence as not capable of being stated in terms of other objects or events. If, for example, it should happen to be noted that effort ensues upon desire and that the effort put forth changes existing conditions, these considerations would then be looked upon as matters wholly external to desire—provided, that is, desire is taken to be original and complete in itself, independent of an observable contextual situation.

When, however, desires are seen to arise only within certain existential contexts (namely, those in which some lack prevents the immediate execution of an active tendency) and when they are seen to function in reference to these contexts in such a way as to make good the existing want, the relation between desire and *valuation* is found to be such as both to make possible, and to require, statement in verifiable propositions. (i) The content and object of desires are seen to depend upon the particular context in which they arise, a matter that in turn depends upon the antecedent state of both personal activity and of surrounding conditions. Desires for food, for example, will hardly be the same if one has eaten five hours or five days previously, nor will they be of the same content in a hovel and a palace or in a nomadic or agricultural group. (ii) Effort, instead of being something that comes after desire, is seen to be of the very essence of the tension involved in desire. For the latter, instead of being merely personal, is an active relation of the organism to the environment (as is obvious in the case of hunger), a factor that makes the difference between genuine desire and mere wish and fantasy. It follows that valuation in its connection with desire is linked to existential situations and that it differs with differences in its existential context. Since its existence depends upon the situation, its adequacy depends upon its adaptation to the needs and demands imposed by the situation. Since the situation is open to observation, and since the consequences of effort-behavior as observed determine the adaptation, the adequacy of a given desire can be stated in propositions. The propositions are capable of empirical test because the connection that exists between a given desire and the conditions with reference to which it functions are ascertained by means of these observations.

The word 'interest' suggests in a forcible way the active connection between personal activity and the conditions that must be taken into account in the theory of valuation. Even in etymology it indicates something in which both a person and surrounding conditions participate in intimate connection with one another. In naming this something that occurs between them it names a transaction. It points to an activity which takes effect through the mediation of external conditions. When we think, for example, of the interest of any particular group, say the bankers' interest, the trade-union interest, or the interest of a political machine, we think not of mere states of mind but of the group as a pressure group having organized channels in which it directs action to obtain and make secure conditions that will produce specified consequences. Similarly in the case of singular persons, when a court recognizes an individual as having an interest in some matter, it recognizes that he has certain claims whose enforcement will affect an existential issue or outcome. Whenever a person has an interest in something, he has a stake in the course of events and in their final issue—a stake which leads him to take action to bring into existence a particular result rather than some other one.

It follows from the facts here adduced that the view which connects valuation (and "values") with desires and interest is but a starting-point. It is indeterminate in its bearing upon the theory of valuation until the nature of interest and desire has been analyzed, and until a method has been established for determining the constituents of desires and interests in their concrete particular occurrence. Practically all the fallacies in the theories that connect valuation with desire result from taking "desire" at large. For example, when it is said (quite correctly) that "values *spring from* the immediate and inexplicable reaction of vital impulse and from the irrational part of our nature," what is actually stated is that vital impulses are a *causal condition* of the existence of desires. When "vital impulse" is given the only interpretation which is empirically verifiable (that of an organic biological tendency), the fact that an "irrational" factor is the causal condition of valuations proves that valuations have their roots *in an existence* which, like any existence *taken in itself*, is *a*-rational. Correctly interpreted, the statement is thus a reminder that organic tendencies are existences which are connected with other existences (the word 'irrational' adds nothing to "*existence*" as such) and hence are observable. But the sentence cited is often interpreted to mean that vital impulses *are* valuations—an interpretation which is incompatible with the view which connects valuations with desires and interests, and which, by parity of logic, would justify the statement that trees are seeds since they "spring from" seeds. Vital impulses are doubtless conditions *sine qua non* for the existence of desires and interests. But the latter include foreseen consequences along with ideas in the form of signs of the measures (involving expenditure of energy) required to bring the ends into existence. When valuation is identified with the activity of desire or interest, its identification with vital impulse is denied. For its identification with the latter would lead to the absurdity of making every organic activity of every kind an act of valuation, since there is none that does not involve some "vital impulse."

The view that "a value is any object of any interest" must also be taken with great caution. On its face it places all interests on exactly the same level. But, when interests are examined in their concrete makeup in relation to their place in some situation, it is plain that everything depends upon the objects involved in them. This in turn depends upon the care with which the needs of existing situations have been looked into and upon the care with which the ability of a proposed act to satisfy or fulfil just those needs has been examined. That all interests stand on the same footing with respect to their function as valuators is contradicted by observation of even the most ordinary of everyday experiences. It may be said that an interest in burglary and its fruits confers value upon certain objects. But the valuations of the burglar and the policeman are not identical, any more than the interest in the fruits of productive work institutes the same values as does the interest of the burglar in the pursuit of his calling—as is evident in the action of a judge when stolen goods are brought before him for disposition. Since interests occur in definite existential contexts and not at large in a void, and since these contexts are situations within the life-activity of a person or group, interests are so linked with one another that the valuation-capacity of any one is a function of the set to which it belongs. The notion that a value is equally any object of any interest can be maintained only upon a view that completely isolates them from one another—a view that is so removed from readily observed facts that its existence can be explained only as a corollary of the introspectionist psychology which holds that desires and interests are but "feelings" instead of modes of behavior.

IV. Propositions of Appraisal

Since desires and interests are activities which take place in the world and which have effects in the world, they are observable in themselves and in connection with their observed effects. It might seem then as if, upon any theory that relates valuation with desire and interest, we had now come within sight of our goal—the discovery of valuation-propositions. Propositions *about* valuations have, indeed, been shown to be possible. But they are valuation-propositions only in the sense in which propositions about potatoes are potato-propositions. They are propositions about matters-of-fact. The fact that these occurrences happen to be valuations does not make the propositions valuation-propositions in any distinctive sense. Nevertheless, the fact that such matter-of-fact propositions can be made is of importance. For, unless they exist, it is doubly absurd to suppose that valuation-propositions in a *distinctive* sense can exist. It has also been shown that the subject matter of personal activities forms no theoretical barrier to institution of matter-of-fact propositions, for the behavior of human beings is open to observation. While there are practical obstacles to the establishment of valid general propositions about such behavior (i.e., about the relations of its constituent acts), its conditions and effects may be investigated. Propositions about valuations made in terms of their conditions and consequences delimit the problem as to existence of valuation-propositions in a *distinctive* sense. Are propositions about existent valuations themselves capable of being appraised, and can the appraisal when made enter into the constitution of further valuations? That a mother prizes or holds dear her child, we have seen, may be determined by observation; and the conditions and effects of different kinds of prizing or caring for may, in theory, be compared and contrasted with one another. In case the final outcome is to show that some kinds of acts of prizing are *better* than others, valuation-acts are themselves evaluated, and the evaluation may modify further direct acts of prizing. If this condition is satisfied, then propositions about valuations that actually take place become the subject matter of valuations in a distinctive sense, that is, a sense that marks them off both from propositions of physics and from historical propositions about what human beings have in fact done.

We are brought thus to the problem of the nature of appraisal or evaluation which, as we saw, is one of the two recognized significations of 'valuation.' Take such an elementary appraisal proposition as "This plot of ground is worth $200 a front foot." It is different in form from the proposition, "It has a frontage of 200 feet." The latter sentence states a matter of accomplished fact. The former sentence states a rule for determination of an act to be performed, its reference being to the future and not to something already accomplished or done. If stated in the context in which a tax-assessor operates, it states a regulative condition for levying a tax against the owner; if stated by the owner to a real estate dealer, it sets forth a regulative condition to be observed by the latter in offering the property for sale. The future act or state is not set forth as a prediction of what will happen but as something which *shall* or *should* happen. Thus the proposition may be said to lay down a norm, but "norm" must be understood simply in the sense of a condition *to be* conformed to in definite forms of future action. That rules are all but omnipresent in every mode of human relationship is too obvious to require argument. They are in no way confined to activities to which the name 'moral' is applied. Every recurrent form of activity, in the arts and professions, develops rules as to the best way in which to accomplish the ends in view. Such rules are used as criteria or "norms" for judging the value of proposed modes of behavior. The existence of rules for valuation of modes of behavior in different fields as wise or unwise, economical or extravagant, effective or futile, cannot be denied. The problem concerns not their existence as general propositions (since every rule of action is general) but whether they express only custom, convention, tradition, or are capable of stating relations between things as means and other things as consequences, which relations are themselves grounded in empirically ascertained and tested existential relations such as are usually termed those of cause and effect.

In the case of some crafts, arts, and technologies, there can be no doubt which of these alternatives is correct. The medical art, for example, is approaching a state in which many of the rules laid down for a patient by a physician as to what it is *better* for him to do, not merely in the way of medicaments but of diet and habits of life, are based upon experimentally ascertained principles of chemistry and physics. When engineers say that certain materials subjected to certain technical operations are *required* if a bridge capable of supporting certain loads is to be built over the Hudson River at a certain point, their advice does not represent their personal opinions or whims but is backed by acknowledged physical laws. It is commonly believed that such devices as radios and automobiles have been greatly improved (bettered) since they were first invented, and that the betterment in the relation of means to consequences is due to more adequate scientific knowledge of underlying physical principles. The argument does not demand the belief that the influence of custom and convention is entirely eliminated. It is enough that such cases show that it is possible for rules of appraisal or evaluation to rest upon scientifically warranted physical generalizations and that the ratio of rules of this type to those expressing mere customary habits is on the increase.

In medicine a quack may cite a number of alleged cures as evidential ground for taking the remedies he offers. Only a little examination is needed to show in what definite respects the procedures he recommends differ from those said to be "good" or to be "required" by competent physicians. There is, for example, no analysis of the cases presented as evidence to show that they are actually like the disease for the cure of which the remedy is urged; and there is no analysis to show that the recoveries which are said (rather than proved) to have taken place were in fact due to taking the medicine in question rather than to any one of an indefinite number of other causes. Everything is asserted wholesale with no analytic control of conditions. Furthermore, the first requirement of scientific procedure—namely, full publicity as to materials and processes—is lacking. The sole justification for citing these familiar facts is that their contrast with competent medical practice shows the extent to which the rules of procedure in the latter art have the warrant of tested empirical propositions. Appraisals of courses of action as better and worse, more and less serviceable, are as experimentally justified as are nonvaluative propositions about impersonal subject matter. In advanced engineering technologies propositions that state the *proper* courses of action to be adopted are evidently grounded in generalizations of physical and chemical science; they are often referred to as *applied* science. Nevertheless, propositions which lay down rules for procedures as being fit and good, as distinct from those that are inept and bad, are different in form from the scientific propositions upon which they rest. For they are rules for the use, in and by human activity, of scientific generalizations as means for accomplishing certain desired and intended ends.

Examination of these appraisals discloses that they have to do with things as they sustain to each other the relation of *means to ends or consequences*. Wherever there is an appraisal involving a rule as to better or as to needed action, there is an end to be reached: the appraisal is a valuation of things with respect to their serviceability or needfulness. If we take the examples given earlier, it is evident that real estate is appraised for the purpose of levying taxes or fixing a selling price; that medicinal treatments are appraised with reference to the end of effecting recovery of health; that materials and techniques are valued with respect to the building of bridges, radios, motorcars, etc. If a bird builds its nest by what is called pure "instinct," it does not have to appraise materials and processes with respect to their fitness for an end. But if the result—the nest—is contemplated as an object of desire, then either there is the most arbitrary kind of trial-and-error operations or there is consideration of the fitness and usefulness of materials and processes to bring the desired object into existence. And this process of weighing obviously involves comparison of different materials and operations as alternative possible means. In every case, except those of sheer "instinct" and complete trial and error, there are involved observation of actual materials and estimate of their potential force in production of a particular result. There is always some observation of the *outcome attained* in comparison and contrast with that intended, such that the comparison throws light upon the actual fitness of the things employed as means. It thus makes possible a better judgment in the future as to their fitness and usefulness. On the basis of such observations certain modes of conduct are adjudged silly, imprudent, or unwise, and other modes of conduct sensible, prudent, or wise, the discrimination being made upon the basis of the validity of the estimates reached about the relation of things as means to the end or consequence actually reached.

The standing objection raised against this view of valuation is that it applies only to things *as means*, while propositions that are genuine valuations apply to things as *ends*. This point will be shortly considered at length. But it may be noted here that ends are appraised in the same evaluations in which things as means are weighed. For example, an end suggests itself. But, when things are weighed as means toward that end, it is found that it will take too much time or too great an expenditure of energy to achieve it, or that, if it were attained, it would bring with it certain accompanying inconveniences and the promise of future troubles. It is then appraised and rejected as a "bad" end.

The conclusions reached may be summarized as follows: (1) There are propositions which are not merely about valuations that have actually occurred (about, i.e., prizings, desires, and interests that have taken place in the past) but which describe and define certain things as good, fit, or proper in a definite existential relation: these propositions, moreover, are *generalizations*, since they form rules for the proper use of materials. (2) The existential relation in question is that of means-ends or means-consequences. (3) These propositions in their generalized form may rest upon scientifically warranted empirical propositions and are themselves capable of being tested by observation of results actually attained as compared with those intended.

The objection brought against the view just set forth is that it fails to distinguish between things that are good and right in and of themselves, immediately, intrinsically, and things that are simply good *for* something else. In other words, the latter are useful for attaining the things which have, so it is said, value in and of themselves, since they are prized for their own sake and not as means to something else. This distinction between two different meanings of 'good' (and 'right') is, it is claimed, so crucial for the whole theory of valuation and values that failure to make the distinction destroys the validity of the conclusions that have been set forth. This objection definitely puts before us for consideration the question of the relations to each other of the categories of *means* and *end*. In terms of the dual meaning of 'valuation' already mentioned, the question of the relation of *prizing* and *appraising* to one another is explicitly raised. For, according to the objection, appraising applies only to *means*, while prizing applies to things that are *ends*, so that a difference must be recognized between valuation in its full pregnant sense and evaluation as a secondary and derived affair.

Let the connection between prizing and valuation be admitted and also the connection between desire (and interest) and prizing. The problem as to the relation between appraisal of things as means and prizing of things as ends then takes the following form: Are desires and interests ('likings,' if one prefers that word), which directly effect an institution of end-values, independent of the appraisal of things as means or are they intimately influenced by this appraisal? If a person, for example, finds after due investigation that an immense amount of effort is required to procure the conditions that are the means required for realization of a desire (including perhaps sacrifice of other end-values that might be obtained by the same expenditure of effort), does that fact react to modify his original desire and hence, by definition, his valuation? A survey of what takes place in any deliberate activity provides an affirmative answer to this question. For what is deliberation except weighing of various alternative desires (and hence end-values) in terms of the conditions that are the means of their execution, and which, as means, determine the consequences actually arrived at? There can be no control of the operation of foreseeing consequences (and hence of forming ends-in-view) save in terms of conditions that operate as the causal conditions of their attainment. The proposition in which any object adopted as an end-in-view is statable (or explicitly stated) is *warranted* in just the degree to which existing conditions have been surveyed and appraised in their capacity as means. The sole alternative to this statement is that no deliberation whatsoever occurs, no ends-in-view are formed, but a person acts directly upon whatever impulse happens to present itself.

Any survey of the experiences in which ends-in-view are

formed, and in which earlier impulsive tendencies are shaped through deliberation into a *chosen* desire, reveals that the object finally valued as an end to be reached is determined in its concrete makeup by appraisal of existing conditions as means. However, the habit of completely separating the conceptions of ends from that of means is so ingrained because of a long philosophical tradition that further discussion is required.

1. The common assumption that there is a sharp separation between things, on the one hand, as useful or helpful, and, on the other hand, as *intrinsically* good, and hence that there exists a separation between propositions as to what is expedient, prudent, or advisable and what is inherently desirable, does not, in any case, state a *self-evident* truth. The fact that such words as 'prudent,' 'sensible,' and 'expedient,' in the long run, or after survey of all conditions, merge so readily into the word 'wise' suggests (though, of course, it does not prove) that ends framed in separation from consideration of things as means are foolish to the point of irrationality.

2. Common sense regards some desires and interests as shortsighted, "blind," and others, in contrast, as enlightened, farsighted. It does not for a moment lump all desires and interests together as having the same status with respect to end-values. Discrimination between their respective shortsightedness and farsightedness is made precisely on the ground of whether the object of a given desire is viewed as, in turn, itself a conditioning means of further consequences. Instead of taking a laudatory view of "immediate" desires and valuations, common sense treats refusal to mediate as the very essence of short-view judgment. For treating the end as *merely* immediate and exclusively final is equivalent to refusal to consider what will happen after and because a particular end is reached.

3. The words 'inherent,' 'intrinsic,' and 'immediate' are used ambiguously, so that a fallacious conclusion is reached. Any quality or property that actually belongs to any object or event is properly said to be immediate, inherent, or intrinsic. The fallacy consists in interpreting what is designated by these terms as out of relation to anything else and hence as absolute.

For example, *means* are by definition relational, mediated, and mediating, since they are intermediate between an existing situation and a situation that is to be brought into existence by their use. But the relational character of the *things* that are employed as means does not prevent the things from having their own immediate qualities. In case the things in question are prized and cared for, then, according to the theory that connects the property of value with prizing, they necessarily have an immediate quality of value. The notion that, when means and instruments are valued, the value-qualities which result are only instrumental is hardly more than a bad pun. There is nothing in the nature of prizing or desiring to prevent their being directed to things which are means, and there is nothing in the nature of means to militate against their being desired and prized. In empirical fact, the measure of the value a person attaches to a given end is not what he *says* about its preciousness but the care he devotes to obtaining and using the *means* without which it cannot be attained. No case of notable achievement can be cited in any field (save as a matter of sheer accident) in which the persons who brought about the end did not give loving care to the instruments and agencies of its production. The dependence of ends attained upon means employed is such that the statement just made reduces in fact to a tautology. Lack of desire and interest are proved by neglect of, and indifference to, required means. As soon as an attitude of desire and interest has been developed, then, because without full-hearted attention an end which is professedly prized will not be attained, the desire and interest in question automatically attach themselves to whatever other things are seen to be required means of attaining the end.

The considerations that apply to 'immediate' apply also to 'intrinsic' and 'inherent.' A quality, including that of value, is inherent if it actually belongs to something, and the question of whether or not it belongs is one of *fact* and not a question that can be decided by dialectical manipulation of the concept of inherency. If one has an ardent desire to obtain certain things as means, then the quality of value belongs to, or inheres in, those things. For the time being, producing or obtaining those means *is* the end-in-view. The notion that only that which is out of relation to everything else can justly be called *inherent* is not only itself absurd but is contradicted by the very theory that connects the value of objects as ends with desire and interest, for this view expressly makes the value of the end-object relational, so that, if the inherent is identified with the nonrelational, there are, according to this view, no inherent values at all. On the other hand, if it is the fact that the quality exists in this case, because that to which it belongs is conditioned by a relation, then the relational character of means cannot be brought forward as evidence that their value is not inherent. The same considerations apply to the terms 'intrinsic' and 'extrinsic' as applied to value-qualities. Strictly speaking, the phrase 'extrinsic value' involves a contradiction in terms. Relational properties do not lose their intrinsic quality of being just what they are because their coming into being is *caused* by something 'extrinsic.' The theory that such is the case would terminate logically in the view that there are no intrinsic qualities whatever, since it can be shown that such intrinsic qualities as *red*, *sweet*, *hard*, etc., are causally conditioned as to their occurrence. The trouble, once more, is that a dialectic of concepts has taken the place of examination of actual empirical facts. The extreme instance of the view that to be intrinsic is to be out of any relation is found in those writers who hold that, since values *are* intrinsic, they cannot depend upon *any* relation whatever, and certainly not upon a relation to human beings. Hence this school attacks those who connect value-properties with desire and interest on exactly the same ground that the latter equate the distinction between the values of means and ends with the distinction between instrumental and intrinsic values. The views of this extreme nonnaturalistic school may, accordingly, be regarded as a definite exposure of what happens when an analysis of the abstract concept of 'intrinsicalness' is substituted for analysis of empirical occurrences.

The more overtly and emphatically the valuation of objects as ends is connected with desire and interest, the more evident it should be that, since desire and interest are ineffectual save as they co-operatively interact with environing conditions, valuation of desire and interest, as means correlated with other means, is the sole condition for valid appraisal of objects as ends. If the lesson were learned that the object of scientific knowledge is *in any case* an ascertained correlation of changes, it would be seen, beyond the possibility of denial, that anything taken *as end* is in its own content or constituents a correlation of the energies, personal and extra-personal, which operate as means. An end as an *actual* consequence, as an existing outcome, is, like any other occurrence which is scientifically analyzed, nothing but the interaction of the conditions that bring it to pass. Hence it follows necessarily that the *idea* of the object of desire and interest, the *end-in-view* as distinct from the end or outcome actually effected, is warranted in the precise degree in which it is formed in terms of these operative conditions.

4. The chief weakness of current theories of valuation which relate the latter to desire and interest is due to failure to make an empirical analysis of concrete desires and interests as they actually exist. When such an analysis is made, certain relevant considerations at once present themselves.

(i) Desires are subject to frustration and interests are subject to defeat. The likelihood of the occurrence of failure in attaining desired ends is in direct ratio to failure to form desire and interest (and the objects they involve) on the basis of conditions that operate either as obstacles (negatively valued) or as positive resources. The difference between reasonable and unreasonable desires and interests is precisely the difference between those which arise casually and are not reconstituted through consideration of the conditions that will actually decide the outcome and those which are formed on the basis of existing liabilities and potential resources. That desires as they first present themselves are the product of a mechanism consisting of native organic tendencies and acquired habits is an undeniable fact. All growth in maturity consists in *not* immediately giving way to such tendencies but in remaking them in their first manifestation through consideration of the consequences they will occasion *if* they are acted upon—an operation which is equivalent to judging or evaluating them as means operating in connection with extra-personal conditions as also means. Theories of valuation which relate it to desire and interest cannot both eat their cake and have it. They cannot continually oscillate between a view of desire and interest that identifies the latter with impulses just as they happen to occur (as products of organic mechanisms) and a view of desire as a modification of a raw impulse through foresight of its outcome; the latter alone being desire, the whole difference between impulse and desire is made by the presence in desire of an end-in-view, of objects *as* foreseen consequences. The foresight will be dependable in the degree in which it is constituted by examination of the conditions that will in fact decide the outcome. If it seems that this point is being hammered in too insistently, it is because the issue at stake is nothing other and nothing less than the possibility of distinctive valuation-propositions. For it cannot be denied that propositions having evidential warrant and experimental test are possible in the case of evaluation of things as means. Hence it follows that, if these propositions enter into the formation of the interests and desires which are valuations of ends, the latter are thereby constituted the subject matter of authentic empirical affirmations and denials.

(ii) We commonly speak of "learning from experience" and the "maturity" of an individual or a group. What do we mean by such expressions? At the very least, we mean that in the history of individual persons and of the human race there takes place a change from original, comparatively unreflective, impulses and hard-and-fast habits to desires and interests that incorporate the results of critical inquiry. When this process is examined, it is seen to take place chiefly on the basis of careful observation of differences found between desired and proposed ends (ends-*in-view*) and attained ends or actual consequences. Agreement between what is wanted and anticipated and what is actually obtained confirms the selection of conditions which operate as means to the desired end; discrepancies, which are experienced as frustrations and defeats, lead to an inquiry to discover the causes of failure. This inquiry consists of more and more thorough examination of the conditions under which impulses and habits are formed and in which they operate. The result is formation of desires and interests which are what they are through the union of the affective-motor conditions of action with the intellectual or ideational. The latter is there in any case if there is an end-in-view of any sort, no matter how casually formed, while it is adequate in just the degree in which the end is constituted in terms of the conditions of its actualization. For, wherever there is an *end-in-view* of any sort whatever, there is affective-*ideational*-motor activity; or, in terms of the dual meaning of valuation, there is union of prizing and appraising. Observation of results obtained, of *actual* consequences in their agreement with and difference from ends anticipated or held in view, thus provides the conditions by which desires and interests (and hence valuations) are matured and tested. Nothing more contrary to common sense can be imagined than the notion that we are incapable of changing our desires and interests by means of learning what the consequences of acting upon them are, or, as it is sometimes put, of *indulging* them. It should not be necessary to point in evidence to the spoiled child and the adult who cannot "face reality." Yet, as far as valuation and the theory of values are concerned, any theory which isolates valuation of ends from appraisal of means equates the spoiled child and the irresponsible adult to the mature and sane person.

(iii) Every person in the degree in which he is capable of learning from experience draws a distinction between what is desired and what is desirable whenever he engages in formation and choice of competing desires and interests. There is nothing far-fetched or "moralistic" in this statement. The contrast referred to is simply that between the object of a desire as it first presents itself (because of the existing mechanism of impulses and habits) and the object of desire which emerges as a revision of the first-appearing impulse, after the latter is critically judged in reference to the conditions which will decide the actual result. The "desirable," or the object which *should* be desired (valued), does not descend out of the a priori blue nor descend as an imperative from a moral Mount Sinai. It presents itself because past experience has shown that hasty action upon uncriticized desire leads to defeat and possibly to catastrophe. The "desirable" as distinct from the "desired" does not then designate something at large or a priori. It points to the difference between the operation and consequences of unexamined impulses and those of desires and interests that are the product of investigation of conditions and consequences. Social conditions and pressures are part of the conditions that affect the execution of desires. Hence they have to be taken into account in framing ends in terms of available means. But the distinction between the "is" in the sense of the object of a casually emerging desire and the "should be" of a desire framed in relation to actual conditions is a distinction which in any case is bound to offer itself as human beings grow in maturity and part with the childish disposition to "indulge" every impulse as it arises.

Desires and interests are, as we have seen, themselves causal conditions of results. As such they are potential means and have to be appraised as such. This statement is but a restatement of points already made. But it is worth making because it forcibly indicates how far away some of the theoretical views of valuation are from practical common-sense attitudes and beliefs. There is an indefinite number of proverbial sayings which in effect set forth the necessity of not treating desires and interests as final in their first appearance but of treating them as means—that is, of appraising them and forming objects or ends-in-view on the ground of what consequences they will tend to produce in practice. "Look before you leap"; "Act in haste, repent at leisure"; "A stitch in time saves nine"; "When angry count ten"; "Do not put your hand to the plow until the cost has been counted"—are but a few of the many maxims. They are summed up in the old saying, "*Respice finem*"—a saying which marks the difference between simply *having* an end-in-view for which *any* desires suffices, and *looking*, examining, to make sure that the consequences that will actually result are such as will be actually prized and valued when they occur. Only the exigencies of a preconceived theory (in all probability one seriously infected by the conclusions of an uncritically accepted "subjectivistic" psychology) will ignore the concrete differences that are made in the content of "likings" and "prizings," and of desires and interests, by evaluating them in their respective causal capacities when they are taken as means.

V. Ends and Values

It has been remarked more than once that the source of the trouble with theories which relate value to desire and interest, and then proceed to make a sharp division between prizing and appraisal, between ends and means, is the failure to make an empirical investigation of the actual conditions under which desires and interests arise and function, and in which end-objects, ends-in-view, acquire their actual contents. Such an analysis will now be undertaken.

When we inquire into the actual emergence of desire and its object and the value-property ascribed to the latter (instead of merely manipulating dialectically the general concept of desire), it is as plain as anything can be that desires arise only when "there is something the matter," when there is some "trouble" in an existing situation. When analyzed, this "something the matter" is found to spring from the fact that there is something lacking, wanting, in the existing situation as it stands, an absence which produces conflict in the elements that do exist. When things are going completely smoothly, desires do not arise, and there is no occasion to project ends-in-view, for "going smoothly" signifies that there is no need for effort and struggle. It suffices to let things take their "natural" course. There is no occasion to investigate what it would be better to have happen in the future, and hence no projection of an end-object.

Now vital impulses and acquired habits often operate without the intervention of an end-in-view or a purpose. When someone finds that his foot has been stepped on, he is likely to react with a push to get rid of the offending element. He does not stop to form a definite desire and set up an end to be reached. A man who has started walking may continue walking from force of an acquired habit without continually interrupting his course of action to inquire what object is to be obtained at the next step. These rudimentary examples are typical of much of human activity. Behavior is often so direct that no desires and ends intervene and no valuations take place. Only the requirements of a preconceived theory will lead to the conclusion that a hungry animal seeks food because it has formed an idea of an end-object to be reached, or because it has evaluated that object in terms of a desire. Organic tensions suffice to keep the animal going until it has found the material that relieves the tension. But if and when *desire* and *an end-in-view* intervene between the occurrence of a vital impulse or a habitual tendency and the execution of an activity, then the impulse or tendency is to some degree modified and transformed: a statement which is purely tautological, since the occurrence of a desire related to an end-in-view *is* a transformation of a prior impulse or routine habit. It is only in such cases that valuation occurs. This fact, as we have seen, is of much greater importance than it might at first sight seem to be in connection with the theory which relates valuation to desire and interest,[1] for it proves that valuation takes place only when there is something the matter; when there is some trouble to be done away with, some need, lack, or privation to be made good, some conflict of tendencies to be resolved by means of changing existing conditions. This fact in turn proves that there is present an intellectual factor—a factor of inquiry—whenever there is valuation, for the end-in-view is formed and projected as that which, if acted upon, will supply the existing need or lack and resolve the existing conflict. It follows from this that the difference in different desires and their correlative ends-in-view depends upon two things. The first is the adequacy with which inquiry into the lacks and conflicts of the existing situa-

[1] Cf. pp. 29 ff., above.

tion has been carried on. The second is the adequacy of the inquiry into the likelihood that the particular end-in-view which is set up will, if acted upon, actually fill the existing need, satisfy the requirements constituted by what is needed, and do away with conflict by directing activity so as to institute a unified state of affairs.

The case is empirically and dialectically so simple that it would be extremely difficult to understand why it has become so confused in discussion were it not for the influence of irrelevant theoretical preconceptions drawn in part from introspectionist psychology and in part from metaphysics. Empirically, there are two alternatives. Action may take place with or without an end-in-view. In the latter case, there is overt action with no intermediate valuation; a vital impulse or settled habit reacts directly to some immediate sensory stimulation. In case an end-in-view exists and is valued, or exists in relation to a desire or an interest, the (motor) activity engaged in is, tautologically, mediated by the anticipation of the consequences which *as a foreseen end* enter into the makeup of the desire or interest. Now, as has been so often repeated, things can be anticipated or foreseen *as ends* or outcomes only in terms of the conditions by which they are brought into existence. It is simply impossible to have an end-in-view or to anticipate the consequences of any proposed line of action save upon the basis of some, however slight, consideration of the means by which it can be brought into existence. Otherwise, there is no genuine desire but an idle fantasy, a futile wish. That vital impulses and acquired habits are capable of expending themselves in the channels of daydreaming and building castles in the air is unfortunately true. But by description the contents of dreams and air castles are *not* ends-in-view, and what makes them fantasies is precisely the fact that they are *not* formed in terms of actual conditions serving as means of their actualization. *Propositions in which things (acts and materials) are appraised as means enter necessarily into desires and interests that determine end-values.* Hence the importance of the inquiries that result in the appraisal of things as means.

The case is so clear that, instead of arguing it directly, it will prove more profitable to consider how it is that there has grown up the belief that there are such things as ends having value apart from valuation of the means by which they are reached.

1. The mentalistic psychology which operates "to reduce" affective-motor activities to mere *feelings* has also operated in the interpretations assigned to *ends-in-view*, *purposes*, and *aims*. Instead of being treated as anticipations of consequences of the same order as a prediction of future events and, in any case, as depending for their contents and validity upon such predictions, they have been treated as merely mental states; for, when they are so taken (and only then), ends, needs, and satisfactions are affected in a way that distorts the whole theory of valuation. An end, aim, or purpose as a *mental* state *is* independent of the biological and physical means by which it can be realized. The want, lack, or privation which exists wherever there is desire is then interpreted as a mere state of "mind" instead of as something lacking or absent *in the situation*—something that must be supplied if the empirical situation is to be complete. In its latter sense, the needful or required is that which is *existentially necessary* if an end-in-view is to be brought into actual existence. *What* is needed cannot in this case be told by examination of a state of mind but only by examination of actual conditions. With respect to interpretation of "satisfaction" there is an obvious difference between it as a state of mind and as fulfilment of conditions, i.e., as something that meets the conditions imposed by the conjoint potentialities and lacks of the situation in which desire arises and functions. Satisfaction of desire signifies that the lack, characteristic of the situation evoking desire, has been so met that the means used make sufficient, in the most literal sense, the conditions for accomplishing the end. Because of the subjectivistic interpretation of end, need, and satisfaction, the verbally correct statement that valuation is a *relation* between a personal attitude and extra-personal things—a relation which, moreover, includes a motor (and hence physical) element—is so construed

as to involve separation of means and end, of appraisal and prizing. A "value" is then affirmed to be a "feeling"—a feeling which is not, apparently, the feeling of anything but itself. If it were said that a "value" is *felt*, the statement *might* be interpreted to signify that a certain existing relation between a personal motor attitude and extra-personal environing conditions is a matter of direct experience.

2. The shift of ground between valuation as *desire-interest* and as *enjoyment* introduces further confusion in theory. The shift is facilitated because in fact there exist both enjoyments of things directly possessed *without* desire and effort and enjoyments of things that are possessed only *because* of activity put forth to obtain the conditions required to satisfy desire. In the latter case, the enjoyment is in functional relation to desire or interest, and there is no violation of the definition of valuation in terms of desire-interest. But since the same *word*, 'enjoyment,' is applied also to gratifications that arise quite independently of prior desire and attendant effort, the ground is shifted so that "valuing" is identified with any and every state of enjoyment no matter how it comes about—including gratifications obtained in the most casual and accidental manner, "accidental" in the sense of coming about apart from desire and intent. Take, for example, the gratification of learning that one has been left a fortune by an unknown relative. There is *enjoyment*. But if valuation is defined in terms of desire and interest, there is no valuation, and in so far no "value," the latter coming into being only when there arises some desire as to what shall be done with the money and some question as to formation of an end-in-view. The two kinds of enjoyment are thus not only different but their respective bearings upon the theory of valuation are incompatible with each other, since one is connected with direct possession and the other is conditioned upon prior lack of possession—the very case in which desire enters.

For sake of emphasis, let us repeat the point in a slightly varied illustration. Consider the case of a man gratified by the unexpected receipt of a sum of money, say money picked up while he is walking on the street, an act having nothing to do with his purpose and desire at the moment he is performing it. If values are connected with desire in such a way that the connection is involved in their definition, there is, so far, no valuation. The latter begins when the finder begins to consider *how* he shall prize and care for the money. Shall he prize it, for example, as a means of satisfying certain wants he has previously been unable to satisfy, or shall be prize it as something held in trust until the owner is found? In either case, there is, by definition, an act of valuation. But it is clear that the property of value is attached in the two cases to very different objects. Of course, the uses to which money is put, the ends-in-view which it will serve, are fairly standardized, and in so far the instance just cited is not especially well chosen. But take the case of a child who has found a bright smooth stone. His sense of touch and of sight is gratified. But there is no valuation because no desire and no end-in-view, until the question arises of what shall be done with it; until the child *treasures* what he has accidentally hit upon. The moment he begins to prize and care for it he puts it to some use and thereby employs it as a *means* to some end, and, depending upon his maturity, he estimates or values it *in that relation*, or as means to end.

The confusion that occurs in theory when shift is made from valuation related to desire and interest, to "enjoyment" independent of any relation to desire and interest is facilitated by the fact that attainment of the objectives of desire and interest (of valuation) is itself enjoyed. The nub of the confusion consists in isolating enjoyment from the conditions under which it occurs. Yet the enjoyment that is the consequence of fulfilment of a desire and realization of an interest is what it is because of satisfaction or making good of a need or lack—a satisfaction conditioned by effort directed by the idea of something as an end-in-view. In this sense "enjoyment" involves inherent connection with *lack* of possession; while, in the other sense, the "enjoyment" is that of sheer possession. Lack of possession and possession are tautologically incompatible. Moreover, it is a common experience that the object of desire when attained is *not* enjoyed, so common that there are proverbial sayings to the effect that enjoyment is in the seeking rather than in the obtaining. It is not necessary to take these sayings literally to be aware that the occurrences in question prove the existence of the difference between value as connected with desire and value as mere enjoyment. Finally, as matter of daily experience, enjoyments provide the primary material of *problems* of valuation. Quite independently of any "moral" issues, people continually ask themselves whether a given enjoyment is worth while or whether the conditions involved in its production are such as to make it a costly indulgence.

Reference was made earlier to the confusion in theory which results when "values" are *defined* in terms of vital impulses. (The ground offered is that the latter are conditions of the existence of values in the sense that they "spring from" vital impulse.) In the text from which the passage was quoted there occurs in close connection the following: "The ideal of rationality is itself as arbitrary, as much dependent upon the needs of a finite organization, as any other ideal." Implicit in this passage are two extraordinary conceptions. One of them is that an ideal is arbitrary if it is causally conditioned by actual existences and is relevant to actual needs of human beings. This conception is extraordinary because naturally it would be supposed that an ideal is arbitrary in the degree in which it is *not* connected with things which exist and is not related to concrete existential requirements. The other astounding conception is that the ideal of rationality is "arbitrary" because it is so conditioned. One would suppose it to be peculiarly true of the ideal of rationality that it is to be judged as to its reasonableness (versus its arbitrariness) on the ground of its function, of what it does, not on the ground of its origin. If rationality as an ideal or generalized end-in-view serves to direct conduct so that things experienced in consequence of conduct so directed are more reasonable in the concrete, nothing more can be asked of it. Both of the implied conceptions are so extraordinary that they can be understood only on the ground of some unexpressed preconceptions. As far as one can judge, these preconceptions are (i) that an ideal *ought* to be independent of existence, that is, a priori. The reference to the origin of ideals in vital impulses is in fact an effective criticism of this a priori view. But it provides a ground for calling ideas arbitrary only if the a priori view is accepted. (ii) The other preconception would seem to be an acceptance of the view that there are or ought to be "ends-in-themselves"; that is to say, ends or ideals that are not also means, which, as we have already seen, is precisely what an ideal is, if it is judged and valued in terms of its function. The sole way of arriving at the conclusion that a generalized end-in-view or ideal is arbitrary because of existential and empirical origin is by first laying down as an ultimate criterion that an end should also *not* be a means. The whole passage and the views of which it is a typical and influential manifestation is redolent of the survival of belief in "ends-in-themselves" as the solely and finally legitimate kind of ends.

VI. The Continuum of Ends-Means

Those who have read and enjoyed Charles Lamb's essay on the origin of roast pork have probably not been conscious that their enjoyment of its absurdity was due to perception of the absurdity of any "end" which is set up apart from the means by which it is to be attained and apart from its own further function as means. Nor is it probable that Lamb himself wrote the story as a deliberate travesty of the theories that make such a separation. Nonetheless, that is the whole point of the tale. The story, it will be remembered, is that roast pork was first enjoyed when a house in which pigs were confined was accidentally burned down. While searching in the ruins, the owners touched the pigs that had been roasted in the fire and scorched their fingers. Impulsively bringing their fingers to their mouths to cool them, they experienced a new taste. Enjoying the taste, they henceforth set themselves to building houses, inclosing pigs in them, and then burning the houses down. Now, if ends-in-view are what they are entirely apart from means, and have their value independently of valuation of means, there is nothing absurd, nothing ridiculous, in this procedure, for the end attained, the *de facto* termination, *was* eating and enjoying roast pork, and that was just the end desired. Only when the end attained is estimated in terms of the means employed—the building and burning-down of houses in comparison with other available means by which the desired result in view might be attained—is there anything absurd or unreasonable about the method employed.

The story has a direct bearing upon another point, the meaning of 'intrinsic.' *Enjoyment* of the taste of roast pork may be said to be immediate, although even so the enjoyment would be a somewhat troubled one, for those who have memory, by the thought of the needless cost at which it was obtained. But to pass from immediacy of enjoyment to something called "intrinsic value" is a leap for which there is no ground. The *value* of enjoyment of an object *as* an attained end is a value of something which in being an end, an outcome, stands in relation to the means of which it is the consequence. Hence if the object in question is prized *as* an end or "final" value, it is valued *in this relation* or as mediated. The first time roast pork was enjoyed, it was *not* an end-value, since by description it was not the result of desire, foresight, and intent. Upon subsequent occasions it was, by description, the outcome of prior foresight, desire, and effort, and hence occupied the position of an end-in-view. There are occasions in which previous effort enhances enjoyment of what is attained. But there are also many occasions in which persons find that, when they have attained something as an end, they have paid too high a price in effort and in sacrifice of other ends. In such situations *enjoyment* of the end attained is itself *valued*, for it is not taken in its immediacy but in terms of its cost—a fact fatal to its being regarded as "an end-in-itself," a self-contradictory term in any case.

The story throws a flood of light upon what is usually meant by the maxim "the end justifies the means" and also upon the popular objection to it. Applied in this case, it would mean that the value of the attained end, the eating of roast pork, was such as to warrant the price paid in the means by which it was attained—destruction of dwelling-houses and sacrifice of the values to which they contribute. The conception involved in the maxim that "the end justifies the means" is basically the same as that in the notion of ends-in-themselves; indeed, from a historical point of view, it is the fruit of the latter, for only the conception that certain things are ends-in-themselves can warrant the belief that the relation of ends-means is unilateral, proceeding exclusively from end to means. When the maxim is compared with empirically ascertained facts, it is equivalent to holding one of two views, both of which are incompatible with the facts. One of the views is that only the specially selected "end" held in view will actually be brought into existence by the means used, something miraculously intervening to prevent the means employed from having their other usual effects; the other (and more probable) view is that, as compared with the importance of the selected and uniquely prized end, other consequences may be completely ignored and brushed aside no matter how intrinsically obnoxious they are. This arbitrary selection of some one part of the attained consequences as *the* end and hence as the warrant of means used (no matter how objectionable are their *other* consequences) is the fruit of holding that *it*, as *the* end, is an end-in-itself, and hence possessed of "value" irrespective of all its existential relations. And this notion is inherent in *every* view that assumes that "ends" can be valued apart from appraisal of the things used as means in attaining them. The sole alternative to the view that *the* end is an arbitrarily selected part of actual consequences which *as* "the end" then justifies the use of means irrespective of the other consequences they produce, is that desires, ends-in-view, and consequences achieved be valued in turn as means of further consequences. The maxim referred to, under the guise of saying that ends, in the sense of actual consequences, provide the warrant for means employed—a correct position—actually says that some fragment of these actual consequences—a fragment arbitrarily selected because the heart has been set upon it—authorizes the use of means to obtain *it*, without the need of foreseeing and weighing other ends as consequences of the means used. It thus discloses in a striking manner the fallacy involved in the position that ends have value independent of appraisal of means involved and independent of their own further causal efficacy.

We are thus brought back to a point already set forth. In all the physical sciences (using 'physical' here as a synonym for *nonhuman*) it is now taken for granted that all "effects" are also "causes," or, stated more accurately, that nothing happens which is *final* in the sense that it is not part of an ongoing stream of events. If this principle, with the accompanying discrediting of belief in objects that are ends but not means, is employed in dealing with distinctive human phenomena, it necessarily follows that the distinction between ends and means is temporal and relational. Every condition that has to be brought into existence in order to serve as means is, *in that connection*, an object of desire and an end-in-view, while the end actually reached is a means to future ends as well as a test of valuations previously made. Since the end attained is a condition of further existential occurrences, it must be appraised as a potential obstacle and potential resource. If the notion of some objects as ends-in-themselves were abandoned, not merely in words but in all practical implications, human beings would for the first time in history be in a position to frame ends-in-view and form desires on the basis of empirically grounded propositions of the temporal relations of events to one another.

At any given time an adult person in a social group has certain ends which are so standardized by custom that they are taken for granted without examination, so that the only problems arising concern the best means for attaining them. In one group money-making would be such an end; in another group, possession of political power; in another group, advancement of scientific knowledge; in still another group, military prowess, etc. But such ends in any case are (i) more or less blank frameworks where the nominal "end" sets limits within which definite ends will fall, the latter being determined by appraisal of things as means; while (ii) as far as they simply express habits that have become established without critical examination of the relation of means and ends, they do not provide a model for a theory of valuation to follow. If a person moved by an experience of intense cold, which is highly objectionable, should momentarily judge it worth while to get warm by burning his house down, all that saves him from an act determined by a "compulsion neurosis" is the intellectual realization of what other consequences would ensue with the loss of his house. It is not necessarily a sign of insanity (as in the case cited) to isolate some event projected as an end out of the context of a world of moving changes in which it will in fact take place. But it is at least a sign of immaturity when an individual fails to view his end as also a moving condition of further consequences, thereby treating it as *final* in the sense in which 'final' signifies that the course of events has come to a complete stop. Human beings do indulge in such arrests. But to treat them as models for forming a theory of ends is to substitute a manipulation of ideas, abstracted from the contexts in which they arise and function, for the conclusions of observation of concrete facts. It is a sign either of insanity, immaturity, indurated routine, or of a fanaticism that is a mixture of all three.

Generalized ideas of ends and values undoubtedly exist. They exist not only as expressions of habit and as uncritical and probably invalid ideas but also in the same way as valid general ideas arise in any subject. Similar situations recur; desires and interests are carried over from one situation to another and progressively consolidated. A schedule of general ends results, the involved values being "abstract" in the sense of not being directly connected with any particular existing case but not in the sense of independence of all empirically existent cases. As with general ideas in the conduct of any natural science, these general ideas are used as intellectual instrumentalities in judgment of particular cases as the latter arise; they are, in effect, tools that direct and facilitate examination of things in the concrete while they are also developed and tested by the results of their application in these cases. Just as the natural sciences began a course of sure development when the dialectic of concepts ceased to be employed to arrive at conclusions about existential affairs and was employed instead as a means of arriving at a hypothesis fruitfully applicable to particulars, so it will be with the theory of human activities and relations. There is irony in the fact that the very continuity of experienced activities which enables general ideas of value to function as rules for evaluation of particular desires and ends should have become the source of a belief that desires, by the bare fact of their occurrence, confer value upon objects as ends, entirely independent of their contexts in the continuum of activities.

In this connection there is danger that the idea of "finality" be manipulated in a way analogous to the manipulation of the concepts of "immediacy" and "intrinsic" previously remarked upon. A value is *final* in the sense that it represents the conclusion of a process of analytic appraisals of conditions operating in a concrete case, the conditions including impulses and desires on one side and external conditions on the other. Any conclusion reached by an inquiry that is taken to warrant the conclusion is "final" for that case. "Final" here has logical force. The quality or property of value that is correlated with the *last* desire formed in the process of valuation is, tautologically, ultimate for that particular situation. It applies, however, to a specifiable temporal *means-end relation* and not to something which is an end per se. There is a fundamental difference between a final property or quality and the property or quality of finality.

The objection always brought against the view set forth is that, according to it, valuation activities and judgments are involved in a hopeless *regressus ad infinitum*. If, so it is said, there is no end which is not in turn a means, foresight has no place at which it can stop, and no end-in-view can be formed except by the most arbitrary of acts—an act so arbitrary that it mocks the claim of being a genuine valuation-proposition.

This objection brings us back to the conditions under which desires take shape and foreseen consequences are projected as ends to be reached. These conditions are those of need, deficit, and conflict. Apart from a condition of tension between a person and environing conditions there is, as we have seen, no occasion for evocation of desire for something else; there is nothing to induce the formation of an end, much less the formation of one end rather than any other out of the indefinite number of ends theoretically possible. Control of transformation of active tendencies into a desire in which a particular end-in-view is incorporated, is exercised by the needs or privations of an actual situation as its requirements are disclosed to observation. The "value" of different ends that suggest themselves is estimated or measured by the capacity they exhibit to guide action in making good, *satisfying*, in its literal sense, existing lacks. Here is the factor which cuts short the process of foreseeing and weighing ends-in-view in their function as means. Sufficient unto the day is the evil thereof and sufficient also is the *good* of that which does away with the existing evil. Sufficient because it is the means of instituting a complete situation or an integrated set of conditions.

Two illustrations will be given. A physician has to determine the value of various courses of action and their results in the case of a particular patient. He forms ends-in-view having the value that justifies their adoption, on the ground of what his examination discloses is the "matter" or "trouble" with the patient. He estimates the worth of what he undertakes on the ground of its capacity to produce a condition in which these troubles will not exist, in which, as it is ordinarily put, the patient will be "restored to health." He does not have an idea of health as an absolute end-in-itself, an absolute good by which to determine what to do. On the contrary, he forms his general idea of health as an end and a good (value) for the patient on the ground of what his techniques of examination have shown to be the troubles from which patients suffer and the means by which they are overcome. There is no need to deny that a general and abstract conception of health finally develops. But it is the outcome of a great number of definite, empirical inquiries, not an a priori preconditioning "standard" for carrying on inquiries.

The other illustration is more general. In all inquiry, even the most completely scientific, what is proposed as a conclusion (the end-in-view in that inquiry) is evaluated as to its worth on the ground of its ability to resolve the *problem* presented by the conditions under investigation. There is no a priori standard for determining the value of a proposed solution in concrete cases. A hypothetical possible solution, as an end-in-view, is used as a methodological means to direct further observations and experiments. Either it performs the function of resolution of a problem for the sake of which it is adopted and tried or it does not. Experience has shown that problems for the most part fall into certain recurrent kinds so that there are general principles which, it is believed, proposed solutions must satisfy in a particular case. There thus develops a sort of framework of conditions to be satisfied—a framework of reference which operates in an *empirically* regulative way in given cases. We may even say that it operates as an "a priori" principle, but in exactly the same sense in which rules for the conduct of a technological art are both empirically antecedent and controlling in a given case of the art. While there is no a priori standard of health with which the actual state of human beings can be compared so as to determine whether they are well or ill, or in what respect they are ill, there have developed, out of past experience, certain criteria which are operatively applicable in new cases as they arise. Ends-in-view are appraised or valued as *good* or *bad* on the ground of their serviceability in the direction of behavior dealing with states of affairs found to be objectionable because of some lack or conflict in them. They are appraised as fit or unfit, proper or improper, *right* or *wrong*, on the ground of their *requiredness* in accomplishing this end.

Considering the all but omnipresence of troubles and "evils" in human experience (evils in the sense of deficiencies, failures, and frustrations), and considering the amount of time that has been spent explaining them away, theories of human activity have been strangely oblivious of the concrete function troubles are capable of exercising when they are taken as *problems* whose conditions and consequences are explored with a view to finding methods of solution. The two instances just cited, the progress of medical art and of scientific inquiry, are most instructive on this point. As long as actual events were supposed to be judged by comparison with some absolute end-value as a standard and norm, no sure progress was made. When standards of health and of satisfaction of conditions of knowledge were conceived in terms of analytic observation of existing conditions, disclosing a trouble statable in a problem, criteria of judging were progressively self-corrective through the very process of use in observation to locate the source of the trouble and to indicate the effective means of dealing with it. These means form the content of the specific end-in-view, not some abstract standard or ideal.

This emphasis upon the function of needs and conflicts as the controlling factor in institution of ends and values does not signify that the latter are themselves negative in content and import. While they are framed with reference to a negative factor, deficit, want, privation, and conflict, their function is positive, and the resolution effected by performance of their function is positive. To attempt to gain an end *directly* is to put into operation the very conditions that are the source of the experienced trouble, thereby strengthening them and at most changing the outward form in which they manifest themselves. Ends-in-view framed with a negative *reference* (i.e., to some trouble or problem) are means which inhibit the operation of conditions producing the obnoxious result; they enable positive conditions to operate as resources and thereby to effect a result which is, in the highest possible sense, positive in content. The content of the end as an object *held in view* is intellectual or methodological; the content of the attained outcome or the end *as consequence* is existential. It is positive in the degree in which it marks the doing-away of the need and conflict that evoked the *end-in-view.* The negative factor operates as a condition of forming the appropriate *idea* of an end; the idea when acted upon determines a positive outcome.

The attained end or consequence is always an organization of activities, where organization is a co-ordination of all activities which enter as factors. The *end-in-view* is that particular activity which operates as a co-ordinating factor of all other subactivities involved. Recognition of the end as a co-ordination or unified organization of activities, and of the end-in-view as the special activity which is the means of effecting this co-ordination, does away with any appearance of paradox that seems to be attached to the idea of a temporal continuum of activities in which each successive stage is equally end and means. The *form* of an attained end or consequence is always the same: that of adequate co-ordination. The content or involved matter of each successive result differs from that of its predecessors; for, while it is a *reinstatement* of a unified ongoing action, after a period of interruption through conflict and need, it is also an *enactment* of a new state of affairs. It has the qualities and properties appropriate to its being the consummatory resolution of a previous state of activity in which there was a peculiar need, desire, and end-in-view. In the continuous temporal process of organizing activities into a co-ordinated and co-ordinating unity, a constituent activity is both an end and a means: an end, in so far as it is temporally and relatively a close; a means, in so far as it provides a condition to be taken into account in further activity.

Instead of there being anything strange or paradoxical in the existence of situations in which means are constituents of the very end-objects they have helped to bring into existence, such situations occur whenever behavior succeeds in intelligent projection of ends-in-view that direct activity to resolution of the antecedent trouble. The cases in which ends and means fall apart are the abnormal ones, the ones which deviate from activity which is intelligently conducted. Wherever, for example, there is sheer drudgery, there is separation of the required and necessary means from both the end-in-view and the end attained. Wherever, on the other side, there is a so-called "ideal" which is utopian and a matter of fantasy, the same separation occurs, now from the side of the so-called *end*. Means that do not become constituent elements of the very ends or consequences they produce form what are called "necessary evils," their "necessity" being relative to the existing state of knowledge and art. They are comparable to scaffoldings that had to be later torn down, but which were necessary in erection of buildings until elevators were introduced. The latter remained for use in the building erected and were employed as means of transporting materials that in turn became an integral part of the building. Results or consequences which at one time were necessarily waste products in the production of the particular thing desired were utilized in the light of the development of human experience and intelligence as means for further desired consequences. The generalized ideal and standard of economy-efficiency which operates in every advanced art and technology is equivalent, upon analysis, to the conception of means that are constituents of ends attained and of ends that are usable as means to further ends.

It must also be noted that *activity* and *activities*, as these words are employed in the foregoing account, involve, like any actual behavior, existential materials, as breathing involves air; walking, the earth; buying and selling, commodities; inquiry, things investigated, etc. No human activity operates in a vacuum; it acts in the world and has materials upon which and through which it produces results. On the other hand, no material—air, water, metal, wood, etc.—is *means* save as it is employed in some human activity to accomplish something. When "organization of activities" is mentioned, it always includes within itself organization of the materials existing in the world in which we live. That organization which is the "final" value for each concrete situation of valuation thus forms part of the existential conditions that have to be taken into account in further formation of desires and interests or valuations. In the degree in which a particular valuation is invalid because of inconsiderate shortsighted investigation of things in their relation of means-end, difficulties are put in the way of subsequent reasonable valuations. To the degree in which desires and interests are formed after critical survey of the conditions which as means determine the actual outcome, the more smoothly continuous become subsequent activities, for consequences attained are then such as are evaluated more readily as means in the continuum of action.

VII. Theory of Valuation as Outline of a Program

Because of the confusion which affects current discussion of the problem of valuation, the analysis undertaken in the present study has been obliged to concern itself to a considerable extent with tracking the confusion to its source. This is necessary in order that empirical inquiry into facts which are taken for granted by common sense may be freed from irrelevant and confusing associations. The more important conclusions may be summarized as follows.

1. Even if "value-expressions" were ejaculatory and such as to influence the conduct of other persons, genuine propositions about such expressions would be possible. We could investigate whether or not they had the effect intended; and further examination would be able to discover the differential conditions of the cases that were successful in obtaining the intended outcome and those that were not. It is useful to discriminate between linguistic expressions which are "emotive" and those which are "scientific." Nevertheless, even if the former said nothing whatever, they would, like other natural events, be capable of becoming the subject matter of "scientific" propositions as a result of a examination of their conditions and effects.

2. Another view connects valuation and value-expressions with desires and interests. Since desire and interest are behavioral phenomena (involving at the very least a "motor" aspect), the valuations they produce are capable of being investigated as to *their* respective conditions and results. Valuations are empirically observable patterns of behavior and may be studied as such. The propositions that result are *about* valuations but are not of themselves value-propositions in any sense marking them off from other matter-of-fact propositions.

3. Value-propositions of the distinctive sort exist whenever things are appraised as to their suitability and serviceability as means, for such propositions are not about things or events that have occurred or that already exist (although they cannot be validly instituted apart from propositions of the kind mentioned in the previous sentence), but are about things *to be*

brought into existence. Moreover, while they are logically conditioned upon matter-of-fact predictions, they are more than simple predictions, for the things in question are such as will *not* take place, under the given circumstances, except through the intervention of some personal act. The difference is similar to that between a proposition predicting that in *any* case a certain eclipse will take place and a proposition that the eclipse will be seen or experienced by certain human beings in case the latter intervene to perform certain actions. While valuation-propositions as appraisals of means occur in all arts and technologies and are grounded in strictly physical propositions (as in advanced engineering technologies), nevertheless they are distinct from the latter in that they inherently involve the means-end relationship.

4. Wherever there are desires, there are *ends-in-view*, not simply effects produced as in the case of sheer impulse, appetite, and routine habit. Ends-in-view as anticipated results reacting upon a given desire are *ideational* by definition or tautologically. The involved foresight, forecast or anticipation is warranted, like any other intellectual inferent factor, in the degree in which it is based upon propositions that are conclusions of adequate observational activities. Any given desire is what it is in its actual content or "object" *because* of its ideational constituents. Sheer impulse or appetite may be described as affective-motor; but any theory that connects valuation with desire and interest by that very fact connects valuation with behavior which is affective-*ideational*-motor. This fact proves the *possibility* of the existence of distinctive valuation-propositions. In view of the role played by ends-in-view in directing the activities that contribute either to the realization or to the frustration of desire, the *necessity* for valuation-propositions is proved if desires are to be intelligent, and purposes are to be other than shortsighted and irrational.

5. The required appraisal of desires and ends-in-view, as means of the activities by which actual results are produced, is dependent upon observation of consequences attained when they are compared and contrasted with the content of ends-in-

view. Careless, inconsiderate action is that which foregoes the inquiry that determines the points of agreement and disagreement between the desire actually formed (and hence the valuation actually made) and the things brought into existence by acting upon it. Since desire and valuation of objects proposed as ends are inherently connected, and since desire and ends-in-view need to be appraised as means to ends (an appraisal made on the basis of warranted physical generalizations) the valuation of ends-in-view is tested by consequences that actually ensue. It is verified to the degree in which there is agreement upon results. Failure to agree, in case deviations are carefully observed, is not mere failure but provides the means for improving the formation of later desires and ends-in-view.

The net outcome is (i) that the problem of valuation in general as well as in particular cases concerns things that sustain to one another the relation of means-ends; that (ii) ends are determinable only on the ground of the means that are involved in bringing them about; and that (iii) desires and interests must themselves be evaluated as means in their interaction with external or environing conditions. Ends-in-view, as distinct from ends as accomplished results, themselves function as directive means; or, in ordinary language, as *plans*. Desires, interests, and environing conditions as means are modes of action, and hence are to be conceived in terms of energies which are capable of reduction to homogeneous and comparable terms. Co-ordination or organizations of energies, proceeding from the two sources of the organism and the environment, are thus both means and attained result or "end" in all cases of valuation, the two kinds of energy being theoretically (if not as yet completely so in practice) capable of statement in terms of physical units.

The conclusions stated do not constitute a complete theory of valuation. They do, however, state the conditions which such a theory must satisfy. An actual theory can be completed only when inquiries into things sustaining the relation of ends-means have been systematically conducted and their results brought to bear upon the formation of desires and ends. For the theory of valuation is itself an intellectual or methodological means and as such can be developed and perfected only in and by use. Since that use does not now exist in any adequate way, the theoretical consideration advanced and conclusions reached outline a program to be undertaken, rather than a complete theory. The undertaking can be carried out only by regulated guidance of the formation of interests and purposes in the concrete. The prime condition of this undertaking (in contrast with the current theory of the relation of valuation to desire and interest) is recognition that desire and interest are not given ready-made at the outset, and a fortiori are not, as they may at first appear, starting-points, original data, or premisses of any theory of valuation, for desire always emerges within a prior system of activities or interrelated energies. It arises within a *field* when the field is disrupted or is menaced with disruption, when conflict introduces the tension of need or threatens to introduce it. An interest represents not just a desire but a set of interrelated desires which have been found in experience to produce, because of their connection with one another, a definite order in the processes of continuing behavior.

The test of the existence of a valuation and the nature of the latter is actual behavior as that is subject to observation. Is the existing field of activities (including environing conditions) *accepted*, where "acceptance" consists in effort to maintain it against adverse conditions? Or is it *rejected*, where "rejection" consists of effort to get rid of it and to produce another behavioral field? And in the latter case, what is the actual field to which, as an end, desire-efforts (or the organization of desire-efforts constituting an interest) are directed? Determination of this field as an objective of behavior determines *what* is valued. Until there is actual or threatened shock and disturbance of a situation, there is a green light to go ahead in immediate act—overt action. There is no need, no desire, and no valuation, just as where there is no doubt, there is no cause of inquiry. Just as the problem which evokes inquiry is related to an empirical situation in which the problem presents itself, so desire and the projection of ends as consequences to be reached are relative to a concrete situation and to its need for transformation. The burden of proof lies, so to speak, on occurrence of conditions that are impeding, obstructive, and that introduce conflict and need. Examination of the situation in respect to the conditions that constitute lack and need and thus serve as positive means for formation of an attainable end or outcome, is the method by which warranted (required and effective) desires and ends-in-view are formed: by which, in short, valuation takes place.

The confusions and mistakes in existing theories, which have produced the need for the previous prolonged analysis, arise very largely from taking desire and interest as original instead of in the contextual situations in which they arise. When they are so taken, they become ultimate in relation to valuation. Being taken, so to speak, at large, there is nothing by which we can empirically check or test them. If desire were of this original nature, if it were independent of the structure and requirements of some concrete empirical situation and hence had no function to perform with reference to an existential situation, then insistence upon the necessity of an ideational or intellectual factor in every desire and the consequent necessity for fulfilment of the empirical conditions of its validity would be as superfluous and irrelevant as critics have said it is. The insistence might then be, what it has been called, a "moral" bias springing from an interest in the "reform" of individuals and society. But since in empirical fact there are no desires and interests apart from some field of activities in which they occur and in which they function, either as poor or as good means, the insistence in question is simply and wholly in the interest of a correct empirical account of what actually exists as over against what turns out to be, when examined, a dialectical manipulation of *concepts* of desire and interest at large, a procedure which is all that is possible when desire is taken in isolation from its existential context.

It is a common occurrence in the history of theories that an error at one extreme calls out a complementary error at the other extreme. The type of theory just considered isolates desires as sources of valuation from any existential context and hence from any possibility of intellectual control of their contents and objectives. It thereby renders valuation an arbitrary matter. It says in effect that any desire is just as "good" as any other in respect to the value it institutes. Since desires—and their organization into interests—are the sources of human action, this view, if it were systematically acted upon, would produce disordered behavior to the point of complete chaos. The fact that in spite of conflicts, and unnecessary conflicts, there is not complete disorder is proof that actually some degree of intellectual respect for existing conditions and consequences does operate as a control factor in formation of desires and valuations. However, the implications of the theory in the direction of intellectual and practical disorder are such as to evoke a contrary theory, one, however, which has the same fundamental postulate of the isolation of valuation from concrete empirical situations, their potentialities, and their requirements. This is the theory of "ends-in-themselves" as ultimate standards of all valuation—a theory which denies implicitly or explicitly that desires have anything to do with "final values" unless and until they are subjected to the external control of a priori absolute ends as standards and ideals for their valuation. This theory, in its endeavor to escape from the frying pan of disordered valuations, jumps into the fire of absolutism. It confers the simulation of final and complete rational authority upon certain interests of certain persons or groups at the expense of all others: a view which, in turn, because of the consequences it entails, strengthens the notion that no intellectual and empirically reasonable control of desires, and hence of valuations and value-properties, is possible. The seesaw between theories which by definition are not empirically testable (since they are a priori) and professed empirical theories that unwittingly substitute conclusions derived from the bare *concept* of desire for the results of observation of desires in the concrete is thus kept up. The astonishing thing about the a priori theory (astonishing if the history of philosophical thought be omitted from the survey) is its complete neglect of the fact that valuations are constant phenomena of human behavior, personal and associated, and are capable of rectification and development by use of the resources provided by knowledge of physical relations.

VIII. Valuation and the Conditions of Social Theory

We are thus brought to the problem which, as was shown in the opening section of this study, is back of the present interest in the problem of valuation and values, namely, the possibility of genuine and grounded propositions about the purposes, plans, measures, and policies which influence human activity whenever the latter is other than merely impulsive or routine. A theory of valuation *as* theory can only set forth the conditions which a method of formation of desires and interests must observe in concrete situations. The problem of the existence of such a method is all one with the problem of the possibility of genuine propositions which have as their subject matter the intelligent conduct of human activities, whether personal or associated. The view that value in the sense of *good* is inherently connected with that which promotes, furthers, assists, a course of activity, and that value in the sense of *right* is inherently connected with that which is needed, required, in the maintenance of a course of activity, is not in itself novel. Indeed, it is suggested by the very etymology of the word *value*, associated as it is with the words 'avail,' 'valor,' 'valid,' and 'invalid.' What the foregoing discussion has added to the idea is proof that if, and *only* if, valuation is taken in this sense, are empirically grounded propositions about desires and interests as sources of valuations possible—such propositions being grounded in the degree in which they employ scientific physical generalizations as means of forming propositions about activities which are correlated as ends-means. The resulting general propositions provide rules for valuation of the aims, purposes, plans, and policies that direct intelligent human activity. They are not rules in the sense that they enable us to tell directly, or upon bare inspection, the values of given particular ends (a foolish quest that underlies the belief in a priori values as ideals and standards); they are rules of methodic procedure in the conduct of the investigations that determine the respective conditions and consequences of various modes of behavior. It does not purport to solve the problems of valuation in and of itself; it does claim to state conditions that inquiry must satisfy if these problems are to be resolved, and to serve in this way as a leading principle in conduct of such inquiries.

I. Valuations exist in fact and are capable of empirical observation so that propositions about them are empirically verifiable. What individuals and groups hold dear or prize and the grounds upon which they prize them are capable, in principle, of ascertainment, no matter how great the *practical* difficulties in the way. But, upon the whole, in the past values have been determined by customs, which are then commended because they favor some special interest, the commendation being attended with coercion or exhortation or with a mixture of both. The practical difficulties in the way of scientific inquiry into valuations are great, so great that they are readily mistaken for inherent theoretical obstacles. Moreover, such knowledge as does exist about valuations is far from organized, to say nothing about its being adequate. The notion that valuations do not exist in empirical fact and that therefore value-conceptions have to be imported from a source outside experience is one of the most curious beliefs the mind of man has ever entertained. Human beings are continuously engaged in valuations. The latter supply the primary material for operations of further valuations and for the general theory of valuation.

Knowledge of these valuations does not of itself, as we have seen, provide valuation-propositions; it is rather of the nature of historical and cultural-anthropological knowledge. But such factual knowledge is a *sine qua non* of ability to formulate valuation-propositions. This statement only involves recognition that past experience, when properly analyzed and ordered, is the sole guide we have in future experience. An individual within the limits of his personal experience revises his desires and purposes as he becomes aware of the consequences they have produced in the past. This knowledge is what enables him to foresee probable consequences of his prospective activities and to direct his conduct accordingly. The ability to form valid propositions about the relation of present desires and purposes to future consequences depends in turn upon ability to analyze these present desires and purposes into their constituent elements. When they are taken in gross, foresight is correspondingly coarse and indefinite. The history of science shows that power of prediction has increased *pari passu* with analysis of gross qualitative events into elementary constituents. Now, in the absence of adequate and organized knowledge of human valuations as occurrences that have taken place, it is a fortiori impossible that there be valid propositions formulating new valuations in terms of consequences of specified causal conditions. On account of the continuity of human activities, personal and associated, the import of present valuations cannot be validly stated until they are placed in the perspective of the past valuation-events with which they are continuous. Without this perception, the future perspective, i.e., the consequences of present and new valuations, is indefinite. In the degree in which existing desires and interests (and hence valuations) can be judged in their connection with past conditions, they are seen in a context which enables them to be revaluated on the ground of evidence capable of observation and empirical test.

Suppose, for example, that it be ascertained that a particular set of current valuations have, as their antecedent historical conditions, the interest of a small group or special class in maintaining certain exclusive privileges and advantages, and that this maintenance has the effect of limiting both the range of the desires of others and their capacity to actualize them. Is it not obvious that this knowledge of conditions and consequences would surely lead to revaluation of the desires and ends that had been assumed to be authoritative sources of valuation? Not that such revaluation would of necessity take effect immediately. But, when valuations that exist at a given time are found to lack the support they have previously been supposed to have, they exist in a context that is highly adverse to their continued maintenance. In the long run the effect is similar to a warier attitude that develops toward certain bodies of water as the result of knowledge that these bodies of water contain disease germs. If, on the other hand, investigation shows that a given set of existing valuations, including the rules for their enforcement, be such as to release individual potentialities of desire and interest, and does so in a way that contributes to mutual reinforcement of the desires and interests of all members of a group, it is impossible for this knowledge not to serve as a bulwark of the particular set of valuations in question, and to induce intensified effort to sustain them in existence.

II. These considerations lead to the central question: What are the conditions that have to be met so that knowledge of past and existing valuations becomes an instrumentality of valuation in formation of new desires and interests—of desires and interests that the test of experience show to be best worth fostering? It is clear upon our view that no abstract theory of valuation can be put side by side, so to speak, with existing valuations as the standard for judging them.

The answer is that improved valuation must grow out of existing valuations, subjected to critical methods of investigation that bring them into systematic relations with one another. Admitting that these valuations are largely and probably, in the main, defective, it might at first sight seem as if the idea that improvement would spring from bringing them into connection with one another is like recommending that one lift himself by his bootstraps. But such an impression arises only because of failure to consider how they actually may be brought into relation with one another, namely, by examination of their respective conditions and consequences. Only by following this path will they be reduced to such homogeneous terms that they are comparable with one another.

This method, in fact, simply carries over to human or social phenomena the methods that have proved successful in dealing with the subject matter of physics and chemistry. In these fields before the rise of modern science there was a mass of facts which were isolated and seemingly independent of one another. Systematic advance dates from the time when conceptions that formed the content of theory were derived from the phenomena themselves and were then employed as hypotheses for relating together the otherwise separate matters-of-fact. When, for example, ordinary drinking water is operatively regarded as H_2O what has happened is that water is related to an immense number of other phenomena so that inferences and predictions are indefinitely expanded and, at the same time, made subject to empirical tests. In the field of human activities there are at present an immense number of facts of desires and purposes existing in rather complete isolation from one another. But there are no hypotheses of the same empirical order which are capable of relating them to one another so that the resulting propositions will serve as methodic controls of the formation of future desires and purposes, and, thereby, of new valuations. The material is ample. But the means for bringing its constituents into such connections that fruit is borne are lacking. This lack of means for bringing actual valuations into relation with one another is partly the cause and partly the effect of belief in standards and ideals of value that lie outside ("above" is the usual term) actual valuations. It is cause in so far as some method of control of desires and purposes is such an important desideratum that in the absence of an empirical method, *any* conception that seems to satisfy the need is grasped at. It is the effect in that a priori theories, once they are formed and have obtained prestige, serve to conceal the necessity for concrete methods of relating valuations and, by so doing, provide intellectual instruments for placing impulses and desires in a context where the very place they occupy affects their evaluation.

However, the difficulties that stand in the way are, in the main, practical. They are supplied by traditions, customs, and institutions which persist without being subjected to a systematic empirical investigation and which constitute the most influential source of further desires and ends. This is supplemented by a priori theories serving, upon the whole, to "rationalize" these desires and ends so as to give them apparent intellectual status and prestige. Hence it is worth while to note that the same obstacles once existed in the subject matters now ruled by scientific methods. Take, as an outstanding example, the difficulties experienced in getting a hearing for the Copernican astronomy a few centuries ago. Traditional and customary beliefs which were sanctioned and maintained by powerful institutions regarded the new scientific ideas as a menace. Nevertheless, the methods which yielded propositions verifiable in terms of actual observations and experimental evidence maintained themselves, widened their range, and gained continually in influence.

The propositions which have resulted and which now form the substantial content of physics, of chemistry, and, to a growing extent, of biology, provide the very means by which the change which is required can be introduced into beliefs and ideas purporting to deal with human and social phenomena. Until natural science had attained to something approaching its present estate, a grounded empirical theory of valuation, capable of serving in turn as a method of regulating the production of new valuations, was out of the question. Desires and interests produce consequences only when the activities in which they are expressed take effect in the environment by interacting with physical conditions. As long as there was no adequate knowledge of physical conditions and no well-grounded propositions regarding their relations to one another (no known "laws"), the kind of forecast of the consequences of alternative desires and purposes involved in their evaluation was impossible. When we note how recently—in comparison with the length of time man has existed on earth—the arts and technologies employed in strictly physical affairs have had scientific support, the backward conditions of the arts connected with the social and political affairs of men provides no ground for surprise.

Psychological science is now in much the same state in which astronomy, physics, and chemistry were when they first emerged as genuinely experimental sciences, yet without such a science systematic theoretical control of valuation is impossible; for without competent psychological knowledge the force of the human factors which interact with environing nonhuman conditions to produce consequences cannot be estimated. This statement is purely truistic, since knowledge of the human conditions *is* psychological science. For over a century, moreover, the ideas central to what passed for psychological knowledge were such as actually obstructed that foresight of consequences which is required to control the formation of ends-in-view. For when psychological subject matter was taken to form a psychical or mentalistic realm set over against the physical environment, inquiry, such as it was, was deflected into the metaphysical problem of the possibility of interaction between the mental and the physical and away from the problem central in evaluation, namely, that of discovering the concrete interactions between human behavior and environing conditions which determine the actual consequences of desires and purposes. A grounded theory of the phenomena of human behavior is as much a prerequisite of a theory of valuation as is a theory of the behavior of physical (in the sense of nonhuman) things. The development of a science of the phenomena of living creatures was an unqualified prerequisite of the development of a sound psychology. Until biology supplied the material facts which lie between the nonhuman and the human, the apparent traits of the latter were so different from those of the former that the doctrine of a complete gulf between the two seemed to be the only plausible one. The missing link in the chain of knowledge that terminates in grounded valuation propositions is the biological. As that link is in process of forging, we may expect the time soon to arrive in which the obstacles to development of an empirical theory of valuation will be those of habits and traditions that flow from institutional and class interests rather than from intellectual deficiencies.

Need for a theory of human relations in terms of a sociology which might perhaps instructively be named cultural anthropology is a further condition of the development of a theory of valuation as an effective instrumentality, for human organisms live in a cultural environment. There is no desire and no interest which, in its distinction from raw impulse and strictly organic appetite, is not what it is because of transformation effected in the latter by their interaction with the cultural environment. When current theories are examined which, quite properly, relate valuation with desires and interests, nothing is more striking than their neglect—so extensive as to be systematic—of the role of cultural conditions and institutions in the shaping of desires and ends and thereby of valuations. This neglect is perhaps the most convincing evidence that can be had of the substitution of dialectical manipulation of the concept of desire for investigation of desires and valuations as concretely existent facts. Furthermore, the notion that an adequate theory of human behavior—including particularly the phenomena of desire and purpose—can be formed by considering individuals apart from the cultural setting in which they live, move, and have their being—a theory which may justly be called metaphysical individualism—has united with the metaphysical belief in a mentalistic realm to keep valuation-phenomena in subjection to unexamined traditions, conventions, and institutionalized customs.[2] The separation alleged to exist between the "world of facts" and the "realm of values" will disappear from human beliefs only as valuation-phenomena are seen to have their immediate source in biological modes of behavior and to owe their concrete content to the influence of cultural conditions.

The hard-and-fast impassible line which is supposed by some to exist between "emotive" and "scientific" language is a reflex of the gap which now exists between the intellectual and

[2] The statement, sometimes made, that metaphysical sentences are "meaningless" usually fails to take account of the fact that culturally speaking they are very far from being devoid of meaning, in the sense of having significant cultural effects. Indeed, they are so far from being meaningless in this respect that there is no short dialectic cut to their elimination, since the latter can be accomplished only by concrete applications of scientific method which modify cultural conditions. The view that sentences having a nonempirical reference are meaningless, is sound in the sense that what they purport or pretend to mean cannot be given intelligibility, and this fact is presumably what is intended by those who hold this view. Interpreted as symptoms or signs of actually existent conditions, they may be and usually are highly significant, and the most effective criticism of them is disclosure of the conditions of which they are evidential.

the emotional in human relations and activities. The split which exists in present social life between ideas and emotions, especially between ideas that have *scientific* warrant and uncontrolled emotions that dominate practice, the split between the affectional and the cognitive, is probably one of the chief sources of the maladjustments and unendurable strains from which the world is suffering. I doubt if an adequate explanation upon the psychological side of the rise of dictatorships can be found which does not take account of the fact that the strain produced by separation of the intellectual and the emotional is so intolerable that human beings are willing to pay almost any price for the semblance of even its temporary annihilation. We are living in a period in which emotional loyalties and attachments are centered on objects that no longer command that intellectual loyalty which has the sanction of the methods which attain valid conclusions in scientific inquiry, while ideas that have their origin in the rationale of inquiry have not as yet succeeded in acquiring the force that only emotional ardor provides. The *practical* problem that has to be faced is the establishment of cultural conditions that will support the kinds of behavior in which emotions and ideas, desires and appraisals, are integrated.

If, then, discussion in the earlier sections of this study seems to have placed chief emphasis upon the importance of valid *ideas* in formation of the desires and interests which are the sources of valuation, and to have centered attention chiefly upon the possibility and the necessity of control of this ideational factor by empirically warranted matters-of-fact, it is because the *empirical* (as distinct from a priori) theory of valuation is currently stated in terms of desire as emotional in isolation from the ideational. In fact and in net outcome, the previous discussion does not point in the least to supersession of the emotive by the intellectual. Its only and complete import is the need for their integration in behavior—behavior in which, according to common speech, the head and the heart work together, in which, to use more technical language, prizing and appraising unite in direction of action. That growth of knowledge of the physical—in the sense of the nonpersonal—has limited the range of freedom of human action in relation to such things as light, heat, electricity, etc., is so absurd in view of what has actually taken place that no one holds it. The operation of desire in producing the valuations that influence human action will also be liberated when they, too, are ordered by verifiable propositions regarding matters-of-fact.

The chief *practical* problem with which the present *Encyclopedia* is concerned, the unification of science, may justly be said to center here, for at the present time the widest gap in knowledge is that which exists between humanistic and nonhumanistic subjects. The breach will disappear, the gap be filled, and science be manifest as an operating unity in fact and not merely in idea when the conclusions of impersonal nonhumanistic science are employed in guiding the course of distinctively human behavior, that, namely, which is influenced by emotion and desire in the framing of means and ends; for desire, having ends-in-view, and hence involving valuations, is the characteristic that marks off human from nonhuman behavior. On the other side, the science that is put to distinctively human use is that in which warranted ideas about the nonhuman world are integrated with emotion as human traits. In this integration not only is science itself *a* value (since it is the expression and the fulfilment of a special human desire and interest) but it is the supreme means of the valid determination of all valuations in all aspects of human and social life.

Selected Bibliography

AYER, A. J. *Language, Truth and Logic.* New York, 1936.

DEWEY, JOHN. *Essays in Experimental Logic.* Pp. 349–89. Chicago, 1916.

———. *Experience and Nature.* "Lectures upon the Paul Carus Foundation, First Series." 1st ed., Chicago, 1925; 2d ed., New York, 1929.

———. *Human Nature and Conduct.* New York, 1922.

———. *Logical Conditions of a Scientific Treatment of Morality.* Chicago, 1903. Reprinted from *The Decennial Publications of the University of Chicago, First Series,* III, 115–39.

———. *The Quest for Certainty.* New York, 1929.

———. *Art as Experience.* New York, 1934.

DEWEY, JOHN, and TUFTS, J. H. *Ethics.* Rev. ed. New York, 1932.

DEWEY, JOHN, *et al. Creative Intelligence.* New York, 1917.

JOERGENSEN, J. "Imperatives and Logic," *Erkenntnis,* VII (1938), 288–96.

KALLEN, H. "Value and Existence in Philosophy, Art and Religion," in *Creative Intelligence,* ed. JOHN DEWEY *et al.* New York, 1917.

KÖHLER, W. *Place of Value in a World of Fact.* New York, 1938.

KRAFT, VIKTOR. *Die Grundlagen einer wissenschaftlichen Wertlehre.* Vienna, 1937.

LAIRD, JOHN. *The Idea of Value.* Cambridge, 1929.

MEAD, G. H. "Scientific Method and the Moral Sciences," *International Journal of Ethics,* XXXIII (1923), 229–47.

MOORE, G. E. *Principia ethica.* London, 1903.

NEURATH, OTTO. *Empirische Soziologie; der wissenschaftliche Gehalt der Geschichte und Nationalökonomie.* Vienna, 1931.

PELL, O. A. H. *Value Theory and Criticism.* New York, 1930.

PERRY, RALPH BARTON. *General Theory of Value.* New York, 1926. Also articles in the *International Journal of Ethics* (1931), *Journal of Philosophy* (1931), and *Philosophical Review* (1932).

PRALL, DAVID, W. "A Study in the Theory of Value," *University of California Publications in Philosophy,* III, No. 2 (1918), 179–290.

———. "In Defense of a 'Worthless' Theory of Value," *Journal of Philosophy,* XX (1923), 128–37.

REID, JOHN. *A Theory of Value.* New York, 1938.

RUSSELL, B. *Philosophical Essays.* New York, 1910.

SCHLICK, MORITZ. *Fragen der Ethik.* Vienna, 1930. English trans., *Problems of Ethics.* New York, 1939.

STUART, HENRY WALDGRAVE. "Valuation as Logical Process," in *Studies in Logical Theory,* ed. JOHN DEWEY *et al. The Decennial Publications of the University of Chicago,* Vol. XI. Chicago, 1903.

The Technique of Theory Construction

Joseph H. Woodger

The Technique of Theory Construction

Contents:

The Technique of Theory Construction

J. H. Woodger

I. The Purpose of This Monograph

I am writing this contribution to the *Encyclopedia* because I believe that a wider diffusion of knowledge about modern discoveries relating to logic will promote the unification of science and have other beneficial effects on scientific theory. I believe it will do this for the following three principal reasons.

1. The utilization of such knowledge in the construction of scientific theories would help to remove those tendencies toward *dis*unity which owe their existence merely to the ambiguities and other defects of natural languages. Disunity which issues from a genuine difference of opinion may be a sign of life and growth and in no way a matter of regret. If unity is to mean no more than a sterile uniformity of opinion, it would hardly be worth a moment's consideration. Let there be as much variety in choice of foundations and as much fertility in the throwing-up of new hypotheses as possible. But, if the maximum benefit is to be derived from such abundance, it is essential that alternative theories should be in some way easily and exactly comparable through the use of a common means of communication. The kind of unity that is required is, therefore, unity of language, or at least the availability of a means of expressing alternative hypotheses in a manner which will make their mutual relations as clear as possible.

But the possibility of unification through a unitary language as here contemplated is not to be achieved by merely choosing a single language in the ordinary sense of that word, i.e., a natural language like English or French, or even an artificial language like Esperanto. Such languages have certain characteristics which render them unfit in the last resort for establishing the kind of unification which is here under discussion. Moreover, these characteristics are in many respects the very ones which render them such admirable vehicles of *literary* style and expression. The richness of their vocabularies and the arbitrariness of their syntactical rules militate against their suitability for scientific purposes by rendering them difficult of control. For this reason, as will be seen later, the process of calculation is not possible in a natural language.

But the natural languages have a further and even more serious disadvantage. Because we learn them during the most impressionable period of our lives, they become to such an extent part of ourselves that we come to use them without ever being aware of their conventional and arbitrary character, and thus of certain of their properties which are least admirable from the point of view of science. Conventions which are so deeply imbedded in our nature have the force of moral and religious principles. They are extremely difficult to face and criticize frankly, and departures from custom in such matters are likely to arouse antagonism. A study of recent advances in knowledge relating to logic will give us courage to experiment with language. During the last thirty years a great deal of such experimenting has been going on, and the results reached are well worth the attention of those interested in the theoretical side of natural science. But they are still only available for the most part in technical treatises and journals.

These researches are the outcome of the union of two tendencies which began about the middle of the nineteenth century, namely, a tendency toward the mathematization of logic, on the one hand, and a tendency toward greater logical rigor and a more critical attention to foundations in mathematics, on the other.

The historical development of mathematics presents a most extraordinary contrast with that of logic. Whereas the former has been uninterrupted during some two thousand years, the latter has slumbered during the same period until its awakening in the nineteenth century by the publication in 1847 of George Boole's first paper on the algebra of logic. It is not surprising, therefore, that the value of mathematics in natural science should by now be recognized while that of logic still remains to be explored. Moreover, the particular course taken by the development of mathematics has been to a large extent determined by the requirements of one science, namely, physics. These circumstances have been responsible for two important consequences. First, the linguistic difficulties already referred to are not so acute in physics as in the biological sciences. At the same time they exist, because physics and mathematics do not coincide, and in consequence a part of physics remains which has to be stated in a natural language. Second, in so far as the *direction* of the development of mathematics has been given a bias by the special requirements of the problems of physics (and of the particular mode of treating those problems which happens to have been followed), in so far it may, in some of its most fully developed branches, be unsuitable for the use of other sciences, e.g., for the biological sciences.

It is thus necessary that we should free ourselves not only from the conventions of natural language but also from the accidental restrictions of traditional mathematics, i.e., the mathematics which has arisen to meet the needs of physics. We require a technique for constructing new kinds of mathematics as well as for constructing new languages for scientific purposes. Indeed these two techniques, if not identical, are extremely closely related.

2. Second, the application of a knowledge of modern discoveries relating to logic to the construction of scientific theories would help to compensate to some extent for the undesirable effects of the inevitable specialization which attends the progress of every branch of scientific investigation. To show how this could happen will be one of the objects of the present work.

3. Third, a wider diffusion of a knowledge of modern discoveries relating to logic would dispel a good deal of the prevailing confusion (which shows itself from time to time in controversies which disfigure the pages of *Nature*) concerning topics which are usually included under the title 'methodology', such as the relation of mathematics to natural science, "the logic of

science", the relation of science to metaphysics, and kindred subjects.

For these reasons the present work attempts to make available to the scientific reader some of the results reached by the logical and metalogical investigations of the last thirty years from the point of view of their utilization in the construction of clearer, more concise, better organized, and so more controllable theories.

There are two ways in which a book with such aims could be written. A general account of the main results achieved might be given and accompanied by simple illustrations. This method would be suitable as an introduction intended for the reader for whom these subjects are a primary interest. But it would involve the use and explanation of much technical terminology; it would be very abstract, and the examples would tend to have an air of isolation and triviality. For these reasons it would be unsuitable as an exposition addressed to men of science. Alternatively, we might try to inculcate these general principles not by means of an abstract exposition, not in fact by *talking* about them at all, but by *exhibiting* them in use by presenting a detailed example of a scientific theory which embodies them. This is the method which will be adopted in the following pages. Even this method is not without its difficulties. If for the sake of simplicity and brevity we choose a simple example, the reader may get the impression that we are being extremely pedantic about things which are obvious and trivial. If we choose an example that is more lengthy and complicated in the hope of impressing the reader and escaping the charge of triviality, we run the risk of losing his attention in a maze of unavoidable details. For our present purposes an illustrative theory must be sufficiently extensive to provide a connected illustration of all the general methodological principles concerned in theory construction. At the same time it should require little or no technical knowledge of the subject matter of the theory, in order that it should not be necessary to divert the reader's attention too much from the principles used in constructing the theory to explaining the ideas contained in the theory itself. In this example, moreover, we shall try to be as explicit and precise as possible—not because such a degree of explicitness and precision is necessary for the treatment of the simple topics with which the theory deals, but because we wish to use this theory to show how such a degree of explicitness and precision is to be attained. It must also be emphasized that this specimen theory is *not* offered as an example of an ideal to which it is suggested that all scientific theories should conform but again only as an illustration of the general principles which are to be discussed when it has been given. As will be explained later, these general principles can be utilized to various degrees and in a variety of ways. In order to make the fullest use of the specimen theory, it must be constructed not from the standpoint of its scientific utility in the science to which it belongs but from that of its completeness for illustrative purposes.

It is not to be expected that a reader who is quite unfamiliar with these topics will be able to understand this monograph thoroughly from a single reading. Some parts (and these, unfortunately, must come early in the exposition) will only be fully understood when their use at later stages has been seen. Accordingly, the reader is recommended at the first reading not to delay too much over difficult points but to read on until they are clarified by later developments. On a second reading such difficulties will be found to have disappeared or will vanish by reference to later parts of the monograph which the reader will by then have read and be able to find. What reward can a reader hope for who studies this monograph attentively in this way? He will have gained at firsthand an insight into logic and into the nature of scientific theory which deserves to form part of a scientific education and to be more widespread among men of science than appears to be the case at present. He will be in a position to extend and deepen his knowledge of these topics by reading more abstract and general treatises, and to apply it to the particular scientific problems in which he is interested.

It is hoped that the specimen theory to be presented will fulfil the requirements set forth above. But, before we begin to expound it, some preliminary explanation is necessary.

II. Preliminary Explanations

(i) *Distinction between theory and metatheory.*—In the first place it is most important to keep constantly in mind the distinction between a theory T which deals with a certain subject matter (in the way in which a particular biological theory deals with some aspect of animals or plants) and a theory which has the theory T itself as *its* subject matter. The theory which has a given theory T as its subject matter is called the metatheory of T. This distinction is very important because, as we shall see, it is the absence of an explicitly formulated metatheory which distinguishes a theory using a natural language from one having the special scientific properties of which we are now in search. Our first task, therefore, will be to state the metatheory of the biological theory which will be offered as an example.

Next we must understand a further distinction within the metatheory. Corresponding to the two aspects—structure and meaning—of T, its metatheory will have two subdivisions. The first—called the *syntax* or morphology of T—is concerned solely with the structure of T. Any written language (and we are concerned only with theories written down) consists of ink (or other) marks of various shapes and sizes arranged (usually) in horizontal lines on the paper and read from left to right. This is the purely structural aspect of language which forms the province of syntax. But, in addition to its purely structural aspect, the elements of a language have a reference to something other than themselves by virtue of which the language functions as a means of communication. Names and designatory expressions denote objects. By means of other expressions—called statements—we are able to assert that so and so is the case and this assertion may be true or false. Such notions as denoting, truth, and falsehood all involve the relation of the signs or words of the given theory T to the subject matter of T, and the branch of metatheory which contains notions involving this relation is called *semantics.*

In order to speak about a language, we obviously require names for the words and statements of that language. In the present instance we shall have recourse to the device of inclosing a word or statement in single quotation marks when we want to speak not about what the word denotes nor about what the statement asserts but about the word or statement itself. In the syntax of a theory *T* the words of *T* will occur in single quotes only; in the semantic of *T* they will occur both with and without single quotes. For example:

'Cats eat meat'

is a biological statement.

' 'Cats' consists of four letters'

is a correctly formulated syntactical statement, and

' 'Cats' denotes cats' and ' 'Cats eat meat' is true'

are semantic statements. But the following would involve a confusion of theory and metatheory because they would not be meaningful expressions in either language:

'Cats consists of four letters'

' 'Cats' eat meat'.

The syntax of a given theory *T* will consist of statements which: (1) enumerate and classify the signs or words of *T;* (2) provide rules which determine when an expression (i.e., a linear combination of one or more signs or words of *T*) is to be regarded as significant, and when a significant expression is a *statement* of *T;* and (3) provide rules which determine when a given statement of *T* is a *consequence* of (or follows from) one or more other statements of *T*. These rules are called the rules of statement-transformation in *T*.

All these tasks fall entirely within the province of syntax, because they can be carried out without reference to meaning. When we turn to questions of application and testing, however, we at once require reference to the subject matter and so enter the province of semantics. One task for semantics is the elucidation of the meaning of the words or signs of *T* with the help of a language already known to the reader. With other tasks of this branch of metatheory the present work will not deal.

From now onward we shall use the letter '*T*' to denote the particular specimen theory we are about to construct and the syntax of which will first be given in English.

(ii) *The classification of signs.*—Before we give a list of the words or signs of *T*, we must settle a number of questions. First, we have to decide whether *T* is to be a word language or a symbolic language. Symbolic languages have a very great deal to recommend them, but they have the great disadvantage that, except in the case of mathematics and chemistry, most people are unaccustomed to them. On the other hand, a pure word language which used English words but not English grammar, not only would sound uncouth but would be extremely cumbersome and tedious if at the same time it were to be precise. We shall in the present instance adopt a compromise. We shall use a language partly of words and partly of symbols, and we shall use 'signs' to include both.

Signs can be divided into two major groups: (1) constants and (2) variables. Constants are signs which have a fixed meaning, and variables have no meaning at all. Their use is to mark places into which constants can be inserted, as will be explained later. Constants can be divided into (*a*) logical constants and (*b*) subject-matter, or descriptive, constants. *T* will contain only three primitive subject-matter constants. These are: '**P**', '**T**', and '**cell**'. '**P**' will denote the relation *part of*, '**T**' the relation *before in time*, and '**cell**' will denote the class of cells (in the biological sense). (Subject-matter constants will always be printed in boldface type.) All the remaining signs will be either logical constants, variables, or subject-matter constants which can be defined with the help of the foregoing three primitives, logical constants, and variables.

For syntactical purposes it is also convenient to classify constants into (*a*) proper constants (e.g., '**cell**') and (*b*) functors. Functors are expressions which, while being themselves neither designations nor statements, form either designatory expressions or statements when combined (in accordance with the syntactical rules) with other signs. They can be divided into designatory or sentential functors according to whether designatory expressions or statements are combined with them, and each of these divisions can be divided further into name-forming and statement-forming functors, according to whether they yield designatory expressions or statements. This classification will be used and illustrated below in formulating the syntax of the theory *T*.[1]

In order to explain the use of logical constants and variables, a little must first be said about statement construction in the theory *T*. The simplest kind of statement will consist of only three signs. For example:

'Chelsea **P** London'

would be such a statement (meaning: Chelsea is part of London), although not one which could be formulated in *T*, because *T* does not contain the words 'Chelsea' and 'London'. These are individual names, and *T* contains no such names but only *individual variables*. This is because we shall not have occasion to make *particular* statements about things being parts of one another but only *general* statements, and this can be done most conveniently by means of variables. Individual variables are signs which can have designations of individuals substituted for them, and as variables of this kind we shall use the letters '*u*', '*v*', '*w*', '*x*', '*y*', and '*z*'. The subject-matter constants '**P**' and '**T**' will appear accompanied by such variables in the following way:

'*x***P***y*', '*u***T***v*' .

The meaning and use of such expressions will be explained later. The foregoing are both instances of the same kind of simple statement, namely, one consisting of a relation-sign with one individual variable to the left and one to the right of it. '**P**' and

[1] The word 'functor' (due to Professor T. Kotarbinski) is used in the foregoing sense by the Warsaw school of logicians. The word is used by R. Carnap in a more restricted sense.

'**T**' are thus examples of statement-forming designatory functors.

A second and different kind of simple statement consists of the logical constant 'ϵ' preceded by an individual variable and followed by the word '**cell**', thus:

'$x\epsilon$**cell**' ,

which can be read 'x is a cell'. The sign 'ϵ' is thus used to express membership of a class.

(iii) *Operations on statements.*—From such unitary statements composite ones are constructed by means of five other logical constants. Thus we can construct:

'not (x**P**y)' ,

which will mean: x is not part of y; and the *negation* of any other statement can be formed by inclosing it in parentheses and writing 'not' in front of it. Similarly, we construct the *conjunction* (also called logical product) of two statements by inclosing them in parentheses and writing 'and' between them. Thus

'(x**P**y) and (x**T**y)'

will mean: x is part of y and before y in time.

The terms 'and' and 'not' will presumably be understood without further explanation. The meanings of the remaining three logical constants of this group will be clear if careful attention is paid to their definitions, since they can all be defined by means of 'not' and 'and'. That is to say, we can construct statements (containing only 'not', 'and', and variables) which can be substituted for statements containing the remaining three constants of this group and variables. Thus, using 'p' and 'q' as variables for which any statements of T can be substituted, we can define 'p or q' as meaning

not ((not (p)) and (not (q))) ,

i.e., 'p or q' is to mean: not both not (p) and not (q). Hence, when we assert 'p or q', we are asserting 'either p or q or both

p and q'. (There is another 'or' which excludes the assertion of 'both p and q', but this will not be used.)

The next logical constant is easily definable but somewhat difficult to represent in a word language on account of the confusion which is likely to become associated with the usual ways of formulating it. But, if the reader will take the precaution of attaching no meaning to the verbal formulation beyond that required by the definition, such difficulties will not arise. This constant can be represented by the words 'if' and 'then' each followed by a statement, or by the single word 'implies' with one statement to the left and one to the right of it. The word 'implies' is so ambiguous in English that it is perhaps more liable to generate confusion than 'if . . . then - - -', but it has the great advantage of consisting of only a single word *between* two statements and thus falls into line with 'and' and 'or'. We define 'p implies q' as being substitutable for

not ((p) and (not (q))) .

If we substitute 'x**T**y' for 'p' and 'x**P**y' for 'q' in this, we see that

'x**T**y implies x**P**y'

will mean: it is not the case that x is before y in time and x is not part of y. Alternatively, 'p implies q' may be read 'if p then q'. But, whichever reading is preferred, the interpretation is that fixed by the foregoing definition and *no other*. Expressed with the help of 'not' and 'or', 'p implies q' may be written

'(not (p)) or (q)' .

(In passing it should be noted that whether the statements here given as examples are true or false is a matter of complete indifference from the present standpoint. Because we are here concerned only with the syntax of T, and the notions of truth and falsity are entirely outside the province of syntax. A false statement serves the purpose of illustration of sentential structure just as well as a true one.)

The last constant of this set is usually represented in words by 'if, and only if,' or by 'is equivalent to', but we shall abbreviate this to 'equiv'. This is used to join two statements 'p' and 'q' when it is the case that p implies q and q implies p, in the sense of 'implies' above explained. It is, therefore, usually defined as being substitutable for

'((p implies q) and (q implies p))' .

It is important to notice that 'p implies q' and 'p equiv q' are not statements *about* statements any more than 'p and q' and 'p or q' are. Otherwise they would belong not to T but to the metatheory of T. Moreover, 'implies' as it is used here is not functioning as a transitive verb but as a conjunction. All these four logical constants are signs with the help of which we join statements belonging to T to form other statements belonging to T. They can accordingly be called *sentential operation-signs*. Just as in arithmetic we can join two number-signs by the sign of multiplication to form a new number-sign which denotes a number called the product of the two numbers denoted by the first two number-signs, so in the case of statements we can join two together by 'and' to form a new statement which is called their logical product. This new statement is clearly not a statement about statements but about whatever its two constituent statements are about. And this is also the case when we join two statements by 'implies' or by 'equiv' in the sense explained above. The logical constant 'not' differs from the other four sentential operation-signs in operating upon *one* statement instead of two. Negation is accordingly called a unitary operation and the others binary operations.

Letters such as 'p' and 'q' which can have statements substituted for them will be called *sentential variables*. Formulas like 'p and q' consisting of sentential variables combined by means of sentential operation-signs are called *truth-functions*. They are given this name because, when statements are substituted for their variables, the truth of the resulting compound statement depends only on the truth or falsity of the statements substituted. Thus 'p and q' yields a true statement only if a true statement is substituted for 'p' and a true one for 'q'; 'p or q' yields a true one so long as at least one of the two variables is replaced by a true statement; and any substitution in 'p implies q' yields a true statement so long as we do not substitute a true one for 'p' and at the same time a false one for 'q'. Thus, of the following four examples of substitutions in 'p implies q', the first three are true statements and the fourth alone is false:

(1) (2 + 2 = 4) implies (Rome is in Italy)

(2) (2 + 2 = 5) implies (Rome is in Italy)

(3) (2 + 2 = 5) implies (Rome is in France)

(4) (2 + 2 = 4) implies (Rome is in France)

Finally, the truth-function 'p equiv q' yields a true statement if *both* variables are replaced by true statements or if *both* are replaced by false ones.

But it is also possible to construct formulas of this kind which yield true statements for *any* substitution we care to make for the variables they contain. Such formulas are used, in conjunction with the rules of statement-transformation which were mentioned above (p. 7), for deriving statements from other statements in the process of proof. For example:

(1) (((p) and (q)) implies (r)) implies ((q) implies ((p) implies (r)))

is such a formula. With its help we can obtain

(2) ($y\mathbf{T}z$) implies (($x\mathbf{T}y$) implies ($x\mathbf{T}z$))

from

(3) (($x\mathbf{T}y$) and ($y\mathbf{T}z$)) implies ($x\mathbf{T}z$)

in the following way. We first make use of a rule governing the process of substitution which enables us to substitute '$x\mathbf{T}y$' for 'p', '$y\mathbf{T}z$' for 'q', and '$x\mathbf{T}z$' for 'r' in (1), by which means we obtain

(4) ((($x\mathbf{T}y$) and ($y\mathbf{T}z$)) implies ($x\mathbf{T}z$)) implies (($y\mathbf{T}z$) implies (($x\mathbf{T}y$) implies ($x\mathbf{T}z$))) .

We next make use of a second rule which states that, if A is a statement of T and 'A implies B' is a statement of T, then B is a consequence of these two statements. For it will be noticed that (4) consists of (3) followed by 'implies' followed by (2). According to the second rule, therefore, (2) is a consequence of (4) and (3). Now (4) has been obtained from a universally valid formula in accordance with the first rule; (3) is adopted by the author as a postulate of the theory T. The rules have been so framed that they yield true statements when they are applied to true statements. Accordingly, anyone who agrees with the author in admitting (3) as a true statement of T is also committed to admitting (2).

The system of all such generally valid statements constructed with sentential variables and operation-signs is called the calculus of statements. In the first part of the theory T we shall give a list of those statements belonging to it which will be needed for the derivation of consequences in T. For the demonstration that they are universally valid and for the axiomatic exposition of this calculus the reader must be referred to works upon logic.

(iv) *Quantifiers.*—We must now consider a little more closely the interpretation of statements containing variables. An algebraical identity such as

$$`x + y = y + x'$$

is understood to hold good for *any* number-signs which we care to substitute for its 'x' and 'y'. In the same way such a formula as

$$`(x\mathbf{T}y \text{ and } y\mathbf{T}z) \text{ implies } (x\mathbf{T}z)'$$

is meaningless unless we are told whether it is to hold for *all* or only for *some* substitutions of the three variables it contains. When no indication is given, it is to be understood that *all* is meant, just as in the case of the algebraical identity. Thus the foregoing statement will mean: whatever thing is before another thing in time, and this other thing is before a third thing, then the first will also be before the third thing. But sometimes it is necessary explicitly to state whether all or some is meant. This is especially important when negation is involved. For example: suppose '$x \epsilon$**cell**' is to be translated into: 'For every x, x is a cell' (i.e., into: 'everything is a cell'), then the interpretation of 'not ($x \epsilon$**cell**)' might be either

(1) it is not the case that everything is a cell

or

(2) for every x, x is not a cell (i.e., nothing is a cell) .

These are evidently quite different statements which must accordingly be distinguished, and for this purpose we need a new logical constant. In those cases where it is necessary we shall write

(3) '(All x) ($x \epsilon$**cell**)'

for: 'For each and every x, x is a cell' or 'everything is a cell'. Then

(4) 'not ((All x) ($x \epsilon$**cell**))'

will mean: it is not the case that everything is a cell, (corresponding to case (1)). And we shall write

(5) '(All x) (not ($x \epsilon$**cell**))'

for: 'For each and every x, it is not the case that x is a cell', i.e., for 'nothing is a cell' (corresponding to case (2)).

Now it will be noticed that (4) is equivalent to saying

(6) 'Something is not a cell'

and is the negation of (3). We can, therefore, make use of this fact in order to define a logical constant for expressing 'for some x' or 'there is at least one x such that' as follows:

'(Some x) ($x \epsilon a$)'

is defined as being equivalent to and so substitutable for

'not ((All x) (not ($x\epsilon\alpha$)))'.

'(Some x) ($x\epsilon\alpha$)' is thus defined in such a way that it means: it is not the case that for every x, x is not a member of α. In this definition we have, of course, used a variable after the 'ϵ' instead of the constant '**cell**'. This is a new kind of variable which we shall call a class variable because it can have class-names substituted for it. The letters 'α', 'β', 'γ', and 'δ' will be used as class variables when the classes concerned are classes of individuals. Constant class-signs will always consist of lower-case letters.

Variables occurring in statements which are not preceded by '(All . . .)' or by '(Some . . .)' containing a variable of the same shape and size are called *free* variables, otherwise they are called *bound* variables. For example, in

'(All x) (x**T**y)'

'x' is bound and 'y' is free. The two new logical constants here introduced are called *quantifiers*. '(All . . .)' is called the universal and '(Some . . .)' the existential quantifier. When used with statements containing relation-signs, when two variables are involved, they will be written as follows:

'(All x) ((All y) (x**P**y))',

which will mean: for every x and every y, x is part of y. And

'(Some x) ((Some y) (x**P**y))',

which will mean: there is an x and a y such that x is part of y. But, as we here have two individual variables, other combinations will now be possible which did not occur in the case of classes. In fact, including the two above, we can have six combinations which must be carefully distinguished:

(1) (All x) ((All y) (x**P**y)) (everything is part of everything)

(2) (All x) ((Some y) (x**P**y)) (everything is part of something)

(3) (All y) ((Some x) ($x\mathbf{P}y$)) (everything has a part)

(4) (Some x) ((All y) ($x\mathbf{P}y$)) (something is part of everything)

(5) (Some y) ((All x) ($x\mathbf{P}y$)) (there is something of which everything is a part)

(6) (Some x) ((Some y) ($x\mathbf{P}y$)) (something is part of something)

It will be noticed that the members of the pairs (2) and (5) and (3) and (4) differ only in the order in which the quantifiers are written, and yet the meaning of the two members of each pair is quite different. On the other hand, the order in which the quantifiers are written in cases (1) and (6) is indifferent.

As variables for which relation-signs like '**P**' and '**T**' can be substituted we shall use the capital letters 'P', 'Q', 'R', and 'S' when the relations concerned are relations between individuals. These will be called *relation variables*. Relation-signs will consist either of a single capital letter or of two or more letters of which the first is a capital. In the theory T the letter '**S**' will be used to denote a certain relation between an individual and a class. The variables used for relations of this type will be 'T', 'U', 'X', and 'Y'.

(v) *Identity.*—To express the identity of x and y, we shall write '$x = y$'. It must be understood that '$=$' is used to denote *strict* identity, not merely likeness or equality with respect to *some* properties. If $x = y$, then everything which can be said about x can also be said about y; and, in consequence, 'y' can be substituted for 'x' in any statement in which 'x' occurs and vice versa.

With the help of the logical constants we now possess we can formulate the statement:

'There is exactly one x such that $x \epsilon a$'.

This can be done by joining the following two statements by 'and':

(1) 'There is at least one x such that $x \epsilon a$'.

(2) 'There is at most one x such that $x \epsilon a$'.

For (1) we already have '(Some x) ($x \epsilon a$)', and for (2) we can now write

$$\text{'(All } x\text{) ((All } y\text{) ((} x \epsilon a \text{ and } y \epsilon a\text{) implies (} y = x\text{)))'},$$

which means: for every x and every y, if x is a member of a and y is a member of a, then y is identical with x. The result of combining these two statements by means of 'and' is a statement which means: there is a member of a, and everything which is a member of a is identical with it.

(vi) *Definitions.*—The constants we have now introduced as primitive or undefined ones, namely, the subject-matter constants '**P**', '**T**', and '**cell**', and the logical constants 'ϵ', 'not', 'and', '(All . . .)', and '=', together with the parentheses '('and')', and the five kinds of variables are all that are absolutely indispensable in order to formulate all the statements of T. Consequently, every statement of T must consist of a combination of some of these signs, or of signs defined with their help. Definitions are not strictly necessary, but they introduce an enormous simplification by enabling us to use a short expression containing the new defined sign in the place of a long combination of the undefined ones. It is obviously better to write 'p or q' instead of 'not ((not p) and (not q))'. Moreover, being defined, the meaning of 'p or q' is fixed, provided the meaning of 'not' and of 'and' is fixed.

In the specimen theory T, in order to avoid the very considerable complications involved in formulating syntactical rules for the construction of definitions, the place of definitions will be taken by statements which are not structurally distinguished from postulates. They will consist of two statements joined by 'equiv', the one on the left of this sign containing the new sign, that on the right containing only signs which have already been introduced. Since two statements joined by 'equiv' can be substituted for each other wherever they occur, such "verbal postulates" will serve the purpose of definitions without involving syntactical complications. This procedure has the disadvantage that we cannot introduce a new sign into the theory without changing it, because each new sign involves an addition to the postulates of the theory. But this will be no obstacle in the present instance where the theory is being constructed purely for illustrative purposes in an introductory exposition.

(vii) *Classes and relations.*—By means of the logical constants which have now been explained it is possible to develop two more logical calculuses in which individual variables no longer appear and in which we operate directly with class-signs and relation-signs. These are called the calculus of classes and the calculus of relations, respectively. Thus, instead of writing

'(All x) (($x \epsilon \alpha$) implies ($x \epsilon \beta$))' ,

we can write

'α is included in β' ,

which is usually symbolized by

'$\alpha \subset \beta$' .

Similarly, in the case of relation-statements, instead of

'(All x) ((All y) ((xRy) implies (xSy))'

we can write

'$R \subset S$' .

For example: suppose R is the relation of parent to child and S that of teacher to pupil, then '$R \subset S$' would mean: every person who is parent of another is also teacher of that other.

Lists of statements belonging to the elementary parts of these calculuses which will be used in the theory T will be given in Part I of the theory. These two calculuses, especially that of relations, are particularly important from the standpoint of natural science. With their help an enormous amount of unnecessary verbiage could be swept away from scientific theories, and a properly constructed scientific language should enable us to make use of them.

To facilitate the formulation of certain of the syntactical rules (especially those relating to the construction of statements), it is necessary to arrange certain of the signs of T in classes called *types*. In the list about to be given, the type to be assigned to each sign is indicated by a type-index. The purpose of these type-indices will be understood from their use in formulating the rules of statement construction. Although the syntax of T may seem complicated when read before the theory itself has been studied, it is in fact quite simple to understand and apply when its use in the theory has been seen and the meanings of the signs enumerated have been learned from the definitional postulates of the theory. The reader should therefore not dwell too long on difficulties which on a first reading appear in this part of the work.

III. The Syntax of the Specimen Theory T[2]

List of Signs Used in Constructing the Expressions of T and Rules for Their Use

1. Variables

   1.1 Sentential variables: 'p', 'q', 'r', 's'.

   1.2 Individual variables: 'u', 'v', 'w', 'x', 'y', 'z'.

   1.3 Class variables: 'α', 'β', 'γ', 'δ'.

   1.4 Relation variables: 'P', 'Q', 'R', 'S'.

   1.5 Relation variables: 'T', 'U', 'X', 'Y'.

Variables under 1.2 will be called variables of type $*$, those under 1.3 will be of type $(*)$, those under 1.4 of type $(*, *)$, and those under 1.5 of type $(*, (*))$.

2. Constants

   2.1 Statement-forming sentential functors
   'not', 'and', 'or', 'implies', 'equiv', 'All', 'Some'.

   2.2 Systematically ambiguous statement-forming functors
   'ϵ', '$=$', '$\neq$', '$\subset$'.

[2] I am greatly indebted to Dr. A. Mostowski for his generous assistance with this section.

2.3 Constants of fixed type

2.31 Proper constants

(∗) ‘Λ’, **‘mom’**, **‘cell’**.

((∗, ∗)) ‘sym’, ‘asym’, ‘trans’, ‘1→many’, ‘many→1’.

2.32 Statement-forming designatory functors

(∗, (∗)) **‘.S.’**

(∗, ∗) **‘.P.’**, **‘.Pp.’**, **‘.T.’**, **‘.C.’**, **‘.Z.’**, **‘.Sl.’**, **‘.B.’**, **‘.E.’**, **‘.D.’**, **‘.F.’**

2.33 Name-forming designatory functors

In the parentheses on the left of each line are given two type-indices separated by a semicolon. On the right of the semicolon is given the type-index of the sign or signs which can be inserted in place of the dot or dots associated with the functor. On the left of the semicolon is given the type-index of the resulting expression.

((∗); ∗) ‘ι‘.’

((∗); (∗), (∗)) ‘.&.’, ‘.or.’

((∗); (∗)) ‘—.’

((∗, ∗); (∗, ∗)) ‘—.’, ‘.$^{-1}$’

((∗, ∗); (∗, ∗), (∗, ∗)) ‘.&.’, ‘.or.’, ‘.—.’

((∗); (∗, ∗)) ‘Dom‘.’, ‘Cnvdom‘.’

((∗); (∗, ∗), ∗) ‘.s‘.’

((∗, ∗); (∗, ∗), (∗, ∗)) ‘.|.’

(∗; (∗)) **‘S‘.’**

(∗; ∗) **‘B‘.’**, **‘E‘.’**

3. Rules for the Construction of (Meaningful) Expressions

3.1 *Rules for the construction of designatory expressions*

3.11 The variables under 1.2, 1.3, 1.4, and 1.5 and the proper constants under 2.31 are called designatory expressions of the 0-th (zero-) level.

3.12 Assuming that designatory expressions of the 0, 1, 2, . . . , kth level have been defined and a type-index has been assigned to each, a designatory expression of the $k + 1$-th level is one which results if we replace the dot in any sign of 2.33 by bracketed designatory expressions of $\leq k$th level with the corresponding type-indices. The expression so arising has the type indicated by the sign standing to the left of the semicolon.

3.13 An expression A is called a designatory expression if there is a number k such that A is a designatory expression of the kth level.

3.2 *Rules for the construction of statements*

3.21 If t is the type of a designatory expression A and (t) the type of a designatory expression B, then we shall say that the type of B is one higher than the type of A.

3.22 An expression A is called a statement of the 0-th level if either (1) it is a sentential variable (see 1.1) or (2) it has the form $(e_1)e_2(e_3)$, where e_1, e_2, e_3 satisfy one of the following conditions:

3.221 e_2 is 'ϵ', and e_1 and e_3 are designatory expressions, the type of e_3 being one higher than the type of e_1.

3.222 e_2 is '$=$' or '$\neq$' or '$\subset$' and e_1 and e_3 are designatory expressions of the same type.

3.223 e_2 is '**S**', e_1 is a designatory expression of type $*$, and e_3 is one of type $(*)$.

3.224 e_2 is one of the expressions 2.32 of type $(*, *)$ and e_1 and e_3 are designatory expressions of type $*$.

3.23 Assuming that statements of the 0,1,2, . . . , kth level have been defined, we call an expression A a statement of the $k + 1$-th level if there are statements B,C of at most the kth level and a variable v which is not a sentential variable, and A is identical with one of the following expressions: 'not (B)', '(B) and (C)', '(B) or (C)', '(B) implies (C)', '(B) equiv (C)', '(All v) (B)', '(Some v) (B)'.

3.24 An expression A is called a statement if there is a number k such that A is a statement of the kth level.

4. Rules of Statement-Transformation in T

4.1 *Rule of substitution*

If A is a statement containing the free variable v, and if B results from A by replacing v wherever it occurs in A by:

4.11 a statement if v is a sentential variable;

4.12 a designatory expression of the same type as v if v is not a sentential variable,

then B is a consequence of A provided always that no free variable of the expression inserted becomes bound after the insertion.

4.2 *Rule of abruption*

If A is a statement of T and 'A implies B' is a statement of T, then B is a consequence of these two statements.

4.3 *Rule for the use of equivalences*

If A is a statement of T, B is a statement of T and a part of A, and C is any statement of T which is equivalent to B, and D is the result of substituting C for B in A, then D is a consequence of A and the statement which asserts the equivalence of B with C.

4.4 *Rule for the use of identities*

If A is a statement of T containing an expression e_1, and e_2 is an expression such that e_1 followed by '=' followed by e_2 is a statement B of T, and C is the statement obtained from A by inserting e_2 in the place of e_1 in one or more (not necessarily all) places of its occurrence in A, then C is a consequence of A and B.

4.5 *Rules for the introduction and removal of quantifiers*

4.51 'A implies B' is a consequence of 'A implies (All v) (B)'.

4.52 'A implies B' is a consequence of '(Some v) (A) implies B'.

4.53 If v is not free in A, then 'A implies (All v) (B)' is a consequence of 'A implies B'.

4.54 If v is not free in A, then '(Some v) (B) implies A' is a consequence of 'B implies A'.

ILLUSTRATIONS OF THE FOREGOING RULES

3.11 Examples of designatory expressions of zero-level: 'x', 'u', 'α', 'P', 'T', '**cell**', 'sym'.
Examples of designatory expressions of the first level: 'ι' (u)', '(α) & (**cell**)', '—(α)', '—(P)', '(P) or (Q)', 'Dom' (R)', '(**P**)s' (x)', '(R) | (S)', '**S**'(α)', '**B**'(x)'.
Examples of designatory expressions of the second level: '(Dom' (R)) & ((α) or (β))', '—(ι' (x))', '((R) | (S))' (x)'.

3.22 Examples of statements of zero-level: 'p', '(u) ϵ (**cell**)', '(**P**) ϵ (trans)', '(x) = (y)', '(α) = (β)', '(P) $\neq$ (Q)', '(R) $\subset$ (S)', '(z) **S** (α)', '(u) **P** (v)', '(x) **T** (y)'.
Examples of statements of the first level: '(p) implies (q)', 'not ((u) ϵ (**cell**))', '(All x) ((x) **T** (y))'.
Examples of statements of the second level: '((p) implies (q)) or ((p) implies (r))', '(All x) ((Some y) ((x) **T** (y)))'.

4.11 '((x) **T** (y)) implies (q)' is a consequence of '(p) implies (q)'

4.12 '(**cell**) $\neq$ (Λ)' is a consequence of '(α) $\neq$ (Λ)'

4.2 'not (p)' is a consequence of
'(p) implies (not (p))' and
'((p) implies (not (p))) implies (not (p))'

4.3 '((x) R (y)) or ((x) S (y))' is a consequence of '((x) R (y)) or ((x) Q (y))' and '((x) Q (y)) equiv ((x) S (y))'

4.4 '(x) R (z)' is a consequence of '(x) R (y)' and '(y) = (z)'.

4.51 ‘(p) implies $((x) \epsilon (a))$’ is a consequence of
‘(p) implies (All x) $((x) \epsilon (a))$’

4.52 ‘$((x) \epsilon (a))$ implies (q)’ is a consequence of
‘(Some x) $((x) \epsilon (a))$ implies (q)’

4.53 ‘(p) implies (All x) $((x)\ R\ (y))$’ is a consequence of
‘(p) implies $((x)\ R\ (y))$’

4.54 ‘(Some x) $((x)\ R\ (y))$ implies (q)’ is a consequence of
‘$((x)\ R\ (y))$ implies (q)’

In order to avoid too great a multiplication of parentheses, we shall also make use of the following:

5.1 *Rules for the omission of parentheses*

5.11 Designatory expressions replacing dots in the signs of 2.33 in accordance with 3.12 will not be bracketed if they are of zero-level.

5.12 The parentheses around ‘e_1’ and ‘e_3’ in 3.22 will be omitted if e_1 and e_3 are of zero-level.

5.13 The parentheses around ‘B’ and ‘C’ in 3.23 will be omitted when B and C are statements of zero-level, except in those cases in which ‘(B)’ is immediately preceded by a quantifier.

5.14 Parentheses around any statement which is not part of another statement may be omitted.

Examples.—We can now write: ‘P or Q’ instead of ‘(P) or (Q)’, ‘Dom‘ R’ instead of ‘Dom‘ (R)’, ‘$u\epsilon$**cell**’ instead of ‘$(u) \epsilon$ (**cell**)’; ‘p implies q’ instead of ‘(p) implies (q)’ and ‘(All x) (x**T**y)’ instead of ‘(All x) $((x)$ **T** $(y))$’.

The reader may at this stage omit reading Part I of the theory T and pass directly to Part II, using Part I only for reference purposes. The statements of Part I are only used for the purpose of deriving consequences in Part II. It will, however, be desirable to read through Sections 0.4 and 0.5 of Part I because these contain explanations of notational devices which are used frequently in Part II. It must be emphasized that

Part I consists only of such extracts from the logical calculuses as are actually used in Part II. No attempt is made to give anything like a complete or systematic exposition of these calculuses. On the contrary, every effort has been made to restrict the statements of Part I to the barest minimum. For a systematic treatment of these topics the reader is referred to works upon logic.

IV. The Theory *T*, Part I

Section 0.1

List of Statements from the Calculus of Statements

0.101 (p or q) equiv (not ((not p) and (not q)))
0.102 (p implies q) equiv (not (p and (not q)))
0.103 (p equiv q) equiv ((p implies q) and (q implies p))
0.111 (q) implies ((p) implies (q))
0.112 p implies (p or q)
0.113 (p implies (not p)) implies (not p)
0.1141 (p and q) implies p
0.1142 (p and q) implies q
0.115 (p and q) implies (p or q)
0.116 ((p implies q) and (q implies r)) implies (p implies r)
0.117 ((p implies q) and (r implies s)) implies ((p and r) implies (q and s))
0.118 ((p implies (q or r)) and (p implies (not r))) implies (p implies q)
0.119 (p implies (q and r)) implies (p implies q)
0.121 p equiv (not (not p))
0.122 p equiv (p and p)
0.123 (p equiv q) equiv (q equiv p)
0.124 (p and q) equiv (q and p)
0.125 (p or q) equiv (q or p)
0.126 (p implies q) equiv ((not q) implies (not p))
0.127 ((p and q) implies r) equiv ((p and (not r)) implies (not q))
0.128 p implies p
0.131 ((p implies q) and (p implies r)) equiv (p implies (q and r))
0.132 ((q implies p) and (r implies p)) equiv ((q or r) implies p)
0.133 (p equiv q) equiv ((not p) equiv (not q))

0.134 ((p equiv q) and (q equiv r)) implies (p equiv r)
0.135 (p equiv q) implies (p implies q)
0.136 ((p and q) implies r) equiv (p implies (q implies r))
0.137 (p implies (q implies r)) equiv (q implies (p implies r))
0.138 (p) implies ((q) implies ((p) and (q)))
0.141 ((p implies q) and ((r and q) implies s)) implies (p implies (r implies s))
0.142 ((p implies q) and ((q and r) implies s)) implies ((p and r) implies s)
0.143 ((p implies (r implies s)) and (q implies r)) implies (p implies (q implies s))
0.144 ((p implies q) and ((p and q) implies r)) implies (p implies r)
0.145 ((p implies (q implies r)) and (q implies s)) implies (p implies (q implies (r and s)))

Section 0.2

Statements from the Calculus of Statements with Quantifiers

0.201 (Some x) ($x\epsilon\alpha$) equiv (not ((All x) (not ($x\epsilon\alpha$))))
0.202 (Some x) (xRy) equiv (not ((All x) (not (xRy))))
0.211 (All x) ($x\epsilon\alpha$) implies $x\epsilon\alpha$
0.212 $x\epsilon\alpha$ implies (Some x) ($x\epsilon\alpha$)
0.213 (All x) ((All y) (xRy)) implies xRy
0.214 xRy implies (Some x) ((Some y) (xRy))
0.215 (xRy and xSz) implies (Some u) (uRy and uSz)
0.216 (xRy and ySz) implies (Some u) (xRu and uSz)

Section 0.3

Statements from the Theory of Identity

0.311 $x = y$ equiv (All α) ($x\epsilon\alpha$ implies $y\epsilon\alpha$)
0.312 $x \neq y$ equiv (not ($x = y$))
0.313 $x = x$
0.314 $x = y$ implies $y = x$
0.315 ($x = y$ and $y = z$) implies $x = z$

Section 0.4

Statements from the Calculus of Classes

0.411 $\alpha \subset \beta$ equiv (All x) ($x\epsilon\alpha$ implies $x\epsilon\beta$)
0.412 $\alpha = \beta$ equiv ($\alpha \subset \beta$ and $\beta \subset \alpha$)

0.413 $(x\epsilon\ (\alpha\ \&\ \beta)\ )$ equiv $(x\epsilon\alpha$ and $x\epsilon\beta)$
0.414 $(x\epsilon\ (-\alpha)\ )$ equiv (not $(x\epsilon\alpha)$)
0.415 $\alpha = \Lambda$ equiv (not (Some x) $(x\epsilon\alpha)$)
('Λ' thus denotes the null or empty class; cf. '0' in arithmetic)
0.416 $\alpha \neq \Lambda$ equiv (Some x) $(x\epsilon\alpha)$
0.417 $(\alpha\ \&\ \beta) \subset \alpha$
0.418 $(\alpha\ \&\ \beta) \subset \beta$

Section 0.5

Statements from the Calculus of Relations

0.511 $R \subset S$ equiv (All x) ((All y) (xRy implies xSy))

A relation R is said to be included in a relation S if, and only if, every pair x,y such that x has the relation R to y is also such that x has the relation S to y (see the example on p. 19).

0.512 $xR\ \&\ Sy$ equiv (xRy and xSy)

'$R\ \&\ S$' denotes the logical product of R and S, i.e., the relation which holds between x and y when x stands in both R and S to y.

0.513 xR or Sy equiv (xRy or xSy)

The relation in which x stands to y when either R or S holds between x and y is called the logical sum of R and S.

0.514 $R = S$ equiv ($R \subset S$ and $S \subset R$)

As in the case of classes two relations are said to be identical if, and only if, they mutually include each other.

0.515 $x{-}R\ y$ equiv (not (xRy))

This is the analogue for relations of 0.414.

0.516 $xR{-}Sy$ equiv $xR\ \&\ {-}Sy$

This merely serves to eliminate '&'.

0.517 $xR^{-1}y$ equiv yRx

R^{-1} is called the converse of R and is the relation in which x stands to y when y stands in R to x, e.g., '$x\mathbf{P}^{-1}y$' will mean that x has y as a part.

0.518 $R = \dot{\Lambda}$ equiv (not (Some x) ((Some y) (xRy))) (cf. 0.415)
0.521 ((All z) (zRy equiv $z = x$)) implies ($x = R'y$)

'$R'y$' may be read '*the* R of y' because it denotes the unique individual which stands in the relation R to y, e.g., 'the father of y'. This notation is very frequently used. But the expression '$R'y$' is only significant when one such unique individual exists and this will be the case if the condition expressed by the equivalence to the left of the word 'implies' in 0.521 is satisfied. The corresponding condition for relations belonging to the type ($*$, ($*$)) is given in the next statement.

0.522 ((All z) ($zT\alpha$ equiv $z = x$)) implies ($x = T'\alpha$)

In the next statement a further use of this notation is introduced, namely, in connection with a relation-sign obtained by writing a small 's' after any given relation-sign, to denote the *class* of things standing in that relation to a given individual. For example, from '**P**' which denotes the relation of a part to that of which it is a part, so that '$x\mathbf{P}y$' means x is part of y, we construct the sign '$\mathbf{P}s$' which denotes the relation of a class to an individual when that class is the class of all the parts of that individual, i.e., '$\alpha\mathbf{P}sy$' will mean α is the class of all the parts of y. Using the notation introduced in 0.521, we can write '$\mathbf{P}s'y$' to denote the class of all the parts of y. In this case no special condition is necessary for significance because a class sign is always significant even when the class has no members and so is identical with Λ.

0.523 $x\epsilon(Rs'y)$ equiv xRy

A special use of this notation is introduced by

0.524 $y\epsilon(\iota'x)$ equiv $y = x$

'$\iota'x$' denotes the class of which x is the only member. Here the Greek letter 'ι' is used as a relation sign representing '=' followed by 's' as in 0.523. '$\iota'x$' thus denotes the class of things which are identical with x.

0.525 $x\epsilon$(Dom' R) equiv (Some y) (xRy)

'Dom' R' (read: 'the domain of R') denotes the class of things which stand in R to something, e.g., Dom' **P** is the class of things which are parts of something.

0.526 $x\epsilon$(Cnvdom' R) equiv (Some y) (yRx)

x is a member of the converse domain of R if, and only if, there is something y which stands in R to it. 'Cnvdom' **P**' will therefore denote the class of things which have parts.

0.531 $R\epsilon$sym equiv (All x) ((All y) (xRy implies yRx))

R is a symmetrical relation if, and only if, whenever we have xRy we also have yRx, i.e., if, using 0.511 and 0.517, we have $R \subset R^{-1}$.

0.532 $R\epsilon$asym equiv (All x) ((All y) (xRy implies (not (yRx)))

R is an asymmetrical relation if, and only if, xRy always excludes yRx.

0.533 $R\epsilon$trans equiv (All x) ((All y) ((All z) ((xRy and yRz) implies xRz)))

R is a transitive relation if, and only if, whenever we have xRy and yRz we also have xRz, e.g., if x**P**y and y**P**z we always have x**P**z, so that **P** is transitive.

0.534 $R\epsilon 1 \rightarrow$many equiv (All x) ((All y) ((All z) ((xRy and zRy) implies $x = z$)))

R is a one-many relation if, and only if, whenever we have xRy and zRy we also have $x = z$.

0.535 $R\epsilon$many$\rightarrow 1$ equiv (All x) ((All y) ((All z) (xRy and xRz) implies $y = z$)))

R is many-one if, and only if, whenever we have xRy and xRz we also have $y = z$.

0.536 ($R\epsilon 1 \rightarrow$many and xRy) implies ($x = R$'y)

If R is a one-many relation and x stands in R to y, then x must be the one and only one term which stands in R to y.

From 0.534 it will be seen that, if $R\epsilon 1\rightarrow$many and xRy, the condition for the significance of 'R'y' laid down in 0.521 is satisfied.

0.541 $xR \mid Sy$ equiv (Some z) (xRz and zSy)

$R \mid S$ is called the relative product of R and S and is the relation in which x stands to y when there is a z such that xRz and zSy. For example, we shall have $x\mathbf{P} \mid \mathbf{T}y$ if x is part of some thing z which is before y in time.

The numbered statements of Part II of T which follows immediately will be found to fall into two groups: (1) those which have the word 'Postulate' written after them and (2) the remaining numbered statements. The word 'Postulate' belongs not to T but to the metatheory of T. It serves to indicate to the reader which statements the author regards as true statements for reasons which are not stated in the theory itself. Among the postulates some have '(d)' written after them. These are the definitional postulates since they are playing the part of definitions, i.e., statements which serve merely to fix the meaning of signs additional to the three primitives '**P**', '**T**', and '**cell**'.

By repeated applications of the transformation rules 4.1 to 4.54 of T to the postulates of Part II, and with the help of the statements of Part I, all the statements of group (2) mentioned above are derivable. The statements of this group are thus all *consequences* of the statements of group (1) in the sense of 'consequence' determined by the rules given in the syntax of T, provided, of course, that no mistake has been made in performing the transformations.

In the case of statements of Sections 1 and 2 of Part II the steps by which consequences are derived are not given in detail. (They are given [for some of these statements] in Appen. E of the author's *Axiomatic Method in Biology*.) But after each statement the numbers of the preceding statements of Part II upon which the derivation depends are given. In Section 3 the derivations are given in full. A careful perusal of a few of these will suffice to illustrate the use of the rules of statement transformation and the part played by the statements of Part I in the derivation of consequences in Part II.

For the convenience of the reader each statement of Part II is immediately followed by a translation into English. But care must be exercised in using these translations. They must be regarded only as aids to reading the precisely formulated statements of the theory T, *not* as alternatives or substitutes for the latter, because such translations frequently involve the use of familiar English words in some special and restricted sense. For example, the word 'sum' has many meanings in English, but its use in the translations of statements of T is throughout restricted to one precise and definite meaning by the definitional postulate (1.1) which stands at the beginning of Section 1. In using these translations it is, therefore, essential to keep such restrictions of meaning in mind.

This will be a suitable place in which to say a little more in explanation of the two primitive signs '**P**' and '**T**'. It must be understood that, when we say '$x\mathbf{P}y$' means that x is part of y, the range of things which 'x' and 'y' can represent is not restricted to purely spatial things but includes four-dimensional time-extended things. Thus 'y' might represent the whole life of a man from birth to death and 'x' the same man from, say, his twenty-first to his twenty-second birthday. In the sense of 'part of' denoted here by '**P**' we should then have x part of y. In fact, the things whose designations belong to the type $*$ (i.e., individuals) in the theory T are to be thought of always as time-extended things in the first instance. Purely spatial things (momentary things) are introduced later. Cells, for example (in the biological sense), are here conceived as time-extended things with a beginning and end in time, such beginnings and endings being called 'time-slices' of the whole time-extended cell.

When we say '$x\mathbf{T}y$' means that x is before y in time, this might be understood to mean that *every* part of x precedes y in time, or that x ends at the moment when y begins so that, although the end of x and the beginning of y are coincident in time, all other parts of x wholly precede the parts of y in time.

In the theory T when *either* of these states of affairs holds for a given pair x and y of things, we shall say that $x\mathbf{T}y$. The accompanying diagram will perhaps help to make clear what is meant:

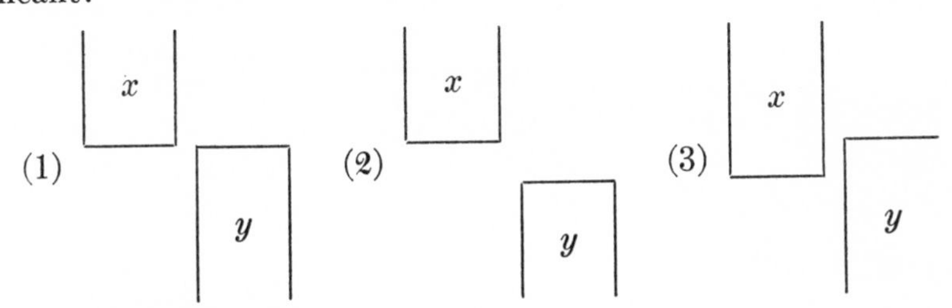

In cases (1) and (2) we have $x\mathbf{T}y$, in case (3) we have x—$\mathbf{T}\, y$, i.e., not $(x\mathbf{T}y)$. The time direction is supposed to be from the top to the bottom of the page.

V. The Theory *T*, Part II

Section 1

1.1 $x\mathbf{S}\alpha$ equiv (($\alpha \subset (\mathbf{P}s\text{'}x)$) and (All y) ($y\mathbf{P}x$ implies (Some z) ($z\epsilon\alpha$ and (($(\mathbf{P}s\text{'}z)$ & $(\mathbf{P}s\text{'}y)$) $\neq \Lambda$))))

Postulate (d)

A thing x is a sum of a class α of things if, and only if, every member of α is a part of x, and if for every thing y which is a part of x there exists a member z of α which has a part in common with y.

1.11 $\mathbf{P}\epsilon$ trans — Postulate

$\mathbf{P}$ is a transitive relation, i.e., if a thing is a part of another thing, and the latter is part of a third thing, then the first thing is also part of the third thing.

1.12 $y\mathbf{S}(\iota\text{'}x)$ implies $y = x$ — Postulate

If a thing y is a sum of a class of things which has a thing x as its only member, then y and x are identical (see 0.524).

1.13 $\alpha \neq \Lambda$ implies (Some y) ($y\mathbf{S}\alpha$) — Postulate

If a class of things is not empty, then there exists a thing which is its sum.

Consequences

1.22 $x = y$ implies $x\mathbf{P}y$ [1.1, 1.12, 1.13

If a thing x is identical with a thing y, then x is part of y (i.e., with the foregoing postulates a thing will be part of itself).

1.231 ($x\mathbf{P}y$ and $y\mathbf{P}x$) equiv $x = y$ [1.1, 1.11, 1.12, 1.22

If a thing x is part of a thing y and y is part of x, then x is identical with y; and, if x is identical with y, then x is part of y and y is part of x.

1.25 $x = (\mathbf{S}\text{‘}(\mathbf{P}s\text{‘}x))$ [1.13, 1.22, 1.1, 1.11, 1.12

A thing is identical with the sum of its parts. (As expressed in English this statement is often disputed in bio-theoretical discussions. But, owing to the ambiguity of ‘identical’ and ‘sum’ in English, the precise point at issue is never quite clear in such disputes. In the theory T, ‘=’ and ‘**S**’ have precise meanings.)

1.3 $\alpha \neq \Lambda$ implies (Some x) $(x = (\mathbf{S}\text{‘}\alpha))$ [1.11, 1.1, 1.13, 1.12

If a class α of things is not empty, then there exists a thing which is *the* (one and only one) sum of α.

1.32 $z\epsilon\alpha$ implies $(z\mathbf{P}(\mathbf{S}\text{‘}\alpha))$ [1.3, 1.1

If a thing is a member of a class of things, then it is part of the sum of that class.

1.33 $(((\mathbf{P}s\text{‘}x) \& (\mathbf{P}s\text{‘}y)) = \Lambda)$ implies $(x—\mathbf{P}—\mathbf{P}^{-1}y)$ [1.22

If two things have no part in common, then neither is part of the other.

1.331 $\alpha \neq \Lambda$ implies $(((\mathbf{S}\text{‘}\alpha)\mathbf{P}x)$ equiv $(\alpha \subset (\mathbf{P}s\text{‘}x)))$ [1.11, 1.1, 1.3, 1.12, 1.13

If a class of things is not empty, then the sum of that class is part of a thing x if, and only if, the class is contained in the class of parts of x.

1.4 $x\mathbf{Pp}y$ equiv ($x\mathbf{P}y$ and $x \neq y$) Postulate (d)

x is a proper part of y if, and only if, x is part of y and x is not identical with y.

1.41 **Pp** ϵ trans [1.4, 1.11, 1.231, 1.22

If a thing is a proper part of another thing and the latter is a proper part of a third thing, then the first thing is also a proper part of the third thing.

SECTION 2

2.1 x ϵ **mom** equiv x**T**x Postulate (d)

x is a momentary thing if, and only if, x stands in **T** to itself.

2.11 **T** ϵ trans Postulate

T is a transitive relation.

2.12 ($x\epsilon$**mom** and $y\epsilon$**mom**) implies (x**T**y or y**T**x) Postulate

Of any two momentary things, one stands in **T** to the other.

2.13 x**T**y equiv
(All u) ((All v) ((($u\epsilon$**mom** and u**P**x) and ($v\epsilon$**mom** and v**P**y)) implies u**T**v)) Postulate

A thing x stands in **T** to a thing y if, and only if, every momentary part of x stands in **T** to every momentary part of y.

Consequences

2.221 (**P** | **T** = **T**) and (**T** = **T** | **P**$^{-1}$) [1.11, 2.13, 1.22

The relative product of **P** with **T** is identical with **T**, and **T** is identical with the relative product of **T** with the converse of **P**; i.e., if a thing x is part of a thing y which stands in **T** to a thing z, then x stands in **T** to z; and, if x stands in **T** to z, then there is a thing u to which x stands in **T** of which z is a part. (Note the conciseness achieved by the use of the calculus of relations.)

2.24 ($\alpha \neq \Lambda$ and $\beta \neq \Lambda$) implies
(((**S'**α) **T** (**S'**β)) equiv (All x) ((All y) (($x\epsilon\alpha$ and $y\epsilon\beta$) implies x**T**y))) [2.1, 2.11, 2.13, 2.221, 1.3, 1.1, 1.22

If α is not empty and β is not empty, then the sum of α stands in **T** to the sum of β if, and only if, every member of α stands in **T** to every member of β.

2.241 $\alpha \neq \Lambda$ implies $(((\mathbf{S}`\alpha)\mathbf{T}y)$ equiv $(\alpha \subset (\mathbf{T}s`y)))$
[2.24, 1.13, 1.12

If α is not empty, then the sum of α stands in **T** to y if, and only if, every member of α stands in **T** to y.

2.242 $\alpha \neq \Lambda$ implies $((y\mathbf{T}(\mathbf{S}`\alpha))$ equiv $(\alpha \subset (\mathbf{T}^{-1}s`y)))$
[2.24, 1.13, 1.12

If α is not empty, then y stands in **T** to the sum of α if, and only if, every member of α stands in the converse of **T** to y.

2.35 **mom** = (Dom` $(\mathbf{T} \mathbin{\&} \mathbf{T}^{-1})$) [2.1, 2.11

The class of momentary things is identical with the domain of the logical product of **T** with its converse, i.e., x is a momentary thing if, and only if, there exists a thing y such that x stands in **T** to y and in the converse of **T** to y.

2.36 $((\mathbf{S}`\alpha)\epsilon\mathbf{mom})$ equiv
$(\alpha \neq \Lambda$ and (All x) ((All y) (($x\epsilon\alpha$ and $y\epsilon\alpha$) implies $x\mathbf{T}y$)))
[2.1, 2.24

The sum of a class α is a momentary thing if, and only if, α is not empty and every member of it stands in **T** to every other member of it.

2.5 $(x\mathbf{C}y)$ equiv $(x\mathbf{T} \mathbin{\&} \mathbf{T}^{-1}y)$ Postulate (*d*)

A thing x coincides in time with a thing y if, and only if, x stands in **T** to y and y stands in **T** to x.

2.52 $\mathbf{C} \in$ (sym & trans) [2.5, 2.11

Coincidence in time is a symmetrical and transitive relation; i.e., if a thing x is coincident in time with a thing y, then y is coincident in time with x; and, if x is coincident in time with y and y with z, then x is coincident in time with z.

2.53 $x\epsilon\mathbf{mom}$ equiv $((\mathbf{P}s`x) \subset (\mathbf{C}s`x))$ [2.1, 2.221, 2.5, 1.22

x is a momentary thing if, and only if, all the parts of x are coincident in time with x.

2.531 (($x\epsilon$**mom** and $y\epsilon$**mom**) and ((**P**s'x) & (**P**s'y)) $\neq \Lambda$) implies x**C**y [2.53, 2.52

If two momentary things have a part in common, then they are coincident in time.

2.54 ((**S**'α)ϵ**mom**) equiv (Some x) ($x\epsilon\alpha$ and ($\alpha \subset$ (**C**s'x))) [2.1, 2.24, 2.5, 2.52

The sum of a class α is a momentary thing if, and only if, there is a thing x which is a member of α and such that α is contained in the class of things which are coincident in time with x.

2.6 x**Z**y equiv (($x\epsilon$**mom** and $y\epsilon$**mom**) and x—**T**^{-1}y) Postulate(d)

x stands in the relation **Z** to y if, and only if, x and y are momentary things and y does not stand in **T** to x. 'x**Z**y' thus means that x and y are momentary things and x is wholly before y in time (cf. 2.12).

2.622 **Z** ϵ (asym & trans) [2.6, 2.12, 2.11

The relation **Z** is asymmetrical and transitive (cf. 0.532).

2.65 ($x\epsilon$**mom** and $y\epsilon$**mom**) implies (x**T**y equiv (x**C** or **Z**y)) [2.12, 2.5, 2.6

If x and y are momentary things, then x stands in **T** to y if, and only if, x is coincident in time with y or is wholly before y in time.

2.652 (**C** $\subset$—**Z**) and (**C** $\subset$—**Z**$^{-1}$) [2.5, 2.6

If a thing x is coincident in time with a thing y, then x is not wholly before y and y is not wholly before x.

2.653 (((x(**Z** | (**P**$^{-1}$)) or (**Z** | **P**)y) and $y\epsilon$**mom**) or ((x ((**P**$^{-1}$) | **Z**) or (**P** | **Z**)y) and $x\epsilon$**mom**)) implies x**Z**y [2.6, 2.531, 2.221, 1.22, 2.5

If a thing x is wholly before some thing z in time of which a thing y is a momentary part or which is part of a momentary thing y, or if x is a momentary thing and some thing z is part of x or x is part of z, and z is wholly before y in time, then x is wholly before y in time, and both are momentary.

2.654 $\mathbf{Z} \subset (-\mathbf{P}-\mathbf{P}^{-1})$ [2.6, 2.652, 2.531, 1.33

If a thing x is wholly before a thing y in time and both are momentary, then neither is part of the other.

2.66 $(\mathbf{C} \mid \mathbf{Z} = \mathbf{Z})$ and $(\mathbf{Z} = \mathbf{Z} \mid \mathbf{C})$ [2.5, 2.6, 2.35, 2.11, 2.1

The relative product of **C** with **Z** is identical with **Z**, and **Z** is identical with the relative product of **Z** with **C** (see 0.541 and 0.514).

2.8 $x\mathbf{Sl}y$ equiv $(x\epsilon\,(\mathbf{mom}\ \&\ (\mathbf{P}s'y)\,)\,)$ and
$(\,(\,(\mathbf{C}s'x)\ \&\ (\mathbf{P}s'y)\,) \subset (\mathbf{P}s'x)\,)\,)$ Postulate (d)

A thing x is a time-slice of a thing y ($x\mathbf{Sl}y$) if, and only if, x is momentary and a part of y, and all things which are coincident in time with x and parts of y are also parts of x.

2.83 $(x\mathbf{Sl}z$ and $y\mathbf{Sl}z)$ implies
$(\,(x\mathbf{C}y$ equiv $x\mathbf{P}y)$ and $(x\mathbf{P}y$ equiv $y\mathbf{P}x)$ and
$(y\mathbf{P}x$ equiv $(\,(\,(\mathbf{P}s'x)\ \&\ (\mathbf{P}s'y)\,) \neq \Lambda)\,)$ and
$(y\mathbf{P}x$ equiv $x = y)\,)$ [2.8, 2.52, 1.231, 1.33, 2.531

If x and y are both time-slices of a third thing z, then x is coincident in time with y if, and only if, x is part of y; and x is part of y if, and only if, y is part of x; and y is part of x if, and only if, x and y have parts in common, and y is part of x if, and only if, x is identical with y.

2.84 $(x\epsilon\,(\mathbf{mom}\ \&\ (\mathbf{P}s'z)\,)\,)$ implies $(\,(\,(\mathbf{S}'\,(\,(\mathbf{C}s'x)\ \&\ (\mathbf{P}s'z)\,)\,)\,\mathbf{Sl}z$ and $(\mathbf{S}'\,(\,(\mathbf{C}s'x)\ \&\ (\mathbf{P}s'z)\,)\,)\,\mathbf{C}x)$ and (All y) $(\,(y\mathbf{Sl}z$ and $y\mathbf{C}x)$ implies $(y = (\mathbf{S}'\,(\,(\mathbf{C}s'x)\ \&\ (\mathbf{P}s'z)\,)\,)\,)\,)\,)$
[2.8, 2.83, 2.54, 2.53, 1.3, 1.32, 1.331

If x is a momentary thing and part of z, then the sum of the class of things which are coincident in time with x and parts of z is the only thing which is a time-slice of z and coincident in time with x.

2.901 $x\mathbf{B}y$ equiv $(x\mathbf{Sl}y$ and (not (Some z) $(\,(z\mathbf{Sl}y$ and $z\mathbf{T}x)$
and $z \neq x)\,)\,)$ Postulate (d)

A thing x is a beginning slice of a thing y if, and only if, x is a time-slice of y and there is no thing z such that z is a time-slice of y, z stands in **T** to x and is not identical with x.

2.902 $x\mathbf{E}y$ equiv ($x\mathbf{SI}y$ and (not (Some z) (($z\mathbf{SI}y$ and $z\mathbf{T}^{-1}x$) and $z \neq x$))) Postulate (d)

A thing x is an end slice of a thing y if, and only if, x is a time-slice of y and there is no thing z such that z is a time-slice of y, x stands in **T** to z and z is not identical with x.

2.911 ($x\mathbf{SI}$ & $\mathbf{T}y$) implies (not (Some z) (($z\mathbf{SI}y$ and $z\mathbf{T}x$) and $z \neq x$)) [2.8, 2.221, 2.5, 2.83

If x is a time-slice of y and stands in **T** to it, then there is no z such that z is a time-slice of y and stands in **T** to x and is not identical with x.

2.912 $\mathbf{B} \subset \mathbf{T}$
[2.1, 2.5, 1.3, 2.84, 2.901, 2.12, 2.8, 2.221, 2.11, 2.13

If a thing is a beginning slice of anything, then it stands in **T** to it.

2.913 $\mathbf{B} = (\mathbf{SI}$ & $\mathbf{T})$ [2.911, 2.901, 2.912

If a thing is a beginning slice of anything, then it is a time-slice of it and stands in **T** to it, and if a thing is a time-slice of anything and stands in **T** to it, then it is a beginning slice of it.

2.914 $\mathbf{B} \in 1 \rightarrow$many [2.901, 2.8, 2.12

B is a one-many relation, i.e., if a thing is a beginning slice of another thing, it is the only beginning slice or the first time-slice of that thing.

2.921 ($x\mathbf{SI}$ & ($\mathbf{T}^{-1}$) y) implies (not (Some z) (($z\mathbf{SI}y$ and $z\mathbf{T}^{-1}x$) and $z \neq x$)) [2.8, 2.221, 2.5, 2.83

If x is a time-slice of y and y stands in **T** to x, then there is no z such that z is a time-slice of y and x stands in **T** to z and z is not identical with x.

2.922 $\mathbf{E} \subset (\mathbf{T}^{-1})$
[2.1, 2.5, 1.3, 2.84, 2.902, 2.12, 2.8, 2.221, 2.11, 2.13

If a thing is an end slice of another thing, then the latter stands in **T** to it.

2.923 $\mathbf{E} = (\mathbf{Sl} \,\&\, (\mathbf{T}^{-1}))$ [2.921, 2.902, 2.922

A thing is an end slice of another thing if, and only if, it is a time-slice of that thing and that thing stands in **T** to it.

2.924 $\mathbf{E} \in 1 \rightarrow \text{many}$ [2.902, 2.8, 2.12

E is a one-many relation, i.e., if a thing x is an end slice of another thing, then it is the only end slice, or the last time-slice, of that thing.

SECTION 3

3.11 $x \epsilon \mathbf{cell}$ implies $(((\mathbf{P}s\text{‘}x) \,\&\, (\mathbf{T}s\text{‘}x)) \neq \Lambda)$ Postulate

If a thing x is a cell, then there are parts of x which stand in **T** to x.

3.12 $x \epsilon \mathbf{cell}$ implies $(((\mathbf{P}s\text{‘}x) \,\&\, (\mathbf{T}^{-1}s\text{‘}x)) \neq \Lambda)$ Postulate

If a thing x is a cell, then there are parts of x to which x stands in **T**.

3.13 $(\mathbf{cell} \,\&\, \mathbf{mom}) = \Lambda$ Postulate

No cell is a momentary thing.

3.14 $((x \epsilon \mathbf{cell}$ and $y \epsilon \mathbf{cell})$ and $(x \neq y$ and $((\mathbf{P}s\text{‘}x) \,\&\, (\mathbf{P}s\text{‘}y)) \neq \Lambda))$ implies $(((\mathbf{B}\text{‘}x)\ \mathbf{Pp}$ or $(\mathbf{Pp}^{-1})\ (\mathbf{E}\text{‘}y))$ or $((\mathbf{E}\text{‘}x)\ \mathbf{Pp}$ or $(\mathbf{Pp}^{-1})\ (\mathbf{B}\text{‘}y)))$ Postulate

If two distinct cells have a part in common, then the first time-slice of one is a proper part of the last time-slice of the other, or the last time-slice of one is a proper part of the first time-slice of the other.

Consequences

3.21 $x \epsilon \mathbf{cell}$ implies $((\mathbf{S}\text{‘}((\mathbf{P}s\text{‘}x) \,\&\, (\mathbf{T}s\text{‘}x))) \epsilon (\mathbf{mom} \,\&\, (\mathbf{P}s\text{‘}x)))$

If a thing x is a cell, then the sum of the parts of x which stand in **T** to x is a momentary thing and a part of x.

Derivation: (This derivation is given and explained in full in order to illustrate the use of the derivation or transformation rules and the part played by the formulas of Part I in derivations. At the same time, the derivation would be inordinately long if every step were written down in full. The later steps will therefore be progressively shortened when the procedure has been sufficiently illustrated by the earlier ones. It will be noted that, in obtaining each step in a derivation, we remain entirely within the province of syntax, and it is therefore only necessary to attend to the form of the statements involved.)

In accordance with Rule 4.12 we first put 'z' for 'x', '$(\mathbf{P}s\text{‘}x)$' for 'α', and '$(\mathbf{T}s\text{‘}x)$' for 'β' in 0.413 and obtain

(1) $(z\epsilon\ ((\mathbf{P}s\text{‘}x)\ \&\ (\mathbf{T}s\text{‘}x)))$ equiv $(z\epsilon\ (\mathbf{P}s\text{‘}x)$ and $z\epsilon\ (\mathbf{T}s\text{‘}x))$.

Using Rule 4.11, we substitute '$z\epsilon\ (\mathbf{P}s\text{‘}x)$' for '$p$' and '$z\epsilon\ (\mathbf{T}s\text{‘}x)$' for '$q$' in 0.1142 and get

(2) $(z\epsilon\ (\mathbf{P}s\text{‘}x)$ and $z\epsilon\ (\mathbf{T}s\text{‘}x))$ implies $z\epsilon\ (\mathbf{T}s\text{‘}x)$.

Applying Rule 4.3 to (1) and (2), we have

(3) $(z\epsilon\ ((\mathbf{P}s\text{‘}x)\ \&\ (\mathbf{T}s\text{‘}x)))$ implies $z\epsilon\ (\mathbf{T}s\text{‘}x)$.

Substituting 'z' for 'x', 'x' for 'y', and '$\mathbf{T}$' for 'R' in 0.523 in accordance with Rule 4.12 gives us

(4) $z\epsilon\ (\mathbf{T}s\text{‘}x)$ equiv $z\mathbf{T}x$.

By the use of Rule 4.3 from (3) and (4) we obtain

(5) $(z\epsilon\ ((\mathbf{P}s\text{‘}x)\ \&\ (\mathbf{T}s\text{‘}x)))$ implies $z\mathbf{T}x$.

Putting 'y' for 'z' in (1), according to Rule 4.12, yields

(6) $(y\epsilon\ ((\mathbf{P}s\text{‘}x)\ \&\ (\mathbf{T}s\text{‘}x)))$ equiv $(y\epsilon\ (\mathbf{P}s\text{‘}x)$ and $y\epsilon\ (\mathbf{T}s\text{‘}x))$.

Again using Rule 4.11, we put '$y\epsilon\ (\mathbf{P}s\text{‘}x)$' for '$p$' and '$y\epsilon\ (\mathbf{T}s\text{‘}x)$' for '$q$' in 0.1141, and we have

(7) $(y\epsilon\ (\mathbf{P}s\text{‘}x)$ and $y\epsilon\ (\mathbf{T}s\text{‘}x))$ implies $y\epsilon\ (\mathbf{P}s\text{‘}x)$.

And from (6) and (7) by Rule 4.3 we get

(8) $(y\epsilon\ ((\mathbf{P}s\text{‘}x)\ \&\ (\mathbf{T}s\text{‘}x)))$ implies $y\epsilon\ (\mathbf{P}s\text{‘}x)$.

Substituting 'y' for 'x' and 'x' for 'y' and '$\mathbf{P}$' for 'R' in 0.523 (by Rule 4.12), we obtain

(9) $y\epsilon\ (\mathbf{P}s\text{‘}x)$ equiv $y\mathbf{P}x$.

Again, putting '$\mathbf{P}$' for 'R' (by Rule 4.12) in 0.517 gives

(10) $x\mathbf{P}^{-1}y$ equiv $y\mathbf{P}x$.

From (9) and (10) by Rule 4.3 we get

(11) $y \epsilon (\mathbf{P} s`x)$ equiv $x\mathbf{P}^{-1}y$.

And from (11) and (8), also by Rule 4.3, we obtain

(12) $(y \epsilon ((\mathbf{P} s`x) \& (\mathbf{T} s`x)))$ implies $x\mathbf{P}^{-1}y$.

In order to combine (5) and (12) into a single statement, we proceed as follows. We first substitute '$(z \epsilon ((\mathbf{P} s`x) \& (\mathbf{T} s`x)))$' for '$p$' and '$(y \epsilon ((\mathbf{P} s`x) \& (\mathbf{T} s`x)))$' for '$r$', '$z\mathbf{T}x$' for '$q$', and '$x\mathbf{P}^{-1}y$' for '$s$' in 0.117 in accordance with Rule 4.11 and get

(13) (($(z \epsilon ((\mathbf{P} s`x) \& (\mathbf{T} s`x)))$ implies $z\mathbf{T}x$) and
($(y \epsilon ((\mathbf{P} s`x) \& (\mathbf{T} s`x)))$ implies $x\mathbf{P}^{-1}y$)) implies
(($(z \epsilon ((\mathbf{P} s`x) \& (\mathbf{T} s`x)))$ and $(y \epsilon ((\mathbf{P} s`x) \& (\mathbf{T} s`x)))$)
implies ($z\mathbf{T}x$ and $x\mathbf{P}^{-1}y$)) .

We must next obtain the conjunction of (5) and (12) in order to apply Rule 4.2 to this conjunction and (13). This we do by first substituting (5) for 'p' and (12) for 'q' in 0.138, obtaining,

(14) ($(z \epsilon ((\mathbf{P} s`x) \& (\mathbf{T} s`x)))$ implies $z\mathbf{T}x$) implies
(($(y \epsilon ((\mathbf{P} s`x) \& (\mathbf{T} s`x)))$ implies $x\mathbf{P}^{-1}y$) implies
(($(z \epsilon ((\mathbf{P} s`x) \& (\mathbf{T} s`x)))$ implies $z\mathbf{T}x$) and
($(y \epsilon ((\mathbf{P} s`x) \& (\mathbf{T} s`x)))$ implies $x\mathbf{P}^{-1}y$))) .

By Rule 4.2 from (5) and (14) we then obtain

(15) ($(y \epsilon ((\mathbf{P} s`x) \& (\mathbf{T} s`x)))$ implies $x\mathbf{P}^{-1}y$) implies
(($(z \epsilon ((\mathbf{P} s`x) \& (\mathbf{T} s`x)))$ implies $z\mathbf{T}x$) and
($(y \epsilon ((\mathbf{P} s`x) \& (\mathbf{T} s`x)))$ implies $x\mathbf{P}^{-1}y$)) .

And by Rule 4.2 again we finally get from (12) and (15)

(16) ($(z \epsilon ((\mathbf{P} s`x) \& (\mathbf{T} s`x)))$ implies $z\mathbf{T}x$) and
($(y \epsilon ((\mathbf{P} s`x) \& (\mathbf{T} s`x)))$ implies $x\mathbf{P}^{-1}y$) .

And, by applying Rule 4.2 to (16) and (13), we obtain

(17) ($(z \epsilon ((\mathbf{P} s`x) \& (\mathbf{T} s`x)))$ and $(y \epsilon ((\mathbf{P} s`x) \& (\mathbf{T} s`x)))$)
implies ($z\mathbf{T}x$ and $x\mathbf{P}^{-1}y$) ,

which is the desired combination of (5) and (12). We now substitute 'z' for 'x', 'x' for 'z', '$\mathbf{T}$' for 'R', and '$\mathbf{P}^{-1}$' for 'S' in 0.541, obtaining,

(18) $z\mathbf{T} \mid \mathbf{P}^{-1}y$ equiv (Some x) ($z\mathbf{T}x$ and $x\mathbf{P}^{-1}y$) .

In 0.216 we substitute 'z' for 'x', 'x' for 'y', 'y' for 'z', 'x' for 'u', '**T**' for 'R', and '$\mathbf{P}^{-1}$' for 'S' and obtain

(19) $(z\mathbf{T}x \text{ and } x\mathbf{P}^{-1}y) \text{ implies (Some } x) (z\mathbf{T}x \text{ and } x\mathbf{P}^{-1}y)$.

Applying Rule 4.3 to (18) and (19), we get

(20) $(z\mathbf{T}x \text{ and } x\mathbf{P}^{-1}y) \text{ implies } z\mathbf{T} \mid \mathbf{P}^{-1}y$.

And, by applying Rule 4.4 to (20) and 2.221, we have

(21) $(z\mathbf{T}x \text{ and } x\mathbf{P}^{-1}y) \text{ implies } z\mathbf{T}y$.

By substitutions in 0.116 (analogous to those made in 0.117 in the derivation of (13)) and then by applying Rule 4.2 to the resulting statement and the conjunction of (17) and (21) we can, in the same way in which we obtained (17), also obtain

(22) $((z\epsilon ((\mathbf{P}s`x) \;\&\; (\mathbf{T}s`x))) \text{ and } (y\epsilon ((\mathbf{P}s`x) \;\&\; (\mathbf{T}s`x)))) \text{ implies } z\mathbf{T}y$.

With the help of 0.123, 0.124, 0.135, 0.136 and Rules 4.1 and 4.3, theorem 2.36 can be transformed into

(23) $(\text{All } x) ((\text{All } y) ((x\epsilon\alpha \text{ and } y\epsilon\alpha) \text{ implies } x\mathbf{T}y)) \text{ implies } (\alpha \neq \Lambda \text{ implies } ((\mathbf{S}`\alpha)\epsilon\mathbf{mom}))$.

Substituting 'z' for 'x' and '$((\mathbf{P}s`x) \;\&\; (\mathbf{T}s`x))$' for 'α' in (23) gives

(24) $(\text{All } z) ((\text{All } y) ((z\epsilon ((\mathbf{P}s`x) \;\&\; (\mathbf{T}s`x)) \text{ and } y\epsilon ((\mathbf{P}s`x) \;\&\; (\mathbf{T}s`x))) \text{ implies } z\mathbf{T}y)) \text{ implies } (((\mathbf{P}s`x) \;\&\; (\mathbf{T}s`x)) \neq \Lambda \text{ implies } ((\mathbf{S}` ( (\mathbf{P}s`x) \;\&\; (\mathbf{T}s`x))) \epsilon\mathbf{mom}))$.

To introduce universal quantifiers in front of (22), we proceed as follows. We first substitute (22) for 'q' in 0.111 and obtain

(25) $(((z\epsilon ((\mathbf{P}s`x) \;\&\; (\mathbf{T}s`x))) \text{ and } (y\epsilon ((\mathbf{P}s`x) \;\&\; (\mathbf{T}s`x)))) \text{ implies } z\mathbf{T}y) \text{ implies } ((p) \text{ implies } (((z\epsilon ((\mathbf{P}s`x) \;\&\; (\mathbf{T}s`x))) \text{ and } (y\epsilon ((\mathbf{P}s`x) \;\&\; (\mathbf{T}s`x)))) \text{ implies } z\mathbf{T}y))$.

By Rule 4.2 from (22) and (25) we obtain

(26) $(p) \text{ implies } (((z\epsilon ((\mathbf{P}s`x) \;\&\; (\mathbf{T}s`x))) \text{ and } (y\epsilon ((\mathbf{P}s`x) \;\&\; (\mathbf{T}s`x)))) \text{ implies } z\mathbf{T}y)$.

We now apply Rule 4.53 to (26) at the same time substituting in it for 'p' any logical formula belonging to the calculus of statements, e.g., 0.128, and obtain

$$(27)\quad (p \text{ implies } p) \text{ implies } (\text{All } y)\ (\ (\ (z\epsilon\ (\ (\mathbf{P}s`x)\ \&\ (\mathbf{P}s`x)\ )\ )\ \text{and } (y\epsilon\ (\ (\mathbf{P}s`x)\ \&\ (\mathbf{T}s`x)\ )\ )\ )\ \text{implies } z\mathbf{T}y)\ .$$

By Rule 4.2 from 0.128 and (27) we can now get

$$(28)\quad (\text{All } y)\ (\ (\ (z\epsilon\ (\ (\mathbf{P}s`x)\ \&\ (\mathbf{T}s`x)\ )\ )\ \text{and } (y\epsilon\ (\ (\mathbf{P}s`x)\ \&\ (\mathbf{T}s`x)\ )\ )\ )\ \text{implies } z\mathbf{T}y)\ .$$

By a repetition of the same process we can introduce '(All z)' and so obtain

$$(29)\quad (\text{All } z)\ (\ (\text{All } y)\ (\ (\ (z\epsilon\ (\ (\mathbf{P}s`x)\ \&\ (\mathbf{T}s`x)\ )\ )\ \text{and } (y\epsilon\ (\ (\mathbf{P}s`x)\ \&\ (\mathbf{T}s`x)\ )\ )\ )\ \text{implies } z\mathbf{T}y)\ )\ .$$

By Rule 4.2 from (29) and (24) we now get

$$(30)\quad (\ (\mathbf{P}s`x)\ \&\ (\mathbf{T}s`x)\ ) \neq \Lambda \text{ implies } (\ (\mathbf{S}`\ (\ (\mathbf{P}s`x)\ \&\ (\mathbf{T}s`x)\ )\ )\ \epsilon\mathbf{mom})\ .$$

From 3.11 and (30) with the help of Rule 4.11 and 0.116 (as in obtaining (22)) and of Rule 5.13 we obtain

$$(31)\quad x\epsilon\mathbf{cell} \text{ implies } (\mathbf{S}`\ (\ (\mathbf{P}s`x)\ \&\ (\mathbf{T}s`x)\ )\ )\ \epsilon\mathbf{mom}\ .$$

Substituting '($(\mathbf{P}s`x)$ & $(\mathbf{T}s`x)$)' for 'α' in 1.331, we have

$$(32)\ (\ (\mathbf{P}s`x)\ \&\ (\mathbf{T}s`x)\ ) \neq \Lambda \text{ implies } (\ (\ (\mathbf{S}`\ (\ (\mathbf{P}s`x)\ \&\ (\mathbf{T}s`x)\ )\ )\ \mathbf{P}x) \text{ equiv } (\ (\ (\mathbf{P}s`x)\ \&\ (\mathbf{T}s`x)\ ) \subset (\mathbf{P}s`x)\ )\ )\ .$$

From 3.11 and (32) we can, with the help of 0.116, obtain

$$(33)\quad x\epsilon\mathbf{cell} \text{ implies } (\ (\ (\mathbf{S}`\ (\ (\mathbf{P}s`x)\ \&\ (\mathbf{T}s`x)\ )\ )\mathbf{P}x) \text{ equiv } (\ (\ (\mathbf{P}s`x)\ \&\ (\mathbf{T}s`x)\ ) \subset (\mathbf{P}s`x)\ )\ )\ .$$

With the help of 0.116, 0.123, 0.135, and 0.137 this may be transformed into

$$(34)\quad (\ (\mathbf{P}s`x)\ \&\ (\mathbf{T}s`x)\ ) \subset (\mathbf{P}s`x) \text{ implies } (x\epsilon\mathbf{cell} \text{ implies } (\mathbf{S}`\ (\ (\mathbf{P}s`x)\ \&\ (\mathbf{T}s`x)\ )\ )\ \ \mathbf{P}x)\ .$$

Substituting '$(\mathbf{P}s`x)$' for '$\alpha$' and '$(\mathbf{T}s`x)$' for 'β' in 0.417, we have

$$(35)\quad (\ (\mathbf{P}s`x)\ \&\ (\mathbf{T}s`x)\ ) \subset (\mathbf{P}s`x)\ .$$

From (35) and (34) we then obtain, by the use of Rule 4.2,

(36) $\quad x \epsilon \mathbf{cell}$ implies $(\mathbf{S}{'}\ (\ (\mathbf{P}s{'}x)\ \&\ (\mathbf{T}s{'}x)\ )\ )\ \mathbf{P}x$.

From (31) and (36) with the help of 0.131 we get

(37) $\quad x \epsilon \mathbf{cell}$ implies $(\ (\mathbf{S}{'}\ (\ (\mathbf{P}s{'}x)\ \&\ (\mathbf{T}s{'}x)\ )\ )\ \epsilon \mathbf{mom}$ and
$(\mathbf{S}{'}\ (\ (\mathbf{P}s{'}x)\ \&\ (\mathbf{T}s{'}x)\ )\ )\ \mathbf{P}x)$.

And from (37) with the help of 0.413 and 0.523 we finally get

(38) $x \epsilon \mathbf{cell}$ implies $(\mathbf{S}{'}\ (\ (\mathbf{P}s{'}x)\ \&\ (\mathbf{T}s{'}x)\ )\ )\ \epsilon\ (\mathbf{mom}\ \&\ (\mathbf{P}s{'}x)\ )$,

which is the statement to be derived, and is thus seen to be derivable from 2.221, 2.36, 3.11, and 1.331 by means of transformations performed with the help of 0.413, 0.1142, 0.523, 0.1141, 0.517, 0.117, 0.138, 0.541, 0.216, 0.116, 0.123, 0.124, 0.135, 0.136, 0.111, 0.128, 0.137, and 0.131 from Part I and the syntactical rules: 4.11, 4.12, 4.2, 4.3, 4.4, 4.53, and 5.13.

3.22 $\quad x \epsilon \mathbf{cell}$ implies $(\ (\mathbf{S}{'}\ (\ (\mathbf{P}s{'}x)\ \&\ (\mathbf{T}s{'}x)\ )\ )\ \mathbf{T}x)$

If a thing x is a cell, then the sum of the parts of x which stand in $\mathbf{T}$ to x itself stands in $\mathbf{T}$ to x.

Derivation:

In accordance with Rule 4.12, we substitute '$(\ (\mathbf{P}s{'}x)\ \&\ (\mathbf{T}s{'}x)\ )$' for '$\alpha$' and '$x$' for '$y$' in 2.241 and obtain

(1) $\quad (\ (\mathbf{P}s{'}x)\ \&\ (\mathbf{T}s{'}x)\ ) \neq \Lambda$ implies
$(\ (\ (\mathbf{S}{'}\ (\ (\mathbf{P}s{'}x)\ \&\ (\mathbf{T}s{'}x)\ )\ )\ \mathbf{T}x)$ equiv
$(\ (\ (\mathbf{P}s{'}x)\ \&\ (\mathbf{T}s{'}x)\ ) \subset (\mathbf{T}s{'}x)\ )\ )$.

By Rule 4.11 we substitute '$x \epsilon \mathbf{cell}$' for 'p', '$(\ (\mathbf{P}s{'}x)\ \&\ (\mathbf{T}s{'}x)\ ) \neq \Lambda$' for '$q$', and '$(\ (\ (\mathbf{S}{'}\ (\ (\mathbf{P}s{'}x)\ \&\ (\mathbf{T}s{'}x)\ )\ )\ \mathbf{T}x)$ equiv $(\ (\ (\mathbf{P}s{'}x)\ \&\ (\mathbf{T}s{'}x)\ ) \subset (\mathbf{T}s{'}x)\ )\ )$' for '$r$' in 0.116 and get

(2) $\quad (\ (x \epsilon \mathbf{cell}$ implies $(\ (\mathbf{P}s{'}x)\ \&\ (\mathbf{T}s{'}x)\ ) \neq \Lambda)$ and
$(\ (\ (\mathbf{P}s{'}x)\ \&\ (\mathbf{T}s{'}x)\ ) \neq \Lambda$ implies
$(\ (\ (\mathbf{S}{'}\ (\ (\mathbf{P}s{'}x)\ \&\ (\mathbf{T}s{'}x)\ )\ )\ \mathbf{T}x)$ equiv
$(\ (\ (\mathbf{P}s{'}x)\ \&\ (\mathbf{T}s{'}x)\ ) \subset (\mathbf{T}s{'}x)\ )\ )\ )\ )$ implies
$(x \epsilon \mathbf{cell}$ implies $(\ (\ (\mathbf{S}{'}\ (\ (\mathbf{P}s{'}x)\ \&\ (\mathbf{T}s{'}x)\ )\ )\ \mathbf{T}x)$ equiv
$(\ (\ (\mathbf{P}s{'}x)\ \&\ (\mathbf{T}s{'}x)\ ) \subset (\mathbf{T}s{'}x)\ )\ )\ )$.

Applying Rule 4.2 to (2) and the conjunction of 3.11 with (1), we get

(3) $x \epsilon$**cell** implies
$$(((\mathbf{S}`\ ( (\mathbf{P}s`x)\ \&\ (\mathbf{T}s`x) ) )\ \mathbf{T}x) \text{ equiv } ( ( (\mathbf{P}s`x)\ \&\ (\mathbf{T}s`x) ) \subset (\mathbf{T}s`x))) .$$

With the help of 0.123 (by a procedure of substitution analogous to that employed in obtaining (3)) we can from (3) obtain

(4) $x \epsilon$**cell** implies
$$((((\mathbf{P}s`x)\ \&\ (\mathbf{T}s`x)) \subset (\mathbf{T}s`x) ) \text{ equiv } ( (\mathbf{S}`\ ((\mathbf{P}s`x)\ \&\ (\mathbf{T}s`x)))\ \mathbf{T}x)) .$$

From (4) with the help of 0.135 we then get

(5) $x \epsilon$**cell** implies
$$((((\mathbf{P}s`x)\ \&\ (\mathbf{T}s`x)) \subset (\mathbf{T}s`x) ) \text{ implies } ( (\mathbf{S}`\ ((\mathbf{P}s`x)\ \&\ (\mathbf{T}s`x)))\ \mathbf{T}x)) .$$

With the help of 0.137 and Rule 4.3, (5) is transformed into

(6)
$$((\mathbf{P}s`x)\ \&\ (\mathbf{T}s`x)) \subset (\mathbf{T}s`x) \text{ implies } (x\epsilon\mathbf{cell} \text{ implies } ( (\mathbf{S}`\ ((\mathbf{P}s`x)\ \&\ (\mathbf{T}s`x)))\ \mathbf{T}x)) .$$

We now substitute '$(\mathbf{P}s`x)$' for '$\alpha$' and '$(\mathbf{T}s`x)$' for 'β' in 0.418 and get

(7)
$$((\mathbf{P}s`x)\ \&\ (\mathbf{T}s`x)) \subset (\mathbf{T}s`x) .$$

Applying Rule 4.2 to (7) and (6), we finally reach (using Rule 5.14)

(8)
$$x\epsilon\mathbf{cell} \text{ implies } ((\mathbf{S}`\ ( (\mathbf{P}s`x)\ \&\ (\mathbf{T}s`x)))\ \mathbf{T}x) ,$$

which is the statement required. In the remaining derivations the method of obtaining each step will not be described in detail, but the statements upon which the transformations depend will be quoted by number at the end of each line.

3.23 $x\epsilon$**cell** implies
$$((\mathbf{C}s`\ (\mathbf{S}`\ ((\mathbf{P}s`x)\ \&\ (\mathbf{T}s`x))))\ \& \ (\mathbf{P}s`x) ) \subset (\mathbf{P}s`\ (\mathbf{S}`\ ( (\mathbf{P}s`x)\ \&\ (\mathbf{T}s`x))))$$

If a thing x is a cell, then the class of things which are both parts of x and coincident in time with the sum of the parts of x which stand in **T** to x is included in the parts of the sum of the parts of x which stand in **T** to x.

Derivation:

(1) $y\epsilon\,((\mathbf{C}s‘\,(\mathbf{S}‘a))\ \&\ (\mathbf{P}s‘x))$ implies $y\mathbf{C}\,(\mathbf{S}‘a)$ [0.523, 0.413, 0.1141

(2) $y\mathbf{C}\,(\mathbf{S}‘a)$ equiv $y\mathbf{T}\&\mathbf{T}^{-1}\,(\mathbf{S}‘a)$ [2.5

(3) $y\mathbf{T}\,\&\,\mathbf{T}^{-1}\,(\mathbf{S}‘a)$ implies $y\mathbf{T}(\mathbf{S}‘a)$ [0.512, 0.1141

(4) $y\mathbf{C}\,(\mathbf{S}‘a)$ implies $y\mathbf{T}(\mathbf{S}‘a)$ [Rule 4.3, (2) (3)

(5) $y\epsilon\,((\mathbf{C}s‘\,(\mathbf{S}‘a))\ \&\ (\mathbf{P}s‘x))$ implies $y\mathbf{T}\,(\mathbf{S}‘a)$ [(1) (4), 0.116

(6) $y\mathbf{C}\,(\mathbf{S}‘\,((\mathbf{P}s‘x)\ \&\ (\mathbf{T}s‘x)))$ implies
$y\mathbf{T}\,(\mathbf{S}‘\,((\mathbf{P}s‘x)\ \&\ (\mathbf{T}s‘x)))$ [Rule 4.1, (4)

(7) $(y\mathbf{T}\,(\mathbf{S}‘\,((\mathbf{P}s‘x)\ \&\ (\mathbf{T}s‘x)))$ and
$(\mathbf{S}‘\,((\mathbf{P}s‘x)\ \&\ (\mathbf{T}s‘x)))\,\mathbf{T}x)$ implies
$y\mathbf{T}x$ [Rule 4.1, 2.11, 0.533

(8) $x\epsilon\mathbf{cell}$ implies $(y\mathbf{T}\,(\mathbf{S}‘\,((\mathbf{P}s‘x)\ \&\ (\mathbf{T}s‘x)))$ implies
$y\mathbf{T}x)$ [3.22, (7), 0.141

(9) $y\epsilon\,((\mathbf{C}s‘\,(\mathbf{S}‘\,((\mathbf{P}s‘x)\ \&\ (\mathbf{T}s‘x))))\ \&\ (\mathbf{P}s‘x))$ implies
$y\mathbf{T}\,(\mathbf{S}‘\,((\mathbf{P}s‘x)\ \&\ (\mathbf{T}s‘x)))$ [Rule 4.12, (5)

(10) $x\epsilon\mathbf{cell}$ implies
$(y\epsilon\,((\mathbf{C}s‘\,(\mathbf{S}‘\,((\mathbf{P}s‘x)\ \&\ (\mathbf{T}s‘x))))\ \&\ (\mathbf{P}s‘x))$ implies
$y\mathbf{T}x)$ [(9) (8), 0.143

(11) $y\epsilon\,((\mathbf{C}s‘\,(\mathbf{S}‘\,((\mathbf{P}s‘x)\ \&\ (\mathbf{T}s‘x))))\ \&\ (\mathbf{P}s‘x))$ implies
$y\mathbf{P}x$ [0.413, 0.1142, 0.523

(12) $x\epsilon\mathbf{cell}$ implies
$(y\epsilon\,((\mathbf{C}s‘\,(\mathbf{S}‘\,((\mathbf{P}s‘x)\ \&\ (\mathbf{T}s‘x))))\ \&\ (\mathbf{P}s‘x))$ implies
$(y\mathbf{T}x$ and $y\mathbf{P}x))$ [(10) (11), 0.145

(13) $(y\mathbf{T}x$ and $y\mathbf{P}x)$ implies $(y\mathbf{P}x$ and $y\mathbf{T}x)$ [0.124, 0.135

(14) $(y\mathbf{P}x$ and $y\mathbf{T}x)$ implies $y\epsilon\,((\mathbf{P}s‘x)\ \&\ (\mathbf{T}s‘x))$
[0.523, 0.413, 0.123, 0.135

(15) $y\epsilon\,((\mathbf{P}s‘x)\ \&\ (\mathbf{T}s‘x))$ implies $y\mathbf{P}\,(\mathbf{S}‘\,((\mathbf{P}s‘x)\ \&\ (\mathbf{T}s‘x)))$
[1.32

(16) $y\mathbf{P}\,(\mathbf{S}‘\,((\mathbf{P}s‘x)\ \&\ (\mathbf{T}s‘x)))$ equiv $y\epsilon\,(\mathbf{P}s‘\,(\mathbf{S}‘\,((\mathbf{P}s‘x)\ \&$
$(\mathbf{T}s‘x))))$ [0.523

(17) $x\epsilon\mathbf{cell}$ implies

$(y\epsilon\,((\mathbf{C}s‘\,(\mathbf{S}‘\,((\mathbf{P}s‘x)\ \&\ (\mathbf{T}s‘x))))\ \&\ (\mathbf{P}s‘x))$ implies
$y\epsilon\,(\mathbf{P}s‘\,(\mathbf{S}‘\,((\mathbf{P}s‘x)\ \&\ (\mathbf{T}s‘x)))))$

[(12) to (16), 0.116

(18) $x\epsilon\mathbf{cell}$ implies

$(\text{All } y)\ (y\epsilon\,((\mathbf{C}s‘\,(\mathbf{S}‘\,((\mathbf{P}s‘x)\ \&\ (\mathbf{T}s‘x))))\ \&\ (\mathbf{P}s‘x))$
implies $y\epsilon\,(\mathbf{P}s‘\,(\mathbf{S}‘\,((\mathbf{P}s‘x)\ \&\ (\mathbf{T}s‘x)))))$ [(17) Rule 4.53

(19) $x\epsilon\mathbf{cell}$ implies

$((\mathbf{C}s‘\,(\mathbf{S}‘\,((\mathbf{P}s‘x)\ \&\ (\mathbf{T}s‘x))))\ \&$
$(\mathbf{P}s‘x))\subset(\mathbf{P}s‘\,(\mathbf{S}‘\,((\mathbf{P}s‘x)\ \&\ (\mathbf{T}s‘x))))$ [(18), 0.411

3.24 $x\epsilon\mathbf{cell}$ implies $(\mathbf{S}‘\,((\mathbf{P}s‘x)\ \&\ (\mathbf{T}s‘x)))\ \mathbf{Sl}x$

If a thing x is a cell, then the sum of the parts of x which stand in $\mathbf{T}$ to x is a time-slice of x.

Derivation:

(1) $x\epsilon\mathbf{cell}$ implies

$((\mathbf{S}‘\,((\mathbf{P}s‘x)\ \&\ (\mathbf{T}s‘x)))\ \epsilon\ (\mathbf{mom}\ \&\ (\mathbf{P}s‘x))$ and
$((\mathbf{C}s‘\,(\mathbf{S}‘\,((\mathbf{P}s‘x)\ \&\ (\mathbf{T}s‘x))))\ \&(\mathbf{P}s‘x))\subset$
$(\mathbf{P}s‘\,(\mathbf{S}‘\,((\mathbf{P}s‘x)\ \&\ (\mathbf{T}s‘x)))))$ [3.21, 3.23, 0.131

(2) $(\mathbf{S}‘\,((\mathbf{P}s‘x)\ \&\ (\mathbf{T}s‘x)))\ \mathbf{Sl}x$ equiv

$((\mathbf{S}‘\,((\mathbf{P}s‘x)\ \&\ (\mathbf{T}s‘x)))\ \epsilon\ (\mathbf{mom}\ \&\ (\mathbf{P}s‘x))$ and
$((\mathbf{C}s‘\,(\mathbf{S}‘\,((\mathbf{P}s‘x)\ \&\ (\mathbf{T}s‘x))))\ \&\ (\mathbf{P}s‘x))\subset$
$(\mathbf{P}s‘\,(\mathbf{S}‘\,((\mathbf{P}s‘x)\ \&\ (\mathbf{T}s‘x)))))$ [Rule 4.12, 2.8

(3) $x\epsilon\mathbf{cell}$ implies $(\mathbf{S}‘\,((\mathbf{P}s‘x)\ \&\ (\mathbf{T}s‘x)))\ \mathbf{Sl}x$ [Rule 4.3, (1) (2)

3.25 $x\epsilon\mathbf{cell}$ implies $(\mathbf{S}‘\,((\mathbf{P}s‘x)\ \&\ (\mathbf{T}s‘x))) = \mathbf{B}‘x$.

If a thing x is a cell, then the sum of the class of parts of x which stand in $\mathbf{T}$ to x is identical with the first time-slice of x.

Derivation:

(1) $x\epsilon\mathbf{cell}$ implies

$((\mathbf{S}‘\,((\mathbf{P}s‘x)\ \&\ (\mathbf{T}s‘x)))\ \mathbf{Sl}x$ and
$(\mathbf{S}‘\,((\mathbf{P}s‘x)\ \&\ (\mathbf{T}s‘x)))\ \mathbf{T}x)$ [3.22, 3.24, 0.131

(2) $x\epsilon\mathbf{cell}$ implies $(\mathbf{S}‘\,((\mathbf{P}s‘x)\ \&\ (\mathbf{T}s‘x)))\ \mathbf{Sl}\ \&\ \mathbf{T}x$ [(1), 0.512

(3) $x\epsilon$**cell** implies (**S**‘ ((**P**s‘x) & (**T**s‘x))) **B**x [(2), 2.913

(4) (**B**ϵ1→many and (**S**‘ ((**P**s‘x) & (**T**s‘x))) **B**x) implies
(**S**‘ ((**P**s‘x) & (**T**s‘x))) = **B**‘x [0.536

(5) **B**ϵ1→many implies
($x\epsilon$**cell** implies (**S**‘ ((**P**s‘x) & (**T**s‘x))) = **B**‘x
[(3) (4), 0.141, 0.137

(6) $x\epsilon$**cell** implies (**S**‘ ((**P**s‘x) & (**T**s‘x))) = **B**‘x
[2.914, (5), Rule 4.2

3.26 $x\epsilon$**cell** implies (((**B**‘x) ϵ**mom** and
(**B**‘x) **P**x) and ((**B**‘x) **T**x and (**B**‘x) **Sl**x))

If a thing x is a cell, then the first time-slice of x is a momentary thing and a part of x and stands in **T** to x and is a time-slice of x.

Derivation: 3.26 is easily derivable from 3.25, 3.21, 3.22, and 3.24 with the help of 0.131, Rule 4.4, 0.116, 0.413, and 0.523.

(The next six statements are analogous to the last six, but using 3.12 and 2.902 instead of 3.11 and 2.901. Accordingly, only the reference numbers of the statements of Part II which are required for their derivation are given, not the derivations themselves.)

3.31 $x\epsilon$**cell** implies (**S**‘ ((**P**s‘x) & (**T**^{-1}s‘x))) ϵ (**mom** & (**P**s‘x))
[3.12, 2.221, 2.36, 1.331, 1.3

If a thing x is a cell, then the sum of the class of those parts of x to which x stands in **T** is momentary and a part of x.

3.32 $x\epsilon$**cell** implies x**T** (**S**‘ ((**P**s‘x) & (**T**^{-1}s‘x)) [3.12, 2.242

If a thing x is a cell, then it stands in **T** to the sum of the class of those parts of x to which x stands in **T**.

3.33 $x\epsilon$**cell** implies ((**C**s‘ (**S**‘ ((**P**s‘x) & (**T**^{-1}s‘x)))) & (**P**s‘x)) ⊂
(**P**s‘ (**S**‘ ((**P**s‘x) & (**T**^{-1}s‘x)))) [3.32, 2.5, 2.11, 1.32

If a thing x is a cell, then the class of things which are both parts of x and coincident in time with the sum of those parts of x to which x stands in **T** is included in the class of things which are parts of the sum of the parts of x to which x stands in **T**.

3.34 $x\epsilon$**cell** implies (**S**' ((**P**s'x) & (**T**^{-1}s'x))) **Sl**x [3.31, 3.33, 2.8

If a thing x is a cell, then the sum of the class of parts of x to which x stands in **T** is a time-slice of x.

3.35 $x\epsilon$**cell** implies (**S**' ((**P**s'x) & (**T**^{-1}s'x))) = **E**'x
[3.34, 3.32, 2.923, 2.924

If a thing x is a cell, then the sum of the class of parts of x to which x stands in **T** is identical with the last time-slice of x.

3.36 $x\epsilon$**cell** implies (((**E**'x) ϵ**mom** and (**E**'x) **P**x) and (**x**T(**E**'x) and (**E**'x)**Sl**x)) [3.35, 3.31, 3.32, 3.34

If a thing x is a cell, then its last time-slice is momentary and a part of x, x stands in **T** to it, and it is a time-slice of x.

3.41 $x\epsilon$**cell** implies (**B**'x) $\neq$ (**E**'x)

If a thing x is a cell, then its first time-slice is not identical with its last time-slice.

Derivation:

(1) $x\epsilon$**cell** implies (x**T** (**E**'x) and (**B**'x) **T**x)
[3.36, 3.26, 0.119, 0.131

(2) ($x\epsilon$**cell** and (**B**'x) = (**E**'x)) implies (x**T** (**E**'x) and (**E**'x) **T**x) [(1), Rule 4.4, 0.142

(3) (x**T** (**E**'x) and (**E**'x) **T**x) implies x**T**x [2.11, 0.533

(4) x**T**x equiv $x\epsilon$**mom** [2.1

(5) $x\epsilon$**cell** and (**B**'x) = (**E**'x)) implies $x\epsilon$**mom**
[(2) (3), 0.116, (4), Rule 4.3

(6) ($x\epsilon$**cell** and $x\epsilon$—**mom**) implies (**B**'x) $\neq$ (**E**'x)
[0.127, (5), 0.312, 0.414

(7) $x\epsilon$**cell** implies (**B**'x) $\neq$ (**E**'x) [(6), 3.13, 0.415, 0.414, 0.144

3.42 $x\epsilon$**cell** implies (**B**'x) **T**(**E**'x)

If a thing is a cell, then its first time-slice stands in **T** to its last time-slice.

Derivation: 3.42 is easily derivable from 3.26, 3.36, and 2.11 with the help of 0.119, 0.131, 0.533 and 0.116.

3.43 $x\epsilon$**cell** implies (**B**'x) **Z** (**E**'x)

If a thing is a cell, then its first time-slice is wholly before its last time-slice in time.

Derivation:

(1) $x\epsilon$**cell** implies ((**B**'x) ϵ**mom** and (**E**'x) ϵ**mom**) [3.26, 3.36, 0.119, 0.131

(2) ((**B**'x) ϵ**mom** and (**E**'x) ϵ**mom**) implies ((**B**'x) **T** (**E**'x) equiv ((**B**'x) **C** or **Z** (**E**'x))) [2.65

(3) $x\epsilon$**cell** implies ((**B**'x) **T** (**E**'x) equiv ((**B**'x) **C** or **Z** (**E**'x))) [(1) (2), 0.116

(4) $x\epsilon$**cell** implies ((**B**'x) **Sl**x and (**E**'x) **Sl**x) [3.26, 3.36, 0.119, 0.131

(5) ((**B**'x) **Sl**x and (**E**'x) **Sl**x) implies ((**B**'x) = (**E**'x) equiv (**B**'x) **C** (**E**'x)) [2.83, 0.134

(6) $x\epsilon$**cell** implies ((**B**'x) = (**E**'x) equiv (**B**'x) **C** (**E**'x)) [(4) (5), 0.116

(7) $x\epsilon$**cell** implies (((**B**'x) **T** (**E**'x) equiv ((**B**'x) **C** or **Z** (**E**'x)) and ((**B**'x) = (**E**'x) equiv (**B**'x) **C** (**E**'x))) [(3) (6), 0.131

(8) $x\epsilon$**cell** implies ((**B**'x) **T** (**E**'x) equiv ((**B**'x) = (**E**'x) or (**B**'x) **Z** (**E**'x))) [(7), Rule 4.3, 0.513

(9) $x\epsilon$**cell** implies ((**B**'x) = (**E**'x) or (**B**'x) **Z** (**E**'x)) [3.42, (8), 0.131, Rule 4.3

(10) $x\epsilon$**cell** implies (**B**'x) **Z** (**E**'x) [3.41, (9), 0.118

3.44 (($x\epsilon$**cell** and $y\epsilon$**cell**) and ((**B**'x) **C** (**B**'y) or (**E**'x) **C** (**E**'y))) implies (**B**'x)—**P**—**P**$^{-1}$(**E**'y)

If two cells have their first time-slices coincident in time or their last time-slices coincident in time, then the first time-slice of the one cannot be part of the last of the other, nor the last of one part of the first of the other.

Derivation:

(1) (($x \epsilon$ **cell** and $y \epsilon$ **cell**) and (**B**'x) **C** (**B**'y)) implies ((**B**'x) **C** (**B**'y) and (**B**'y) **Z** (**E**'y)) [3.43, 0.114, 0.117, 0.124, 0.122

(2) ((**B**'x) **C** (**B**'y) and (**B**'y) **Z** (**E**'y)) implies (**B**'x) **C** | **Z** (**E**'y) [0.541, 0.216

(3) (**B**'x) **C** | **Z** (**E**'y) implies (**B**'x) **Z** (**E**'y) [2.66, Rule 4.4

(4) (**B**'x) **Z** (**E**'y) implies (**B**'x)—**P**—$\mathbf{P}^{-1}$(**E**'y) [2.654, 0.511

(5) (($x \epsilon$ **cell** and $y \epsilon$ **cell**) and (**B**'x) **C** (**B**'y)) implies (**B**'x)—**P**—$\mathbf{P}^{-1}$(**E**'y) [(1) (2) (3) (4), 0.116

(6) (($x \epsilon$ **cell** and $y \epsilon$ **cell**) and (**E**'x) **C** (**E**'y)) implies ((**B**'x) **Z** (**E**'x) and (**E**'x) **C** (**E**'y)) [3.43, 0.117, 0.124, 0.122, 0.114

(7) (($x \epsilon$ **cell** and $y \epsilon$ **cell**) and (**E**'x) **C** (**E**'y)) implies (**B**'x) **Z** | **C** (**E**'y) [(6), 0.541, 0.116

(8) (($x \epsilon$ **cell** and $y \epsilon$ **cell**) and (**E**'x) **C** (**E**'y)) implies (**B**'x) **Z** (**E**'y) [(7), 2.66, Rule 4.4, 0.116

(9) (($x \epsilon$ **cell** and $y \epsilon$ **cell**) and (**E**'x) **C** (**E**'y)) implies (**B**'x)—**P**—$\mathbf{P}^{-1}$(**E**'y) [(8), 2.654, 0.511, 0.116

(10) (($x \epsilon$ **cell** and $y \epsilon$ **cell**) and ((**B**'x) **C** (**B**'y) or (**E**'x) **C** (**E**'y))) implies (**B**'x)—**P**—$\mathbf{P}^{-1}$(**E**'y) [(5) (9), 0.136, 0.131, 0.132

3.45 (($x \epsilon$ **cell** and $y \epsilon$ **cell**) and (((**B**'x) **C** (**B**'y) or (**E**'x) **C** (**E**'y)) and ((**P**s'x) & (**P**s'y)) $\neq \Lambda$)) implies $x = y$

If x is a cell and y is a cell and the first time-slice of x is coincident in time with the first time-slice of y, or the last of x is coincident with the last of y, and if there are parts of x which are also parts of y, then x is identical with y.

Derivation:

(1) (**B**'x)—**P**—$\mathbf{P}^{-1}$(**E**'y) equiv (not ((**B**'x) **P** or $\mathbf{P}^{-1}$(**E**'y))) [0.516, 0.512, 0.121, 0.101, 0.513, 0.133

(2) (($x\epsilon$**cell** and $y\epsilon$**cell**) and ((**B'**x) **C** (**B'**y) or (**E'**x) **C** (**E'**y)))
implies (not ((**B'**x) **P** or **P**$^{-1}$(**E'**y))) [3.44, (1), Rule 4.3

(3) (($y\epsilon$**cell** and $x\epsilon$**cell**) and ((**B'**y) **C** (**B'**x) or (**E'**y) **C** (**E'**x)))
implies (not ((**B'**y) **P** or **P**$^{-1}$(**E'**x))) [(2), Rule 4.12

(4) (not ((**B'**y) **P** or **P**$^{-1}$(**E'**x))) equiv (not ((**E'**x) **P** or
P^{-1}(**B'**y))) [0.513, 0.125, 0.517

(5) (($x\epsilon$**cell** and $y\epsilon$**cell**) and ((**B'**x) **C** (**B'**y) or (**E'**x) **C** (**E'**y)))
equiv (($y\epsilon$**cell** and $x\epsilon$**cell**) and ((**B'**y) **C** (**B'**x) or
(**E'**y) **C** (**E'**x))) [2.52, 0.124, Rule 4.3

(6) (($x\epsilon$**cell** and $y\epsilon$**cell**) and ((**B'**x) **C** (**B'**y) or (**E'**x) **C** (**E'**y)))
implies ((not ((**B'**x) **P** or **P**$^{-1}$(**E'**y))) and
(not ((**E'**x) **P** or **P**$^{-1}$(**B'**y))))
[(2) (3) (4) (5), 0.131, Rule 4.3

(7) ((not ((**B'**x) **P** or **P**$^{-1}$(**E'**y))) and (not ((**E'**x) **P** or
P^{-1}(**B'**y)))) equiv (not (((**B'**x) **P** or **P**$^{-1}$(**E'**y)) or
((**E'**x) **P** or **P**$^{-1}$(**B'**y)))) [0.121, 0.101

(8) (not (((**B'**x) **P** or **P**$^{-1}$(**E'**y)) or ((**E'**x) **P** or
P^{-1}(**B'**y)))) implies (not (((**B'**x) **Pp** or **Pp**$^{-1}$(**E'**y)) or
((**E'**x) **Pp** or **Pp**$^{-1}$(**B'**y))))
[0.112, 0.101, 0.121, 0.312, 1.4, 0.133

(9) (($x\epsilon$**cell** and $y\epsilon$**cell**) and (not (((**B'**x) **Pp** or **Pp**$^{-1}$(**E'**y)) or
((**E'**x) **Pp** or **Pp**$^{-1}$(**B'**y)))) and ((**P**s'x) & (**P**s'y)) $\neq \Lambda$)
implies $x = y$ [3.14, 0.124, 0.127

(10) 3.45 [(6) (7), Rule 4.3, (8), 0.116; (9), 0.142, 0.124, 0.122

3.46 ($x\epsilon$**cell** and $y\epsilon$**cell**) implies (($x = y$ equiv (**B'**x) = (**B'**y))
and ($x = y$ equiv (**E'**x) = (**E'**y)))

If x is a cell and y is a cell, then x is identical with y if, and only if, the first time-slice of x is identical with the first of y, and if, and only if, the last time-slice of x is identical with the last of y.

Derivation:

(1) ($x\epsilon$**cell** and $y\epsilon$**cell**) implies
((**S'** ((**P**s'x) & (**T**s'x))) = (**B'**x) and (**S'** ((**P**s'y) &
(**T**s'y))) = (**B'**y)) [3.25, 0.117

(2) ($x = y$ and (**S'** ((**P**s'y) & (**T**s'y))) = (**B'**y)) implies
(**S'** ((**P**s'x) & (**T**s'x))) = (**B'**y) [Rule 4.4

(3) $((\mathbf{S}`((\mathbf{P}s`x) \,\&\, (\mathbf{T}s`x))) = (\mathbf{B}`x)$ and
$(\mathbf{S}`((\mathbf{P}s`x) \,\&\, (\mathbf{T}s`x))) = (\mathbf{B}`y))$ implies $(\mathbf{B}`x) = (\mathbf{B}`y)$
[Rule 4.4

(4) $((x \epsilon \mathbf{cell}$ and $y \epsilon \mathbf{cell})$ and $x = y)$ implies
$(\mathbf{B}`x) = (\mathbf{B}`y)$
[0.119, 0.124, 0.142, 0.117, 0.122, 0.116, (1) (2) (3)

(5) $(x \epsilon \mathbf{cell}$ and $y \epsilon \mathbf{cell})$ implies $(x = y$ implies $(\mathbf{B}`x) = (\mathbf{B}`y))$
[(4), 0.136

(6) $((x \epsilon \mathbf{cell}$ and $y \epsilon \mathbf{cell})$ and $(\mathbf{B}`x) = (\mathbf{B}`y))$ implies
$((\mathbf{B}`x)\, \mathbf{P} x$ and $(\mathbf{B}`x)\, \mathbf{P} y)$ [3.26

(7) $((\mathbf{B}`x)\, \mathbf{P} x$ and $(\mathbf{B}`x)\, \mathbf{P} y)$ implies
$((\mathbf{P}s`x) \,\&\, (\mathbf{P}s`y)) \neq \Lambda$ [0.215, 0.523, 0.413, 0.416

(8) $((x \epsilon \mathbf{cell}$ and $y \epsilon \mathbf{cell})$ and $(\mathbf{B}`x) = (\mathbf{B}`y))$ implies
$((\mathbf{B}`x)\, \mathbf{P}\, (\mathbf{B}`x)$ and $(\mathbf{B}`x)\, \mathbf{P}\, (\mathbf{B}`y)$ [1.22

(9) $((\mathbf{B}`x)\, \mathbf{P}\, (\mathbf{B}`x)$ and $(\mathbf{B}`x)\, \mathbf{P}\, (\mathbf{B}`y))$ implies
$(\mathbf{P}s`\,(\mathbf{B}`x)) \,\&\, (\mathbf{P}s`\,(\mathbf{B}`y)) \neq \Lambda$
[0.217, 0.523, 0.413, 0.416

(10) $(x \epsilon \mathbf{cell}$ and $y \epsilon \mathbf{cell})$ implies $((\mathbf{B}`x) \,\epsilon \mathbf{mom}$ and $(\mathbf{B}`y) \,\epsilon \mathbf{mom})$
[3.26, 0.119, 0.117

(11) $((x \epsilon \mathbf{cell}$ and $y \epsilon \mathbf{cell})$ and $(\mathbf{B}`x) = (\mathbf{B}`y))$ implies
$(((\mathbf{B}`x) \,\epsilon \mathbf{mom}$ and $(\mathbf{B}`y) \,\epsilon \mathbf{mom})$ and $((\mathbf{P}s`\,(\mathbf{B}`x)) \,\&$
$(\mathbf{P}s`\,(\mathbf{B}`y))) \neq \Lambda)$ [(8) (9) (10), 0.116

(12) $((x \epsilon \mathbf{cell}$ and $y \epsilon \mathbf{cell})$ and $(\mathbf{B}`x) = (\mathbf{B}`y))$ implies
$(\mathbf{B}`x)\, \mathbf{C}\, (\mathbf{B}`y)$ [(11),2.531, 0.116

(13) $((x \epsilon \mathbf{cell}$ and $y \epsilon \mathbf{cell})$ and $(\mathbf{B}`x) = (\mathbf{B}`y))$ implies
$x = y$ [(6) (7), 0.116, (12), 3.45, 0.116

(14) $(x \epsilon \mathbf{cell}$ and $y \epsilon \mathbf{cell})$ implies $((\mathbf{B}`x) = (\mathbf{B}`y)$ implies $x = y)$
[(13), 0.136

(15) $(x \epsilon \mathbf{cell}$ and $y \epsilon \mathbf{cell})$ implies $((\mathbf{B}`x) = (\mathbf{B}`y)$ equiv $x = y)$
[(5) (14), 0.103, 0.131

By an exactly analogous derivation, using 3.35 and 3.36 instead of 3.25 and 3.26, we get

(16) $(x \epsilon \mathbf{cell}$ and $y \epsilon \mathbf{cell})$ implies $((\mathbf{E}`x) = (\mathbf{E}`y)$ equiv $x = y)$.

And from (15) and (16) with the help of 0.131 we get 3.46.

3.5 $x \mathbf{D} y$ equiv $((x \epsilon \mathbf{cell}$ and $y \epsilon \mathbf{cell})$ and $(\mathbf{B}`y)\, \mathbf{Pp}\, (\mathbf{E}`x))$
Postulate (*d*)

A thing y will be said to be derived immediately by division from a thing x if, and only if, x and y are cells and the first time-slice of y is a proper part of the last time-slice of x.

3.51 $x\mathbf{D}y$ implies (($\mathbf{B}$'x) $\mathbf{Z}$ ($\mathbf{B}$'y) and ($x\mathbf{T}y$ and $x \neq y$))

If a cell y is derived immediately by division from a cell x, then the first time-slice of x stands in **Z** to the first of y, x stands in **T** to y and is not identical with it.

Derivation:

(1) $x\mathbf{D}y$ implies ($\mathbf{B}$'y) $\mathbf{Pp}$ ($\mathbf{E}$'x) [3.5, 0.119

(2) $x\mathbf{D}y$ implies ($\mathbf{B}$'y) $\mathbf{P}$ ($\mathbf{E}$'x) [(1), 1.4, 0.116

(3) $x\mathbf{D}y$ implies ($\mathbf{E}$'x) $\mathbf{P}^{-1}$($\mathbf{B}$'y) [(2), 0.517, 0.116

(4) $x\mathbf{D}y$ implies (($\mathbf{B}$'x) $\mathbf{Z}$ ($\mathbf{E}$'x) and ($\mathbf{E}$'x) $\mathbf{P}^{-1}$($\mathbf{B}$'y)) [3.5, 3.43, (3), 0.131

(5) (($\mathbf{B}$'x) $\mathbf{Z}$ ($\mathbf{E}$'x) and ($\mathbf{E}$'x) $\mathbf{P}^{-1}$($\mathbf{B}$'y)) implies ($\mathbf{B}$'x) $\mathbf{Z} \mid \mathbf{P}^{-1}$ ($\mathbf{B}$'y) [0.216, 0.541, 0.116

(6) $x\mathbf{D}y$ implies ($\mathbf{B}$'x) $\mathbf{Z} \mid \mathbf{P}^{-1}$ ($\mathbf{B}$'y) [(4) (5), 0.116

(7) $x\mathbf{D}y$ implies ($\mathbf{B}$'y) $\epsilon\mathbf{mom}$ [3.5, 0.119, 3.26, 0.116

(8) $x\mathbf{D}y$ implies ($\mathbf{B}$'x) $\mathbf{Z}$ ($\mathbf{B}$'y) [(6) (7), 2.653, 0.131

(9) $x\mathbf{D}y$ implies ($x\mathbf{T}$ ($\mathbf{E}$'x) and ($\mathbf{E}$'x) $\mathbf{P}^{-1}$($\mathbf{B}$'y)) [3.5, 0.119, 3.36, (3), 0.131

(10) ($x\mathbf{T}$ ($\mathbf{E}$'x) and ($\mathbf{E}$'x) $\mathbf{P}^{-1}$($\mathbf{B}$'y)) implies $x\mathbf{T} \mid \mathbf{P}^{-1}$($\mathbf{B}$'$y$) [0.216, 0.541, 0.116

(11) $x\mathbf{D}y$ implies $x\mathbf{T} \mid \mathbf{P}^{-1}$($\mathbf{B}$'$y$) [(9) (10), 0.116

(12) $x\mathbf{D}y$ implies $x\mathbf{T}$ ($\mathbf{B}$'y) [(11), 2.221, Rule 4.4

(13) $x\mathbf{D}y$ implies ($\mathbf{B}$'y) $\mathbf{T}y$ [3.5, 0.119, 3.26, 0.116

(14) $x\mathbf{D}y$ implies ($x\mathbf{T}$ ($\mathbf{B}$'y) and ($\mathbf{B}$'y) $\mathbf{T}y$) [(12) (13), 0.131

(15) $x\mathbf{D}y$ implies $x\mathbf{T}y$ [(14), 2.11, 0.533, 0.116

(16) $x\mathbf{D}y$ implies ($\mathbf{B}$'x)—$\mathbf{P}$—$\mathbf{P}^{-1}$($\mathbf{B}$'y) [(8), 2.654, 0.511, 0.116

(17) $x\mathbf{D}y$ implies ($\mathbf{B}$'x) $\neq$ ($\mathbf{B}$'y) [1.231, (16), 0.516, 0.512, 0.115, 0.116, 0.133

(18) $x\mathbf{D}y$ implies $x \neq y$
[3.46, 0.119, 0.133, 3.5, 0.116, (17), 0.123, 0.135, 0.136, 0.144

(19) $x\mathbf{D}y$ implies (($\mathbf{B}$'x) $\mathbf{Z}$ ($\mathbf{B}$'y) and ($x\mathbf{T}y$ and $x \neq y$))
[(8) (15) (18), 0.131

3.52 $\mathbf{D} \epsilon$ (1→many & asym)

The relation in which one cell stands to another when the latter is derived immediately by division from the former is a one-many and asymmetrical relation.

Derivation:

(1) ($x\mathbf{D}y$ and $z\mathbf{D}y$) implies (($\mathbf{B}$'y) $\mathbf{P}$ ($\mathbf{E}$'x) and ($\mathbf{B}$'y) $\mathbf{P}$ ($\mathbf{E}$'z))
[3.5, 1.4, 0.119

(2) (($\mathbf{B}$'y) $\mathbf{P}$ ($\mathbf{E}$'x) and ($\mathbf{B}$'y) $\mathbf{P}$ ($\mathbf{E}$'z)) implies (Some u) ($u\mathbf{P}$ ($\mathbf{E}$'x) and $u\mathbf{P}$ ($\mathbf{E}$'z))
[0.215

(3) (Some u) ($u\mathbf{P}$ ($\mathbf{E}$'x) and $u\mathbf{P}$ ($\mathbf{E}$'z)) implies (Some u) ($u\epsilon$ ($\mathbf{P}s$' ($\mathbf{E}$'x)) and $u\epsilon$ ($\mathbf{P}s$' ($\mathbf{E}$'z)))
[0.523

(4) (Some u) ($u\epsilon$ ($\mathbf{P}s$' ($\mathbf{E}$'x)) and $u\epsilon$ ($\mathbf{P}s$' ($\mathbf{E}$'z))) equiv (($\mathbf{P}s$' ($\mathbf{E}$'x)) & ($\mathbf{P}s$' ($\mathbf{E}$'z))) $\neq \Lambda$
[0.413, 0.416

(5) (($x\epsilon$**cell** and $z\epsilon$**cell**) and (($\mathbf{P}s$' ($\mathbf{E}$'x)) & ($\mathbf{P}s$' ($\mathbf{E}$'z))) $\neq \Lambda$) implies (($\mathbf{P}s$'x) & ($\mathbf{P}s$'z)) $\neq \Lambda$
[3.36, 0.119, 0.416, 1.11, 0.533, 0.217

(6) ($x\mathbf{D}y$ and $z\mathbf{D}y$) implies (($\mathbf{P}s$'x) & ($\mathbf{P}s$'z)) $\neq \Lambda$
[(1)–(4), 3.5, 0.131, (5), 0.116

(7) ($x\mathbf{D}y$ and $z\mathbf{D}y$) implies ($\mathbf{E}$'x) $\mathbf{C}$ ($\mathbf{E}$'z)
[3.5, 0.119, 3.36, 2.531, (1)–(4), 0.116, 0.131

(8) ($x\mathbf{D}y$ and $z\mathbf{D}y$) implies $x = z$ [(6), 3.5, (7), 0.131, 3.45

(9) (All x) ((All y) ((All z) (($x\mathbf{D}y$ and $z\mathbf{D}y$) implies $x = z$)))
[(8), Rule 4.53

(10) $\mathbf{D} \epsilon$ 1→many [(9), 0.534

(11) $x\mathbf{D}y$ implies ($\mathbf{B}$'x) $\mathbf{Z}$ ($\mathbf{B}$'y) [3.51, 0.119

(12) ($\mathbf{B}$'x) $\mathbf{Z}$ ($\mathbf{B}$'y) implies (not (($\mathbf{B}$'y) $\mathbf{Z}$ ($\mathbf{B}$'x))) [2.622, 0.532

(13) (not (($\mathbf{B}$'y) $\mathbf{Z}$ ($\mathbf{B}$'x))) implies (not ($y\mathbf{D}x$))
[(11), Rule 4.12, 0.126

(14) $x\mathbf{D}y$ implies (not $(y\mathbf{D}x)$) [(11) (12) (13), 0.116

(15) (All x) ((All y) ($x\mathbf{D}y$ implies (not $(y\mathbf{D}x)$))) [(14), Rule 4.53

(16) $\mathbf{D} \in$ asym [(15), 0.532

(17) $\mathbf{D} \in$ (1→many & asym) [(10) (16), 0.413

3.6 $x\mathbf{F}y$ equiv (($x\in$**cell** and $y\in$**cell**) and ($\mathbf{E}$'x) **Pp** ($\mathbf{B}$'y))
Postulate (*d*)

A thing y arises immediately through the fusion of a thing x with some other thing if, and only if, x and y are cells and the last time-slice of x is a proper part of the first time-slice of y.

(The next two statements are the analogues for the relation **F** of 3.51 and 3.52. Their derivations are closely analogous to those of the latter two statements and are therefore omitted.)

3.61 $x\mathbf{F}y$ implies (($\mathbf{B}$'x) **Z** ($\mathbf{B}$'y) and ($x\mathbf{T}y$ and $x \neq y$))

3.62 $\mathbf{F} \in$ (many→1 & asym)

3.71 ((Dom' **D**) & (Dom' **F**)) = Λ

No cell both divides and fuses with another cell.

Derivation:

(1) ($x\mathbf{D}y$ and $x\mathbf{F}z$) implies
(($\mathbf{B}$'y) **Pp** ($\mathbf{E}$'x) and ($\mathbf{E}$'x) **Pp** ($\mathbf{B}$'z))
[3.5, 3.6, 0.119, 0.117

(2) ($x\mathbf{D}y$ and $x\mathbf{F}z$) implies ($\mathbf{B}$'y) **Pp** ($\mathbf{B}$'z) [(1), 1.41, 0.533

(3) ($x\mathbf{D}y$ and $x\mathbf{F}z$) implies (($\mathbf{B}$'y) **P** ($\mathbf{B}$'z) and ($\mathbf{B}$'y) $\neq$ ($\mathbf{B}$'z))
[(2), 1.4, Rule 4.3

(4) ($x\mathbf{D}y$ and $x\mathbf{F}z$) implies ((($\mathbf{B}$'y) $\in$**mom** and ($\mathbf{B}$'z) $\in$**mom**) and
((**P**s' ($\mathbf{B}$'y)) & (**P**s' ($\mathbf{B}$'z))) $\neq \Lambda$
[3.5, 3.6, 3.26, (3), 0.131, 1.22, 0.214, 0.313

(5) ($x\mathbf{D}y$ and $x\mathbf{F}z$) implies ($\mathbf{B}$'y) **C** ($\mathbf{B}$'z) [(4), 2.531, 0.116

(6) ($z\mathbf{D}y$ and $x\mathbf{F}z$) implies (($y\in$**cell** and $z\in$**cell**) and
((**P**s'y) & (**P**s'z)) $\neq \Lambda$) [3.5, 3.6, (4), 1.11, 3.26

(7) ($x\mathbf{D}y$ and $x\mathbf{F}z$) implies $y = z$ [(5) (6), 3.45

(8) ($x\mathbf{D}y$ and $x\mathbf{F}z$) implies $y \neq z$ [(3), 0.119, 3.5, 3.6, 3.46

(9) ($x\mathbf{D}y$ and $x\mathbf{F}z$) implies (not ($x\mathbf{D}y$ and $x\mathbf{F}z$))
[(7) (8), 0.126, 0.312, 0.116

(10) not ($x\mathbf{D}y$ and $x\mathbf{F}z$) [(9), 0.113, Rule 4.2

(11) (All x) (not ((Some y) ($x\mathbf{D}y$) and (Some z) ($x\mathbf{F}z$)))
[(10), Rules 4.53, 4.54

(12) ((Dom‘ **D**) & (Dom‘ **F**)) = Λ equiv
(All x) (not ((Some y) ($x\mathbf{D}y$) and (Some z) ($x\mathbf{F}z$)))
[0.415, 0.413, 0.525

(13) ((Dom‘ **D**) & (Dom‘ **F**)) = Λ [(11) (12), 0.135, Rule 4.2

3.72 ((Cnvdom‘ **D**) & (Cnvdom‘ **F**)) = Λ

No cell arises both by division and by fusion.

(The derivation of 3.72 is closely analogous to that of 3.71.)

3.73 (**D** & **F**) = $\dot{\Lambda}$

There are no cells x and y such that x stands in **D** and in **F** to y.

Derivation:

(1) $x\mathbf{D}$ & $\mathbf{F}y$ implies ($x\mathbf{D}y$ and $x\mathbf{F}y$) [0.512

(2) ($x\mathbf{D}y$ and $x\mathbf{F}y$) implies ($x\epsilon$ (Dom‘ **D**) and $x\epsilon$ (Dom‘ **F**))
[0.525, Rule 4.54

(3) ($x\epsilon$ (Dom‘ **D**) and $x\epsilon$ (Dom‘ **F**)) implies $x\epsilon$ ((Dom‘ **D**)) &
(Dom‘ **F**)) [0.413

(4) $x\epsilon$ ((Dom‘ **D**) & (Dom‘ **F**)) implies ((Dom‘ **D**) & (Dom‘ **F**))
$\neq$ Λ [Rule 4.54, 0.416

(5) $x\mathbf{D}$ & $\mathbf{F}y$ implies ((Dom‘ **D**) & (Dom‘ **F**)) $\neq$ Λ
[(1) (2) (3), 0.116

(6) ((Dom‘ **D**) & (Dom‘ **F**)) = Λ implies (not ($x\mathbf{D}$ & $\mathbf{F}y$))
[(5), 0.126, 0.312, 0.121

(7) (not ($x\mathbf{D}$ & $\mathbf{F}y$)) implies (All x) ((All y) (not ($x\mathbf{D}$ & $\mathbf{F}y$)))
[Rule 4.53

(8) (All x) ((All y) (not ($x\mathbf{D}$ & $\mathbf{F}y$))) equiv
(not (Some x) ((Some y) ($x\mathbf{D}$ & $\mathbf{F}y$))) [0.202, 0.133

(9) (not (Some x) ((Some y) (x**D** & **F**y))) equiv (**D** & **F**) = $\dot{\Lambda}$ [0.518

(10) ((Dom' **D**) & (Dom' **F**)) = Λ implies (**D** & **F**) = $\dot{\Lambda}$ [(6) (7) (8) (9), 0.116, Rule 4.3

(11) (**D** & **F**) = $\dot{\Lambda}$ [(10), 3.71, Rule 4.2

3.74 x**D** or **F**y equiv (($x\epsilon$**cell** and $y\epsilon$**cell**) and (x**T**y and ((**P**s'x) & (**P**s'y)) $\neq \Lambda$))

The necessary and sufficient condition that a thing x stands in **D** or **F** to a thing y is that x and y are cells, x stands in **T** to y, and x and y have a part in common.

Derivation:

(1) x**D** or **F**y implies(($x\epsilon$**cell** and $y\epsilon$**cell**) and x**T**y) [3.5, 3.6, 3.51, 3.61

(2) x**D** or **F**y implies ((**B**'y) **P** (**E**'x) or (**E**'x) **P** (**B**'y)) [3.5, 3.6, 1.4

(3) ((**B**'y) **P** (**E**'x) or (**E**'x) **P** (**B**'y)) implies ((**P**s'x) & (**P**s'y)) $\neq \Lambda$ [1.22, 1.11, 2.8, 2.901, 2.902

(4) x**D** or **F**y implies ((**P**s'x) & (**P**s'y)) $\neq \Lambda$ [(2) (3)

(5) x**D** or **F**y implies (($x\epsilon$**cell** and $y\epsilon$**cell**) and (x**T**y and ((**P**s'x) & (**P**s'y)) $\neq \Lambda$)) [(1) (4)

(6) (($x\epsilon$**cell** and $y\epsilon$**cell**) and x**T**y) implies (((**E**'x) ϵ**mom** and (**B**'y) ϵ**mom**) and ((**B**'x) **Z** (**E**'x) and (**E**'x) **T** (**B**'y))) [3.26, 3.36, 2.13, 3.43

(7) (((**E**'x) ϵ**mom** and (**B**'y) ϵ**mom**) and (**E**'x) **T** (**B**'y)) implies (**E**'x) **C** or **Z** (**B**'y) [2.65

(8) ((**B**'x) **Z** (**E**'x) and (**E**'x) **C** or **Z** (**B**'y)) implies (**B**'x) **Z** (**B**'y) [2.622, 2.66

(9) (**B**'x) **Z** (**B**'y) implies (**B**'x) $\neq$ (**B**'y) [2.654, 1.231

(10) (($x\epsilon$**cell** and $y\epsilon$**cell**) and (**B**'x) $\neq$ (**B**'y)) implies $x \neq y$ [3.46

(11) (($x\epsilon$**cell** and $y\epsilon$**cell**) and x**T**y) implies ((**B**'x) **Z** (**B**'y) and $x \neq y$) [(6)–(10)

(12) $((x\epsilon\mathbf{cell} \text{ and } y\epsilon\mathbf{cell}) \text{ and } (x \neq y \text{ and } ((\mathbf{P}s‘x) \,\&\, (\mathbf{P}s‘y)) \neq \Lambda))$
$\text{implies } ((\mathbf{B}‘x)\ \mathbf{Pp} \text{ or } \mathbf{Pp}^{-1}(\mathbf{E}‘y) \text{ or } (\mathbf{E}‘x)\ \mathbf{Pp} \text{ or } \mathbf{Pp}^{-1}(\mathbf{B}‘y))$
[3.14

(13) $y\epsilon\mathbf{cell} \text{ implies } (\mathbf{B}‘y)\ \mathbf{Z}\ (\mathbf{E}‘y)$ [3.43

(14) $((\mathbf{B}‘x)\ \mathbf{Z}\ (\mathbf{B}‘y) \text{ and } (\mathbf{B}‘y)\ \mathbf{Z}\ (\mathbf{E}‘y)) \text{ implies } (\mathbf{B}‘x)\ \mathbf{Z}\ (\mathbf{E}‘y)$
[2.622

(15) $(\mathbf{B}‘x)\ \mathbf{Z}\ (\mathbf{E}‘y) \text{ implies } (\text{not } ((\mathbf{B}‘x)\ \mathbf{Pp} \text{ or } \mathbf{Pp}^{-1}(\mathbf{E}‘y)))$
[2.654, 1.4

(16) $((x\epsilon\mathbf{cell} \text{ and } y\epsilon\mathbf{cell}) \text{ and } (x\mathbf{T}y \text{ and } ((\mathbf{P}s‘x) \,\&\, (\mathbf{P}s‘y)) \neq \Lambda))$
$\text{implies } (\mathbf{E}‘x)\ \mathbf{Pp} \text{ or } \mathbf{Pp}^{-1}(\mathbf{B}‘y)$ [(11)–(15)

(17) $((x\epsilon\mathbf{cell} \text{ and } y\epsilon\mathbf{cell}) \text{ and } (x\mathbf{T}y \text{ and } ((\mathbf{P}s‘x) \,\&\, (\mathbf{P}s‘y)) \neq \Lambda))$
$\text{implies } x\mathbf{D} \text{ or } \mathbf{F}y$ [(16), 3.5, 3.6

(18) $x\mathbf{D} \text{ or } \mathbf{F}y \text{ equiv } ((x\epsilon\mathbf{cell} \text{ and } y\epsilon\mathbf{cell}) \text{ and } (x\mathbf{T}y \text{ and }$
$((\mathbf{P}s‘x) \,\&\, (\mathbf{P}s‘y)) \neq \Lambda))$ [(5) (17)

Statements 3.52, 3.62, 3.71, and 3.72 are the most important theorems of the theory *T* from the point of view of the theory of cells considered only as things related by division and fusion. In these four theorems the properties of **D** and **F** are summarized. They form the basis for an extensive theory of relations between cells and of relations between such relations. They have here been shown to be consequences of the simple postulates for ‘**P**’, ‘**T**’, and ‘**cell**’ with which the three sections of Part II began. The further developments of the theory need not be pursued here. The interested reader will find them in Section 2 of the author’s *Axiomatic Method in Biology* (pp. 68–83). What has now been given fully suffices to illustrate the general principles to be discussed in the remaining pages of this monograph.

On the accompanying chart (Fig. 1) are depicted certain relations between some of the statements of Section 3 and those of previous sections which are used in their derivations.

VI. General Principles Illustrated by the Theory *T*

(i) *The process of generalization within the theory T.*—We began the theory *T* with ‘**P**’ and its postulates; we then introduced ‘**T**’ with postulates connecting it with ‘**P**’; and, finally,

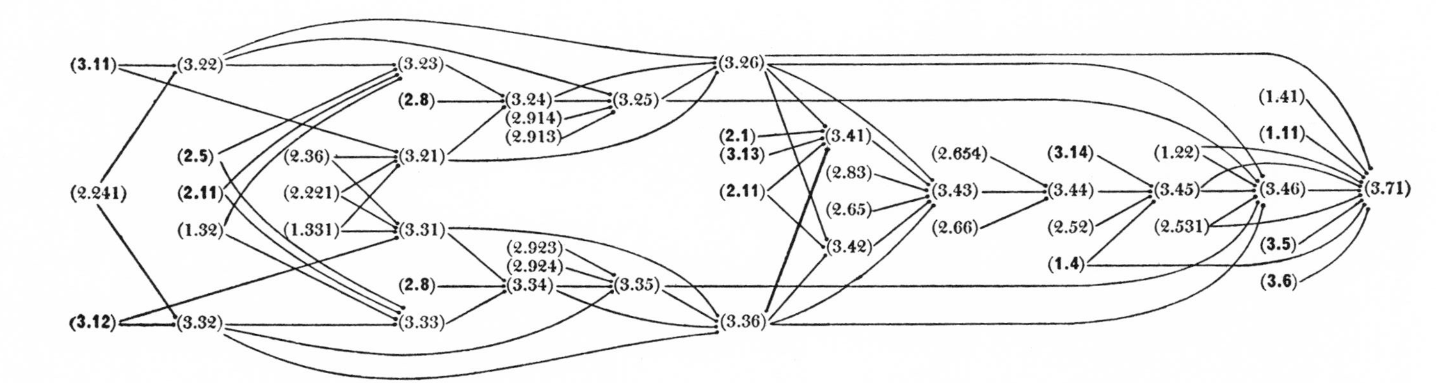

FIG. 1.—Chart showing relations between statements involved in the derivations of some of the statements of Section 3. A line running from one number-sign and ending in a dot on another indicates that the statement denoted by the former sign is used in the derivation of that denoted by the latter. A statement is a consequence of the conjunction of all the statements from the number-signs of which lines run to its number-sign. The number-signs of postulates are printed in boldface type.

we introduced '**cell**'. But the four postulates containing this biological constant are in no way specifically biological because what is said in them about cells is stated exclusively with the help of '**P**' and '**T**'. Moreover, the first three of these postulates (3.11, 3.12, and 3.13) are satisfied by a great many classes other than **cell**. We can, in fact, substitute for '**cell**' in these postulates any class-name denoting a class of things which are not momentary and have a beginning and an end in time, and all the consequences of these postulates in Section 3 will hold for this class. We could, therefore, have proceeded much more generally by putting a class variable 'α' in the place of '**cell**' in these theorems and stating them in the form of consequences of the *hypothesis* that α satisfied one or more of the three postulates 3.11, 3.12, and 3.13. For example, 3.22 could have been written in the form

$$(x\epsilon\alpha \text{ implies } ((\mathbf{P}s`x) \ \& \ (\mathbf{T}s`x)) \neq \Lambda) \text{ implies} \\ (x\epsilon\alpha \text{ implies } (\mathbf{S}` ( (\mathbf{P}s`x) \ \& \ (\mathbf{T}s`x))) \mathbf{T}x) \,.$$

Or, we could have introduced a new class-sign, say '**be**', to denote the class of things which are not momentary and have a beginning and an end in time, defining it with the help of the three postulates as follows:

$$x\epsilon\mathbf{be} \text{ equiv } ((((\mathbf{P}s`x) \ \& \ (\mathbf{T}s`x)) \neq \Lambda \text{ and} \\ ((\mathbf{P}s`x) \ \& \ (\mathbf{T}^{-1}s`x)) \neq \Lambda) \text{ and } x\epsilon\text{—}\mathbf{mom}) \,.$$

3.22 could then be written in the form

$$x\epsilon\mathbf{be} \text{ implies } (\mathbf{S}` ( (\mathbf{P}s`x) \ \& \ (\mathbf{T}s`x))) \mathbf{T}x \,.$$

And the three postulates could be replaced by

'**cell** $\subset$ **be**'.

It will further be noticed that the fourth postulate containing '**cell**' (3.14) is also satisfied by other classes. We could, for example, perfectly well replace '**cell**' by 'nucleus' or 'chromosome' in 3.14. This postulate is, in fact, satisfied by any class of things which, as we often express it, reproduce by division.

Actually, 3.14 covers more than this because it also includes origin by fusion, a property more frequently and regularly exhibited by cells than by members of other classes. But, before we can generalize this postulate, certain complications must be considered. In the first place, although the four postulates here given suffice for the derivation of the consequences which follow them on the foregoing pages, they do not suffice to characterize fully the division and fusion relations. For example, if we have $x\mathbf{D}y$, there is always at least one z such that $x\mathbf{D}z$ and $z\epsilon$**cell** and $z \neq y$; and, if we have $x\mathbf{F}y$, there is at least one z such that $z\mathbf{F}y$ and $z\epsilon$**cell** and $z \neq x$. But these statements are not derivable from the four postulates here given for '**cell**'. Second, when division in the biological sense occurs, the thing which divides and the things into which it divides are, in an important sense, all things of the same sort; e.g., a cell divides into cells, a nucleus into nuclei, a chromosome into chromosomes, etc. The same thing is also seen in the division of whole organisms in the process of budding. Were this not so, we should not speak of *reproduction* by division. Corresponding remarks are true also of the process of fusion in the biological sense.

If, then, we wish to generalize postulate 3.14 and to define a more general concept of which '**cell**' is only a special case—in the way suggested above for the first three postulates containing '**cell**'—these two complications must be dealt with: the first by introducing further conditions, and the second by defining not a class of individuals but a class of classes. If we use '**df**' to denote this class of classes, the definition would be as follows:

$$\begin{aligned}
\alpha\epsilon\mathbf{df} \text{ equiv } & ((\alpha \subset \mathbf{be} \text{ and} \\
& (((x\epsilon\alpha \text{ and } y\epsilon\alpha) \text{ and } (x \neq y \text{ and } ((\mathbf{P}s\text{'}x)\ \& \\
& \quad (\mathbf{P}s\text{'}y)) \neq \Lambda)) \text{ implies } ((\mathbf{B}\text{'}x)\ \mathbf{P} \text{ or } \mathbf{P}^{-1}(\mathbf{E}\text{'}y) \text{ or} \\
& \quad (\mathbf{E}\text{'}x)\ \mathbf{P} \text{ or } \mathbf{P}^{-1}(\mathbf{B}\text{'}y)))) \text{ and} \\
& ((((x\epsilon\alpha \text{ and } y\epsilon\alpha) \text{ and } (\mathbf{B}\text{'}x)\ \mathbf{P}\ (\mathbf{E}\text{'}y)) \text{ implies} \\
& \quad (\text{Some } z)((z\epsilon\alpha \text{ and } z \neq x) \text{ and } (\mathbf{B}\text{'}z)\ \mathbf{P}\ (\mathbf{E}\text{'}y))) \text{ and} \\
& (((x\epsilon\alpha \text{ and } y\epsilon\alpha) \text{ and } (\mathbf{E}\text{'}x)\ \mathbf{P}\ (\mathbf{B}\text{'}y)) \text{ implies} \\
& \quad (\text{Some } z)((z\epsilon\alpha \text{ and } z \neq x) \text{ and } (\mathbf{E}\text{'}z)\ \mathbf{P}\ (\mathbf{B}\text{'}y))))).
\end{aligned}$$

Of the four conditions prescribed by this definition, the first corresponds to postulates 3.11, 3.12, and 3.13; the second corresponds to postulate 3.14; and the last two cover the first requirement explained in the preceding paragraph (with these additional conditions '**P**' may be used instead of '**Pp**' in stating the second condition). In the generalized theory the four postulates for '**cell**' would therefore be covered, and more than covered, by

$$\text{'}\mathbf{cell} \in \mathbf{df}\text{'} \, .$$

It is then quite easy to give definitions of relations of a more general character corresponding to the definitions of '**D**' and '**F**' which here denote relations holding only between cells. This procedure would also facilitate the introduction of new undefined signs denoting other classes which are also members of **df**. We should then have a general theory of things which are not momentary, have a beginning and an end in time, and exhibit the processes of division and fusion in the biological sense. Moreover, if it is only among organisms and their parts that we find classes of things which satisfy the definition above given of '**df**', then this might provide a basis for a definition of 'biology' itself, without the use of a single specifically biological concept.

The condition that a given member α of **df** agrees with the doctrine of abiogenesis (which is not provided for in the postulates of the theory T) could be formulated as follows:

$$\alpha \epsilon \mathbf{df} \text{ and } (x \epsilon \alpha \text{ implies (Some } y) \; (\; (y \epsilon \alpha \text{ and } y\mathbf{T}x) \text{ and } \\ (\; (\mathbf{P}s\text{'}x) \; \& \; (\mathbf{P}s\text{'}y) \;) \neq \Lambda) \;) \, .$$

It will be noticed that this process of generalization consists in essentials of introducing a variable in the place of the subject-matter constant '**cell**' and of giving the consequences of the postulates a hypothetical form. This process could obviously be extended to the remaining two subject-matter constants '**P**' and '**T**'. But, if this were done, the theory would no longer contain a subject-matter constant, and we should then no longer have a theory belonging to natural science but a purely abstract one belonging to logic or mathematics. This, then, illustrates the fact that we pass from natural science to logic and mathematics by replacing *all* the subject-matter constants in theories belonging to the former by variables of the appropriate types, and expressing the theorems as implications with the transformed postulates on the left and the transformed consequences on the right of the word 'implies' (as was done in the case of 3.22 in the example on p. 62). This is a process of *complete* generalization. On the other hand, when we *use* statements belonging to logic or mathematics (as we did in deriving consequences from the postulates of the theory *T*), we proceed by a reverse process of particularization—substituting subject-matter constants for certain of the variables which they contain.

All that remains of a theory after *complete* generalization is its *logical form*, and it is this logical form which determines the relation of premiss to consequence in a scientific theory, not the particular subject-matter constants which may be substituted for its variables. On the other hand, it is our knowledge of the subject matter which guides us in determining the logical form of the theory, because we choose our undefined signs, and frame our definitions, with a view to the precise denotation of the things about which we wish to speak, and we choose our postulates in such a way that their consequences will agree with our observations regarding the subject matter. The importance of logical form in the organization of scientific theories is obscured and easily overlooked when they are expressed in a natural language. By the use of a properly constructed scientific language it is clearly brought to light.

(ii) *Explanation of some technical terms.*—Before we proceed, a little must be said in explanation of four technical terms which have not been used in the foregoing pages.

By the *formalization* of a scientific theory is meant the process of constructing its metatheory, the most important tasks being, as we have seen, (i) enumerating and elucidating the undefined signs, (ii) establishing rules of statement construction, and (iii) establishing rules of statement transformation. A theory may be said to be completely formalized when its metatheory is completely and explicitly stated. A formalized theory has the advantage of possessing, in its syntactical rules, *objective criteria* for deciding when an expression is meaningful and when one statement is a consequence of others—questions which in a natural language sometimes give rise to disputes and misunderstandings because they are only decided by subjective considerations. Formalization has proved to be specially important for dealing with philosophical problems where linguistic pitfalls are liable to be numerous and difficult to discover. By this method, for example, A. Tarski has succeeded in formulating for the first time an adequate definition of truth (one of the chief problems of semantics) for certain classes of languages.

A theory is said to be *axiomatized* when it possesses a set of primitive or undefined concepts with the help of which all its remaining concepts can be defined, and a set of primitive statements or postulates from which all the remaining statements can be derived as consequences. This can be done, as it was done, for example, by Euclid, without formalization, leaving the consequence-relation to be determined by common-sense considerations which are never explicitly stated; or it can be done more precisely and objectively on the basis of a previous formalization.

The ability to enumerate such a set of primitive concepts, and to choose such a set of postulates, presupposes that the concepts and statements of the theory have been subjected to a prior process of analysis. This process (only the outcome of which appears explicitly in the theory itself) is here called *logical analysis*.

By *symbolization* is meant the use of single letters or other printed shapes or of small groups of such in the place of ordinary words. But it also includes the use of variables, an aspect of statement-construction which is unusual in natural languages and which has other merits than those resulting from mere abbreviation (see below).

It is plain that these four processes are to a large extent independent of one another and, in consequence, that the use of one does not necessarily involve the use of all the others. Our specimen theory T is an example of a theory axiomatized on the basis of a complete formalization and contains the outcome of the logical analysis of a number of scientific concepts as well as furnishing a basis for the logical analysis of others. The system (**P**, **T**, **U**, etc.) given in the author's *Axiomatic Method in Biology* is an example of a completely axiomatized and symbolized theory without formalization. The theory *T* is only partly symbolized. Symbolization is in no way theoretically essential, but its merits have long been recognized in mathematics and chemistry, and it is all but indispensable in practice in a theory in which calculation (i.e., the derivation of consequences by the application of rules of statement-transformation) is to be performed. The derivation of consequences in Part II of the theory *T* would have been intolerably long and extremely difficult to follow had we used no symbolization. They would have been still more abbreviated and correspondingly easy to survey had we used a complete symbolization.

The use of variables is a special instance of symbolization which deserves a little further consideration. It will be noticed that in the English translations of some of the statements of Part II of *T* we were able to avoid the use of variables without loss of precision. In other cases this was not done, because without taking advantage of this convenient device an accurate English translation would have been possible only at the cost of intolerable circumlocution and prolixity. In further illustration of these facts we may consider the following examples.

Suppose we wish to state that the relation of being earlier than or before in time (in the sense we have denoted by '**T**') is transitive, but without using the technical logical predicate 'transitive'. In a word language this can be done as follows:

(*a*) If any thing is earlier than another thing, and the latter is earlier than a third thing, then the first is also earlier than the third thing.

By the use of individual variables this becomes:

(*b*) For every x, y, and z, if x is earlier than y, and y is earlier than z, then x is earlier than z.

Thus the use of variables enables us to eliminate such words as 'another', 'the latter', 'the first', 'a third', etc., without ambiguity, and reduces the length of the statement by about one-half. The use of a single relation-sign '**T**' in the place of 'is earlier than' prepares the way for the use of the calculus of relations and reduces the statement to a single line:

(*c*) For every x, y, and z, if $x\mathbf{T}y$ and $y\mathbf{T}z$, then $x\mathbf{T}z$.

In the notation of the theory T this was expressed by

(*d*) (All x) ((All y) ((All z) (($x\mathbf{T}y$ and $y\mathbf{T}z$) implies $x\mathbf{T}z$))) .

In the notation of *Principia mathematica* 'All' is omitted from the quantifiers, and the logical constants 'and' and 'implies' are symbolized by '.' and '$\supset$', respectively, so that we reach a complete symbolization:

(*e*) $$(x, y, z) : x\mathbf{T}y \;.\; y\mathbf{T}z \;.\supset .\; x\mathbf{T}z \;.$$

(Here also some of the parentheses are replaced by dots.) Finally, we reach the highest degree of brevity by formulating the statement by means of '**T**' and signs belonging to the calculus of relations (see 0.541):

(*f*) $$\mathbf{T} \mid \mathbf{T} \subset \mathbf{T} \,.$$

This last formulation owes its brevity to the fact that, by the use of constants belonging to the calculus of relations, we are able to eliminate individual variables.

These examples, then, illustrate successive stages in symbolization, and the advantages of the use of variables even in a word language. In the theory T the reader will easily find examples of statements which would be extremely difficult to state precisely in a word language without the use of variables. The examples (*c*) to (*f*) also show the convenience of using single letters like '**T**' (or short combinations like '**SI**') for subject-matter relation-signs, on account both of their brevity and of the manner in which they prepare the way for the use of signs belonging to the calculus of relations. These devices might all be employed with advantage in natural science quite apart from formalization and axiomatization.

In the four processes above described—formalization, axiomatization, logical analysis, and symbolization—the investigations of these topics during the present century have provided natural science with a technique for dealing with all that concerns the formal or structural aspect of theory construction.

(iii) *Relations between Parts I and II of T.*—The chief difference between Parts I and II is that Part I contains only logical constants and variables, whereas Part II contains subject-matter constants in addition to these. It is further distinguished from Part I by containing no sentential variables. It is because it contains the subject-matter constants that Part II is a piece of natural science. In fact, from the standpoint of natural science the theory would seem to be contained wholly within the bounds of Part II. The example shows that the role of Part I is primarily to furnish a basis for the successive steps in the derivations of the statements called consequences from those called postulates in Part II. Ordinarily, when a scientific theory is stated, nothing corresponding to Part I is given. Nevertheless, such a Part I is implicitly contained in every scientific theory because it is tacitly assumed that the author and his readers will agree about the transformations of statements, without explicitly stating rules (in a syntax) or citing universally valid formulas (in a Part I) for performing them. Even where arithmetical or higher mathematical operations are involved it would obviously lead to needless repetition if every scientific theory which needed them were preceded by a Part I in which all the principles applied in its Part II were stated. It is rightly assumed that every serious reader either will already be familiar with them or will be able to find them in appropriate books. In the present instance it was necessary to give an explicitly formulated Part I in order to illustrate this point and because *T* was intended to furnish an example of a completely axiomatized theory and the logical formulas involved are not so universally understood among men of science as are the principles of arithmetic. Arithmetic is a very ancient science and has been taught in schools for centuries; the new technique of theory construction is as young as the art of flying and is not taught in schools at all.

(iv) *Summary of principles illustrated and merits of the technique.*—

1. Every scientific theory implicitly contains two parts—a Part I and a Part II—the former only being explicitly stated when it involves some special and unfamiliar branch of mathematics, or is stated only by references to books.

2. Part I in a scientific theory contains no subject-matter constants and is used for transforming statements belonging to Part II which do contain such constants into others also belonging to Part II.

3. If we suppose the Part I of each scientific theory to be so constructed as to contain the whole of logic and mathematics (e.g., by stating all their basic postulates, as was attempted, for example, in *Principia mathematica*), then it is clear that logic and mathematics can be regarded as *that which is common to all scientific theories* and contains no subject-matter constants.

4. Modern logic has now been so far developed that it enables us to mathematicize the whole of a scientific theory and not only the numerical or quantitative part. It furnishes us with calculuses which enable us to perform complicated transformations with precision upon statements which contain no signs belonging to traditional mathematics. (This is especially well illustrated by the theory of relations, although only the most elementary part of that theory has been used in the theory *T*.)

5. Two scientific theories are only distinguished from each other by the subject-matter constants which occur in them, and/or by the postulates in which those constants occur.

6. Subject-matter constants are either individual names (e.g., 'Venus' in astronomy, 'Paris' in geography), or class-signs like '**cell**', or relation-signs like '**P**' and '**T**' (although they need not always be two-termed relation-signs as these are).

7. Every formalized theory is accompanied by an explicitly formulated metatheory. If it were necessary to draw up a distinct set of syntactical rules for each scientific theory, formaliza-

tion would be a very complicated process. But, in fact, this is not necessary. We need only enumerate and explain the primitive subject-matter constants. The rules of statement construction and transformation and all that concerns variables and logical constants could be formulated quite generally and so be applicable to any theory.

8. A great simplification and unification would be achieved in natural scientific theory: (i) by the use of variables and especially by the use of a common system of notation by which to distinguish variables belonging to the different logical types; (ii) by the use of a common system of notation for the logical constants; and (iii) by the use of a common system of abbreviations for the subject-matter constants (class-signs being easily distinguished from relation-signs).

9. The reforms suggested in the last paragraph would provide not only a common basis for stating scientific theories but also a basis for stating them in such a form that the logical and mathematical calculuses could at once be applied to them.

10. Scientific methodology belongs to metatheory. Its business is to compare and criticize scientific theories both from the point of view of their internal structure and from that of their success in practical application. Formalization, or at least axiomatization, is in the long run an essential preliminary to the successful pursuit of methodology. Only when two theories are formalized or axiomatized can we properly compare them, because only then are the essentials upon which a comparison rests laid bare—only then do they possess a definite structure. The neglect of this precaution is chiefly responsible for the misunderstandings, confusion, and barrenness which frequently infect methodological discussions. Formalization also helps us to avoid confusion between questions of fact and questions of linguistic convention.

11. Although it is by intuition or common sense that we ordinarily think and discover new hypotheses, the use of a logical technique can help intuition in two ways: (i) by providing a check which enables us to determine whether a theorem arrived at by intuition is in fact a consequence of our assumptions or not

and (ii) by providing intuition with a guide rope in complicated and unfamiliar regions where, without some such aid, it could not penetrate or would easily go astray. Intuition and calculation are both necessary for science. Neither is infallible. Together they compensate for each other's defects.

12. As far as technical knowledge about theory construction is concerned, there seems to be no obstacle at the present day (as there was at the time of Leibniz) to the construction of a universal symbolic (or semi-symbolic) language (or languages) for scientific purposes. The use of such a language would bring with it all the advantages above enumerated. Owing to its simplicity, universality, conciseness, and structural affinity with the logical calculuses, the use and diffusion of such a language would make the results of one branch of science more easily comprehensible to workers in other branches than is the case at present, and so would overcome to some extent the consequences of specialization which are now aggravated by linguistic factors.

13. A wider diffusion of knowledge concerning the general linguistic principles discussed in this monograph could have effects of greater scope than the improvement of scientific theories. It could help to make people language-conscious and so to render them less vulnerable to the attacks of propaganda.

(v) *Requirements and directions for further progress.*—The degree to which a given science might profit by the use of such a technique as is here outlined will depend upon the degree to which theories have been developed in it by intuitive methods. But, supposing a suitable subject to be chosen, two things are necessary in order to axiomatize a theory: a fund of the relevant scientific knowledge and an understanding of the technique of theorizing. Important results with this method have so far been reached only in logic and mathematics. This is not because *only* such subjects can be treated in this way (our example shows that this is not the case) but because it has happened that the people who have understood the technique of theorizing have also been mathematicians or logicians or both. What is needed, therefore, is a wider distribution of a knowledge of this technique among men of science, or a greater degree of collaboration between logicians armed with the new knowledge and those engaged in constructing scientific theories.

There are two directions in which this technique can be exercised: (1) First, we can begin with a few very general concepts and construct systems embodying them, as has now been done to such a great extent in logic and mathematics. This is the procedure exemplified also by the specimen theory T in a non-mathematical sphere. Such a system can then be enlarged by introducing new undefined concepts one at a time, determining their postulates, connecting them with those already introduced, deriving consequences, and defining as many other concepts as possible with their help. In other words, in this direction we begin with what are taken, provisionally at least, for foundations in order to discover how much can be done with them. (2) Second, we can take some special and restricted theory, near the growing-point of some branch of science, and axiomatize it in the hope that by so doing we shall lay bare its logical form, see its relation to other theories, and find ways of further extending or otherwise improving it. When a number of such studies have been completed for related theories, they can gradually be united into more and more comprehensive systems.

Only when much more work along these lines has been done will it be possible to form a rational estimate of the contribution which this new technique of theory construction is likely to make toward the further development of scientific theory. At the dawn of mathematical history it would have been impossible to foresee what part it was destined to play in natural science. Today it is futile to dogmatize about the possibilities of applied logic. What it can do can only be discovered by trial.

VII. Toward the Removal of Misunderstandings

In view of the persistence of misunderstandings regarding the relation of logic to natural science (as revealed, for example, by remarks in reviews of the author's *Axiomatic Method in Biology*), it seems to be desirable to devote a few lines to an attempt to answer specific objections which have been raised against the use of axiomatization in the natural sciences. Anyone who has attentively studied the example furnished by the theory T will have no difficulty in meeting such objections, but the following additional remarks may be useful.

Speaking generally, the misunderstandings in question seem to be persistent remnants of the Baconian revolt against scholasticism. This revolt seems (as is often the way with revolts) to have swung from one extreme to another—from too much occupation with formalism to a failure to understand at all the place of logical form in scientific theory. The whole history of the part played by mathematics in physical theory, and the corresponding backwardness of those sciences which have not been able to utilize mathematics, provides a warning commentary on such an attitude. Is not this ancient controversy by now a dead issue? Should we not turn from it to equip ourselves with a knowledge of what is meant by logic and metalogic *at the present day* and with some understanding of the discoveries in these subjects since the time of Boole in the last century? So much for misunderstandings in general. A few specific examples will now be considered.

In the first place it seems to be felt in some quarters that the deliberate use of a technique of theorizing involves (in the case of biology) "fitting the facts of life" into some rigid predetermined scheme. Nothing could be farther from the truth. Far from making facts conform to a scheme (which in any case would be impossible), we deliberately *construct* the theoretical system in such a way that it will as faithfully represent the facts as possible. It is true that such schemes cannot profess to be exhaustive of the "riches of nature". But surely no serious student of science ever expects them to be. Moreover, what is true in this respect of theories constructed according to an explicit technique is no less true of those expressed in a natural language. Indeed the more clearly, explicitly, and unequivocally a theory says what it does say, the more easily any errors or inadequacies it may contain will be detected and corrected. Not only may scientific theories be very abstract in the sense of only covering a selected set of facts; they may even be realized only by "ideal" objects (cf. physics). Nevertheless, if such theories cover "real" objects to a sufficiently close degree of approximation, they may still be useful in practice.

Second, it has been said that "we cannot get any new information about the cell by considering our definition of it" (the suggestion being that in mathematical sciences, e.g., in geometry, new information can be obtained in this way). But the role of definitions is not primarily to provide new information. They might rather be said to be the *receptacles* of present information with the aid of which such information is tested. As we have seen, definitions play the part of postulates, and if the consequences of a postulate, when confronted by observations, are falsified, the postulate must be altered until its consequences conform with what is observed. Only in this indirect way could a definition be expected to lead to new discoveries, although the working-out of its consequences might reveal properties of the thing defined which were not anticipated when the definition was first formulated. In any case no such thing as our definition of the cell exists at present. If it did, biological science would be very much farther advanced than it is. To frame an *adequate* definition of 'cell' would require far more knowledge and theoretical insight than we possess at present or are likely to possess for many years to come.

Third, it has been argued that logic and mathematics are deductive sciences, that biology is an inductive science, and that therefore any attempt to use the axiomatic method (or method of postulates) in biology is a perverse procedure destined to failure by the very nature of biological science, since the axiomatic method is only appropriate to deductive sciences. This argument is convincing only so long as we neglect to examine its premisses, and it falls to pieces after a little reflection in the light of modern knowledge.

In the first place, our theory T has shown that the axiomatic method is applicable in its full rigor to a topic containing physical and biological concepts but (in the extent to which it is here developed) using neither metrical nor arithmetical ones. In the second place, there is no sharp distinction in actual fact between deductive and inductive sciences. The so-called deductive sciences have been constructed in the first instance in the particular way in which they have been constructed precisely in order that they may have the applications which they do have. The postulates of Euclidean geometry, for example, have been chosen so that their consequences agree with our spatial intuitions and can be used as guides in the operations of determining heights, distances, areas, etc. With this starting-point other choices of postulates have suggested themselves, having different consequences and so generating new systems with either no application at present or more sophisticated ones than those of Euclidean geometry. On the other hand, the so-called inductive sciences all use deductive procedures, although with varying degrees of explicitness and extensiveness. Roughly speaking, the inductive procedure consists in making a guess (working hypothesis), deriving its consequences, and then testing these against observation. The derivation of the consequences of hypotheses may involve some existing branch of mathematics, as in physics, or it may require nothing more than the simpler transformations which are the stock-in-trade of everyday life, as is usually the case in the biological sciences. In both cases, whether we give it that name or not, we are using a deductive procedure. The more extensive cultivation of the axiomatic method may some day help us to do this in situations where existing branches of mathematics and the resources of everyday language fail us.

In conclusion we may say that a great part of scientific activity consists in accurately formulating reliable generalizations, i.e., statements which contain at least one free variable or at least one universal quantifier, and which have continued, for a sufficiently long time during which their consequences have been adequately tested, not to be falsified by observations. The question of the use of a technique of theory construction arises as soon as a number of such generalizations have been established and it becomes necessary to *order* them in a system. There is presumably no question that ordering them in some way is desirable. A railway timetable cut up into strips which are shaken up and thoroughly mixed together in a bag is much less useful than one in the more usual book form, if only because the information it contains is so much more easily accessible in the latter form than in the former. At the same time (provided the cutting-up has been carefully done) this information is still the same in the book as it is in the bag. The mere ordering of statements does not of itself create new information. Granting, then, that our generalizations are to be ordered when we have accumulated a number of them, the next question is: What *ordering relation* are we to use for attaining this end?

When we adopt the so-called deductive method, we use the *consequence-relation* as our ordering relation among the statements of our theory (cf. Fig. 1, p. 61). In the case of the specimen theory T this relation was explicitly established by the syntactical rules 4.1 to 4.5. It has already been pointed out that this consequence-relation is used to some extent even in the elementary stages of the so-called inductive sciences. All that we do when we use the axiomatic method is to use this relation explicitly and systematically. Even so it is usually employed intuitively without the formulation of transformation rules. This is the case even in ordinary mathematics. The final step, which we have called formalization, and which involves such formulations, is a recent invention which need only be resorted to for dealing with special problems. We adopted it in the theory T not so much because it was necessary but in order to illustrate the procedure fully. Nevertheless, there seems to be no reason why a formalized language should not be constructed for general scientific use to which the special signs for the various branches of science could be added as required.

Judging by the example of the more theoretically advanced sciences, and by the general principles adduced here, there seems every reason to hope for good results from an extensive use of the axiomatic method and symbolic languages in the biological sciences. But, again judging by the history of physics, important results are not to be expected in a short time, nor without the co-operative efforts of many workers. Here the words of Ernst Schröder may be recalled: ‘ Freilich darf man die Ernte nicht schon während der Aussaat fordern, und am wenigsten da, wo Bäume gepflanzt werden."

VIII. Epilogue

Toward the end of his life, after all his greater works had been published, Francis Bacon wrote a little fantasy called the *New Atlantis*. It is given the form of a narrative told by a returned traveler describing a visit to a lost island in the Pacific Ocean. Bacon uses this narrative in order to depict a future state of scientific development. The central feature of the community described was Salomon's House, which has since been regarded as an anticipation of the Royal Society. Two features of this institution are especially interesting to the modern reader. First, it is clear from the description that Bacon anticipated an equal development of the biological with the physical sciences. But, whereas the achievements of physical science at the present day are far in advance of anything that Bacon foresaw, those of biology are as far behind. We read, for example, of achievements in what we should now call genetics, ecology, and physiological morphology which are very much in advance of anything to which those sciences have attained at the present day. It seems clear from this that Bacon did not correctly anticipate the *order* in which the sciences were destined to develop.

Second, the modern reader is struck by the contrast between the simplicity of the theoretical equipment which accompanied all this practical achievement and the degree of mathematical complexity which has been demanded by the actual development of physics, no hint of which is to be found in Bacon. Bacon describes what he calls the employments and offices of the fellows of Salomon's House in some detail. First, there were twelve Merchants of Light who visited foreign countries secretly and brought back books, abstracts, and "patterns of experiments" from them. Three Depredators collected experiments from books. Three Mystery Men collected experiments used in the practice of trades and mechanical arts. It was the duty of three Pioneers or Miners to "try new experiments, such as themselves think good". The results obtained by the foregoing were tabulated by three Compilers, and the possibilities of making practical applications of them were studied by three Dowry Men or Benefactors. To three Lamps was given the task of considering the labors of their colleagues with a view to directing new experiments "of a higher light, more penetrating into Nature than the former". The experiments so suggested were performed and reported upon by three Inoculators. Finally, there were three Interpreters of Nature whose business it was to "raise the former discoveries by experiments into greater observations, axioms and aphorisms".

In the light of what has happened during the three hundred years since the publication of the *New Atlantis* I believe a modern visitor to Salomon's House would find considerable changes in this organization. He would find that the Merchants of Light and the Dowry Men are still used, although in greater numbers. But the work of Depredators and Mystery Men has long been completed or superseded. Pioneers are no longer distinguished from Inoculators. The chief theoretical work is done, as before, by the Lamps and the Interpreters of Nature, although no distinction is now made between them, and very many more than six are needed. Every actively growing subject, however small, has at least one Lamp. His duty is, on the basis of a thorough knowledge of the topic for which he has chosen to be responsible, *to conceive new hypotheses.* When he has discovered a hypothesis which he considers worthy of testing, the Lamp writes it down on a card in the universal symbolic notation which has long been adopted and sends it to a Calculator—a category of worker not provided for by Bacon. The Calculator operates a gigantic calculating machine, like our machines but of very much wider scope, being capable of working out the consequences of any hypothesis which can be formulated in the universal notation. On receipt of a new hypothesis, the Calculator sets up his machine in accordance with the present state of the theory in the branch of science concerned, but incorporating the new hypothesis. He then sets the machine in motion. The consequences of the new hypothesis (in conjunction with the remaining postulates of the theory) are then shot out of the machine printed on cards. These cards are handed at once to another set of workers called Sorters, who act as apprentices to the Lamps and form a reserve from which new Lamps are recruited from time to time. Their duty is to cast out those cards which bear trivial or already well-known consequences and to keep only those which are new. The latter are then stamped by the Sorter with the number of the Lamp to which they belong and the number of the laboratory in which they are to be tested, and dispatched to the appropriate Inoculator. On receiving such a card, the Inoculator sets up the necessary apparatus, performs the experiment, and writes the result on the card, which is then returned to the Lamp whose number it bears. When the Lamp has received all the results he requires, he holds a meeting with his Sorters at which these results are discussed and further changes in the theory are proposed and considered. The holding of such meetings is advertised, and other Lamps and their Sorters are permitted to attend. (Such meetings are also open to Inoculators, but these rarely attend unless they wish to become Sorters.)

The duty of the Compilers is to maintain a card index of the postulates, and another of the fundamental concepts, of all scientific theories in accordance with the reports published from time to time by the Lamps. To the Compilers is also assigned the important task of studying the mutual relations between these theories and bringing the results of such studies to the notice of the Lamps concerned.

Little importance is attached to divisions between the sciences. The differences between them are not artificially exaggerated by notational peculiarities. The workers of Salomon's House, having long ago appreciated the importance of logical form, have so constructed their universal notation that the purely mechanical part of their theoretical work can be done by machines and the theoretical workers are set free to devote their energies to the important task of devising new hypotheses. Most honored among them are those whose minds are most fertile in this direction. In this way they have long ago achieved a real unification of science.

Selected Bibliography

The beginner is recommended to read first Tarski's *Einführung in die mathematische Logik* (Wien, 1937), which provides a good introduction to general principles without symbolism. As a good introduction to the theory of classes and relations and the notation of *Principia mathematica*, with numerous suggestions for applications, Carnap's *Abriss der Logistik* (Wien, 1929) is recommended. A good account of the calculus of statements will be found in Hilbert and Ackermann's *Grundzüge der theoretischen Logik* (2d ed.; Berlin, 1938). The student can then read the numbered sections of Volume I, and those dealing with series in Volume II, of Whitehead and Russell's *Principia mathematica* (Cambridge, 1925). Examples of applications of the new logical ideas will be found in the book by Hempel and Oppenheim, *Der Typusbegriff im Lichte der neuen Logik* (Leiden, 1936), dealing with the classification of psychological types, and in the article by Grelling and Oppenheim, "Der Gestaltbegriff im Lichte der neuen Logik," in *Erkenntnis*, Volume VII, Heft 3 (1938). Suggestions for applications to genetics and embryology will be found in the author's *Axiomatic Method in Biology* (Cambridge, 1937) and for applications to physiological and psychological theories in a brief article by the author in the above-mentioned volume of *Erkenntnis*. The reader should also consult Volume I, No. 2 (*Foundations of the Theory of Signs*) and No. 3 (*Foundations of Logic and Mathematics*) of this *Encyclopedia*.

Methodology of Mathematical Economics and Econometrics

Gerhard Tintner

Methodology of Mathematical Economics and Econometrics

Contents

Methodology of Mathematical Economics and Econometrics

Gerhard Tintner

Preface

In this essay I have attempted to present some methodological problems of mathematical economics and econometrics, and also of operations research (econometrics of enterprise), to a public which does not necessarily consist of economists. Hence, it has been impossible to treat the whole field, and the selection of topics may be found to be one-sided.

Mathematical economics is the use of mathematics in the construction of economic models. Econometrics may be defined as the utilization of mathematics, economics, and statistics in an effort to evaluate economic models empirically with the help of concrete data and to investigate the empirical support of certain economic theories.

I also have to apologize for sometimes presenting the results of my own research. My excuse is that I am, of course, more familiar with these particular problems and hence able to discuss them more adequately than the results established by others. I concede, however, that a diligent search might have discovered more suitable examples. But the literature is so extensive and has become so specialized that more time and effort would have been expended than I felt I could spare.

It seems to me that it would be fruitless to engage in long philosophical discussions of some of the methodological issues which have bedeviled the enormous literature in this field. Instead, I have tried to present various views on those questions together with some simple examples—a method which should enable the reader to form his own opinions on such questions as the suitability of mathematics, the possibility of measurement in economics, the various views on probability, some aspects of economic prediction, and so on. If the reader desires more information, he is referred to the literature cited.

All these questions, in my opinion, are far from settled, and this essay might be considered a progress report which endeavors to present the contemporary status of some of the more important methodological problems. I hope that it will stimulate interest among those who are not economists, especially mathematicians and natural scientists, in the problem of modern mathematical economics and econometrics. A valuable survey of almost the entire field of mathematical economics may be found in the two books of R. G. D. Allen (1949, 1963). For a simple introduction to the mathematical and statistical methods used, I might refer the reader to my own books (Tintner 1952*b*, 1960*c*, 1962) and to Kemeny, Snell, and Thompson (1957), which also contain a great many practical applications of econometric methods to concrete economic data.

I am much obliged to the National Science Foundation, Washington, D.C., for financial support. I also want to thank Aurelius Morgner, Department of Economics, University of Southern California, for bibliographical help.

I thank Rudolf Carnap and Charles Morris for helpful advice and criticism of a first version of this work. Also, I would like to thank a number of my friends and colleagues for critical remarks: M. Bronfenbrenner (Pittsburgh), W. W. Cooper (Pittsburgh), C. R. Fayette (Lyons), E. Fels (Munich), N. Georgescu-Roegen (Nashville, Tennessee), T. Haavelmo (Oslo), B. Higgins (Austin, Texas), C. Humphrey (Los Angeles), T. W. Hutchison (Birmingham, England), K. Kimura (Nagoya), F. Machlup (Princeton), C. McConnell (Los Angeles), R. N. Narayanan (New Haven), J. B. Nugent (Los Angeles), J. Pfanzagl (Cologne), H. Wold (Uppsala). I am obliged to R. Watson (Los Angeles) for computational help.

G. T.

LOS ANGELES

I. Introduction

This monograph deals with some of the fundamental methodological problems of mathematical economics and econometrics. We start with a definition of economics. But the word of Cairnes (1875, p. 148), one of the outstanding older writers on economic methodology, should be remembered: "Definitions in the present state of economic science should be regarded as provisional only, and may be expected to need constant revision and modification with the progress of economic knowledge." Economics has changed much in the past, and continues to change very fast. Hence, the words of Cairnes are as true now as when they were written.

Economics has been defined as the science which studies human behavior as a relationship between ends and scarce means which have alternative uses (Robbins 1949, p. 16) or, on the other hand, as the science of administration of scarce resources in human society (Lange, 1953). A more recent definition by the same author (Lange 1963, p. 1) reads: "Political economy, or social economy, is the study of social laws governing the production and distribution of the material means of satisfying human needs."

Economics has a long history, dating back to Plato and Aristotle (Schumpeter 1954) or even earlier (Bernadelli 1961). It would be useless to deny that the bulk of the results of theoretical economics has been achieved without mathematical means. Mathematical economics, which also has a long history, has only recently come into prominence.

In theoretical economics we construct fundamental models which we try to apply to concrete economic problems. The necessity of economic theory, which was denied by the historical school in Germany and by the institutionalist school in the United States, is now almost universally recognized. This, however, is not true in regard to the use of the method of mathematics in economics (L. von Mises 1949, pp. 347 ff., 697 ff.,

706 ff.; see also Georgescu-Roegen 1966, p. 49). The repudiation of the anti-theoretical schools of German historicism and American institutionalism should, however, not imply that the study of economic history and of economic institutions is useless and should be neglected. Also, it need not be claimed that mathematical economics and econometrics are the *only* methods for the study of problems in economics (Tintner 1952, p. 13).

No one can deny that economics holds a special position among the social sciences. It has perhaps reached a degree of scientific maturity that is still lacking in many of the other social sciences. Nevertheless, the scientific achievement of economics is not yet comparable to that of modern physics or genetics. Georgescu-Roegen (1965) attributes this limited success of economics to the nature of many economic models, which are constructed after the example of classical physics. The implied conditions of measurability and linearity are not always fulfilled. There is much to this criticism, and we shall come back to some of these problems. It remains to be seen whether stochastic models of the economy will be more successful.

Some of the fundamental difficulties of modern mathematical economics have also been pointed out by Georgescu-Roegen (1966, pp. 49 ff.). In economics we frequently, perhaps always, deal with phenomena which are qualitative and which we can quantify only imperfectly. The qualitative residual shows itself then in non-linearities of the relations between quantified phenomena.

I believe, however, that some (but not all) of our difficulties in mathematical economics (especially in utility theory and the theory of choice) are of our own making. Modern physics has shown that the number of ultimate particles (whatever these may be) is large but finite. Similarly, whereas subjective time may be continuous, any measured clock time is also discontinuous. Many of the difficulties connected with set and measure theory disappear when we deal with finite aggregates. Hence we should properly use only difference equations, and those we use in mathematical economics and elsewhere are convenient idealizations. (Kemeny, Snell, and Thompson 1957).

First, let us examine a spirited criticism of modern economics which emphasizes some of its difficulties. Schoeffler (1955, pp. 17 ff.) in a critical study points to many weaknesses existing in contemporary economics. Mechanical behavior models, like the theory of the firm, may be called artificial mechanization. Batteries of homogeneous active agents – for example, a competitive industry – may lead to artificial simplification. The empirical behavior equation – for example, the consumption function (consumption explained by income) – is an artificial generalization. Classification and description by types – for example, classification of firms into industries – may be called artificial systematization. Mapping of structural elements of the world – for example, production surfaces (production explained in terms of various factors of production, such as capital and labor) – may be artificial fixation. Times series analysis, like analysis into trend, business cycle, and seasonal components, may imply artificial factorization. Endogenous models – for example, certain business cycle models which explain the cycle without assuming outside influences – may lead to artificial closure. Semi-endogenous models, which also include exogenous variables – those which influence the system but are not influenced by it – may encompass artificial semiclosure. Partial consideration of variables – the tendency to consider only economic or perhaps closely related effects of policy measures – may be artificial isolation. Assumptions verified by their logical consequences – for example, when the economist tests his hypotheses not directly by opinion surveys but with the help of available statistical series – may lead to artificial indirectness.

I believe that few economists would deny the cogency of much of this criticism. It shows, however, only that, considerable as the achievements of contemporary economics are, they are still far from complete.

The success of econometrics depends, of course, upon the availability of good data (Morgenstern 1963), but our models also need improvements. Until the theoretical problem of oligopoly (a few sellers in a market with many consumers) is solved, there is little hope that we can achieve somewhat realistic results in a country like the United States, where oligopoly in many important markets is the prevailing form of organization. Oligopoly is a situation different from free competition (many sellers and buyers) and monopoly (one seller and many buyers).

Since there are few sellers and many buyers in an oligopolistic market (Baran and Sweezy, 1966; Shonfield 1965), even the economic models of statics are seen to be deficient. Economic statics is defined by Hicks (1946, p. 115):

> I call economic statics those parts of economic theory where we do not trouble about dating; economic dynamics those parts where every quantity must be dated. For example, in economic statics we think of an entrepreneur employing such-and-such quantities of factors and producing with their aid such-and-such quantities of products, but we do not ask when the factors are employed and when the products come to be ready.

If time is introduced, the difficulties multiply. An interesting discussion of the semantic confusion in economics is that of Machlup (1963). Papandreou (1958) investigates the structure of economics from the point view of modern mathematical logic and semantics. His analysis stresses the theoretical shortcomings of many economic theories and models.

The progress of economics has been slow because of the influence of ideological bias and some ancient metaphysical ideas which have long been discarded in the more mature natural sciences. Ideology is well defined by Robinson (1962, p. 8): "What then are the criteria of an ideological proposition opposed to a scientific one? First, that if an ideological proposition is treated in a logical manner, it either dissolves into a completely meaningless noise or turns out to be a circular argument." (See also Topitsch 1958, 1961.) Perhaps it should be recognized that "ideology" is a very ill-defined and much abused word of uncertain meanings (Quine 1960).

It is certainly understandable (even if deplorable) that in the past and in the present, on the left and on the right, economists have been very much influenced by the ideological struggles of their time and have sometimes illegitimately presented value judgments as scientic truth. This is perhaps unavoidable. Nobody will be concerned with economics who is not vitally interested in social matters and is not looking for solutions of burning social problems. Until recently, with few exceptions, the intellectual level of discussion of economic matters and the standards in this field were not high. More gifted intellectuals were (and perhaps still are) attracted to the natural sciences and mathematics. These fields, it must be confessed, are certainly intellectually much more attractive, and the danger of purely ideological influences is perhaps smaller there.

Topitsch (1958, pp. 235 ff.; 1961) has emphasized the importance of the idea of natural right in classical political economy (Adam Smith and his school). The same is also true of Schumpeter (1954, pp. 122 ff.). It cannot be denied that traces of this empty idea are still to be found in contemporary economic writings. Topitsch shows how since antiquity social phenomena and organizations (e.g., government, the state) have been used to explain natural phenomena. They are interpreted as "natural laws," and the natural laws are then used in turn to explain other social phenomena. In the course of history they have become empty formulas and have been used to justify conservative, liberal, and socialist ideas.

Both the Physiocrats and the classical British school of economists base themselves largely upon this empty concept of natural law. Hence early economic writings are almost completely normative. In spite of this ideological orientation, there are of course important scientific contributions made by the physiocratic and classical school of economics. This ideological influence also explains in part the famous "invisible hand" of Adam Smith, which makes selfish ends promote the common good (Viner 1927; see also Wicksell 1934, pp. 72 ff.) We should not forget, however, as has been emphasized by Robbins (1961), that the English classical economists were not *uncritical* advocates of laissez-faire. This is true even of Hume, Adam Smith, Bentham, Malthus, Ricardo, and of the writings of J. S. Mill, which constitute in a sense the great synthesis of the classical system. In disagreement with Schumpeter (1954), who stressed the novelty and undeniable contributions of the marginal revolution, especially the Austrian school, I would like to emphasize the continuity of the classical tradition in economics, stressed by Marshall (1948). It is particularly evident in the development of economics after the death of Schumpeter. Economists who deal with the problems of economic development are rediscovering many ideas of the English classical school of economics, especially those of Malthus and Ricardo. (Baumol 1959). Hence, perhaps it is also not surprising that economists working in the field of economic development make use of Marxist models (Morishima 1954, Robinson 1949). It has also been pointed out recently by Chipman (1965) that the English classical writers anticipated many of the results which were later established by modern mathematical economists with the help of advanced mathematical methods.

In the social sciences, just as much as in the natural sciences, we must distinguish between the often ideological and metaphysical motivations of a given thinker and his real scientific contributions. The scientific contributions of Newton do not depend fundamentally upon his peculiar theology and metaphysics, even if they are perhaps partly motivated by these ideas. But it must be confessed that the task of separating ideology and scientific contribution of many of the most famous economists is much more difficult. It is true that economics, like all social sciences, deals with human action. This fact, however, should not prevent us from trying to apply the fundamental scientific methods used elsewhere. The difficulty should not be denied, as has been recently emphasized by Georgescu-Roegen (1966, pp. 3–132). But why has economics not yet been more successful in these endeavors? One source of our difficulty, emphasized by Morgenstern (1963), is the deficiency in much of economic statistics. But the strictly theoretical difficulties are also very great. We find in recent economic literature an undue emphasis on purely static models of a competitive economy. These static models are such that the quantities involved are undated. Free competition prevails in markets with many buyers and sellers. Static competitive models are plainly insufficient for application to economic reality (which is essentially non-static and also involves many elements of monopoly, oligopoly, and so on).

The work on dynamic models which has flourished in recent years (Morishima 1964) shows that the fundamental theoretical difficulties connected with economic development have not been mastered. It is doubtful if the application of the theory of stochastic processes, which has been advocated, will be more successful (Sengupta and Tintner 1963, 1964; Mukherjee *et al.* 1964; Tintner and Patel 1966).

One weakness of present-day economics, which has been emphasized by Georgescu-Roegen (1960), consists in the fact that almost all modern economics is strongly related to the analysis of the competitive phase of modern capitalism. This prevents economists from coming to grips with pre-capitalist structures, which are still important in underdeveloped countries (T. W. Schultz 1964; Georgescu-Roegen 1960). But also, in spite of some progress in the analysis of non-competitive capitalist structures (J. Robinson 1938; E. H. Chamberlin 1948; Fellner 1959), these ideas have never been really well integrated into a comprehensive system of economic equilibrium. Bowley (1924) is almost the only modern writer who has at least tried to include monopoly (one seller, many buyers) and bilateral monopoly (one seller, one buyer) into a system of general economic equilibrium. The outstanding problem in mature capitalist economies is, however, really oligopoly (few sellers, many buyers) and similar structures (Shubik 1959; Baran and Sweezy 1966). Another difficulty with economics is what Schumpeter (1954) has called the "Ricardian vice." He says of Ricardo (pp. 472–73):

> His interest was in the clear-cut result of direct, practical significance. In order to get this he cut this general system to pieces, bundled up as large parts as possible, and put them in cold storage – so that as many things as possible should be frozen and "given." He then piled one simplifying assumption upon another, until, having really settled everything by these assumptions, he was left with only a few aggregative variables between which, given these assumptions, he set up simple, one-way relations so that, in the end, the desired results emerged almost as tautologies.

This Ricardian vice was, unfortunately, not confined to Ricardo and his immediate disciples. Schumpeter (p. 1171) attributes it with some justice even to Keynes.

The case of Marx is extremely interesting. Topitsch (1961, p. 252) convincingly shows the way in which ancient gnostic and even cabalistic speculations were transmitted to Hegel and from Hegel to Marx. The main idea is the concept of dialectics (Popper 1963*b*) with its scheme of thesis, antithesis, and synthesis, which again in its long historical career became a perfectly empty formula, useful for any purpose. This method of analysis, used by Hegel for the justification of the existing Prussian state (Popper 1957), was used by the Hegelian Marx for the purposes of the revolutionary socialist movement. Again, in spite of the fact that a large portion of the thought of Marx is ideological (or at least motivated by ideology), he is by no means a negligible figure in the history of economics, as Schumpeter (1951*a*) has shown (see Sievers 1962). Recently Georgescu-Roegen (1966, pp. 3–132) has argued that "dialectical" concepts are useful and may even be indispensable in economics. There can be no doubt that he is right in stating that many concepts and ideas in economics, especially in economic policy, are imprecise and vague, in some sense pre-scientific. But since the concrete problems are pressing, we must by necessity use imperfect and not well-delimitated concepts in our discussion. The introduction of dialectial concepts should be welcomed as perhaps the beginning of a serious discussion between Marxist and "standard" (modern non-Marxist) economists (see also Bronfenbrenner 1965).

Whatever his ideological and metaphysical motivations, in economics Marx appears as one of the most important members of the British classical school of economics, extending from Adam Smith to John Stuart Mill (Schumpeter 1951*a*). His thinking must be understood in terms of the fundamental conceptions of this school, which, in spite of progress in many fields, are still largely the fundamental ideas of modern economics (Lange 1963). It is no wonder that Marx, as an economic thinker, now enjoys a certain renaissance and that as conservative an economist as Schumpeter is deeply indebted to Marx. In recent years, in view of the understandable concern of contemporary economists over problems of economic development, one of the most important subjects of the classics — "magnificent dynamics" (Baumol 1959; Hicks 1965) — has brought back certain Marxist concepts which may prove fruitful in our endeavor to understand the process of economic development (J. Robinson 1949, 1962; Morishima 1964; Bronfenbrenner 1965).

Concerning ideological bias, we might well agree with Schumpeter (1951*b*, pp. 280–81):

> There is little comfort in postulating, as has been done sometimes, the existence of detached minds that are immune to ideological bias and ex hypothesis able to overcome it. . . . There is more comfort in the observation that no economic ideology lasts forever and that, with a likelihood that approximates certainly we eventually grow out of each . . . But this still leaves us with the result, some ideology will always be with us, and so, I feel convinced, it will.

That ideological influences are recognized even among Marxists is shown by Lange (1963, pp. 338–9): "At one period in the building of socialism political economy was fettered by dogmatism and by a tendency to transform science into apologetics. This was connected with Stalin's system of the 'cult of personality.'" In an extremely interesting book (Baran and Sweezy 1966), two American Marxists have given a penetrating (but one-sided) analysis of the economy of the United States. This book is especially remarkable for discarding many preconceptions of Marx (and Lenin) and using some ideas and techniques of 'modern' or 'standard' (non-Marxist) economics. May we hope that this example will be followed by Russian, Chinese, and other economists in Communist countries? May we also hope that at least some economists in the non-Communist world will follow this example in the opposite direction and try to learn something from Marxist economics and the experience of planning in the Communist world?

We cannot do better than quote one of the outstanding Marxist economists of our time on the subject of the objectivity of economics:

> Our conclusion about the objectivity of economic science may seem startling. Economists are rather notorious for being unable to reach agreement and for being divided into opposing "schools of thought", "orthodox, and unorthodox," "bourgeois" and "socialist," and many others. The existence of profound disagreement among economists, however, does not refute our thesis about the objectivity of economics as a science. The disagreements can all be traced to one or more of the following sources:
> 1) Disagreements about social objectives. This is the most frequent source of disagreement, but acts as such only as long as it is implicit and unrecognized. If the social objectives are stated explicitely, the disagreement disappears. For any given set of social objectives and with given assumptions as to empirical conditions, conclusions are drawn with interpersonal validity by the rules of logic and verification. (2) Disagreements about facts. Such disagreements can always be resolved by further observation and study of empirical material. Frequently, however, the empirical data necessary to resolve the disagreement are unavailable. In such cases the issue remains unsettled. The conclusion that the issue cannot be settled with the data available has interpersonal validity. Agreement is reached to withhold judgment. (3) Failure to abide by the rules of logic, of identification and verification. The disagreement can be removed by correct application of the rules. (Lange 1953, p. 749.)

In order to bridge the gap between theoretical concepts and empirical observations, it is necessary to have a procedure of identification, which contains rules establishing the correspondence between the two. This concept of identification introduced by Lange has nothing to do with identification in econometrics discussed below.

There should be no doubt of the fundamental methodological unity of the social and natural sciences:

> There are no other methods or aims in the social and cultural sciences than exist in the natural sciences: observation, description, measurement, statistics, the discovery of explanatory laws and theories – more difficult of achievement in the former than in the latter – are the basic procedures. The role of sympathetic "understanding" or "empathy" as a practical guide is certainly not to be minimized, but its results, if they are to be scientifically valid, are subject to the very same objective tests as are the results of inorganic science. . . . To what extent sociology, economics or history are capable of discovering reliable laws on some

level of concept formation is an empirical question and therefore cannot be decided *a priori* on logical grounds. (Feigl 1949, p. 22).

As Carnap (1938) has pointed out, the procedures of social and hence economics are fundamentally the same as in the natural sciences (see also Morris 1938, 1946).

Prices, and quantities sold, interest rates, and the like are all quantitative concepts. Prices are measured in monetary units — for example, dollars. Quantities are measured in pounds, kilograms, or number of items, and so on. Interest rates are given in per cent. For investigations of a whole economy, it is frequently necessary to construct index numbers — for example, a cost-of-living index, various price indexes which represent prices in sectors of the total economy (for example, an index of prices of producers' goods). These problems involve the difficulties of aggregation, which will be discussed below. The economic magnitudes can be observed, and indeed it is the task of economic statistics to give the economist quantitative information about them. Censuses give more or less complete information and sample surveys give information derived by the methods of modern sampling.

But the psychological dispositions of consumers, entrepreneurs, and so on are also of great importance. Consider a farmer who produces a commodity which has a definite period of production. Then his rational action at the time of starting the production process will depend not on the existing price of the commodity in question but on the price he anticipates to prevail when the process of production is complete. Here again, sample surveys will give us more information about anticipated prices, business conditions, and the like (Theil 1961; Katona *et al.* 1954).

For emphasis on the unity of scientific methods in the natural and social sciences we quote Popper (1957, pp. 130 ff.):

. . . I am going to propose a doctrine of the unity of method; that is to say, the view, that all theoretical or generalizing sciences make use of the same method, whether they are natural or social sciences. . . . I do not intend to assert that there are no differences whatever between the methods of the theoretical sciences of nature and of society; such differ-

> ences clearly exist. . . . But I agree with Comte and Mill and with many others, such as C. Menger, that the methods in the two fields are fundamentally the same. . . . The methods always consist in offering deductive casual explanations, and in testing them (by way of predictions). This has sometimes been called the hypothetical-deductive method, or more often the method of hypothesis, for it does not achieve absolute certainty for any of the scientific statements which it tests; rather, these statements always retain the character of tentative hypotheses, even though their character of tentativeness may cease to be obvious after they have passed a great number of severe tests.

Popper (1957, pp. 136–37) maintains the unity of method in the social and natural sciences against Hayek (1952, p. 140) and even claims that social science is less complicated than physics.

The subject of human action in economics is sometimes called praxeology. It has made much progress in recent years, and many methods connected with it will be discussed below. It is here that the application of modern mathematical methods has been most successful, especially in connection with operations research, the econometrics of enterprise. It is interesting that both the extreme proponent of laissez-faire L. von Mises and the Marxist Lange define praxeology in similar terms. Von Mises (1949, p. 39) says: "The real thing, which is the subject matter of praxeology, human action, stems from the same source as human reasoning. . . . Praxeology conveys exact and precise knowledge of real things." Lange (1963, pp. 188–89) says: "In view of the fact that rationality of action is now a feature of many fields of human activity, there arises the problem of discovering what is that that is common to all fields of rational activity. This has led to the general study of rational activity, *praxeology*." Lange goes on to include operations research, cybernetics, decision theory, and the marginal calculus in the field of praxeology.

Praxeological methods (econometrics, operations research, cybernetics, programing, and so on) have been used in the United States mainly for military planning and rational planning of private enterprise (Churchman *et al.*, 1957; Dantzig 1963; Vadja 1961; Heady and Candler 1958; Holt *et al.* 1960).

In communist countries they have been used for central planning (O. Lange 1959; Kantorovich 1963). The planning models of certain West European countries (Theil 1961, 1964; Massé 1959; Lesourne 1960) and India (Mahalanobis 1955) deal with mixed economic systems, which in a sense are intermediate between free enterprise and pure collectivism.

II. Mathematical Economics

Economics uses the "logical-deductive" method and derives conclusions from certain fundamental assumptions or axioms, such as rationality and profit maximization. The status of these fundamental postulates is somewhat in doubt (Machlup 1955). We owe to Koopmans (1957, pp. 132 ff.) an interesting discussion of the postulational structure of economic theory. He criticized the ideas of Robbins (1949) and Friedmann (1953), two economists who were similar in their political views – both were ardent advocates of extreme laissez-faire. But Lord Robbins had great confidence in the introspectively established postulate of a preference ordering. He believed that it was possible to derive from this ordering practically the whole of economic theory, perhaps including certain policy recommendations. In contrast Friedmann, who would agree with the position Robbins held on policy, insisted on the view that the fundamental postulates of economics were irrelevant and that only their consequences were testable. The assumptions need not be realistic.

Against Robbins' views, Koopmans maintains that the assumption of a complete and invariable preference ordering contradicts some well-known facts in actual economic choice. Against Friedmann, Koopmans insists that direct verification of the postulates by their consequences is hardly ever possible in economics by experimentation. The indirect verifications involve a lengthy and perhaps uncertain chain of reasoning. Koopmans (1957, p. 142) says: "The theories that have become dear to us can very well stand by themselves as an impressive and highly valuable system of deductive thought, erected on a few premises that seem to be well chosen first approximations to a complicated reality."

A more liberal point of view about the status of economic theory is that of Blaug (1962, p. 606):

> A "theory" is not to be condemned merely because is as yet untestable; not even if it is so formed as to preclude testing, provided it draws

attention to a significant problem and provides a framework for its discussion from which a testable implication may some day emerge. It cannot be denied that many so-called "theories" in economics have no substantive content and serve merely as filing systems for organizing empirical information. To demand the removal of all heuristic postulates and theorems in the desire to press the principle of verifiability to the limit is to proscribe further research in many branches of economics. It is perfectly true that economists have often deceived themselves – and their readers – by engaging in what Leontief called "implicit theorizing," presenting tautologies in the guise of substantive contributions to economic knowledge. But the remedy for this practice is clarification of purpose, not radical and possibly premature surgery.

The use of mathematics in economics has been criticized not only by anti-theoretical schools like the German historical school and the American institutionalists but also by some writers who emphasize the use of economic theory (L. von Mises 1949; Stigler 1949). Against these views we might quote the opinion of a recognized specialist on the methodology of economics: "On the whole, we arrive at the conclusions, first, that political economy involves conceptions of a mathematical nature requiring to be analyzed in a mathematical spirit; and secondly, that there are certain departments of the science in which valuable aid may be derived from the actual employment of symbolical or diagrammatical methods." (J. N. Keynes 1955, p. 267). One of the founders (Jevons 1911, p. xxiii) of the modern marginal utility theory has this to say: "I hold then, that to argue mathematically, whether correctly or incorrectly, constitutes no real differentia as regards writers on the theory of economics. But it is one thing to argue and another thing to understand and to recognize explicitely the method of argument."

Norbert Wiener (1964, p. 90), whose work in cybernetics has also been most stimulating in economics, is critical: "The mathematics that the social scientists employ and the mathematical physics that they use as their models are the mathematics and mathematical physics of 1850. An econometrician will develop an elaborate and ingenuous theory of demand and supply, inventories and unemployment, and the like, with a relative or total indifference to the methods by which these elusive quantities are observed or measured." This statement by one of the greatest mathematicians of our time cannot be dismissed lightly, since it points to certain definite weaknesses in contemporary mathematical economics and econometrics which no doubt exist. It is perhaps too early to say whether the explicit introduction of stochastic processes into the treatment of economic problems will yield more reliable results (Tintner, Sengupta, and Thomas 1966).

We should deal here with some of the objections against mathematical economics and econometrics by the eminent Austrian economist L. von Mises (1949, pp. 347–54, 374–76, 697–98, 706–11). According to von Mises, the numerical results of econometrics lack universal validity, are essentially historical, and always refer to a given country and a given (past) time period.

We must concede that economics has not yet derived universal laws and constants like physics. Perhaps the only possible exception is the controversial Pareto distribution of incomes (Pareto 1927; Davis 1941, pp. 28 ff.). Discussion of this subject has recently been resumed (Mandelbrot 1960; Steindl 1965). Pareto stated that personal income (at least for higher incomes) follows the law

$$N = Ax^{-B}; \tag{1}$$

where N is the number of persons receiving income x or more; A and B are constants, and B is about $+1.5$. Now it is possible to derive this law from a specific stochastic process (Champernowne 1953; Mandelbrot 1960; Simon 1957, pp. 145–64; Klein 1962, pp. 140 ff.; Steindl 1965). The matter cannot yet be considered settled, but some empirical investigations contained in the literature give us hope that the Pareto law (perhaps in a slightly more complicated form) might be valid at least as a somewhat crude approximation of many economic phenomena.

But are the specific results of econometric research useless, since they evidently refer to a given country and a given time period? The Swedish econometrician Wold (Wold and Jureén

1953, pp. 307 ff.) has derived demand functions for a number of consumer's goods for the period 1921–39. A demand function explains the quantity demanded of a given commodity in terms of prices and money income. He compares predictions based upon these demand functions with actual consumption of the commodities in Sweden in 1950, as shown in Table 1.

TABLE 1

WOLD'S COMPARISON OF PREDICTIONS OF DEMAND FUNCTIONS AND OBSERVED COMMODITY CONSUMPTION
(Percentage Change)

Commodity	Predicted	Observed
Milk and cream.	+7	+2
Consumer milk	+1	+6
Butter and margarine	+11	+26
Butter.	+7	+28
Margarine	+16	+23
Cheese.	+13	+12
Eggs.	+6	+42
Meat	+1	+3
Meat (excluding pork).	—3	—13
Pork.	+6	+20
Wheat and rye flour.	—11	—10
Sugar (refined)	+12	+11
Potatoes.	—7	—6

This econometric investigation is based upon the following version of the theory of demand: The quantity demanded for a given good is influenced by the price of this commodity and income. The quantity demanded is consumption per head of population; the price is the real price—that is, the money price divided by a consumer price index. Also, income is real income per head—that is, money income divided by the consumers price index.

To appreciate the comparison, it should be realized that Sweden rationed many foodstuffs during the war, and rationing was not abolished until 1949. The prediction depends upon relationships based on family surveys and market statistics collected in Sweden. It is derived from a model which assumes that only price and income changes influence consumption.

A similar analysis is by Fox (1958). He investigates price changes of 30 agricultural commodities, based upon an econometric analysis with data taken from statistics in the United States 1922–41. These examples show, however, only that predictions based upon econometric methods are sometimes successful. The econometric method does not guarantee that this will always be the case. If, for example, the relationships investigated show a change in time, it is ideally up to the economist to construct a "dynamic" theory which will explain the very change, but it should not be denied that our dynamic theories are still insufficient.

Another use of econometrics is the "verification" of economic theories. May I here refer to one of my own investigations. It is well known that the celebrated general theory of Keynes (1936), which in a sense has revolutionized economics and was important for the economic policy of various countries, assumes that the (static) supply function of labor depends upon money wages and not upon real wages – that is, money wages divided by the cost of living index. By the supply function of labor, we mean the relationship between the amount of labor offered or supplied as a function of wages, either money wages or real wages. The supply of labor is here measured as the sum of the number of workers employed plus the number of insured unemployed. The wage index is computed as the weighted average of sixteen industries. This gives us an index of money wages. Real wages are computed as the ratio between money wages and a cost of living index. I have investigated this question empirically, using data for British industry 1920–38 (Tintner 1952, pp. 143 ff.). The result of the econometric investigation using various mathematical models is as follows. A statistical test of the fitted relations indicates that it is probable that the supply of British industrial labor in the interwar period depended upon real wages rather than upon money wages alone. This result is of course dependent upon a number of assumptions. The model must be at least approximately valid; the errors must be approximately normally distributed, independent over time, and so on.

The investigation yields an estimate of the elasticity of British labor supply in relation to real wages. This elasticity is estimated as −0.19. Under the assumptions mentioned, it is statistically significant. Hence, if *ceteris paribus* the real wage of industrial labor in Great Britain increases by 1 per cent, we might expect that the supply of labor will decrease by about 0.2 per cent. The negative estimate of the elasticity of supply for labor agrees with economic theory (Bowley 1924, p. 40) and also with empirical results for the United States (Mosback 1959). This negative elasticity of the supply of labor can be explained in the following way. As real wages increase, there will at the same time be a decrease of the number of people who are willing to supply labor. This means that, for example, the number of working wives will decrease, some children will continue their schooling instead of trying to enter the labor market, and so on. This example should be carefully interpreted. It perhaps tells us something about the conditions of the supply of industrial labor in Great Britain during the period investigated, but nothing (except by analogy) about the character of the supply of labor in other countries and in other periods. Nevertheless, it could be used to tentatively question the Keynesian assumption that the demand for labor depends upon money wages and not upon real wages. If the results of the investigation are somewhat reliable, at least the *universal* validity of this fundamental assumption must be questioned.

Problems of Measurability

It has been maintained that some important economic magnitudes cannot be measured (Painlevé 1960; Georgescu-Roegen 1966, pp. 114 ff.). It is evident that many important economic concepts (consumption, production, labor, interest rates, prices) are quantitative. For instance, consumption of specific commodities is given in terms of pounds and number of items. Production of various commodities is again in quantitative form – pounds, number of items. Labor can be measured in terms of days or hours worked. Interest rates are expressed in percentages. Prices appear in dollars. There is perhaps only one important exception which appears in modern but not in classical (and Marxian) economics – satisfaction or utility (Alt 1936;

Pfanzagl 1959). The problem of anticipations and expectations will be discussed below.

Let us consider the static theory of choice (Tintner 1955). Assume that a given individual is faced with the choice among three combinations of goods and services, which we denote by A, B, and C. Without the use of mathematics, the adherents of the Austrian school establish that a rational individual will act as follows. If he does not prefer A to B and also not B to C, he will not prefer A to C. We follow the example of Arrow (1963), Stone (1951), and von Wright (1963*a*) and introduce a relation R for the given individual. This relation is defined in the following manner. XRY means that the individual in question does not prefer X to Y. Our proposition can be reformulated in terms of mathematical logic. From the two propositions ARB and BRC, ARC follows. The relation R is transitive. Now we use the functional calculus. Let $U(X)$ be the satisfaction or utility of the individual derived from the combination of goods and services X. Then we deduce that if $U(A) \leqq U(B)$ and also $U(B) \leqq U(C)$, it follows that $U(A) \leqq U(C)$. Following Pareto (1927), it can also be shown that utility or satisfaction need not be measurable. An ordinal scale of utility is sufficient. It is easy to see that the three formulations given above are logically equivalent. Hence, non-mathematical economic theoreticians have no reason to reject the last two formulations if they are willing to accept the first.

This short discussion is not presented as a very realistic model of economic choice. It neglects the possibility that utility might be a multidimensional concept (Georgescu-Roegen 1954). This question is discussed below. Also, it neglects the possibility of the existence of a psychological threshold, which might invalidate the transitivity of the concept of indifference (Georgescu-Roegen 1950). For instance, an individual may say he is indifferent between X and Y and also between Y and Z. But still, when confronted by a choice between X and Z, he may prefer Z because of existing psychological thresholds. The problem of measurability in economics has been recently discussed in a most penetrating analysis by Georgescu-Roegen (1965), who especially shows its importance in the theory of production.

Much of the modern discussion of the pure theory of choice is carried on in terms of revealed preference. This is a strictly behavioristic point of view in choice theory, in which we try to discover the underlying structure of choice of a given individual from his overt actions of choice. (Samuelson 1938, 1947, 1948; Houthakker 1950). This point of view has been criticized by Georgescu-Roegen (1950), who has pointed out that the existence of psychological thresholds may make the recognition of revealed preference (e.g., transitivity of choice) impossible.

Thanks to a method proposed by Wald (1940), we are able to approximate the static utility function or indifference system. By a utility function we mean a function which measures utility or satisfaction derived from various goods and services consumed. An indifference surface indicates combinations of various goods and services which are such that they give the same utility or satisfaction. These ideas are important in the modern theory of demand, because they enable us to explain consistently the quantity demanded of a given commodity or service as a function of the given prices of all commodities and services and of money income. Wald uses quadratic utility functions as approximations. An Engel curve is a relationship between the quantity of a commodity or service consumed and money income, assuming prices being constant. The Engel curves – relations between consumption of a given commodity by an individual and money income of the same individual with constant prices – are linear if the utility function is quadratic. J. A. Nordin (Tintner 1952, pp. 60 ff.) uses data from statistics in the United States 1935–6 and 1941. His first sample includes 300,000 families; his last 3,060. In this highly aggregated model, x is an index of consumption of food and y an index of consumption of non-food – all other items. It is assumed that tastes and preferences of American families have been approximately constant during the period involved; utility is taken as a one-dimensional concept. There is also no consideration of stochastic problems, and the existence of a psychological threshold is ignored. Further, no attention is paid to saving and other dynamic factors which might influence consumption. Finally, there is no consideration of the possible interrelationship of the demand for a given good between individual consumers (Tintner 1946, 1960*b*).

The utility function derived here is an average utility function for the United States. The sample surveys give us numerical information about expenditure on various items – for example, expenditures on food. By dividing food expenditures by an index of food prices, we derive an index of the quantity of food consumed. We also have information about expenditure on all other items. Dividing these expenditure figures by an index of all other prices except food, we derive an index of consumption of non-food.

Let x be consumption of food and y be consumption of all other commodities and services. As an empirical approximation, the following utility function for the American economy is

$$U = -0.000890x^2 + 0.022401xy + 0.008353y^2 + 104.572144x + 96.68771y. \tag{2}$$

This utility function is supposed to represent approximately the average American's satisfaction derived and choice between food and all other commodities. Making $U = k$, where k is a constant, we might derive the indifference curves, which show combinations of x (food consumed) and y (all other things consumed) between which the typical American individual is indifferent. Naturally, these indifference curves are only crude approximations and subject to the limitations of the analysis pointed out above. At best, they might give us some idea of the behavior of consumers in the region covered by the data. It is known from the modern theory of indifference curves that we might substitute for this utility function U a non-decreasing function of U: thus

$$V = f(U) \tag{3}$$

as long as

$$dV/dU > 0. \tag{4}$$

This fact emphasizes the ordinal character of the utility function. This example, based by necessity upon an extremely aggregated model, should be interpreted with care. It shows that the "classical" concept of static utility theory can be implemented or at least illustrated by the use of empirical data. The idea of utility is not completely empty, nor does it have to rely on the dubious merits of "introspection" alone. The demand function for food shows the relationship between the quantity of food demanded as a function of the price of food, the price of all other goods, and money income. The demand function for non-food shows the relationship between the quantity of all items other than food consumed as a function of a price index of food, a price index for all other commodities, and money income. The utility function may be used, for example, for the computation of demand functions (see Tintner 1952, p. 61), and these might potentially be useful in policy. The derived functions make it possible also to test indirectly the goodness of approximation of the underlying estimations (for a related method see Afriat 1967).

It should, however, be mentioned that according to Georgescu-Roegen (1954) the very existence of indifference surfaces might be doubted. Following Aristotle and some ideas of the early Austrian writers, this implies the existence of a hierarchy of wants. The more urgent wants will be satisfied first, and less urgent wants only after the satisfaction of the most urgent ones. Georgescu-Roegen (1954) considers as an example the choice between butter and margarine. First, the desire for food (calories) will be fulfilled, then the desire of taste, finally the desire for entertainment. These three form a hierarchical order for choice and give rise to a lexicographical ordering.

One of the most interesting developments of recent years is the theory of *measurable* utility by von Neumann and Morgenstern (1944). The concept of utility has a long history which cannot be presented here (Stigler 1950). It might only be mentioned that Bernouilli (1730) investigated a specific form of measurable utility, characteristically in connection with a problem in probability theory (K. Menger 1934*a*). The three founders of modern utility theory—C. Menger, Walras, and Jevons—also considered utility measurable. Since Pareto (1927), however, it is recognized that measurable utility is not necessary for the purposes of static economics in order to explain choice in consumption.

TABLE 2

MARSCHAK'S PRESENTATION OF INDIVIDUAL CHOICE AND RESULTANT CONSEQUENCE

Acts	Consequence	
	State of World (s_1)	State of World (s_2)
a_1	b	b
a_2	a	c

Marschak's (1964) concept of individual choice and resultant consequence utilizes some of the ideas of Ramsey (1928) and Savage (1954). Consider an individual who has the choice between two acts a_1 and a_2 and who is faced with two possible states of the world s_1 and s_2. Table 2 shows the choices open to him. If the state of the world is s_1, the consequence of his action will be b if he chooses a_1 and a if he chooses a_2. If the state of the world is s_2, his action a_1 will have consequence b and action a_2 will result in consequence c. There is complete uncertainty about the state of the world. Assume that a is preferred to b and b to c. Let $u(x)$ be the utility of x. To fix an arbitrary scale of utility, we assume that

$$u(c)=0 \text{ and } u(a)=1, \tag{5}$$

and it follows that

$$0 \leqq u(b) \leqq 1. \tag{6}$$

Now assume that the individual has a subjective probability p for the state of the world s_1 and the subjective probability $1-p$ for the state s_2 $(0 \leqq p \leqq 1)$. We assume that the individual evaluates his actions by the principle of the mathematical expectation of utility. The mathematical expectation is the weighted arithmetic mean with the probabilities as weights. We have for action a_2 the mathematical expectation of utility

$$u(a_2)=u(a)p+u(c)\ (1-p)=p; \qquad (7)$$

and similarly, for action a_1 the mathematical expectation of utility

$$u(a_1)=u(b)p+u(b)\ (1-p)=u(b). \qquad (8)$$

There should be a subjective probability p_0 for the individual at which he is *indifferent* between actions a_1 and a_2. Thus the utility of b is p_0.

$$u(c)=0\,u(b)=p_0\,u(a)=1. \qquad (9)$$

Hence we have now assigned measurable utility to a, b, and c. These utilities are unique up to a linear transformation. The zero point and the scale of utility can be assigned arbitrarily.

The above example shows how we may construct a measurable utility function, which is determinate except for origin and scale. Any linear function $V = a+bU$ with a and b constant and b positive is equivalent. This example shows also the importance of the axiom of expected utility maximization and the relation between measurable utility and subjective probability.

An interesting theory of utility which tries to explain gambling and the buying of insurance has been proposed by Friedman and Savage (1948). For experimental determination of measurable utility, see Mosteller and Nogee (1951) and Davidson, Siegel, and Suppes (1957). The whole theory has been criticized by Allais (1953) from the point of view of the psychological assumption involved. The von Neumann-Morgenstern concept of measurable utility has been criticized, because utility of gambling and love of danger are excluded (Graff 1957, p. 36; Marschak 1950). This is in many ways a serious shortcoming but may be remedied in time. Nevertheless, the idea of measurable utility represents a great advance in economics and has been used extensively in modern statistics, especially in decision theory (Wald 1950; Blackwell and Girshick 1954) and the personal, subjective, or Bayesian approach to statistics (Savage 1954, 1962).

Statics

Economic statics is defined by Hicks (1946) as the theory of an economic system in which time does not enter, where the variables (prices, quantities produced and consumed, and so on) are not dated. Much of present-day economics is still statics, especially the great system of static equilibrium of Walras. This is perhaps one of the main weaknesses of present-day economics.

One of the most important uses of economic statics is the comparison of two distinct static systems. This is called comparative statics. It has the advantage that frequently important economic conclusions can be derived from very simple assumptions (Samuelson 1947). As an example of comparative statics, consider the position of a simple monopolist in face of a tax (Samuelson 1947, pp. 15–16). Monopoly exists if in a market there is one seller and many buyers. Let x be the output of the monopolist, $f(x)$ the profits before the tax – that is, the difference between total revenue (price times quantity demanded) and total production cost – and t the (constant) tax rate per unit. Hence, we have for the profit after tax

$$P = f(x) - tx. \qquad (10)$$

This is maximized if

$$dP/dx = f'(x) - t = 0 \qquad (11)$$

and

$$d^2P/dx^2 = f''(x) < 0. \qquad (12)$$

We assume that the function $f(x)$ has first and second derivates, denoted by $f'(x)$ and $f''(x)$. To investigate the influence of the tax, we differentiate (11) with respect to the parameter t and obtain

$$f''(x)\ (dx/dt) - 1 = 0. \tag{13}$$

Hence,

$$dx/dt = 1/f''(x) < 0 \tag{14}$$

because of (12). Output will decline with increase of the tax rate t.

As Kaufmann (1944, pp. 218 ff.) points out, to determine the monopoly price and optimum level of output, the monopolist must know (a) his total cost function, (b) the demand function for the product, and (c) the maximum value of the profit function. This might require perfect foresight. Whereas (a) and (b) are synthetic (empirical) propositions, (c) is analytic. These assumptions show that conclusions drawn from comparative statics are frequently of limited validity in practical applications. Nevertheless, since they are relatively easy to obtain from rather simple models, such methods are popular among economists, who do not always realize the severe limitations of the conclusions and do not hesitate to apply the results sometimes to concrete questions of policy.

As an example of the more modern linear methods used in mathematical economics and operations research, let us present a simple instance of linear programing, a method discovered by Kantorovich (1939, 1963) and Dantzig (1949, 1963). Mathematically, the problem of linear programing consists in maximizing (or minimizing) a linear form, subject to linear inequalities and the condition that the solutions might not be negative (Vajda 1961).

Consider the situation of a typical farm in Hancock County, Iowa, during the period 1928–52 (Tintner 1960*a*). There are two products: corn (x_1) and flax (x_2). We analyze the situation in the short run (Heady and Candler 1958); hence, we can neglect fixed costs – that is, costs which are independent of the amounts produced, since in the short run they are incurred anyway (Tintner 1960*a*). We consider production in the short run – that is, in a situation where the amount of various factors of production (land, labor, capital) is fixed. Consider land, for instance. In the short run the amount of land is given for the farmer, but he does not have to use his total amount of land and can leave part of it uncultivated. In the long run he may sell some of his land or buy more land. Similar considerations also hold for other factors of production.

Further, we assume constant coefficients of production. This means that the amount of any product (e.g., corn) is proportional to the inputs used in the production of this commodity (proportional to the amounts of land and capital). This assumption must be considered as a great simplification of the real conditions of production. In the short run the amounts of the factors of production used cannot be increased. Also we assume that the farm produces under static conditions. The price of a bushel of corn is $1.56 and that of a bushel of flax is $3.81. The "objective function" — the short run profits the farmer wants to maximize — is

$$f = 1.56x_1 + 3.81x_2. \tag{15}$$

For the conditions of production in the short run we again make the simplest possible assumptions — fixed coefficients of production. We assume that outputs are approximately proportional to inputs. The sample survey tells us that it takes an average of 0.022740 acres of land to produce a bushel of corn and an average of 0.09244 acres of land to produce a bushel of flax. The typical farm we are investigating has 148 acres of land. Similarly, it takes an average $0.317720 of capital to produce a bushel of corn, and on the average of $0.969500 of capital to produce a bushel of flax. The average capital available for the farm is $1,800.

Again for the sake of simplicity, we neglect other factors of production apart from land and capital — for example, labor. Now in the short run the farmer can use only the land (148 acres) and the capital ($1,800) he actually has available. But he is not obliged to utilize all the land and capital he has. In the long run, however, he might sell some land or borrow more capital. The conditions of production in our simple example are in the short run

$$0.022740x_1 + 0.092440x_2 \leqq 148 \qquad (16)$$

and

$$0.317720x_1 + 0.969500x_2 \leqq 1800.$$

To these conditions of production in the short run we must also add the condition that it is impossible to produce negative amounts of corn and flax

$$x_1 \geqq 0, \ x_2 \geqq 0. \qquad (17)$$

The solution to the problem of finding the maximum of short term profits (15) under the stated conditions of short term production may be found by the simplex method of Dantzig (1951, 1963). The results are: To maximize profits the farmer ought to produce $x_1 = 5{,}365.366$ bushels of corn and $x_2 = 0$ bushels of flax. Then his optimal profit will be $f = \$8{,}837.971$. This maximum is achieved if the farmer uses all his available capital ($1,800) but only 128.83 acres of the total available 148 acres of land. It is a remarkable mathematical fact that to each maximum problem in linear programming there exists a dual minimum problem. This establishes the formal relation between linear programming and the von Neumann and Morgenstern (1944) theory of two person zero sum games. This dual also has a very interesting economic interpretation.

Again considering our example, the farmer in question will have to establish certain accounting (book keeping) or shadow prices for the two factors of production used. Let u_1 be the shadow price for an acre of land and u_2 the shadow price for each dollar of capital. It should be emphasized that these are merely accounting or shadow prices. They express the rational valuation of units of factors of production (land and capital) for the farmer in the short run and are not necessarily identical with market prices.

Since the typical farm possesses in the short run 148 acres of land and $1,800 of capital, the farmer will try to *minimize*:

$$g = 148u_1 + 1{,}800u_2 \tag{18}$$

This expression under our assumptions is the book keeping or accounting cost of the farm enterprise in the short run.

The inequalities imposed are that for each activity (bushels of corn and flax produced) the *imputed cost* (using the accounting prices) must be at least as great as the net price of the activity (price of a bushel of corn or flax): thus

$$0.022740u_1 + 0.317720u_2 \geqq 1.56 \tag{19}$$

and

$$0.092440u_1 + 0.969500u_2 \geqq 3.81.$$

The last condition says that the accounting prices are not negative:

$$u_1 \geqq 0,\ u_2 \geqq 0. \tag{20}$$

The solution of this minimum problem is that the imputed price of land is $u_1 = 0$. Land is for the farmer in question a free good, like air. This is shown by the fact that he did not use all the land available (148 acres) but only 128.83 acres. Hence, it would not cost him anything to use more land. The high imputed price of capital, $u_2 = 4.91$, is explained by the scarcity of capital. The total imputed cost $g =$ $8,837.971. Hence, we see that the dual minimum problem has the same solution (minimum value of imputed cost) as the original maximum problem (maximum of short run profits). If the factors of production are correctly evaluated then they exhaust the profits, and no extra profits are made in equilibrium.

In evaluating this example, the following fundamental assumptions used should be kept in mind: The model is static, but the data for its verification are taken from a dynamic economy. We assume pure and perfect competition; that is, the farmer cannot in any way influence the prices at which he sells his products. We investigate production in the short run; that is, the farmer cannot increase or decrease the amounts of the factors of production (land and capital) available to him. Also, for the sake of simplicity we distinguish only two factors of production. In a more realistic investigation labor should be introduced and various types of labor, capital, and land distinguished. The assumption of constant coefficients of production is a very strong one. The output of a given commodity (corn or flax) is strictly proportional to the inputs (land and capital). This is perhaps the most convenient available economic model of production but a great simplification of reality. Finally, we assume that the farmer only tries to maximize his short run profit or equivalently tries to minimize his accounting cost in the short run.

In spite of these severe limitations, the method of linear programming has been applied with some success to concrete economic problems. It should be pointed out that certain generalizations are possible. We might generalize the method by non-linear programming in which neither the objective function nor the inequalities need to be linear (Kuhn and Tucker 1951). When it is necessary for the solutions to be integers (Baumol 1961, pp. 148 ff.), we must use the method of integer programming. The method can also be generalized to deal with dynamic problems – that is, production over time, which leads to dynamic programming (Bellman 1957). Finally, we may introduce probability considerations into a linear program. The methods of stochastic programming will be discussed below (Charnes and Cooper 1959; Moeseke 1965).

One of the most interesting developments in mathematical economics was the introduction of game theory by von Neumann and Morgenstern (1944). Here for the first time we find a mathematical model which is not borrowed from the models of classical (deterministic) physics but from the theory of games of strategy (see also Luce and Raiffa 1957). Games of chance (e.g., roulette, dice) have played a very important part in the development of the theory of probability. But the theory of games is entirely different and has actually very little to do

TABLE 3

GAIN MATRIX OF A (LOSS MATRIX OF B) IN A TWO PERSON ZERO SUM GAME

STRATEGIES OF A	STRATEGIES OF B			ROW MINIMA
	B_1	B_2	B_3	
A_1	21	11	31	11
A_2	32	0	4	0
COLUMN MAXIMA	32	11	31	

with games of chance (e.g., throwing dice, roulette). It treats, on the contrary, games of strategy (e.g., poker, bridge, or chess) where each participant pursues his aim intelligently. Chance plays a very minor part in this theory, but it is not entirely absent (as in chance moves). Consider a simple example of a two person zero sum game (Tintner 1957). There are two players, A and B. Assume that A plays against B. What A wins B loses, and vice versa (zero sum). The totality of all possible moves of the game by A is called a strategy. Strategies are defined by von Neumann and Morgenstern (1944, p. 79) as follows:

Imagine now that each player . . . instead of making each decision as the necessity for it arises, makes up his mind in advance for all possible contingencies; i.e. that the player . . . begins to play with a complete plan: a plan which specifies what choices he will make in every possible situation, for every possible actual information which he may possess at that moment in conformity with the pattern of information which the rules of the game provide for him in that case. We call such a plan a *strategy*.

Assume that A has the strategies A_1, A_2 and B the strategies B_1, B_2, B_3. In Table 3 it can be seen that the gains of A are at the same time the losses of B. If A uses his strategy A_1, he will gain 21, 11, or 31 according to whether B uses B_1, B_2, or B_3. Since A knows that B is an intelligent opponent and that A's gains are B's losses, B will minimize his loss (which is A's gain). Hence A can only count on winning the *minimum* of the first row, 11, if he uses his strategy A_1. Suppose A uses his strategy A_2. Then, if B uses B_1, B_2, or B_3, he will gain 32, 0, or 4. For the same reasons as before, he can only count on the minimum of the second row, 0.

From the point of view of B the table represents losses. Suppose he uses his strategy B_1. Then, according to whether A uses A_1 or A_2, B may count on a loss of 21 or 32. But since B's loss is A's gain, he can only expect the *maximum* loss, 32, if he uses B_1. Similarly, for his remaining strategies B_2 and B_3 he has to take account of the fact that he must expect for each strategy the maximum loss – 11 for B_2 and 31 for B_3.

An equilibrium exists if the desire of A to maximize his minimum gain for each strategy coincides with the aim of B to minimize his maximum loss for each of his own strategies. This

TABLE 4

GAIN MATRIX OF A (LOSS MATRIX OF B) IN A TWO PERSON ZERO SUM GAME WITHOUT MINIMAX

STRATEGIES OF A	STRATEGIES OF B			ROW MINIMA
	B_1	B_2	B_3	
A_1	9	10	11	9
A_2	11	10	9	9
A_3	12	10	8	8
COLUMN MAXIMA	12	10	11	

equilibrium solution is a minimax or saddle point. In our simple example it is evidently 11. Hence, A will use A_1, and B will utilize B_2. This combination of strategies make sure that A will gain at least 11, and B will not lose more than 11.

Since the gain and loss matrix is arbitrary, we might well ask if there is always an equilibrium. It is of course easy to construct matrices which have no minimax. Consider, for instance, a case where A and B have 3 strategies (Table 4). It is evident that in this case no minimax exists. But if we change the problem slightly, we may consider the situation where A and B play the game not just once but many times. Suppose that A uses the strategy A_1 with a probability p_1, the strategy A_2 with probability p_2, and A_3 with probability p_3. Also, B uses strategy B_1 with probability q_1, B_2 with probability q_2, and B_3 with probability q_3. In a long series of games played by A and B the average gain of A (loss of B) will be the mathematical expectation

E, which is the weighted arithmetic mean of the gains or losses with the probabilities as weights. Thus,

$$E = 9p_1q_1 + 10p_1q_2 + 11p_1q_3 + 11p_2q_1 + 10p_2q_2 + 9p_2q_3 + 12p_3q_1 + 10p_3q_2 + 8p_3q_3. \tag{21}$$

Assume that A tries to maximize and B tries to minimize the mathematical expectation (21). Then A has the choice of two probability distributions: $p_1 = 2/3$, $p_2 = 0$, $p_3 = 1/3$; and alternatively, $p_1 = 1/2$, $p_2 = 1/2$, $p_3 = 0$. B has to choose the probability distribution: $q_1 = 0$, $q_2 = 1$, $q_3 = 0$, or $q_1 = ½$, $q_1 = 0$ $q_3 = ½$. If A and B choose their strategies with the indicated probabilities, the mathematical expectation of the gain of A (loss of B) is $E = 10$ — this is to say that by choosing the given probabilities, A can make sure to gain at least 10 in a long series of games, and B can make sure to lose not more than 10 in a long series.

Ideas based on the theory of games have been very important in modern mathematical economics in connection with the theory of measurable utility, decision theory, and the theory of static general equilibrium systems under free competition. The theory has not yet been quite successful in connection with problems of market organization, which are in a sense between free competition (many buyers and sellers) and monopoly (one seller) or monopsony (one buyer) (Shubik 1959). The problem of market organizations which are neither purely competitive (a great number of buyers and sellers) nor monopoly (one seller) nor monopsony (one buyer) has stubbornly resisted theoretical analysis. These are problems of oligopoly (a few sellers), oligopsony (a few buyers), bilateral monopoly (one seller and one buyer), and so on. Game theory has made a valuable contribution by pointing out that the main problem is the formation of *coalitions* (Shubik 1959).

We shall discuss a bargaining model, using bilateral monopoly (one seller, one buyer) as an example. The problem of bilateral monopoly is very important in the labor market, where, for example, a labor union faces a single monopolistic enterprise or a cartelized or trustified industry. An interesting baring model has been proposed by Harsanyi (1962) on the basis of a theory by Nash (1953), which has its origin in Zeuthen's ideas (1933). Assume that two people bargain, whose utility we denote by u_1 and u_2. We define a prospect space, the set of all utilities the two bargainers can obtain by a joint strategy. Let the conflict point C represent the utility they would obtain if they could not reach agreement. Then a maximization of the product,

$$[u_1(S) - u_1(C)] \cdot [u_2(S) - u_2(C)], \qquad (22)$$

with respect to S over the whole prospect space will determine the solution. As Harsanyi (1962, p. 449) points out, this solution of the bargaining problem gives intuitively attractive results:

> In general, the Nash solution assigns to any party a larger payoff: 1. the larger the party's willingness to risk a conflict rather than making concessions to his opponent. 2. the smaller the other party's willingness to risk a conflict. 3. the larger the damage that in the case of a conflict the first party could cause to the second party, at a given cost to himself. 4. the smaller the damage that the second party could cause to the first party at a given cost to himself. (See also Harsanyi 1966.)

The following discussion of the existence and uniqueness of the Walras-Cassel system of competitive static equilibrium is an adaptation of an earlier discussion by Wald (1951) and by Dorfmann, Samuelson, and Solow (1958, p. 346 ff.). A more detailed investigation is found in Debreu (1959).

We assume constant coefficients of production: a_{ij} is the amount of factor i used in producing the commodity j. Let $x_1, x_2, \ldots, x_n$ be the amounts of the n commodities produced, $r_1, r_2, \ldots, r_m$ the amounts of the m resources or factors of production utilized. The total amount of factor i used in the production of commodity j is then $a_{ij}x_j$. The total amount of the factor i used in the production of all commodities in the economy is $a_{i1}x_1 + a_{i2}x_2 + \ldots + a_{in}x_n$. But evidently this cannot be greater than the total amount of the factor i available, r_i. Hence we obtain the system of inequalities

$$\begin{aligned} a_{11}x_1 + a_{12}x_2 + \ . \ . \ . \ + a_{1n}x_n &\leqq r_1 \\ a_{21}x_1 + a_{22}x_2 + \ . \ . \ . \ + a_{2n}x_n &\leqq r_2 \\ \cdots\cdots\cdots\cdots\cdots\cdots\cdots\cdots\cdots\cdots \\ a_{m1}x_1 + a_{m2}x_2 + \ . \ . \ . \ + a_{mn}x_n &\leqq r_m. \end{aligned} \tag{23}$$

Inequalities are introduced because some of the resources might be redundant and then become free goods (i.e., their prices are zero).

Let $p_1, p_2, \ . \ . \ . \ , p_n$ be the prices of the n final goods and $v_1, v_2, \ . \ . \ . \ , v_m$ be the prices of the m factors of production. We assume that there are market demand equations for all final commodities, which (in principle) depend on all the prices

$$\begin{aligned} x_1 &= F_1 \ (p_1, p_2, \ . \ . \ . \ , p_n, v_1, v_2, \ . \ . \ . \ , v_m) \\ x_2 &= F_2 \ (p_1, p_2, \ . \ . \ . \ , p_n, v_1, v_2, \ . \ . \ . \ , v_m) \\ \cdots\cdots&\cdots\cdots\cdots\cdots\cdots\cdots\cdots\cdots \\ x_n &= F_n \ (p_1, p_2, \ . \ . \ . \ , p_n, v_1, v_2, \ . \ . \ . \ , v_m). \end{aligned} \tag{24}$$

These demand functions are homogeneous of degree zero in the prices; that is, if *all* prices are multiplied by a positive constant L, the quantities demanded are not changed:

$$\begin{aligned} F_i \ (Lp_1, Lp_2, \ . \ . \ . \ , Lp_n, Lv_1, Lv_2, \ . \ . \ . \ , Lv_m) \\ = F_i \ (p_1, p_2, \ . \ . \ . \ , p_n, v_1, v_2, \ . \ . \ . \ , v_m). \end{aligned}$$

The supply functions of the factors of production are

$$\begin{aligned} r_1 &= G_1 \ (p_1, p_2, \ . \ . \ . \ , p_n, v_1, v_2, \ . \ . \ . \ , v_m) \\ r_2 &= G_2 \ (p_1, p_2, \ . \ . \ . \ , p_n, v_1, v^2, \ . \ . \ . \ , v_m) \\ \cdots\cdots&\cdots\cdots\cdots\cdots\cdots\cdots\cdots\cdots \\ r_m &= G_m \ (p_1, p_2, \ . \ . \ . \ , p_n, v_1, v_2, \ . \ . \ . \ , v_m). \end{aligned} \tag{26}$$

These functions are also homogeneous of degree zero in the prices. If *all* prices are multiplied by the same positive constant, the quantities supplied are not changed.

The cost of factor i in the production of commodity j is $a_{ij}v_i$. The total cost per unit of producing commodity j is under our assumptions $a_{1j}v_1 + a_{2j}v_2 + \dots + a_{mj}v_m$. But this unit cost cannot be smaller than the price of the commodity p_j. In long run static competitive equilibrium the price of each commodity cannot be greater than the unit costs:

$$\begin{aligned} a_{11}v_1 + a_{21}v_2 + \dots + a_{m1}v_m &\geqq p_1 \\ a_{12}v_1 + a_{22}v_2 + \dots + a_{m2}v_m &\geqq p_2 \\ \dots\dots\dots\dots\dots\dots\dots\dots \\ a_{1n}v_1 + a_{2n}v_2 + \dots + a_{mn}v_m &\geqq p_n. \end{aligned} \tag{27}$$

If the strict inequality sign holds in any of the equations, then the unit cost of producing this commodity is greater than its price, and the corresponding output of the commodity in question is zero. If the cost of production is greater than the price of a commodity, the commodity is not produced.

By methods related to linear programming, the existence of a system of solutions for the quantities of the final commodities $x_1, x_2, \dots, x_n$, their prices $p_1, p_2, \dots, p_n$, the quantities $r_1, r_2, \dots, r_m$, and the prices of the factors of production $v_1, v_2, \dots, v_m$ can be proved; and these results are economically meaningful since none of these quantities is negative. If some rather strong assumptions are also made about the demand functions, it can be shown that the solutions are unique.

For the sake of simplicity this model of general static competitive economic equilibrium has been presented in a highly aggregative form. It can be shown, however, that individual demand functions for all commodities and individual supply functions of original services (labor) for all individuals in the system can be determined from the modern theory of choice. Similarly, the theory of production can be used to find the demand for all factors of production and the supply of all final products by competitive firms (Hicks 1946). Under our assumptions of atomistic independence it would then be possible to simply add these individual demand and supply functions, which form the basis of the system actually investigated. Morgenstern (1947) has pointed out that this procedure is not permissible if we deviate from our assumptions.

Competition has been defined by Moore (1929) as involving the following principles: every economic factor seeks and obtains a maximum net income; there is but one price of the commodities of the same quality in the same market; the influence of the product of any one producer upon the price per unit of the total product is negligible; the output of any one producer is negligible as compared with the total output; each producer regulates the amount of his output without regard to the effects of his act upon the conduct of his competitors. These conditions should warn us that the application of the model described above to a concrete economic system would be very hazardous.

There has been a great deal of discussion of the stability of solutions of a static competitive economic system. The discussion has been conveniently summarized by Morishima (1964); but since the assumptions of the theory are unrealistic, it seems to be only remotely connected with a truly dynamic theory of economics.

It might be mentioned that several authors (Koopmans 1957; Debreu 1959; Malinvaud 1953; Kuenne 1963) have extended this model in a number of directions to include individual households and consumption, and also individual competitive firms. The model could be generalized to include production and consumption over time and even certain aspects of uncertainty (Arrow and Debreu 1959). We should in all honesty, however, remember some of the fundamental restrictions which cannot be removed: All commodities and services are indefinitely divisible, and the problem of indivisibilities has, up to now, resisted analysis (Lerner 1944; Koopmans and Beckman 1957; Hurwicz 1960). Also, no elements of monopoly and similar phenomena can be included.

As a small example of the empirical analysis of a static system, I would like to present a model of the Portuguese economy in the year 1957 (Tintner and Murteira 1960; Tintner 1965). This economy is divided into 4 sectors: (1) all enterprises; (2) government; (3) foreign trade; (4) households. Such a highly aggregated system, of course, is not very useful, but Leontief (1951) and others have calculated systems with many more sectors. With data taken from national accounting in Portugal, we can compute the constant coefficients of production. The assumption is that in a first approximation the Portuguese economy works under conditions of constant coefficients of production — that is to say, we assume that the output of each sector is strictly proportional to the inputs coming from other sectors. We also assume perfect competition.

Let X_1 be the net output of the enterprise sector, X_2 the value of government services, and X_3 the value of exports. The net output of each sector is the output absorbed by the other sectors. Designate the demand of private consumers for products of the enterprise sector by y_1, the demand of consumers for government services by y_2, and their demand for imports by y_3. The demand of the private consumers is assumed to be given.

With our assumptions we derive the following system:

$$
\begin{aligned}
X_1 &= 1.276y_1 + 0.635y_2 + 1.000y_3 \\
X_2 &= 0.084y_1 + 1.042y_2 + 0.184y_3 \\
X_3 &= 0.323y_1 + 0.161y_2 + 1.253y_3.
\end{aligned}
\tag{28}
$$

This system shows the linear dependence of the net output of each sector (X_1, X_2, X_3) upon the autonomous demand of consumers in Portugal for the goods and services of the three sectors (y_1, y_2, y_3).

The interpretation of these results is as follows: Assume that our hypotheses (constant coefficients of production and free competition) hold, at least approximately, in the Portuguese economy. Assume further that the demand for products and services of the enterprise sector (y_1) alone increases by 1 escudo.

We may then expect that the net product of the enterprise sector will increase by 1.276 escudos, the net value of government services by 0.084 escudos, the value of exports by 0.323 escudos. Assume now that *ceteris paribus* the demand of households for government services (y_2) increases by 1 escudo. The effects on the net output of the various sectors is as follows: The net output of enterprises must increase by 0.635 escudos, the value of government services by 1.042 escudos, the value of exports by 0.161 escudos. Finally, if *ceteris paribus* the demand of households for imports (y_3) increases by 1 escudo, the effects are as follows: We might expect the net product of enterprises to increase by 1 escudo, the value of government services to increase by 0.184 escudos, the value of exports by 1.253 escudos.

This analysis of the Portuguese economy should be considered only as an example which demonstrates the general methodology of input-output analysis. For a serious investigation of a given economic system the economy must of course be divided into many sectors. Apart from this great simplification, there are certain other hypotheses which underlie the analysis: We assume constant coefficients of production: that is, in each sector outputs are strictly proportional to inputs. Also, free and perfect competition is assumed. Capital and stocks are here neglected. No attention is paid to monetary phenomena. The final demand of the families for the goods and services of the various sectors is assumed to be given. It should perhaps be pointed out that all these assumptions are not very realistic for the Portuguese economy. Systems of this kind are called input-output systems (Leontief systems). They are of course more useful if they contain more sectors than our small model. They can also be generalized to include capital and other phenomena (Leontief 1951; Dorfman, Samuelson, and Solow 1958) and they play a certain role in the theoretical analysis of dynamic phenomena (Morishima 1964). The development of input-output systems (Leontief 1951), which has been exemplified above, belongs to the most successful and interesting advances of econometrics. Many such empirical systems with a great many sectors have been constructed in various countries (Chenery and Clark 1959). They are very useful for short term prediction and may also be combined with other econometric models, which are not necessarily linear, for the study of a national economy like the recent extensive model of the United States (Duesenberry *et al* 1966).

Non-Static Systems

The philosopher Northrop (1947, p. 235) makes an important contribution with his greatly detailed examination of what he calls "classical economic science." He comes to the conclusion that on this basis no economic dynamics is possible. This conclusion is certainly justified insofar as the systems investigated are by their very assumption static. In order to make them dynamic (i.e., in order to introduce time) special additional assumptions are needed, which are mostly concerned with the theory of anticipation of relevant economic quantities (Hicks 1946; Tintner 1942*a-c*, 1941) — for example, anticipated prices (G. G. Granger 1955, p. 88 ff., 1960).

All economic phenomena which are not static we call non-static. Following the ideas of Knight (1933), non-static economic phenomena are classified as follows: dynamics — single valued anticipations; risk — existence of a single known probability distribution of anticipations; uncertainty — existence of several probability distributions of anticipations, perhaps connected by an a priori probability distribution.

In Hicks (1946) view of dynamics anticipations are single valued — that is, we assume that faced with a decision about present behavior which involves the future, the individual households and firms have unique and single valued anticipations (Tintner 1941, 1942*a-c*). This is, of course, only a limiting case. Consider a farmer who has to plan his crop. He must take into account the anticipated future price of the commodity he produces at the time when the production will be finished.

As an example, consider an empirical investigation of the production of pork in Austria, 1948–55 (Tintner 1960*c*, p. 85). Denote the logarithm of the quantity of pork offered in the year

t by X_{1t} and the logarithm of the price of pork in the year t by X_{2t}. The estimate of the supply function of pork in Austria is

$$X_{1t} = 0.81 + 0.74X_{2t-1}. \tag{29}$$

The assumption here is that the anticipated price of pork is equal to the existing price of pork at the time of the start of production. This equation gives us an estimate of the elasticity of the supply of pork—namely, 0.74. This has to be interpreted in this way: Assume that the price of pork increases by 1 per cent in a given year. We might then predict an increase of the supply of pork in the next year of about ¾ per cent. Dynamic problems have been studied by many economists and econometricians, especially Roos (1934) and Evans (1930). There are also related studies of business cycle phenomena—for example, Kalecki (1935). In recent years the problem of economic growth has come into the foreground, especially in connection with the underdeveloped countries (Baumol 1959; Higgins 1959).

Friedman (1957) has computed a "dynamic" consumption function for the United States 1905–1951 (war years omitted). Let C_t be consumption and R_t disposable income in the year t. The empirical relation is

$$C_t = 0.58R_t + 0.32R_{t-1} + 53. \tag{30}$$

The short term marginal propensity to consume is 0.58 and the long term marginal propensity is $0.58 + 0.32 = 0.90$—that is, if disposable income (income received by consumers) increases by \$1.00, in the short run consumption will increase by \$0.58 and in the long run by \$0.90. A more complicated model gives

$$\begin{aligned} C_t = 0.29R_t + 0.19R_{t-1} + 0.13R_{t-2} \\ + 0.09R_{t-3} + 0.06R_{t-4} + 0.04R_{t-5} + \dots - 4. \end{aligned} \tag{31}$$

In this more comprehensive model the short term propensity to consume is 0.29, and the long term propensity is 0.88. This is of course a very highly aggregated model and some of its assumptions must be taken into account: There is no change in the consumption habits of the persons in the economy involved during the period investigated. Consumption depends *only* upon disposable income and anticipated disposable income, which is approximated as a linear function of the disposable income in the past. More complicated and more truly dynamic models which make consumption depend upon the highest experienced income of the past are by Modigliani (1949) and Duesenberry (1949).

It is undeniable that ideas about economic development which have been discussed in the recent past (Harrod 1952; Higgins 1959; Domar 1957; von Neumann 1945) are closely related to concepts of the classical school of economics and of Marx (Lange 1963). This is particularly apparent in the recent book of Morishima (1964, pp. 136 ff.) which introduces Marxist ideas explicitly.

One of the most interesting and most discussed models of economic development is the one proposed by von Neumann (1945). In Morishima's interpretation (1964, pp. 131 ff.) consumption of goods takes place only through the processes of production, which includes necessities of life consumed by the workers. Wages are at the subsistence level, and all capitalist income is reinvested. This model can be illustrated by an example in which only two goods are produced and three processes of production available, although the model was given originally for any number of goods and processes.

Let c_{ij} by the input coefficient of process i for good j. This includes the quantity of good j technologically necessary per unit of process i, and also the minimum of good j necessary to persuade people employed in process i to work. The output coefficient b_{ij} is the quantity of good j produced in process i. The coefficients are assumed to be constants. Each process has unit duration (e.g., one year). Processes of longer duration than one time unit are broken down into a number of processes with unit duration.

Let $f(t)$ be the interest factor in period t, that is, one plus the rate of interest (expressed in decimals). Thus in this notation if the rate of interest is 5 per cent, then $f(t) = 1.05$. Also, $q_1(t)$, $q_2(t)$, $q_3(t)$ are the intensities of the processes of production (e.g., number of bushels of corn produced, etc.). $P_1(t)$ and $P_2(t)$ are the prices of the two commodities. Since competitive equilibrium is assumed, there cannot be any processes which yield a return *greater* than the prevailing rate of interest, because under perfect competition extra profits would attract competitors to use the same process, and the prices of factors of production would rise. Hence symbolically,

$$b_{11}P_1(t+1) + b_{12}P_2(t+1) \leqq f(t)\,[c_{11}P_1(t) + c_{12}P_2(t)] \tag{32}$$

and

$$b_{21}P_1(t+1) + b_{22}P_2(t+1) \leqq f(t)\,[c_{21}P_1(t) + c_{22}P_2(t)]$$

and

$$b_{31}P_1(t+1) + b_{32}P_2(t+1) \leqq f(t)\,[c_{31}P_1(t) + c_{32}P_2(t)].$$

The first inequality refers to the first process. It shows that the output of this process evaluated at the prices prevailing at time $t+1$ cannot be greater than the cost of this process, evaluated at the prices at time t and multiplied by the interest factor, $f(t)$. If the inequality sign holds, the process will not be used; then $q_1(t)$ is equal to zero. Similar considerations hold for the second and third commodity. Hence it follows that in equilibrium

$$\begin{aligned} &q_1(t)b_{11}P_1(t+1) + q_1(t)b_{12}P_2(t+1) \\ &+q_2(t)b_{21}P_1(t+1) + q_2(t)b_{22}P_2(t+1) \\ &+ q_3(t)b_{31}P_1(t+1) + q_3(t)b_{32}P_2(t+1) = \\ &f(t)\,[q_1(t)c_{11}P_1(t) + q_1(t)c_{12}P_2(t) + q_2(t)c_{21}P_1(t) + \\ &q_2(t)c_{22}P_2(t) + q_3(t)c_{31}P_1(\mathrm{t}) + q_3(t)c_{32}P_2(t)]. \end{aligned} \tag{33}$$

The left hand side shows the total value of the production at time t evaluated with the prices at time $t+1$ when the production process is finished. The right hand side is the total cost of production, evaluated at prices at time t, times the interest factor. These two quantities must be equal for equilibrium.

Since one cannot consume more of a given good than is available from the production of the preceding period, we have the inequalities,

$$q_1(t-1)b_{11}+q_2(t-1)b_{21}+q_3(t-1)b_{31} \geqq q_1(t)c_{11}+q_2(t)c_{21}+q_3(t)c_{31} \qquad (34)$$

and

$$q_1(t-1)b_{12}+q_2(t-1)b_{22}+q_3(t-1)b_{32} \geqq q_1(t)c_{12}+q_2(t)c_{22}+q_3(t)c_{32}.$$

The first inequality refers to the first commodity. On the left hand side we have the total amount of this good produced at period $t-1$. This must be greater than or equal to the quantity of this good necessary as an input at period t. If the inequality sign actually holds, the good is a free good and its price $P_1(t)=0$. We argue similarly for the second commodity thus,

$$q_1(t-1)b_{11}P_1(t)+q_1(t-1)b_{12}P_2(t)+q_2(t-1) b_{21}P_1(t)+q_2(t-1)b_{22}P_2(t)+q_3(t-1)b_{31}P_1(t)+ q_3(t-1)b_{32}P_2(t)= q_1(t)c_{11}P_1(t)+q_1(t)c_{12}P_2(t)+q_2(t)c_{21}P_1(t)+ q_2(t)c_{22}P_2(t)+q_3(t)c_{31}P_1(t)+q_3(t)c_{32}P_2(t). \qquad (35)$$

This equation tells us that under our assumption of free competition the total value of the output of the economy in period $t-1$ evaluated at prices at period t (left side of the equation) must be exactly equal to the total value of the input necessary at period t (right side of the equation).

We will have *balanced growth* if the prices are constant:

$$P_1(t)=P_1(t+1),\ P_2(t)=P_2(t+1). \qquad (36)$$

If the interest factor is constant,

$$f(t)=c. \qquad (37)$$

If intensities of processes in subsequent periods are proportional,

$$q(t) = aq(t-1). \tag{38}$$

Here a is constant, equal to one plus the rate of balanced growth, $a=c$. This implies exponential growth, or growth in a geometric series.

Von Neumann assumed that all the input and output coefficients are non-negative, but he also assumed, rather unrealistically, that every good is involved as input or output in every process. In that case the system yields unique solutions for c and a, and the interest and growth factor are equal. There may be, however, several solutions for the intensities $q_i(t)$ and the prices $P_j(t)$.

In a generalization of the model Kemeny, Morgenstern, and Thompson (1956) make additional assumptions: The total value of all goods produced must be positive, every good can be produced by some processes, and every process uses some inputs. Under these conditions, but without the unrealistic assumption of von Neumann, they are able to prove the existence but not the uniqueness of the growth and interest factors, which are again equal.

The von Neumann model leads to exponential growth—that is, growth in a geometric series. If some further quite restrictive assumptions are made (Morishima 1964, pp. 154 ff.; Radner 1961), then it can be proved that the following holds: If an economy starts from any quite arbitrary situation and wishes to reach a certain final situation, also almost as arbitrary, it will in the long run be best to follow von Neumann—that is, the behavior of the corresponding von Neumann model (turnpike theorem). This theorem, in spite of very restrictive assumptions, is useful for problems connected with the development of underdeveloped countries.

The Ramsey (1928) model in the form given by Stone (1962) illustrates dynamic models in the narrow sense that we present. (See also Allen 1949, pp. 536 ff.) This highly aggregated model assumes the existence of a production function;

$$Y = f(L, K); \tag{39}$$

where Y is a product in general, L labor, and K capital stock. In this model, like in many more modern models of economic development, we reduce the whole economy to one single sector. All labor utilized is aggregated into an index of total labor used in the economy. Also, the value of all capital goods (factories, machinery, land) used in the economy is represented by a single index of capital goods. Similarly, it is assumed that all consumption goods used by various individuals in the economy can be aggregated into a single index of consumption. The satisfaction enjoyed by consumption will be represented by a single, average utility function: $\partial Y/\partial L$ is the marginal product of labor—that is, the increment in the total product (Y) if labor (L) increases by a small unit. Similarly, $\partial Y/\partial K$ is the marginal product of capital—that is, the increment in total product (Y) resulting if capital stock (K) increases by a small unit. The time derivative of the capital stock $K: I = dK/dt$ is investment. Let C be consumption. Then by definition

$$C = Y - I = Y - dK/dt \tag{40}$$

Consumption is the difference between total product and investment. Let the utility of consumption be

$$M = h(C). \tag{41}$$

Utility is assumed to be measurable. The derivative of this function $h'(C)$ is the marginal utility—that is, the additional utility of a small increment of consumption. The disutility of labor is

$$N = g(L). \tag{42}$$

Again disutility is measurable. The derivative of this function $g'(L)$ is marginal disutility—that is, the incremental disutility of a small increase in labor. Hence, total utility is

$$U = M - N. \tag{43}$$

Total utility is utility of consumption minus disutility of labor. To maximize,

$$W = \int_{t_1}^{t_2} U \, dt. \tag{44}$$

This is the sum (integral) of total utility between t_1 and t_2. By using the methods of the calculus of variations (Allen 1949, pp. 521 ff.) we derive the following results:

$$\partial Y/\partial L = g'(L)/h'(C). \tag{45}$$

At all times the marginal product of labor must equal the ratio of the marginal disutility of labor to the marginal utility of consumption.

$$\partial Y/\partial K = \frac{dh'(C)/dt.}{h'(C)} \tag{46}$$

At all times the marginal product of capital must be equal to the proportionate rate of decrease of the marginal utility of consumption over time. This purely theoretical model throws some light on a problem which is of great importance in connection with economic development, especially of underdeveloped countries: What is (in a sense) an optimal allocation between consumption and investment?

By abandoning the somewhat unreal but convenient assumption of single value anticipations, we have a case of *risk* (Tintner 1941, 1942*a-c*). Here we assume not a unique value of anticipated conditions (prices etc.), but probability distributions of anticipations. At least these probability distributions, however, are known.

As an example, consider again the situation of an Iowa farm (Tintner 1960*a*). We assume that the farmer uses only two factors of production, land and capital, since labor is abundant. With data from the period 1938–52, we estimate the probability distribution of the input coefficients of land and capital in the production of corn and flax. We then determine by numerical methods the approximate probability distribution of the short run net profit of the farmer. This is a problem of *stochastic programming.* The input coefficients are assumed to be normally and independently distributed. By numerical methods we may approximate the probability distribution of short term profits (thus in our example the arithmetic mean of profits is $11,081).

TABLE 5

ARITHMETIC MEAN OF PROFITS ACCORDING TO PROPORTIONAL USE OF *Land* VS. *Capital* FOR CORN AND FLAX
(In Dollars)

PROPORTION OF CAPITAL USED FOR PRODUCTION OF CORN (Rest in Flax)	PROPORTION OF LAND USED FOR PRODUCTION OF CORN (Rest in Flax)		
	None	One-Half	All
None	5,704	5,168	0
One-Half	4,082	7,008	4,945
All	0	5,075	8,472

This is called the *passive* approach. It may be used for comparing, for example, the probability distribution of profits for a farm in Iowa with those of a farm in California.

More important perhaps is the *active* approach in which the decision variables are the proportions of the factors of production (land and capital) assigned to growing various crops (corn and flax). In Table 5 the arithmetic mean of the estimated probability distribution of short term profits for various possible allocations of the factors of production (land and capital) is shown. The entries in this table must be interpreted in the following way: Assume that the farmer uses no land for corn (hence all the land for flax). He also utilizes no capital for corn and all the capital for flax. Then, his average profit in the short run will be $5,704; but if he divides both land and capital evenly between the two commodities corn and flax, he will receive on the average $7,008. It is easily seen from this table that the best policy for the farmer is the following: He should devote all his land and all his capital to corn, none to flax. Then, under our very simplified conditions his average short term profit will be $8,472, which is the highest he can obtain.

We should mention again that of course some very unrealistic assumptions underly our analysis. The probability distribution of our input coefficients is assumed known; actually it is estimated from past experience. Very crude numerical methods have been utilized to estimate the mean values of profits found in the table. The enormous amount of computations involved, even for a simple example, makes this method of stochastic

TABLE 6

DECISIONS UNDER UNCERTAINTY—*Individual* VS. *Nature*

STRATEGIES OF THE INDIVIDUAL (Values of a)	STRATEGIES OF NATURE (Values of x)				ROW MINIMA	ROW MAXIMA
	0	1	2	3		
0	0	0	0	0	0	0
1	—0.75	0.25	0.25	0.25	—0.75	0.25
2	—1.50	—0.50	0.50	0.50	—1.50	0.50
3	—2.25	—1.25	—0.25	0.75	—2.25	0.75
COLUMN MAXIMA	0	0.25	0.50	0.75		

linear programming impractical for the study of important and realistic empirical problems. (See also Sengupta, Tintner, Morrison 1963.)

The case of risk concerns a situation in which the relevant probability distributions are known. This means in practice that we are dealing with a stable situation (unchanging tastes and technology) and that we have ample experience in the past to estimate accurately the relevant probability distributions. If this is not the case, if the underlying probability distributions are not known, we deal with a case of *uncertainty*.

Consider now the general problem of decisions under uncertainty (Tintner 1959, 1966). This is also called "games against nature" (Milnor 1964). We utilize a game theoretical setup and assume that the strategies of nature are $x = 0, 1, 2, 3$ and that the possible strategies of the individual playing against nature are the actions $a = 0, 1, 2, 3$. (See Table 6.) If the individual chooses the action a = 1, he will get − 0.75 if nature plays the strategy x = 0 and 0.25 if nature is in the state x = 1, 2 or 3.

According to the *minimax criterion* of Wald (1950; see also Davidson *et al.* 1957) the individual has to treat nature as if he was playing a two person zero sum game. He has always to expect the worst. This is expressed in our table as the minimum for each row – that is, for each strategy of the individual. It is sensible under such conditions that the individual will maximize the row minimum – that is, choose the largest figure among the row minima. Hence, in our case the individual will choose action a = 0, since this is the maximum of all the figures in the column of row minima. This procedure has been criticized from the point of view of personal subjective probability by Savage (1954, pp. 200 ff.; see also Good 1950).

TABLE 7

SAVAGE'S MATRIX APPLIED TO GAMES AGAINST NATURE

STRATEGIES OF THE INDIVIDUAL (Values of a)	STRATEGIES OF NATURE (Values of x)				ROW MINIMA
	0	1	2	3	
0	0.00	−0.25	−0.50	−0.75	−0.75
1	−0.75	0.00	−0.25	−0.50	−0.75
2	−1.50	−0.75	0.00	−0.25	−1.50
3	−2.25	−1.50	−0.75	0.00	−2.25

The *regret* matrix of Savage (1954 pp. 163 ff.) is formed by considering the regret of the acting individual as the loss occurring between the actual result for a given strategy and the result which could be obtained if the state of nature was known. By deducting the column maxima from each figure in its given column of Table 6 we obtain the regret matrix of Savage (see Table 7). By the application of the minimax principle to the regret matrix, the individual will choose the maximum from among the row minima. Thus, in our case, he is free to choose $a = 0$ or $a = 1$.

Another criterion has been given by Hurwicz (Luce and Raifa 1957, pp. 282 ff.). Consider an individual who is influenced by the worst (row minimum) and the best (row maximum) that could happen for each of his strategies. His criterion is a weighted arithmetic mean of the maximum and minimum (see Table 8). The weights chosen may be regarded as measures of the optimism or pessimism of the individual. For instance, if he gives equal weight to the best and worst for each strategy (0.5min + 0.5max), he will choose $a = 0$ as the best strategy. On

TABLE 8

MEASURE OF INDIVIDUAL OPTIMISM OR PESSIMISM

Strategy of the Individual (Value of a)	Row Minimum (Table 6)	Row Maximum (Table 6)	0.5 Minimum + 0.5 Maximum	0.1 Minimum + 0.9 Maximum
0	0.00	0.00	0.00	0.00
1	—0.75	0.25	—0.25	0.15
2	—1.50	0.50	—0.50	0.30
3	—2.25	0.75	—0.75	0.45

TABLE 9

LAPLACE'S CRITERION

Strategy of the Individual (Value of a)	Expected Profit
0.	0
1.	0
2.	—0.25
3.	—0.75

the other hand, if he is more optimistic and acts according to 0.1min + 0.9max, he will choose $a = 3$.

Another method is Laplace's criterion. Consider all 4 strategies of nature equally probable and compute the mathematical expectation for each strategy (see Table 9). In our example the individual who wants to maximize the expected profit (computed with the help of the Laplace assumption) will have the choice between strategies $a = 0$ and $a = 1$. Other possibilities are the assignment of a priori probabilities with the help of the theory of logical probability of Carnap (1950).

Here I want to show how one version of the probability theory of Carnap might be used in the face of complete uncertainty to construct a decision model. We are dealing with a perishable commodity and also an entirely new commodity where no past experience is possible. Consider for instance a man who contemplates constructing a rocket for commercial travel to the moon (Tintner 1959, 1960d, 1966). This is a simple version of an inventory problem. How big a rocket should he construct?

Assume that there are only two customers, C_1 and C_2. Each

TABLE 10

Application of Carnap's Theory of Logical Probability
(Number of Tickets)

State Description	Customer's Action		A Priori Probability
	C_1	C_2	
1	0	0	1/6
2	0	1	1/12
3	1	0	1/12
4	0	2	1/12
5	2	0	1/12
6	1	1	1/6
7	1	2	1/12
8	2	1	1/12
9	2	2	1/6

of these might buy 0, 1 or 2 tickets. In this universe Table 10 shows us all that could happen. Each line in this table is a *state description.* In the first line none of the customers buys a ticket. In the second the first customer buys none, but the second buys one. In the third C_1 buys one, and the C_2 buys none. In the fourth the C_1 buys no ticket, but the second customer buys two, and so on.

Now, according to Carnap's theory, we must treat the two individuals on a par. The entrepreneur does not care who buys the tickets but only how many are sold. Hence we see that certain state descriptions may be obtained by permutation of the two individuals. These are the *structure descriptions*, which

TABLE 11

GAIN OF THE ENTREPRENEUR
(In Money Units)

NUMBER OF SEATS SOLD (x)	A PRIORI PROBABILITY (Px)	GAIN PER NUMBER OF PLACES (Values of a)				
		0	1	2	3	4
0	1/6	0	—1	—2	—3	—4
1	1/6	0	1	0	—1	—2
2	1/3	0	1	2	1	0
3	1/6	0	1	2	3	2
4	1/6	0	1	2	3	4
MATHEMATICAL EXPECTATION		0	2/3	1	2/3	0

are classes of equivalent state descriptions. In table 10, structure descriptions are separated by horizontal lines. Hence, state descriptions 2 and 3 form one structure description, 4 and 5 constitute another structure description, and 7 and 8 still another structure description. State descriptions 1, 6, and 9 each form a structure description. According to one version of Carnap's probability theory (see Table 10) we give each structure description the same a priori probability. Since there are 6 structure descriptions, each receives probability 1/6; within each structure description each state description receives the same probability. The a priori probabilities of the state descriptions are indicated in the last column of the table.

Now assume that it costs 1 money unit to construct one place in the rocket and that each ticket may be sold for 2 money units. Hence, if a rocket of a places is available and x places are sold, the profit is given by

$$P = 2x - a, \quad 0 \leqq x \leqq a, \tag{47}$$

and

$$P = a \quad , \qquad x > a.$$

For various sizes of the rocket (a) and for various numbers of places sold (x) we obtain Table 11.

The a priori probabilities are taken from Table 10. Assume

that $a = 2$ (i.e., a rocket is constructed which has 2 places) and $x = 0$ seats are sold; then the gain of the entrepreneur is -2. If $x = 1$ seat is sold, the gain is 0. If $x = 2$, 3, or 4 seats are sold, the gain is 2. By using Carnap's probabilities in order to compute the mathematical expectation (weighted arithmetic mean of gains and losses with the a priori probabilities as weights), we see that it will, under the given circumstances, be most profitable to construct a rocket which has $a = 2$ places. In this case the average gain, 1 money unit, is the highest.

The theory of action under uncertainty has received much stimulation through ideas borrowed from game theory (von Neumann and Morgenstern 1944), statistical decision theory (Wald 1950), and the theory of personal (subjective) probability (Savage 1954; Schlaiffer 1959).

III. Econometrics

Probability

Econometrics is an important special method for the evaluation of mathematical economic models in numerical terms and for the verification of economic theories; it uses the methods of modern statistics for this purpose. The importance of statistics for economics has been long recognized: "In connection with the process . . . of verification and the discovery of disturbing causes, or (to express the same idea differently) the discovery of the minor influences affecting economic phenomena, we find the proper place of statistics in economic reasoning" (Cairnes 1875, p. 85). "The deductive science of Economics must be verified and rendered useful by the purely empirical science of Statistics" (Jevons 1911, p. 22).

Modern statistics, which is based upon the theory of probability, may be described as applied probability. Probability (Nagel 1939) has been defined in various ways. The *classical* definition of probability defines it as the ratio between the number of favorable and the number of equally likely cases. Consider a die with six faces and the probability of throwing an ace. There is one favorable case and six equally likely cases. Hence the a priori probability is 1/6, according to the classical definition.

This definition is unsatisfactory, especially since it is in many cases very difficult, or even impossible, to determine what the equally likely cases are. A purely empirical definition of probability defines it as related to the *relative frequency* of events in a long series of trials. It may be conceived as the limit of this relative frequency as the number of trials increases indefinitely (R. von Mises 1951). Most statisticians follow Fisher, who defined probability as the relative frequency in an infinite hypothetical population. Kolmogoroff (1933) defines probability axiomatically in terms of modern set and measure theory

(Cramér 1946, pp. 137 ff.). To illustrate the axiomatic system of the probability calculus, take the following axioms of Fisz (1963, pp. 12–13).

Assume a set of elementary events: to every random event there corresponds a certain number, called its probability, which lies between zero and one; the probability of the sure event equals one; the probability of the alternative of the finite or denumerable infinite number of pairwise exclusive events is the sum of the probabilities of these events.

From these axioms follow a number of theorems: if a set of events exhausts the set of elementary events, their probability is one (certainty); the sum of the probabilities of any event and its complement is one; the probability of the impossible event is zero, and so on. This abstract concept of probability is then identified with relative frequency in certain random experiments.

Because of its great importance in recent discussions, we should be cognizant of *Bayes' theorem* (Lindley 1965, vol. 1, p. 20). Let B be some event and denote its probability by $p(B)$. Let $A_1, A_2, \ldots, A_n$ be a set of exclusive and exhaustive events and denote the probability of A_i by $p(A_i)$. Furthermore, let $p(A_i/B)$ be the probability of A_i, given B, and $p(B/A_i)$ be the probability of B, given A_i. These are conditional probabilities. Then we have

$$p(A_i/B) = p(B/A_i)p(A_i)/p(B/A_1)p(A_1) + p(B/A_2)p(A_2) + \ldots + p(B/A_n)p(A_n). \tag{48}$$

To illustrate Bayesian methods, we take an example from Good (1965, pp. 12 ff.). Consider a simple sample where the result of N independent trials is either a success or a failure. Suppose we have obtained r successes and s failures $(r+s=N)$. Let p be the a priori probability of a success and $1-p$ be the a priori probability of a failure. Assume that the sequence of trials is permutable (De Finetti 1937) – that is, a success does not depend upon the previous sequence of successes and failures. Good suggests for the estimation of p (a priori probability of success) a beta distribution, which is proportional to $p^a(1-p)^b$. The mathematical expectation of a success is, given r successes and s failures in a sample of N independent trials,

$$\mathrm{E}=\frac{a+r+1}{a+b+N+2}. \tag{49}$$

This should be compared with Laplace's law of succession, which gives

$$L=\frac{r+1}{N+2}, \tag{50}$$

and with the maximum likelihood estimate of classical statistics,

$$M=r/N. \tag{51}$$

In contrast to the concept of statistical probability, related to relative frequency, one school of theoreticians considers a *logical concept* of probability, which is frequently called subjective or personal probability. The most important writers in this field are J. M. Keynes (1948), Ramsey (1928), Jeffreys (1948), De Finetti (1937), and, most recently, Savage (1954). See also Kryburg and Smockler (1964); Good (1965). Here probability is conceived as a logical concept, the rational degree of belief in a proposition based upon a certain amount of evidence. The "subjective" concept of probability, the degree of belief, has recently achieved some prominence (Savage 1962, Lindley 1965). The so-called Bayesian approach in statistics is essentially based upon it.

If probability is defined as being subjective then this concept is really a subject of psychology. If the subjective concept of probability is interpreted as ideal rational behavior, it seems that it is related to the logical concept of probability as developed by Carnap (1962).

The philosopher Carnap (1950; 1952) has pointed out that we should really distinguish between these two ideas of probability (the frequency concept and the logical concept) and that both are useful. In many scientific endeavors we utilize probability in the statistical sense as a quantity related to relative frequency, an empirical concept (Good 1965). On the other hand, it is also necessary to talk about the probability (in the logical sense) of a proposition or theory. Carnap (1962) and Kemeny (1963) have tried to lay the foundations for such a concept, which they call *degree of confirmation.* (See also Fels 1963; Tintner 1949.) A solution which would be practically useful in a science like physics or economics is still to be found. We can illustrate this concept by an example of Kemeny's (1963). We must have a certain "language" (in the semantic sense). Up to now the theory has only been developed for languages containing a finite number (or denumerable infinite number) of predicates. The full language of physics (or economics) cannot be included in the analysis. A model or interpretation of a logistic system must be defined. The number of models is infinite, and the domain of the individuals must be finite. If e is given as the sentence which stands for the evidence and h as the sentence which stands for the hypothesis, we can denote the degree of confirmation of the hypothesis h given the evidence e by $c(h,e)$.

The following conditions are imposed upon the degree of confirmation C:

*CA*1: C must define a system of betting which is coherent (fair) – (a) $0 \leq c(h,e) \leq 1$; (b) if h and e are logically equivalent to h' and e', then $c(h,e) = c(h',e')$; (c) if e logically implies h, then $c(h,e) = 1$; (d) if e logically implies the negation of (h and h'), then $c(h \text{ or } h',e) = c(h,e) + c(h',e)$; (e) $c(h \text{ and } e',e) = c(e',e) \cdot c(h,e \text{ and } e')$ where e and e' are not self contradictory.

In case we require strict coherence (fairness) (c) becomes $c(h,e) = 1$, if and only if e logically implies h. The concept of coherence (fairness) is defined as follows: Let $c(h,e) = r$, then the odds of a bet are $r/(1-r)$. In a coherent (fair) betting system (Ramsey 1928, De Finetti 1937) no profits are possible. Also the gambler can not be sure if he will win or lose. The case is then strict coherence (fairness)· (f) $c(h,e) = m(h \text{ and } e)/m(e)$.

Here m is a measure function. Equivalent sentences have the same measure. If h and h' are mutually exclusive, (d) becomes $m(h \text{ or } h') = m(h) + m(h')$. The set of models is disjoint, and the measure is additive. The measure of the set of all models is one, and the measure of the empty set is zero. Given a finite set and a measure which is additive: the measure of the universal set is one; of the empty set is zero, each measure is nonnegative; if the measure is defined over two subsets, it is also defined for all complements, sums, and products, and then it can be extended to all subsets. Hence only weights (measures) have to be assigned to models. If strict coherence is required, then $m(h) = 1$, only if h is analytic.

CA2: $c(h,e)$ is to depend only on the proposition expressed in h and e. We may select a minimal language, which allows for finiteness.

CA3: Constants which are logically alike can be treated a priori alike. This is the requirement of empiricism and is a version of the principle of indifference.

CA4: The definition of c must enable us to *learn from experience.* If we have a series of evidences consisting of more and more confirming instances, the c values must increase.

CA5: We need consider only that part of e which is relevant to h.

These conditions *CA*1–5 are sufficient to solve a dice problem. The minimal language is the set of properties: P_1, P_2, P_3, P_4, P_5, P_6, (outcomes of the throw of the die). The number of throws (individuals) must be finite. The language L_n assigns one property P_i to each individual (throw). We then have 6^n models. How are the weights assigned? Let us say that the hypothesis is h_i; this means that the next throw will be P_i, and the evidence e^k states the outcome of the k previous throws. We have n individual constants a_i (throws). A sentence, $P_{i_1}a_1 \,.\, P_{i_2}a_2 \,.\,.\,.\, P_{i_n}a_n$, is a state description.

By using Carnap's theory (1952), Kemeny comes to this important conclusion: That the degree of confirmation of the hypothesis h_i that the outcome of the next throw of the die will be P_i based upon the evidence e^k, that among k previous throws there have been k' throws with the result P_i and $k-k'$ throws with other results, depends only upon one single parameter L:

$$c(h_i, e^k) = \frac{k' + L/6}{k + L}. \tag{52}$$

L may be any non negative real number and may be interpreted as an index of caution. Consider the case where $k' = k$ – that is, where all previous throws are favorable. Then when $c = 1/2$, $k = (2/3)L$. This shows us after how many trials, all of them favorable, we can bet even money that the next throw will be P_i. If $L = 3$, we would do so after only 2 throws and after 200 consecutive successes if $L = 300$. The formula (52) may be written

$$c(h_i, e^k) = \frac{(k'/k) + (L/6k)}{1 + (L/k)}. \tag{53}$$

and we see that

$$\lim_{k \to \infty} c(h_i, e^k) = k'/k \tag{54}$$

For a long series of trials the "logical" factor L drops out, and the probability that the next throw will be a success is the relative frequency of successes, k'/k. This corresponds to the maximum likelihood estimate of classical statistics.

Carnap also points out that L may be considered a measure of our belief of how evenly we choose our alternatives. The less homogeneous the world is, the larger the optimal value of L. If $L = 0$, we have unbiased estimation in the statistical sense. A statistical estimate is unbiased, if its mean value (mathematical expectation, weighted arithmetic mean with the probabilities as weights) is equal to the parameter estimated in the unknown population. It would be best in a completely homogeneous world. If all k's of the k throws are aces, the probability of aces is one. Strict coherence (fairness) forbids this value.

Another interesting value is $L = \infty$, and $c = 1/6$ in Wittgenstein's method. Here we assign all state descriptions equal probability. Hence, the degree of confirmation is independent of past experience and forbidden by *CA*4, but it would be optimal in a world where individuals (throws) are evenly distributed among the 6 alternatives.

As a slight generalization, consider the problem of a *K*-faced die or a family of *K* (instead of 6) properties $P_1, P_2, \ldots, P_K$. Our formula becomes

$$c(h_i, e^k) = \frac{k' + L/K}{k + L}. \qquad (52')$$

Now, assume that $K = \infty$ — that is, a die with infinite number of faces.

$$c(h_i, e^k) = (k'/k)/\ 1 + (L/k) \qquad (53')$$

Again, what is the relation between *L* and *K*, if $k' = k$ — that is, if only favorable instances are observed? We have $L = k$ when $c = 1/2$ — that is, if $L = 1$, we will make an even bet that aces will turn up if only one ace has been observed. If $L = 10$, we will make an even bet on aces if we have 10 observations, all aces, and so on.

Laplace's rule of succession,

$$c(h_i, e^k) = (k' + 1)/(k + 2), \qquad (54')$$

must be restricted to simple alternatives — $K = 2$ to be consistent. This leads then to $L = 2$.

The problem of several families of predicates brings up some novel features of analogies (Carnap and Stegmueller 1958, p. 251 ff.) — for example, if balls of different colors and size are drawn from an urn, the combination of the two predicates, color and size, must be taken into account. This introduces another parameter. This problem is related to Nagel's (1939) problem of varieties of instances. Later work by Carnap (1963) connects his ideas with the subjective probability of Savage (1954).

It should be emphasized that the controversies about the foundations of probability are mostly of importance for the *interpretation* of probability statements. All writers in this field agree as far as the concrete laws of probability are concerned.

One branch of probability theory which has been very much developed in our time is the theory of *stochastic processes.* A stochastic process is a family of random variables depending upon a parameter (Fisz 1963, p. 272). There are many interesting applications in the fields of natural science but also in operations research. The applications in economics proper are still somewhat tentative. The use of the theory of stochastic processes in economics makes it possible to meet Hayek's objection (1952 pp. 61 ff.) that statistics cannot be applied to economics because our data must be samples of an unchanging population (see also F. M. Fisher 1960).

A stochastic process is stationary if its characteristics depend upon time differences and not upon absolute time. A stationary time series hence cannot have a trend; the variance and higher moments cannot change with time. For applications of stochastic processes to economics see Cootner (1964), Granger and Hatakana (1964).

An interesting application of a simple type of a stochastic process has been given by Blumen *et al.* (1955; see also Kemeny and Snell 1960, p. 199). It is based upon empirical data taken from social security records. The authors investigate empirically the transition probabilities of change between given occupational groups. Let $f(x, t\colon y, s)$ be the conditional probability (or probability density) when a random variable will have the value x at time t if it had the value y at time $s < t$: then this conditional probability is a transition probability. They find a remarkable agreement between the theoretically computed ultimate probability distribution and the empirical distribution. Similar methods have been used by Orcutt *et al.* (1961) for the analysis of various demographic problems—labor force, debt, and liquid asset behavior, and so on. (See also Kemeny and Snell [1962].)

We can take another example from operations research, which is, after all, nothing else but the econometrics of enterprise. This enterprise may be private or public. As a matter of fact, some of the best work in this field has been done by French econometricians in close connection with experiences in nationalized industries (Massé 1959; Lesourne 1960).

Consider a simple *queuing* process (Lindley 1965, vol. 1, p. 187). Denote the probability that at time t there are n customers in the queue (including those who are being served) by $p_n(t)$. In a small time interval Δt the probability that another customer will join the queue is approximately $L(\Delta t)$, and the probability that a customer will leave the queue is approximately $K(\Delta t)$. These probabilities are independent; L and K are constants. As time passes, the queuing process can be shown (under certain conditions) to settle down to a stationary situation which is quite independent of the initial situation. In the limit (i.e., after a reasonably long lapse of time) we get limiting probabilities. From these simple assumptions we may compute the limiting probability distribution, which is given by

$$P_r = R^r P_0, \tag{55}$$

where $R = L/K < 1$ and $P_0 = 1 - R$. This is the limiting probability that the server is not occupied. The limiting distribution is a geometric distribution, and R is the traffic intensity. With average size of the queue expressed as $R/(1-R)$, the expected waiting time of a customer is in the limit $R/K(1-R)$, and the expected busy time is $(K-L)^{-1}$. More complicated queuing processes have also been investigated. The methods are applicable to machine interference, and similar methods can be used on problems of dams and storage (inventories).

Another example is the stochastic theory of *economic development* (Sengupta and Tintner 1963, 1964; Tintner and Thomas 1963). By following the theory of *economic stages* by Rostow and others, we may approximate economic development by a simple type of stochastic process in each stage of economic development. This method has been applied to the data on British industry compiled by Hoffman. The whole series covers the period 1700 to 1939. The nature of the stochastic process at each stage, assuming that it changes from stage to stage was estimated. If only two types of regimes are recognized, the change is estimated to have occurred in 1834; with three regimes we have an estimated change in 1791 and 1869. If we assume the existence of four types of economic development, the change occurred, according to our estimation, in 1777, 1820, and 1869.

In a recent paper (Tintner and Patel 1966) we have tried to apply the idea of a *lognormal diffusion process* to the theory of economic development. A random variable (in our case real national income) is lognormally distributed, if its *logarithm* follows the familiar bell-shaped normal distribution (Aitcheson and Brown 1957). By making some assumptions about the nature of the transition probabilities, we can derive a lognormal distribution where the logarithms of real national income are normally distributed, but the arithmetic mean and variance (a measure of dispersion) of the logarithm of national income are linear functions of time. Then, national income itself has an exponential trend. This theory is verified and parameters of the process are estimated with the help of the method of maximum likelihood on the basis of Indian data, 1948–61. As a tentative application to policy, we also assume that government expenditure influences the transition probabilities in a simple way. Then the trend of national income becomes a function of total government expenditure. On the basis of this model we make predictions about the trend of Indian national income, assuming various plausible developments for future government expenditure.

Dynamic programming (Bellman 1957) is also most promising for the treatment of dynamic economic problems (Boulding 1950; Holt, Modigliani, Muth, and Simon 1960).

Models of this type make it possible to rationally treat problems of planning production, employment, and inventories for large private corporations. Also, there is the possibility of utilizing similar models in economic planning for a whole economy.

Statistics

Probability theory (and with it statistics, which is applied probability) finds itself in some predicaments today. There is still a large number of statisticians who cling to the traditional conception of probability, defined as being related to empirical relative frequency in the long run or in large (theoretically infinite) samples. There is, however, a growing movement toward the Bayesian point of view (Schlaiffer 1959). Traditional statisticians are reluctant to use this theorem, since from a classical point of view the conditions for its application are hardly ever present. The Bayesian point of view is connected with a revival of the notion of personal probability (Kryburg and Smockler, 1954) denoting the subjective belief of the investigator. If this belief is truly personal or subjective, it would make probability theory and statistics a subject for psychological investigation. This is also suggested by the large use of introspection and experimentation in this field. If, on the other hand, personal probability signifies the rational point of view then perhaps it must be interpreted in a sense like the Carnap's (1962) theory of the degree of confirmation.

The movement is still largely influenced by the purely pragmatic point of view of decision theory (Wald 1950, games against nature). The justification of the decision outlook in industrial and perhaps, some other applications cannot be denied. On the other hand, the particular method of reaching decisions (e.g., by the minimax principle), may appear more doubtful. But how can decision theory be applied in the field of scientific investigation, where the losses connected with various types of errors can hardly be taken for granted (R. A. Fisher 1956, pp. 99 ff.)? Also, in the field of econometrics the application of decision theory appears doubtful in view of the possible failure of welfare economics to produce on acceptable welfare function. Here, as in many other fields of application, the appropriateness of the decision theoretic point of view is questionable.

Because of its great importance for econometrics, we must make a more detailed examination of modern statistics. Since modern statistics is firmly based upon probability theory, a strict distinction must be made between the concrete empirical sample and the universe or population from which it is taken.

The fundamental ideas of modern statistics are applicable in econometrics, but they cannot be applied blindly or by analogy.

In many applications (agricultural experimentation, industrial experimentation, biological, and especially genetic applications) carefully planned experiments are possible. This is not true for economics, however, except in a very few cases. Hence, while in ordinary econometric work the fundamental ideas of modern statistics must be used, the specific methods useful with designed experiments (e.g., analysis of variance) are not necessarily applicable.

We use sampling methods to get information about a given universe or population. A random sample must be taken in such a way that each item in the population has the same chance of being chosen. More refined methods of sampling, like stratified sampling or systematic sampling are also useful for the collection of economic data.

To sample a numerical characteristic of a given population, say the expenditure on food by the urban population of Austria, care must be taken that every family has the same chance of being chosen as a member of the sample. This may be achieved by numbering all families and then by choosing the families in the sample with the help of a table of random numbers.

The fundamental problem of statistics is: given a sample, how can we derive inferences about an unknown population? A complete census would give us information about the numerical characteristic of which we are interested (the expenditure on food by the urban population of Austria); but since such a census is costly and frequently not feasible, we must use methods of sampling. The most fundamental method is estimation.

To estimate a characteristic of the unknown population (the average expenditure on food by the urban Austrian population), we compute a function of the observed values of our *sample*—for example, the arithmetic mean or median in the sample. To sample expenditure on food, let $x_1, x_2, \ldots, x_n$ be the food expenditure of the n families in a random sample. Then the sample mean $M = (x_1 + x_2 + \ldots + x_n)/n$ is the average food expenditure of the families in the sample. The median is computed as follows: The items are ordered according to size. In odd numbered samples items in the middle constitute the median, and in an even numbered sample the median is the arithmetic mean of the two items in the middle. What properties of statistical estimates are desirable so that we may choose between various possibilities?

We want the empirical estimate to be consistent. An estimator is consistent if it converges in probability to the parameter of the population estimated (Wilks 1962, p. 351). By convergence in probability we mean the following (Wilks 1962, p. 99): Consider a set of random variables (e.g., our estimators). The event that the value of our estimator deviates from the parameter in the population by more than a given (small) number is investigated. If the probability of this event tends to zero, as the sample becomes larger and larger, we have convergence in probability. If we sample from a normal population, which has a bell-shaped symmetrical probability density, for example, the arithmetic mean of a random sample is a consistent estimate of the mathematical expectation—that is, of the mean value of the unknown normal population. If we assume that the food expenditure by the Austrian urban population has a normal distribution, then the arithmetic mean of food expenditure of a random sample of families is a consistent estimate of the population mean—that is, the true (unknown) mean of food expenditure by the population of urban Austrian families.

Another property is *unbiasedness* of estimation (Cramér 1946, pp. 351 ff.). The mean value (mathematical expectation) of the estimate is equal to the estimated parameter in the population. Practically, this means that if many estimates of a given sample size are available and we compute the arithmetic mean of the estimates, the chances are that this general mean will deviate from the true value of the estimated parameter in the unknown population by a small number. The property of unbiasedness has been criticized by Carnap (1952) and also from a Bayesian point of view by Lindley (1965, vol. 1, pp. 130 ff.).

Under certain circumstances, the unbiased estimate (e.g., the arithmetic mean of the sample) will itself have a tendency to be normally distributed, at least in large samples. Such a distribution can be characterized by its mean value and by its variance, which measures the dispersion about the mean. A consistent estimate is *efficient* if (for a given sample size) its variance is as small as possible. Assume that we have a (large) random sample of n items from a normal population. The variance of this population is σ^2. The variance of the arithmetic mean of our random sample is σ^2/n. The sample mean is a consistent estimate of the population mean and is also unbiased: its own mathematical expectation is equal to the estimated parameter in the population.

To compute the sample median, the sample is ordered according to size and (for an odd n) the item in the middle or (for an even n) the arithmetic mean of the two items in the center is chosen. Again, this is a consistent estimate. It tends to be normally distributed for large samples. The variance of the sample median (for large samples) is $1.57\,\sigma^2/n$. Hence the sample mean is a more efficient estimate. Assume we have a sample of $n = 100$ items; then we must have a sample of $n = 157$ to achieve the same accuracy for the sample median as for the arithmetic mean, if we measure accuracy by the variance.

Sufficiency (Wilks 1962, p. 351) is a more complicated characteristic of statistical estimates. A sufficient estimate has the property that the conditional distribution of any other estimator, given the sufficient estimate, is independent of the estimated parameter. The arithmetic mean of a sample taken from a normal population with unknown mathematical expectation but known variance is sufficient. The distribution of any other estimator (e.g., the sample median), given a knowledge of the sufficient estimate (sample mean), is independent of the estimated parameter (mathematical expectation in the population). Sufficient statistics are of particular importance in the Bayesian approach (Savage 1962, Lindley 1965, vol. 2, p. 46 ff.)

The most important procedure of estimation in econometrics is R. A. Fisher's method of *maximum likelihood*. It is based upon the idea of choosing from the infinite number of possible values of the unknown parameter in the population the one which has, in view of the given sample, the highest chance to have produced the empirical sample. Unbiased estimates correspond to the arithmetic mean and maximum likelihood estimates to the mode. The mode is the value which has the highest probability or probability density. Under some conditions the results of maximum likelihood estimation are consistent and efficient for large samples; on the other hand, although they are not generally unbiased, they have the desirable property of being invariant against (non-singular) transformations, which is not in general true of unbiased estimates. For instance the maximum likelihood estimate of the standard deviation in the population, which is the square root of the population variance, will be the square root of the maximum likelihood estimate of the variance. The unbiased estimate of the standard deviation, however, is not, in general, the square root of the unbiased estimate of the variance.

In the modern Bayesian approach the method of maximum likelihood is justified as a convenient approximation (Savage 1954, pp. 224 ff.; Lindley 1965, vol. 2, pp. 128 ff.). For example, from a stratified sample of urban families in Austria (1954–5), consisting of 7,019 families, I have estimated the income elasticity of food expenditure by the method of maximum likelihood (Tintner 1960*c*, p. 36). The estimated income elasticity is 0.46. By making certain assumptions about the distribution of the elasticity in the unknown population, I chose, as my estimate, the population which had the maximum likelihood to produce the given sample. This figure should be interpreted as follows: If *ceteris paribus* (especially with unchanging prices and tastes) the income of the urban population in Austria increases by 1 per cent, we may expect an increase in expenditure on food by about 0.46 per cent—that is, among all possible unknown populations characterized by different income elasticities the one with the given income elasticity is the most likely to have produced the empirical sample.

Another important method of estimation which yields unbiased estimates is that of *least squares*. Here an estimate is found by making a minimum of the sum of squares of the deviation of the observations in the sample from the estimated value. The results of maximum likelihood estimation and the estimation by the method of least squares are frequently (but not always) identical. In our example, for instance, the estimate of the income elasticity as 0.46 per cent is also the estimate obtained by the method of least squares.

In econometrics, as in other applications of statistics, we frequently desire not only single estimates (point estimates) but also regions into which the estimated parameter may fall with given probability (e.g., 95 per cent). There are two methods, which in most cases will lead to the same numerical results, but have different interpretations (Fraser 1961). In our example, *fiducial limits* for the elasticity with 95 per cent probability are 0.49 and 0.44. The fiducial probability of 95 per cent leads to the conclusion that the unknown universe of the urban population in Austria from which our sample is taken will react as follows, *ceteris paribus*, to an increase of income by 1 per cent: Food expenditure will increase by not less than 0.44 per cent and not more than 0.49 per cent (Tintner 1960*c*, p. 38). A similar interpretation might be given from the Bayesian point of view (Lindley 1965, vol. 2, p. 15).

Fiducial probability is a concept related to logical probability and cannot be interpreted in the frequency sense; hence many statisticians object to it. In contrast to Fisher, Neymann (1938) considers one single population with a single unknown parameter to be estimated; for example, our income elasticity in the unknown universe of the urban population of Austria. He contemplates, however, a statistician who makes estimates very frequently and computes limits for his estimates—here called *confidence limits*. These limits are random variables since they are functions of the items in the sample which are themselves random variables. The limits will sometimes include the estimated (constant) parameter of the population (the true income elasticity)—sometimes not. It is these limits, which are random or chance variables. Suppose the statistician fixes the limits in such a fashion that they will include the unknown parameter in 95 per cent of all cases and the parameter will fall outside the limits in only 5 per cent of all cases. Then, we will obtain exactly the same limits as before, but their meaning is different.

From a strictly Bayesian point of view we might derive different limits (Schlaiffer 1959, pp. 661 ff.).

Our limits, now interpreted as confidence limits, show that there is only one unknown population, characterized by the unknown true parameter in the universe (income elasticity). If we determine limits by methods that have a probability of 95 per cent (now interpreted in the frequency sense) of including the parameter in a long series of trials, the limits are 0.44 and 0.49.

Another important idea is R. A. Fisher's concept of *test of significance*. Here we test a hypothesis called the null hypothesis (e.g., the hypothesis that the income elasticity in the unknown population is zero). What are the chances that in the unknown universe as income increases there will be *no* reaction as far as food expenditure is concerned? To test this hypothesis we must arbitrarily fix a level of significance — that is, a probability which is such that the null hypothesis will be rejected if the sample could have been obtained with a smaller probability. Assume a level of significance of 1 per cent — that is, we will reject all null hypothesis with a probability of less than 1 per cent. On the other hand, we will not reject null hypotheses which could have arisen with a probability of more than 1 per cent. It is unlikely, on the basis of our sample, that the urban population in Austria will not change its food expenditure if income increases. Thus, the test of our null hypothesis gives us a probability which is much smaller than 1 per cent or even than 0.1 per cent; hence, we may reject the null hypothesis (Tintner 1960*c*, p. 40). Significance tests have been treated by Lindley (1965, vol. 2, p. 58 ff.) from a Bayesian point of view.

There is a divergence among statisticians about testing hypotheses. Neymann (1938) and E. S. Pearson not only consider the probability of rejecting a true hypothesis (level of significance — called errors of the first kind, Lehmann 1959) but also consider the existence of rival hypotheses. The error committed in not rejecting a false hypothesis is called an error of the second kind. The method recommended by Neymann and Pearson and adopted by most statisticians is to keep the probability of committing an error of the first kind (rejecting a true hypothesis) on a given level (level of significance, e.g. 1 per cent) and then to construct the test in such a way as to minimize the chance of not rejecting a false hypothesis (committing an error of the second kind: Kendall and Stuart 1961, pp. 161 ff.).

Here all probabilities are interpreted in the frequency sense. Let us test in our example the hypothesis that in the unknown population the income elasticity for food expenditure by the urban population in Austria is 1. Our hypothesis then states that in the population, as income increases by a given percentage, the urban Austrians will react by increasing food expenditure by the same percentage. By constructing a test in the manner of Neymann and Pearson, we find that this hypothesis must be rejected at the level of significance of 1 per cent (Tintner 1960*c*, pp. 41–42.). It is unlikely that the urban population in Austria will react in the indicated manner to a percentage increase in income, *ceteris paribus*, by increasing food expenditure by the same percentage.

Neymann and Pearson's method may also be defended from the point of view of decision theory. Wald (1950) has generalized this method to a general method of decision functions (Lehmann 1959; Braithwaite 1953, pp. 196 ff.). Assume that there are various actions possible (rejection of a hypothesis, nonrejection, or continuing the experiment) and that we know the losses which all the actions incur. Then we may compute the probabilities connected with each action and find the mathematical expectation of the loss. This risk function should then be minimized. This idea is especially important in connection with industrial experimentation and some methods of operations research. The Bayesian approach (De Finetti 1937; Savage 1954; 1962; Lindley 1965; Kryburg and Smockler 1964) bases itself firmly upon the personalist or subjective concept of probability. In a sense these ideas are related to decision theory, and many of the classical ideas of statistics, described above, appear questionable from this point of view.

Methodology of Econometrics

Econometrics, the result of a certain outlook on the role of economics, consists of the application of mathematical statistics to economic data to lend empirical support to the models constructed by mathematical economics and to obtain numerical results (Tintner 1952, p. 3; 1960*c*, p. 1).

The models which are utilized in econometrics depend upon the special scope of the investigation. This, in turn, will be mostly influenced by the policy purposes which the econometrician has in mind. Although there may be partial equilibrium models in econometrics (e.g., the investigation of a given market in isolation), models of general equilibrium, which try to analyze the interaction of many economic variables in a given national economy, are more important. Also, the models can be classified into static models, where time plays no particular part, and dynamic models, where time enters in a significant fashion into the analysis. Such models will in general involve lagged values of the economic variables. For instance, a dynamic model of a given economy will involve not only contemporary prices but also prices of previous years.

Mathematical economics in its modern form develops models for individual households and firms. From this point of view it is already useful for applications in market research and operations research (the econometrics of the enterprise). It is also possible to combine these results and study, under certain (not always very realistic) assumptions, questions which pertain to a total economy. If we want to study these problems econometrically (i.e., with the help of statistical methods), we need aggregates or index numbers for practical reasons. The number of households and firms is too great to permit the study of even the simplest relations between all of them, even with the help of the biggest existing (or easily imaginable) electronic computers. The problem of the appropriate construction of aggregates or index numbers is one of the most difficult but evidently fundamental one in econometrics. Stone (1947) and Tintner (1952, pp. 102 ff.) have proposed purely statistical criteria, which are based upon Hotelling's idea of principal components—a method closely related to factor analysis.

The practical interest in aggregation in econometric work based upon the theories of Keynes (1936) which are essentially macro-economic (i.e., deal with aggregates) stimulated much important theoretical work, especially by American and French mathematical economists (see Green 1964). The problem was further considered by Theil (1954), who, however, dealt only with linear aggregation and introduced statistical as well as theoretical considerations. He has shown the difficulty of perfect aggregation, and he and his school have followed certain ideas of Frisch and Wald (see Pfanzagl 1955) and used statistical methods for the construction of aggregates or index numbers which have certain optimal properties (Theil 1954; 1962).

Prais and Houttakker (1955, p. 13 ff.) have considered lognormal distributions in the theory of aggregation. A variable is said to be lognormally distributed if its logarithm follows the well known bell-shaped normal or Gaussian distribution. For instance, let y be consumption and x income. The individual consumption functions are

$$y_i = a + b \log x_i \tag{59}$$

for individual or family $i(i = 1, 2, \ldots, n)$. Also, a and b are constants. Total consumption may be written in the form

$$y = na + bn \log x^*, \tag{60}$$

if we assume that income (x) is lognormally distributed and x^* is the *geometric* mean of the individual incomes x_1. Let $x_1, x_2, \ldots, x_n$ be the income of n families. The geometric mean (defined only for positive incomes) is

$$x^* = \sqrt[n]{x_1, x_2, \ldots, x_n}. \tag{61}$$

For an application of information theory to aggregation see Theil (1967).

One fundamental problem in econometrics is the division

of variables into two classes. Endogenous variables are economic variables whose interaction determines the economic system; for example, quantities sold and bought, interest rates, and so on. The number of endogenous variables must normally be equal to the number of equations in the system in question. In addition, there are the predetermined variables—variables which influence the system but are not influenced by it (exogenous variables like the weather and lagged values of the endogenous variables like past prices).

After constructing a model, the econometrician has to make certain assumptions about the stochastic nature of the equations involved. Broadly speaking, he has to evaluate errors in the variables (akin to errors of observations, Morgenstern 1963) and errors in the equations (i.e., the effect of economic variables which are left out of given equations and are replaced by stochastic variables). Other assumptions of a statistical or probability nature have also to be made; for example, independence of the stochastic variables over time.

Before statistical estimation of the postulated relationships can be discussed, the problem of *identification* must be considered. Simon (1957, p. 10 ff.) sees this from the general point of view of an analysis of causal relations as the following: Given the model and the stochastic assumptions involved, is it possible (in principle) to estimate the individual relationships (or even individual parameters) in the hypothetical case where we have infinite samples?

To illustrate the particular idea of causal relationships used in econometrics, we present the following example taken from Simon (1957; see also Wold 1965). Denote an index measuring the favorableness of the weather for growing wheat by x_1, the size of the wheat crop by x_2, and the price of wheat by x_3. The system, assumed to be linear for the sake of simplicity, can thus be stated as

$$a_{11}x_1 = a_{10}, \tag{62}$$

$$a_{21}x_1 + a_{22}x_2 = a_{20}, \tag{63}$$

and

$$a_{32}x_2 + a_{33}x_3 = a_{30}. \tag{64}$$

Here a_{ij} are constants. No stochastic terms are introduced. This model might be stated in the form: x_1 (state of the weather) is the direct cause of x_2 (wheat crop); and x_2 (wheat crop) is the direct cause of x_3 (price of wheat). It should be noted that this causal relation is established by observing which variables appear in a given equation. This idea of causal relations has been much discussed in recent years (Liu 1960; Strotz and Wold 1960).

As an example, consider a Marshallian market of partial equilibrium. Denote the price of commodity in question by P and the quantity exchanged by Q. Let a_{ij} be some constants to be estimated and assume as a first approximation that the relations involved are linear. The demand function is

$$a_{10} + a_{11}P + a_{12}Q = u_1, \qquad (65)$$

and the supply function is

$$a_{20} + a_{21}P + a_{22}Q = u_2. \qquad (66)$$

The stochastic variables u_1 and u_2 are errors in the equations. In the demand function (65) we consider a linear relationship between the price and the quantity of the given commodity, but there are other variables which enter into the demand relation; for example, income, prices of complementary and competing commodities, and so on. If the individual influences of all these neglected variables are small and irregular, we may replace them all by the stochastic variable u_1. Similarly the supply equation (66) is conceived in first approximation as a linear relationship between the price and the quantity of the commodity in question, but there are other factors which influence the supply of the commodity and are here omitted (technological conditions, cost conditions, prices of the factors of production, etc.). All these omitted variables are replaced by the stochastic variable u_2.

Now, consider the identification of the demand function (65) without making assumptions about the random variables u_1

and u_2. Our system consists of two endogenous variables P and Q, and thus also of two linear equations which in equilibrium determine these two variables. There are no predetermined variables.

Multiply the first equation (65), the demand equation, by an arbitrary constant K_1 and the second equation (66), the supply equation, by an arbitrary constant K_2. This linear combination of the demand and the supply equation (which is economically meaningless) is

$$K_1a_{10} + K_2a_{20} + (K_1a_{11} + K_2a_{21})P + (K_1a_{12} + K_2a_{22})Q = K_1u_1 + K_2u_2. \tag{67}$$

Can we distinguish this meaningless linear combination of the demand and the supply function (67) from the true demand function (65)? Evidently, this is not possible within our model. One must remember that all we know from (65) about the demand function is that it is a linear relation between price (P) and quantity (Q) with a random variable u_1. Since the linear combination (67) is of the same form, we have a linear combination of price and quantity with random variables $K_1u_1 + K_2u_2$. Hence we say that in our model the demand function (65) is not identified. Koopmans (see Hood and Koopmans 1953, pp. 27 ff.) has given rules for identification, and a statistical test for identification has also been supplied by Basmann (1960).

Having satisfied ourselves that a given relationship in the model is identified, there arises the problem of statistical estimation. A number of methods have been proposed in this field. The *full maximum likelihood method* is in almost all cases so complicated that one has to use numerical approximation. If we use the full maximum likelihood method, we maximize the probability of obtaining the given sample, taking all restrictions implied by the model into account. As a substitute, the *limited information method* of Anderson and Rubin (1949) or the *two stage least squares method* of Theil (1961, pp. 334 ff.) and Basmann (1957) sometimes can be used.

In estimating a single equation by the method of limited information, we neglect all the properties of the total system except the following: Which endogenous and predetermined variables enter into the equation which is to be estimated and which predetermined variables are in the system but not in the equation?

If we use the two stage least square method we first compute the reduced form equations—that is, we estimate by the method of least squares the linear relations between each one of the endogenous variables in the equation to be estimated and all the predetermined variables, both those in the equation and also the predetermined variables in the system but not in the equation. Then we utilize the estimated values of the endogenous variables in the equation to estimate the original equation, again by the method of least squares.

As an example, let us consider a model proposed by Tintner (1952, pp. 166 ff.) and treated by various methods of estimation by Goldberger (1964, pp. 328 ff., 344 ff.). Denote the quantity of meat consumed per capita by Q, a price index of meat by P, disposable income per capita by Y, unit cost of meat processing by C, and an index of cost of agricultural production by K. The data are taken from United States agricultural statistics, 1919–41. For simplicity all variables are deviations from their means so that constant terms can be omitted in the equations. The model is

$$Q = a_1 P + a_2 Y + u_1. \tag{68}$$

This is the demand equation for meat. It assumes that in the first approximation the demand for meat depends linearly upon the price of meat and disposable income. The variable u_1 is a random variable and stands for other economic factors which influence the demand for meat (e.g., prices of other commodities). These are omitted and replaced by the stochastic variable u_1. Also a_1 and a_2 are constants to be estimated. The supply equation is

$$Q = b_1 P + b_2 C + b_3 K + u_2. \tag{69}$$

Again, the supply of meat is assumed in first approximation to depend in a linear fashion upon the price of meat, the cost of meat processing, and the cost of agricultural production. The random variable u_2 stands for all other factors which influence the supply of meat (*e.g.*, the weather). Here our aim is to estimate the constants b_1, b_2, b_3. Note that we have a system of two linear relations and two endogenous variables—Q the quantity of meat consumed and P the price of meat. These are simultaneously determined in equilibrium through the interaction of our two equations. We also have, however, 3 variables which are assumed to be predetermined. They influence the market of meat but are not influenced by it, under our assumptions. They are Y disposable income, C cost of processing meat, K cost of agricultural production. It should be emphasized that in this model, as in many other econometric models, the predetermined variables are not really exogenous because part of the disposable income Y will be generated in the meat market. The variables are assumed only to be predetermined for the purpose of studying this particular market.

The reduced form equations give the linear relations between each endogenous variable (Q and P) and all the predetermined variables (Y, C, and K). They have been estimated by the method of least squares:

$$q = -0.030839Y - 0.344626C + 0.613660K \qquad (70)$$

and

$$p = 0.096075Y + 0.208343C - 0.033902K.$$

Since we have just enough conditions for the identification of the supply equation (69), we may estimate it by the *indirect method of least squares.* We simply eliminate from the system (70) the variable Y, which does not appear in the supply equation (69). We obtain an estimate of the supply equation:

$$Q = -0.32P - 0.28C + 0.60K. \qquad (71)$$

For the demand equation (68) we have more conditions than are necessary for identification, hence, it is *overidentified.*

The simplest method of elimination (indirect least squares) is not applicable. By applying the method of two stage least squares (Theil 1961, pp. 334 ff.; Basmann (1957), we use the estimated value of p from (70) in equation (69) and compute the following estimate for the supply function for meat:

$$q = -1.58p + 0.20Y. \tag{72}$$

Again, this has been obtained by the method of least squares. On the other hand, the limited information method of Anderson and Rubin (1949) gives a different estimate:

$$Q = 4.85P + 0.53Y. \tag{73}$$

Johnston (1963, p. 269) has computed a *consumption function* for the United Kingdoms, 1948–58. With C_t as consumption and Y_t as disposable income and by using the method of limited information, he obtained

$$C_t = 3{,}841 + 0.4923Y_t; \tag{74}$$

and with the application of two stage least squares the result was

$$C_t = 2{,}041 + 0.6029Y_t. \tag{75}$$

The marginal propensity to consume is estimated in the first case as about 1/2 and in the second case as about 3/5. The results of various econometric estimation methods are frequently quite different since they are based upon different theoretical assumptions. Since the small sample properties of these estimates are not yet sufficiently explored, it is difficult to say which method is preferable (Basmann 1960; Johnston 1963, pp. 275 ff.).

A convenient method of analysis are the *recursive* systems of Wold (1953, pp. 28 ff.), which arise if the matrix of contemporaneous endogenous variables becomes triangular. These ideas can be illustrated by a simple example from Tintner's (1960, pp. 252 ff.) Austrian data, 1948–55. Denote the logarithm of the quantity by X_{1t} and the logarithm of the price of pork in the year t by X_{2t}. We assume that the supply of pork depends upon the price of the year before, but that the demand is a relation between the price and the quantity of pork in the same year. Then, if the errors or deviations of the two equations are independent, we may estimate them in this form by the classical method of least squares: thus,

$$X_{1t} = 0.81 + 0.74X_{2t-1} \tag{76}$$

and

$$X_{2t} = 0.80 - 0.15X_{1t}. \tag{77}$$

The first of these equations (76) is the supply function, and the second (77) is the demand function. If the price of pork is known in a given year, we may compute the supply for the next year from (76); and by using the result in equation (77), we compute the price for the subsequent year, and so on. The system (76) and (77) forms a *causal chain.*

F. M. Fisher (1961) has established proximity theorems and shows that the customary assumptions for identification of simultaneous equations system also hold as approximations if the a priori restrictions are close approximations, omitted variables have small coefficients, endogenous variables have negligible effect on the assumed exogenous variables, and endogenous variables have negligible indirect effect on the assumed exogenous variables. An interesting method for estimation of simultaneous econometric equations has been proposed by F. M. Fisher (1966). He starts with the idea of block recursive systems, which is in fact a generalization of Wold's recursive systems (1965), illustrated above. Then he makes some reasonable assumptions about the errors in the equations. The estimation is effected by the use of instrumental variables (Malinvaud 1963, pp. 352 ff.; Christ 1966, pp. 402 ff.). Instrumental variables are uncorrelated with the disturbances and are chosen by taking the casual relations with the variables in the equation to be estimated into account. Fisher gives a method of classifying these instrumental variables and choosing them in the most advantageous way. In our discussion of the estimation of econometric models we have neglected all problems not connected with errors in the equations. Even there, difficult problems of estimation occur if the model includes lagged values of the endogenous variables and also if the errors in the equations are not independent over time but show autocorrelations. Since economic data are rarely as accurate as one is prone to assume (Morgenstern 1963), we will invariably meet with errors in the variables, akin to errors of observations. These errors constitute a difficult problem in the estimation of econometric relations. They may also give rise to the problem of multicollinearity (Frisch 1934; Tintner 1952, pp. 121 ff.) — the situation where (approximate) linear relations exist between the "independent" variables in a least squares problem.

Various methods are available to deal with the problems arising when autocorrelations or serial correlations between the errors or deviations exist (Tintner 1952, part 3). Autocorrelation measures the relationship between lagged values of the same random variable, and serial correlation measures the relationship between the values of one random variable and the lagged value of another random variable. Not much is known about the treatment of these problems in small samples, but for stationary time series (i.e., time series without trend) some modern methods of spectral analysis (Hannan 1960; Granger and Hatakana 1964) may be applicable.

Prediction

Prediction in economics is not easy. An amusing list of wrong predictions by some great economists of the past is contained in Zweig (1950, p. 25 ff.). It includes Adam Smith, Malthus, Senior, Say, Ricardo, Sismondi, Mill, Marx, Keynes. It seems that the besetting sin of economists is the inclination to project existing situations and tendencies of their own time and society into the future. It is ultimately the goal of econometrics to achieve reliable predictions. Such predictions are of great scientific interest, as tentative verifications of economic models and theories.

But they are also of importance for applications to economic policy.

It cannot be denied that up to now prediction with the help of econometric methods has not been too successful. This point of view has been stated forcefully by Machlup (1955, p. 19):

> Where the economist's prediction is *conditional*, that is, based upon specific conditions, but where it is not possible to check the fulfilment of all the conditions stipulated, the underlying theory cannot be disconfirmed, whatever the outcome observed. Nor is it possible to disconfirm a theory where the prediction is made with a *stated probability* value less than 100%; for, if an event is predicted with, say, 70% probability, any kind of outcome is consistent with the prediction. Only if the same "case" were to occur hundreds of times could we verify the stated probability by the frequency of "hits" and "misses." This does not mean complete frustration of all attempts to verify our economic theories. But it does mean that the tests of most our theories will be more nearly of the character of *illustrations* than of verifications of the kind possible in relation with repeatable controlled experiments or with recurring fully-identified situations.

An excellent survey of forecasting is contained in Theil (1961 pp. 49 ff.). He utilizes American, Dutch, Norwegian, Swedish, Danish, Canadian, and West German forecasts, both private and public. He comes to the conclusion that forecasts almost invariably underestimate change. This is particularly important for questions of policy. There it is necessary to predict the exogenous variables — that is, the variables which influence the economic system but are not influenced by it. If biased forecasts are used for policy purposes, they will naturally lead to losses in welfare.

It has sometimes been maintained that predictions in economics are impossible in principle since these very predictions, if they are believed by the acting economic subjects, will falsify the original predictions. Grunberg and Modigliani (1954; see also Simon 1957, pp. 79 ff.) have investigated the old problem of prediction in economics and the question of whether public prediction can influence the eventual outcome of the economic process. They come to the following conclusion:

It has been shown that, provided that correct private prediction is possible, correct public prediction is also conceptually possible. Two possibilities may be distinguished (1) The public prediction does not affect the course of events because the agents are indifferent to or incapable of reacting to the public prediction. In this case, correct public prediction coincides with correct private prediction. (2) Agents react to public prediction, and their reaction alters the course of events. The reaction can conceptually be known and taken into account. It has been shown that boundedness of the variables of the predictive system and the continuity over the relevant interval of the functions relating the variables to each other are sufficient, though not necessary conditions for the existence of correct public predictions. These conditions were found to be normally fulfilled in the world about which predictive statements are to be made. The argument of this paper establishes the falsity of the proposition that the agents' reaction to public prediction necessarily falsifies all such predictions and that, therefore, social scientists may never hope to predict both publicly and correctly. But it demonstrates no more than that correct public prediction is possible if the possibility of correct private prediction is accepted. About the possibility of private prediction it has nothing to say. So, in the end, the major difficulties of prediction in the domain of social prediction turn out to be those of private prediction.

We (Hohenbalken and Tintner 1962) have constructed a small static model for the United States, using national income data for the period 1948–60. The endogenous variables are those which are simultaneously determined by the interaction of the equations of the system. Here there are only 5 endogenous variables: C is personal nominal consumption expressed in current dollars, which, represents the total value of all goods and services consumed. Y is the nominal gross national product, which represents the total value of all commodities produced. P is the price index of the gross national product, which shows the general movement of the price level. D is total employment, the sum of total civilian, agricultural and non-agricultural employment—that is, the number of people employed in all civilian occupations. X is real national income, national income expressed in constant dollar values derived by dividing nominal national income (Y) by an appropriate price index (P).

The system, of course, is also influenced by other (exogenous) variables: N is the population of the United States. G is nominal public consumption, the total consumption of federal and state as well as local authorities. I, which is nominal gross asset formation, expresses private investment. L, nominal increases in stocks, is the total change in the value of inventories. E is nominal exports and M nominal imports, expressing the dollar value of international trade. W is nominal yearly wage per worker (hourly earnings in manufacturing times weekly average of hours worked in manufacturing times 52); Time is t. It should be emphasized that some of these variables are not really exogenous to the functioning of the American economic system, but they have been assumed to be exogenous and would become endogenous in a larger dynamic system like the one of Klein and Goldberger (1957; see also Duesenberry *et al.* 1965). Our purpose was to construct a small model which would be immediately comparable to other similar models of Western European countries and Canada (von Hohenbalken and Tintner 1962).

By using the method of simultaneous equations, we derived the following estimates of the equations in our system:

$$C_t/N_tP_t = 687.9 + 0.336\, Y_t/N_tP_t. \tag{78}$$

This is a Keynesian type consumption function. It relates real consumption per head ($C_t/N_t\, P_t$) to real national income per head (Y_t/N_tP_t) linearly. Our estimate of the marginal propensity to consume is 0.336. This means that if real national income per capita increases by \$1.00, real per capita consumption may be expected to increase by about \$0.34. The next equation is simply a bookkeeping definition of nominal national income:

$$Y_t = C_t + G_t + I_t + L_t + E_t - M_t. \tag{79}$$

We have the obvious definition of real national income:

$$X_t = Y_t/P_t. \tag{80}$$

Real national income is nominal national income divided by a suitable index of the price level.

The demand for labor is determined by the marginal productivity of labor:

$$dX_t/dD_t = W_t/P_t. \tag{81}$$

This follows from the theory of production under free competition and neglects the monopolistic elements in the American economy. Under free competition it can be shown that if a firm maximizes its profits with given prices and wages, the maximum of profits implies the condition that the marginal productivity of each factor of production is equal to the ratio between the price of this factor and the price of the product. To understand our concept of marginal productivity consider, *ceteris paribus*, a small increase in the amount of one factor used; the the corresponding increase of the product will be the marginal productivity. The last equation is a primitive production function of the Cobb-Douglas type. Such functions are linear in the logarithm of the product and the factors of production. It has been estimated by the method of simultaneous equations by using the assumption of free competition. Here again we neglect the monopolistic elements and, for example, the existence of trade unions thus,

$$\log X_t = 6.652 + 0.630 \log D_t. \tag{82}$$

It should also be noted that in this static model capital has been neglected. The estimated coefficient 0.63 is an elasticity, thus if employment in the United States increases by 1 per cent we may expect an increase of production of about 0.63 per cent.

To utilize this non-linear model for problems of economic policy (Tinbergen 1954; Theil 1961), we have considered the exogenous variables as policy variables and asked the following question: Given an isolated autonomous increase in an exogenous variable by 1%, what will be the simultaneous percentage change in the endogenous variables? (See Table 12.) Assume that population (N) increases by 1 per cent by immigration, for example. Then we may expect the following reaction of the endogenous variables: the price level (P) will increase by about 1/5 per cent; nominal income (Y) will increase by more than 1/2 per cent; real national income (*X*) will increase by about 1/3 percent; employment (*D*) will increase by about 1/2 per cent; and nominal consumption (*C*) will increase about 9/10 per cent. Now consider an increase of nominal (money) wages (*W*) by 1 per cent. Such an increase may be the result of deliberate wage policy. We may expect that the price level will increase by about 3/4 per cent and nominal income by about

TABLE 12

Percentage Change in Endogenous Variables with an Isolated Autonomous Increase in the Exogenous Variables

Endogenous Variable	Exogenous Variable						
	N	*W*	*G*	*I*	*L*	*E*	*M*
P	0.21	0.76	0.12	0.11	0.01	0.04	—0.03
Y	0.56	0.35	0.32	0.30	0.02	0.10	—0.08
X	0.35	—0.41	0.20	0.19	0.01	0.06	—0.05
D	0.56	—0.65	0.31	0.30	0.02	0.10	—0.08
C	0.87	0.55	0.22	0.21	0.01	0.07	—0.06

1/3 per cent; but real national income will decrease by about 4/10 per cent and employment by almost 2/3 per cent and nominal consumption will increase by about 1/2 per cent. What are the effects of an increase in public consumption (G) by 1 per cent? The price level will increase by more than 1/10 per cent, nominal national income by about 1/3 per cent, real national income by 1/5 per cent, employment by about 3/10 per cent, and nominal consumption by about 1/5 per cent. The consequences of an increase of nominal investment (*I*) by 1 per cent effected, for example, by tax policy are as follows: The price level will increase by about 1/10 per cent, nominal national income by 3/10 per cent, real national income by about 1/5 per cent, employment by 3/10 per cent, and nominal consumption by about 1/5 per cent. The effects of changes in stocks (*L*), imports (*M*), and exports (*E*) are very small. This shows that these variables are not very suitable for use in influencing the endogenous variables by public policy.

For the estimation of our model, data from the period 1948–60 have been used. The model is very simple, and we cannot expect great accuracy in prediction. Nevertheless, I have ventured to use the elasticities in the above table to predict the changes in the United States economy for 1963–64. Table 13 gives the actual and computed changes. Whereas there is, of course, no great accuracy in the prediction, the sign of the change and the order of magnitude has been correctly predicted. This shows that even such a small and simple model is perhaps not completely useless. We have presented this model only as an example. Much larger models are necessary to achieve predictions which are potentially useful in applications to economic policy.

TABLE 13

United States Econimic Change, 1963/64—Predicted and Actual
(Per Cent)

Endogenous Variable	Actual	Predicted
Price level	1.1	4.5
Nominal income	5.9	3.1
Real income	4.8	1.3
Employment	4.4	2.1
Nominal consumption	5.6	5.4

One of the most elaborate models is the Klein-Goldberger (1955) dynamic model for the United States, 1928–52. It contains 20 endogenous variables, 20 equations, and 15 predetermined variables. Christ (1956) has applied the model for forecasts to 1953 and 1954. Apart from forecasts of investment and the price level, the forecasts of such important economic variables as national income, consumption, and employment are successful in the sense that the error of forecast is about 1 per cent, and the errors for forecasts of investment and prices are 10 and 5 per cent.

IV. Welfare Economics and Economic Policy

The Value Problem

In contrast to natural science, especially physics, economics is much more closely related to problems of applications, and its history is possibly more akin to the history of medicine. In any case it is perhaps only during the present century that "pure" or positive economics has been studied for its own sake. This is not to say, of course, that analytical propositions have not been developed by economists primarily interested in practical problems, as Schumpeter (1954) has so brilliantly shown. The orientation towards policy, especially, explains the type of economic problems which were treated by economists at any given time. But the fact that many (perhaps all) great economists were interested in problems of policy, passionately desired the adoption of certain policy measures, and perhaps even created ideologies does not mean that they did not try to apply scientific methods objectively. If we say that economics (pure or positive) should be *value free*, we mean that the economist should try to investigate objectively and in an unprejudiced way the economic phenomena. These by necessity, however, include certain valuations—for example, given preferences of the consumers. But as indicated earlier, problems of choice may also be investigated objectively by methods of mathematical economics and econometrics.

Many writers (M. Weber 1922, Weber and Topitsch 1952) have emphasized that economics should be value free. The distinction between positive and policy oriented economies is discussed by Hutchinson (1964). He especially stresses the shortcomings of modern welfare economics and also the fact that the consequences of economic policy measures must be considered in the total social framework and not as isolated economic events. Such an admonition is much more needed today than at the time of J. S. Mill when economics was less specialized and less divorced from other social sciences. The fundamental neutrality of scientific economics has been emphasized by Cairnes (1875, p. 20): "In the first place, then, you will remark that, as thus conceived, Political Economy stands apart from all particular systems of social or industrial existence. It has nothing to do with laissez faire any more than with communism." We may well agree with the view expressed by Albert (1964, p. 396):

> 1. In a certain sense *no science* can be value free: All sciences have a value foundation; they are influenced by valuations. 2. In a certain sense, *social sciences* cannot be value-free; they must analyze valuations in the realm of objects. 3. In a certain sense, *any science* can be value-free: no science is in need of value judgments within the context of its propositions. There is no value freedom in the absolute sense. But the methodological principle of value freedom can be maintained almost in the sense of Max Weber.

It seems obvious that we should distinguish between positive economics and economic policy which has to be based upon normative ideas. It is equally apparent that this distinction has rarely been made in the past and is indeed difficult to maintain.

Problems of Ethics

The problems of economic policy, like all political problems, involve economics in questions of ethics and social philosophy. This preoccupation can be traced back to Plato and Aristotle. It is particularly obvious in the writings of the English classical economists who were under the potent influence of utilitarian philosophy. This influence has persisted to this day. The efforts of Pareto and his modern followers and of the writers in decision theory to emancipate economics from utilitarian ethics form an important part of the background of much of the modern literature in this field. Unfortunately, logical positivists have, with a few exceptions, up to now somewhat neglected the study of ethics and politics. It seems clear, however, that methods of logical analysis are potentially important for the study of ethics (Menger 1934*b*; von Wright 1951 and 1963; Kraft 1951; Becker 1952). It is obvious that these contributions might be useful in the study of welfare economics and economic policy.

The application of *modal logic* to problems of ethics makes a study of the consistency of ethical systems possible. We present the following examples from Wright (1951, pp. 39 ff.): If doing what we ought to do commits us to do something else, then this new act is also something we ought to do. Doing the permitted can never commit us to do the forbidden. If doing something commits us to do the forbidden, we are forbidden to do the first thing. An act, which commits us to a choice between forbidden alternatives is forbidden. It is logically impossible to be obliged to choose between forbidden alternatives. Our commitments are not affected by our other obligations. If failure to commit an act commits us to perform it, then this act is obligatory. Churchman (1961) argues from a thoroughly pragmatic point of view. Basing himself upon many of the recent advances in operations research he tries to construct a science of ethics.

For the study of economic policy we might proceed in the manner of Weldon (1953). He analyses the vocabulary of politics essentially from a positivist point of view (see also von Wright 1963). Real essences, absolute standards, and the geometrical method should be recognized as illusions. Classical political theory (this includes utilitarianism and Marxism) are based upon certain metaphysical suppositions and hence are largely quite meaningless. Freedom, for instance, should be recognized as involving two concepts: free from and free to—for example, the industrial worker in a Western community at the present time, as compared to the time of Marx, is less free from interference from trade unions and government agencies, but he is more free to acquire a high living standard. It is quite senseless and not at all helpful to talk about political "foundations." This holds for democratic foundations (laissez faire liberalism), Hegelian foundations (conservativism), Marxist foundations (socialism and communism), and philosopher kings (dictatorship). The elimination of metaphysical ideas in this field, as in many other fields, would clarify the discussion of important issues. This point of view is similar to Popper's proposition regarding piecemeal engineering (Popper 1950, pp. 154 ff.).

Weldon, as well as Topitsch (1961, pp. 271 ff.), emphasizes that this critical point of view applied to traditional politics does not necessarily lead to skepticism. Extreme subjectivism and relativism in political and ethical matters is not implied. These important problems and their solution are more than mere matters of taste, but the search for absolute standards of conduct in politics cannot be justified. There are no objective standards in politics comparable to standards in physics. Judgments about ethics and politics are typically more subject to error than propositions in the natural sciences, but they are not meaningless. Appraisal statements are simply not very accurate. We may conclude that for this reason policies should always be experimental, and the danger of developing vested interest in given policies should be minimized.

Political institutions cannot always, or even usually, be completely appraised in the language of means and ends, and the same holds in aesthetics. Appraisals are empirical judgments made by individuals. There remains a *personal* component in the appraisal of political and social institutions. I would be tempted to agree with Weldon (p. 176) that we should judge such institutions according to the following criteria: Does the political system under consideration censor the reading of those who are subject to it and impose restrictions on teaching? Does it maintain that any political or other principles are immutable and therefore beyond criticism? Does it impose restrictions on the intercourse of its members with those who live under different systems? Do the rulers of the associations which have these institutions find most of their supporters among the illiterates, the uneducated, and the superstitious?

There is, of course, the danger that with the adoption of a thoroughly empirical point of view the social sciences may degenerate into conformist endeavors (Topitisch 1961; Mills 1959; Hoffmann 1961). The academic situation in the United States at the present time, where large corporations and government departments provide much employment to social scientists and also where a great deal of research in the social sciences is being supported by the government, points in this direction. Piecemeal engineering and piecemeal social technology as proposed by Popper (1950, pp. 154 ff.) may (against the wish of its originator) degenerate into a defense of the status quo. Intellectuals may become (conscious or unconscious) social managers of the existing system and their endeavors may not be much more than its ideological defense.

One may also emphasize, however, that the pressing problems of our times, the prevention of war and the economic development of the underdeveloped countries (Boulding 1964), are such that it is very doubtful if the application of piecemeal engineering would always be very helpful. For instance, the economic development of India necessitates perhaps an almost complete revolution of existing values and institutions. The transition of the retarded Indian peasant society to modern capitalism would be just as revolutionary as the transition to a Marxist dictatorship or to a mixed system, as is being attempted today. It is not easy to see how methods of slow reform can be very helpful in this connection. After all, the transition from feudalism to the modern society, which made the rise of modern science possible, was also revolutionary and painful, as Popper has repeatedly emphasized.

Welfare Economics

A tradition of doctrines of welfare economics, which is based upon the philosophy of the utilitarians (e.g., Bentham, J. S. Mill), many of whom made important contributions to positive economics, has grown up and persisted in the writings of many economists. The somewhat naive ideas of Bentham about "the greatest happiness of the greatest number" were refined and presented in very plausible form by Marshall (1948) and Pigou (1920). For an application of these ideas to Indian problems see Tintner and Patel (1966). Pareto (1927), for philosophical and political reasons, was opposed to the utilitarian point of view and proposed another method which is still in the center of discussion. The Pareto optimum is defined as the state in which no individual can be made better off without making some individual worse off. Looked at as a prescription of economic policy this amounts, of course, to a defense of the status quo.

Bergson (1948, 1954, 1966) makes a most important contribution to welfare economics. He introduces a *welfare function* which depends upon the utilities or satisfactions of *all* individuals in a given economy, but the welfare function which he proposes has been criticized by Arrow (1963; see also Rothenberg 1961 and Scitovsky 1964.)

By starting from purely individualistic and atomistic assumptions, Arrow showed that if there are at least three alternatives which the members of a society are free to order in any arbitrary way, then a social ordering must be either imposed or dictatorial. The assumption here is that choices are comparable and transitive. The welfare function is conceived in such a fashion that individual preferences are taken into account and irrelevent alternatives play no part. Dictatorship means that the choices of a single individual are decisive. Imposition means that social welfare is independent of individual choices. For a criticism of welfare economics from the extreme laissez-faire point of view see Rothbard (1956). As has been pointed out (Albert 1964), welfare economics has never really been a firm basis of economic policy (see also Robbins 1961, 1963).

The modern welfare economics considers economic phenomena in isolation. In this form modern welfare economics is decidedly inferior to the theories of the utilitarian philosophers and economists (Bentham and J. S. Mill), who considered economics as only one of the aspects of the desirable social order. Much of welfare economics was conceived as a defense of unrestricted laissez-faire. This involves, among other assumptions, the idea of consumers sovereignty—that is, each consumer is assumed to act in isolation, independently of all other consumers, to maximize his satisfaction. Whereas there is not much objection to this idea as a first approximation in the explanation of choice under static conditions, it becomes most questionable if it is made the basis of policy considerations in welfare economics. Is it really possible to consider the consumer as an independent, atomistic, and sovereign being, who ultimately determines by his choice the totality of economic phenomena? This contradicts an old idea, expressed by Aristotle, that man is "a political animal." A society of isolated atomistic individuals, as conceptualized in much of modern welfare economics, is hardly imaginable (Tintner 1946, 1960*b*).

The questionable nature of the assumption of consumer's sovereignty which is still often maintained as a foundation of welfare economics, can be indicated by an example. In an unpublished dissertation R. L. Basmann has investigated the demand for tobacco and tobacco products in the United States, 1926–45 (Tintner 1962, p. 216). He finds that the elasticity of demand for tobacco per head of population with respect to the real cost of advertising for tobacco is 0.085. If the cost of advertising for tobacco and tobacco products increases by 1 per cent, the demand for these products will increase by a little less than 1/10 per cent. This example shows how tastes and preferences of consumers are actually influenced by advertising (Chamberlin 1948). Hence it seems impossible to consider the consumer as completely sovereign and independent.

The following statement is pertinent:

> A combination of the critical examination of social life and its institutions with the aid of theoretical thinking and of practical attempts to reform them by application of theoretically founded technological systems — in short, a combination of social criticism and social technology as the basis of politics — this would be the realization of the idea of approximation in social life, of approximation to a state for which there is just as little a criterion as for truth in the realm of knowledge (Albert 1964, p. 408–9).

The criticism of welfare economics by Albert (1964) is certainly not to be neglected. Perhaps we may, however, tentatively retain this idea and reform it in the direction indicated in an earlier article (Tintner 1946): take collective wants into account (Galbraith 1958, 1967), make the theory dynamic by not neglecting historical factors, perhaps introduce stochastic (i.e., probability) considerations to deal with irrational elements, and pay serious attention to external economies and diseconomies in consumption (Baumol 1965). The last point involves the fact that the individuals in the society do not really live as isolated atoms but interact, and that their preferences and tastes may be at least in part determined by this interaction.

We want to illustrate the contribution of modern mathematical economics to welfare problems by an example. The method of linear programming can be applied to short term production in general. It can make a contribution to the interesting problem of whether rational calculation and decentralized decision is possible in a static collectivist or planned economy (L. von Mises 1951). By generalizing older contributions by Pareto (1927), Barone (1935), Lange (1948), and Lerner (1944), Koopmans has shown the following: If an economic system which is essentially a generalization of the linear programming model presented above is assumed, he again proves the proposition that under free competition no firm makes any profit. But what about a collectivistic economy? Would all decisions have to be made by the central planning board? Koopmans (1951, pp. 33 ff.) proves that even under collectivism decentralized decisions are possible. Accounting or shadow prices are used in the same fashion as illustrated above. Assume also that there is a helmsman (central planning board), a custodian for each commodity, and a manager for each activity. A social optimum which corresponds exactly to the competitive optimum will be reached if in a collectivist economy the following rules are imposed: For the helmsman: choose a set of positive prices for all final commodities and inform the custodian of each commodity of this price. For the custodians: buy and sell your commodity from and to managers at a single price announced to all managers; buy all that is offered at this price; sell all that is demanded at the given price to the limit of the availability of your commodity. For all custodians of final commodities: announce the prices set on your commodity by the helmsman to managers. For all custodians of intermediary commodities: announce a tentative price on your commodity; if the demand by the managers falls short of the supply by managers, lower the prices; if demand exceeds supply, raise the price. For all custodians of primary commodities: regard the available inflow from nature as part of the supply of the commodity and then follow the rules for custodians of intermediary commodities; do not announce a price lower than zero but accept a demand below supply at zero price, if necessary. For all managers: do not engage in activities with negative (shadow) profits but maintain activities of zero profitability at a constant level and expand activities with positive profits by increasing orders for the necessary inputs with the custodians of the pertinent commodities and by offers of the outputs in question to the custodians of the commodities concerned.

This is certainly an interesting contribution of mathematical economics to one of the most important problems of our times: The theoretical possibility of collectivist economic planning. It is, however, important to realize that the result is subject to severe limitations. The theory is completely static and tells us nothing about the more important economic problems which involve time. These are, for example, economic fluctuations or economic growth and are hardly negligible in practice. Even within the subject of static economics it should be realized that the result applies to a severely idealized model of a competitive or collectivist economy. No indivisibilies are allowed. All coefficients of production are constant. There are no increasing or decreasing returns to scale. Thus if all inputs are increased in a fixed proportion, the output will also increase in the same proportion. The solution is optimal in the Pareto sense (production under unrestricted ideal free competition): no more of any final commodity can be produced, under the given assumptions, without diminishing the production of some other final commodity.

Even within the compass of the model the question of incentives is not treated. Evidently, the incentive of a capitalist entrepreneur to maximize money profits in a competitive market economy and the incentive of a manager in a collectivist economy to maximize bookkeeping profits are psychologically different. Problems of this nature are probably more adequately treated by methods of social psychology or sociology than by economics. It should be realized, however, that, perhaps in practice, there is not much difference between the manager of a large capitalist corporation and the manager of a state trust or a nationalized enterprise. It is likely that the problems faced by these managers are very similar (Schumpeter 1950), but these problems are quite different from the case analyzed by Koopmans.

It is deplorable that mathematical economists and econometricians have not occupied themselves more with one of the most important problems of our time – the question of the best economic regime, capitalism or socialism. Tinbergen's (1959, pp. 264 ff.) interesting article is one exception, which presents a promising beginning but does not come to very impressive conclusions (see also Marschak 1954).

Against this pessimistic view of welfare economics, one may maintain that the fundamental idea underlying it is a sound one, but like so many other ideas in economics it is difficult to formulate precisely and without ambiguity.

Theil's (1964) application, based upon Tinbergen's ideas (1954), of econometric methods to policy questions is without a doubt a most important contribution. These ideas are the more remarkable because the author was closely connected with government planning in the Netherlands. First, a quadratic welfare function is constructed as a rough approximation. This is based upon considerations related to the modern theory of welfare economics (Arrow 1963; Harsanyi 1955). The analogy to committee decisions is made, and a method of compromising different viewpoints is suggested. The measurable utility theory of von Neumann and Morgenstern (1944) plays a most important part. The expected value of utility is to be maximized, and certainty equivalents are introduced in order to deal with risk situations. A probability distribution is replaced by the mean value of the variable in question.

In retrospect, these ideas are applied to the economic policy of the United States 1933–6, using the Klein's model (1950). Both the static and the dynamic version of the theory are considered – that is, planning over time. There is also an application to cost minimization of a paint factory by production and employment scheduling (Holt *et al.* 1960). Finally, the theory is applied to questions of economic policy in the Netherlands 1957–9. Rothenberg (1961), who compares the essence of social decisions in a "going society" to actual choice as made in a typical family, also comes to similar conclusions. Hence, perhaps the idea of a social welfare function might be retained and even used successfully in planning for large capitalist enterprises or a whole economy.

It cannot, however, be denied that planning in Western (Theil 1961, 1964, 1966) and underdeveloped countries (Mahalanobis 1955; Schultz 1954; Tintner 1960*a*; Myrdal 1960) is profoundly nationalistic (Hayek 1960, p. 405). This fact in itself creates great difficulties (Robinson 1962). Except for the weaknesses of existing international organizations (Myrdal 1960), there is no reason why similar methods could not be used for international planning. It would be well for both the followers of laissez-faire capitalism and of Marxism to remember that the founders of these ideas were strong internationalists. This is as true of Karl Marx as it is of Adam Smith and David Ricardo (Robbins 1963, pp. 134 ff.). Extreme nationalism, whose growth was not anticipated by any of the writers named, has already in our century lead to two destructive world wars and seems likely to result in an even more destructive (and perhaps fatal) third war.

Welfare considerations can surely only give a partial answer to problems of economic policy. For one thing, as Albert (1964) emphasizes, economic welfare is only a part of general welfare and perhaps in some cases not even the most important part; consider, for example, the modernization of agriculture in India—surely one of the most important aspects of economic development in this country (Georgescu-Roegen 1960; Schultz 1964). In the Indian villages modernization may break up the traditional Indian society and have a number of disagreeable consequences, which cannot be overlooked in any general program.

Bibliography

Afriat, S. *The Cost of Living Index. Essays in Mathematical Economics. In Honour of O. Morgenstern*, ed. M. Shubik, pp. 335–66. Princeton: Princeton University Press, 1967.

Aitcheson, J., and Brown, J. A. C. *The Lognormal Distribution.* Cambridge: Cambridge University Press, 1957.

Albert, A. *Social Science and Moral Philosophy*, ed. M. Bunge, pp. 385–409, 1964.

Allais, M. Le Comportement de l'homme rationel devant le risque; critique des postulates et axiomes de l'école americaine, *Econometrica* 21 (1953), 503–46.

Allen, R. G. D. *Mathematical Analysis for Economists.* London: Macmillan, 1949.

———. *Mathematical Economics.* 2d ed., London: Macmillan, 1963.

Alt, F. Ueber die Messbarkeit des Nutzens, *Zeitschrift für Nationaloekonomie* 7 (1939), 161.

Anderson, T. W., and Rubin, H. Estimation of the parameters of a single equation in a complete system of stochastic equations, *Annals of Mathematical Statistics* 20 (1949), 46–63.

Arrow, K. J. *Social Choice and Individual Values.* 2d ed., New York: Wiley, 1963.

Arrow, J. K., and Debreu, G. Existence of an equilibrium for a competitive economy, *Econometrica* 22 (1954), 265–90.

Baran, P. A., and Sweezy, P. M. *Monopoly Capital.* New York: Monthly Review Press, 1966.

Barone, E. *The Ministry of Production in a Collectivist Economy*, ed. F. A. von Hayek, pp. 245–90.

Basmann, R. L. A generalized classical method of linear estimation of coefficients in a structural equation, *Econometrica* 25 (1957), 77–83.

———. On finite sample distribution of generalized classical linear identifiability tests statistics, *Journal American Statistical Association* 55, (1960), 650–59.

Baumol, W. *Economic Dynamics.* 2d ed., New York: Macmillan, 1959.

———. *Economic Theory and Operations Analysis.* Englewood Cliffs, N.J.: Prentice Hall, 1961.

———. *Welfare Economics and the Theory of the State.* 2d ed., London: G. Bell, 1965.

Becker, P. *Untersuchungen über den Modalkalkuel.* Meisenheim: Westkulturverlag, 1952.

Bellmann, R. *Dynamic Programming.* Princeton: Princeton University Press, 1957.

Bergson, A. *Socialist Economics*, ed. H. S. Ellis, pp. 412–48, 1948.

———. On the concept of social welfare, *Quarterly Journal of Economics* 58 (1954), 233–52.

———. *Essays in Normative Economics.* Cambridge, Mass.: Harvard University Press, 1966.

BERNADELLI, H. The origins of modern economic theory, *Economic Record* 37 (1961), 320–38.

BERNOUILLI, D. Exposition of a New Theory in the Measurement of Risk [1730], tr. L. Sommer, *Econometrica* 22 (1954), 23–36.

BLACKWELL, D. and GIRSHICK, M. A. *Theory of Games and Statistical Decisions.* New York: Wiley, 1954.

BLAUG, M. *Economic Theory in Retrospect.* Homewood, Ill.: Irwin, 1962.

BLUMEN, I., KOGAN, L. M., and MCCARTHY, P. J. The industrial mobility of labor as a probabilistic process. *Cornell Studies in Industrial and Labor Relations* 7 (1955).

BOULDING, K. E. *A Reconstruction of Economics.* New York: Wiley, 1950.

———. *The Meaning of the 20th Century.* New York: Harper and Row, 1964.

BOWLEY, A. L. *Mathematical Groundwork of Economics.* Oxford: Clarendon, 1924.

BRAITHWAITE, R. B. *Scientific Explanation.* Cambridge: Cambridge University Press, 1953.

BRONFENBRENNER, M. Das Kapital for modern man, *Science and Society* 29 (1965), 419–38.

BUNGE, M. (ed.) *The Critical Approach to Science and Philosophy.* Glencoe, Ill.: Free Press, 1964.

CAIRNES, J. E. *Character and Logical Method of Political Economy.* London: Macmillan, 1875.

CARNAP, R. Testability and meaning, *Philosophy of Science* 3 (1936), 419–71.

———. Logical foundations of the unity of science, *International Encyclopedia of Unified Science* Vol. 1, No. 1, Chicago: University of Chicago Press, 1938.

———. Foundations of logic and mathematics, *International Encyclopedia of Unified Science* Vol. 1, No. 3, Chicago: University of Chicago Press, 1939.

———. *Logical Foundations of Probability.* Chicago: University of Chicago Press, 1950.

———. *The Continuum of Inductive Methods.* Chicago: University of Chicago Press, 1952.

———. *The Aim of Inductive Logic*, ed. A. Nagel, P. Suppes, and A. Tarski, pp. 303–18, 1962.

———. *My Basic Conceptions of Probability and Induction*, ed. P. A. Schlipp, pp. 966–72, 1963.

CARNAP, R., and STEGMUELLER, W. *Induktive Logik und Wahrscheinlichkeit.* Vienna: Springer-Verlag, 1958.

CHAMBERLIN, E. H. *The Theory of Monopolistic Competition.* 6th ed. Cambridge, Mass.: Harvard University Press, 1948.

Feigl, H., and Sellars, W. (ed.) *Readings in Philosophical Analysis.* New York: Appleton-Century-Crafts, 1949.

Fellner, W. *Competition among the Few.* New York: Knopf, 1949.

Fels, E. M. About probability-like measures for entire theories, *Metrika* 7 (1963), 1–22.

Finetti, B. de. La Prévision; ses lois logiques, ses sources subjectives, *Annals Institut Henri Poincaré* 7 (1937), 1–68.

Fisher, F. M. On the analyis of history and independence of the social sciences, *Philosophy of Science* 27 (1960), 147–58.

———. On the cost of approximate specification in simultaneous equations estimation, *Econometrica* 29 (1961), 139–70.

———. *Dynamic Structure and Estimation in Economy Wide Econometric Models*, ed. J. S. Duesenberry *et al.* pp. 589–637, 1966.

Fisher, R. A. *Statistical Methods and Scientific Inference.* New York: Hafner, 1956.

Fisz, M. *Probability Theory and Mathematical Statistics.* 3d ed., New York: Wiley, 1963.

Fox, K. *Economic Analysis for Public Policy.* Ames, Iowa: Iowa State University Press, 1958.

Fraser, D. A. S. On fiducial inference, *Annals of Mathematical Statistics* 32 (1961), 661–76.

Friedman, M. *Essays in Positive Economics.* Chicago: University of Chicago Press, 1953.

———. *A Theory on the Consumption Function.* Princeton: Princeton University Press, 1957.

Friedman, M., and Savage, L. J. The utility analysis of choices involving risk, *Journal of Political Economy* 56 (1948) 279–304.

Frisch, R. *Statistical Confluence Analysis by Means of Complete Regression Systems.* Oslo: Universitetets Økonomiske Institutt, 1934.

Galbraith, J. K. *The Affluent Society.* Boston: Houghton Mifflin, 1958.

———. *The New Industrial State.* Boston: Houghton Mifflin, 1967.

Georgescu-Roegen, N. The theory of choice and constancy of economic laws, *Quarterly Journal of Economics* 54 (1950), 125–38.

———. Choice, expectations and measurability, *Quarterly Journal of Economics* 68 (1954), 503–34.

———. Economic theory and agrarian economics, *Oxford Economic Papers* 12 (1960), 1–40.

———. *Measure, quality, and optimal size.* ed. C. R. Rao pp. 231–56, 1965.

———. *Analytical Economics.* Cambridge, Mass.: Harvard University Press, 1966.

Goldberger, A. S. *Econometric Theory.* New York: Wiley, 1964.

Good, I. J. *Probability and the Weighing of Evidence.* London: Griffin, 1950.

———. *The Estimation of Probabilities.* Cambridge, Mass.: Mass. Inst. Technol. Press, 1965.

Graff, J. de V. *Theoretical Welfare Economics.* Cambridge: Cambridge University Press, 1957.

Champernowne, D. G. A model of income distribution, *Economic Journal,* 63 (1953), 318.

Charnes, A., and Cooper, W. W. Chance-constrained programming, *Management Science* 6 (1959), 134–48.

Chenery, H., and Clark, P. *Interindustry Economics.* New York: Wiley, 1959.

Chipmann, J. S. A survey of the theory of international trade, *Econometrica* 33 (1965), 477–519 and 685–760.

Christ, C. Aggregative economic models, *American Economic Review* 46 (1956), 385–408.

———. *Econometric Models and Methods.* New York: Wiley, 1966.

Churchman, C. W., Ackoff, R. L., and Arnoff, E. L. *Introduction to Operations Research.* New York: Wiley, 1957.

Churchman, C. W. *Prediction and Optimal Decision.* Englewood Cliffs, N.J.: Prentice Hall, 1961.

Cootner, P. H. (ed.) *The Random Character of Stock Market Prices.* Cambridge, Mass.: Mass. Inst. Technol. Press, 1964.

Cramér, H. *Mathematical Methods of Statistics.* Princeton: Princeton University Press, 1946.

Dantzig, G. B. Programming of interdependent activities, *Econometrica* 17 (1949), 200–211.

———. Maximisation of a linear function of variables subject to linear inequalities. ed. T. C. Koopmann, pp. 339–47, 1951.

———. *Linear Programming and Extensions.* Princeton: Princeton University Press, 1963.

Davidson, D. S., Siegel, S., and Suppes, P. *Decision Making.* Stanford, Cal.: Stanford University Press, 1957.

Davis, H. T. *Theory of Econometrics.* Bloomington, Ind.: Principia Press, 1941.

Debreu, J. *Theory of Value.* New York: Wiley, 1959.

Domar, E. *Essays in the Theory of Economic Growth.* New York: Oxford, 1957.

Duesenberry, J. S. *Income, Saving and the Theory of Consumer Behavior.* Cambridge, Mass.: Harvard University Press, 1949.

Duesenberry, J. S., Fromm, G., Klein, L. A., Kuh, E., (eds.) *The Brookings Quarterly Economic Model for the United States.* Chicago: Rand McNally 1966.

Domar, E. *Essays in the Theory of Economic Growth.* New York: Oxford University Press, 1957.

Dorfman, R., Samuelson, P. A., and Solow, R. N. *Linear Programming and Economic Analysis.* New York: McGraw-Hill, 1958.

Ellis, H. S. (ed.) *A Survey of Contemporary Economics.* Philadelphia: Blakiston, 1948.

Evans, G. C. *Mathematical Introduction to Economics.* New York: 1930.

Feigl, H., and Brodbeck, M. (ed.) *Readings in the Philosophy of Science.* New York: Appleton-Century-Crafts, 1953.

GRANGER, C. W. J., and HATAKANA, M. *Spectral Analysis of Economic Time Series.* Princeton: Princeton University Press, 1964.
GRANGER, G. G. *Méthodologie Économique.* Paris: Presses Universitaires de France, 1955.
———. *Pensée formelle et science de l'homme.* Paris: Aubier, 1960.
GREEN, H. A. J. *Aggregation in Economic Analysis.* Princeton: Princeton University Press, 1964.
GRUNBERG, E., and MODIGLIANI, F. The predictability of social events, *Journal of Political Economy* 62 (1954), 465–78.
HAAVELMO, T. The probability approach to econometrics, *Econometrica* 12 (1944), suppl.
———. *Studies in the Theory of Economic Evolution.* Amsterdam: North-Holland, 1954.
HALEY, B. F. (ed.) *A Survey of Contemporary Economics.* Vol. 2, Homewood, Ill.: Irwin, 1952.
HANNAN, E. J. *Time Series Analysis.* London: Methuen, 1960.
HARROD, F. R. *Towards a Dynamic Economics.* New York: Macmillan, 1952.
HARSANYI, J. C. Cardinal welfare, individualistic ethics, and interpersonal comparison of utility, *Journal of Political Economy.* 63 (1955), 309–21.
———. *Models for the Analysis of Balance of Power in Society,* ed. E. Nagel, P. Suppes, A. Tarski, pp. 442–62, 1962.
———. A general theory of rational behavior in game situations, *Econometrica* 34 (1966), 613–34.
HAYEK, F. A. (ed.) *Collectivist Economic Planning.* London: Routledge & Kegan Paul, 1935.
———. *The Counterrevolution of Science.* Glencoe, Ill.: Free Press, 1952.
———. *The Constitution of Liberty.* Chicago: University of Chicago Press, 1960.
HEADY, E. O., and CANDLER, W. *Linear Programming Methods.* Ames, Iowa: Iowa State University Press, 1958.
HICKS, J. R. *Value and Capital.* 2d ed., Oxford: Clarendon Press, 1946.
———. *Capital and Growth.* Oxford: Oxford University Press, 1965.
HIGGINS, B. *Economic Development.* New York: Norton, 1959.
HOFMANN, W. *Gesellschaftslehre als Ordnungsmacht.* Berlin: Duncker and Humbolt, 1961.
HOHENBALKEN, B. VON, and TINTNER, G. Econometric models for the OEEC member countries, the United States and Canada: Their application to economic policy, *Weltwirtschaftliches Archiv* 89 (1962), 29–86.
HOLT, C. C. MODIGLIANI, F., MUTH, J. F., and SIMON, H. A. *Planning, Production, Inventories and Work Force.* Englewood Cliffs, N.J.: Prentice Hall, 1960.
HOOD, W. C., and KOOPMANS, T. (eds.) *Studies in Econometric Method.* New York: Wiley, 1953.
HOUTHAKKER, H. S. Revealed preference and the utility function, *Economica* 17 (1950), 159–74.
HURWICZ, L. Conditions for economic efficiency of centralized and decen-

tralized structures. ed. G. Grossman, *Value and Plan.* Berkeley: University of California Press, 1960.

Hutchison, T. W. *Positive Economics and Policy Objectives.* Cambridge, Mass.: Harvard University Press, 1964.

Jeffreys, H. *Theory of Probability.* Oxford: Clarendon, 1948.

Jevons, W. S. *The Theory of Political Economy.* London: Macmillan, 1911.

Johnston, J. *Econometric Methods.* New York: McGraw-Hill, 1963.

Kalecki, M. A macrodynamic theory of business cycles, *Econometrica* 3 (1935), 327–44.

Kantorovich, L. V. Mathematical methods in the organization and planning of production (1939), *Management Science* 6 (1960), 366–422.

———. *Calcul économique et utilisation des ressources.* Paris: Dunod, 1963.

Katona, G., Klein, L. R., Lansing, J. B., and Morgan, J. N., *Contributions of Survey Methods to Economics.* New York: Columbia University Press, 1954.

Kaufmann, F. *Methodology of Social Sciences.* New York: Oxford University Press, 1944.

Kemeny, J. G. *Carnap's Theory of Probability and Induction.* ed. P. A. Schilpp, pp. 711–38, 1963.

Kemeny, J. G., Morgenstern, O., and Thompson, G. L. A generalization of the von Neumann model of an expanding economy, *Econometrica* 24 (1956), 115–35.

Kemeny, J. G., and Snell, J. L. *Finite Markov Chains.* Princeton, N.J.: Van Nostrand, 1960.

———. *Mathematical Models in the Social Sciences.* Boston: Ginn, 1962.

Kemeny, J. G., Snell, J. L., and Thompson, G. L. *Introduction to Finite Mathematics.* New York: Prentice Hall, 1957.

Kendall, M. G., and Stuart, A. *The Advanced Theory of Statistics.* Vols. 1 and 2. New York: Hafner, 1958 and 1961.

Keynes, J. M. *The General Theory of Employment, Interest and Money.* London: Macmillan, 1936.

———. *A Treatise on Probability.* London: Macmillan, 1948.

Keynes, J. N. *Scope and Method of Political Economy,* 4th ed. New York: Keller, 1955.

Klein, L. R. *Economic Fluctuations in the United States, 1921–1941.* New York: Wiley, 1950.

———. *An Introduction to Econometrics.* Englewood Cliffs, N.J.: Prentice Hall, 1962.

Klein, L. R., and Goldberger, A. S. *An Econometric Model for the United States, 1929–1952.* Amsterdam: North-Holland, 1957.

Knight, F. H. *Risk, Uncertainty and Profit.* (reprint.) London: London School of Economics, 1933.

Kolmogoroff, A. *Grundbegriffe der Wahrscheinlichkeitsrechnug* Berlin: Springer-Verlag, 1933.

Koopmans, T. C. (ed.) *Activity Analysis of Production and Allocation.* New York: Wiley, 1951.

KOOPMANS, T. C. *Three Essays on the State of Economic Science.* New York. McGraw-Hill, 1957.
KOOPMANS, T. C., and BECKMANN, M. J. Assignment problems and the location of economic activities, *Econometrica* 25 (1957) 53–76.
KRAFT, V. *Grundlagen einer wissenschaftlichen Wertlehre* Vienna: Springer-Verlag, 1951.
KRYBURG, H. E., and SMOCKLER, H. E. *Studies in Subjective Probability.* New York: Wiley, 1964.
KUENNE, R. E. *Theory of General Equilibrium.* Princeton: Princeton University Press, 1963.
KUHN, H., and TUCKER, A., Nonlinear programming. In *Proceedings of the Second Berkeley Symposium on Mathematical Statistics and Probability,* ed. J. Neyman. Berkeley: University of California Press, 1951.
LANGE, O. *On the Economic Theory of Socialism,* ed. E. B. Lippincott, pp. 55–141, 1948.
———. *The Scope and Method of Economics.* ed. H. Feigl and M. Brodbeck pp. 744–56, 1953.
———. *Introduction to Econometrics.* London: Pergamon, 1959.
———. *Political Economy,* Vol. 1. New York: Macmillan, 1963.
LEHMANN, E. L. *Testing Statistical Hypotheses.* New York: Wiley, 1959.
LEONTIEF, W. W. *The Structure of the American Economy 1919–1939* New York: Oxford University Press, 1951.
LERNER, A. P. *The Economics of Control.* New York: Macmillan, 1944.
LESOURNE, J. *Technique économique et gestion industrielle.* Paris: Dunod, 1960.
LINDLEY, D. V. *Introduction to Probability and Statistics.* Vols. 1 and 2. Cambridge: Cambridge University Press, 1965.
LIPPINCOTT, E. B. (ed.) *On the Economic Theory of Socialism.* Minneapolis: University of Minnesota Press, 1948.
LIU, T. C. Underidentification, structural estimation and forecasting, *Econometrica* 28 (1960), 855–65.
LUCE, R. D., and RAIFFA, H. *Games and Decision.* New York: Wiley, 1957.
MACHLUP, F. The problem of verification in economics, *Southern Economic Journal* 22 (1955), 1–21.
———. *Essays on Economic Semantics.* Englewood Cliffs, N.J.: Prentice Hall, 1963.
MAHALANOBIS, P. C. The approach of operational research to planning, *Sankhya* 16 (1955), 3–130.
Malinvaud, E. Capital accumulation and efficient allocation of resources, *Econometrica* 21 (1953), 233–68.
———. *Méthodes statistiques de l'économetrie.* Paris: Dunod, 1964.
MANDELBROT, B. The Pareto-Levy Law and the distribution of income, *International Economic Review* 1 (1960), 79–106.
MARSCHAK, J. Rational behavior, uncertain prospect, and measurable utility, *Econometrica* 18 (1950), 111–41.
———. Economic measurement for policy and prediction. In *Studies in*

Econometric Method, ed. W. C. Hood, and T. C. Koopmans, pp. 1–26. New York: Wiley, 1953.
———. Towards an economic theory of organization and information. ed. Thrall, Davis, and Coombs, In *Decision Processes*, pp. 187–220. New York: Wiley, 1954.
———. *Scaling of Utility and Probability*, ed. M. Shubik, pp. 95–109, 1964.
MARSHALL, A. *Principles of Economics*. 8th ed. New York: Macmillan, 1948.
MASSÉ, P. *Le Choix des investissements*. Paris: Dunod, 1959.
MENGER, K. Das Unsicherheitsmoment in der Wertlehre. *Zeitschrift für Nationaloekonomie*. 5 (1934), 459–85.
———. *Moral, Wille und Weltgstaltung*. Vienna: Springer-Verlag, 1934*b*.
MILLS, C. W. *The Sociological Imagination*. New York: Oxford University Press, 1949.
MILNOR, J. *Games against Nature*, ed. M. Shubik, pp. 120–34, 1964.
MISES, L. VON. *Human Action*. New Haven: Yale University Press, 1949.
———. *Socialism*. New Haven: Yale University Press, 1951.
MISES, R. VON. *Wahrscheinlichkeit, Statistik und Wahrheit*. 3d ed. Vienna: Springer-Verlag, 1951.
MODIGLIANI, F. Fluctuations in the savings-income rate, *Studies in Income and Wealth* 6 (1949), New York: National Bureau of Economic Research, 1949.
MOESEKE, P. VAN. Stochastic linear programming, *Yale Economics Essays* 5 (1965), 197–254.
MOORE, H. L. *Synthetic Economics*. New York: Macmillan, 1929.
MORGENSTERN, O. Demand Theory Reconsidered, *Quarterly Journal of Economics* 62 (1947), 165–201.
———. *On the Accuracy of Economic Observations*. 2d ed. Princeton: Princeton University Press, 1963.
MORISHIMA, M. *Equilibrium, Stability and Growth*. Oxford: Clarendon, 1964.
MORRIS, C. Foundations of the theory of signs, *International Encyclopedia of Unified Science*, Vol. 1, No. 2. Chicago: University of Chicago Press, 1938.
———. *Signs, Language, and Behavior*. New York: Prentice Hall, 1946.
MOSBAEK, E. Fitting a static supply and demand function for labor, *Weltwirtschaftliches Archiv* 82 (1959), 133–46.
MOSTELLER, F., and NOGEE, P. An experimental measurement of utility, *Journal of Political Economy* 59 (1951), 371–404.
MUKHERJEE, V., TINTNER, G., NARAYANAN, R. A generalized Poisson process for the explanation of economic development, *Arthaniti* 8 (1964), 156–64.
MYRDAL, G. *Beyond the Welfare State*. New Haven: Yale University Press, 1960.
NAGEL, E. Principles of the theory of probability, *International Encyclopedia of Unified Science*, Vol. 1, No. 6. Chicago: University of Chicago Press, 1939.

NAGEL, E., SUPPES, P., TARSKI, A. (eds.) *Logic, Methodology and Philosophy of Science*. Stanford: Stanford University Press, 1962.

NASH, J. F. The bargaining problem, *Econometrica* 18 (1950) 155–62.

———. Two person cooperative games, *Econometrica* 21 (1953), 128–40.

NEUMANN, J. VON. A model of general economic equilibrium, *Review of Economic Studies* 13 (1945), 1–9.

NEUMANN, J. VON, and MORGENSTERN, O. *Theory of Games and Economic Behavior*. Princeton: Princeton University Press, 1944.

NEYMAN, J. *Lectures and Conferences on Mathematical Statistics*. Washington: Graduate School Department of Agriculture, 1938.

NORTHROP, F. S. C. *The Logic of the Sciences and the Humanities*. New York: Macmillan, 1947.

ORCUTT, G. H., GREENBERGER, M., KORBEL, J., and RIVLIN, A. M. *Microanalysis of Socioeconomic Systems*. New York: Harper & Row, 1961.

PAINLEVÉ, P. The place of mathematical reasoning in economics, ed. L. Sommer. In *Essays in European Economic Thought*. pp. 120–32. Princeton: Van Nostrand, 1960.

PAPANDREOU, A. G. *Economics as a Science*. Chicago: Lippincott, 1958.

PARETO, V. *Manual d'économie politique*. Paris: M. Giard, 1927.

PFANZAGL, J. Zur Geschichte der Theorie des Lebenshaltungsindices, *Statistische Vierteljahrsschrift* 7 (1955), 1–52.

———. *Die axiomatischen Grundlagen einer allgemeinen Theorie des Messens*. Würzburg: Physica Verlag, 1959.

PIGOU, A. C. *The Economics of Welfare*. London: Macmillan, 1920.

POPPER, K. R. *The Open Society and Its Enemies*. Princeton: Princeton University Press, 1950.

———. *The Poverty of Historicism*. New York: Harper & Row, 1957.

———. *The Logic of Scientific Discovery*. New York: Science Editions, 1961.

———. *Conjectures and Refutations*. London: Routledge & Kegan Paul, 1963*a*.

———. What is dialectic? (In 1963*a*), pp. 312–35, 1963*b*.

———. (1963C): Prediction and Prophecy in the Social Sciences. (In 1963*a*), pp. 336–46, 1963*c*.

PRAIS, S. J., and HOUTHAKKER, H. S., *The Analysis of Family Budgets*. Cambridge: Cambridge University Press, 1955.

QUINE, W. V. *Word and Object*. New York: Wiley, 1960.

RADNER, R. Paths of economic growth that are optimal with regard only to final states: A turnpike theorem, *Review of Economic Studies* 28 (1961), 98–104.

RAO, C. R. (ed.) *Essays in Econometrics and Planning*. Oxford: Pergamon, 1965.

RAMSEY, F. P. A. Mathematical Theory of Saving, *Economic Journal* 36 (1928), 543–59.

———. *The Foundations of Mathematics*. London: Routledge & Kegan Paul, 1931.

ROBBINS, L. *An Essay on the Nature and Significance of Economic Science*. 2d ed. London: Macmillan, 1949.

———. *The Theory of Economic Policy*. London: Macmillan, 1961.

———. *Politics and Economics*. London: Macmillan, 1963.

ROBINSON, J. *Economics of Imperfect Competition*. London: Macmillan, 1938.

———. *An Essay in Marxian Economics*. London: Macmillan, 1949.

———. *Economic Philosophy*. London: Penguin, 1962.

ROOS, C. F. *Dynamic Economics*. Bloomington, Ind.: Principia, 1934.

ROTHBARD, M. N. *Towards a Reconstruction of Utility and Welfare Economics*, ed. M. Sennholz, pp. 224–62, 1956.

ROTHENBERG, J. *The Measurement of Social Value*. Englewood Cliffs, N.J.: Prentice Hall, 1961.

SAMUELSON, P. A. A note on the pure theory of consumers behavior, *Economica* 5 (1938), 61–71 and 353–54.

———. *Foundations of Economic Analysis*. Cambridge, Mass.: Harvard University Press, 1947.

———. Consumption theory in terms of revealed preference, *Economica* 15 (1948), 243–53.

———. *The Collected Scientific Papers of Paul A. Samuelson*, Vols. 1 and 2. Cambridge, Mass.: Mass. Inst. Technol. Press, 1966.

SAVAGE, L. J. *Foundations of Statistics*. New York: Wiley, 1954.

SAVAGE, L. J. *The Foundations of Statistical Inference*. London: Methuen, 1962.

SCHLAIFFER, R. *Probability and Statistics for Business Decisions*. New York: McGraw-Hill, 1959.

SCHILPP, P. A. (ed.) *The Philosophy of Rudolf Carnap*. La Salle, Ill.: Open Court, 1963.

SCHOEFFLER, S. *The Failures of Economics*. Cambridge, Mass.: Harvard University Press, 1955.

SCHULTZ, T. W. *Transforming Traditional Agriculture*. New Haven: Yale University Press, 1964.

SCHUMPETER, J. A. *Ten Great Economists*. New York: Oxford, 1951*a*.

———. *Essays*. Cambridge, Mass.: Addison-Wesley, 1951*b*.

———. *History of Economic Analysis*. New York: Oxford, 1954.

———. *Capitalism, Socialism and Democracy*, 3d ed. New York: Harper, 1950.

SCITOVSKI, T. *Welfare and Growth*. Stanford: Stanford University Press, 1964.

SENGUPTA, J. K., and TINTNER, G. An approach to a stochastic theory of economic development with applications, *Problems of Economic Dynamics and Planning: Essays in honour of M. Kalecki*, pp. 373–93. Warsaw: Polish Scientific Publishers, 1964.

———. Some aspects of trend in the aggregative models of economic growth, *Kyklos* 16 (1963), 47–61.

SENGUPTA, J. K., TINTNER, G., and MORRISON, B., Stochastic linear programming with applications to economic models, *Economica* 30 (1963), 262–76.

SENNHOLZ, M. (ed.) *Freedom and Free Enterprise*. Princeton: Von Nostrand, 1956.

SHONFIELD, A. *Modern Capitalism.* New York: Oxford University Press, 1965.
SHUBIK, M. *Strategy and Market Structure.* New York: Wiley, 1959.
SHUBIK, M. (ed.). *Game Theory and Related Approaches to Social Behavior.* New York: Wiley, 1964.
SIEVERS, A. M. *Revolution, Evolution and the Economic Order.* Englewood Cliffs, N.J.: Prentice Hall, 1962.
SIMON, H. A. *Models of Man.* New York: Wiley, 1957.
SOLOMON, H. (ed.) *Mathematical Thinking in the Measurement of Behavior.* Glencoe, Ill. Free Press, 1960.
STEINDL, J. *Random Processes and the Growth of Firms.* New York: Hafner, 1965.
STIGLER, G. J. *Five Lectures on Economic Problems.* London: Longmans Green, 1949.
———. The development of utility theory, *Journal of Political Economy* 58 (1950), 307–27, 373–96.
STONE, R. On the interdependence of blocks of transactions, *Journal Royal Statistical Society* Suppl. 8 (1947), 1.
———. *The Role of Measurement in Economics.* Cambridge: Cambridge University Press, 1951.
———. Three models of economic growth. ed. E. Nagel, P. Suppes, and A. Tarski, pp. 494–506, 1962.
STROTZ, R. H., and WOLD, H. Recursive *vs.* nonrecursive systems: An attempt of a synthesis, *Econometrica* 28 (1960), 417–27.
THEIL, H. *Linear Aggregation of Economic Relations.* Amsterdam: North-Holland, 1954.
———. *Economic Forecasts and Policy,* 2d ed., Amsterdam: North-Holland, 1961.
———. *Alternative Approaches to the Aggregation Problem,* ed. E. Nagel, P. Suppes, and A. Tarski, pp. 507–27, 1962.
———. *Studies in Mathematical and Managerial Economics.* Amsterdam: North-Holland, 1964.
———. *Applied Economic Forecasting.* Amsterdam: North-Holland, 1966.
———. *Economics and Information Theory.* Amsterdam: North-Holland, 1967.
THRALL, R. M., COOMBS, C. H., and DAVIS, R. L. (eds.) *Decision Processes.* New York: Wiley, 1954.
TINBERGEN, J. *On the Theory of Economic Policy.* Amsterdam: North-Holland, 1954.
———. *Selected Papers.* Amsterdam: North-Holland, 1959.
TINTNER, G. The theory of choice under subjective risk and uncertainty, *Econometrica* 9 (1941), 298–304.
———. The theory of production under nonstatic conditions, *Journal of Political Economy* 50 (1942*a*), 645–67.
———. A contribution to the non-static theory of choice, *Quarterly Journal of Economics* 56 (1942*b*), 274–306.
———. A contribution to the nonstatic theory of production, ed. Lange,

O., McIntyre, F., and Yntema, Th., *Studies in Mathematical Economics and Econometrics.* Chicago: University of Chicago Press, 1942*c*.

———. A note on welfare economics, *Econometrica* 14 (1946), 69–78.

———. Foundations of probability and statistical inference, *Journal of the Royal Statistical Society* Ser. A, 112 (1949), 251–79.

———. *Econometrics* New York: Wiley, 1952.

———. The definition of econometrics, *Econometrica* 21 (1953*a*), 31–40.

———. *Mathematics and Statistics for Economists.* New York: Rinehart, 1953*b*.

———. Einige Grundprobleme der Oekonometrie, *Zeitschrift für die gesamte Staatswissenschaft* 111 (1955), 601–10.

———. Game theory, linear programming and input-output analysis, *Zeitschrift für Nationaloekonomie* 17 (1957), 1–38.

———. The application of decision theory to a simple inventory problem, *Trabajos de Estadistica* 10 (1959), 239–47.

———. A note on stochastic linear programming, *Econometrica* 28 (1960*a*), 490–95.

———. External economies in consumption. *Essays in Economics and Econometrics*, pp. 107–112. Chapel Hill: University of North Carolina Press, 1960*b*.

———. *Handbuch der Oekonometrie.* Berlin: Springer-Verlag, 1960*c*.

———. Eine Anwendung der Wahrscheinlichkeitstheory von Carnap auf ein Problem der Unternehmungsforschung, *Unternehmungsforschung* 4 (1960*d*), 164–70.

———. The use of stochastic linear programming in planning, *Indian Economic Review* 5 (1960*e*), 159–67.

———. *Mathématiques et statistiques pour les économistes.* Paris: Dunod, 1962.

———. Lineare Programme und Input-Output Analyse, *Statistische Hefte* 5 (1965), 50–55.

———. Modern decision theory. *Journal of the Indian Society of Agricultural Research Statistics* 18 (1966), 82–98.

Tintner, G., and Murteira, B. Un modelo input-output simplificado para a economia portuguesa. *Colectanea de Estudos*, No. 8. Lisbon: Centro de Estudos de Estadistica Economica, 1960.

Tintner, G., and Patel, R. C. A long-normal diffusion process applied to the economic development of India, *Indian Economic Journal* 13 (1966), 465–75.

Tintner, G., and Murteira, B. Um modelo input-output simplificado para *nal of Farm Economics* 48 (1966), 704–10.

Tintner, G., and Pátel, R. C. A log-normal diffusion process applied to Theory of Stochastic Processes to Economic Development. The Theory and Design of Economic Development, ed. I. Adelman and E. Thorbeck, pp. 99–110. Baltimore: Johns Hopkins, 1966.

Tintner, G., and Thomas, E. J., Un modèle stochastique de développement économique avec application à l'industrie anglaise *Revue d'Economie Politique* 73 (1963), 278–80.

Topitsch, E. *Vom Ursprung und Ende der Metaphysik*, Vienna: Springer-Verlag, 1958.
———. *Sozialphilosophie zwischen Ideologie und Wissenschaft*. Neuwied: Luchterhand, 1961.
Vadja, S. *Mathematical Programming*. Reading, Mass.: Addison-Wesley, 1961.
Viner, J., Adam Smith and laissez-faire, *Journal of Political Economy* 35 (1927), 198–232.
Wald, A. The approximate determination of indifference systems by means of Engel curves, *Econometrica* 8 (1940), 144–75.
———. *Statistical Decision Functions*. New York: Wiley, 1950.
———. On some systems of equations of mathematical economics, *Econometrica* 19 (1951), 368–403.
Weber, M. *Gesammelte Aufsätze zur Wissenschaftslehre*. Tübingen: Mohr, 1922.
Weber, W., and Topitsch, E. Das Wertfreiheitsproblem seit Max Weber, *Zeitschrift für Nationaloekonomie* 13 (1952), 199.
Weldon, T. D. *The Vocabulary of Politics*. London: Penguin, 1953.
Wicksell, K. *Lectures on Political Economy*, Vol. 1. London: Routledge & Kegan Paul, 1934.
Wiener, N. *God and Golem*. Cambridge, Mass.: Mass. Inst. Technol. Press, 1964.
Whittle, P. *Prediction and Regulation*. London: English Universities Press, 1963.
Wilks, S. S. *Mathematical Statistics*. New York: Wiley, 1962.
Wold, H. *A Letter report to Professor P. C. Mahalanobis*, Rao, R. C., ed., pp. 309–328, 1965.
Wold, H., and Jureén, L. *Demand Analysis*. New York: Wiley, 1953.
Wright, G. H. von *An Essay in Modal Logic*. Amsterdam: North-Holland, 1951.
———. *The Logic of Preference*. Edinburgh: Edinburgh University Press, 1963*a*.
———. *The Varieties of Goodness*. London: Routledge & Kegan Paul, 1963*b*.
Zeuthen, F. *Problems of Monopoly and Economic Welfare*. London: Routledge & Kegan Paul, 1933.
Zweig, F. *Economic Ideas*. Englewood Cliffs, N.J.: Prentice Hall, 1950.

Fundamentals of Concept Formation in Empirical Science

Carl G. Hempel

Fundamentals of Concept Formation in Empirical Science

Contents:

Fundamentals of Concept Formation in Empirical Science

Carl G. Hempel

1. Introduction

Empirical science has two major objectives: to describe particular phenomena in the world of our experience and to establish general principles by means of which they can be explained and predicted. The explanatory and predictive principles of a scientific discipline are stated in its hypothetical generalizations and its theories; they characterize general patterns or regularities to which the individual phenomena conform and by virtue of which their occurrence can be systematically anticipated.

In the initial stages of scientific inquiry, descriptions as well as generalizations are stated in the vocabulary of everyday language. The growth of a scientific discipline, however, always brings with it the development of a system of specialized, more or less abstract, concepts and of a corresponding technical terminology. For what reasons and by what methods are these special concepts introduced and how do they function in scientific theory? These are the central questions which will be examined in this monograph.

It might seem plausible to assume that scientific concepts are always introduced by definition in terms of other concepts, which are already understood. As will be seen, this is by no means generally the case. Nevertheless, definition is an important method of concept formation, and we will therefore begin by surveying, in Chapter I, the fundamental principles of the general theory of definition. Chapter II will analyze the methods, both definitional and nondefinitional, by means of which scientific concepts are introduced. This analysis will lead to a closer examination of the function of concepts in scientific theories and will show that concept formation and theory formation in science are so closely interrelated as to constitute virtually two different aspects of the same procedure. Chapter III, finally, will be concerned with a study of qualitative and quantitative concepts and methods in empirical science.

We shall use, in this study, some of the concepts and techniques of modern logic and occasionally also a modicum of symbolic notation; these will, however, be explained, so that the main text of this monograph can be understood without any previous knowledge of symbolic logic. Some remarks of a somewhat more technical nature as well as points of detail and bibliographic references have been included in the notes at the end.[1]

I. Principles of Definition

2. On Nominal Definition

The word 'definition' has come to be used in several different senses. For a brief survey of the major meanings of the term, we choose as our point of departure the familiar distinction made in traditional logic between "nominal" and "real" definition. A real definition is conceived of as a statement of the "essential characteristics" of some entity, as when man is defined as a rational animal or a chair as a separate movable seat for one person. A nominal definition, on the other hand, is a convention which merely introduces an alternative—and usually abbreviatory—notation for a given linguistic expression, in the manner of the stipulation

(2.1) Let the word 'tiglon' be short for (i.e., synonymous with) the phrase 'offspring of a male tiger and a female lion'

In the present section we will discuss nominal definition; in the following one, real definition and its significance for scientific inquiry.

A *nominal definition* may be characterized as a stipulation to the effect that a specified expression, the *definiendum*, is to be synonymous with a certain other expression, the *definiens*, whose meaning is already determined. A nominal definition may therefore be put into the form

(2.2) Let the expression E_2 be synonymous with the expression E_1

This form is exemplified by the definition of the popular neologism 'tiglon' in (2.1) and by the following definitions of scientific terms:

(2.3) Let the term 'Americium' be synonymous with the phrase 'the element having 95 nuclear protons'

(2.4) Let the term 'antibiotic' be synonymous with (and thus short for) the expression 'bacteriostatic or bactericidal chemical agent produced by living organisms'

If a nominal definition is written in the form (2.2), it clearly speaks about certain linguistic expressions, which constitute its definiendum and its definiens; hence, it has to contain names for them. One simple and widely used method of forming a name for an expression is to put the expression between single quotation marks. This device is illustrated in the preceding examples and will frequently be used throughout this monograph.

There exists, however, an alternative way of formulating definitions, which dispenses with quotation marks, and which we will occasionally use. In its alternative form, the definition (2.3) would appear as follows:

(2.5) Americium $=_{Df}$ the element with 95 nuclear protons

The notation '$=_{Df}$' may be read 'is, by definition, to equal in meaning', or briefly, 'equals by definition'; it may be viewed as stipulating the synonymy of the expressions flanking it. Here are two additional illustrations of this manner of stating nominal definitions:

(2.6)
$$\text{the cephalic index of person } x =_{Df} 100\,\frac{\text{maximum skull breadth of person } x}{\text{maximum skull length of person } x}$$

(2.7) x is dolichocephalic $=_{Df}$ x is a person with a cephalic index not exceeding 75

All these definitions are of the form

(2.8) $\text{———} =_{Df} \cdots\cdots$

with the definiendum expression appearing to the left, and the definiens expression to the right of the symbol of definitional equality.

According to the account given so far, a nominal definition introduces, or defines, a new *expression*. But it is sometimes expeditious and indeed quite customary to describe the function of nominal definition in an alternative manner: We may say that a nominal definition singles out a certain *concept*, i.e., a nonlinguistic entity such as a property, a class, a relation, a function, or the like, and, for convenient reference, lays down a special name for it. Thus conceived, the definition (2.5) singles out a certain property, namely, that of being the chemical element whose atoms have 95 nuclear protons, and gives it a brief name. This second characterization is quite compatible with the first, and it elucidates the sense in which—as is often said—a nominal definition defines a *concept* (as distinguished from the expression naming it). Henceforth, we will permit ourselves to speak of definition, and later more generally of introduction, both in regard to expressions and in regard to concepts; the definition (2.6), for example, will be alternatively said to define the expression 'cephalic index of person x' or the concept of cephalic index of a person.

The expression defined by a nominal definition need not consist of just one single word or symbol, as it does in (2.5); it may instead be a compound phrase, as in (2.6) and (2.7). In particular, if the expression to be introduced is to be used only in certain specific linguistic contexts, then it is sufficient to provide synonyms for those contexts rather than for the new term in isolation. A definition which introduces a symbol s by providing synonyms for certain expressions containing s, but not for s itself, is called a *contextual definition*. Thus, e.g., when the term 'dolichocephalic' is to be used only in contexts of the form 'so-and-so is dolichocephalic', then it suffices to provide means for eliminating the term from those contexts; such means are provided by (2.7), which is a contextual definition.

The idea that the definiendum expression of an adequate nominal definition must consist only of the "new" term to be introduced is a misconception which is perhaps related to the doctrine of classical logic that every definition must be stated in terms of *genus proximum* and *differentia specifica*, as in the definition

(2.9) minor $=_{Df}$ person less than 21 years of age

This definition characterizes, in effect, the class of minors as that subclass of the genus, persons, whose members have the specific characteristic of being less than 21 years old; in other words, the class of minors is defined as the logical product (the intersection) of the class of persons and the class of beings less than 21 years of age.

The doctrine that every definition must have this form is still widely accepted in elementary textbooks of logic, and it sometimes seriously hampers the adequate formulation of definitions—both nominal and "real"—in scientific writing and in dictionaries.[2] Actually, that doctrine is unjustifiable for several reasons. First, a definition by genus and differentia characterizes a class or a property as the logical product of two other classes or properties; hence this type of definition is inapplicable when the definiendum is not a class or a property but, say, a relation or a function. Consider, for example, the following contextual definition of the relation, harder than, for minerals:

(2.10) x is harder than y $=_{Df}$ x scratches y, but y does not scratch x

or consider the contextual definition of the average density of a body—which is an example of what, in logic, is called a function:

(2.11) $$\text{average density of } x =_{Df} \frac{\text{mass of } x \text{ in grams}}{\text{volume of } x \text{ in cc.}}$$

In cases of this sort the traditional requirement is obviously inapplicable. And it is worth noting here that the majority of terms used in contemporary science are relation or function terms rather than class or property terms; in particular, all the terms representing metrical magnitudes are function terms and thus have a form which altogether precludes a definition by genus and differentia. Historically speaking, the genus-and-differentia rule reflects the fact that traditional logic has been concerned almost exclusively with class or property concepts—a limitation which renders it incapable of providing a logical analysis of modern science.

But even for class or property concepts the traditional form of definition is not always required. Thus, e.g., a property might be defined as the logical sum of certain other properties rather than as a product. This is illustrated by the following definition, which is perfectly legitimate yet states neither genus nor differentia for the definiendum:

(2.12) Scandinavian $=_{Df}$ Dane or Norwegian or Swede or Icelander

The genus-and-differentia form is therefore neither necessary nor sufficient for an adequate definition. Actually, the nominal definition of a term has to satisfy only one basic requirement: it must enable us to eliminate that term, from any context in which it can grammatically occur, in favor of other expressions, whose meaning is already understood. In principle, therefore, signs introduced by nominal definition can be dispensed with altogether: "To define a sign is to show how to avoid it."[3]

3. On "Real" Definition

A "real" definition, according to traditional logic, is not a stipulation determining the meaning of some expression but a statement of the "essential nature" or the "essential attributes" of some entity. The notion of essential nature, however, is so vague as to render this characterization useless for the purposes of rigorous inquiry. Yet it is often possible to reinterpret the quest for real definition in a manner which requires no reference to "essential natures" or "essential attributes," namely, as a search either for an empirical explanation of some phenomenon or for a meaning analysis. Thus, e.g., the familiar pronouncement that biology cannot as yet give us a definition of life clearly is not meant to deny the possibility of laying down some nominal definition for the term 'life'. Rather, it assumes that the term 'life' (or, alternatively, 'living organism') has a reasonably definite meaning, which we understand at least intuitively; and it asserts, in effect, that at present it is not possible to state, in a nontrivial manner, explicit and general criteria of life, i.e., conditions which are satisfied by just those phenomena which are instances of life according to the customary meaning of the term. A real definition of life would then consist in an equivalence sentence of the form

(3.1a) x is a living organism if and only if x satisfies condition C

or, in abbreviatory symbolization:

(3.1b) $$Lx \equiv Cx$$

Here, 'C' is short for an expression indicating a more or less complex set of conditions which together are necessary and sufficient for life. One set of conditions of this kind is suggested by Hutchinson[4] in the following passage:

> It is first essential to understand what is meant by a living organism. The necessary and sufficient condition for an object to be recognizable as a living organism, and so to be the subject of biological investigation, is that it be a discrete mass of matter, with a definite boundary, undergoing continual interchange of material with its surroundings without manifest alteration of properties over short periods of time, and, as ascertained either by direct observation or by analogy with other objects of the same class, originating by some process of division or fractionation from one or two pre-existing objects of the same kind. The criterion of continual interchange of material may be termed the *metabolic criterion*, that of origin from a pre-existing object of the same class, the *reproductive criterion*.

If we represent the characteristic of being a discrete mass with a definite boundary by 'D' and the metabolic and reproductive criteria by 'M' and 'R', respectively, then Hutchinson's characterization of life may be written thus:

(3.2) $$Lx \equiv Dx \cdot Mx \cdot Rx$$

i.e., a thing x is a living organism if and only if x has the characteristic of being a discrete mass, etc., and x satisfies the metabolic criterion, and x satisfies the reproductive criterion.

As the quoted passage shows, this equivalence is not offered as a convention concerning the use of the term 'living' but rather as an assertion which claims to be true. How can an assertion of this kind be validated? Two possibilities present themselves:

The expression on the right-hand side of (3.2) might be claimed to be synonymous with the phrase 'x is a living organism'. In this case, the "real" definition (3.2) purports to characterize the meaning of the term 'living organism'; it constitutes what we shall call a *meaning analysis*, or an *analytic definition*, of that term (or, in an alternative locution, of the concept of living organism). Its validation thus requires solely a reflection upon the meanings of its constituent expressions and no empirical investigation of the characteristics of living organisms.

On the other hand, the "real" definition (3.2) might be intended to assert, not that the phrase 'x is a living organism' has the same meaning as the expression on the right, but rather that, as a matter of empirical fact, the three conditions D, M, and R are satisfied simultaneously by those and only those objects which are also living things. The sentence (3.2) would then have the character of an empirical law, and its validation would require reference to empirical evidence concerning the characteristics of living beings. In this case, (3.2) represents what we shall call an *empirical analysis* of the property of being a living organism.

It is not quite clear in which of these senses the quoted passage was actually intended; the first sentence suggests that a meaning analysis was aimed at.

Empirical analysis and meaning analysis differ from each other and from nominal definition. Empirical analysis is concerned not with linguistic expressions and their meanings but with empirical phenomena: it states characteristics which are, as a matter of empirical fact, both necessary and sufficient for the realization of the phenomenon under analysis. Usually, a sentence expressing an empirical analysis will have the character of a general law, as when air is characterized as a mixture, in specified proportions, of oxygen, nitrogen, and inert gases. Empirical analysis in terms of general laws is a special case of scientific *explanation*, which is aimed at the subsumption of empirical phenomena under general laws or theories.

Nominal definition and meaning analysis, on the other hand, deal with the meanings of linguistic expressions. But whereas a nominal definition introduces a "new" expression and gives it meaning by stipulation, an analytic definition is concerned with an expression which is already in use—let us call it the *analysandum expression* or, briefly, the *analysandum*—and makes its meaning explicit by providing a synonymous expression, the *analysans*, which, of course, has to be previously understood. Dictionaries for a natural language are intended to provide analytic definitions for the words of that language; frequently, however, they supplement their meaning analyses by factual information about the subject matter at hand, as when, under the heading 'chlorine', a chemical characterization of the substance is supplemented by mentioning its use in various industrial processes.

According to the conception here outlined, an analytic definition is a statement which is true or false according as its analysans is, or is not, synonymous with its analysandum. Evidently, this conception of analytic definition presupposes a language whose expressions have precisely determined meanings—so that any two of its expressions can be said either to be, or not to be, synonymous. This condition is met, however, at best by certain artificial languages and surely is not generally satisfied by natural languages. Indeed, to determine the meaning of an expression in a given natural language as used by a specified linguistic community, one would have to ascertain the conditions under which the members of the community use—or, better, are disposed to use—the expression in question. Thus, e.g., to ascertain the meaning of the word 'hat' in contemporary English as spoken in the United States, we would have to determine to what kinds of objects—no matter whether they actually occur or not—the word 'hat' would be applied according to contemporary American usage. In this sense the conception of an analysis of "the" meaning of a given expression presupposes that the conditions of its application are (1) well determined for every user of the language and are (2) the same for all users during the period of time under consideration. We shall refer to these two presuppositions as the conditions of *determinacy* and of (personal and interpersonal) *uniformity of usage.* Clearly, neither of them is fully satisfied by any natural language. For even if we disregard ambiguity, as exhibited by such words as 'field' and 'group', each of which has several distinct meanings, there remain the phenomena of vagueness (lack of determinacy) and of inconsistency of usage.[5] Thus, e.g., the term 'hat' is vague; i.e., various kinds of objects can be described or actually produced in regard to which one would be undecided whether to apply the term or not. In addition, the usage of the term exhibits certain inconsistencies both among different users and even for the same user of contemporary American English; i.e., instances can be described or actually produced of such a kind that different users, or even the same user at different times, will pass different judgments as to whether the term applies to those instances.

These considerations apply to the analysandum as well as to the analysans of an analytic definition in a natural language. Hence, the idea of a true analytic definition, i.e., one in which the meaning of the analysans is the same as that of the analysandum, rests on an untenable assumption. However, in many cases, there exists, for an expression in a natural language, a class of contexts in which its usage is practically uniform (for the word 'hat' this class would consist of all those contexts in which practically everybody would apply the term and of those in which practically none would); analytic definitions within a natural language might, therefore, be qualified as at least more or less adequate according to the extent to which uniform usage of the analysandum coincides with that of the analysans. When subsequently we speak of, or state, analytic definitions for expressions in a natural language, we will accordingly mean characterizations of approximately uniform patterns of usage.

Meaning analysis, or analytic definition, in the purely descriptive sense considered so far has to be distinguished from another procedure, which is likewise adumbrated in the vague traditional notion of real definition. This procedure is often called logical analysis or rational reconstruction, but we will refer to it, following Carnap's proposal, as *explication*.[6] Explication is concerned with expressions whose meaning in conversational language or even in scientific discourse is more or less vague (such as 'truth', 'probability', 'number', 'cause', 'law', 'explanation'—to mention some typical objects of explicatory study) and aims at giving those expressions a new and precisely determined meaning, so as to render them more suitable for clear and rigorous discourse on the subject matter at hand. The Frege-Russell theory of arithmetic and Tarski's semantical definition of truth are outstanding examples of explication.[7] The definitions proposed in these theories are not arrived at simply by an analysis of customary meanings. To be sure, the considerations leading to the precise definitions are guided initially by reference to customary scientific or conversational usage; but eventually the issues which call for clarification become so subtle that a study of prevailing usage can no longer shed any light upon them. Hence, the assignment of precise meanings to the terms under explication becomes a matter of judicious synthesis, of rational reconstruction, rather than of merely descriptive analysis: An explication sentence does not simply exhibit the commonly accepted meaning of the expression under study but rather proposes a specified new and precise meaning for it.

Explications, having the nature of proposals, cannot be qualified as being either true or false. Yet they are by no means a matter of arbitrary convention, for they have to satisfy two major requirements: First, the explicative reinterpretation of a term, or—as is often the case—of a set of related terms, must permit us to reformulate, in sentences of a syntactically precise form, at least a large part of what is customarily expressed by means of the terms under consideration. Second, it should be possible to develop, in terms of the reconstructed concepts, a comprehensive, rigorous, and sound theoretical system. Thus, e.g., the Frege-Russell reconstruction of arithmetic gives a clear and uniform meaning to the arithmetical terms both in purely mathematical contexts, such as '$7 + 5 = 12$', and in their application to counting, as in the sentence 'The Sun has 9 major planets' (a purely axiomatic development of arithmetic would not accomplish this); and the proposed reconstruction provides a basis for the deductive development of pure arithmetic in such a way that all the familiar arithmetical principles can be proved.

Explication is not restricted to logical and mathematical concepts, however. Thus, e.g., the notions of purposiveness and of adaptive behavior, whose vagueness has fostered much obscure or inconclusive argumentation about the specific characteristics of biological phenomena, have become the objects of systematic explicatory efforts.[8] Again, the basic objective of the search for a "definition" of life is a precise and theoretically fruitful explication, or reconstruction, of the concept. Similarly, the controversy over whether a satisfactory definition of personality is attainable in purely psychological terms or requires reference to a cultural setting[9] centers around the question whether a sound explicatory or predictive theory of personality is possible without the use of sociocultural parameters; thus, the problem is one of explication.

An explication of a given set of terms, then, combines essential aspects of meaning analysis and of empirical analysis. Taking its departure from the customary meanings of the terms, explication aims at reducing the limitations, ambiguities, and inconsistencies of their ordinary usage by propounding a reinterpretation intended to enhance the clarity and precision of their meanings as well as their ability to function in hypotheses and theories with explanatory and predictive force. Thus understood, an explication cannot be qualified simply as true or false; but it may be adjudged more or less adequate according to the extent to which it attains its objectives.

In conclusion, let us note an important but frequently neglected requirement, which applies to analytic definitions and explications as well as to nominal definitions; we will call it the *requirement of syntactical determinacy:* A definition has to indicate the syntactical status, or, briefly, the syntax, of the expression it explicates or defines; i.e., it has to make clear the logical form of the contexts in which the term is to be used. Thus, e.g., the word 'husband' can occur in contexts of two different forms, namely, 'x is a husband of y' and 'x is a husband'. In the first type of context, which is illustrated by the sentence 'Prince Albert was the husband of Queen Victoria', the word 'husband' is used as a *relation term:* It has to be supplemented by two expressions referring to individuals if it is to form a sentence. In contexts of the second kind, such as 'John Smith is a husband', the word is used as a *property term,* requiring supplementation by only one individual name to form a sentence. Some standard English dictionaries, however, define the term 'husband' only by such phrases as 'man married to woman', which provide no explicit indication of its syntax but suggest the use of the word exclusively as a property term applying to married men; this disregards the relational use of the term, which is actually by far the more frequent. Similarly, the dictionary explication of 'twin' as 'being one of two children born at a birth' clearly suggests use of the word as a property term, i.e., in contexts of the form 'x is a twin', which is actually quite rare, and disregards its prevalent relational use in contexts of the form 'x and y are twins'. This shortcoming of many explications reflects the influence of classical logic with its insistence on construing all sentences as being of the subject-predicate type, which requires the interpretation of all predicates as property terms. Attempts to remedy this situation are likely to be impeded by the clumsiness of adequate formulations in English, which could, however, be considerably reduced by the use of variables. Thus, in a somewhat schematized form, an entry in the dictionary might read:

husband. (1) x is a h. of y: x is a male person, and x is married to y; (2) x is a h.: x is a male person who is married to some y.[10]

Nominal definitions have to satisfy the same requirement: Certainly, a term has not been defined if not even its syntax has been specified. In the definitions given in section 2, this condition is met by formulating the definiens in a way which unambiguously reflects its syntactical status; in some of them variables are used for greater clarity. Similarly the definition of a term such as 'force' in physics has to show that the term may occur in sentences of the form 'The force acting upon point P at time t equals vector f'. By way of contrast, consider now the concept of vital force or entelechy as adduced by neovitalists in an effort to explain certain biological phenomena which they consider as inaccessible in principle to any explanation by physicochemical theories. The term 'vital force' is used so loosely that not even its syntax is shown; no clear indication is given of whether it is to represent a property or a scalar or a vectorial magnitude, etc.; nor whether it is to be assigned to organisms, to biological processes, or to yet something else. The term is therefore unsuited for the formulation of even a moderately precise hypothesis or theory; consequently, it cannot possess the explanatory power ascribed to it.

A good illustration of the importance of syntactical determinacy is provided by the concept of probability. The definitions given in older textbooks, which speak of "the probability of an event" and thus present probabilities as numerical characteristics of individual events, overlook or conceal the fact that probabilities are relative to, and change with, some reference class (in the case of the statistical concept of probability) or some specific information (in the case of the logical concept of probability) and thus are numerical functions not of one but of two arguments. Disregard of this point is the source of various "paradoxes" of probability, in which "the same event" is shown to possess different probabilities, which actually result from a tacit shift in the reference system.

4. Nominal Definition within Theoretical Systems

Nominal definition plays its most important role in the formulation of scientific theories. In the present section we will consider the fundamental logical principles governing its use for this purpose.[11]

By the total vocabulary of a theory T let us understand the class of all the words or other signs which occur in the sentences of T. The total vocabulary of any scientific theory contains certain terms which belong to the vocabulary of logic and mathematics, such as 'not', 'and', 'or', 'if . . . then ___', 'all', 'some', etc., or their symbolic equivalents; symbols for numbers as well as for operations on them and relations between them; and, finally, variables or equivalent verbal expressions. The balance of the terms in the total vocabulary of a theory T will be called the *extra-logical vocabulary*, or, briefly, the *vocabulary*, of T. Apart from a few exceptions which serve mainly illustrative purposes, we shall discuss here only the definition of extra-logical terms in scientific theories.

While many terms in the vocabulary of a theory may be defined by means of others, this is not possible for all of them, short of an infinite regress, in which the process of defining a term would never come to an end, or a definitional circle, in which certain terms would be defined, mediately or immediately, by means of themselves. Definitional circles are actually encountered in dictionaries, where one may find 'parent' defined by 'father or mother', then, 'father' in turn by 'male parent' and 'mother' by 'female parent'. This is unobjectionable for the type of analytic definition intended by dictionaries; in the context of nominal definition within scientific theories, however, such circularity is inadmissible because it defeats the purpose of nominal definition, namely, to introduce convenient notations which, at any time, can be eliminated in favor of the defining expressions. Infinite definitional regress evidently has to be barred for the same reason.

Thus the vocabulary of a theory falls into two classes: the *defined terms*, i.e., those which are introduced by definition in terms of other expressions of the vocabulary, and the so-called *primitive terms*, or *primitives*, by means of which all other terms of the theoretical vocabulary are ultimately defined. The primitives themselves, while not defined within the theory, may nevertheless have specific meanings assigned to them. Methods of effecting such interpretation of the primitives will be considered later.

By way of illustration let us define a set of words which might be used in a theory of family relationships. As primitives, we choose the words 'male' and 'child'. The former will be used as a property term, i.e., in contexts of the form 'x is a male', or, briefly, 'Male x'; the latter will serve as a relation term, i.e., in contexts of the form 'x is a child of y', or, briefly, 'x Child y'. In formulating our definitions, we use, besides the dot symbol of conjunction, also the denial sign '$\sim$' (to be read 'it is not the case that'), and the notation for existential quantification—'$(Ez)(.....)$' stands for 'there is at least one entity z such that .....'. As our universe of discourse, i.e., the totality of objects under consideration, we choose the class of human beings. Now we lay down the following definitions:

(4.1a) x Parent y $=_{Df}$ y Child x
(4.1b) x Father y $=_{Df}$ Male $x \cdot x$ Parent y
(4.1c) x Mother y $=_{Df}$ x Parent $y \cdot \sim x$ Father y
(4.1d) x Grandparent y $=_{Df}$ $(Ez)(x$ Parent $z \cdot z$ Parent $y)$
(4.1e) x Grandmother y $=_{Df}$ $\sim$Male $x \cdot x$ Grandparent y

As the symbol '$=_{Df}$' indicates, these sentences are to be understood as nominal definitions, even though the meanings they assign to the definienda are those of ordinary usage if the primitives are taken in their customary meanings. Thus, e.g., the definition (4.1a) may be paraphrased as stipulating that 'x is a parent of y' is to be synonymous with 'y is a child of x', while (4.1d) lays down the convention that 'x is a grandparent of y' is to mean the same as 'x is a parent of someone, z, who is a parent of y'.

Let us note in passing that not one of these definitions has the genus-and-differentia form and, furthermore, that two of them, (4.1c) and (4.1e), are couched in what classical logic would call "negative terms." The traditional injunction against definition in negative terms[12] has no theoretical justification; indeed, it is highly questionable whether any precise meaning can be given to the very distinction of positive and negative concepts which it presupposes.

As the formulas (4.1) illustrate, each defined term in a theory is connected with the primitives by a "chain" of one or more definitions. This makes it possible to eliminate any occurrence of a defined term in favor of expressions in which all extra-logical symbols are primitives. Thus, 'parent' is eliminable directly in favor of 'child' by virtue of (4.1a); elimination of the word 'father' requires the use of two definitions; and the term 'grandmother' is linked to the definitional basis by a chain of three definitions, by virtue of which the phrase 'x Grandmother y' can always be replaced by the following expression, which contains only primitives: '$\sim$Male $x \cdot (Ez)(z$ Child $x \cdot y$ Child $z)$'. Generally, a *definition chain* for a term t, based on a given class of primitives, is a finite ordered set of definitions; in each of these any term occurring in the definiens either is one of the given primitives or has been defined in one of the preceding definitions of the chain; and the definiendum of the last definition is the term t.

Since any expression introduced by definition—i.e., by a single definition sentence or a chain of them—can be eliminated in favor of primitives, nominal definition, at least theoretically, can be entirely dispensed with: Everything that can be said with the help of defined terms can be said also by means of primitives alone. But even in our simple illustration the abbreviatory notations introduced by definition afford a noticeable convenience; and, in the complex theoretical systems of logic, mathematics, and empirical science, definitions are practically indispensable; for the formulation of those theories exclusively in terms of primitives would become so involved as to be unintelligible. Thus, not even the moderately advanced scientific disciplines could be understood—let alone actually have been developed—without extensive use of nominal definition.

The nominal definitions in a scientific theory are subject to one fundamental requirement, which we have mentioned repeatedly: They must permit the elimination of all defined terms in favor of primitives. More fully, this requirement may be stated thus:

Requirement of univocal eliminability of defined expressions:

For every sentence S containing defined expressions, there must exist an essentially unique expansion in primitive terms, i.e., a sentence S' which satisfies the following conditions: (1) S' contains no defined term; (2)S' and S are deducible from one another with the help of the definition chains for the defined expressions occurring in S; (3) if S'' is another sentence which, in the sense of (2), is definitionally equivalent with S, then S' and S'' are logically deducible from each other and thus logically equivalent.

Thus, e.g., by virtue of the definition system (4.1), the phrase 'x Father y' has such alternative expansions as 'Male $x \cdot y$ Child x' and 'y Child $x \cdot$ Male x', but these are mutually deducible by virtue of the principles of formal logic.

The requirement of univocal eliminability has important consequences. First of all, it evidently precludes the possibility of giving two different definitions for the same term, an error which is usually avoided in practice but which could easily introduce contradictions if allowed to pass. In addition, a definition system which satisfies the requirement of univocal eliminability is noncircular, for any circularity would clearly preclude complete eliminability of defined terms.

It is sometimes held that nominal definitions, in contradistinction to "real" definitions, are arbitrary and may be chosen completely as we please. In reference to nominal definition in science, this characterization is apt to be more misleading than enlightening. For, in science, concepts are chosen with a view to functioning in fruitful theories, and this imposes definite limitations on the arbitrariness of definition, as will be pointed out in some detail in a later section. Furthermore, a nominal definition must not give rise to contradictions. As a consequence of this obvious requirement, the introduction of certain kinds of nominal definition into a given theoretical system is permissible only on condition that an appropriate nondefinitional sentence, which might be called its *justificatory sentence,* has been previously established. Thus, e.g., in Hilbert's axiomatization of Euclidean geometry, the line segment determined by two points, P_1 and P_2, is defined, in effect, as the class of points between P_1 and P_2 on the straight line through P_1 and P_2.[13] This definition evidently presupposes that through any two points there exists exactly one straight line; and it is permissible only because this presupposition can be proved in Hilbert's system and thus can function as justificatory sentence for the definition.

An illustration which was proposed and analyzed by Peano[14] shows well how disregard of the need for a justificatory theorem may engender contradictions. Consider the following definition of a "question-mark operation" for rational numbers:

$$\frac{x}{y} ? \frac{z}{u} =_{Df} \frac{x+z}{y+u}.$$

By virtue of this definition, we have:

$$\frac{1}{2} ? \frac{2}{3} = \frac{3}{5} \quad \text{and} \quad \frac{2}{4} ? \frac{2}{3} = \frac{4}{7};$$

but since $\frac{1}{2} = \frac{2}{4}$, it follows that $\frac{3}{5} = \frac{4}{7}$, which introduces a contradiction into arithmetic. Now, the given definition purports to introduce a question-mark operation as a unique function of its two arguments, i.e., in such a manner that to each couple of rational numbers—no matter in what particular form they are symbolized—it assigns exactly one rational number, which is to be regarded as the result of applying the question-mark operation to them. Clearly, this assumption is presupposed in deriving two incompatible results from the definition. But a definition purporting to introduce a unique function is acceptable only if accompanied by a justificatory theorem establishing this uniqueness—a requirement, which, in the case at hand, obviously cannot be met.

Consider now an illustration from the field of empirical science: The definition of the melting point of a given chemically homogeneous substance as the temperature at which the substance melts is permissible only if it has been previously established that all samples of that substance melt at the same temperature independently of other factors, such as pressure. Actually, the second of these conditions is not strictly satisfied, and in cases where the variation with pressure is marked a relativized concept of melting point at a specified pressure has to be used. Similar observations apply to the definitions of the density, specific heat, boiling point, specific resistance, and thermal conductivity of a substance, as well as to the definition of many other concepts in empirical science.

Nominal definitions in empirical science, then, are not entirely arbitrary, and in many cases they even require legitimation by a properly established justificatory sentence.

II. Methods of Concept Formation in Science

5. The Vocabulary of Science: Technical Terms and Observation Terms

Empirical science, we noted earlier, does not aim simply at a description of particular events: it looks for general principles which permit their explanation and prediction. And if a scientific discipline entirely lacks such principles, then it cannot establish any connections between different phenomena: it is unable to foresee future occurrences, and whatever knowledge it offers permits of no technological application, for all such application requires principles which predict what particular effects would occur if we brought about certain specified changes in a given system. It is, therefore, of paramount importance for science to develop a system of concepts which is suited for the formulation of general explanatory and predictive principles.

The vocabulary of everyday discourse, which science has to use at least initially, does permit the statement of generalizations, such as that any unsupported body will fall to the ground; that wood floats on water but that any metal sinks in it; that all crows are black; that men are more intellectual than women; etc. But such generalizations in everyday terms tend to have various shortcomings: (1) their constituent terms will often lack precision and uniformity of usage (as in the case of 'unsupported body', 'intellectual', etc.), and, as a consequence, the resulting statement will have no clear and precise meaning; (2) some of the generalizations are of very limited scope (as, for example, the statement dealing only with crows) and thus have small predictive and explanatory power (compare in this respect the generalization about floating in water with the general statement of Archimedes' principle); (3) general principles couched in everyday terms usually have "exceptions," as is clearly illustrated by our examples.

In order to attain theories of great precision, wide scope, and high empirical confirmation, science has therefore evolved, in its different branches, comprehensive systems of special concepts, referred to by technical terms. Many of those concepts are highly abstract and bear little resemblance to the concrete concepts we use to describe the phenomena of our everyday experience. Actually, however, certain connections must obtain between these two classes of concepts; for science is ultimately intended to systematize the data of our experience, and this is possible only if scientific principles, even when couched in the most esoteric terms, have a bearing upon, and thus are conceptually connected with, statements reporting in "experiential terms" available in everyday language what has been established by immediate observation. Consequently, there will exist certain connections between the technical terms of empirical science and the experiential vocabulary; in fact, only by virtue of such connections can the technical terms of science have any empirical content. Much of the discussion in the present Chapter II will concern the nature of those connections. Before we can turn to this topic, however, we have to clarify somewhat more the notion of experiential term.

The experiential vocabulary is to be used in describing the kind of data which are usually said to be obtainable by direct experience and which serve to test scientific theories or hypotheses. Such experiential data might be conceived of as being sensations, perceptions, and similar phenomena of immediate experience; or else they might be construed as consisting in simple physical phenomena which are accessible to direct observation, such as the coincidence of the pointer of an instrument with a numbered mark on a dial; a change of color in a test substance or in the skin of a patient; the clicking of an amplifier connected with a Geiger counter; etc. The first of these two conceptions of experiential data calls for a phenomenological vocabulary, which might contain such expressions as 'blue-perception', 'looking brighter than' (applicable to areas of a visual field, not to physical objects), 'sour-taste-sensation', 'headachy feeling', etc. The second conception requires, for the description of experiential data, a set of terms signifying certain directly observable characteristics of physical objects, i.e., properties or relations whose presence or absence in a given case can be intersubjectively ascertained, under suitable circumstances, by direct observation. A vocabulary of this kind might include such terms as 'hard', 'liquid', 'blue', 'coincident with', 'contiguous with', etc., all of which are meant here to designate intersubjectively ascertainable attributes of physical objects. For brevity, we will refer to such attributes as *observables*, and to the terms naming them as *observation terms*.

A phenomenalistic conception will appeal to those who hold that the data of our immediate phenomenal experience must constitute the ultimate testing ground for all empirical knowledge; but it has at least two major disadvantages: first, while many epistemologists have favored this view, no one has ever developed in a precise manner a linguistic framework for the use of phenomenalistic terms;[15] and, second, as has been pointed out by Popper,[16] the use of observation reports couched in phenomenalistic language would seriously interfere with the intended objectivity of scientific knowledge: The latter requires that all statements of empirical science be capable of test by reference to evidence which is public, i.e., which can be secured by different observers and does not depend essentially on the observer. To this end, data which are to serve as scientific evidence should be described by means of terms whose use by scientific observers is marked by a high degree of determinacy and uniformity in the sense explained in section 3. These considerations strongly favor the second conception mentioned above, and we will therefore assume, henceforth, especially in the context of illustrations, that the vocabulary used in science for the description of experiential evidence consists of observation terms. Nevertheless, the basic general ideas of the following discussion can readily be transferred to the case of an experiential vocabulary of the phenomenalistic kind.

6. Definition vs. Reduction to an Experiential Basis

We now turn to a consideration of the connections between the technical terms of science and its observational vocabulary —connections which, as we noted, must exist if the technical terms are to have empirical content. Since the scientist has to introduce all his special terms on the basis of his observational vocabulary, the conjecture suggests itself that the former are defined in terms of the latter. Whether this is the case or not cannot be ascertained, however, by simply examining the writings and the pronouncements of scientists; for most presentations of science fail to state explicitly just what terms are taken to be defined and what others function as primitives. In general, only definitions of special importance will be stated, others will be tacitly taken for granted. Furthermore, the primitive terms of one presentation may be among the defined ones of another, and the formulations offered by different authors may involve various divergences and inconsistencies. The task of analyzing the logical relations among scientific terms is, therefore, one of rational reconstruction as characterized in section 3. Its ultimate objective is the construction of a language which is governed by well-determined rules, and in which all the statements of empirical science can be formulated. For the purposes of this monograph, it is not necessary to enter into the details of the complex problem—which is far from a complete solution—of how a rational reconstruction of the entire system of scientific concepts might be effected; it will suffice here to consider certain fundamental aspects of such a reconstruction.

The conjecture mentioned in the preceding paragraph may now be restated thus: Any term in the vocabulary of empirical science is definable by means of observation terms; i.e., it is possible to carry out a rational reconstruction of the language of science in such a way that all primitive terms are observation terms and all other terms are defined by means of them.

This view is characteristic of the earlier forms of positivism and empiricism, and we shall call it the *narrower thesis of empiricism*. According to it, any scientific statement, however abstract, could be transformed, by virtue of the definitions of its constituent technical terms, into an equivalent statement couched exclusively in observation terms: Science would really deal solely with observables. It might well be mentioned here that among contemporary psychologists this thesis has been intensively discussed in reference to the technical terms of psychology; much of the discussion has been concerned with the question whether the so-called intervening variables of learning theory are, or should be, completely definable in terms of directly observable characteristics of the stimulus and response situations.[17]

Despite its apparent plausibility, the narrower empiricist thesis does not stand up under closer scrutiny. There are at least two kinds of terms which raise difficulties: disposition terms, for which the correctness of the thesis is at least problematic, and quantitative terms, to which it surely does not apply. We will now discuss the status of disposition terms, leaving an examination of quantitative terms for the next section.

The property term 'magnetic' is an example of a disposition term: it designates, not a directly observable characteristic, but rather a disposition, on the part of some physical objects, to display specific reactions (such as attracting small iron objects) under certain specifiable circumstances (such as the presence of small iron objects in the vicinity). The vocabulary of empirical science abounds in disposition terms, such as 'elastic', 'conductor of heat', 'fissionable', 'catalyzer', 'phototropic', 'recessive trait', 'vasoconstrictor', 'introvert', 'somatotonic', 'matriarchate'; the following comments on the term 'magnetic' can be readily transferred to any one of them.

Since an object may be magnetic at one time and nonmagnetic at another, the word 'magnetic' will occur in contexts of the form '(object) x is magnetic at (time) t', and a contextual definition (cf. sec. 2) with this expression as definiendum has to be sought. The following formulation—which is deliberately oversimplified in matters of physical detail—might suggest itself:

(6.1) x is magnetic at $t =_{Df}$ if, at t, a small iron object is close to x, then it moves toward x

But the conditional form of the definiens, while clearly reflecting the status of the definiendum as a disposition concept, gives rise to irksome problems.[18] In formal logic the phrase 'if . . . then ___' is usually construed in the sense of material implication, i.e., as being synonymous with 'either not . . . or also ___'; accordingly, the definiens of (6.1) would be satisfied by an object x not only if x was actually magnetic at time t but also if x was not magnetic but no small iron object happened to be near x at time t.

This shows that if sentences of the form illustrated by (6.1) are to serve as definitions for disposition terms, the 'if . . . then ___' clause in the definiens requires a different interpretation, whose import may be suggested by using the subjunctive mood:

(6.2) x is magnetic at $t =_{Df}$ if, at t, a small iron object should be close to x, then that object would move toward x.

Surely, the subjunctive conditional phrase cannot be interpreted in the sense of the material conditional; but before it can be accepted as providing an adequate formulation for the definition of disposition terms, the meaning of the phrase 'if . . . then ___' in subjunctive clauses would have to be made explicit. This is a problem of great interest and importance, since the formulation of so-called counterfactual conditionals and of general laws in science calls for the use of 'if . . . then ___' in the same sense; but despite considerable analytic efforts and significant partial results, no fully satisfactory explication seems available at present,[19] and the formulation (6.2) represents a program rather than a solution.

An alternative way of avoiding the shortcomings of (6.1) has been suggested, and developed in detail, by Carnap.[20] It consists in construing disposition terms as introduced, not by definition, but by a more general procedure, which he calls *reduction*, it amounts to partial, or conditional, definition and includes the standard procedure of explicit definition as a special case.

We will briefly explain this idea by reference to the simplest form of reduction, effected by means of so-called bilateral reduction sentences. A bilateral reduction sentence introducing a property term 'Q' has the form

(6.3) $$P_1x \supset (Qx \equiv P_2x)$$

Here, 'P_1x' and 'P_2x' symbolize certain characteristics which an object x may have; these may be more or less complex but must be stated in terms which are already understood.

In a somewhat loose paraphrase, which however suggests the scientific use of such sentences, (6.3) may be restated thus:

(6.31) If an object x has characteristic P_1 (e.g., x is subjected to specified test conditions or to some specified stimulus), then the attribute Q is to be assigned to x if and only if x shows the characteristic (i.e., the reaction, or the mode of response) P_2

Now the idea that was to be conveyed by (6.1) may be restated in the following reduction sentence:

(6.4) If a small iron object is close to x at t, then x is magnetic at t if and only if that object moves toward x at t

In reduction sentences, the phrase 'if . . . then ——' is always construed as synonymous with 'not . . . or ——', and 'if and only if' is understood analogously; yet the difficulty encountered by (6.1) does not arise for (6.4): If no small iron object is close to x at t, then the whole statement (6.4) is true of x, but we cannot infer that x is magnetic at t.

A reduction sentence offers no complete definition for the term it introduces, but only a partial, or conditional, determination of its meaning; it assigns meaning to the "new" term only for its application to objects which satisfy specific "test conditions." Thus, e.g., (6.4) determines the meaning of 'magnetic at t' only in reference to objects which meet the test condition of being close to some small iron body at t; it provides no interpretation for a sentence such as 'object x is now magnetic, but there is no iron whatever in its vicinity'. Hence, terms introduced by reduction sentences cannot generally be eliminated in favor of primitives. There is one exception to this rule: If the expression, 'P_1x', in (6.3) is analytic, i.e., is satisfied with logical necessity by any object x whatever (which is the case, for example, if 'P_1x' stands for 'x is green or not green'), then the bilateral reduction sentence is equivalent to the explicit definition '$Qx \equiv P_2x$'; hence, it fully specifies the meaning of 'Qx' and permits its elimination from any context. This shows that reduction is actually a generalization of definition. To put the matter in a different way, which will be useful later: A set of reduction sentences for a concept Q lays down a necessary condition for Q and a sufficient one; but, in general, the two are not identical. A definition of Q, on the other hand, specifies, in the definiens, a condition which is both necessary and sufficient for Q.

The indeterminacy in the meaning of a term introduced by a reduction sentence may be decreased by laying down additional reduction sentences for it which refer to different test conditions. Thus, e.g., if the concept of electric current had been introduced previously, (6.4) might be supplemented by the additional reduction sentence:

(6.5) If x moves through a closed wire loop at t, then x is magnetic at t if and only if an electric current flows in the loop at t

The sentences (6.4) and (6.5) together provide criteria of application for the word 'magnetic' in reference to any object that satisfies the test condition of at least one of them. But, since the two conditions are not exhaustive of all logical possibilities, the meaning of the word is still unspecified for many conceivable cases. On the other hand, the test conditions clearly are not logically exclusive: both may be satisfied by one and the same object; and for objects of this kind the two sentences imply a specific assertion, namely: Any physical object which is near some small iron body and moves through a closed wire loop will generate a current in the loop if and only if it attracts the iron body. But this statement surely is not just a stipulation concerning the use of a new term—in fact, it does not contain the new term, 'magnetic', at all; rather, it expresses an empirical law. Hence, while a single reduction sentence may be viewed simply as laying down a notational convention for the use of the term it introduces, this is no longer possible for a set of two or more reduction sentences concerning the same term, because such a set implies, as a rule, certain statements which have the character of empirical laws; such a set cannot, therefore, be used in science unless there is evidence to support the laws in question.

To summarize: An attempt to construe disposition terms as introduced by definition in terms of observables encounters the difficulties illustrated by reference to (6.1). These can be avoided by introducing disposition terms by sets of reduction sentences. But this method has two peculiar features: (1) In general, a set of reduction sentences for a given term does not have the sole function of a notational convention; rather, it also asserts, by implication, certain empirical statements. Sets of reduction sentences combine in a peculiar way the functions of concept formation and of theory formation. (2) In general, a set of reduction sentences determines the meaning of the introduced term only partially.

Now, as was noted in section 4, even an explicit nominal definition may imply a nondefinitional 'justificatory' statement which has to be established antecedently if the definition is to be acceptable; thus, the first characteristic of introduction by reduction sentences has its analogue in the case of definition. But this is not true of the second characteristic; and it might seem that the partial indeterminacy of meaning of terms introduced by reduction sentences is too high a price to pay for a method which avoids the shortcomings of definitions such as (6.1). It may be well, therefore, to suggest that this second characteristic of reduction sentences does justice to what appears to be an important characteristic of the more fruitful technical terms of science; let us call it their *openness of meaning*. The concepts of magnetization, of temperature, of gravitational field, for example, were introduced to serve as crystallization points for the formulation of explanatory and predictive principles. Since the latter are to bear upon phenomena accessible to direct observation, there must be "operational" criteria of application for their constitutive terms, i.e., criteria expressible in terms of observables. Reduction sentences make it possible to formulate such criteria. But precisely in the case of theoretically fruitful concepts, we want to permit, and indeed count on, the possibility that they may enter into further general principles, which will connect them with additional variables and will thus provide new criteria of application for them. We would deprive ourselves of these potentialities if we insisted on introducing the technical concepts of science by full definition in terms of observables.[21]

7. Theoretical Constructs and Their Interpretation

A second group of terms which fail to bear out the narrower thesis of empiricism are the metrical terms, which represent numerically measurable quantities such as length, mass, temperature, electric charge, etc. The term 'length', for example, is used in contexts of the form 'the length of the distance between points u and v is r cm.', or briefly

(7.1) $$\text{length}\ (u, v) = r$$

Similarly, the term 'mass' occurs in contexts of the form 'the mass of physical body x is s grams', or briefly

(7.2) $$\text{mass}\ (x) = s$$

In the hypotheses and theories of physics, these concepts are used in such a way that their values—r or s, respectively—may equal any nonnegative number. Thus, e.g., in Newton's general law of gravitation, which expresses the force of the gravitational attraction between two physical bodies as a function of their masses and their distance, all these magnitudes are allowed to take any positive real-number value. The concept of length, therefore—and similarly that of mass, and any other metrical concept whose range of values includes some interval of the real-number system—provides for the theoretical distinction of an infinity of different possible cases, each of them corresponding to one of the permissible real-number values. If, therefore, the concept of length were fully definable in terms of observables, then it would be possible to state, purely in terms of observables, the meaning of the phrase 'length $(u, v) = r$' for each of the permissible values of r. But this cannot be done, as we will now argue in two steps.

First, suppose that we try to define the characteristic of having a length of r cm. as tantamount to some specific combination (expressible by means of 'and', 'or', 'not', etc.) of observable attributes. (In effect, this restricts the definiens to a molecular sentence in which all predicates are observation terms.) This is surely not feasible for every theoretically permissible value of r. For in view of the limits of discrimination in direct observation, there will be altogether only a finite, though large, number of observable characteristics; hence, the number of different complexes that can be formed out of them will be finite as well, whereas the number of theoretically permissible r-values is infinite. Hence, the assignment of a numerical r-value of length (or of any other measurable quantity) to a given object cannot always be construed as definitionally equivalent to attributing to that object some specific complex of observable characteristics.

Let us try next, therefore, to construe the assignment of a specified r-value to a given object as equivalent to a statement about that object which can be expressed by means of observation terms and logical terms alone. The latter may now include not only 'and', 'or', 'not', etc., but also the expressions 'all', 'some', 'the class of all things satisfying such and such a condition', etc. But even if definition in terms of observables is construed in this broad sense, the total number of defining expressions that can be formed from the finite vocabulary available is only denumerably infinite, whereas the class of all theoretically permissible r-values has the power of the continuum. Hence, a full definition of metrical terms by means of observables is not possible.

It might be replied, in a pragmatist or extreme operationist vein, that a theoretical difference which makes no observable difference is no significant difference at all and that therefore no metrical concept in science should be allowed to take as its value just any real number within some specified interval. But compliance with this rule would make it impossible to use the concepts and principles of higher mathematics in the formulation and application of scientific theories. If, for example, we were to allow only a discrete set of values for length and for temporal duration, then the concepts of limit, derivative, and integral would be unavailable, and it would consequently be impossible to introduce the concepts of instantaneous speed and acceleration and to formulate the theory of motion. Similarly, all the formulations in terms of real and complex functions and in terms of differential equations, which are so characteristic of the theoretically most powerful branches of empirical science, would be barred. The retort that all those concepts and principles are "mere fictions to which nothing corresponds in experience" is, in effect, simply a restatement of the fact that theoretical constructs cannot be definitionally eliminated exclusively in favor of observation terms. But it is precisely these "fictitious" concepts rather than those fully definable by observables which enable science to interpret and organize the data of direct observation by means of a coherent and comprehensive system which permits explanation and prediction. Hence, rather than exclude those fruitful concepts on the ground that they are not experientially definable, we will have to inquire what nondefinitional methods might be suited for their introduction and experiential interpretation.

Do reduction sentences provide such a method? The conjecture has indeed been set forth in more recent empiricist writings that every term of empirical science can be introduced, on the basis of observation terms, by means of a suitable set of reduction sentences.[22] Let us call this assertion the *liberalized thesis of empiricism.*

But even for this thesis difficulties arise in the case of metrical terms. For, as was noted in section 6, a set of reduction sentences for a term t lays down a necessary and a (usually different) sufficient condition for the application of t. Hence, suitable reduction sentences for the phrase 'length $(u, v) = r$' would have to specify, for every theoretically permissible value of r, a necessary and a sufficient condition, couched in terms of observables, for an interval (u, v) having a length of exactly r cm.[23] But it is not even possible to formulate all the requisite sufficient conditions; for this would mean the establishment, for every possible value of r, of a purely observational criterion whose satisfaction by a given interval (u, v) would entail that the interval was exactly r cm. long. That a complete set of such criteria cannot exist is readily seen by an argument analogous to the one presented earlier in this section in reference to the limits of full definability in observational terms.

The metrical concepts in their theoretical use belong to the larger class of *theoretical constructs*, i.e., the often highly abstract terms used in the advanced stages of scientific theory formation, such as 'mass', 'mass point', 'rigid body', 'force', etc., in classical mechanics; 'absolute temperature', 'pressure', 'volume', 'Carnot process', etc., in classical thermodynamics; and 'electron', 'proton', 'ψ function', etc., in quantum mechanics. Terms of this kind are not introduced by definitions or reduction chains based on observables; in fact, they are not introduced by any piecemeal process of assigning meaning to them individually. Rather, the constructs used in a theory are introduced jointly, as it were, by setting up a theoretical system formulated in terms of them and by giving this system an experiential interpretation, which in turn confers empirical meaning on the theoretical constructs. Let us look at this procedure more closely.

Although in actual scientific practice the processes of framing a theoretical structure and of interpreting it are not always sharply separated, since the intended interpretation usually guides the constructions of the theoretician, it is possible and indeed desirable, for the purposes of logical clarification, to separate the two steps conceptually.

A theoretical system may then be conceived as an uninterpreted theory in axiomatic form, which is characterized by (1) a specified set of primitive terms; these are not defined within the theory, and all other extra-logical terms of the theory are obtained from them by nominal definition; (2) a set of postulates—we will alternatively call them primitive, or basic, hypotheses; other sentences of the theory are obtained from them by logical deduction.[24]

As an example of a well-axiomatized theory which is of fundamental importance for science, consider Euclidean geometry. Its development as "pure geometry," i.e., as an uninterpreted axiomatic system, is logically quite independent of its interpretation in physics and its use in navigation, surveying, etc. In Hilbert's axiomatization,[25] the primitives of the theory are the terms 'point', 'straight line', 'plane', 'incident on' (signifying a relation between a point and a line), 'between' (signifying a relation between points on a line), 'lies in' (signifying a relation between a point and a plane), and two further terms, for congruence among line segments and among angles, respectively. All other terms, such as 'parallel', 'angle', 'triangle', 'circle', are defined by means of the primitives: the term 'parallel', for example, can be introduced by the following contextual definition:

(7.3) x is parallel to y $=_{Df}$ x and y are straight lines; there exists a plane in which both x and y lie; but there exists no point which is incident on both x and y

The postulates include such sentences as these: For any two points there exists at least one, and at most one, straight line on which both are incident; between any two points incident on a straight line there exists another point which is incident on that line; etc. From the postulates, the other propositions of Euclidean geometry are obtained by logical deduction. Such proof establishes the propositions as theorems of pure mathematical geometry; it does not, however, certify their validity for use in physical theory and its applications, such as the determination of distances between physical bodies by means of triangulation, or the computation of the volume of a spherical object from the length of its diameter. For no specific meaning is assigned, in pure geometry, to the primitives of the theory[26] (and, consequently, none to the defined terms either); hence, pure geometry does not express any assertions about the spatial properties and relations of objects in the physical world.

A physical geometry, i.e., a theory which deals with the spatial aspects of physical phenomena, is obtained from a system of pure geometry by giving the primitives a specific interpretation in physical terms. Thus, e.g., to obtain the physical counterpart of pure Euclidean geometry, points may be interpreted as approximated by small physical objects, i.e., objects whose sizes are negligible compared to their mutual distances (they might be pinpoints, the intersections of cross-hairs, etc., or, for astronomical purposes, entire stars or even galactic systems); a straight line may be construed as the path of a light ray in a homogeneous medium; congruence of intervals as a physical relation characterizable in terms of coincidences of rigid rods; etc. This interpretation turns the postulates and theorems of pure geometry into statements of physics, and the question of their factual correctness now permits—and, indeed, requires—empirical tests. One of these is the measurement made by Gauss of the angle-sum in a triangle formed by light rays, to ascertain whether it equals two right angles, as asserted by physical geometry in its Euclidean form. If the evidence obtained by suitable methods is unfavorable, the Euclidean form of geometry may well be replaced by some non-Euclidean version which, in combination with the rest of physical theory, is in better accord with observational findings. In fact, just this has occurred in the general theory of relativity.[27]

In a similar manner, any other scientific theory may be conceived of as consisting of an uninterpreted, deductively developed system and of an interpretation which confers empirical import upon the terms and sentences of the latter.[28] The term to which the interpretation directly assigns an empirical content either may be primitives of the theory, as in the geometrical example discussed before, or may be defined terms of the theoretical system. Thus, e.g., in a logical reconstruction of chemistry, the different elements might be defined by primitives referring to certain characteristics of their atomic structure; then, the terms 'hydrogen', 'helium', etc., thus defined might be given an empirical interpretation by reference to certain gross physical and chemical characteristics typical of the different elements. Such an interpretation of certain defined terms of a system confers mediately, as it were, some empirical content also upon the primitives of the system, which have received no direct empirical interpretation. This procedure appears well suited also for Woodger's axiomatization of biology,[29] in which certain defined concepts, such as division and fusion of cells, permit of a more direct empirical interpretation than some of the primitives of the system.

An adequate empirical interpretation turns a theoretical system into a testable theory: The hypotheses whose constituent terms have been interpreted become capable of test by reference to observable phenomena. Frequently the interpreted hypotheses will be derivative hypotheses of the theory; but their confirmation or disconfirmation by empirical data will then mediately strengthen or weaken also the primitive hypotheses from which they were derived. Thus, for example, the primitive hypotheses of the kinetic theory of heat concern the mechanical behavior of the micro-particles constituting a gas; hence, they are not capable of direct test. But they are indirectly testable because they entail derivative hypotheses which can be formulated in certain defined terms that have been interpreted by means of such "macroscopic observables" as the temperature and the pressure of a gas.[30]

The double function of such interpretation of defined terms—to indirectly confer empirical content upon the primitives of the theory and to render its basic hypotheses capable of test—is illustrated also by those hypotheses in physics or chemistry which refer to the value of some magnitude at a space-time point, such as the instantaneous speed and acceleration of a particle; or the density, pressure, and temperature of a substance at a certain point: none of these magnitudes is capable of direct observation, none of these hypotheses permits of direct test. The connection with the level of possible experimental or observational findings is established by defining, with the help of mathematical integration, certain derived concepts, such as those of average speed and acceleration in a certain time interval, or of average density in a certain spatial region, and by interpreting these in terms of more or less directly observable phenomena.

A scientific theory might therefore be likened to a complex spatial network: Its terms are represented by the knots, while the threads connecting the latter correspond, in part, to the definitions and, in part, to the fundamental and derivative hypotheses included in the theory. The whole system floats, as it were, above the plane of observation and is anchored to it by rules of interpretation. These might be viewed as strings which are not part of the network but link certain points of the latter with specific places in the plane of observation. By virtue of those interpretive connections, the network can function as a scientific theory: From certain observational data, we may ascend, via an interpretive string, to some point in the theoretical network, thence proceed, via definitions and hypotheses, to other points, from which another interpretive string permits a descent to the plane of observation.

In this manner an interpreted theory makes it possible to infer the occurrence of certain phenomena which can be described in observational terms and which may belong to the past or the future, on the basis of other such phenomena, whose occurrence has been previously ascertained. But the theoretical apparatus which provides these predictive and postdictive bridges from observational data to potential observational findings cannot, in general, be formulated in terms of observables alone. The entire history of scientific endeavor appears to show that in our world comprehensive, simple, and dependable principles for the explanation and prediction of observable phenomena cannot be obtained by merely summarizing and inductively generalizing observational findings. A hypothetico-deductive-observational procedure is called for and is indeed followed in the more advanced branches of empirical science: Guided by his knowledge of observational data, the scientist has to invent a set of concepts—theoretical constructs, which lack immediate experiential significance, a system of hypotheses couched in terms of them, and an interpretation for the resulting theoretical network; and all this in a manner which will establish explanatory and predictive connections between the data of direct observation.

Is it possible to specify a generally applicable form in which the interpretive statements for a scientific theory can be expressed? Let us note, to begin with, that those statements are not, in general, tantamount to full definitions in terms of observables. We will state the reasons by reference to the physical interpretation of geometrical terms. First, some of the expressions used in the interpretation, such as 'light ray in a homogeneous medium', are not observation terms but at best disposition terms which can be partially defined through observables by means of chains of reduction sentences.[31] Second, even if all the terms used in interpreting geometry were accepted as observation terms, the interpretive statements still would not express conditions which are both necessary and sufficient for the interpreted terms; hence, they would not have the import of definitions. If, e.g., it were necessary and sufficient for a physical point to be identical with, or at least at the same place as, a pinpoint or an intersection of cross-hairs or the like, then many propositions of geometry would be clearly false in their physical interpretation; among them, for example, the theorem that between any two points on a straight line there are infinitely many other points. Actually, no geometrical theory is rejected in physics for reasons of this type; rather, it is understood that comparatively small physical bodies constitute only approximations of points in the sense of physical geometry. The term 'point' as used in theoretical physics is a construct and does not denote any objects that are accessible to direct observation.

But an interpretation is not even generally tantamount to a set of reduction sentences: The interpretation of a theoretical term may well make use of expressions which are introducible by means of a set of reduction sentences based on observation predicates; but those expressions again will, as a rule, be used not to specify necessary and sufficient conditions for the theoretical term in question but only to provide a partial assignment of empirical content to it. Consider the case mentioned before of a reconstruction of chemistry in which the elements are theoretically defined in terms of their atomic structure and empirically interpreted by reference to their gross physical and chemical characteristics: Some of the latter, such as solubility in specified solvents, malleability, chemical affinity, etc., have the character of disposition concepts rather than of observables, and, furthermore, the interpretation is applicable only to sufficiently large amounts of the substance in question, so that surely only a partial interpretation of the theoretical terms is achieved.

As a consequence, an interpreted scientific theory cannot be equivalently translated into a system of sentences whose constituent extra-logical terms are all either observation predicates or obtainable from observation predicates by means of reduction sentences; and a fortiori no scientific theory is equivalent to a finite or infinite class of sentences describing potential experiences.

Considerations of the kind here surveyed have led some authors to the opinion that the rules of interpretation for a scientific theory cannot be stated in precise terms at all, that they will always have to remain somewhat vague.[32] Others have suggested that the interpretation of theoretical terms may have to be put into the form of probability statements.[33] Thus, e.g., the key terms of psychoanalytic theory (which, to be sure, has never been stated quite explicitly and precisely and for which no axiomatization is available) receive an empirical interpretation by reference to free associations, reports on dreams, slips of tongue, pen, and memory, and other more or less directly ascertainable aspects of overt behavior. But a cautious reconstruction will have to treat observable clues of these kinds not as strictly necessary or as strictly sufficient conditions for certain hypothetical states or processes such as oedipal fixation or regression or transference, but rather as "indicators" which are tied to those hypothetical states by probability relations. Thus, the rules of interpretation for a psychoanalytic concept *C* might specify the probabilities of specified observable symptoms occurring in individuals who have the nonobservable characteristic *C*, and conversely the probabilities for the presence of *C* if such and such observable symptoms are present. Generally, a theoretical system might be said to have been given an empirical interpretation if rules of confirmation have been laid down in such a way that for every sentence *S* of the theory and for every evidence sentence *E* that can be formulated in terms of observation predicates (no matter whether it is factually true or false) the rules determine (1) to what extent *E* confirms *S*, or what probability *E* confers upon *S;* and conversely (2) what probability *S* gives to *E*.

This conception of interpretation is at present largely a program, however. Its realization requires the development of an adequate theory of the probability of hypotheses. Important steps in this direction have been taken, but the results are still the object of controversy.[34] A discussion of details would take us too far beyond the scope of our present survey.

8. Empirical and Systematic Import of Scientific Terms; Remarks on Operationism

A theoretical system without empirical interpretation is incapable of test and thus cannot constitute a theory of empirical phenomena; we shall say of its terms as well as of its concepts that they lack *empirical import.*

Neovitalism, for example, provides no interpretation for its key term, 'entelechy', or for terms definable by means of it, nor does it offer an indirect interpretation by formulating a system of general laws and definitions which connect the term 'entelechy' with other, interpreted, terms of the theory. Consequently, the concept of entelechy cannot serve the explanatory purpose for which it was intended; for a concept can have explanatory power only in the context of an interpreted theory. Thus, e.g., to say that the regularities of planetary motion can be explained by means of the concept of universal gravitation is an elliptic way of asserting that those regularities are explainable by means of the formal *theory* of gravitation, together with the customary interpretation of its terms.

Another illustration is provided by the use of the term 'purpose' in some teleological accounts of biological phenomena. Thus, when a certain form of mimicry is said to have the purpose of preserving a given species by protecting its members from their natural enemies, no direct or indirect interpretation of 'purpose' in terms of observables is provided; i.e., no criteria are laid down by means of which it is possible to test assertions about the purposes of biological phenomena and to decide, let us say, between the view just mentioned and the alternative opinion that mimicry has the purpose of bringing aesthetic variety into the animate world. The assertion that mimicry actually does protect the members of a given species to some extent is not suited as interpretation for the former view, for then the alternative idea would have to be construed as stating that mimicry actually does effect aesthetic variety; and thus both claims would turn out to be true—a result which surely does not agree with the spirit and the intentions of teleological arguments. Thus, when used in contexts such as these, the term 'purpose' and the statements in which it occurs lack empirical import; they cannot provide any theoretical understanding of the phenomena in question. This use of teleological terminology may be characterized as an illegitimate transfer from contexts of similar grammatical form, for which, however, an empirical interpretation is available, such as 'the safety valves of steam engines have the purpose of preventing explosions'. This sentence can be interpreted by reference to the intentions and beliefs of the designer, which can be ascertained, at least under favorable conditions, by various observational methods. No doubt it is the close linguistic analogy to cases of the latter type which creates the illusion of empirical import in teleological arguments of the kind considered before.

Insistence that no term of science can be significant unless it possesses an empirical interpretation is the basic tenet of the operationist school of thought, which has its origin in the methodological work of the physicist P. W. Bridgman,[35] and which has exerted a great influence also in psychology and the social sciences.

The basic idea of operationism is "the demand that the concepts or terms used in the description of experience be framed in terms of operations which can be unequivocally performed";[36] in other words, the requirement that there must exist, for the terms of empirical science, criteria of application couched in terms of observational or experimental procedure. The operational criteria for the application of the term 'length', for example, would consist in appropriate rules for the measurement of length. But the idea should not be restricted to quantitative terms. Thus, the operational criteria of application for the term 'diphtheria' might be formulated in terms of the various symptoms of diphtheria; these would include not only such symptoms as are ascertainable by the "operation" of directly observing the patient but also the results of bacteriological and other tests which call for such "operations" as the use of microscopes and the application of staining techniques.

The statement of such criteria of application for a given term is often referred to as operational definition. This terminology is misleading, however. For, first of all, the term 'operational' would seem to preclude any criteria which require simply direct observation without any manipulation; and, as our last example illustrates, this would mean an unwarranted limitation. And, second, insistence that every scientific concept be *defined* in "operational" terms is unduly restrictive: as we tried to show in the preceding two sections, it would disqualify, among others, the most powerful theoretical constructs.

An attempt is sometimes made to reconcile insistence on a specific operational interpretation for all scientific terms with the endorsement of highly abstract theoretical constructs by representing the introduction of the latter as involving, in addition to "physical" operations, also "mental," "verbal," and "paper-and-pencil" operations.[37] This idea is then extended even to purely mathematical concepts, which are said to be defined in terms of mental operations. Such a view, however, fails to distinguish the systematic from the psychological aspects of concept formation. Thus, e.g., the definition of mathematical terms requires no reference to mental operations, although mental operations are involved in the psychological processes of defining and using mathematical terms. Moreover, mental operations are involved as well in the use of terms such as 'entelechy' or 'absolute simultaneity of spatially separated events', which the operationist criterion is intended to rule out as devoid of empirical import. Hence, to countenance "mental operations" in the criteria of application for scientific terms is to open a back door to all the concepts which operationism was originally designed to bar from the vocabulary of science.

An alternative to the reliance on "mental operations" and to the conception of operational criteria as generally providing full *definitions* is suggested by reflection upon the rationale of the operationist approach, i.e., the consideration that scientific terms are to function in statements which are capable of objective test by reference to data furnished by direct observation. "Operational definition" was conceived as a means to insure their suitability for this purpose by providing criteria for the test of the statements in which the terms occur. Thus, e.g., if the expression 'mineral x is harder than mineral y' is "operationally defined" by 'a sharp point of a sample of mineral x scratches a smooth surface of a sample of mineral y', then criteria have been established for the intersubjective test of comparative judgments about hardness. But the testability of a theory does not require full definition of its constituent concepts in terms of observables: a partial empirical interpretation will suffice. This suggests a broadening of the concept of operational definition: In its widest sense, we may construe an operational "definition" of a set of terms, $t_1, t_2, \ldots, t_n$ in a scientific theory as an interpretation, by reference to observables, of $t_1, t_2, \ldots, t_n$ or of other expressions which are connected with $t_1, t_2, \ldots, t_n$ by definitions within the theory. Thus, e.g., those operational criteria of diphtheria which refer to more or less directly observable symptoms might be viewed as a partial interpretation of the term 'diphtheria' itself, while the criteria referring to microscopic findings afford a partial interpretation of such terms as 'Krebs-Loeffler bacillus', which are definitionally connected with the term 'diphtheria' within bacteriological theory. In the broadened interpretation here suggested, the basic principle of operationism is just another formulation of the empiricist requirement of testability[38] for the theories of empirical science. Indeed, there exists a close correspondence between the ways in which those two ideas have been gradually liberalized. The earlier insistence that each statement of empirical science should be fully verifiable or falsifiable by means of observational evidence has been modified in two respects: (1) by the recognition that a scientific hypothesis cannot, as a rule, be tested in isolation but only in combination with other statements—the test of a bacteriological hypothesis by means of staining techniques and microscopes, for example, presupposes various hypotheses of mechanics, optics, and chemistry—so that the criterion of testability has to be applied to comprehensive systems of hypotheses rather than to single statements; and (2) by replacing the overly rigid standard of complete verifiability or falsifiability by the more liberal requirement that a system of hypotheses must be capable of being more or less highly confirmed by observational evidence. Analogously, the idea that each scientific term should be capable of definition in terms of observables has to be broadened by (1) application to a whole system of terms, connected by laws and definitions within a theoretical network, and (2) replacement of the requirement of complete definition by that of partial interpretation, as discussed in the preceding section.

As was noted in section 5, science strives for objectivity in the sense that its statements are to be capable of public test with results that do not vary essentially with the tester. This requirement makes it imperative that the vocabulary used in the interpretation of scientific terms have a high determinacy and uniformity of usage. This consideration accounts for the tendency to replace such experiential criteria as the direct comparison of two adjoining plane surfaces in regard to brightness (as used in photometry) or the reliance on specific smells, tastes, and visual appearances, or on the soapy feel of a lye in chemical work, by the use of instruments; this reduces the requisite observational evidence to statements describing spatial and temporal coincidences or specific pointer readings—a reduction which affords a considerable increase in pragmatic determinacy.

Another manifestation of the quest for scientific objectivity is the concern of psychologists and social scientists with the so-called *reliability* of their "operational definitions." Consider, for example, the social status scale propounded by Chapin.[39] Its purpose is to rate American homes in regard to their social status on the basis of readily obtainable information concerning the presence of such items as hardwood floors, radios, etc., and their condition of repair, as well as the general cleanliness and orderliness of the living-room. Each possible finding on these different counts is assigned a certain positive or negative integer as its weight, and the social status rating of a home is then determined by a specified arithmetical procedure from the weights assigned to it by a trained investigator. Evidently, the objectivity of the ratings thus obtained depends upon the determinacy and uniformity with which different investigators apply the basic criteria determining the scale. The reliability of the scale is intended to express this consistency of usage in numerical terms. Chapin used two different measures of reliability: (1) the correlation between the scores obtained by one observer on two sucessive visits to the same set of homes (this reflects what we have called determinacy of usage), and (2) the correlation between the scores obtained by two different observers for the same set of homes (this reflects uniformity of usage).[40] While the reliability thus obtained was high, it was not perfect, especially when determined in the second manner. The inter-observer differences here reflected seem to be largely attributable to the fact that, while most of the rating criteria are purely descriptive in character, there is at least one which requires the investigator to express in numerical terms his general impression of good taste in the appearance of the living-room: If that impression is bizarre, inharmonious, or offensive, for example, the score on this count is −4. It is to be expected that valuational expressions have a smaller uniformity of usage than descriptive ones; their use in specifying the meanings of scientific terms will tend to conflict, therefore, with the requirement of uniform intersubjective testability for the statements of science.

Another illustration of this point is provided by Ogburn's hypothesis of cultural lag.[41] In rough summary, this hypothesis asserts that certain aspects of the nonmaterial culture of a social group are dependent on, and adapt themselves to, the prevailing material culture; so that a change in the latter (for example, excessive depletion of forests) will bring about a corresponding change in the adaptive culture (for example, enactment of forest conservation laws). However, between the occurrence of corresponding changes there is a temporal lag of varying length. An estimate of the lag in any particular case requires a determination not only of the time when a change in the adaptive culture did take place in response to a material change but also of the time when it "should" have taken place to prevent serious "maladjustment." But as Ogburn himself has candidly pointed out, "one's notion of adaptation in some cases depends somewhat on one's attitude towards life, one's idea of progress, or one's religious beliefs."[42] This is exactly why hypotheses employing such valuational notions are not capable of uniform intersubjective test and thus lack that objectivity which is an indispensable prerequisite of scientific formulations.[43] The factual information which the hypothesis of cultural lag is intended to convey needs therefore restating in a nonvaluational terminology.[44] Analogous considerations seem to be applicable to the functional theory of culture,[45] which proposes to account for social institutions and cultural change by reference to specific social needs they satisfy. Some of the concepts used for this purpose appear to be valuational, others teleological in character; in order to ascertain the empirical import of functional analyses, it is therefore essential that those concepts be given an interpretation in nonnormative terms.

Notwithstanding the soundness of the insistence on operational interpretations for scientific terms, it must not be forgotten that good scientific constructs must also possess *theoretical,* or *systematic, import;* i.e., they must permit the establishment of explanatory and predictive principles in the form of general laws or theories. Loosely speaking, the systematic import of a set of theoretical terms is determined by the scope, the degree of factual confirmation, and the formal simplicity of the general principles in which they function.[46]

In the theoretically advanced stages of science these two aspects of concept formation are inseparably connected; for, as we saw, the interpretation of a system of constructs presupposes a network of theoretical statements in which those constructs occur. In the initial stages of research, however, which are characterized by a largely observational vocabulary and by a low level of generalization, it is possible to separate the questions of empirical and of systematic import; and to do so explicitly may be helpful for a clarification of some rather important methodological issues.

Concepts with empirical import can be readily defined in any number, but most of them will be of no use for systematic purposes. Thus, we might define the hage of a person as the product of his height in millimeters and his age in years. This definition is operationally adequate and the term 'hage' thus introduced would have relatively high precision and uniformity of usage; but it lacks theoretical import, for we have no general laws connecting the hage of a person with other characteristics.

The significance of systematic import is well illustrated by the various attempts which have been made to divide human beings into different types according to physical or psychological characteristics. There are many different ways in which such typological classifications can be achieved by means of "operationally significant" criteria, so that the corresponding type concepts undeniably have empirical import. Yet the resulting systems differ considerably in scientific fruitfulness; for, in typological as well as in any other classificatory systems, it is essential that the defining characteristics of each of the different classes should be empirically associated with a large number of other attributes, so that the members of any one of the different classes will exhibit clusters of empirically correlated features. In typological systems, for example, one of the prevalent objectives is the delimitation of different types by means of physical characteristics which are closely correlated both with other physical attributes and with certain groups of psychological traits. Classifications which meet this requirement are often called *natural classifications*, in contradistinction to so-called *artificial classifications*, which are characterized by the absence of regular connections between the defining characteristics and others which are logically independent of them. This distinction will be examined more fully in section 9.

Recent typological theories, especially that of Sheldon,[47] allow for gradations in the various physical and mental traits they deal with. This yields a theoretical arrangement of individuals in an ordering framework reminiscent of a co-ordinate system instead of a classification with sharp boundary lines. A fruitful ordering schema is one in which the traits whose gradations determine the order have a high correlation with bundles of other physical or psychological characteristics.[48]

In the contemporary methodological literature of psychology and the social sciences, the need for "operational definitions" is often emphasized to the neglect of the requirement of systematic import, and occasionally the impression is given that the most promising way of furthering the growth of sociology as a scientific discipline is to create a large supply of "operationally defined" terms of high determinacy and uniformity of usage,[49] leaving it to subsequent research to discover whether these terms lend themselves to the formulation of fruitful theoretical principles. But concept formation in science cannot be separated from theoretical considerations; indeed, it is precisely the discovery of concept systems with theoretical import which advances scientific understanding; and such discovery requires scientific inventiveness and cannot be replaced by the—certainly indispensable, but also definitely insufficient—operationist or empiricist requirement of empirical import alone.

In the literature of psychology and sociology the opinion is sometimes expressed or implied that when a term of conversational language—such as 'manual dexterity', 'introversion', 'submissiveness', 'intelligence', 'social status', etc.—is taken over into the scientific vocabulary and given a more precise interpretation by reference to specified tests or similar criteria, then it is essential that the latter be "*valid*" in the sense of providing a correct characterization of the feature to which the term refers in ordinary usage. But if that term has so far been used solely in prescientific discourse, then the only way of ascertaining whether a proposed set of precise criteria affords a "valid" gauge of the characteristic in question is to determine to what extent the objects satisfying the criteria coincide with those to whom the characteristic in question would be assigned in prescientific usage. And, indeed, various authors have adopted, as an index of the validity of a proposed testing procedure for a psychological or social characteristic, the correlation between the test scores of a group of subjects and the ratings which those subjects were given, for the characteristic in question, by acquaintances who judged them "intuitively," i.e., who applied the term under consideration according to its prescientific usage. But the "intuitive" use of those terms in conversational language lacks both determinacy and uniformity. It is therefore unwarranted to consider them as referring to clearly delimited and unambiguously specifiable characteristics and to seek "correct" or "valid" indicators or tests for the presence or absence of the latter.

It should be added, however, that in the contemporary psychological and sociological literature, at least two other concepts of validity are used, which are not subject to this kind of criticism.[50] These concepts are applied especially to numerical scales (as determined by specified criteria) for such attributes as the pitch of a tone or the intelligence of an individual. In effect, one or both of the following criteria are applied to such scales (not being equivalent, they determine two different concepts of validity): (1) A high correlation of the given scale with other scales designed to represent the same attribute, or with some other previously adopted criterion; (2) ability of the attribute which is "operationally defined" by the criteria for the given scale to enter into fruitful and simple theoretical connections with other characteristics in the same area of investigation.[51] Validity in the first sense is a relative matter: A scale, or the graded characteristic represented by it, can be assigned a specific validity only in reference to certain previously accepted criteria or test scales; and its validity may be high in reference to one such set, low in reference to another. This conception of validity does not, therefore, seem to be of great theoretical importance; its major significance lies perhaps in the fact that it adumbrates, by aiming at correlation with other scaled characteristics, the requirement of systematic import; and it is again this requirement which clearly constitutes the rationale for the second conception of validity.

To summarize and amplify: In its exploratory pretheoretical research, science will often have to avail itself of the vocabulary of conversational language with all its imperfections; but in the course of its growth it has to modify its conceptual apparatus so as to enhance the theoretical import of the resulting system and the precision and uniformity of its interpretation —without being hampered by the consideration of preserving and explicating the prescientific usage of conventional terms taken over into its vocabulary. Physics, for example, would not have attained its present theoretical strength if it had insisted on using such terms as 'force', 'energy', 'field', 'heat', etc., in a manner that was "valid" in the sense discussed above.[52] At present, in fact, the connection between the technical and the prescientific meaning of theoretical terms has become quite tenuous in many instances, but the gain achieved by this "alienation" has been an enormous increase in the scope, simplicity, and experiential confirmation of scientific theories. Indeed, it is largely a matter of historical accident, and partly one of convenience, that terms of conversational English are used in the formulation of abstract theories; specially created words or symbols—as they are in fact used to some extent in all theoretical disciplines—would serve the same theoretical purpose and might offer the practical advantage of forestalling various misconceptions to which the use of familiar conversational terms gives rise.[53]

An interest in theoretical import increasingly influences concept formation also in psychology and sociology; and it is to be expected that if a system of concepts is advanced within these disciplines which has a clear empirical interpretation and considerable theoretical strength, it will be adopted even if it should differ radically from the conventional concepts and cut across the groupings and patterns established by them. In fact, such changes in the conceptual framework are clearly reflected in some recent theoretical developments. One of these is the attempt to isolate, through factor analysis,[54] systems of primary mental characteristics. The primary factors evolved by this method do not, as a rule, correspond closely to the psychological characteristics commonly distinguished in prescientific discourse. But this is irrelevant for the strictly systematic objective of the procedure, namely, the representation of numerous psychological traits, each of which is characterized by specific tests, as specific compounds—more exactly, as linear functions—of a relatively small number of independent primary factors.[55] While in factor analysis, emphasis is placed largely on a concept system which affords economy of descriptive representation, there are other theoretical trends which are aimed mainly at the construction of systems with explanatory and predictive power; these trends are illustrated, among others, by recent behavioristic theories of learning[56] and by the psychoanalytic and related approaches to certain aspects of personality and of culture.

III. Some Basic Types of Concept Formation in Science

9. Classification

The preceding chapter dealt with general logical and methodological problems of concept formation in empirical science. In the present chapter we will examine three widely used specific types of scientific concept formation, namely, the procedures of classification, of nonmetrical ordering, and of measurement. In this section we briefly consider classification.

Generally speaking, a classification of the objects in a given domain D (such as numbers, plane figures, chemical compounds, galactic systems, bacteria, human societies, etc.) is effected by laying down a set of two or more criteria such that every element of D satisfies exactly one of those criteria. Each criterion determines a certain class, namely, the class of all objects in D which satisfy the criterion. And if indeed each object in D satisfies exactly one of the criteria, then the classes thus determined are mutually exclusive, and they are jointly exhaustive of D.

Thus, e.g., one customary anthropometric classification of human skulls is based on the following criteria, C_1 to C_5:

(9.1) C_1: The cephalic index, $c(x)$, of skull x is 75 or less; or, $c(x) \leq 75$; C_2: $75 < c(x) \leq 77.6$; C_3: $77.6 < c(x) \leq 80$; C_4: $80 < c(x) \leq 83$; C_5: $83 < c(x)$

The properties determined by these criteria are referred to, respectively, as (1) dolichocephaly, (2) subdolichocephaly, (3) mesaticephaly, (4) subbrachycephaly, and (5) brachycephaly.

In this case the requirements of exclusiveness and exhaustiveness are satisfied simply as a logical consequence of the determining criteria; for, by definition, the cephalix index of any skull is a positive number, and every positive number falls into exactly one of the five intervals referred to by the criteria (9.1). An analogous observation applies, in particular, to all those dichotomous classifications which are determined by some property concept and its denial, such as the division of integers into those which are and those which are not integral multiples of 2, of chemical compounds into organic and inorganic, of bacteria into Gram-positive and Gram-negative.

Of greater significance for empirical science, however, is the case where at least one of the conditions of exclusiveness and exhaustiveness is satisfied not simply as a logical consequence of the determining criteria but as a matter of empirical fact; for this indicates an empirical law and thus confers some measure of systematic import upon the classificatory concepts involved. Thus, e.g., the division of humans into male and female on the basis of primary sex characteristics or of the animal kingdom into various species, and the classification of crystals as developed in crystallography, are not logically exhaustive; hence, to the extent that they are factually so, they possess some systematic import by virtue of laws to the effect that every object in the domain under consideration satisfies one of the determining criteria.

As was mentioned earlier, a distinction is frequently made between *natural* and *artificial classifications*. The former are sometimes said to be based on essential characteristics of the things under investigation and to group together objects which possess fundamental similarities, whereas the latter are viewed as groupings determined by superficial resemblances or external criteria. Thus, e.g., the taxonomic division of plants or of animals into orders, families, genera, and species by reference to phylogenetic criteria would be considered as determining a natural classification; their division into several weight classes according to the average weight of the fully grown specimens would usually be regarded as artificial.

But the notion of essential characteristic invoked here is too obscure to be acceptable as the determining criterion for this classification of classifications. Indeed, it seems impossible to speak significantly of the essential characteristics of an individual thing. Surely, no examination of a given object could establish any of its characteristics as essential; and the customary interpretation of an essential characteristic of a thing as one without which the thing would not be what it is would obviously qualify every characteristic of a thing as essential and would thus render the concept trivial.

The idea of essential characteristic can be given a clearer meaning when it is used in reference to kinds or classes of objects rather than individual objects, as when chemical transmutation is said to be an essential characteristic of radioactive elements or metabolic activity an essential element of living organisms. Statements of this sort have the form 'Attribute Q is an essential characteristic of things of kind P', which may be construed as asserting that the characteristic Q is invariably associated with the characteristic P, i.e., that the sentence 'Whatever has the characteristic P also has the characteristic Q' is true either on logical grounds or as a matter of empirical fact. But this concept of essentiality can be applied only relatively to some given characteristic P; it does not justify the notion that the objects in a given field of inquiry can be individually described as to their essential characteristics and can then be divided into groups forming a natural classification.

The rational core of the distinction between natural and artificial classifications is suggested by the consideration that in so-called natural classifications the determining characteristics are associated, universally or in a high percentage of all cases, with other characteristics, of which they are logically independent. Thus, the two groups of primary sex characteristics determining the division of humans into male and female are each associated —by general law or by statistical correlation—with a host of concomitant characteristics; this makes it psychologically quite understandable that the classification should have been viewed as one "really existing in nature"—as contrasted with an "artificial" division of humans according to the first letter in their given names, or even according to whether their weight does or does not exceed fifty pounds.

Again, the taxonomic categories of genus, species, etc., as used in biology, determine classes whose elements share various biological characteristics other than those defining the classes in question; frequently, the groupings they establish also reflect relations of phylogenetic descent. Thus, the concepts by means of which biology seeks to establish a "natural system" are definitely chosen with a view to attaining systematic, and not merely descriptive, import: "The devising of a classification is, to some extent, as practical a task as the identification of specimens, but at the same time it involves more speculation and theorizing,"[57] for a "natural system is . . . one which enables us to make the maximum number of prophecies and deductions."[58] Similarly, the determining characteristics in the classification of crystals according to the number, relative length, and angles of inclination of their axes are empirically associated with a variety of other physical and chemical characteristics. Analogous observations apply to the ingenious arrangement of the chemical elements according to the periodic system, whose governing principles enabled Mendeleev to predict the existence of several elements then missing in the system and to anticipate with great accuracy a number of their physical and chemical properties.

If a natural classification is thus construed as one whose defining characteristics have a high systematic import, then evidently the distinction between natural and artificial classifications becomes a matter of degree; furthermore, the extent to which a proposed classification is systematically fruitful and thus natural has to be ascertained by empirical investigation; and, finally, a particular classification may well prove "natural" for the purposes of biology; others may be fruitful for psychology or sociology,[59] etc., and each of them would presumably be of little use in some of these contexts.

10. Classificatory vs. Comparative and Quantitative Concepts

A classificatory concept represents, as we saw, a characteristic which any object in the domain under consideration must either have or lack; if its meaning is precise, it divides the domain into two classes, separated by a sharp boundary line. Science uses concepts of this kind largely, though not exclusively, for the description of observational findings and for the formulation of initial, and often crude, empirical generalizations. But with growing emphasis on a more subtle and theoretically fruitful conceptual apparatus, classificatory concepts tend to be replaced by other types, which make it possible to deal with characteristics capable of gradations. In contrast to the "either . . . or" character of classificatory concepts, these alternative types allow for a "more or less": each of them provides for a gradual transition from cases where the characteristic it represents is nearly or entirely absent to others where it is very marked. There are two major types of such concepts which are used in science: comparative and quantitative concepts.[60] These will now be described briefly, in preparation for a closer analysis in the following sections.

The idea of more or less of a certain attribute may be expressed in quantitative terms, as when the classificatory distinction of hot, cold, etc., is replaced by the concept of temperature in degrees centigrade. Concepts such as length in centimeters, temporal duration in seconds, temperature in degrees centigrade, etc., will be called *quantitative* or *metrical* concepts or, briefly, *quantities:* they attribute to each item in their domain of applicability a certain real number,[61] the value of the quantity for that item. In addition to these so-called scalar quantities, whose values are single numbers, there exist other metrical concepts, each of whose values is a set of several numbers; among these are vectors, such as velocity, acceleration, force, etc. The basic problems of measurement concern the introduction of scalars, and we will therefore limit our discussion of quantitative concepts to these.

But conceptual distinctions as to more or less may also be made without any use of numerical values, as when the classificatory terms 'hard', 'soft', etc., are replaced, in mineralogy, by the expressions 'x is as hard as y' and 'x is less hard than y'—both of which are defined by means of the scratch test. These two concepts allow for a comparison of any two pieces of mineral in regard to their relative hardness, and they thus determine an ordering of all pieces of mineral according to increasing hardness; but they do not introduce or presuppose any numerical measure of hardness. We will say that they determine a comparative concept of hardness, and we will generally refer to a concept based on criteria of the kind here illustrated as a *comparative concept.*

It has often been held that the transition from classificatory to the more elastic comparative and quantitative concepts in science has been necessary because the objects and events in the world we live in simply do not exhibit the rigid boundary lines called for by classificatory schemata, but present instead continuous transitions from one variety to another through a series of intermediate forms. Thus, e.g., the distinctions between long and short, hot and cold, liquid and solid, living and dead, male and female, etc., all appear, upon closer consideration, to be of a more-or-less character, and thus not to determine neat classifications. But this way of stating the matter is, at least, misleading. In principle, every one of the distinctions just mentioned can be dealt with in terms of classificatory schemata, simply by stipulating certain precise boundary lines. Thus, e.g., we might, by definitional fiat, qualify the interval between two points as long or short according as, when put alongside some arbitrarily chosen standard, the given interval does or does not extend beyond the latter. This criterion determines a dichotomous division of intervals according to length. And by means of several different standard intervals, we may exhaustively divide all intervals into any not too large finite number of "length classes," each of them having clearly specified boundaries.

However—and this observation suggests a more adequate way of stating the crucial point—suitably chosen comparative or quantitative concepts will often prove so considerably superior for the purposes of scientific description and systematization that they will seem to reflect the very nature of the subject matter under study, whereas the use of classificatory categories will seem an artificial imposition. We now turn to a brief survey of the major advantages offered by those nonclassificatory concepts.[62] Some of the characteristics included in our list will be found exclusively or predominantly in concepts of the quantitative type.

a) By means of ordering or metrical concepts, it is often possible to differentiate among instances which are lumped together in a given classification; in this sense a system of quantitative terms provides a greater descriptive flexibility and subtlety. Thus, e.g., the essentially classificatory wind scale of Beaufort distinguished twelve wind strengths: calm, light air, light breeze, gentle breeze, moderate breeze, fresh breeze, etc., which are defined by such criteria as smoke rising vertically, ripples on the surface of the water, white caps on the waves, etc. A corresponding quantitative concept is that of wind speed in miles per hour, which plainly permits subtler differentiations and which, in addition, covers all possible instances, whereas the

Beaufort classes are not necessarily mutually exclusive and exhaustive of all possibilities. Now, the descriptive subtlety of a classificatory schema may be enhanced by the construction of narrower subclasses; but this possibility does not change the basic fact that the number of distinctions must remain limited. Besides, such subdivision requires the introduction of new terms for the various cases to be distinguished—an inconvenience which is avoided by metrical concepts.

b) A characterization of several items by means of a quantitative concept shows their relative position in the order represented by that concept; thus, a wind of 30 miles per hour is stronger than one of 18 miles per hour. Qualitative characterizations, such as 'gentle breeze' and 'moderate breeze', indicate no such relationship. This advantage of quantitative concepts is closely related to that obtained by the use of numerals rather than proper names in naming streets and houses: numerals indicate spatial relationships which are not reflected by proper names.

c) Greater descriptive flexibility also makes for greater flexibility in the formulation of general laws. Thus, e.g., by means of classificatory terms, we might formulate laws such as this: "When iron is warm, it is gray; when it gets hot, it turns red; and when it gets very hot, it turns white," whereas with the help of ordering terms of the metrical type it is possible to formulate vastly more subtle and precise laws which express the energy of radiation in different wave lengths as a mathematical function of the temperature.

d) The introduction of metrical terms makes possible an extensive application of the concepts and theories of higher mathematics: General laws can be expressed in the form of functional relationships between different quantities. This is in fact the standard form of general laws in the theoretically most advanced branches of empirical science; and mathematical methods such as those of the calculus can be used in applying scientific theories formulated in terms of mathematical functions to concrete situations, for purposes of test, prediction, or explanation.

In the following sections we will examine methods for the introduction of comparative and then of quantitative concepts. The introduction of a scalar quantitative concept will also be referred to as the determination of a *scale of measurement* or of a *metrical scale*.

Following Campbell, we distinguish two major ways of introducing scalar quantities: *fundamental measurement*, which presupposes no prior metrical scales, and *derived measurement*, i.e., the determination of one metrical concept by means of others, such as the definition of density in terms of mass and volume, or of certain anthropological indices as specified functions of the distances between certain reference points in the human body.

The most important—and perhaps the only—type of fundamental measurement used in the physical sciences is illustrated by the fundamental measurement of mass, length, temporal duration, and a number of other quantities. It consists of two steps: first, the specification of a comparative concept, which determines a nonmetrical order; and, second, the metrization of that order by the introduction of numerical values. In the next two sections these two steps will be analyzed in some detail. As our basic illustration we choose the fundamental measurement of mass by means of a balance.[63] This procedure is applicable only to bodies of "medium" size; let D_1 be the class of those bodies. The first, nonmetrical, stage of fundamental measurement will be discussed in section 11; the metrization of the resulting orders will be examined in section 12.[64]

11. Comparative Concepts and Nonmetrical Orders

To establish a comparative concept of mass for the class D_1 of medium-sized bodies is to specify criteria which determine for any two objects in D_1 whether they have the same mass, and, if not, which of them has the smaller one. Similarly, a comparative concept of hardness for the class D_2 of mineral objects is determined by criteria which specify, for any two elements of D_2, whether they are of equal hardness, and, if not, which of them is less hard. By means of these criteria it must be possible to arrange the elements of the given domain in a serial kind of order, in which an object *precedes* another if it has a smaller mass, hardness, etc., than another, whereas objects of equal mass, hardness, etc., *coincide*, i.e., share the same place.

To generalize now: A comparative concept with the domain of application D is introduced by specifying criteria of coincidence and precedence for the elements of D in regard to the characteristic to be represented by the concept; the relations C of coincidence and P of precedence must be so chosen as to arrange the elements of D in a quasi-serial order, i.e., in an array that is serial except that several elements may occupy the same place in it. This means that C and P have to meet certain conditions, which will now be stated, and which will provide a precise definition of the concept of quasi-serial order. In the following formulations, x, y, and z are meant to be any elements of D, i.e., any of the objects to which the comparative concept characterized by C and P is to be applicable.

(11.1a) C is transitive; i.e., whenever x stands in C to y and y stands in C to z, then x stands in C to z; or briefly: $(xCy \cdot yCz) \supset xCz$

(11.1b) C is symmetric; i.e., whenever x stands in C to y, then y stands in C to x; or briefly: $xCy \supset yCx$

(11.1c) C is reflexive; i.e., any object x stands in C to itself; or briefly: xCx

(11.1d) P is transitive

(11.1e) If x stands in C to y, then x does not stand in P to y; or briefly: $xCy \supset \sim xPy$. If this condition is satisfied, we will say that P is C-irreflexive

(11.1f) If x does not stand in C to y, then x stands in P to y or y stands in P to x; or briefly: $\sim xCy \supset (xPy \vee yPx)$. If this condition is satisfied, we will say that P is C-connected

The need for these requirements is intuitively clear, and it will presently be illustrated by reference to the comparative concept of mass. Here we note only that the last two conditions jointly amount to the requirement that any two elements of D

must be comparable in regard to the attribute under consideration; i.e., they must either have it to the same extent, or one must have it to a lesser extent than the other. This clarifies further the strict meaning of the idea of comparative concept. We now define:

(11.2) Two relations, C and P, determine a *comparative concept*, or a *quasi-series*, for the elements of a class D if, within D, C is transitive, symmetric, and reflexive, and P is transitive, C-irreflexive, and C-connected[65]

We now return to our illustration. In formulating specific criteria for this case, we will use two abbreviatory phrases: of any two objects, x and y, in D_1, we will say that x *outweighs* y if, when the objects are placed into opposite pans of a balance in a vacuum, x sinks and y rises; and we will say that x *balances* y if under the conditions described the balance remains in equilibrium.

A quasi-serial order for the elements of D according to increasing mass may now be determined by the following stipulations, in which x, y, z, are always assumed to belong to D_1:

(11.3a) x coincides with y, or x has the same mass as y, if and only if x is either identical with y or balances y

(11.3b) x precedes y, or x has less mass than y, if and only if y outweighs x

The two relations thus defined satisfy the requirements for quasi-serial order: Coincidence is reflexive and symmetrical simply as a consequence of the definition (11.3a); similarly, precedence is C-irreflexive as a consequence of our definitions. The satisfaction of the remaining requirements is a matter of empirical fact. Thus, e.g., the two relations are transitive by virtue of two general laws: Whenever a body x outweighs y and y outweighs z, then x outweighs z, and analogously for the relation representing coincidence. These two general statements do not simply follow from our definitions; in fact, the second of them holds only with certain limitations imposed by such factors as the sensitivity of the balance employed.[66]

The criteria laid down in (11.3) thus determine a nonmetrical order. They enable us to compare any two objects in D_1 in regard to mass; they do not assign, to the individual elements of D_1, numerical values as measures of their masses.

In mineralogy a comparative concept of hardness is defined by reference to the scratch test: A mineral x is called harder than another mineral, y, if a sharp point of x scratches a smooth surface of y; x and y are said to be of equal hardness if neither scratches the other. These criteria, however, are not entirely adequate for the determination of a quasi-series, for the relation of scratching is not strictly transitive.[67]

The rank orders which play a certain role in the initial stages of ordering concept formation in psychology and sociology are likewise of a nonmetrical character. A number of objects—frequently persons—are ordered by means of some criterion, which may be simply the intuitive judgment of one or more observers; the sequence in which the given objects—let there be n of them—are arranged by the criterion is indicated by assigning to them the integers from 1 to n. Any other monotonically increasing sequence of numbers—no matter whether integers or not—would serve the same purpose, since the rank numbers thus assigned have merely the function of ordinal numbers, not of measures. The purely ordinal character of ranking is reflected also by the fact that the rank assigned to an object will depend not only on that object but also on the group within which it is ranked. Thus, the number assigned to a student in ranking his class according to height, say, will depend on the other members of the class; the number representing a measure of his height will not. The need to change rank numbers when a new element is added to the group is sometimes avoided by inserting fractional ranks between the integral ones. Thus, e.g., in the hardness scale of Mohs, certain standard minerals are assigned the integers from 1 to 10 in such a way that a larger number indicates greater hardness in the sense of the scratch test. Now, this test places lead between the standard minerals talc, of hardness 1, and gypsum, of hardness 2; and instead of assigning to lead the rank number 2 and increasing that of all the harder standard minerals by 1, lead is assigned the hardness 1.5. Generally, a substance which the scratch test places between two standard minerals is assigned the arithmetic mean of the neighboring integral values. This method is imperfect, since it may assign the same number to substances which differ in hardness according to the scratch test. Similarly, in Sheldon's typologies of physique and of temperament,[68] positions intermediate to those marked by the integral values 1, 2, . . . , 7 in any one of the three typological components are indicated by so-called half numbers, i.e., by adding $\frac{1}{2}$ to the smaller integer.

When the intuitive judgment of some "qualified" observers is used as a criterion in comparing different individuals in regard to some psychological characteristic, then those observers play a role analogous to that of the balance in the definition of the comparative concept of mass. The use of human instruments of comparison has various disadvantages, however, over the use of nonorganismic devices such as scales, yardsticks, thermometers, etc.: Interobserver agreement is often far from perfect, so that the "yardstick" provided by one observer cannot be duplicated; besides, even one and the same observer may show inconsistent responses. In addition, it appears that most of the concepts defined by reference to the responses of human instruments are of very limited theoretical import—they do not give rise to precise and comprehensive generalizations.

12. Fundamental Measurement

In this section we examine first fundamental measurement as used in physics; the concept of mass again serves as an illustration. Subsequently, alternative types of fundamental measurement are considered briefly.

Fundamental measurement in physics is accomplished by laying down a quasi-serial order and then metricizing it by a particular procedure, which will be explained shortly. What we mean by metricizing a quasi-series is stated in the following definition:

(12.1) Let C and P be two relations which determine a quasi-serial order for a class D. We will say that this order has been metricized if criteria have been specified which assign to each element x of D exactly one real number, $s(x)$, in such a manner that the following conditions are satisfied for all elements x, y of D:

(12.1a) If xCy, then $s(x) = s(y)$

(12.1b) If xPy, then $s(x) < s(y)$

Any function s which assigns to every element x of D exactly one real-number value, $s(x)$, will be said to constitute a *quantitative* or *metrical concept*, or briefly a *quantity* (with the domain of application D); and if s meets the conditions just specified, we will say that it *accords with* the given quasi-series. It follows readily, in view of (11.2), that if s satisfies the conditions (12.1a) and (12.1b), then it also satisfies their converses, i.e.,

(12.1c) If $s(x) = s(y)$, then xCy

(12.1d) If $s(x) < s(y)$, then xPy

In the fundamental measurement of mass, a metrical concept which accords with the comparative one specified in (11.3) may be introduced through the following stipulations, in which '$m(x)$' stands for 'the mass, in grams, of object x':

(12.2a) If x balances y, then let $m(x) = m(y)$

(12.2b) If a physical body z consists of two bodies, x and y, which have no part in common and together exhaust z, then the m-value of z is to be the sum of the m-values of x and y. Under the specified conditions let us call z a *join* of x and y and let us symbolize this by writing: $z = xjy$. Now our stipulation can be put thus: $m(xjy) = m(x) + m(y)$

(12.2c) A specific object k, the International Prototype Kilogram, is to serve as a standard and is to be assigned the m-value 1,000; i.e., $m(k) = 1{,}000$

Now, by (12.2a), any object that balances the standard k has the m-value 1,000; by (12.2b), any object that balances a total of n such "copies" of k has an m-value of $1{,}000n$; each one of two objects that balance each other and jointly balance k has the m-value 500; etc. In this manner it is possible—within limits set by the sensitivity of balances—to give an "operational interpretation" to integral and fractional multiples of the m-value 1,000 assigned to the standard k. Because of the limited sensitivity of balances, the rational values thus interpreted suffice to assign a mass value, $m(x)$, to any object x in the domain D_1; thus, the objective of fundamental measurement for mass has been attained.[69] Let us note that the conditions (12.1a) and (12.1b) are both satisfied; the former is enforced through the stipulation (12.2a), whereas the validity of the latter reflects a physical law: If x precedes y in the sense of (11.3b), i.e., if y outweighs x, then, as a matter of physical fact, y can be balanced by a combination of x with a suitable additional object z. But then, by (12.2b), $m(y) = m(x) + m(z)$; hence,[70] $m(x) < m(y)$.

The fundamental measurement of certain other magnitudes, such as length, and electrical resistance, exhibits the same logical structure: A quasi-serial order for the domain D in question is defined by means of two appropriately chosen relations, P and C, and this order is then metricized. The crucial phase in determining a scalar function s that accords with the given quasi-series consists, in each of those cases of fundamental measurement, in selecting some specific mode of combining any two objects of D into a new object, which again belongs to D; and in stipulating that the s-value of the combination of x and y is always to be the sum of the s-values of the components.

For a somewhat closer study of this process, let us choose, as a nonspecific symbol for any such mode of combination, a small circle (rather than the often-used plus sign, which is apt to be confusing). Thus, '$x \circ y$' will designate the object obtained by combining, in some specified way, the objects x and y. The particular mode of combination varies from case to case; in the fundamental measurement of mass, e.g., it is the operation j of forming an object, xjy, which has x and y as parts; this can be done, in particular, by joining x and y or by placing them together. In the fundamental measurement of length the basic mode of combination consists in placing intervals marked on rigid bodies end to end along a straight line.

We now state the stipulations by which, in fundamental physical measurement, a quasi-series, determined by two relations, C and P, is metricized with the help of some mode of combination, $\circ$; again, x and y are to be any elements of D.

(12.3a) If xCy, then $s(x) = s(y)$

(12.3b) $s(x \circ y) = s(x) + s(y)$

(12.3c) Some particular element of D, say b, is chosen as a standard and is assigned some positive rational number r as its s-value: $s(b) = r$

To give another illustration of the third point: In the fundamental measurement of length in centimeters, b is the interval determined by two marks on the International Prototype Meter; r is chosen as 100.

Now, just as the relations C and P must meet specific conditions if they are to determine a quasi-series, so the operation $\circ$, together with C and P, has to satisfy certain requirements if the assignment of s-values specified by (12.3) is to be unambiguous and in accordance with the standards laid down in (12.1). We will refer to these requirements as the *conditions of extensiveness*. In effect, they demand that the operation $\circ$ together with the relations C and P obey a set of rules which are analogous to certain laws which are satisfied by the arithmetical operation $+$, together with the arithmetical relations $=$ and $<$ (less than) among positive numbers. To consider a specific example: By (12.3b), we have $s(x \circ y) = s(x) + s(y)$, and $s(y \circ x) = s(y) + s(x)$; hence, by virtue of the commutative law of addition, which states that $a + b = b + a$, we have $s(x \circ y) = s(y \circ x)$. If, therefore, (12.1c) is to hold, we have to require that for any x and y in D,

(12.4a) $x \circ y \ C \ y \circ x$

This, then, is one of the conditions of extensiveness. Because of its strict formal analogy to the commutative law of addition, we shall say that (12.4a) requires the operation o to be commutative relatively to C.

In a similar fashion the following further conditions of extensiveness are seen to be necessary.[71] (They are understood to apply to any elements x, y, z, . . . , of D. The horseshoe again symbolizes the conditional, '. . . ⊃ ——' meaning 'if . . . then ——'; the dot serves as the sign of conjunction and thus stands for 'and'; finally, the existential quantifier, '(Ez)' is to be read 'there exists a thing z [in D] such that'.)

(12.4b) $x \circ (y \circ z)\ C\ (x \circ y) \circ z$ (o must be associative relatively to C)

(12.4c) $(xCy \cdot uCv) \supset x \circ u\ C\ y \circ v$

(12.4d) $(xCy \cdot uPv) \supset x \circ u\ P\ y \circ v$

(12.4e) $(xPy \cdot uPv) \supset x \circ u\ P\ y \circ v$

(12.4f) $xCy \supset x\ P\ y \circ z$

(12.4g) $xPy \supset (Ez)(y\ C\ x \circ z)$

The arithmetic analogues of these conditions are obtained by replacing 'o', 'C', 'P', by '+', '=', '<', respectively. The resulting formulas are readily seen to be true for all positive numbers. Condition (12.4f) serves to exclude what might be called zero-elements (in the case of mass, for example, a zero-element would be a body whose combination with another body does not yield an object with greater mass than the latter); while such elements often are useful for theoretical purposes, their admission on the level of fundamental measurement would introduce complications.

The conditions stated so far are sufficient to make sure that the stipulations (12.3) assign no more than one s-value to any one element of D and that the s-values thus assigned satisfy the requirements (12.1a) and (12.1b). They are not sufficient, however, to guarantee that the stipulations (12.3) assign some s-value to every object in D. If, for example, D should contain no two elements that stand in C to each other, then clearly the procedure of fundamental measurement specified in (12.3) would assign an s-value to no other element of D than that chosen as a standard. The kind of further requirement that $\circ$, C, and P have to meet is suggested by the circumstance that fundamental measurement gives rise to rational s-values exclusively. Let us state this point more explicitly:

(12.5) The assignment, by the rules (12.3), of an s-value to an object x in D is based on finding a certain number, n, of other objects, $y_1, y_2, \ldots, y_n$, in F such that

(12.5a) any two of the y's stand in C to each other

(12.5b) joining a suitable number, say, m, of the y's by the operation $\circ$ yields an object that stands in C to the standard b

(12.5c) by joining a suitable number, say k, of the y's, it is possible to obtain an object that stands in C to x

If these conditions are satisfied for a given object x, then the stipulations (12.3) yield $s(x) = (k \cdot r)/m$, which is a rational number. (By virtue of requiring a count of the number of intermediary standards, represented by the y's, which "make up" the standard b and the object x, respectively, the type of fundamental measurement here considered may be said to reduce ultimately to counting.)

If, for a given object x in D, there exists a set of intermediary standard objects $y_1, y_2, \ldots, y_n$ which satisfy the conditions listed in (12.5), we will say that x is *commensurable* with b in D on the basis of C and $\circ$. If x is not thus commensurable with b, then the rules of fundamental measurement assign no s-value to it at all. To insure, therefore, the assignment, required in (12.1) of an s-value to every x in D, we would have to supplement the requirements (12.4) by the following addition:

(12.6) *Condition of commensurability:* Every x in D is commensurable with b on the basis of C and $\circ$

Now, the limitations of observational discrimination preclude the possibility that this condition should ever be found to be violated in the fundamental measurement of physical quantities. Nevertheless, theoretical considerations strongly militate against its acceptance; for it restricts the possible values of those quantities to rational numbers, whereas it is of great importance for physical theory that irrational values be permitted as well. We will return to this point in the next section.

Condition (12.6), therefore, is imcompatible with physical theory and has to be abandoned. This has the consequence, however, that the stipulations for fundamental measurement as here considered do not guarantee the assignment of an s-value to each element of D and hence do not provide a full definition of the quantity s. Rather, they have to be viewed as a partial interpretation,[72] by reference to observables, of the expression '$s(x)$', which itself has the status of a theoretical construct. In leading to this result, our discussion corroborates and amplifies the ideas presented in section 7; it also makes clear that even the fundamental measurement of physical quantities requires the satisfaction of various general conditions, namely, those for quasi-serial order and for extensiveness. And while, depending on the particular choice of D, C, P, $\circ$, and b, some of these conditions may be true just by definition, others will have the character of empirical laws. Hence, fundamental measurement is not just a matter of laying down certain operational rules: it must go hand in hand with the establishment of general laws and theories.

Fundamental measurement based on stipulations of the kind discussed so far, while probably the only type of nonderivative measurement used in physics, is not the only type that is conceivable. Psychology, for example, has developed procedures of a quite different character for the fundamental measurement of various characteristics. As an example,[73] let us briefly consider a method, developed by S. S. Stevens and his associates,[74] for the measurement of pitch, which is an attribute of tones and must be distinguished from the frequency of the corresponding vibrations in a physical medium. The scale in question is obtained by means of "fractionation" experiments, in which a subject is presented with a pair of pure tones of fixed frequencies and then divides the pitch interval determined by them into what he perceives as four phenomenally equal parts, with the help of an apparatus which permits the production of pure tones of any frequency between the two given ones. This type of experiment was performed repeatedly with several subjects and for different pairs of tones whose frequency ranges partly overlapped. The responses obtained from different observers were found to be in satisfactory agreement, and for different frequency intervals they dovetailed so as to determine one uniform scale. A certain pure tone was then chosen as standard and was assigned the arbitrary value of 1,000 mels as a measure of its pitch; and another pure tone, which marks the limit of pitch perception, was assigned zero pitch. Through these stipulations it was possible to assign, in a reasonably univocal fashion, numerical pitch values to the pure tones with frequencies of between 20 and 12,000.

The procedure just described differs plainly from the method analyzed before; in particular, it nowhere relies on any mode of combination. On the other hand, it presupposes no other scale of measurement (theoretically not even that of frequency); it has therefore to be qualified as fundamental measurement.[75]

There are indications that the measure of pitch thus determined may have theoretical import: It correlates with the scales obtained through somewhat different procedures (such as having the subject produce a tone which sounds "halt as high" as a given other tone); and—what is more significant—there is evidence which suggests the existence of a physiological counterpart of the pitch scale: The distance in mels between pure tones seems to be proportional to the linear distances between those regions on the basilar membrane in the inner ear which are made to vibrate most vigorously by those tones; a relationship which does not obtain between the distances in frequency between pure tones and the corresponding basilar areas of maximal excitation.

13. Derived Measurement

By derived measurement we understand the determination of a metrical scale by means of criteria which presuppose at least one previous scale of measurement. It will prove helpful to distinguish between derived measurement by stipulation and derived measurement by law.

The former consists in defining a "new" quantity by means of others, which are already available; it is illustrated by the definition of the average speed of a point during a certain period of time as the quotient of the distance covered and the length of the period of time. Derived measurement by law, on the other hand, does not introduce a "new" quantity but rather an alternative way of measuring one that has been previously introduced. This is accomplished by the discovery of some law which represents the magnitude in question as a mathematical function of other quantities, for which methods of measurement have likewise been laid down previously. Thus, certain laws of physics make it possible to use sound or radar echoes for measuring spatial distances by the measurement of time lapses. Other instances are the measurement of altitude by barometer, of temperature by means of a thermocouple, and of specific gravity by hydrometer, as well as the trigonometric methods used in astronomy and other disciplines in determining the distances of inaccessible points as functions of other distances which are amenable to more direct measurement; these latter methods are based, in particular, on the laws of physical geometry.

While fundamental measurement gives rise to rational values only, its combination, in derived measurement, with general laws or theories, calls for the admission of irrational values as well. Thus, e.g., when direct measurement has yielded the length 10 for the sides of a square, geometry demands that its diagonal be assigned the irrational number $\sqrt{200}$ as its length, although fundamental measurement could never establish or disprove this assignment.[76] Similarly, the law that the period t of a mathematical pendulum is related to its length l by the formula $t = 2\pi\sqrt{l/g}$ requires irrational and even transcendental values for the periods of some pendulums, although again fundamental measurement cannot prove or disprove such an assignment.

As we have seen, the rules of fundamental measurement determine the value of a quantity only for objects of a certain intermediate range, to which we referred as constituting a domain D. In the fundamental measurement of mass, for instance, D consists of those physical bodies which are capable of being weighed on a balance; in the fundamental measurement of length, D is the class of all physical distances capable of direct measurement by means of yardsticks. But the use of such terms as 'mass' and 'length' in physical theory extends far beyond this domain: physics ascribes a mass to the sun and a length to the distance between the solar system and the Andromeda nebula; it determines the masses of submicroscopic particles and the wave lengths of X-rays; and none of these values is obtainable by fundamental measurement.

Thus, also for this reason, the rules for the fundamental measurement of a quantity s do not completely define s, i.e., they do not determine the value of s for every possible case of its theoretically meaningful application. To give an interpretation to 's' outside the original domain D, an extension of those rules is called for. The same remark applies to many quantities for which rules of indirect measurement have been laid down, such as, say, the concept of temperature as interpreted by reference to mercury as a thermometric substance.

One important type of procedure for extending the rules of measurement for a given quantity s consists in combining the methods of derived measurement by law and by stipulation. Suppose, for example, that a scale for the measurement of temperature is originally determined (through derived measurement by stipulation) by reference to a mercury thermometer. Then the concept of temperature is interpreted only for substances which fall within the range between the melting point and the freezing point of mercury. Now, it is an empirical law that, within this range, the temperature of a body of gas under constant pressure can be represented as a specific mathematical function, f, of its volume: $T = f(v)$. This law provides a possibility for the indirect measurement of temperature of a substance by means of a "gas thermometer," i.e., by determining the volume v which a certain standard body of gas assumes under a specified pressure when brought in contact with that substance;

the temperature will then be $f(v)$. Evidently, the formula '$T = f(v)$' represents an empirical law only within the range of the mercury thermometer, since, for other cases, 'T' has received no interpretation. But it is possible to extend its domain of validity by fiat, namely, by stipulating that, outside the range of the mercury thermometer, the formula '$T = f(v)$' is to serve as a definition—or, rather, as a partial interpretation—of the concept of temperature; i.e., that the temperature of a substance outside the range of the mercury thermometer is to be set equal to $f(v)$, where v is the volume which a certain standard body of gas assumes when brought in contact with the substance. Thus, the gas thermometer can now be used for the derived measurement of temperature in a much larger domain than that covered by the mercury thermometer; within the range of the latter, the use of the gas thermometer represents indirect measurement by law; outside that range, indirect measurement by stipulation.

The same type of procedure is used for further extensions of the temperature scale and for the extension of other metrical scales as well. Thus, e.g., the use of trigonometric methods for the determination of certain astronomical distances and the reliance on gravitational phenomena in determining the masses of astronomical bodies may be viewed as extensions of the rules for the fundamental measurement of length and of mass and as providing indirect measurement by law within the original range of the latter and by stipulation outside that range.

It should be noted, however, that this brief account of the interpretation of metrical terms in science as a process of piecemeal stipulative extension of low-order empirical generalizations is considerably schematized in order to exhibit clearly the basic structure of the process. In practice, the empirical "laws" (such as '$T = f(v)$' above) which form the basis of the process often hold only approximately, and there may be considerable deviations from it, particularly at the ends of the original scale. In such cases the original scale of measurement for the given magnitude may be dropped altogether in favor of the more comprehensive scale of indirect measurement; thus, e.g., the temperature scale determined by the gas thermometer approximately, but not strictly, coincides in the range of the mercury thermometer with the scale determined by the latter; it was therefore used to replace the latter in the interest of securing an interpretation of the concept of temperature which would be unambiguous and which would cover a wider range of cases. The process did not stop there, however. For the purposes of theoretical physics, a thermodynamical scale of temperature was eventually introduced which permitted, in conjunction with other concepts, the formulation of a system of thermodynamics distinguished by its theoretical power and its formal simplicity. The thermodynamic concept of temperature has the status of a theoretical construct; it is introduced, not by reference to any particular thermometric substance, but by laying down, hypothetically, a set of general laws couched in terms of it and some other constructs and by providing a partial empirical interpretation for it or for certain derivative terms. The various methods for the measurement of temperature have to be viewed as partial and approximate interpretations of this theoretical construct.

In a similar fashion considerations of theoretical import and systematic simplicity govern the gradual development of the rules for the measurement of many other quantities in the more advanced branches of empirical science; and frequently there is a complex interplay between the development of theoretical knowledge in a discipline and the criteria used for the interpretation of its metrical terms.

Thus, e.g., the concept of time, or of temporal duration of events, represents a theoretical construct whose empirical interpretation has undergone considerable changes. In principle any periodic process might be chosen for the determination of a time scale; the periodically repeated elementary phases of the process would be said to be of equal duration, and the temporal duration of a given event would be measured—to put it briefly—by determining the number of successive elementary processes which take place during the event. Thus, to give an example which Moritz Schlick used in his lectures, the pulse beat of the Dalai Lama might be chosen as the standard clock; but—apart from its enormous technical inconvenience—this convention would have the consequence that the speed of all physical processes would depend on the state of health of the Dalai Lama; thus, e.g., whenever the latter had a fever and showed what, by customary standards, is called a fast pulse, then such events as one rotation of the earth about its axis or the fall of a rock from a given height would take up more temporal units—and would therefore be said to take place more slowly—than when the Dalai Lama was in good health. This would establish remarkable laws connecting the state of health of the Dalai Lama with all events in the universe—and this by instantaneous action at a distance; but it would preclude the possibility of establishing any laws of the simplicity, scope, and degree of confirmation exhibited by Galileo's, Kepler's, and Newton's laws. The customary choice of the earth's daily rotation about its axis as a standard process recommends itself, among other things, because it does not have such strikingly undesirable consequences and because it does permit the formulation of a large body of comprehensive and relatively simple laws of physical change. Eventually, however, those very laws compel the abandonment of the daily rotation of the earth as the standard process for the measurement of time; for they entail the consequence that tidal friction and other factors slowly decelerate that rotation, so that the choice of the earth as a clock would make the speed of physical phenomena dependent upon the age of the earth and would thus have a similar effect as reliance on the pulse beat of the Dalai Lama. This consideration calls for the use of other types of standard processes, such as electrically induced vibrations of a quartz crystal.

Thus, the principles governing the measurement of time and temperature and similarly of all the other magnitudes referred to in physical theory represent complex and never definitive modifications of initial "operational" criteria; modifications which are determined by the objective of obtaining a theoretical system that is formally simple and has great predictive and explanatory power: Here, as elsewhere in empirical science, concept formation and theory formation go hand in hand.[77]

14. Additivity and Extensiveness

A distinction is frequently made between additive and nonadditive quantities and similarly between extensive and intensive characteristics.[78] The ways in which different authors construe these concept pairs conflict to some extent, and some of the criteria offered to explicate them involve certain difficulties. By reference to our preceding analyses, we will now restate concisely what appears to be the theoretical core of those ideas.

The distinction of additive and nonadditive quantities refers to the existence or nonexistence, for a given quantitative concept, of an operational interpretation for the numerical addition of the s-values of two different objects. In this sense, length is called an additive quantity because the sum of two numerical length-values can be represented as the length of the interval obtained by joining two intervals of the given lengths end to end in a straight line; temperature is said to be nonadditive because there is no operation on two bodies of given temperatures which will produce an object whose temperature equals the sum of the latter. To state this idea more precisely, we first define a relative concept of additivity:

(14.1) A quantity s is additive relatively to a combining operation $\circ$ if $s(x \circ y) = s(x) + s(y)$ whenever x, y, and $x \circ y$ belong to the domain within which s is defined

A quantity may be additive relatively to some mode of combination, nonadditive relatively to others. Thus, e.g., the electric resistance of wires is additive relatively to their arrangement in series, nonadditive relatively to their arrangement in parallel. The reverse holds for the capacitance of condensers. Again, the length of intervals marked off on metal rods is additive relatively to the operation of placing them horizontally end to end along a straight line but not strictly additive if the rods are placed on one another vertically in a straight line; in this case the length of the combination will be somewhat less than the sum of the lengths of the components. However, in prevailing usage the terms 'additive' and 'nonadditive' are not relativized with regard to some specified mode of combination: rather, they serve to qualify given quantities categorically as additive or nonadditive. Can such usage be given a satisfactory explication? Can we not simply call a quantity additive if there exists *some* mode of combination in regard to which the quantity satisfies (14.1)? No; this criterion would classify as additive many quantities which in general usage would be called nonadditive (strictly speaking, any quantity is additive under this rule); thus, e.g., the temperature of gases would be additive, for the conditions (14.1) are satisfied by the operation of mixing two bodies of gas and then heating the mixture until its temperature equals the sum of the initial temperatures of its components. To do justice to the intent of the notion of additivity, we would have to preclude such complicated and "artificial" modes of combination as this and to insist on simple and "natural" ones.[79] In the light of our earlier discussions, these two qualifications have to be understood, not in the psychological sense of intuitive simplicity and familiarity, but in the systematic sense of theoretical simplicity and fruitfulness. This suggests the following explication:

(14.2) A quantity s is additive if there exists a mode of combination $\circ$ such that (1) s is additive relatively to $\circ$ in the sense of (14.1); (2) $\circ$ together with s gives rise to a simple and fruitful theory

We now turn to the distinction of extensive and intensive characteristics. It is intended to divide all attributes permitting distinctions as to more or less into two groups: those which can and those which cannot be metricized by the basic physical method of fundamental measurement. Now, any attribute capable of gradations determines a comparative concept in the sense of section 11, and its amenability to fundamental measurement requires the existence of a mode of combination in regard to which the conditions of extensiveness are satisfied. We therefore define:

(14.3) A comparative concept represented by two relations, C and P, which determine a quasi-series within a class D is extensive relative to a given mode of combination, o, if the conditions (12.4) are all satisfied

But, again, the term 'extensive' is customarily used in a non-relativized form. Could we interpret this usage by calling a comparative concept extensive if there exists *some* mode of combination relative to which the concept is extensive in the sense of (14.3)? No; just as in the case of additivity, this criterion is too weak; for it can be shown that if a comparative concept can at all be metricized in the sense of (12.1)—no matter whether derivatively or by one or another kind of fundamental measurement—then there exists some mode of combination, albeit possibly a rather "artificial" one, which satisfies (14.3); hence any comparative concept of this sort would have to count as extensive, which is quite contrary to customary usage. Considerations analogous to those that led to (14.2) suggest the following explication of the distinction between extensive and intensive characteristics:

(14.4) A comparative concept represented by two relations, C and P, which determine a quasi-series within a class D is extensive (intensive) if there exists some (no) mode of combination o such that (1) C, P, and o jointly satisfy the conditions (12.4) of extensiveness; (2) o together with C and P gives rise to a simple and fruitful theory

In the sense of this explication, the comparative concept of mass characterized by the stipulations (11.3) is extensive in its domain of application, since there exists a "simple" and "natural" mode of combination, namely, the operation j specified in (12.2b), in regard to which the conditions of extensiveness are satisfied. On the other hand, consider the comparative concept of hardness determined by the scratch test, which establishes the ordering relations "as hard as" and "less hard than." No simple and natural mode of combining minerals is known which, jointly with those two relations, satisfies the conditions of extensiveness. (The joining operation j, for example, will not do at all; for the join of two different mineral objects is, as a rule, an inhomogeneous body, to which the scratch test is not applicable.) Hardness as characterized by the scratch test has to be qualified, therefore, as an intensive characteristic.

Obviously, an intensive characteristic as here construed is not capable of fundamental measurement by reference to some mode of combination which is governed by simple theoretical principles; but it may well be amenable to some alternative type of fundamental measurement (as are tonal pitch and many other objects of psychophysical measurement) or to derivative measurement (as are density, temperature, refractive index, etc.).

If the explications which have been propounded in this section do justice to the theoretical intent of the two concept pairs under analysis, then a few simple consequences follow, which we now state in conclusion:

Whether a given quantity is qualified as additive or nonadditive and, similarly, whether an attribute is adjudged extensive or intensive will depend on the theoretical knowledge available. In classical mechanics, for example, mass and the speed of rectilinear motion are additive; in relativistic physics, they are not.

Furthermore, there are no sharp boundary lines separating extensive from intensive and additive from nonadditive concepts. For the criteria on which these distinctions are based invoke the concepts of theoretical simplicity and fruitfulness, both of which are surely capable of gradations.

And, lastly, the very fact that questions of simplicity and systematic import enter into the criteria for the distinction reflects once more the pervasive concern of empirical science: to develop a system of concepts which combines empirical import with theoretical significance.

Notes

Preliminary remark. Abbreviated titles inclosed in brackets refer to the Bibliography at the end. Several of the previously published monographs of the *International Encyclopedia of Unified Science* contain material relevant to the problems of concept formation. For convenience, those monographs will be referred to by abbreviations; 'EI3', for example, indicates Volume I, No. 3.

1. I am gratefully indebted to the John Simon Guggenheim Memorial Foundation, which granted me, for the academic year 1947–48, a fellowship for work on the logic and methodology of scientific concept formation. The present monograph is part of the outcome of that work. I sincerely thank all those who have helped me with critical comments or constructive suggestions; among them, I want to mention especially Professors Rudolf Carnap, Herbert Feigl, Nelson Goodman, and Ernest Nagel, Dr. John C. Cooley, and Mr. Herbert Bohnert.

2. Thus, e.g., the genus-and-differentia rule is explicitly advocated in Hart [Report], which presents the views of a special committee on conceptual integration in the social sciences.

3. Quine [Math. Logic], p. 47.

4. Hutchinson [Biology]. (Quoted, with permission of the editor, from the 1948 copyright of Encyclopaedia Britannica.)

5. The concepts of determinacy and uniformity of usage as well as those of vagueness and inconsistency of usage are relative to some class of individuals using the language in question; they are therefore pragmatic rather than syntactic or semantic in character. On the nature of pragmatics, semantics, and syntax see Carnap [EI3], secs. 1, 2, and 3, and Morris [EI2].

6. See [Log. Found. Prob.], chap. i.

7. For details cf. Russell [Math. Philos.] and Tarski [Truth].

8. Cf., for example, Sommerhoff [Analyt. Biol.], which combines a lucid presentation of the general idea of explication with some useful object lessons in the explication of certain fundamental concepts of biology.

9. A spirited discussion of this issue by a group of psychologists and social scientists may be found in Sargent and Smith [Cult. and Pers.], esp. pp. 31–55. This debate illustrates the importance, for theorizing in psychology and the social sciences, of a clear distinction between the various meanings of "(real) definition."

10. Explicit (though not the fullest possible) use of this mode of formulation is made by Hogben in the presentation of his auxiliary international language, Interglossa. His English translations of Interglossa phrases include such items as these:

habe credito ex Y = owe Y; date credito Y de Z = lend Y (some) Z; X acte A Y = X performs the action A on Y ([Interglossal], pp. 45 and 49).

Similarly, Lasswell and Kaplan, in [Power and Soc.], use variables to indicate the syntax of some of their technical terms. Thus, e.g., power is defined as a triadic relation: "*Power* is participation in the making of decisions: G has power over H with respect to the values K if G participates in the making of decisions affecting the K-policies of H" (*ibid.*, p. 75). Note that this definition is expressed contextually rather than by simulating the genus-and-differentia form, which is strictly inapplicable here.

11. Definitions of the form considered in the present and the preceding sections are often called *explicit definitions*. They state explicitly, in the definiens, an expression which is synonymous with the expression to be defined and in favor of which the latter can always be eliminated. They differ in this respect from the so-called *recursive definitions*, which play an important role in logic and mathematics but are not used in empirical science. For details on the formal aspects of explicit and recursive definition cf. Church [Articles] and Carnap [Syntax]. A critical historical study of various conceptions of definition, together with an examination of the function of definition in mathematics and empirical science, may be found in Dubislav [Definition]. Lewis [Analysis] contains a detailed discussion of definition, with special emphasis on explication. Various important observations on the nature and function of definition are included in Quine [Convention]. Chapter i of Goodman [Appearance] is an excellent study of the use of definition in rational reconstruction. Robinson [Definition] presents a nontechnical discussion of various aspects of definition; this book is not predominantly concerned, however, with definition in science.

12. For a fuller statement, and a partly critical discussion, of this injunction see Eaton [Logic], chap. vii.

13. Hilbert [Grundlagen], § 3.

14. Peano [Définitions].

15. Important contributions to the clarification and partial solution of this problem are contained in Carnap [Aufbau] and in Goodman [Appearance].

16. Cf. [Forschung], secs. 25–30.

17. See, e.g., Woodrow [Laws]; Hull [Int. Var.]; Spence [Theory Construction]; and MacCorquodale and Meehl [Distinction].

18. These were first pointed out by Carnap in [Testability].

19. For details on this problem and its ramifications see Chisholm [Conditional]; Goodman [Counterfactuals]; Hempel and Oppenheim [Explan.]; Lewis [Analysis], chap. vii; Reichenbach [Logic], chap. viii.

20. In [Testability]; a less technical account may be found in Parts III and IV of Carnap's [EI1].

21. It might be objected that these remarks disregard the distinction between (1) *assigning a meaning* to a scientific concept by reference to observables and (2) *discovering an empirical regularity* which connects the previously defined concept with certain observables. Such discovery, it might be argued, though providing new criteria of application for the concept, does not affect its meaning at all. Now, I think it is often useful to make a distinction between questions of meaning and questions of fact; but I have doubts about the possibility of finding precise criteria which would explicate the distinction. For this reason, the presumptive objection does not seem to me decisive. Reasons for this view may be found in White [Analytic] and Quine [Dogmas]. This issue, however, is still the object of considerable controversy, and my remarks in the text are therefore deliberately sketchy.

22. This is, in effect, the liberalized version of physicalism by which Carnap proposed to replace the earlier form, to which we referred as the narrower thesis of empiricism. More specifically, the liberalized thesis asserts that each term of empirical science can be introduced by means of an introductive chain, i.e., an ordered set of reduction sentences analogous to a definition chain; for details see Carnap [Testability], esp. secs. 9, 15, and 16; a less technical synopsis is given in Carnap [EI1], Secs. III and IV.

23. More precisely, an introductive chain as defined by Carnap can be shown to specify, in terms of observables, one necessary and one sufficient condition for the term it introduces. The bilateral reduction sentence (6.3), e.g., which is a simple special case of an introductive chain, specifies that $P_1x \cdot P_2x$ is a sufficient condition for Qx, and $\sim(P_1x \cdot \sim P_2x)$ is a necessary one. These conditions apply quite generally to any object x, no matter whether it meets the test condition P_1x or not. Carnap has not discussed in detail reduction sentences for expressions involving more than one variable; but these are called for by the liberalized physicalistic thesis, and his theory of reduction can readily be transferred to this case. But, in doing so, we obtain the consequence that an introductive chain for 'length $(u, v) = r$' must specify a necessary and a sufficient condition for every sentence of this form, i.e., for every possible set of values r (> 0), u, v. And this clearly cannot be accomplished in terms of observation predicates.

24. For fuller details on the axiomatic method cf., e.g., Tarski [Logic], chaps. vi ff., and Woodger's treatises [Ax. Meth.] and [EII5]. It should be noted that the conception of scientific theories as presented in axiomatized form is an idealization made for purposes of logical clarification and rational reconstruction. Actual attempts to axiomatize theories of empirical science have so far been rare. Apart from different axiomatizations of geometry, the major instances of such efforts include Reichenbach's [Axiomatik], Walker's [Foundations], the work of Woodger just referred to, the axiomatization of the theory of rote learning by Hull and his collaborators (cf. [Rote Learning]), and, in economic theory, axiomatic treatments of such concepts as utility (see, e.g., the axiomatization of utility in von Neumann and Morgenstern [Games], chap. iii and Appendix).

25. See [Grundlagen].

26. It is sometimes held that in an uninterpreted axiomatized theory the axioms, or postulates, themselves constitute "implicit definitions" of the primitives and that, accordingly, the latter mean just what the postulates require them to mean. This view has been eloquently advocated by Schlick (cf. [Erkenntnislehre], sec. 7) and by Reichenbach (see [Raum-Zeit-Lehre], § 14), and, more recently, it has been invoked by Northrop (see, e.g., [Logic], chap. v), and it is reflected also in Margenau's concept of a "constitutive definition" for theoretical constructs (cf. [Reality], chap. xii). This conception of the function of postulates faces certain difficulties, however. According to it, the term 'point' in pure Euclidean geometry means an entity of such a kind that, for any two of them, there exists exactly one straight line on which both are incident, etc. But since the terms 'straight line', 'incident', etc., have no prior meaning assigned to them, this characterization cannot confer any specific meaning on the term 'point'. To put it differently: The conjunction of all the postulates of an axiomatized theory may be construed as a sentential function in which the primitives play the role of variables. But a sentential function cannot well be said to "define" one particular meaning of (i.e., one particular set of values for) the variables it contains unless proof is forthcoming that there exists exactly one set of values for the variables which satisfies the given sentential function. The postulates of an uninterpreted deductive system may well be said, however, to impose limitations upon the possible interpretations for the primitives. Thus, e.g., the postulate that for two points there exists exactly one straight line on which both are incident precludes the interpretation of 'point' by 'person', 'line' by 'club', and 'incident on' by 'member of', for this interpretation would turn the postulate into a false sentence.

For a concise discussion of the notion of implicit definition and its historical roots see

Dubislav [Definition], secs. 28 and 29; on the origin of the idea in Gergonne's work see also Nagel [Geometry], secs. 27–30.

27. For details on pure and physical geometry, and for further references, see Carnap [EI3], secs. 21 and 22; Hempel [Geometry]; and especially Reichenbach [Rise], chap. viii, and the detailed work [Raum-Zeit-Lehre]. On the relativistic treatment of physical geometry, cf. also Finlay-Freundlich [EI8].

28. A distinction between abstract theoretical system and interpretation is made by Campbell, who divides a physical theory into the "hypothesis," i.e., a set of propositions "about some collection of ideas which are characteristic of the theory," and the "dictionary"; the latter relates to some (but not to all) propositions of the hypothesis certain empirical propositions whose truth or falsity can be ascertained independently, and it states that one of these sets of related propositions is true if and only if the other is true ([Physics], p. 122). Note that the argument presented earlier in section 7 casts doubt upon the "if and only if." In a similar vein, Reichenbach speaks of the "coördinative definitions" which, by coördinating physical objects with geometrical concepts, specify the denotations of the latter (cf. [Rise], chap. viii; [Raum-Zeit-Lehre], sec. 4). Carnap, in [EI3], secs. 21–24, analyzes the empirical interpretation of abstract calculi as a semantical procedure. The logical structure of scientific theories and their interpretation has recently been discussed at length also by Northrop ([Logic], esp. chaps. iv–vii, and [Einstein]) and by Margenau ([Reality], chaps. iv, v, and xii); the rules effecting the interpretation are called "epistemic correlations" by Northrop, "rules of correspondence" by Margenau. (Both authors envisage an interpretation in phenomenalistic terms rather than by reference to intersubjectively ascertainable observables.) Einstein's lecture, [Method], contains a lucid discussion of the problem at hand in special reference to theoretical physics.

29. See [Ax. Meth.] and [EII5].

30. Two of the monographs in the *International Encyclopedia of Unified Science* give brief accounts of the fundamental ideas of the kinetic theory: Lenzen [EI5], sec. 15, and Frank [EI7], Part IV. For concise analyses of the hypotheses and interpretations involved in this case see Campbell [Physics], pp. 126–29, and Nagel [Reduction], pp. 104–11.

31. Koch's essay, [Motivation], offers some good examples of the interpretation of theoretical constructs in psychology by means of empirical terms which, in turn, are introduced by chains of reduction sentences based on observation terms.

32. Thus, e.g., A. Wald says on the interpretation of a scientific theory, "In order to apply the theory to real phenomena, we need some rules for establishing the correspondence between the idealized objects of the theory and those of the real world. These rules will always be somewhat vague and can never form a part of the theory itself" ([Statist. Inf.], p. 1).

33. Cf. Kaplan [Def. and Spec.].

34. Cf. Carnap [Log. Found. Prob.], chap. ii; Nagel [EI6]; Reichenbach [Probability], chaps. ix and xi; also see Helmer and Oppenheim [Degree], where a theory of logical probability is developed which differs from Carnap's in various respects.

35. Cf. especially [Modern Physics], [Physical Theory], [Op. An.], and [Concepts]. For a concise and lucid presentation and appraisal of the central ideas of operationism see Feigl [Operationism]. For enlightening comments on operationism in psychology, cf. Bergmann and Spence [Operationism] and Brunswik [EI10], chaps. i and ii.

36. [Op. An.], p. 119. In this article, incidentally, Bridgman points out: "I believe that I myself have never talked of 'operationalism' or 'operationism', but I have a distaste for these grandiloquent words which imply something more philosophic and esoteric than the simple thing that I see" (*ibid.*, p. 114).

37. Cf. Bridgman [Op. An.] and [Concepts].

38. For elementary accounts see Ayer [Language], chap. i and Preface; Pap [Anal. Philos.], chap. xiii. A more advanced treatment may be found in Carnap [Testability]; recent critical surveys in Hempel [Emp. Crit.] and [Cogn. Signif.].

39. [Institutions].

40. In some cases the reliability of a test is intended to be an index of the objective consistency of its different components. For a fuller discussion see Thurstone [Reliability] and Guilford [Methods]. Dodd's article [Op. Def.] contains useful comments on the notion of reliability.

41. Cf. Ogburn [Social Change], Parts IV and V; Ogburn and Nimkoff [Sociology], pp. 881 ff.

42. [Social Change], p. 297.

43. The discussion of cultural lag in Ogburn and Nimkoff [Sociology], pp. 881 ff., takes cognizance of this difficulty and proposes certain qualified formulations; these, however, still contain valuational terms such as "best adjustment to a new culture trait," "harmonious integration of the parts of culture," etc., so that the basic objection raised here remains unaffected.

44. To suggest a direction in which such a restatement might be sought, let us note that lag phenomena are known in the area of the physical sciences as well. Thus, when a jar containing a viscous liquid is tilted, a certain period of time elapses before the liquid has "adjusted" itself to the new position of the jar; again, the changes in a magnetic field and the resultant magnetization of a piece of steel in the field exhibit a temporal lag in that the maxima and minima of the latter "lag behind" those in the intensity of the field; however, a theoretical account of the first phenomenon uses descriptive concepts such as that of equilibrium rather than the normative concept of good adjustment; and, in the second case, the theoretical analysis does not assert that the maxima and minima of induced magnetization "should" occur earlier than they do but simply aims at a description of the temporal lag by means of graphs or mathematical functions. It is conceivable that the empirical information which the hypothesis of cultural lag is intended to convey might be satisfactorily restated in an analogous fashion. On this point see also Lundberg's discussion of cultural lag and related concepts in [Foundations], pp. 521 ff.

45. Cf., e.g., Malinowski [Dynamics] and the searching study of functional analysis presented in Merton [Social Theory], chap. i.

46. A rigorous explication of the notions of scope, confirmation, and formal simplicity presents considerable difficulties; but, for our present purposes, an intuitive understanding of these concepts will suffice. The characteristics of a good theory are concisely surveyed in Nagel [EI6], sec. 8. The concept of confirmation is dealt with in the publications listed in n. 34. Certain aspects of the intriguing notion of simplicity are examined in Popper [Forschung], secs. 31–46, and in Goodman [Appearance], chap. iii. Reichenbach [Experience], § 42, suggests a distinction between descriptive and inductive simplicity.

47. Cf. [Physique] and [Temperament].

48. For fuller details, cf. Hempel and Oppenheim [Typusbegriff], in Sheldon [Physique], chap. v, and Lazarsfeld and Barton [Qual. Meas.].

49. Cf., e.g., Dodd's "S-theory" (expounded most fully in [Dimensions]), which actually is not a sociological theory but a system of terminology and classification whose theoretical import is quite problematic.

50. For illustrations and a more detailed discussion of the concept of validity see Adams [Validity]; Dodd [Op. Def.]; Guilford [Methods], pp. 279 and 421; Thurstone Reliability], sec. 25.

51. The second criterion is applied, e.g., by Stevens and Volkmann (cf. [Pitch]), who claim validity for their pitch scale partly on the ground that it permits fitting into one simple curve the data obtained by several sets of experiments, and who also mention a law of simple mathematical form that appears to connect pitch with a certain anatomical aspect of the hearing process (cf. pp. 68 and 69 of this monograph).

52. This is stressed by Lundberg in his critique of common usage as a standard for the operational interpretation of sociological terms; see [Definitions] and, especially, [Measurement].

53. An entirely different view is expressed by G. W. Allport, who insists that "mathematical or artificial symbols" are not suited to name human traits and that "the attributes of human personality can be depicted only with the aid of common speech, for it alone possesses the requisite flexibility, subtlety, and established intelligibility" ([Personality], p. 340). In a similar vein, Allport and Odbert assert, concerning the naming of traits, "Mathematical symbols cannot be used, for they are utterly foreign to the vital functions with which the psychologist is dealing. Only verbal symbols (ambiguous and troublesome as they are) seem appropriate" ([Trait-Names], p. v). In accordance with this view, the authors hold that "the empirical discovery of traits in individual lives is one problem, that of selecting the most appropriate names for the traits thus discovered is another" (*ibid.*, p. 17).

54. Cf. L. L. Thurstone [Vectors], [Analysis], and [Abilities]; for detailed references to the comprehensive literature see Wolfle [Factor Analysis].

55. At present, the theoretical significance of the primary characteristics evolved by factor analysis appears to reside mainly in their ability to permit a formally simplified, or economical, descriptive representation of human traits; and, indeed, Thurstone repeatedly emphasizes considerations of theoretical parsimony (cf., e.g., [Vectors], pp. 47, 48, 73, 150, 151; [Analysis], p. 333). The simplicity thus achieved is of the kind which Reichenbach calls descriptive and which he distinguishes from inductive simplicity; the latter is closely related to predictive power and thus to theoretical import (cf. [Experience], sec. 42). The predictive aspect is not strongly emphasized in Thurstone's work (cf., e.g., [Analysis], pp. 59 ff.); but there are certain remarks as to the possibility of a genetic substructure underlying the system of factors, which would confer upon factor analysis special significance for the study of mental inheritance and of mental growth (cf. Thurstone [Vectors], pp. 51 and 207; and [Analysis], p. 334). If such connections could be ascertained, they would provide the conceptual systems developed by factor analysis with theoretical import in addition to the systematic advantage of descriptive simplicity. At present, however, few if any suggestions of general laws (of causal or statistical form) can be found in the work on factor analysis—except for the statement of independence of certain primary traits. See also Brunswik's discussion, in [EI10], of the issue at hand.

56. Cf. the discussion in Spence [Learning].

57. Mayr [Systematics], p. 10.

58. Huxley [New Syst.], p. 20. Fuller details on this conception of natural classification in modern taxonomy may be found in Gilmour [Taxonomy], Huxley, *op. cit.*, and Mayr [Systematics].

59. On the methodology of classificatory and related procedures in the social sciences see Lazarsfeld and Barton [Qual. Meas.].

60. This terminology is Carnap's; see [Log. Found. Prob.], secs. 4 and 5, where he distinguishes and compares classificatory, comparative, and quantitative concepts.

61. Not a numeral (i.e., a symbol naming a number), as has been asserted by several authors, including Campbell (cf. [Physics], chap. x; [Measurement], p. 1 *et passim;* and Campbell's contribution to Ferguson [Reports]), Reese ([Measurement]), and Stevens (see, e.g., [Scales] and [Math., Meas., and Ps.]). In [Physics], chap. x, p. 267, however, Campbell declares: "Measurement is the process of assigning numbers to represent qualities." And, indeed, the values of quantitative concepts have to be construed so as to be able to enter into mathematical relationships with each other, such as those expressed by Newton's law of gravitation, the laws for the mathematical pendulum, Boyle's law, etc. They must, therefore, permit multiplication, the extraction of roots, etc.; and all these operations apply to numbers, not to numerals. Similarly, it is impossible to speak significantly of the distance, or difference, of two numerals.

62. This survey follows in part Carnap's discussion in [Begriffsbildung], pp. 51–53.

63. Properly speaking, a balance compares and measures gravitational forces exerted by the earth upon bodies placed in the scales. Since these forces, however, are proportional to the masses of those objects, an "operational definition" of mass can be based on the use of a scale. In a similar vein, Frank ([E17], p. 12) gives an "operational definition" of the mass of a body as the number of grams ascertained by reading its weight on a spring balance at sea level. For a fuller discussion of the point at hand see Campbell [Measurement], chap. iii, and esp. p. 45.

64. For further details on the problems of ordering and measurement discussed in the following sections cf. Campbell's analyses in [Physics], Part II, and in [Measurement]; Carnap [Begriffsbildung]; von Helmholtz [Zählen und Messen]: Hempel and Oppenheim [Typusbegriff] (on comparative concepts); Lenzen [EI5] (esp. secs. 5, 6, and 7, which deal with the measurement of length, time, and weight); Nagel [Measurement] and [Log. of Meas.]; Russell]Principles], Part III; Stevens [Math., Meas., and Ps.]; and Suppes' precise study [Ext. Quant.].

65. In logic, a relation P is said to constitute a series within a class D if within D it is transitive, irreflexive (i.e., for no x in D does xPx hold), and connected (i.e., if not $x = y$ then xPy or yPx). Our concepts of C-irreflexivity, C-connectedness, and quasi-series are generalizations of these ideas and include the latter as the special case where C is the relation of identity. Let us note that, among others, the following theorems are consequences of the conditions specified in (11.2):

(T1) $$(x)(y)\ \{xCy \equiv (u)\ [(xPu \equiv yPu) \cdot (uPx \equiv uPy)]\}$$

(T2) $$(x)(y)(z)\ [(xCy \cdot yPz) \supset xPz]$$

(T1) corresponds closely to a definition of coincidence in terms of precedence as used by Campbell; cf. [Measurement], p. 5.

66. Cf. the discussion of this point in Campbell [Measurement], chap. iii.

67. Cf. Bollenrath [Härte], where an account of more recent alternative comparative and metrical concepts of hardness may also be found.

68. See [Physique] and [Temperament].

69. This account of the fundamental measurement of mass is necessarily schematized with a view to exhibiting the basic logical structure of the process. We have to disregard such considerations as that the equilibrium of a balance carrying a load in each pan may not be disturbed by placing into one of the pans an additional object which is relatively light but whose mass is ascertainable by fundamental measurement. This means that fundamental measurement does not assign exactly one number to every object in D_1. A more detailed discussion of problems of this type may be found in Campbell [Measurement].

70. This "hence" relies on yet another physical fact, namely, that fundamental measurement assigns to any object in D_1 a positive number. This would not be the case but for the stipulation, included in our definitions of balancing and outweighing, that the balance be placed in a vacuum; for, in air, objects such as helium-filled balloons would have to be assigned negative numbers. See Campbell's remarks on this point in [Physics], pp. 319–20, and [Measurement], pp. 37–38.

71. A complete statement of the conditions here referred to does not seem to exist in the literature. The most detailed analysis appears to be Campbell's in [Measurement]. Other important contributions to the problem may be found in Helmholtz [Zählen und Messen]; Hölder [Quantität]; Nagel [Log. of Meas.] and [Measurement]; Suppes [Ext. Quant.]. Concerning the conditions of extensiveness, let us note here that we cannot require '$x o y$' to have a meaning for every x and y in D. Indeed, the modes of combination on which fundamental measurement is based are usually inapplicable when x and y are identical or have common parts. (This is clearly illustrated by the operation j we invoked in the fundamental measurement of mass.) Nor can we require that D be closed under the operation o, i.e., that whenever x and y belong to D then $x o y$ belongs to D again; for—to illustrate by reference to the operation j—the joining of two large bodies each of which can just barely be weighed on a scale will yield an object to which this procedure is no longer applicable.

The conditions laid down under (12.4) should accordingly be construed as applying to all those cases where the combinations mentioned exist and belong to D; and these cases are assumed to form a reasonably comprehensive set.

72. This conception was suggested to me by Professor Carnap.

73. For an account of other methods see Guilford [Methods]; Gulliksen [Paired Comparisons]; Thurstone [Methods].

74. Cf. Stevens and Volkmann [Pitch]; Stevens and Davis [Hearing], esp. chap. iii.

75. A similar method has been used by Stevens and his associates to establish scales of measurement for other attributes of tones; cf., e.g., Stevens [Loudness] and Stevens and Davis [Hearing]. Some authors, including Campbell, do not recognize this procedure as measurement, essentially on the ground that it is not of the extensive type, i.e., it does not rely on any mode of combination for the interpretation of addition. A critical discussion of Stevens' procedure from this point of view will be found in Ferguson [Reports], which includes a contribution by Campbell. For reasons mentioned in the present section, the conception of measurement which underlies this criticism appears unduly narrow; this is argued by Stevens in his article [Scales], which contains his direct reply, and in [Math., Meas., and Ps.]. For further light on this issue see Bergmann and Spence [Psych. Meas.] and Reese [Measurement].

76. For details on this important point, which again reflects the connection between theory formation and concept formation, and which also illustrates the grounds for re-

jecting the commensurability requirement (12.6), cf. von Helmholtz [Zählen und Messen] and Hertz [Comments], esp. p. 103; Campbell [Physics], pp. 310–13, and [Measurement], pp. 24 and 140; and Nagel [Op. An.], pp. 184–89.

77. The interplay of concept formation and theory formation is instructively exhibited in the "method of successive approximations," to whose importance in measurement Lenzen calls attention in [EI5]. This method for the successively more precise interpretation of metrical terms clearly presupposes the availability of certain laws and theories.

78. See, e.g., Cohen and Nagel [Logic], chap. xv, and Bergmann and Spence [Psych. Meas.]; also cf. Duhem [Théorie physique], pp. 177–80, where the basic idea underlying the two distinctions is vividly presented (though not logically analyzed) under the heading "Quantité et qualité."

79. In the context of their lucid analytic study, [Psych. Meas.], Bergmann and Spence have made an attempt to specify a restrictive clause for the admissible types of combination. They stipulate that the operation has to "lie within the dimension" of the quantity under consideration. This formulation is too elusive, however, to provide a solution to the problem. It might well be argued, for example, that our "artificial" operation, relatively to which the temperature of gases is additive, satisfies this requirement; for it involves measurement only in the "dimension" of temperature. For further observations on additivity, see especially Carnap [Begriffsbildung], pp. 32–35.

Bibliography

ADAMS, HENRY, F. [Validity] "Validity, Reliability, and Objectivity," in "Psychological Monographs," XLVII, No. 2 (1936), 329–50.

ALLPORT, GORDON W. [Personality] *Personality: A Psychological Interpretation*. New York, 1937.

ALLPORT, GORDON W., and ODBERT, HENRY S. [Trait-Names] *Trait-Names: A Psycho-lexical Study*. "Psychological Monographs," Vol. XLVII, No. 1 (1936).

AYER, ALFRED J. [Language] *Language, Truth and Logic*. 2d ed. London, 1946.

BERGMANN, GUSTAV, and SPENCE, KENNETH W. [Operationism] "Operationism and Theory in Psychology," *Psychological Review*, XLVIII (1941), 1–14.

———. [Psych. Meas.] "The Logic of Psychophysical Measurement," *Psychological Review*, LI (1944), 1–24.

BOLLENRATH, FRANZ [Härte] "Härte und Härteprüfung," in *Handwörterbuch der Naturwissenschaften*, Vol. V, ed. R. DITLER *et al.* 2d ed. Jena, 1931–35.

BRIDGMAN, P. W. [Modern Physics] *The Logic of Modern Physics*. New York, 1927.

———. [Physical Theory] *The Nature of Physical Theory*. Princeton, 1936.

———. [Op. An.] "Operational Analysis," *Philosophy of Science*, V (1938), 114–31.

———. [Concepts] "The Nature of Some of Our Physical Concepts," *British Journal for the Philosophy of Science*, I, 257–72 (1951); II, 25–44 and 142–60 (1951). Also reprinted as a separate monograph, New York, 1952.

BRUNSWIK, EGON. [EI10] *The Conceptual Framework of Psychology*. EI10. Chicago, 1952.

CAMPBELL, NORMAN R. [Physics] *Physics: The Elements*. Cambridge, England, 1920.

———. [Measurement] *An Account of the Principles of Measurement and Calculation*. London and New York, 1928.

CARNAP, RUDOLF. [Begriffsbildung] *Physikalische Begriffsbildung*. Karlsruhe, 1926.

———. [Aufbau] *Der logische Aufbau der Welt*. Berlin, 1928.

———. [Testability] "Testability and Meaning," *Philosophy of Science*, III (1936), 419–71, and IV (1937), 1–40. Also reprinted as a separate pamphlet (with corrigenda and additional bibliography by the author) by Graduate Philosophy Club, Yale University, 1950.

———. [Syntax] *Logical Syntax of Language*. London, 1937.

———. [EI1] "Logical Foundations of the Unity of Science," in EI1, pp. 42–62. Chicago, 1938. Reprinted in FEIGL and SELLARS [Readings].

———. [EI3] *Foundations of Logic and Mathematics*. EI3. Chicago, 1939.

———. [Log. Found. Prob.] *Logical Foundations of Probability*. Chicago, 1950.

CHAPIN, F. STUART. [Institutions] *Contemporary American Institutions.* New York, 1935.

CHISHOLM, RODERICK M. [Conditional] "The Contrary-to-Fact Conditional," *Mind,* LV (1946), 289–307. Reprinted in FEIGL and SELLARS [Readings].

CHURCH, ALONZO. [Articles] Articles "Definition" and "Recursion, Definition by," in D. D. RUNES (ed.), *The Dictionary of Philosophy.* New York, 1942.

COHEN, MORRIS R., and NAGEL, ERNEST. [Logic] *An Introduction to Logic and Scientific Method.* New York, 1934.

DODD, STUART C. [Dimensions] *Dimensions of Society.* New York, 1942.

———. [Op. Def.] "Operational Definitions Operationally Defined," *American Journal of Sociology,* XLVIII (1942–43), 482–89.

DUBISLAV, WALTER. [Definition] *Die Definition.* 3d ed. Leipzig, 1931.

DUHEM, P. [Théorie physique] *La théorie physique: Son objet et sa structure.* Paris, 1906.

EATON, R. M. [Logic] *General Logic.* New York, 1931.

EINSTEIN, ALBERT. [Method] *On the Method of Theoretical Physics.* Herbert Spencer Lecture. Oxford, 1933.

FEIGL, HERBERT. [Operationism] "Operationism and Scientific Method," *Psychological Review,* LII (1945), 250–59. Reprinted, with some alterations, in FEIGL and SELLARS [Readings].

FEIGL, HERBERT, and SELLARS, WILFRID (eds.). [Readings] *Readings in Philosophical Analysis.* New York, 1949.

FERGUSON, A. [Reports] FERGUSON, A., *et al.*, "Interim Report of Committee Appointed To Consider and Report upon the Possibility of Quantitative Estimates of Sensory Events," *Report of the British Association for the Advancement of Science, 1938,* pp. 277–334; and FERGUSON, A., *et al.*, "Final Report . . . ," *ibid., 1940,* pp. 331–49.

FINLAY-FREUNDLICH, E. [EI8] *Cosmology.* EI8. Chicago, 1951.

FRANK, PHILIPP. [EI7] *Foundations of Physics.* EI7. Chicago, 1946.

GILMOUR, J. S. L. [Taxonomy] "Taxonomy and Philosophy," in JULIAN HUXLEY (ed.), *The New Systematics,* pp. 461–74. Oxford, 1940.

GOODMAN, NELSON. [Counterfactuals] "The Problem of Counterfactual Conditionals," *Journal of Philosophy,* XLIV (1947), 113–28.

———. [Appearance] *The Structure of Appearance.* Cambridge, Mass., 1951.

GUILFORD, J. P. [Methods] *Psychometric Methods.* New York and London, 1936.

GULLIKSEN, H. [Paired Comparisons] "Paired Comparisons and the Logic of Measurement," *Psychological Review,* LIII (1946), 199–213.

HART, HORNELL. [Report] "Some Methods for Improving Sociological Definitions: Abridged Report of the Subcommittee on Definition of Definition of the Committee on Conceptual Integration," *American Sociological Review,* VIII (1943), 333–42.

HELMER, OLAF, and OPPENHEIM, PAUL. [Degree] "A Syntactical Definition of Probability and of Degree of Confirmation," *Journal of Symbolic Logic,* X (1945), 25–60.

HELMHOLTZ, HERMANN VON. [Zählen und Messen] "Zählen und Messen," in VON HELMHOLTZ, *Schriften zur Erkenntnistheorie.* Herausgegeben und erläutert von PAUL HERTZ and MORITZ SCHLICK. Berlin, 1921.

HEMPEL, CARL G. [Geometry] "Geometry and Empirical Science," *American Mathematical Monthly*, LII (1945), 7–17. Reprinted in FEIGL and SELLARS [Readings].

———. [Emp. Crit.] "Problems and Changes in the Empiricist Criterion of Meaning," *Revue internationale de philosophie*, No. 11 (1950), pp. 41–63.

———. [Cogn. Signif.] "The Concept of Cognitive Significance: A Reconsideration," *Proceedings of the American Academy of Arts and Sciences*, LXXX, No. 1 (1951), 61–77.

HEMPEL, CARL G., and OPPENHEIM, PAUL. [Typusbegriff] *Der Typusbegriff im Lichte der neuen Logik*. Leiden, 1936.

———. [Explan.] "Studies in the Logic of Explanation," *Philosophy of Science*, XV (1948), 135–75.

HERTZ, PAUL. [Comments] Comments on counting and measurement included in VON HELMHOLTZ [Zählen und Messen].

HILBERT, DAVID. [Grundlagen] *Grundlagen der Geometrie*. 4th ed. Leipzig, 1913.

HÖLDER, O. [Quantität] "Die Axiome der Quantität und die Lehre vom Mass," *Ber. d. Sächs. Gesellsch. d. Wiss., math.-phys. Klasse*, 1901, pp. 1–64.

HOGBEN, LANCELOT. [Interglossa] *Interglossa*. Penguin Books, 1943.

HULL, C. L. [Behavior] *Principles of Behavior*. New York, 1943.

———. [Int. Var.] "The Problem of Intervening Variables in Behavior Theory," *Psychological Review*, L (1943), 273–91.

HULL, C. L.; HOVLAND, C. I.; ROSS, R. T.; HALL, M.; PERKINS, D. T.; and FITCH, F. B. [Rote Learning] *Mathematico-deductive Theory of Rote Learning*. New Haven, 1940.

HUTCHINSON, G. EVELYN. [Biology] "Biology," *Encyclopaedia Britannica* (1948).

HUXLEY, JULIAN. [New Syst.] "Towards the New Systematics," in JULIAN HUXLEY (ed.), *The New Systematics*, pp. 1–46. Oxford, 1940.

KAPLAN, A. [Def. and Spec.] "Definition and Specification of Meaning," *Journal of Philosophy*, XLIII (1946), 281–88.

KOCH, SIGMUND. [Motivation] "The Logical Character of the Motivation Concept," *Psychological Review*, XLVIII (1941), 15–38 and 127–54.

LASSWELL, HAROLD D., and KAPLAN, ABRAHAM. [Power and Soc.] *Power and Society*. New Haven, 1950.

LAZARSFELD, PAUL F., and BARTON, ALLEN H. [Qual. Meas.] "Qualitative Measurement in the Social Sciences: Classification, Typologies, and Indices," in DANIEL LERNER and HAROLD D. LASSWELL (eds.), *The Policy Sciences*, pp. 155–92. Stanford, Calif., 1951.

LENZEN, VICTOR F. [EI5] *Procedures of Empirical Science*. EI5. Chicago, 1938.

LEWIS, C. I. [Analysis] *An Analysis of Knowledge and Valuation*. La Salle, Ill., 1946.

LUNDBERG, GEORGE A. [Foundations] *Foundations of Sociology*. New York, 1939.

———. [Measurement] "The Measurement of Socioeconomic Status," *American Sociological Review*, V (1940), 29–39.

———. [Definitions] "Operational Definitions in the Social Sciences," *American Journal of Sociology*, XLVII (1941–42), 727–43.

MacCorquodale, Kenneth, and Meehl, Paul E. [Distinction] "On a Distinction between Hypothetical Constructs and Intervening Variables," *Psychological Review*, LV (1948), 95–107.

Malinowski, Bronislaw. [Dynamics] *The Dynamics of Culture Change*. Ed. Phyllis M. Kaberry. New Haven, 1945.

Margenau, Henry. [Reality] *The Nature of Physical Reality*. New York, 1950.

Mayr, Ernst. [Systematics] *Systematics and the Origin of Species*. New York, 1942.

Merton, Robert K. [Social Theory] *Social Theory and Social Structure*. Glencoe, Ill., 1949.

Morris, Charles. [EI2] *Foundations of the Theory of Signs*. EI2. Chicago, 1938.

Nagel, Ernest. [Log. of Meas.] *On the Logic of Measurement*. (Thesis, Columbia University, 1931.) New York, 1930.

———. [Measurement] "Measurement," *Erkenntnis*, II (1931), 313–33.

———. [Geometry] "The Formation of Modern Conceptions of Formal Logic in the Development of Geometry," *Osiris*, VII (1939), 142–224.

———. [EI6] *Principles of the Theory of Probability*. EI6. Chicago, 1939.

———. [Op. An.] "Operational Analysis as an Instrument for the Critique of Linguistic Signs," *Journal of Philosophy*, XXXIX (1942), 177–89.

———. [Reduction] "The Meaning of Reduction in the Natural Sciences," in Robert C. Stauffer (ed.), *Science and Civilization*. Madison, Wis., 1949.

Neumann, John von, and Morgenstern, Oskar. [Games] *Theory of Games and Economic Behavior*. 2d ed. Princeton, 1947.

Northrop, F. S. C. [Logic] *The Logic of the Sciences and the Humanities*. New York, 1947.

———. [Einstein] "Einstein's Conception of Science," in P. A. Schilpp (ed.), *Albert Einstein: Philosopher-Scientist*, pp. 387–408. Evanston, Ill., 1949.

Ogburn, William F. [Social Change] *Social Change*. New York, 1922.

Ogburn, William F., and Nimkoff, Meyer F. [Sociology] *Sociology*. New York, 1940.

Pap, Arthur. [Anal. Philos.] *Elements of Analytic Philosophy*. New York, 1949.

Peano, Giuseppe. [Définitions] "Les Définitions mathématiques," *Bibliothèque du Congrès International de Philosophie* (Paris), III (1901), 279–88.

Popper, Karl. [Forschung] *Logik der Forschung*. Wien, 1935.

Quine, W. V. [Convention] "Truth by Convention," in *Philosophical Essays for A. N. Whitehead*, pp. 90–124. New York, 1936. Reprinted in Feigl and Sellars [Readings].

———. [Math. Logic] *Mathematical Logic*. New York, 1940.

———. [Dogmas] "Two Dogmas of Empiricism," *Philosophical Review*, XL (1951), 20–43.

Bibliography

REESE, THOMAS W. [Measurement] *The Application of the Theory of Physical Measurement to the Measurement of Psychological Magnitudes, with Three Experimental Examples.* "Psychological Monographs," Vol. LV, No. 3 (1943).

REICHENBACH, HANS. [Axiomatik] *Axiomatik der relativistischen Raum-Zeit-Lehre.* Braunschweig, 1924.

———. [Raum-Zeit-Lehre] *Philosophie der Raum-Zeit-Lehre.* Berlin, 1928.

———. [Experience] *Experience and Prediction.* Chicago, 1938.

———. [Quantum Mechanics] *Philosophic Foundations of Quantum Mechanics.* Berkeley and Los Angeles, 1944.

———. [Logic] *Elements of Symbolic Logic.* New York, 1947.

———. [Probability] *Theory of Probability.* Berkeley and Los Angeles, 1949.

———. [Rise] *The Rise of Scientific Philosophy.* Berkeley and Los Angeles, 1951.

ROBINSON, RICHARD. [Definition] *Definition.* Oxford, 1950.

RUSSELL, BERTRAND. [Math. Philos.] *Introduction to Mathematical Philosophy.* 2d ed. London, 1920.

———. [Principles] *Principles of Mathematics.* 2d ed. New York, 1938.

SARGENT, S. STANSFELD, and SMITH, MARIAN W. (eds.). [Cult. and Pers.] *Culture and Personality: Proceedings of an Inter-disciplinary Conference Held under Auspices of the Viking Fund, November 7 and 8, 1947.* New York, 1949.

SCHLICK, MORITZ. [Erkenntnislehre] *Allgemeine Erkenntnislehre.* 2d ed. Berlin, 1925.

SHELDON, W. H. [Physique] *The Varieties of Human Physique.* With the collaboration of S. S. STEVENS and W. B. TUCKER. New York and London, 1940.

———. [Temperament] *The Varieties of Temperament.* With the collaboration of S. S. STEVENS. New York and London, 1945.

SOMMERHOFF, G. [Analyt. Biol.] *Analytical Biology.* London, 1950.

SPENCE, KENNETH W. [Theory Construction] "The Nature of Theory Construction in Contemporary Psychology," *Psychological Review,* LI (1944), 47–68.

———. [Learning] "Theoretical Interpretations of Learning," in S. S. STEVENS (ed.), *Handbook of Experimental Psychology,* pp. 690–729. New York and London, 1951.

STEVENS, S. S. [Loudness] "A Scale for the Measurement of a Psychological Magnitude: Loudness," *Psychological Review,* XLIII (1936), 405–16.

———. [Scales] "On the Theory of Scales of Measurement," *Science,* CIII (1946), 677–80.

———. [Math., Meas., and Ps.] "Mathematics, Measurement, and Psychophysics," in S. S. STEVENS (ed.), *Handbook of Experimental Psychology,* pp. 1–49. New York and London, 1951.

STEVENS, S. S., and DAVIS, H. [Hearing] *Hearing: Its Psychology and Physiology.* New York, 1938.

STEVENS, S. S., and VOLKMANN, J. [Pitch] "The Relation of Pitch to Frequency: A Revised Scale," *American Journal of Psychology,* LIII (1940), 329–53.

SUPPES, PATRICK. [Ext. Quant.] "A Set of Independent Axioms for Extensive Quantities," *Portugaliae Mathematica*, X, Fasc. 4 (1951), 163–72.

TARSKI, ALFRED. [Logic] *Introduction to Logic and to the Methodology of Deductive Sciences*. New York, 1941.

———. [Truth] "The Semantic Conception of Truth," *Philosophy and Phenomenological Research*, IV (1943–44), 341–75. Reprinted in FEIGL and SELLARS [Readings].

THURSTONE, L. L. [Reliability] *The Reliability and Validity of Tests*. Ann Arbor, Mich., 1932.

———. [Vectors] *The Vectors of Mind*. Chicago, 1935.

——— [Abilities] *Primary Mental Abilities*. "Psychometric Monographs," No. 1. Chicago, 1943.

———. [Analysis] *Multiple-Factor Analysis*. Chicago, 1947.

———. [Methods] "Psychophysical Methods," in T. G. ANDREWS (ed.), *Methods of Psychology*, chap. v. New York and London, 1948.

TOLMAN, E. C. [Op. Behav.] "Operational Behaviorism and Current Trends in Psychology," *Proceedings of the Twenty-fifth Anniversary Celebration of the Inauguration of Graduate Study, Los Angeles, University of Southern California, 1936*, pp. 89–103.

WALD, ABRAHAM. [Statist. Inf.] *On the Principles of Statistical Inference*. Notre Dame: University of Notre Dame, 1942.

WALKER, A. G. [Foundations] "Foundations of Relativity: Parts I and II," *Proceedings of the Royal Society, Edinburgh*, LXII (1943–49), 319–35.

WHITE, MORTON G. [Analytic] "The Analytic and the Synthetic: An Untenable Dualism," in S. HOOK (ed.), *John Dewey: Philosopher of Science and of Freedom*, pp. 316–30. New York, 1950.

WOLFLE, DAEL. [Factor Analysis] *Factor Analysis to 1940*. Chicago, 1940.

WOODGER, J. H. [Ax. Meth.] *The Axiomatic Method in Biology*. Cambridge, England, 1937.

———. [EII5] *The Technique of Theory Construction*. EII5. Chicago, 1939.

WOODROW, H. [Laws] "The Problem of General Quantitative Laws in Psychology," *Psychological Bulletin*, XXXIX (1942), 1–27.

The Development of Rationalism and Empiricism

Giorgio de Santillana and Edgar Zilsel

The Development of Rationalism and Empiricism

Contents:

Aspects of Scientific Rationalism in the Nineteenth Century

Giorgio de Santillana

Der Mensch muss bei dem Glauben verharren, dass das Unbegreifliche begreiflich sei; er würde sonst nicht forschen.—GOETHE

I. Introduction

This essay does not attempt the analysis of a doctrine but rather the life-story of an idea. The great systems of Descartes, Spinoza, and Leibniz belong to philosophy. My concern is to see what scientists did with a certain idea that runs through the systems and which belongs to science from the beginning.

"Rationalism" is a dangerous word because it is used with a variety of meanings. Colloquially, it is associated with the free-thinker, which term rather vaguely connotes a gentleman about fifty years of age or more who objects to people's going to church and who strenuously believes in the dictates of reason, whatever that may mean. Technically, in philosophy, it is defined as the belief in the a priori as a source of knowledge of the external world.

The present essay concerns itself with neither of these meanings. The historical meaning of the word 'rationalism' is chiefly connected with the faith of men like Bayle, Condorcet, and Comte in the emancipation of human reason, or of the generation of Buckle in the rise of liberal institutions. Scientific rationalism has very strong links with these points of view. But it is nonetheless quite distinct from them, as I shall point out. Reason in science and reason in society do work together and sometimes appear as one and the same thing, but really they are quite different in essence.

The kind of rationalism that I wish to discuss here is expressed in the simple, logical certainty of Parmenides and Wittgenstein: what is conceivable can happen. Which seems to tie up with the other certainty: that what happens is conceivable.

Galileo stated the rationalist attitude for his own time in words not to be forgotten:

> . . . . The understanding is to be taken two ways, that is *intensive*,[1] or *extensive;* and *extensive*, that is as to the multitude of intelligibles, which are infinite, the understanding of man is as nothing, though he should understand a thousand propositions; for that a thousand in respect of infinity is but as a cypher: but taking the understanding *intensive*, in as much as that term imports intensively, that is, perfectly some propositions, I say that human wisdom understandeth some propositions so perfectly, and is as absolutely certain thereof, as Nature herself; and such are the pure Mathematical sciences, to wit, Geometry and Arithmetick: in which Divine Wisdom knows infinite more propositions, because it knows them all; but I believe that the knowledge of those few comprehended by human understanding equalleth the divine, as to the objective certainty, for that it arriveth to comprehend the necessity thereof, than which there can be no greater certainty.

By mathematics Galileo understood implicitly the science of physics, since the book of nature "is written in mathematical characters": a belief that was later shared by Newton.

"Fortunate Newton," says Einstein, "happy childhood of science! Nature to him was an open book, whose letters he could read without effort. The conceptions which he used to reduce the material of experience to order seemed to flow spontaneously from experience itself, from the beautiful experiments which he ranged in order like playthings and describes with an affectionate wealth of detail."

The faith endured through many changes right into the nineteenth century—one might even say right to Einstein himself. The scientist might have become philosophically more circumspect, technically more critical; but still he kept on the same line, never assuming that he imposed reason on nature or that he was being vouchsafed its revelation; just discovering it as he worked along, and wondering how it happened, notwithstanding Professor Kant; slightly bewildered, yet doubting not in the least, and finding the whole thing rather natural: "Raffiniert ist der Herrgott, aber boshaft ist er nicht."

Any treatment of scientific rationalism, and especially in the nineteenth century, has to be unmethodical. For rationalism is never in the spotlight; it is what makes the life of method but is outside method, a persistent form of thought which creates the inner tension necessary for the progress of method. It is for the scientist what the feeling of beauty is for the artist, and hardly more communicable.

What, then, is rationalism? Surely it must be some coherent set of ideas. But if it is a system at all, it is an implicit one, a "hidden system" like those Helmholtz had to devise for his thermodynamics. Essentially, it is the scientist himself, creative science at work.

The structure of science is a well-belabored subject. There is a sufficiency of *pièces à thèse* concerning it, as well as of theories on a science of science.

This study is an attempt to treat science not from the outside—to view the scientist not as the technocrat, or as the laboratory artisan, or as the man in search of power over nature, or as the intellectual challenged by a difficult problem, or as the organizer building up a "well-made language" for things; but simply as the seeker after truth.

This may be proved historically to have been the case with most of the great personalities of science; and, while social or psychological motivations may provide a fascinating field of inquiry, there is something to be said for an attempt at considering the creative scientific mind in its pristine candor and originality, to look at the scientist as a free being and a free agent.

The scientific mind, as I hope to show, is ever looking for certain forms of interpretation and a certain type of unity. Of course, the conceptualizing side of science is an aspect which finds its natural complement in the operational. The present survey will therefore have to be one sided and to appear almost partisan. Under the growing complexity of the problems of science, with the decay of the original simple clarity, a reorganization of its concepts and its language may have become a prime necessity. But there is a certain "invincible surmise" which hardly pauses at necessities and carries on with its own imaginings beyond even the bounds of the probable.

Many of the workers of science never care to consider the forms of thought and the consequences which the greater minds among them envisage. But when the wind sweeps over the wheat, the ears bend to the gusts without knowing the rhythm of the wind.

Scientific rationalism is not only the wind of that spirit. It is, as we said, a faith and an implicit doctrine. As such, it lasts until well into the nineteenth century; but it cannot continue as a body of ideas when all its philosophical props are knocked out from under it.

Its life-story is therefore really a tragedy; and my attempt to retrace it took naturally the form of a romantic biography, as it were of a late friend.

Since this is only a rapid survey of some aspects of rationalism, I have kept to the historically central field of the physical sciences, and I had to leave out a highly interesting course of developments in mathematics, on one side, and in biology, on the other, as well as all the problems concerning classification.

The really new factor of the nineteenth century—historical rationalism—ought to have been brought in, since without it the subject loses its proper perspective; but then this essay would have become a book. For the same reason I have concentrated on the period of flowering, and I had to leave out the later crisis of rationalist ideas under the combined influence of Maxwell and Mach in physics and of Darwinism in philosophy.

II. The Crisis of the Eighteenth Century

The eighteenth century was rightly termed "the Newtonian era." A triumphal mood is perceptible everywhere in thought, even if the wiser minds of Berkeley, D'Alembert, and Diderot are aware of an unstable foundation. To the thinking world at large the theory of gravitation had provided proof positive that the mechanical view of Nature, the Galilean and Cartesian "new science," was objectively right. It appeared natural to believe that the world possesses a rational structure, that is to say, that reality possesses an organization coincident with the organization of the human intellect, taking this, of course, in the form of mathematical reason. The words of Galileo which we quoted in the introduction may be held as the creed of scientific rationalism. Rationalism is not simply the discovery that reason is a sound way of dealing with the outer world. As at its birth in Greece, so at its rebirth in the "century of genius," reason was not the free and irresponsible play of ideas but a radical and infrangible conviction that in astronomical thought man was in contact with an absolute order of the cosmos; it is truly a living faith in the illimitable power of the mind, but only in so far as it is admitted that mathematical reasoning releases a transcendent source of certainty which is more than our individual analytical power.

The new science is far from being only a method. The system of dynamics is a complete conceptual world, and it was present already as such in Newton's mind. Its conceptions of space, matter, force, and motion were interrelated in a whole; and, as successive epochs tended to formalize the science, they only brought out the element of high abstraction that was implicit and necessary to keep it together. The new experimental sciences that sprung from the Newtonian trunk in the eighteenth century had apparently little to do with mechanics. Yet even Lavoisier's effort, from the start, toward stripping science of its unavowed metaphysics and reducing it to a "well-made language" could not rid him of his personal presuppositions, which appeared to him a matter of sheer common sense, as they had already appeared to Condillac. Lavoisier works on the material arranged by the experimentalists who came before him. Such material inevitably implies representation, schemes, and structures; as he found out, it becomes a system "before you can do anything about it." The attempt is truly to converse in real things, like the academicians of Gulliver's Laputa; but it cannot be carried through. Reliance on the balance is not so commonsensical as it might appear, nor did the contemporary chemists, even men like Black and Priestley, admit it to be so. It is something like "flying by instruments." Lavoisier had avoided the confusion and the difficulties of the Newtonian chemists who had tried to explain by attraction the forces of affinity, but he had held fast to the fundamental idea: to explain things in terms of matter only. Not only does he leave inside his system a substantial caloric, which is actually fire matter, but, the more methodical he is, the more he accepts the essential limitations of Newtonianism. He used to say that only physicists understood what he was after. The whole of his extremely positive thought seems unified, both in what he does and in what he leaves undone, in the Newtonian definition: "All matter is heavy in proportion to its quantity, which remains forever constant." And, yet, the very substantial clarity that brings him on a common ground with his friend Laplace entices him into the forbidden realm of images. It seems natural to him to speak of bodies as made of molecules which do not ever touch each other and are kept apart by varying quantities of caloric fluid; of the gaseous state as element plus caloric. These images are not followed up with deductive rigor; but it is significant that they are found right in the creative region of his thought, where theory and experiment meet, where imagination has to deal with the fundamentals of light, heat, matter, continuity and discontinuity. The system that emerges is born of many disparate motives, of intuition, logical clarity, and operational procedure inextricably interwoven. And the thread that guides him is the Newtonian idea. When Berthollet founded chemical statics, he was consciously taking up another strain of the Newtonian system. So was Franklin, for that matter, when he accepted a principle of conservation and admired, while he refused to worship, that "Wisdom which had made all things by weight and measure." Thus we see that the several sciences which are born toward the end of the eighteenth century are the result not of simple method but rather of a central, unifying inspiration proceeding from celestial mechanics and broken down into usable aspects; and that vision still provides a hidden link that will facilitate their interchanges and fusion later on. The basic images are only apparently clear; they are a mixture of quantity and quality, matter and geometry, flexible for any development. Its rigor is not the Cartesian rigor, for that is already blighted at the root, since Cartesianism insists that space is divisible and then talks of indivisible atoms, which surely cannot be. It is Newtonianism more than logic which holds the key to unity; for, if the unity sought after was still deductive, the unity that is established is really analogical, as Newton had seen it:

> For it's well known, that Bodies act upon one another by the Attractions of Gravity, Magnetism, and Electricity; and these Instances shew the Tenor and Course of Nature, and make it not improbable that there may be more attractive Powers than these. For Nature is very consonant and conformable to her self.[2]

The rationalistic postulates are redefined here in a very significant way, and that is how they are understood by Newton's successors. There never is a "strict induction" but contains a considerable amount of deduction, starting from points chosen analogically.

Strict and universal deduction, of course, is still held to be the ideal. A generation later Laplace still believed in it, even if he perceived the difficulties more clearly:

> We ought to be able to explain simply through the variety of molecular forms all the varieties of attractive forces, and thus bring back to one general law all the phenomena of physics and astronomy. But the impossibility of knowing the shapes of the molecules and their mutual distances makes such explanations vague and useless to the progress of science.

Once research had moved beyond Newton's field of action, the vast dreams of a science unified by the "géomètres," and deduced directly from the law of gravitation, had proved no more easy of achievement that Diderot's dream at the other pole of thought, derived from Leibniz's intuition of the continuum. But we see how strictly Laplace adhered to the original canon of explaining things "par figure et mouvement." Franklin, the champion of induction, imagined the particles of electric fluid grouped in the form of triangles to "explain their compressibility." Schelling was fully justified in pointing out that the Lucretian ideal of science still held—that all its theories that tend to explain quality in substances can be reduced, except for straight analytical formulas, to attempts at "expressing qualities through figure, i.e., substituting some geometric figuration for each primordial quality."[3] Laplace's faith is all the more remarkable if we consider that fully a generation before, D'Alembert, his immediate predecessor, the man who had done most to apply analysis to mechanics, had grown to be the most skeptical of the future of this attempt:

> After having reflected for a long time on this important matter [the resistance of fluids], it has seemed to me that the small progress achieved is due to the fact that we have not yet understood the true principles according to which it must be dealt with. Therefore I sought to apply myself to seeking these principles and the way of applying, if possible, calculation to them. For these two purposes must not be confused, and modern geometricians have not perhaps paid enough attention to this point. It is often the desire to be able to make use of methods of calculation which determines the choice of principles, whereas the principles themselves should first be sought without thinking in advance to bend them forcibly to methods of calculation. Geometry which should only obey physics, when united to the latter, sometimes commands it.[4]

III. The Formula as Transcendent

What D'Alembert seemed to fear took place nevertheless. And, by an irony of history, he was to be among those mainly responsible. Out of the complexity of Newton's thought and background one aspect had been given special importance—the distinction between mathematical description and philosophical explanation. But it was difficult not to give to the mathematical expression, once arrived at, some causal dignity, which should be in harmony with its striking simplicity. Huygens, Euler, Le Sage—indeed, Newton himself[5]—had insisted that a purely mechanical explanation was desirable (like ultramundane corpuscles) which should make the simplicity of the gravitational formula appear irrelevant. But the very success of Newtonianism had been too much for the out-and-out mechanists. Before Newton showed how far a simple formula could go in linking facts which were apparently disconnected, no one could have suspected that this would be possible; when philosophers realized the power of the Formula and foresaw that power extending to electricity and magnetism, an opposite movement set in through which mathematical processes were focused upon at the expense of experiment and observation and also at the expense of philosophical reasoning. D'Alembert's misgivings were ignored by his geometrical colleagues, and the mathematical formula took on something of the prestige of a cause.

This intransigent attitude resulted in an antimathematical reaction on the part of naturalists and thinkers at large, who felt that the world was too deep and varied to be written off in a formula, and one impossible of application at that. Still we should note that, when, in 1747, Clairaut suggested that it might be necessary to add a factor of correction to the Newtonian formula to make it fit the annual motion of the moon's apogee, it was Buffon who protested most loudly, for metaphysical reasons, against the infringement of the simplicity of the universe. Thus, whatever the trends of fashion, the charm of mathematics as "divine philosophy" still held sway; and the discussion of the foundations of mechanics still centered on the principle of least action as the one most likely to provide the key to the plan of the universe.

IV. Mechanics and the Nature of Matter

The most significant effort in this direction is that of Euler. Following his discoveries on the maxima and minima,[6] he wrote:

> There must be a double method for solving mechanical problems: one is the direct method founded on the laws of equilibrium or of motion; but the other one is by knowing which formula must provide a maximum or minimum. The former way proceeds by efficient causes, the latter by final causes: both ways lead to the same solution, and it is such a harmony which convinces us of the truth of the solution, even if each method has to be separately founded on indubitable principles. But it is often very difficult to discover the formula which must be a maximum or minimum, and by which the quantity of action is represented.[7]

At this point Euler reaches again beyond the rather cramped principle of economy of Maupertuis, the range and flexibility of Leibniz's original principle of maxima and minima. Euler's actual discovery was that the differential equations of the motion of a particle are given by the simple requirement that the integral $\int v \cdot ds$ taken over two positions of the particle should be a minimum.[8] But we are not concerned here so much with the importance of the principle in the history of mechanics. The really relevant point lies in the phrase, "both ways lead to the same solution, and it is this harmony which convinces us of the truth of the solution, even if each method has to be separately founded on indubitable principles." What appears here, more than a century after Galileo, is a still quite complex conception of the meaning of truth. It would take a miraculous adjustment and parallel development of the two ways of knowledge in one mind for an investigator to carry out correctly the program contained in that sentence. Actually, when Euler found great difficulties in extending his principle from isolated bodies to systems (which was to be achieved only by Lagrange), he unashamedly sought a short cut: "Since motions of that sort are not easy to reduce to calculation, it will be understood more easily starting from first principles."

It was D'Alembert who brought back strict scientific reason—or at least what he felt to be such:

> The great metaphysical problem has been put recently: are the laws of nature necessary or contingent? To settle our ideas on this question, we must first reduce it to the only reasonable meaning it can possibly have . . . . viz. whether the laws of equilibrium and motion that we observe in nature are different from those that matter would have followed, if abandoned to itself. . . . . Hence this is the way the scientist should follow: first he should try to discover through reason alone which would be the laws of mechanics in matter abandoned to itself; then he should investigate experimentally what are really such laws in the universe. If the two sets of laws be different, he shall conclude that the laws of mechanics, such as those yielded by experiment, are of contingent truth, since they would appear to spring from a particular and express decision of the Supreme Being; if, on the other side, the laws yielded by experiment agree with those that could be deduced by logic alone, he shall conclude that those laws are of necessary truth: which does not mean that the Creator could not have established a wholly different set of laws, but that he did not hold it right to establish other laws than those which resulted from the very existence of matter.[9]

An attitude has been enunciated which is going to be the coherent standpoint of mechanistic philosophy from then on. Both Descartes and Leibniz had seen the laws of nature as a choice among an infinity of possibilities, arising either out of an arbitrary creative act or out of a reasoned determination in favor of the greatest variety of co-possibles. With D'Alembert, the laws of mechanics become necessary, in that they are just those which would arise simply out of the existence of matter left to itself. The divine choice is set back to a moment previous to the existence of matter—in other words, it is bowed out.[10] A new universal deduction is taking shape, with the "existence of matter" as a starting-point. But matter could hardly lend itself to the role of first principle without being lifted from its empirical status by an unnoticed regressive deduction. D'Alembert thought he had cut the ground from under metaphysics; actually, he had replaced the divine will with a not-too-clear entity called the nature of things and was resting his case upon a different, but no less remote, principle of explanation. And what did not follow from the nature of matter had to follow from the principle of sufficient reason.

The explicit recourse to this principle is among the most ancient and permanent characteristics of science. Anaximander and Archimedes made use of it. But it is more of the nature of a limiting principle than of a creative one. It has been exorcised in modern times and given a positive heuristic value in van't Hoff's scheme of stereochemistry and in Curie's principle of symmetry. In these cases it means nothing more than a condition for the construction of models.

If for D'Alembert sufficient reason appeared such a strongly positive principle of knowledge, the reason for this should probably be sought in the social rationalism of the *philosophes*, which, having done away with all metaphysical structures, considered sufficient reason (a "good" abstraction) a norm for human relationships and hence, by an unconscious extension, an active law of nature. Scientific rationalism, however, was quite different from social rationalism; whatever its representatives may have believed, it was not dependent on such ontological abstractions as sufficient reason. Its real, though hidden, life lay rather in the "nature of matter" which to D'Alembert and Laplace appeared to belong to mere common sense. Under an inconspicuous appearance, the notion of matter, like Euler's faith in conspiring levels of truth, carried in itself a latent universe.

V. Formalization of Mechanics

Faith was still firm at the beginning of the nineteenth century. Laplace conceived that a universal mind would be competent to foretell the progress of nature for all eternity if only the masses of all bodies, their positions and initial velocities, were given. A dream—but not an absurd one. Matter presented no problem, and analytical mechanics had provided a scheme of incomparable economy.

On the other hand, as Mach said, no fundamental light can be expected from this branch of mechanics. The choice of a basic principle is only one of convenience. As Gauss pointed out in 1829, apropos of his own principle of least constraint, no essentially new principle could be established in dynamics: it was only a matter of new points of view; and, while a clear perception of the relativeness of the various points of view to one another might be fruitful enough, it left some serious problems unanswered. Mechanics is made by mechanicians, and in the first years of the century a very eminent one, Poinsot, voiced his misgivings apropos the principle of virtual velocities:

> We naturally believed that the science was completed, and that it only remained to seek the demonstration of the principle of virtual velocities. But that quest brought back all the difficulties that we had overcome by the principle itself. That general law, wherein are mingled the vague and unfamiliar ideas of infinity, small movements and perturbations of equilibrium, happened somehow to grow more and more obscure upon examination; and the work of Lagrange supplying nothing clearer than the progress of analysis, we saw plainly that the clouds had appeared to be lifted from the course of mechanics only because they had, so to speak, been gathered at the very origin of that science.
>
> At bottom, a general demonstration of the principle of virtual velocities would be equivalent to the establishment of the whole of mechanics upon a different basis: for the demonstration of a law which embraces a whole science is neither more nor less than the reduction of that science to another law just as general, but evident, or at least more simple than the first, and which, consequently, would render that useless.[11]

Poinsot saw clearly, and he was resigned. But he did not like the situation as much as, perhaps, a Condillac or a Lavoisier might have liked it. Even quite positive minds may not like to be confronted with the full implications of their positiveness. The principle of least action might well have been reduced to a formal status, yet there was still some vague hope that these principles of minimum would turn out to mean *something*. "There is undoubtedly," said Heinrich Hertz later, "a special charm in such suggestions; and Gauss felt a natural delight in giving prominence to it in his beautiful discovery of least constraint. Still, it must be confessed that the charm is that of mystery; we do not really believe that we can solve the enigma of the world by such half-suppressed allusions." Poinsot, like Gauss, still felt the charm of illusions lost and could not be satisfied with the present. He already knew what Mach was to state later: that the principle of virtual velocities had its best possible foundation in the impossibility of perpetual motion, and *that*, of course, could not be merely deduced from the principle of causality. It was a fact and, what is worse, a negative fact. "One fundamental fact," warns Mach, "is not at all more intelligible than another: the choice of fundamental facts is a matter of convenience, history, and custom. . . . . It is a result of a misconception to believe, as people do at the present time, that mechanical facts are more intelligible than others, and that they can provide the foundation for other physical facts."[12] This was written in 1871, but, as we see, the problem had been staring scientists in the face as far back as 1800.

Yet the whole ambition of the mathematical school had been to reduce physical reality to those mechanical, but conspicuously positive, factors for which astronomy provided the paradigm. Newton had entertained strong doubts as to the applicability of his laws to molecular dimensions; but of those doubts nothing had resulted except a riot of arbitrary assumptions and no real progress for knowledge. When Laplace, very much aware of the danger, had succeeded in deducing the laws of capillary action from the gravitational formula, all doubts and vagueness were removed.[13] With Coulomb and Ampère the law was extended to electricity. Its generality appeared necessary by now, and the regressive deduction became institutionalized. As late as 1891 Kundt could still write: "We feel it difficult to step out of this circle of ideas."[14] Science had put herself under the spell of the one formula. With the discovery of Neptune as the result of pure calculation, it was clear that there was recompense for such dedication; and the "triumph of rationalism" was brought home to the thinking world. "Descend from Heaven, Urania," intoned Dr. Whewell. And he could write with simple assurance: "We have now to contemplate the last and most splendid period of the progress of astronomy . . . . the first great example of a wide complex assemblage of phenomena indubitably traced to their single, simple cause: in short, the first example of the formulation of a perfect inductive science."

Although humbly proud, he was a shade puzzled by success: "It is a paradox that experiment should lead to acknowledged universal truths and apparently necessary ones, such as the laws of motion. The solution of that paradox is to be found in the fact that the laws are interpretations of the axioms of causality . . . . our idea of cause provides the form, and experiment the matter of the laws."[15] This is very reassuring, and we see how even Kantianism can become an armchair. Rationality obviously belonged to the scheme of things for our convenience, like many other paraphernalia of the Victorian age. But induction was the work of man; it bore the truly moral hallmark of success.

Once they felt the goal had been reached, people could ascribe it to induction or to deduction, according to their temperament. We shall have to investigate later the theoretical reasons for this permanent equivocation. But it has an important historical aspect. Good progressive minds were no longer taking rationality in the spirit of a solemn revelation or a lighthearted adventure, as they had done in the preceding centuries. Somehow, they took it for granted, and celebrated the soundness of Method, while even in mechanics the more philosophical minds, such as Gauss, Poinsot, Riemann, or Helmholtz, conceived growing doubts as to the substructure. The nineteenth century had brought science, and not simply reason, to the public mind. Truly it was, as it claimed to be, a scientific century. The *idées claires* in Whewell's time were—science as it stood.

Actually, the Kantian armchair of stable forms and categories was getting more and more rickety, and non-Euclidean geometry had strongly contributed to its undermining. It was proved that mind had not one "form" of space only to superpose on matter but a multiplicity of forms, flowing into one another, and that the Euclidean one was a choice and not a necessity.[16] Riemann had seen the whole situation at a glance in the fifties.

As for the "indubitable" assumptions of mechanics, they had reached a state of dangerous abstraction which was revealed in its full extent by relativity. We need only to enumerate the chief ones:

1. Assumption of an objective local time, connected with a closed mechanical system endowed with periodicity (clock). This implies, as Einstein was to show later, an initial definition of space by means of rigid bodies.

2. Introduction of the concept of objective time for happenings over all of space: whereby local time is generalized to become the time of physical theory. To one unaware of the operational point of view, the finite velocity of light causes no difficulty; there is no need for a *Gedankenexperiment*. Empirically, events seen at the same time could well be called contemporary.

3. Concept of the material point: a bodily object which can be satisfactorily described in regard to position and movement as a point with three coordinates.

4. Law of motion for the material point, linking force, mass, and acceleration; which becomes the law of inertia when the components of acceleration vanish, i.e., when the point is sufficiently far from any other point.

5. Law of action between material points. These laws were provided empirically by Newton. But in Newtonian mechanics, derived from astronomy, the space K_0 comes in with a new aspect, not contained in the above-mentioned conception of space, which derives, as we have said, from rigid bodies. In this kind of space, condition No. 4 is not valid for any K_0, but only for such K_0 as we call 'inertial systems.'

Thus the frame of reference takes on an independent physical property which has nothing to do with geometrical space proper. This is not a theoretical refinement; it brought up grave difficulties right away, of which Newton was well aware. It was the problem of the whirling bucket, which appeared hopeless until

Mach suggested that the frame of reference of mechanics might not be "absolute space" but the star system as a whole.

Scientists felt it as a disharmony that this system of abstractions should still be founded on the crudely empirical laws of force, and Laplace insisted that the inverse-square law derived a character of necessity from being linked with the very nature of Euclidean space. But, actually, Newton's law stands out among other laws that might be imagined, by reason of its success alone.

Apart from those laws, there is not much in mechanics that might be called inductive. It has to proceed on the concept of the material point, and indeed on the concept of the unextended center of force, as systematized by Cauchy. And, since material objects almost never get anywhere near being such points, the problem is always of a resolution of the given objects into points and forces between them.[17]

It is natural to suppose those material points invariable, as well as the laws that link them. The result is a completely atomistic picture of matter. Such a picture is one of the fundamental scientific a prioris, as we hope to show in the following section. It has succeeded in providing the needed laws of matter by drawing upon celestial bodies as models of atoms. And now the theory goes back to its origins—but in the rather strange forms of materialism without matter, of causality devoid of substance. Such is the paradox of regressive deduction.

VI. The Atomic Idea

The scientist always searches for laws, but he wants to think in terms of things. These things must be real, that is, he must be able to believe in them. Hartmann once pointed out that the scientist who should set out to describe the world in terms of objects that he knows to be untrue would admittedly belong to the lunatic asylum. Whatever the limitations and conventions attached to the words 'true' or 'real,' it is clear that the thing which is the principle of explanation must command at least provisional belief.[18] Furthermore, as we said before in criticizing Mach's statement, it ought to be a positive and not a negative fact. It makes all the difference between the old words *scientia* and *cognitio.*

Every nascent scientific explanation will, therefore, be realistic and not nominalistic. The great theories of the nineteenth century give ample proof of this. Such a realism is not a mere fleeting overture; it accompanies them throughout the creative stage, and its struggle to survive against contradictions and to emerge from the steadily complicating network of experimental relations closing in on it is the very life-story of the theory.

There is still, in the background of all explanation, the old unsolved identity between "mechanical" and "natural." Whether it be sought in material models or in the general analytical forms of theorems on potential, there is the attempt at a true picture, and at one universally true. After D'Alembert, abstract necessity has lost its hold (see p.760), science goes in for "contingency" or "immanence": the order of nature is supposed to express the characters of the real things which jointly compose the existences to be found in nature.[19] But in the *materia signata* of this order there are still hidden all the metaphysical characteristics that go to make an essence, as Laplace had truly called it. Simplicity, the seal of nature, is impressed on it; and so are the laws of continuity, of analogy, of identity, that are to reconstruct a unity out of the scattered pieces of the puzzle.

Theories may come and go, but this it is that makes the atomic idea a central motive of science. Atomism is the recurrent image that has dominated scientific thought for over two thousand years. If anything were more remarkable than its purely a priori origin, it would be its enduring through the ages as a strict matter of faith, its capacity for drawing new life from seemingly unrelated discoveries, and its ultimate vindication. Yet even in its principle it bears the whole drama of rationalism.

The statement of Democritus that all qualities are illusory and "in truth nothing but atoms and the void" was a decisive stroke of simplification. It showed the way both to a clear explanation of nature and to one which could be mathematized. It was the abstract belief of Democritus, as far as possible from experimental verification, which inspired Galileo, Boyle, Bernouilli, and Newton.

The appeal of the atomic idea lay in its rigorous simplicity. Yet that simplicity was deceptive. With the atom a highly complex unit had entered the stream of thought. On one side, it was a physical entity; on the other, a mathematical one. It carried with it all the implications of the Eleatic view of nature from which Democritus had sprung. Whitehead has justly if paradoxically observed that the Void of Democritus, the Space of Newton, and the God of Leibniz are one and the same thing. It is this hidden character, this concentration of properties, that made the void so terribly embarrassing to the Greek philosophers. The contrary intuitions of metaphysical plenitude and of physical void, both linked with our experience of space as a whole, have a confusing and sometimes exasperating way of exchanging aspects in a philosophical quadrille which lasts up to Leibniz and beyond.[20]

Such a permanent equivocation arises, as always, from unexpressed content. The abstract principles that we tried to outline previously must coexist with a simple faith in substance, and that substance must be matter[21]—such a matter as can be invariable and forever equal to itself, since it must represent eternal and necessary Being itself.[22] Such is the essence of Galilean thought, and such is the stated content of the image. The decisive character of this particular conception of substantialism is the additional requirement that the location of substance should be always indubitable through the changes undergone by a system of bodies. The whole of the measuring technique is directed to ascertaining such locations in the flux of phenomena. In other words, there must be some essential system of reference. This is the hidden postulate of any geometrization of the real. The void of Democritus, although bereft of the quality of Being, is the equivalent in this of Newton's absolute space. They both express an intrinsic necessity of spatial designation or structure, which finds a correct expression at last in general relativity, as the texture of world-lines structuring the continuum.

For twenty-five centuries all this content was packed in a simple image—the atom of Democritus. It is materialized logical simplicity. It inherits the implications of Eleatic unity, but it is essentially a unit, and a unit infinitely repeated. It is substantialism reduced to small change.

Genuine realism has only one epistemological function: it refers all qualities to substance. Substance is real; phenomena are also real. All qualities are illusory which are not possessed directly by the substance itself. Mabilleau, at the close of his history of atomism, is led to a plain atomistic profession of faith: "The progress of science consists in linking the outward manifestations of matter to its internal structure, so as to establish, through the interdependence of both orders, the unity of the law without which there is no true explanation." This is practically an identification of structure and phenomenon. W. Thomson and, after him, Helmholtz both observed that atomism cannot "explain" any properties except those which are attributed gratuitously and a priori to the atom itself. In the seventeenth century the transfer is immediate and naïve. Acids have a prickly taste, so their atoms must be pointed. Still more: "Atoms constitute the immutability of elements, they cause fire to be always fire, water to be always water, and the imperceptible germs that form a man never to form a bird."[23]

The traditional atom, then, contains in itself all that the scientist needs to imagine of reality. It traverses the ages laden both with the physical and with the mathematical possibilities that wait to be developed into an explanation. But here the difficulties begin.

Let us examine them first in a strictly theoretical manner as they came up before experiment forced determinations upon them.

Rigorous and naïve substantialism leaves all the modes of substance undetermined—*in mente materiae,* if we may so put it. Substance is an absolute and ought not to bear determination from any outside factor. Why should such a substance, manifesting itself solely out of its inner power, go twice through the same modes under equal conditions or give rise to the same phenomena? If, to take Voltaire's example, fire is an unrelated absolute, why should it behave always as that which we call 'fire' and not as something else? If Peirce's tychism had been there at the very start, people would have stopped trying to explain natural phenomena. Why should we look for diversity elsewhere than in protean substance itself, and why, in particular, should we look for it in outward agents which cannot influence such an autonomous substance? Unless, indeed, substance has attributes which allow for such action. In the end, if we want to build up a science and not an animism, we are left with only one choice, which is the historical one: the atom must be quite dead, its substance devoid of all spontaneity. Realism starts instinctively from a cosmos "which is itself a god" and ends in the most "thingy" thing it can conceive of. On that thing, endowed with only the primary qualities, the science of numbers and extension can operate. Galileo takes up again where Democritus left off, after many centuries of interlude in metaphysics of the soul. But to have science operate, to bring out constant laws arising from constant relations, we must break down that one Thing into so many replicas of itself. And to have that breakdown not a mere arbitrary action of our mind but an objective reality, there must be a limit to it. Only thus can the real be defined outside of our action on the phenomena. If division could be infinite, substance would become itself a veritable phenomenon and escape our analysis in the illusion of an unseizable composite. We would be sitting on one of the horns of the Eleatic dilemma: the World of Opinion. If, on the other hand, reality were one and intangible, there would be no phenomena. We would find ourselves on the other horn: a world that may be true but does not look real.

Atomism is the one way out toward a science;[24] it is a naïve materialism that has become precise so as to be amenable to mathematics.[25]

In Descartes's kinematical universe atoms are merely extension and shock. The purism of such a scheme allows them to be the subjects of only one law—that of a constant quantity of movement. But even that is so arbitrary, the idea of spontaneous movement is so obviously not contained in that of extension, that the occasionalists find it more proper to situate the laws of shock "outside." Atoms cannot even act on one another. They become a cipher, a mere pretext for the law.

Atomism is stripped to the bone; but it is stripped also of its *raison d'être*. Still, Newton has to accept the principle: the enduringness and permanence through change that distinguishes nature does not appear to him explainable otherwise than by particles "solid, massy, hard, impenetrable," coming together and separating again. But there has to be a reaction against Cartesianism, and the Newtonian reaction leads toward a richer atomic substance, endowed with both geometric figure and dynamic properties, essentially attraction; but also magnetism, electricity, affinity, even light. Thus, though against Newton's better judgment, and with much hesitation on his part, atomism was drifting back toward explanations through inherence. For chemists, such explanations were obvious because they were necessary. We have quoted Laplace's doubts on this point. Notwithstanding his doubts, Laplace was still ready to condemn the undulatory theory of Young and Fresnel because it discarded the particle.

But now, apart from vagueness, much more grave and serious difficulties arose. The finite size and the inner cohesion of the atom had never been a problem to the Cartesians, who thought in terms of "figure and movement." Form and substance went together. The atom's geometrical surface was an integrant part of its being: Cartesianism still thought, although it did not care to realize it, in substantial forms. In the Newtonian scheme, however, the atom was dynamic matter. It had properties; that is, things were supposed to happen inside it. Its cohesion had to be explained, and this meant an inner force tying its parts together. The atom was embarrassingly resolving itself into subatoms, and the question started afresh on that level. One had to attribute some kind of "absolute" hardness to those subatomic particles or to go into a regression *ad infinitum*. If scientists wanted to avoid Aristotelian types of explanation, they had to go to the limit and define the atom as a pure mathematical point, for which action by contact was unthinkable and whose only character was force, acting at a distance. Such was Boscovich's theory, the theory that Cauchy took up and extended as the fundamental picture of mechanics. The astronomical view of nature and the atomic view converged under the requirements of dynamics into a materialism devoid of matter, whose only realities were points linked by forces. It would perhaps be more accurate to say that this offshoot of the atomic idea was absorbed by the astronomic system and became formalized along with mechanics. The atomic idea of matter has culminated again in a paradox: it started with a substance that was a cause and it ends in a cause without substance.

VII. Atomism and Central Forces

Such a paradox is of the essence of the problem we are analyzing. We are here at the heart of the rationalist process, and we can see the form it has taken under the influence of celestial mechanics. After watching developments in the peripheral zone of the budding sciences, we are led by atomism back to the central model of explanation that had been set as the ultimate goal of *scientia*. "Understanding," said Cuvier, "that is, explaining in astronomical terms. . . . ."

The unity of the understanding does not rest any longer in a deduction from abstract principles. It is based on the "essence of matter," whatever that may be, as revealed by astronomy. In other terms, the facts of celestial mechanics, organized into a scheme of matter and force-at-a-distance, are conceived as some kind of a priori. And this is hardly to be wondered at, since that particular tie of geometry with mechanics was established by Newton himself.[26] The inevitable consequence is that, once we step beyond the empirical field of validity, the scheme will not hold together. The structure of matter cannot be simply equated to a solar system; but the scientist is willing to risk any paradox at any scale of dimensions. For he thinks by now that both matter and central force are of the essence of nature.

The unity of deduction had been once replaced, at Newton's suggestion, by analogy. But the uncertain web of analogies spun over physical reality tends to converge, all difficulties notwithstanding, toward the central analogy, and that analogy tends in its turn to become an identity.

Weber, referring modestly to his contribution, says:

> After the general laws of motion had furnished a foundation, there remained in physics mainly the investigation of the laws of interaction of bodies. . . . . For a long time Newton's doctrine of gravitation furnished the leading idea for nearly all theories of electricity and magnetism, till a new clue was gained through Oersted's and Ampère's discoveries regarding the equivalence of closed electrical currents with magnets. This led, first, to the reduction of all magnetic effects to the action of electrical currents; and secondly, to the enunciation of a fundamental law of interaction of two elements of electricity in motion. A third leading idea was that of reducing the interaction of all bodies to that of the mutual action of pairs of bodies.[27]

In fact, Weber had been brought to attribute to electricity something like inertia (and this was the origin of Hertz's researches). But it is only by extreme abstraction[28] that we can speak of electrical masses, be they one or two; of elements of current; of velocities of a something which as yet cannot be clearly defined. As Carl Neumann tried to put it safely: "Electrical matters never exist alone, but only in combination with ponderable matter." Heat was another such stuff, whose particles repelled each other. As to ponderable matter, it had no definite range of dimensions and could well carry within itself the other observed forces. Out of its very uncertainty analogy has fled into identity. An abstract substance, a projection of matter, equal unto itself in all modes: such is now the "nature of matter."[29]

The factual law, in itself a dead datum, provides the starting-point for the twofold imaginative process, converging toward identity and expanding toward generality. It is as involuntary as the rhythm of the heart. Before he knows it, the scientist has taken the decisive step; thus, when the problem of cosmic dimensions comes up in the mind of Laplace, it is settled without a doubt. A model universe, he remarks, "reduced to the smallest space imaginable, would have to offer always the same appearances to its observers."[30] Later, the more cautious Helmholtz was led to state apropos of the laws of electricity: "If we are to consider Weber's law as an elementary law, as an expression of the ultimate cause of phenomena, and not merely as an approximately correct expression of fact within narrow limits, then we must demand that, if applied to objects of the largest imaginable dimensions, it should give results which are physically possible."[31] In the same way Democritus, the earliest of physical rationalists, had said: "There is no reason why there should not be somewhere atoms as large as a world."

But within what range of dimensions is matter still matter? Among the many paradoxes of an infinite universe, there is also this one of the lack of absolute dimensions, which was accepted cheerfully, as it confirmed the recurrent analogy of nature suggested by Newton. But in this way an empirical law is well on its way toward becoming itself an absolute, all the more so in that it has been successively extended to the field of electricity, thence to that of electrochemistry by Berzelius, and, through sheer analytical ingenuity, to that of electromagnetism by Ampère and Weber.

The process of unification may take any of a number of ways toward its goal. Here, having to face a really unresolved multiplicity of forces, it is only natural that it should tend to ignore it and concentrate on the law itself, which is an abstraction in regard to the nongravitational forces, but only in regard to those. We are still within the frame imagined by Newton.[32] The embarrassing pictures of electrical fluids and suchlike having been abstracted out of the way, the theory stabilizes itself on the level of two seemingly intuitive representations—matter and force. The mathematical treatment is a powerful determinant in this, since it provides the accomplished mold of the equations of mechanics. Thus the substantialism inherent in science had found a resting-point: a labile one in the dialectic of rationalism, perhaps, but strong by virtue of its ancient prestige and of the highly rationalist pattern in which fact and theory were welded together. And thus the mechanics of central forces was able to weather the age of positivism and the difficulties connected with its extension to the intra-atomic field. Corrections had to be brought in; but the pattern survived, and without it science would have been utterly without a lead when it tackled the inner structure of the atom. Even today quantum mechanics has to start from the Hamiltonian equations.

When H. A. Lorentz was searching for the electromagnetic laws of the theory of the electron, he started from Maxwell's equations, that experience had shown to be well grounded, and eliminated all the quantities in which the influence of matter is revealed by material constants. He assumed that the "true" or microscopic electromagnetic field obeys these simplified harmonic laws, and he was proved to be right.

By this time the original astronomical idea of law has undergone a last and significant transformation, through which it enters modern physics. Substantialism has been reluctantly cast overboard. The exact laws of nature are linked only to atomic structure, and only from it can they derive their constants.

To recapitulate, dynamic atomism had proved intrinsically unstable. Against the wish and the imagination of its nineteenth-century supporters, it developed into a field theory, where "matter" became a substantial shadow, a counterpart set up for action. The unresolved duality provides both the pattern and the fundamental irrational. "Science," says Helmholtz, "considers the objects of the external world under two types of abstraction: on the one hand according to their mere presence, and independently from their action on other objects or on our organs of sense: and this it calls matter." On the other hand, we can attribute to matter the capacity for action, and then we shall know it only through its action. "Pure matter would be indifferent to the rest of nature, for it would never be able to determine any modification in it or on our organs. Pure force would be something that has to be there and yet is not there, since we designate to-be-there as matter."

VIII. The Embarrassment of Reality

From the point of vantage of these later developments, both in mechanics and in electrodynamics, we can better survey the critical period in which "real" representations try to fight their way through progressive abstraction and the all-embracing theorems of the science of energy.

Toward the middle of the nineteenth century it was obvious that the astronomical image had worked itself into a tangle, while, on the other hand, it was becoming clear that the atomic view had a real physical foundation and must be brought back to a physical image. Maxwell was working out the kinetic theory of gases on the assumption of hard elastic particles and meeting with almost complete success. In his second paper he concluded:

> But who will lead me into that still more hidden and dimmer region where Thought weds Fact—where the mental operations of the mathematician and the physical action of the molecules are seen in their true relation? Does not the way to it pass through the very den of the metaphysician, strewed with the remains of former explorers and abhorred by every man of science?

It did. Certain inevitable discrepancies between theory and observation could not be solved—nor could the a priori element be repressed—until the advent of the quantum theory. Meanwhile, atomism groped its way through the den clad in the protective garb of Kantian criticism,[33] until it emerged safely in modern physics, where the existence of ultimate particles is established, but on a level where operational realism and axiomatic nominalism are scrambled together so as to make them foolproof against the metaphysician. The electron is no longer a coherent image; it is the support for a certain number of more or less co-ordinated behaviors revealed by experimental techniques.

But to come back to the punctiform atom as imagined, let us say, about 1840. We can see how the extremely elusive concept provided by the physics of central forces was the meeting-point of several abstract lines of reasoning and, as such, was simply an intimation to science to go forward toward either a stricter analysis or a more coherent representation.[34]

The atom is postulated as an active unit. It is less the unity of a figure than the unity of a force. Still it had a substance, but no one knew what to do with it. From this mongrel being (which was to reappear in the first representations of the electron) we can either go back to substance or go forward to construct a coherent dynamic theory of matter.

Both these ways were followed, and almost at the same time. The first way, back to the physical atom, was taken by the chemists. The issue was long beclouded by the obstinate struggle of the positivists against what they considered a scientific regression; but, in its heuristic thought, chemistry, since Dalton, lived by simple belief in the substantial atom. It has been said that Dalton invented atomism only to give a convenient form to the law of multiple proportions. Lange and Kirchberger, however, have shown by the documents that Dalton was guided by the fundamental atomistic intuition of Newtonianism in organizing his research. Richter had also been looking at the same time for simple numerical proportions. "But," says Lange, "while Richter promptly jumped from the observed constancies to the most general form of the idea, and concluded that all phenomena in nature are ruled by measure, number and weight, Dalton was looking for a sensible representation, and the atomistic one came to meet him half-way."

The philosophical problems raised by Dalton's simple idea nearly smothered it in its inception. Atomism gives an additive explanation of phenomena, but it appears hardly able to explain the new properties of chemical compounds. It imagines only *Aggregatzustände.* Its implicit postulate is: what is in the whole must necessarily be in the parts—and that is just what, except for weight, the whole of chemical experience seemed to disprove. A new principle of association had to be found, less abstract and more conformable to the complexity of the real, and the first clear step toward it was, quite typically also, the reintroduction of form and figure through stereochemistry. Meanwhile, even the most imaginative were left doubtful. "I must confess," says Faraday early in his career, "I am jealous of the term *atom;* for though it is very easy to talk of atoms, it is very difficult to form a clear idea of their nature, especially when compound bodies are under consideration."[35] But ten years later, in 1844, he added:

> The word atom, which can never be used without involving much that is purely hypothetical, is often *intended* to be used to express a simple fact. . . . . There can be no doubt that the words definite proportions, equivalents, primes, etc. which did and do express fully all the *facts* of what is usually called the atomic theory, were dismissed because they were not expressive enough, and did not say all that was in the mind of him who used the word atom in their stead; they did not express the hypotheses as well as the fact.[36]

With the advent of the kinetic theory of gases, the chemical atom had become practically a certainty; yet it took long years even to give back to Avogadro's law the simple appearance of a fact under which it had been conceived, and as late as 1890 Ostwald could speak of the "rout of atomism."

Boltzmann, in his *Vorlesungen über Gastheorie*, has bitter words against his positivist friends and against the "hands off" attitude of Kirchhoff:

> On the Continent, where the theory of central forces had been generalized into a foundation of the theory of knowledge itself, and therefore a few decades ago the electrical theory of Maxwell was hardly noticed, the provisional character of any special hypothesis was again erected into a general principle, and it was concluded that even the kinetic theory of heat was bound to be discarded in time. . . . . Why should the present generalization [against representations] not prove to be dangerous in its turn?[37]

The reasons for this repugnance are many, and all of them are interesting. First, of course, was the positivist taboo against figurative hypotheses; but, on the nonpositivist side, the issue was followed up half-heartedly, for it was realized by then that the material atom raised more problems than it solved. That is the recurrent drama of rationalism. Once natural philosophers had the atom, it turned out to be not quite what they wanted. The real theoretical hope, as revealed in Prout's hypothesis, had been that the various elements could be shown to be formed of a condensation of the original atom of hydrogen. It was again the Cartesian ideal. Stoechiometry forbade that. The chemical atom turned out to be inconveniently multiple and circumstantial. In other words, it was a little too real. And while imagination struggled to adjust itself to the situation (e.g., with Berthollet's ideas of continuity), the kinetic theory supervened to give the *coup de grâce* to old simple schemes. It had to be admitted that the atomic radii and the masses, which could be calculated for several elements from measurements of energy and momentum in the gases, did by no means suggest constant proportion between mass and volume for all elements. In other words, the idea of a unitary matter stamped out in blocks of different size and shape or welded into aggregates was definitely dead. Each element was clearly made of its own particular stuff of a particular density.

But with this simple pluralism all the old problems that confronted the Newtonian atom came back multiplied by 92, with some fresh problems thrown in.

The rigorous equality of all atoms of a kind again inspired John Herschel to propound a direct intervention of the Creator as the only possible explanation, while all the problems of inherence came up again for the relation of the atom to its physical properties (see p.771).

But there was also the new problem: once we are again on the ground of substantialism, what mechanism can we imagine which would select from the twice infinite set of substantial spheres, of all possible radii and masses, only those few discrete possibilities which correspond to the actual elements? At the same time a way out was dimly perceived. The periodic system, discovered in 1847, was hailed by positivism as a satisfactory classification, but to the scientific imagination it was far more than that: the tantalizing blueprint of an as yet unrevealed unitary structure. The solution was to come only through the quantum theory, but the intimations of it were enough to give new courage to believers in the atom.

These hopes were to be fulfilled: first, through stereochemistry; then, through modern physics, matter was to find its way back to that peculiar elusive combination of substance and geometry required by rationalism.

IX. From Mechanics to Physics

We have shown how the unextended atom of the theory of central forces was already by 1820 a dead issue and how any developments leading to new ideas had to drop one or the other of the assumptions embodied in it. One of the ways, that leading back to substantialism, we have already described. Another way led out analytically toward a mechanics of systems and ultimately toward the formalization of the science as set out previously (see p.766). It simply worked out the dynamic implications of the mixed "astronomical" image. But after having originally claimed empirical matter, under its mechanical aspect, for its province, it found itself dealing with something that was matter by convention only. When Laplace found a rigorous solution of the three-body problem for the special case of an arrangement of the three bodies in an equilateral triangle, he remarked rather wistfully that this was "the only case not possible in nature." An attempt to find help in the no less abstract science of energy provided only auxiliary representations of a strictly fictitious character. Hertz was most philosophically clear-sighted on this point.

Such was the coherent ending of the daring Galilean abstraction of primary qualities and "concreteness," until it was reborn for a time with electrical mass instead of mechanical as subject matter.

The other possible way, and the one marked out by the rationalistic urge, was to go back to reality without yielding on generality: to find a new effective image that could fit the ample range of physical fact.

That range was rapidly extending. The theories of elasticity, heat, optics, and electricity had been organizing their inductive research around certain provisional images, like caloric fluid in heat, emission theory of light, the two-fluid theory of electricity, which had developed mathematical instruments; and these in turn, just as D'Alembert had foreseen, constrained and directed imagination. Essentially, the new developments remained within the frame of phenomenological physics, and Diderot's comments on it (see p.801, n. 51) were still valid. But the progress of mathematical technique provided wonderful bridges between domains, such as Fourier's theorem or the potential theory. The mechanics of deformable bodies set certain assumptions which allowed the science to deal with elasticity and hydrodynamics, and it became the realm of partial differential equations.

The assumptions were, of course, fictional. The mechanics of extended masses presupposed, for instance, continuous variations of density and velocity within the mass. But such concepts were linked with certain original images. Once endowed with the proper analytical apparatus, the images are adopted more boldly; they become isolated starting-points for complete deductions on the Newtonian pattern. Researchers trust their intuition and feel they are dealing with actual experiential material, whereas, of course, they are dealing with fundamental images (not yet quite models[38] in the later sense of the word) presenting the usual combination of abstraction and concreteness and, as such, able to replace the traditional forms of dynamic atomism. These new images now have captured the mind and are being belabored on all sides by imagination.

The general pattern is that of continuity. Continuous media are needed to link pure mechanics not only with the theory of deformable bodies but also with "thermology" (as it was called) and optics, where the wave theory, founded anew by Young and Fresnel, was searching for adequate means of representation. There is scarcely an aspect of physics, from the planning of a steel bridge to the evanescent fringes of color exhibited by a crystal, wherein such a theory of matter does not play its part. It is, indeed, fundamental in its relations to the theory of structures, to the theory of hydromechanics, to the elastic solid theory of light, and to the theory of crystalline media.

The sort of matter imagined now is (apart from incompressible fluids) in the form of rubber bands, jellies, and suchlike substances exhibiting elastic properties and shivering behaviors, capable of shear but not of pressure, set up for the purpose of stretching, sliding, shifting, vibrating.

What is taking on a body again, from so many converging

efforts of the imagination, is that old archetype of the continuum, of unity itself, the space ether.

As in the time of Huygens, Newton, and Euler, it has to be "a very subtil and elastic fluid." It is still half-substantial, still burdened with all the contradictions that smothered its theory in the first instance;[39] but this time it is started on its way by definite heuristic success. Fresnel's ether was able to explain interference and double refraction. Its very success, however, added a new difficulty to those that Newton had perceived already. It vibrated transversally, which made it even harder to visualize. The time had passed for talking vaguely, as Blackmore had done in verse, of "the springy Texture of the Air." The nearest analogon would have been a jelly of surpassing rigidity. The image is not concrete enough to control the imagination, yet it is clear enough to guide research. Rowan Hamilton was able to predict conical refraction in crystals as a consequence of the mathematical fact that Fresnel's wave surface in a biaxial crystal possesses four conical points.

Such success obtained even by a first inadequate theory spurred on researchers, and so hypotheses succeeded each other, the value of which had to be estimated by the fruitfulness of their mathematical consequences. After the adynamical theory of Fresnel came the elastic solid theory of ether developed by Navier, Cauchy, Poisson, and Green, the labile ether theory developed by Cauchy and Kelvin, and the rotational ether theory of MacCullagh. But none of these, while constructing images and models that were often elaborate, could provide a really satisfactory entity, a *construirbare Vorstellung.*

Substantialism as a *caput mortuum* still haunted the theory. Ether was a need; it was a background; it was, as was once said, the substantive of the verb 'to vibrate'; but it remained a shadowy presence. As figurations become difficult and uncertain, the analytical instrument gains predominance and all "truth value" seems to concentrate on the idea of law. The whole of French physics from Lagrange to Poincaré is permeated with a positivist spirit, even before positivism became a doctrine. It is at best a transitional period, since it is easier to profess belief in method only than it is to get rid of all hope of an ultimate substantial representation. We have seen how this process took place even within the astronomical representation (see p.774). But it was to find greater scope on a new level. A science of energy took shape after 1840, and although Faraday's intuitions, Carnot's analogies, and J.R. Mayer's metaphysical generalizations played a large part in its inception, it quickly became formalized and deflected the imagination toward a general theory of equivalence, while the embarrassing substantial substratum of all changes fell back into the shadow.[40] The great organizers of the new theory were Thomson and Helmholtz. The latter especially was to attempt a unification of science on the basis of the conservation of energy. As early as 1847 he roughly sketched out the plan of the work he was to take up twenty years later, aiming at co-ordination of the departments of mechanics, physics, and chemistry. We need not follow up these later developments. But, even here, they go from negative fact to positive figuration. The original principle of unification was the impossibility of perpetual motion. But the inevitable end was that energy was given a completely substantial role in the new scheme; and, when Ostwald in 1890 inveighs against the "materialism" of mechanics, it is simply that he has replaced the old stuff with a new one more to his taste.

It may be said of the science of energy, however, that in its perspicuity, economy, and ample symmetry, in its expansion and in its superordination to phenomena, it provided an almost perfect paradigm for the positivist canon of what a science should be, and such it appeared to the late positivist generation. It is a significant irony that in its beginnings Comte should immediately have taken up the cudgels against it.[41] Deriving his inspiration from the abstract analytical models of Lagrange and Fourier,[42] he proclaimed: "Instead of searching blindly for a sterile unity, as oppressive as it would be chimerical, in the vicious reduction of all phenomena to a single order of laws, the mind shall look at last on the different classes of events as having their own special laws."

To formulate theories concerning the "why" or the "what" of phenomena was for him the cardinal sin, which can be committed only "by minds wholly alien to the true scientific spirit." Hence a number of interdictions which today appear at least weird, among which, specifically, that we should never search for the connections between the various forms of energy.

We cannot dwell here upon the strange and important phenomenon of Comtian positivism. But it would be easy to prove that Comte, too, is a rationalist—in fact, a classical Newtonian rationalist as far as his model of the true science is concerned. The novelty of his case is only that he is trying to apply the old model to the field which alone appears to him essential, that of sociology: all other sciences are merely ancillary and treated as such; but sociology, the all-comprising, is again supposed to provide "un miroir exact de l'ordre extérieur." Nothing is really changed; only the stress is shifted, and the reasons for the shift could be stated correctly only through a discussion of the new "historical" rationalism. That is an ample field which we cannot discuss here. Let us simply remark that the fate of Comte's doctrine is generally that of all systematizations which presuppose a closed group of primary ideas. One might extend to Comte, if one considered his fundamentally analytical and Saint-Simonian background, what Whitehead says of Stuart Mill: "His mentality was limited by his peculiar education, which gave him system before any enjoyment of the relevant experience."

X. The Particle and the Field

To go back to the "open" rationalism of the physicists, which goes on regardless of interdictions: a systematic treatment would mark its progressive absorption of empiricist ideas, and the successful attempts at unification which take place through the science of energy. But we wish simply to point out two fundamental historical facts which are in the strictly rationalist line and precede all formalization.

The first is the evolution of thermodynamics. As Sir J. Larmor says:

> A science of thermodynamics leapt into being in 1824, in a fashion too novel and strange in relation to current trends of ideas to gain recognition at the time, coming from the brain of perhaps the supreme scientific genius of the last century, Sadi Carnot. . . . . But the vast subject instinctively mapped out by him as regards its essential ideas could not become a progressive science until there was a basic theory of matter, on which the primary crude conception of the nature of heat, otherwise unfathomable in any exact sense, could take form in some definite way: and that arrived with sufficient precision only through the formulation of the kinetic theory of gases.[43]

Carnot had proceeded by analogy. His original image was that of heat as a waterfall. And in his short career he managed to foresee the coming form of explanation, as shown by his notebooks published only in 1878. The gap between the general idea of particle and the actual theory was bridged by Maxwell's intuitive imagination, using the technique he had developed in the study of the stability of Saturn's rings, and proceeding from assumptions which appeared sweeping to his puzzled contemporaries. Boltzmann's description of his fundamental paper likens it to a great overture; but then Boltzmann was no mean competitor:

> At first are developed majestically the Variations of Velocities, then from one side enter the Equations of State, from the other the Equations of Motion in a central Field; ever higher sweeps the chaos of Formulae; suddenly are heard the four words: "put $n = 5$." The evil spirit V (the relative velocity of two molecules) vanishes and the dominating figure in the bass is suddenly silent; that which had seemed insuperable being overcome as if by a magic stroke. There is no time to say, why this or why that substitution was made; who cannot sense this should lay the book aside, for Maxwell is no writer of programme music. Result after result is given by the pliant formulae till, as unexpected climax, comes the Heat Equilibrium of a heavy gas; the curtain then drops.

Through sheer acrobatic ingenuity, the particle has been found again: the phenomena of mechanics and heat are brought back to the same concrete reality, and in the end Clausius' analytical expressions, such as entropy, are given an illuminating physical meaning.

This is one of the fundamental facts. The other is the personality of Faraday.

In 1844, Faraday comes to grips with the classical atomic representation, although fully conscious of its power (see p.777). He tries to overcome it. What do we know, he asks, of the atom apart from its force?

> To my mind the nucleus vanishes, and the substance consists in the powers of *m* [the force]. And indeed, what notion can we form of the nucleus independent of its powers? What thought remains on which to hang the imagination of a nucleus independent of the acknowledged force?

This view he pushes to its utmost consequences:

> It would seem to involve necessarily the conclusion that matter fills all space, or at least all space to which gravitation extends; for gravitation is a property of matter dependent on a certain force, and it is this force which constitutes the matter. In that view, matter is not merely mutually penetrable;[44] but each atom extends, so to say, throughout the whole of the solar system, yet always retaining its own centre of force.

Faraday was not as yet constructing his field theory; he was using the classical one, which says nothing more than the theory of central forces. But it was already taking a new shape. Force seemed to him an entity dwelling along the line in which it is exerted. The lines along which gravity acts between the sun and the earth seem figured in his mind as so many elastic strings; indeed, he accepts the assumed instantaneity of gravity as the expression of the "lines of weight." Such views, fruitful in the case of magnetism, barren for his time in the case of gravity, explain his efforts to transform the latter force.

The guiding thread in all of Faraday's endeavor is the search for the underlying unity of the forces of nature. Tyndall has remarked that Faraday's difficulty was at bottom the same as Newton's. But, we might add, this time the phenomena were more complicated, and, since Faraday was not a mathematician, he was not tempted to work his way out in a formula. Since 1834 he had been thinking of the correlation of chemical affinity, heat, electricity, magnetism, gravitation: "Now consider a little more generally the relation of all these powers. We cannot say that any one is the cause of the others, but only that all are connected and due to one common cause. As to the connection, observe the productions of any one from another, or the conversion of one into another." And, after examining the known transformations, he adds: "This relation is probably still more extended and inclusive of aggregation, for as elements change in these relations they change in those. . . . . And even gravitation may perhaps be included." These ideas, and the electrochemical discoveries which accompanied them, prompted Joule to his experiments on the mechanical equivalent of heat. In 1849, Faraday writes again: "The exertions in physical science of late years have been directed to ascertain not merely the natural powers, but the manner in which they are linked together, the universality of each in its action, and their probable *unity in one*."

His certainty of an interconnection of the forces brought him (as shown in his notebooks after 1831) to the discovery of electromagnetic induction; and, through his rigorous and patient translating of experimental effects into a coherent image, the space traversed by the lines of force was drawn into the operational picture, and the electromagnetic field was evolved. At this point the astronomical view of nature had to be definitely discarded. To those steeped in the traditional doctrines there was nothing left but bitter carping about "metaphysics,"[45] or helpless wonderment. Helmholtz, still a young man, pored over Faraday's seemingly "impenetrable" statements and despaired of ever understanding what they meant.[46] So far removed was Faraday's geometric visualization from previous schemes.

But while Faraday organized his discoveries in a work which in patience, rigor, and amplitude remains possibly the greatest monument of experimental research, his dream of the unification of all forces followed him vainly to the end of his life. In 1858, aged sixty-seven, he writes in his notebooks:

15,786. Suppose a relation to exist between gravitation and electricity . . . .—is not likely; nevertheless, try, for less likely things apparently have happened in nature.

15,805. Then we might expect a wonderful opening out of the electrical phenomena.

15,808. Perhaps almost all the varying phenomena of atmospheric heat, electricity, etc. may be referrible to effects of gravitation, and in that respect the latter may prove to be one of the most changeable powers, instead of one of the most unchanged.

15,809. Let the imagination go, guarding it by judgment and principle, but holding it in and directing it by experiment.

15,814. If anything results then we shall have

15,815. An entirely new mode of the excitement of either heat or electricity.

15,816. An entirely new relation of natural forces.

15,817. An analysis of gravitational force.

15,818. A justification of the conservation of force.

The last three points are the rationalist's creed. We are reminded of Diderot's meditation (seep.801,n.51). Beyond the relations and equivalences, Faraday is still searching for the underlying one thing which will allow us to *analyze* gravitation, to *justify* the conservation of force and remove it from the positive status of a mere device of scientific accountancy.

In his quest for unity Faraday was always on the verge of the metaphysical sin, of searching for an abstraction. But what he constructed in the course of his search was his real object, the concrete unity; maybe not so universal as he hoped but leading to universality as none before it; as universal as a theory can be that is born of a creative imagination and structures the complexity of one man's thought; still, for his time the very body of experimental truth: the electromagnetic ether.

The unity thus reached could not last long. Faraday had Newton's imagination but not his mathematics. Faraday and Maxwell together might be said to compose a Newton of the nineteenth century. And Maxwell, much against Faraday's hopes and somewhat against his own, was driven to the use of purely operational models, even to a set of inconsistent ones. When he set out to translate Faraday's ideas into symbols, he tried at first to preserve his master's unity of vision. "Faraday's methods," he writes, "resemble those in which we begin with the whole and arrive at the parts by analysis, while the ordinary mathematical methods were founded on the principle of beginning with the parts and building up the whole by synthesis."

But mathematical analogies drove him far beyond any one model or figuration. He was so guarded in assuming anything about electricity that he postulated little about it beyond its name. This is not enough to enable us to visualize the part it plays in electrical phenomena. For this reason, Maxwell's system was to his contemporaries "a book with seven seals": Helmholtz said that he would be puzzled to explain what an electrical charge was in Maxwell's theory other than the recipient of a symbol. And Hertz drew the conclusion when he said that Maxwell's theory was Maxwell's equations. Thus seemingly ends the last attempt at a real figurative structure.

XI. The Fate of Scientific Rationalism

We go back to what we said before: the search is for objects whose characters develop by necessity into the order of nature as we can see it.

For Descartes and Leibniz physics had been an extension of mathematics. Descartes discarded sensible representation and replaced it with a world which is "the object of speculative geometry." Leibniz, thinking he had found in the analysis of the infinite the means of dealing mathematically with that fugitive multitude of motions which manifests itself in sensible qualities, looked to the order of abstract truths for the discriminating principle between illusory and "well-founded phenomena." A philosophy proceeding thus from the a priori is not only called upon to justify the accord between the intelligible and the sensible but it must prove the very existence of the sensible. Newton's natural philosophy starts from a quite different point of view. The link between mathematics and physics is founded on experiment. And the value of experiment lies in bursting the too narrow frame of metaphysical evidence and in establishing types of relations to which no amount of pure reasoning could possibly have led.

After Newton, rationalism, whether scientific or not, has to admit of something which is "given" beyond any argument. This, for the scientist, is matter. The problem is: What can thought do with it? For as a datum it is opaque and irrelevant; and all that counts is the program of inquiry concerning it, what Galileo called the "ordeal."

From then on thought has to forge its way ahead "in the teeth of stubborn facts," and it is matter which is supposed to provide them out of its own fulness. It is of "the essence of matter," as Laplace would have said, that it should present an ever emergent irrational side to investigation—and, also, that the irrational side should be of one texture with the geometrical truths wrung out of reality. Not merely a shadow side of truth but a principle of opposition ever emerging and ever subdued.

How, then, are we to define the intelligible essence of this texture so that it should fulfil the scientist's myth?

First, it must have precision. That is, we must believe that whatever the growing precision of our measures, the object being measured is still more precise in its determinations, whether it be a body or a wave length.[47] In fact, the whole of physical determinism is based on the faith in a limitless precision at the heart of things.[48] This faith is what Hume had already vainly tried to undermine.

Second, the quantitative characteristics which mark objects in the scientific analysis of nature also define, under conditions marked by similar characteristics, their behavior in a wholly definite and predictable way. This is held to be a law of nature, and certain recurrences are supposed to point to invisible structures of the same type but more universal, such as forces, fields, atoms. These invisible structures then become the principle of order.

Third, the rational intellect is competent to organize sequences which are the actual functional sequences of reality. And when several explanatory constructions can be found, it is a postulate that they can be reduced to one. This is just what Hobbes and D'Alembert had doubted, but the rationalist belief goes even farther. As Einstein said in 1918: "We have always found, that among the several theoretical constructions which could be thought out, one has shown itself as unconditionally superior to all the rest. . . . . The world of sensation determines the theoretical system in a practically universal way, even if we are denied logical access to the foundations of the theory."

There is, no doubt, an inherent contradiction in rationalism, but it is a dynamic contradiction endowed with all the characters of a dialectic. Some kind of substance, of real being, is the object of the quest. Yet, as soon as that substance appears to be reached, it shows itself to be a vain image and a stumbling block to thought.

The physical particle was no sooner found than it raised the problem of what it stood for. Tangible matter as an absolute? A metaphysics of the primary qualities? Surely not, and the simple images of Galilean rationalism had been only a mirage. The molecule had to be resolved into spatial structures; then it was the turn of the atom. The process was carried on with inflexible consequentiality, always in the same direction: toward "the core of things," as Leibniz would have said. Even after rationalism was dead as an explicit faith, after positivism and then Darwinism had knocked all the props from under it, the imagination of physicists went on working in the same line and would not be diverted into "economy of thought." The explanatory space became ever more complex, but its identification with the particle ever more precise, until at last the theory was based upon the structure of the continuum itself.

There is no inductive method that leads us to the fundamental concepts of physics. This is a recognized truth. The theoretical structures imagined are ever more removed from experience. First inertia became an obvious idea, then the absurdity of perpetual motion, then the electromagnetic field, and finally relativity. The process was already well on its way in the nineteenth century; and the lack of recognition of this fact caused considerable delay in the success of the molecular theory and of Maxwell's ideas.

For the rationalist not only is this a fact but it is a "positive" fact in that it confirms his ideal. The ideal is, and ever remains, deduction. If not simple deduction from abstract principles, then a complex deduction from the "essence of matter" as successively determined, but always going both ways: progressing toward consequences, regressing toward elementary ideas and basic structures ever more removed from experience;[49] ever alive, ever altered in the course of research by new intuitions but always seeking to embrace principles and consequences in a frame not only fitting needs but claiming absolute universality and resolved on enough strength to bear not only the present but also the future of the theory.

Hence its insistence on embodying principles of a metaphysical order, like continuity, analogy, and simplicity, the "seal of truth." These principles in turn converge toward identity.[50] But it is not tautological identity. It is a postulated identity, which is of the essence of space and denies the essence of time. The universe is what it is, has been, and shall be. "The elements," says the Critic of Königsberg, "must have that perfection which they derive from their origin"; and even so Newton refuses to conceive of "old, worn particles." Democritus had thought he was cutting the ground from under metaphysical speculation when he said: "Of what is and has been forever one should not seek the cause." But then, of course, he was admitting a "forever," and so did all scientists after him. No sooner was the disturbing fact of the degradation of energy discovered, than fantasy was started on the search for a vaster compensatory mechanism. With identity, necessity, and eternity there always goes an initial perfection, and even recent cosmogonies cannot free themselves from that canon. The unity that comes of it has the essential character of a myth; it implies "absolute" pre-existence of all that we are trying to find.

Thus, identity does not really mean tautology. Meyerson saw a fundamental paradox in the fact that science works at bringing out diversity to annul it then into unity. Historically, the paradox works out as a dialectic. But Meyerson would not even have called it a paradox if he had seen that the unity sought by the rationalist is never formal unity. He is always led and betrayed into abstraction by his analytical instrument but never resigned to it. What he searches for is the very body of truth: a unity in depth, so to speak, in which all things are subsumed without canceling out. It is this that allows a contradiction to exist within the frame without its being stultifying. The contradiction is rather felt to be significant; it points to a complexity within the unity, not to be resolved. Leibniz's "all is exact" and Diderot's "rien de précis en nature" really convey one and the same thing: we are led back to the "difficulty" of the One and the Many, to Parmenides and Heraclitus, as it was in the beginning. A difficulty that could be resolved only in the pure intelligible realm of mathematics, and even there not wholly.

For the scientific rationalist, that slightly imaginary quintessential character, the solution would be the Many-in-the-One. He would speak perhaps of the unity of a total system. But here, again, he finds himself at the heart of his contradiction. He is warmly convinced that he is antimetaphysical. He sets "facts" and not principles at the beginning. But it is a historical fact that in all metaphysics principles have been able to fulfil their assigned role only by being at once homogeneous with facts and superior to them in dignity. For Faraday the electromagnetic field is a reality, as the atom was for Newton. And such reality was for them not phenomenic; as little phenomenic as the "true" or "subtle" elements of the ancients. They would have answered Mach's injunction to stay on empirical ground, and to consider all facts as equivalent, by arguing that that was the way for science to get stuck in a Sargasso Sea of pointer readings and conventional entities. Yet they were as conscientiously empirical as anyone could ever be. Their regressive deduction might be taken for induction at times, even by themselves. After all, were they not looking for some basic reality? Was it their fault if the theoretical frame was forever receding from empirical truth? The imagined structures looked real enough to the mind's eye. Here is the root of that permanent equivocation we mentioned apropos of Whewell's inductivism. With uncanny perception, Hegel saw and used it to denounce science as a vast tautology, which awaited an explaining principle from "elsewhere." In strict logic, he was right. How can principles account for facts, or vice versa, without some bond of communion between them? Their homogeneity is implied by empirical doctrines which claim to infer the principle strictly from the observed fact, just as much as by rationalism which tries to go the inverse way. For in both cases the aim is to prove that this or that fact in nature is basically important with respect to all the rest: whether it be motes in a sunbeam, or the breaking of the waves, or the silent omnipresence of cosmic space. It is only a matter of what you see in and through the thing. The true scientist has an empiricist conscience and a rationalist imagination. He may or may not see the full implications of the latter; here is all the difference between Laplace and Faraday. But so long as he remains true to the initial faith of Galileo and Newton, he is a scientific rationalist. He may at one point, from his own sociological motivations, choose to ignore the process of regressive deduction, and he will appear as an inductivist; or from other sociological and historical motivations, he may decide that what was meant by unity is just unification, and then he will be logically right, and a positivist. Or he may follow his geometrizing instinct to the bitter end, where it meets operationalism, and he will find himself in the magic circle of general relativity. At that point his rationalism will have become relatively meaningless. But only relatively.

For the unity of the rationalist is not simply logical—it is symbolic and creative. His symbol takes on an extended meaning and permeates the whole theory. In so far as he has imagination, the scientist is a metaphysician; and it is yet to be seen how long the imagination may preserve its power after the progress of inquiry has destroyed its metaphysical roots. Technically, the analytical developments of rationalism have led it to meet halfway the rise of logical empiricism, and there is no formal issue that divides the two. Vitally, it may be otherwise.

In the zone of active scientific thought there is an instinctive identification of the image with truth. A "hunch" has to become a belief if we are to work it out, and casualties do not count. It is a faith which may be forever removed from one object to another, but is not easy to destroy. The image is born out of an effort of coherent imagination; it is of one piece with all the other certainties that actuate the mind; and it is as if its truth, according to the ancient pattern, laid claim to multiple and convergent confirmation, for it is of the nature of a truth of inner experience to appear limitless in depth. Like Aristotle and Euler, so the classical scientist does not quite conceive of a simple and univocal necessity. Notwithstanding all precautionary clauses, he expects a confirmation to extend far beyond the range of his own thought and the needs of present theory; he postulates a congruence between "empiricism" and deduction, between "analysis" and "justification." Laplace gave thirteen years of work, from 1773 to 1786, to the proof of the "stability of the universe," feeling that this was necessary to dispel all doubts as to the validity of the Newtonian doctrine. No such proof was needed for the limited time requirements of Cuvier's geology; but the doctrine was expected to confirm that the universe could last forever: in other words, that both the universe and the theory were "right."

The search, to be sure, is for necessary laws, as general as possible. It is a course natural to science, away from mere figuration, toward a higher unitary formalization. But always in the dark faith that formal unity will reveal a value, a specific realm of being.

Conservative thought, as exemplified in Hegel and Comte, tends to reduce unity to abstraction and legality. But science happens not to be conservative by nature.

Since the unity as originally conceived was simple and yet complex, the symbol and epitome of all the forces of nature, the rationalist tends to expand its consequences to the infinite, to weight it with all that has been found and with what has yet to be discovered. He refuses to stop and organize the present, for he already lives in the future. He does not care to arrange what we actually have with a view to convenience or even to the necessities of the organization of thought; for the present is to him a fictitious point, dividing what has been from what will be. Therefore, also, there can be no limitation to his fancy, be it operational, theoretical, or social. The immediate goal is an admitted fiction, but beyond that there is the projection of the infinite, the eagerly awaited unforeseen, which is going to yield the illuminating miracle of unity. To the rationalist mind the miracle is only just good enough.

These are philosophical terms and perforce inadequate; for at the core of scientific thought there seems to be something alien to concepts which requires nondialectical treatment; call it an objective correlative, call it a concrete realization. It is of the nature of science to search not only for intersubjective but for objective reality. Some would define the goal as the search for power through law; such a theory is adventitious, even though it goes back to Bacon, and historically irrelevant. Through history the actual goal is the object, something that the mind's eye can see with the inner certainty that it *is* there; it is the object in turn which imposes its nature on thought and makes all philosophical terminology appear alien and fanciful by comparison. As theories expand in scope, as communities of characters and growing interdependences are discovered, all leading away from any "absolute" being, the real course of thought remains progressive concentration on the actual components of the world, the tension increases between the two levels of theory and reality; and with the tension the strength of the bond that unites them. The knowledge of a theory requires not so much clear-cut general concepts as operational acquaintance. It is the object which calls for that handling by the imagination which creates familiarity and adherence. Even today it is hard to conceive that a mere programmatic will to a coherence should produce a structure of reality; to a researcher it is rather the "feel of the thing" which imposes a coherence upon the representation. So long as the scientist can believe that in the tightening web of his theory the "thing itself" is caught, whatever it may be, so long does he preserve the certainty that he is right, that he has captured the object of his quest. Its nature is ambiguous, like all of reality, but at least it is all there; it may elude him indefinitely but not escape him, and all the limitless reserves of rationality and irrationality that go with the real are still there for him to work on. It is still the presence of the old familiar mystery with which the world faces us, and which was projected, perhaps, by ourselves: strange, consistent, and insoluble; many and, at the same time, essentially one.

Some have said that this need for unity expresses the identity of the self; others, that it stands for economy of thought. Possibly, but this is interpretation, psychologically or sociologically tinged. The need for unity is a historical fact and, as such, unresolved. It is also a symbol of science, and its myth. Science is a finely defined and articulated system of symbols; but the ultimate symbol, that of unity, can have no referent. Rather, one might say it stands for the totality of the knowable and the unknowable. A confusing situation for the scientific mind, but one it cannot escape. For the conflict at the heart of rationalism is the source of its strength, as long as it lasts.

Once the faith is lost, something else has to be found.

Under the relentless pressure of social change, with the growing operationalism of physical theory and the metaphysical devastations attendent on Darwinism, the myth of unity could no longer hold. It had to be replaced by unification. But with that the status of science is changed and also that of the scientist. The mirror of nature that reason had endeavored to build up through the ages is shattered, and we look for the first time straight out into an unknown world.

NOTES

1. The Latin word, meaning "intensively," the same to be said of *extensive*. The passage is taken from the *Dialogue of the World Systems;* the translation is Salusbury's.

2. *Opticks* (4th ed.; London, 1931), Query 31.

3. *Einleitung zu dem Entwurf, etc.*, in *Werke* (1st ser.), III, 295.

4. *Essai d'une nouvelle théorie de la resistance des fluides* (1752).

5. Newton started by looking for a mechanical explanation of gravitation. He imagined an ether which might explain magnetic and electric phenomena as well as those of gravitation and light (Letters to Oldenburg, January 25, 1675–76; to Robert Boyle, February 28, 1678–79, and especially the fourth letter to Bentley). Having later found it more profitable to work out the mathematical consequences of the simple fact of gravitation, he abandoned his "guesses," finding, as MacLaurin says, that "he was not able, from experiment and observation, to give a satisfactory account of this medium and the manner of its operation." Thus it is pragmatic results and not a theoretical conviction that led him to distrust of hypotheses in the *Principia*. Even so,

he could never bring himself to believe in action at a distance, and in the queries of the *Opticks* he reiterated a belief that the force of gravitation, like electricity and magnetism, was due to strains in the rare and subtle medium making up the extension between two bodies. The statement that gravity is as much of an experimental fact as extension and mobility and need not be further explained is of Roger Cotes (that clever heretic, as Maxwell calls him), in his Preface to the second edition of the *Principia*. And even this view which may appear positivistic actually recalls Du Bois Reymond's later "Ignorabimus." In fact, it was advocated and extended by such men as Priestley and Bernouilli. The latter wrote to Euler (February 4, 1744), referring to the ether theory: "Moreover, I believe both that the ether is *gravis versus solem* and the air *versus terram*, and I cannot conceal from you that on these points I am a perfect Newtonian, and I am surprised that you adhere so long to the *principiis Cartesianis;* there is possibly some feeling in the matter. If God had been able to create an *animam* whose nature is unknown to us, He has also been able to impress an *attractionem universalem materiae*, though such is *attractio supra captum*, whereas the *principia Cartesiana* involve always something *contra captum*."

6. The first (and best) formulation of the principle of least action is due to Leibniz: "In a free motion, the action of the moving bodies is usually a maximum or a minimum." As deduced a priori and here formulated, it still lacks a necessary condition for its validity, viz., that energy should remain constant. Maupertuis tried to define action as mass by space and velocity, or as time by *vis viva*.

7. *Proceedings of the Berlin Academy* (1728), p. 151.

8. There were conditions, such as that for the velocity *v* should be substituted its value resulting from the principle of *vis viva* and that the theorem holds only if the principle of *vis viva* holds (and, therefore, it cannot hold for motion in a resisting medium), but Euler expected them only to be temporary limitations—and he was right.

9. *Traité de dynamique* (1758), "Discours préliminaire," p. xxiv.

10. From that time the opposition of "necessary" versus "contingent" (i.e., immanent) will mean only that the laws of nature are either rational truths, demonstrable a priori, or (which was Lazare Carnot's view) empirical truths, founded a posteriori.

11. *Elémens de statique* (10 ed.; Paris, 1861), p. 263. The statement, however, goes back to 1806 and was first published in an essay in the *Journal de l'Ecole polytechnique.*

12. *Geschichte und Wurzel des Satzes von der Erhaltung der Arbeit*, chap. iii.

13. *Mécanique céleste*, IV (1805), 2; cf. also *Supplément*, p. 67.

14. *Die neue Entwicklung der Elektricitätslehre*, p. 35.

15. *Philosophy of the Inductive Sciences*, I, xxvii.

16. It is a moot question whether, space and time being not categories but *Anschauungsformen*, the Kantian critique could not have overcome this difficulty. But for most of its interpreters the system had already become a rigid thing.

17. We have to omit the later attempts (Hertz) at a formalization of mechanics on the basis of "purely mechanical" concepts. They were bound to prove sterile, and they did.

18. This is meant for nineteenth-century science. Modern physical theory has a vastly different aspect. Empiricism has come into its own. Yet even here it might be pointed out that both Planck and Einstein are essentially realists and rationalists of the type we are describing.

19. The postulate is well expressed by Euler: "All modifications taking place in bodies have their cause in the essence and the properties of the bodies themselves."

20. We cannot dwell on this point which ought to be qualified. It involves Plato's conception of space and Aristotle's subsequent developments (cf. *Phys.* iv. 8 and 9).

21. For an unsophisticated statement of the requirement cf. Locke, *Essay on Human Understanding*, Book ii [ch. 27, § 3].

22. "The necessary and the eternal always go together . . . . if something exists of necessity, it is eternal, and if eternal, of necessity" (Aristotle *De gen. et corrupt.* ii. 337^b35, 338^a1).

23. Voltaire, *Dictionnaire philosophique*, art. "Atome."

24. Leibniz, the protagonist of the continuum, is strictly an atomist on the physical plane: "In the explanation of particular phenomena I am an out-and-out partisan of corpuscular philosophy" (Letter to Arnauld, *Philos. Schr.*, II, 58).

25. Cf. Bachelard, *Les Intuitions atomistiques.*

26. Cf. *Principia*, Introd.

27. *Electrodynamische Maassbestimmungen* (1878), p. 645.

28. Having only abstraction to guide him, Weber actually found that the constant in his law had the dimensions of a velocity, that it established the connection between the electromagnetic and the electrostatic system of measurements, and that it was practically the same as that for the propagation of light. Yet he passed it by as a coincidence, to which he could attach no physical meaning, although already in 1846 he was speculating about the role of the intervening medium (*ibid.*, Part I, p. 169).

29. This went on until 1853, when Krönig and Clausius came out with the kinetic theory. For the attempts at clarification of this problem of analogy cf. C. Neumann, *Die Principien der Elektrodynamik* (1868); Weber, *op. cit.;* Helmholtz, *Vorträge und Reden.*

30. *Exposition du système du monde* (6th ed.), p. 440.

31. *Wissensch. Abhandlungen*, I, 658.

32. "It seems to me farther, that these particles have not only a *vis inertiae* . . . . but also that they are moved by certain active Principles such as is that of Gravity, and that which causes Fermentation, and the Cohesion of Bodies. These principles I consider, not as occult Qualities, supposed to result from the specifick Form of Things, but as general Laws of Nature, by which the Things themselves are form'd; their Truth appearing to us by phenomena, though their Causes be not yet discover'd" (*Opticks*, Book iii, p. 1).

33. Given the form of our understanding, says, e.g., Lasswitz, "the science of a given epoch must end—or more exactly begin—with a given group of atomic systems that one may imagine as encased the one within the other, and such a science must start thence to explain all that is to be explained. . . . ." Or, again: "There must be a phenomenal object which in itself should be immutable, impenetrable and very small, and forming the subject of all changes in Nature. But then also we have exhausted the properties to be necessarily attributed to the atom if we study it without regard to relations with other atoms. All other properties of the atom are conditioned by the connexions between atoms" (K. Lasswitz, *Atomistik und Kritizismus*, p. 62).

34. Kant, typically enough, thought that the physics of central forces provided an *ubi consistam* for theory. In the *Metaph. Anfangsgründe der Naturwissenschaft* he builds up matter out of a postulated equilibrium between attractive and repulsive forces which is necessary to explain the distribution of matter through space. Physically, the theory could not hold water without supplementary determinations.

35. *Experimental Researches*, No. 869.

36. *Ibid.*, II, 285.

37. "Einleitung" (ed. 1895).

38. The typical "model" in the sense of a mechanism derives from the "image" of an incompressible fluid with its vortices. MacCullagh was the first to suggest, in 1839, the replacement of elasticity by rotary motion, and in the highly mechanical imagination of W. Thomson and J. M. Rankine this gave rise to the concrete hypothesis of a gyrostatic ether. With four gyroscopes jointed in lozenge fashion one can compose a system which represents the elasticity of a spring and can well explain the rotation of the plane of polarization of light in a magnetic field. This led up to Maxwell's conception of the model as a transitory help for the imagination.

39. Newton imagined ether as "filling all space adequately without leaving any Pores, and by consequence much denser than Quick-Silver or Gold." At the same time it was supposed to be so rare that "it would not offer the slightest alteration to the motion of the planet." As to this last, he reminds us of the strange "subtelty, imponderability and potency" of the electric effluvia. And, furthermore, let the objectors tell him "how the effluvia of a magnet can be so rare and subtile, as to pass through a Plate of Glass without any Resistance or Diminution of their Force, and yet be so potent as to turn a magnetick Needle beyond the Glass?" (*Opticks*, Book iii, Part I, Query 22).

40. The name of "energetics" was suggested by Rankine, in 1855, as "the abstract theory of physical phenomena in general."

41. It is also to be noted that Helmholtz's fundamental paper of 1847 was turned down for publication by Poggendorff and his committee (as Mohr's and Mayer's had been before), because, notwithstanding Joule's experimental material, the theory did not appear to live up to empirical standards. Clausius' memoir of 1855 on the kinetic theory had a similar fate.

42. "There exists a very extensive class of phenomena which are not produced by mechanical forces, but which result solely from the presence and accumulation of heat. This part of natural philosophy cannot be brought under dynamical theories; it has principles peculiar to itself, and is based upon a method similar to that of the other exact sciences" (J. B. J. Fourier, *Théorie analytique de la chaleur*, p. 13).

43. *J. C. Maxwell Commemoration Volume* (1931), p. 88.

44. Faraday, as is known, had the first intuition of wave mechanics. He compares the interpenetration of two atoms to the coalescence of two distinct waves which, though for a moment blended into a single mass, preserve their individuality and afterward separate.

45. Cf. G. B. Airy's letter to J. Barlow, February 26, 1855, and Faraday's reply thereto, in Bence Jones, *Life and Letters of Faraday*, II, 354.

46. Cf. e.g., *Vorlesungen über Theoretische Physik*, "Einleitung."

47. In the first half of the nineteenth century it was still an important philosophical issue to test this precision. Rigor in the equality of acceleration for all matter was seen to mark "the absence of *à peu près*" and the presence of a precise, uniform reality inside the varieties of matter, viz., mass. Bessel and Encke showed this in convergent approximation in their experiments on the pendulum. Laplace's speculations on the stability of the solar system led in the same direction.

48. The idea of an "inherent fuzziness" in nature is quite new and of this century. But Diderot already speaks of a "fumbling" of Nature as she tries to reach her goals (*Interprétation de la nature*, §§ xii, xxxvii), and this order of ideas goes back to Plato.

49. Mention should be made of Enriques's recent theory, which recognizes this fact and tends to save rationalism by what might be called the "infinitesimal a priori." There is an invariant relation, he says, between the voluntary premiss and the clarifying experiment. The premiss is the working hypothesis, which is organized and revised all the time so as to fit the requirements of rationality.

50. "We know full and clearly only one law: that of constancy and uniformity. . . . . When we study things that vary, it is only that we may find what is there uniform and constant. And if we find these constants becoming variable, we must take one more step, but in the same direction: it will be not the same relations, but a combination of them" (Poinsot, *Elémens de statique*).

51. It was Diderot's idea that there must be some force or "affection" of which attraction, elasticity, magnetism, and electricity are only isolated aspects. "Lacking such a center of common correspondence, phenomena will remain isolated; all the discoveries of modern physics will tend to bring them nearer through interposition, without ever uniting them: and even if they did succeed in uniting them, it would make a continuous circle of phenomena where no one could tell which is the first and which is the last. A strange situation, wherein experimental physics through sheer work would have shaped a labyrinth for rational physics to wander about in endlessly; but not an impossible one" (*Interprétation de la nature*, § xlv).

Problems of Empiricism*

Edgar Zilsel

I. The Rise of Experimental Science

1. Experiment and Manual Labor

Modern science is accustomed to decide all questions about reality by experience and experiment. This remarkable attitude is not at all self-evident. It is rather a late achievement in the history of mankind, and its import cannot be fully understood as long as that fact is not realized. We shall start our exposition, therefore, with a retrospective view of the rise of empirical thinking and the experimental method. Though at the beginning of the modern era empirical research proceeded from certain empirical achievements of antiquity, classical empiricism can be omitted in this brief survey. In antiquity the empirical sciences were considerably surpassed in intellectual influence by metaphysics and rhetoric, and empiricists always were but a small minority among ancient philosophers and scientists. It will be sufficient, therefore, to begin with the Middle Ages.

The seats of medieval civilization were not towns, which in the early centuries were rare, but monasteries and castles in the country. The castles and the cultural accomplishments of the knights have little bearing on theoretical thinking. The monasteries, being the centers of medieval scholarship, are more important for the present study. Monks, by the conditions of their lives, are not much disposed to look at the world with open eyes. Inclosed within walls, intrusted with the task of transmitting established doctrines to successors by scholastic instruction, they were compelled to indulge in abstract reasoning and

* This article forms part of a study undertaken for the International Institute of Social Research with the help of grants from the Committee in Aid of Displaced Foreign Scholars, the Rockefeller Foundation, and the Social Science Research Council.

to develop their sagacity. This attitude of mind was later taken over by the universities of the late medieval cities. Up to the thirteenth century the method of investigating that appears to be the most natural one to modern science was practically unknown. When medieval theorists, or theologians, intended to solve a problem, they looked first for relevant passages in the Holy Scripture, the patristic writings, and certain works of Aristotle. Then they compared affirmative and negative statements of colleagues and predecessors and, finally, drew conclusions by means of logical deduction from the premisses collected. It cannot be a surprise, therefore, that the scholastic theory of syllogism is the chief contribution of medieval science.

By the end of the Middle Ages, however, a few scholars, among them Roger Bacon in England and Albertus Magnus in Germany, had begun to understand the importance of experience.[1] Their contemporary, the French nobleman Petrus de Maricourt, even experimented successfully upon magnets. This rise of new scientific methods in the thirteenth century is connected with a fundamental change in society. Towns had gradually grown up, the rural monasteries and castles were losing their social importance, and a new social class—the townsmen—entered upon the stage of history. Money and profit began to rule the lives of nations.

The period of transition from feudalism to early capitalism lasted until the fifteenth century. At its beginning, the craftsmen of the towns, to prevent economic competition, organized themselves in guilds which took care lest the working traditions of the past be broken. But when, with the strengthening of capitalism, competition proved stronger and destroyed the guilds, guild constraints upon working began to crumble. The craftsman who worked exactly as his master and his master's master had done, was surpassed by less conservative competitors; one's own spirit of enterprise, one's own experience and inventive genius, made the successful man. The age of inventions had begun. Some great inventions of this period—the manufacturing of paper, gunpowder, and guns, the mariner's compass, the printing press—are generally known. The invention of the blast furnace and the stamping-mill, the introduction of ventilators and hauling-engines in mining, numerous improvements in construction of weaving-looms, ships, canals, and fortresses are scarcely less important. With the inventions of the fifteenth and sixteenth centuries the technology of the Middle Ages was completely revolutionized. Similar effects were produced by the great geographical discoveries. On new shores animals, plants, and things never seen before were found, which even the most acute monk would not have been able to deduce from his authorities. Authorities and syllogisms had been beaten by experience; a new empirically minded type of man went out to conquer the world.

In virtue of their occupations these men were craftsmen and navigators belonging to the lower ranks of society. They were not esteemed too highly either by the noblemen or by the bankers and rich merchants of plebeian origin. Since the literati, the humanists, who wrote for upper-class publicity were not interested in such plebeian people as craftsmen, the biographies and even the names of most of the inventors are seldom known. There are few detailed and reliable literary reports even on the great discoverers.[2] The craftsmen and sailors themselves were not literary men but rather uneducated people. Their empiricism, very far from being science, was a matter of practical life and casual observation and, for the most part, was lacking in systematic method.

But gradually, especially in Italy, a more systematic technique of empirical research developed among certain groups of superior craftsmen whose professions required more knowledge than did their colleagues'. We are speaking of the artists, the makers of nautical and of musical instruments, and the surgeons. In the Middle Ages the painters, sculptors, and architects had not been distinguished from the whitewashers, stonedressers, and masons. With the decay of their guilds they slowly separated from handicraft and eventually, about the middle of the sixteenth century, became free artists. During the course of this evolution a remarkable group developed within their ranks. We may call them artist-engineers, for they did not only paint pictures and build cathedrals but also constructed lifting-engines, earthworks, canals and sluices, guns and fortresses, discovered new pigments, detected the geometrical laws of perspective, and invented new measuring-tools for engineering and gunnery. Many of them composed diaries and treatises on their experiences and inventions for the use of their colleagues. As they belonged neither to the Latin-speaking university scholars nor to the humanistic literati, but were artisans, they wrote their papers in the vernacular. All of them were already accustomed to making experiments. They were the true forerunners of modern experimental science.[3]

Brunelleschi (1377–1446), the constructor of the cupola of the cathedral of Florence, was the first of these artist-engineers. Among his successors are the bronze-founder Ghiberti (d. 1455), the architect and painter Lione Battista Alberti (1407–72), who—as an exception—had had a classical education, the architect and military engineer Giorgio Martini (1425–1506), Leonardo da Vinci (1452–1519), and, finally, the goldsmith, sculptor, and gun constructor Cellini (1500–1571). The architect and gunnery expert Biringucci (d. 1538) may be mentioned because of his book on metallurgy, *Della pirotechnia*. His paper is the first treatise on chemistry based on sound experience and avoiding any alchemistic superstition. The inventors and constructors of the new clavicembali and harpsichords, and the compass-makers also belonged to the experimenting superior craftsmen. A third group was formed by the surgeons. They dissected animals and often human bodies, whereas the learned physicians seldom dared to do such untidy and sinister work. By their common interest in anatomy there were often connections between artists and surgeons.

Experiment requires manual work, and, therefore, in both antiquity and the modern era its use began in handicraft. In antiquity scientific experimentalists were extremely rare. Since rough work was generally done by slaves, contempt of manual labor formed an obstacle that only the boldest of ancient scholars dared to overcome. A similar obstacle, though by scarcity of slaves less unsurmountable, obstructed the rise of experimental science in the modern era.

The educational system under early capitalism took over from the Middle Ages the distinction between liberal and mechanical arts. In the seven liberal arts (grammar, dialectic, rhetoric, arithmetic, geometry, astronomy, and music) thinking and disputing were alone required; on them alone was the education of well-bred men based. All other arts, as requiring manual work, were considered to be more or less plebeian. There are numerous instances to indicate that up to the sixteenth century even the greatest artists of the Renaissance had to fight against social prejudice. And by the same reason the two components of modern scientific method were kept apart: methodical training of intellect was reserved for university scholars and humanistic writers who belonged, or at least addressed themselves, to the upper class; experiment, and to a certain extent observation, was left to lower-class manual workers. Even the great Leonardo, therefore, was not a true scientist. As he had never learned how to inquire systematically, his results form but a collection of isolated, though sometimes splendid, discoveries. In his diaries he several times discusses problems erroneously which he had solved correctly years before. Gradually, however, the technological revolution transformed society and thinking to such a degree that the social barrier between liberal and mechanical arts began to crumble, and the experimental techniques of the craftsmen were admitted to the ranks of the university scholars. Rational training and manual work were united at last: experimental science was born. This was accomplished about 1600 with Galileo Galilei, Francis Bacon, and William Gilbert. One of the greatest events in the history of mankind had taken place.

The scientific work of Copernicus (d. 1543), being rational not experimental, may be omitted here.

Galileo (1564–1642)[4] got his education at the University of Pisa and for more than twenty years was professor at the universities of Pisa and Padua. His relations to handicraft and technology, however, are often underrated. During his student days there was no mathematical instruction at Pisa.[5] He learned mathematics privately, his tutor, Ostilio Ricci, being an architect and teacher at the Accademia del disegno which had been founded in 1562 by the painter Vasari as something between a modern academy of arts and a technical college.[6] Thus Galileo's first mathematical education was directed by persons who were artist-engineers. As a young professor in Padua he lectured at the university in mathematics and astronomy and gave private instruction on engineering in his home. For his experimental studies he established working-rooms in his house and hired craftsmen as assistants.[7] This was the first university laboratory in history. His scientific research started with studies on pumps, the regulation of rivers, and the construction of fortresses. Ever since his student days he had liked to visit dockyards and arsenals. His first printed publication (1606) described a new measuring-tool for military purposes. Even in his last work of 1638, which initiated modern mechanics, the setting of the dialogue is the arsenal of Venice. His greatest achievement, the discovery of the law of falling bodies, also originated in connection with the contemporary technology. Among the gunnery experts of the period there were many discussions on the shape of the trajectory. Galileo realized that the question could not be answered before the problem of falling bodies was solved. Free falling, however, was too fast to be measured exactly. In order to slow down the movement, Galileo took brass balls, made them roll down an inclined groove, and measured the spaces, times, and velocities. He succeeded in correlating his results by means of a mathematical formula and finally determined the shape of the curve of projection. In Galileo's classical inquiry the two pillars on which modern science is based stand out: experiment and mathematical analysis. The experiments of the craftsmen, alone, would never have issued in science.

Francis Bacon (1561–1626) did not make any important discovery in the natural sciences. His writings abound with magical survivals and errors; he did not understand well the achievements of Copernicus, Galileo, Gilbert, and Harvey; the methodological prescriptions he gave in order to promote empirical research were too pedantic to be of great use to scientists. But he was the very first writer who fully realized the importance of science for human civilization. The scholastics and humanists, he explained, have only repeated sayings of the past. Only in the mechanical arts has knowledge been furthered since antiquity. Bacon, therefore, spoke of craftsmen and mariners with enthusiasm and proclaimed their works as models for the scholars. He is an enthusiastic advocate both of scientific induction and of the ideal of progress. Both ideas are closely connected: they are nothing else than the working method of early capitalistic handicraft seen with the eyes of a philosopher. His whole philosophy is one great attack against the ideals of the seven liberal arts. Bacon himself performed numerous experiments; he died from a cold which he caught while stuffing a dead chicken with snow. Most of his learned contemporaries probably considered that experiment more fitting for a cook or flayer than for a former lord chancellor of England.

The first learned book of the modern era on experimental physics had appeared before Galileo's and Bacon's publications. It was William Gilbert's *De magnete* (1600). Gilbert was physician to Queen Elizabeth. His experimental method originated partly in contemporary metallurgy and mining, partly in the experiments of the retired mariner and compass-maker, Robert Norman.[8]

Why is experiment so essential to empirical science? Mere observation is a passive affair. It means but "wait and see" and often depends on chance. Experiment, on the other hand, is an active method of investigation. The experimenter does not wait until events begin, as it were, to speak for themselves; he systematically asks questions. Moreover, he uses artificial means of producing conditions such that clear answers are likely to be obtained. Such preparations are indispensable in most cases. Natural events are usually compounds of numerous effects produced by different causes, and these can hardly be separately investigated until most of them are eliminated by artificial means. There is, therefore, in all empirical sciences a distinct trend toward experimentation. Sciences in which experiment is not feasible are handicapped. They try to solve their problems by referring to other sciences in which experiments can be performed. Thus meteorology, geology, seismology, and astrophysics make use of laboratory physics and laboratory chemistry. Sociologists and economists attempt to utilize results of psychology. A few modern geologists have even attempted to imitate formation of mountains on a diminished scale in laboratories. To a large extent the poor results of the social sciences might be explained by lack of experiments. The only substitute which, under favorable conditions, can to a certain degree compensate for the lack of laboratory experiments is the use of a great number of observations when carefully compared and worked up by means of statistical methods. Until now, however, experiment has been by far the most efficient empirical method.

It is noteworthy that one of the oldest empirical sciences, astronomy of the solar system, was highly successful without any experiment. This is due to the extraordinary fact that in our solar system superimposed effects belong to very different orders of magnitude and therefore can be separated comparatively easily. The solar system is exceptionally well isolated and the sun surpasses by far all planets in mass. Were the solar system continually bombarded by heavy meteorites or, what is the same, were it passing through a dense star-cluster, and were Jupiter's mass of the same order of magnitude as that of the sun, Copernicus, Kepler, and Newton would not have achieved much.

2. Causal Research: The Mechanical Conception of Nature

The young experimental science was forced to fight hard battles with prescientific thinking. Primitive man does not distinguish exactly between animate and inanimate objects; he apprehends all natural events as if they were manifestations of striving, loving, and hating beings. This animistic conception of nature is predominant in all civilizations without money economy and dominated medieval thinking too. When a comet appeared in the sky or a monster was born, medieval man questioned rather the meaning, the aim, and the purpose of these events than their causes. The scholars did not think very differently either. Certainly "entelechies" and "substantial forms" of Aristotle and the Scholastics are not primitive ideas; they are complicated and highly rational constructs. Yet their animistic kernel has, as it were, only dried up; something like a soul, striving to reach its aims, still glimmers through the rational hull. The same holds of the "occult qualities" that were liked so well by the Scholastics. These could never be observed but were supposed to adhere to most objects and to produce effects by sympathy and antipathy, as if they were little ghosts. Animistic survivals like these were of no use to modern technology and had to be cleared away. Teleological explanations were gradually replaced by causal ones. Purposes of inanimate things, the meaning of natural events, and soul-like powers of physical objects cannot be ascertained by observation. On the other hand, the regular connection of cause and effect is testable by experience and experiment. Moreover, engineers are able to produce the effects they want if they know the cause. Causal explanation, therefore, became the chief aim of experimental science.

The discarding of teleological explanations may be illustrated by two well-known instances. The Scholastics, with Aristotle, explained the falling of bodies by the theory of natural places. Each body was supposed to have its correct place to which it moved when it had been brought to a wrong one. Obviously inanimate bodies were conceived as though they were cattle striving to the accustomed stable. As the theory of natural places did not give any information on the empirical details of falling, it was of no use to the artillery men of the modern era who wished to level their guns correctly. It had to be replaced by Galileo's law of falling bodies and his calculation of the parabola of projection.

The working of suction pumps was explained in the late Middle Ages by the doctrine of *horror vacui*. Water was supposed to rise in pump barrels because nature had an antipathy to empty space. Since the well-diggers of the new era could not calculate from this theory how long they might make their pipes, two pupils of Galileo, Viviani and Toricelli, experimented on pipes filled with mercury and discovered and measured atmospheric pressure.

The causal mode of investigation gave rise to a basic, and previously unknown, conception. In a well-governed state there are laws which are prescribed by the government and, for the most part, observed by the citizens. Lawbreaking is rare and is punished when detected. Let us now suppose the government to be omnipotent and the police to be omniscient. In this case laws would always be observed. The seventeenth century began to compare nature with such a perfect state, ruled by an almighty and omniscient king.[9] Thus the recurrent associations of natural processes were named natural 'laws' by the scientists who investigated them—especially if they had succeeded in expressing the regularities by mathematical formulas. The term 'law' became so common that people soon began to forget that it originated in a metaphor; the idea arose that *all* events, without exception, were subject to natural laws. This deterministic conviction dominated philosophy and science from Descartes, Hobbes, and Spinoza, and from Galileo, Huyghens, and Newton almost up to the present time. It impelled all naturalists to look for more and more natural laws and thus proved to be extremely fruitful.

Nevertheless, we must distinguish the empirical and the metaphysical and theological components in these ideas. The statement that there are regular connections between certain events or qualities undoubtedly is an empirical one, for such connections are observable. The case is different with the assertion that events are connected not only as a matter of fact but that, moreover, some necessity subsists, forcing one event to follow the other. Necessity never is observable; it transcends the province of experience. Metaphysical and theological additions become the more marked when necessity is interpreted as the order of a personal deity or of impersonal nature. But even without drawing in necessity, difficulties emerge and experience is transcended when it is asserted that an observed regularity will *always* hold. Obviously no statement speaking of more than a finite number of objects and containing the term 'all' can be completely verified by observation. All these unempirical components were ascertained and criticized by Hume. Before Hume no scientist and no philosopher conceived natural laws merely as empirically ascertained regularities, and even nowadays the idea of lawful necessity has not entirely vanished.

The logical difficulty just mentioned is repeated on a higher level in the thesis of general determinism. What exactly does it mean to say that regularities hold "everywhere" or, which is the same thing, that there is "no" fact that is not subject to some regularity? In mathematics, if any finite number of cases is given, an equation covering the given cases can always be found. Does general determinism maintain only this analytic proposition or does it maintain more? When the nineteenth-century determinists spoke of regularities and laws, they had in mind the equations of classical mechanics. Yet those equations have since then proved inadequate. Plenty of physical facts became known in the last decades which are covered neither by mechanical equations nor by equations of a similar type. Nevertheless, it is instructive to observe how well determinism turned out—up to the twentieth century at least. Even vague and dubious assertions can render good services to empirical research as a heuristic stimulus.

As we have indicated, determinism for the most part was conceived in a special form. Up to the late nineteenth century nature was interpreted as a gigantic but lawfully functioning mechanism. Since movements, pressing, pushing, and pulling, are the only essential factors in mechanical devices, in nature, as well, all processes and qualities were reduced to such movements. All other qualities, though comprising the greater portion of everyday experience—as colors, sounds, and smells—were not regarded as "real" ones. They were interpreted as "illusions," and no scientific explanation and no natural law was considered to be definitive until it was reduced to laws of mechanics.

The mechanical conception of nature began as early as Galileo. It appeared in almost all philosophers from Descartes to Kant and dominated physics from the beginning of the seventeenth to the end of the nineteenth century. As is generally known, mechanical theories of sound, heat, light, and electromagnetism were constructed in complete detail. Such theories were not completed in chemistry and remained only programs in the provinces of smell and taste.

How can the rise of this remarkable conception be explained? Man himself is, in a certain respect, a mechanical being. All actions by which he influences the world around him consist in movements, in pushing and pulling. From the days when he learns as a baby to control his limbs, he regards that way of reacting as the only natural one. Unless he happens to be a physiologist, the more subtle chemical and electrical processes within his muscles, nerves, and bowels are unknown to him. Small wonder, therefore, that, when he looks at the external world, he takes for granted the movements, pushing, and pulling that he finds there. On the other hand, he feels that all processes differing from his own actions require further explanation and tries to reduce them to mechanical ones. An electric eel, gifted with intelligence, might behave in an analogous manner. Imagine an animal of this kind, not only defending itself and attacking by electric shocks but also attracting prey by electromagnetism, splitting and absorbing its food by electrolysis! If it were a physicist, such an animal would probably look for electromagnetic explanations everywhere. To summarize, then: the mechanical conception of nature is anthropomorphic and interprets natural processes after the pattern of human actions.

Unfortunately, more facts seem to be accounted for by this biological explanation than actually prove to be correct in history. Though men have always had a like biological organization, mechanistic theories are met with only very seldom and in very few civilizations. Mechanistic physics appeared only with ancient atomists and Epicureans and in the period we are just discussing. Obviously, some additional conditions are required if theories of this type are to develop. Apparently, the state of the contemporary technology can give the additional explanation we need. Man of the seventeenth century had outgrown primitive animism and begun to see the world with the eyes of an engineer. But since the only machines with which the period was acquainted were, without exception, mechanisms such as printing presses, weaving-looms, pulleys, and clockwork, it was rather natural to think that the whole world was a mere mechanism as well. This prejudice was shaken only when technology had entirely changed, and nonmechanical engines had become more frequent than mechanical ones. This happened in the late nineteenth century.

The scientific value of the mechanist conception of nature cannot be overlooked. Distances and shapes, pressure and movement, can be measured comparatively easily: that is to say, can be co-ordinated to numbers without great difficulty. On the other hand, qualities such as blue and cold never, themselves, appear among the results of any calculation, and their measurement is considerably more complicated. So reduction of all qualities to mechanical ones enabled the physicists to coordinate numbers to qualities, and this again made it possible to grasp the physical world by mathematics. By the mathematization of physics, the building-up of deductive theories was immensely furthered. We need not point out how philosophical rationalism was stimulated by this development, as these achievements belong with an account of the rational side of knowledge. The heuristic value of the mechanist conception, however, was not slight. Everyone who knows the history of physics also knows how many new empirical facts were discovered and causally explained for the first time by means of mechanistic models. Both the rational and the empirical value of the mechanical conception of nature cannot be doubted. Yet, being based on a prejudice, that conception had its dangers as well. Distinction between a "real" world of mechanics and an "illusory" one of qualities has produced a confusion of concepts that has interfered with the analysis of knowledge and philosophy for almost three centuries. This will be discussed in the next section.

II. The Philosophy of Modern Classical Empiricism

3. The Opposition of Outer and Inner World

After natural science had adopted the experimental method from the craftsmen, it developed rapidly during the seventeenth century. It was engaged in its special questions and, consequently, does not offer many problems to epistemological investigation. The case of contemporary philosophy was different. The most remarkable fact in empiricist philosophy of the classical period might be said to be the tendency gradually to transfer its interest from objects to the subject. By this remarkable tendency it soon became entangled in pseudo-problems. Medieval animism and the Aristotelianism of the Scholastics were scarcely overcome when a new metaphysics developed, originating in the contrast of object and subject, of outer and inner world.

With Francis Bacon (1561–1626), interest in objects still prevailed; subjective components of knowledge are only mentioned in his doctrine of idols. Bacon is convinced that knowledge gives a rather accurate picture of nature if only we take care in avoiding a few errors and prejudices. These fallacies are induced partly by social, partly by individual, conditions by which the judgment of the observer is disturbed. They can, however, be eliminated if attention is called to them. As is generally known, such biases are called 'idols' by Bacon and are discussed by him in an entirely empirical way without involving metaphysics. The beginnings of the new subject-object metaphysics appeared in empiricism with Hobbes (1588–1679). Hobbes was a radical mechanist. Since in his opinion all processes consist in movement, he was faced with the task of explaining the origin of all other qualities that are perceived by us. He tried to master the difficulty by his terministic theory of sensation. In his epistemology sensations are distinguished from objective qualities. Sensations are not at all copies of the physical world but only correspond to physical qualities as symbols or terms do to their objects. Plato had already duplicated the world by opposing the realm of Ideas to the world of phenomena. Platonic Ideas, however, are extremely vague constructs, originating chiefly in certain ethical considerations and in the philosophy of mathematics; they were scarcely designed to help naturalists in interpretation of their scientific observations: the metaphysical background of the Platonic Ideas is obvious. Into natural science and empirical philosophy the two-worlds theory was introduced by Hobbes. Although his terministic theory of sensation contains valuable elements (which were not, however, developed until the time of the modern logic of relations), his distinction between the "real" world of movements and the subjective world of qualities has entangled the philosophy of nature in pseudo-problems for more than two centuries. They are the more serious as they are inevitable implications of the mechanistic interpretation of phenomena. As long as the latter was considered the only possible philosophy of nature, the pseudo-problems connected with the contrast of external world and subjective awareness could scarcely be eliminated.

The decisive step that changed empiricist epistemology into an introspective psychological theory of knowledge was taken by Locke (1632–1704). As generally known, Locke's theory of knowledge reduced all statements, both in everyday life and in science, either to sensation or to reflection. It belongs to the character of his philosophy that he spoke rather of ideas and sensations than of facts, observations, and statements. Modern logical empiricism restricts itself to analyzing the methods by which statements are tested and verified, whereas empiricism of the seventeenth century proceeded psychologically and investigated how ideas are obtained. Locke's polemic against rationalism became a psychological analysis of abstract ideas, on the one hand, and an attack upon innate ideas, on the other. Since innate ideas played a rather prominent part in Descartes's epistemology, this attack is of considerable historical importance in the further development of empiricism. But Locke soon turned to a discussion of the mind of the newborn child and revealed the concept of soul as the metaphysical background of all his psychological analyses. Mechanistic philosophy was considered so self-evident by Locke that its physical implications were not even pointed out by him; it appears only as his well-known distinction of primary and secondary qualities. The metaphysical background of his psychology of knowledge becomes most obvious in his exposition of the idea of substance. Substance to Locke was the "unknown support" of qualities. Although in his opinion that support is, and ever must stay, "unknown to us," he did not hesitate to distinguish material and spiritual substances and to raise the problems of their interdependence. Locke, the empiricist, accepted the metaphysical dualism of Descartes without question and discussed an interdependence, never testable by experience, between constructs that could not be tested either. The outstanding scientific and historical importance of Locke's analyses is generally known; but the metaphysical pseudo-problems, which are connected with them, must not be overlooked.

The contradiction between Locke's empiricist principles and his theory of substance was noticed by Berkeley (1685–1753). Berkeley adopted the principles and the introspective method from his predecessor but rejected Locke's theory of substance. He started with a criticism of Locke's theory of abstract ideas. Since he found by psychological introspection that he was not able to form, for instance, the idea of a color which is neither red nor blue nor green, he rejected Locke's analysis and flatly denied the subsistence of abstract ideas. By leaving differences out of consideration and by marking common properties only, one word can obviously represent quite a group of different objects. Thus abstract ideas may and must be replaced by abstract words. Applying this analysis to the concept of substance, Berkeley concluded that substance is a mere word, void of content. The unknown "support" of qualities, which had been assumed by Locke, although it could never be perceived, Berkeley thought could, and should, be eliminated: matter, being an abstract construct, does not exist at all. Actually there are perceivable qualities only, and these are nothing else than perceptions. Thus Berkeley's well-known equation, *esse = percipi*, resulted.

The only point that strikes us in Berkeley's remarkably consistent argumentation is that he failed to apply his analysis to the concept of mind or soul. Can souls ever be experienced? Are they not abstract supports of perceptions or qualities just as matter is? No doubt it was Bishop Berkeley's religious attitude that prevented him from drawing this dangerous consequence. Avoiding the term 'idea,' he always spoke of the 'notion' of soul and would not admit that it is no less abstract than the idea of matter. Thus he built up his odd metaphysical system, constructing the world out of souls, ideas, and God. The tendency in empiricism both of turning to subjectivism and of getting involved in pseudo-problems has reached its peak in Berkeley.

Why did English empiricism tend to introspective psychology and how could it be invaded by subject-object metaphysics? Both in ancient and in medieval philosophy only slight beginnings of analogous ideas are to be found. The modern dualism of inner and outer world might be correlated with the dualism of soul and body and probably can be explained by the influence on mechanistic physics of theology. Belief in immortal souls and the mechanical conception of nature had never been united in one philosophical system before the seventeenth century. In antiquity the atomists and Epicureans were mechanists, but they did not believe in spiritual substances; Platonists and Stoics, on the other hand, distinguished souls and bodies but were not mechanists. Medieval theologians were not compelled to stress a chasm between spiritual and physical substances, since in their philosophy all physical objects were imbued with more or less soul-like powers; they could content themselves with Aristotelian entelechies which, being forms, were rather connected with than opposed to matter. Actual and radical dualism was introduced into philosophy by Descartes, mechanist and devout Catholic. It is comprehensible that, apart from direct Cartesian influence, similar tendencies invaded English empiricism as soon as it turned to mechanism. In their business of investigating phenomena, natural scientists are faced with the task of separating constant relations, on which all observers can agree, from the variable and unstable aspects which are offered under different conditions or to different observers in a different way. This is the sound basis for the distinction between objects and subjects. Bacon, for instance, did not misuse it. But this separation was misinterpreted as soon as the mechanistic philosophy of physical phenomena separated "real" qualities, such as motions, from merely "apparent" ones, such as colors. The rationalistic mechanist Descartes became guilty of the misinterpretation as did the empiristic mechanist Hobbes. Obviously, it was the rise of mechanistic physics that turned the harmless distinction between subjective and objective components of observation into a dualism of inner and outer world. And it is rather comprehensible that, under the influence of religious tradition, this dualism was more or less identified with the contrast of soul and matter. Thus analysis of experience turned with Locke to introspective psychology—that is, to investigation of soul. Experience became a psychical duplicate of the "real" external world, and pseudo-questions arose: whether both halves of the world actually exist and how it happens that they correspond.

4. David Hume

Locke's and Berkeley's principles were consistently applied to the concept of spiritual substance by Hume (1711–76). As is generally known, he eliminated substantial souls and replaced them by heaps or bundles of impressions. Berkeley's philosophy, which had denied the existence of physical objects but approved the existence of souls, seemed paradoxical to common sense. To critics entangled in the problems and language of the period, Hume's position of rejecting spiritual substances as well as physical ones seemed, of course, even more paradoxical. Actually the paradox was rather diminished by him, for he treated objects and subjects in the same way and thus approached common-sense realism again. His analysis is a most important step toward overcoming the pseudo-problems of a subject-object metaphysics. In Hume's time, however, theological ideas and the mechanistic interpretation of physical phenomena still blocked the way to a complete understanding and elimination of such pseudo-problems. Moreover, his analysis was impaired by his dealing with impressions and ideas rather than with statements. This predilection for elements of knowledge which are too minute to be very useful in analysis had originated with Locke and can be traced in the development of empiricism up to Mach and the nineteenth century. The same phenomenon occurs when Berkeley, for instance, as he often does, turns philosophical criticism to criticism of language. In this case he is discussing words rather than statements. This noteworthy but prejudicial predilection apparently also originated in the psychological attitude of classical empiricism, for statements are too logical to attract the attention of psychologists. Analysis of experience shifted from ideas to statements only when axiomatic methods in mathematics of the late nineteenth century had roused interest in statements and their concatenation, and when non-Euclidean geometry and modern symbolic logic began to influence empiricism. With rationalists and with Kant, on the other hand, statements always had played an important part.

Among the most important achievements in the history of philosophy is Hume's analysis of the concept of causality. With surpassing clarity he showed that we speak of cause and effect if phenomena are connected regularly with one another. Since cause and effect are not linked by logical necessity, even the most common causal statements of everyday life depend entirely on past experiences and never can be inferred a priori. Induction, therefore, differs radically from deductive logic and is based psychologically on custom and belief. Even today the man in the street is inclined to conceive cause as a thing that by its activity produces another thing. Hume's criticism destroyed this idea of active production—that last remainder of primitive animism. Therewith it is implied that not things but processes, qualities, and relations alone are causes. And, by stressing regularity of connection, he adapted the concept of cause to the concept of natural law, the very form in which it is used in any mature empirical science. His theory of causality would not have been possible in a period in which natural laws were not yet known, and it was obviously inspired by Newton's and his followers' successful investigation of physical laws. Moreover, it made feasible the future development of physics. Without Hume's concept of cause, nonmechanistic physics of the late nineteenth and the twentieth centuries scarcely could have come into existence. Among later philosophers, however, his analysis has met with considerable criticism, opposition, and—even worse—complete neglect. As to induction, later philosophy has made little progress beyond Hume's remarks.

During the eighteenth century, empiricism, by emphasizing sensations, turned in France to sensationalism and, with many philosophers, to materialism. On the other hand, empiricist philosophy of the seventeenth and eighteenth centuries developed important methods and results of psychological investigation. The laws of association were discussed again and again; contemporary reports on customs of primitive peoples were used by Locke in his polemics against innate ideas, and the anthropological method was successfully developed and applied to the investigation of religion by Hume; the psychology of people with defective senses was occasionally referred to in Berkeley's *Theory of Vision* and was considerably furthered in Diderot's *Letters on the Blind* (1749) and *Letters on the Deaf-Mute* (1751). For the first time perception of space was investigated psychologically, and the visible, tactual, and muscular components of the experience of space were more or less exactly distinguished in Berkeley's *Theory of Vision*.

III. The Advance of Empirical Science

5. Religious Problems

The last remarks have brought us to that field in the evolution of empiricism which is the most fertile one—the special sciences. Only the general expansion of the domain of empirical research can, however, be discussed here.

The empirical spirit of the modern era is entirely contrary to the spirit of medieval scholasticism and, consequently, was compelled at first to overcome the resistance of theological tradition. The rise of empiricism, therefore, is connected, though not identical, with the spreading of the Enlightenment. The beginnings of religious Enlightenment, that is, the ideas of natural religion (Herbert of Cherbury, 1624) were based rather on rational construction than on empirical investigation of the various religious systems. Little attention was paid by Herbert and his followers to the variety of modes of worship; priests and ceremonies were rather disliked by them. As is generally known, the main theses of natural religion stated the existence of God, immortality of the soul, and reward and punishment in the other world. They were primarily products of abstract reasoning and were based on empirical comparison only in so far as they expressed the components common to the three great monotheistic religions—the only religions well known in this period. Obviously, the articles of natural religion indicate the common content of Judaism, Christianity, and Mohammedanism. Yet the articles were void of emotional content and do not occur, as they were formulated by Herbert and his followers, in any living religious system. The religious attitude of most of the empiricist thinkers of the seventeenth and eighteenth centuries was, however, more or less determined by these rather lean ideas.

It is remarkable that the first empirical contribution to scientific investigation of religion was made in the modern era by one of the most consistent rationalists—by Spinoza. In analyzing the worldly literature of antiquity, the humanists of the Renaissance had already created and developed the methods of philological criticism. Hobbes, in his *Leviathan* (1651), had advanced a few critical remarks on the composition of the Old Testament. Spinoza, however, was the first who, in his *Tractatus theologico-politicus* (1670), dared to apply, consistently, philological methods to the Old Testament. If painstaking philological criticism is considered an empirical achievement, Spinoza's successful attempt to determine the time of composition of parts of the Old Testament may be reckoned with the major advances of empiricist thinking. The next step in this field was taken by Hume. Hume's dissertation on the *Natural*

History of Religion (1753) tried for the first time to give an empirical theory of general religious development and started comparative investigation of primitive religions. Hume, however, knew for the most part only the religious ideas of a few primitive nations, of ancient Greece and Rome, and, of course, the three great monotheistic systems. Actual knowledge of the great religious systems of India and China came to Europe not before the early nineteenth century. In the later nineteenth century the empirical science of religion rose considerably, using the philological and anthropological methods of Spinoza and Hume as well as modern psychological and sociological ones. Obviously, empirical thinking could not have spread in the field of religious and anthropological research were it not for the expansion of world-trade and the colonial system that made the white race acquainted with exotic civilizations. In Locke and Hume these connections had already become visible.

6. The Natural Sciences

The natural sciences are more than two centuries older than the science of religion and comparative anthropology. In the field of astronomy and physics empirical research had reached its first overwhelming successes as early as the seventeenth century, in the period from Galileo to Newton. Chemistry joined the advance in the eighteenth century and later was followed by mineralogy, geology, and meteorology. From the methodological point of view success was obtained in the physical sciences by three means: they did not restrict themselves to mere observation but experimented wherever physical processes could be influenced by technological devices; they investigated the quantitative relations of phenomena; and they considered discovery of natural laws as the most important goal of research. Application of mathematics to the empirical findings and the construction of deductive theories cannot be separated neatly from those empirical methods and scarcely are less important; but, as they belong with the rational side of science, they need not be discussed here.

The biological sciences developed considerably more slowly.

Among them, medicine is the most ancient one. In the modern era medical investigation, as well as physical research, was hampered at first by the social prejudice against manual work. This prejudice was overcome in the late sixteenth century, and dissection came in use in anatomy about the same period as experiment did in physics. As early as the beginning of the seventeenth century Harvey experimented with animals in his embryological research and occasionally used quantitative methods in his investigation of the circulation of the blood. Nevertheless, the practical tasks of medicine affect the emotions of men so intensely that metaphysical, teleological, and even superstitious ideas and traditions were eliminated from medicine more slowly than from any other natural science. In zoölogy and botany quantitative and experimental methods did not come into general use before the nineteenth century. Up to that time, biologists restricted themselves chiefly to observation, noncausal description, and classification. Classification of the material must precede the investigation of causes and laws in all empirical sciences, as in business the stock of goods must be ordered and inventoried before actual trading starts. In the field of physics, as well, solid and liquid bodies, acoustical and optical phenomena, for instance, were distinguished before they were investigated scientifically. Whereas in physics, however, most classifications are elementary and, therefore, can be followed rather soon by causal research, the vaster variety of objects in the fields of zoölogy and botany is much harder to survey. Biological research, as a result, has, during almost three centuries, scarcely passed beyond classification and noncausal description. Moreover, it is far more difficult with plants and animals to produce alterations artificially than it is with physical and chemical objects. For all these reasons experimental methods and causal explanations were adopted by biology rather late. Instead of the concept of cause, prescientific teleological interpretation ruled the biological sciences until the nineteenth century.

The eighteenth century, however, was already aware of the difference between "artificial" and "natural" classification.

Linnaeus, the most eminent biologist of the period, gave both artificial and natural classifications of plants and animals. Since plants are extremely numerous and varied, it is often not easy to determine the species to which some individual plant belongs. In order to facilitate that task, Linnaeus classified the plants by the number of their stamens; he was well aware, however, that this classification was merely artificial and hardly more than a technical device of nomenclature. His natural classifications, on the other hand, aimed at putting together plants or animals in one group which "actually" are connected with one another. Thus natural classification includes the viewpoint of empirically ascertained relationships and, therefore, is the first step in the direction of causal research and the theory of evolution.

Lamarck's, and especially Darwin's, achievements opened entirely new ways to biological investigation. Empirical thinking was furthered by the theory of evolution in four respects. Linnaeus had still considered animal and plant species as absolutely rigid; there are as many species, he explained, as have initially been created by God. By the theory of evolution this rigid immutability was liquified, as it were: 'to be' was replaced by 'to become.' Certainly, the concept of temporal process is not necessarily involved in the concept of natural law. There are in physics nontemporal laws of coexistence as well as laws of temporal change, though the former might be less numerous. Moreover, the latter seem to interest men more intensely; human actions are processes themselves and always aim at influencing the future. At any rate, the precursor of the concept of law—the concept of cause—contains the element of temporal change as an essential component. For that reason interpretation of living beings as subjects of a permanent temporal process was indispensable if description and classification were to be supplied by causal explanation and later by investigation of natural laws. Second, Darwin's theory of natural selection (1859) made the first successful, though yet incomplete, attempt to explain causally the obvious fact that organisms are well adapted to their environments. Since the decline of Aristotelian scholasticism, prescientific teleological traditions had been removed from the field of physics. With Darwin they suffered the first serious blow in biology. Third, Darwin's exposition of the descent of man (1871) destroyed traditional anthropocentric philosophy and thus decidedly helped eliminate obstacles to the empirical investigation of mankind. In anthropology, sociology, psychology, and ethics this influence became conspicuous very soon, even if the animal ancestors of man and natural selection were not discussed at all. Darwin's ideas throughout fitted in with the trend of modern empiricist philosophy. Already, Hobbes, Hume, and a few representatives of the French Enlightenment had, without even thinking of animal ancestors of man, consistently treated all human problems as natural, merely empirical ones. This interpretation was confirmed and enormously spread by Darwinism. And, finally, Darwin's theory called scientific attention to a few characteristic traits which can be found in any evolution—even beyond the province of biology. As it was pointed out by Herbert Spencer, for instance, specialization, or family-tree-like ramification, appears in the historical development of occupations and sciences and in the intellectual development of the human individual as well as in the phylogenetic evolution of organisms. Thus, empirical investigation of society, civilization, and mind was also considerably furthered by Darwin's ideas. On the other hand, however, the scientific value of the concept of evolution was sometimes overestimated. The merely descriptive statement of some "evolutionary" process in society or culture is only a preliminary step by which the solution of the scientific problem is prepared but not given. A final statement of causes and laws cannot be supplied simply by description of some one evolutionary process.

Darwin's theory was based on the practice of cattle-breeders and on comparative observation. He did not perform experiments. This most successful and most exact method of empirical research was applied to biological problems in the investigation of heredity and physiology in the late nineteenth century. Mendel discussed his experiments on heredity (1865) in statistical terms, whereas some physiologists of the nineteenth century transferred the ordinary methods and concepts of physics and chemistry to biology and adapted them to the new problems. Moreover, both modern genetics and modern physiology investigate quantitative relations and are seeking natural laws.

At the end of the nineteenth century teleology seemed to revive again in biology. In the final analysis Darwin's causal explanation of the fitness and adaptation of organisms deals with a statistical phenomenon only. It discusses the survival of the fittest within large groups of plants or animals, but it is not interested in the problem as to how the processes which are characteristic of life are accomplished in individuals. Such processes as the development of the fecundated egg, regeneration of injured organs, and others, were interpreted teleologically by Driesch. Even the Aristotelian concept of entelechy was reintroduced by him—a concept which two centuries before had been eliminated from science after long and arduous intellectual conflicts. The strangest phenomenon in this resurrection is that it is connected with a considerable advance of experimental method. For the aforesaid processes were investigated by Driesch by means of experiments that proved to be most fruitful. They showed that processes in one part of an organism are often influenced not only by conditions in that part but also by other, if not all, parts of the organism. Driesch and the neovitalists assume the influence of *all* parts and speak, therefore, of the efficacy of "wholeness"; wholeness, however, is interpreted by them as a nonspatial, soul-like entelechy which aims at ends. This neovitalistic "explanation" is highly contestable. Nonspatial entelechies are in no way observable, and, consequently, the entelechies of a frog and a hydra cannot be distinguished. It is not clear, therefore, what help they can give in explaining, predicting, and controlling the rather different processes occurring in these different animals. Obviously, entelechies are additional metaphysical ingredients and do not contribute anything to a solution of the empirical problems. The phenomena of regeneration and formation are now actively investigated. They cannot yet be explained satisfactorily, though many partial successes have been achieved. Nevertheless, there is no reason to assume that, in this field, solution of the scientific problems will be achieved by other means than by experiments, causal explanation, and investigation of laws.

7. Psychology

The beginnings of psychology in the modern era appear rather remote from empirical research. Descartes and Spinoza, when discussing human passions, gave theoretical constructions which were intended primarily to support the ethical ideals and theories of each philosopher. Yet it cannot be doubted that their "psychological" studies are based on a good deal of introspection into their own minds and comparative observation of fellow-men. The same methods were used more consistently by Locke, Berkeley, and Hume in their analyses of knowledge. Most psychologists of the seventeenth and eighteenth centuries were highly impressed by the scientific success of Newtonian physics; they were, therefore, seeking psychological laws, and considered the laws of association to be analogous to, and as important as, the laws of mechanics. Such physical analogies become rather conspicuous in the psychological treatises of both the physician Hartley and the chemist Priestley. Morals, also, were again and again investigated psychologically. By application of psychological methods to the problems of ethics, certainly a major advance in empirical thinking was made. On the other hand, in the eighteenth century, psychology still dealt chiefly with knowledge and rather neglected emotions, volitions, and actions.

Psychology in the nineteenth century developed considerably, both as to its subjects and as to its methods. In the psychology of knowledge the constructs of the deductive sciences began to be investigated. It was even assumed that logic could be reduced to psychology. Quite a number of epistemologists were convinced that all problems of logic would be solved if the psychological origins of logical concepts and logical propositions were found out. By their critics their line of thought was called "psychologism." Psychologism in the nineteenth-century epistemology distinctly mirrors the empirical tendencies of the period and even exaggerates them. It obviously committed the same error a chess player would make if he thought that knowledge of the historical and psychological origin of all chess rules could answer the question of which chess problems are solvable and which are not. On the other hand, psychologists with a leaning toward the natural sciences investigated sensations in close connection with physiological research. Thus experimental and quantitative methods were introduced into at least one branch of psychology.

With the decline of the Enlightenment, the irrational sides of the human mind also began to attract attention. In the early nineteenth century German romanticists were already highly interested in passion and emotion and, moreover, paid a great deal of attention to hypnotism, somnambulism, and insanity. Philosophers and psychologists belonging, as did Schelling and Schubert, to the Romantic current of thought, began to deal also with the irrational and abnormal components of mind. These were interpreted, both by writers and by theorists, metaphysically, and even more or less magical ideas frequently occur in the expositions. Again a new and fruitful field of research was discussed at first in prescientific terms, as often happens in the history of science. This initial stage, however, was overcome about the middle of the nineteenth century. Scientific psychology divested itself of magic and metaphysical ideas without resuming the eighteenth century's overestimation of intellect. It turned to voluntarism and investigated emotions, appetites, and instincts empirically, frequently in close connection with biology, physiology, and animal psychology. Psychiatrical research arose, and the comparison of normal and abnormal mind contributed fruitful data to psychology. Investigation of neuroses, and of hysteria especially, added unconscious processes to the objects of psychological research.

The new conception that mind is composed of unconscious as well as conscious elements apparently is exposed to methodological objections.[10] Viewed empirically, mind seems to be equivalent to awareness. Unconscious mental processes, therefore, seem to be a contradiction in terms and to be metaphysical constructs which can never be tested by experience. Yet, these objections do not hold. Psychology of the nineteenth century is, for the most part, based either on introspection or on empathic interpretation of the behavior of fellow-men. In this behavior, however, speaking and intentional communications form but a small part; other, and unintentional, reactions are far more numerous. No psychologist, therefore, bases his scientific assertions only on the words by which his fellow-men describe their awareness but makes use of their physiological reactions—for instance their blushing, their actions, and their omissions of action—as well. Is there any reason to abstain from investigating processes which do produce such reactions, actions, and omissions, though they cannot be described by the individual in whom they happen? Since the individual in whom they occur is not aware of them, it might be objected that they ought to be called physiological processes. Actually, we may hope that in the future they will be explored by physiological means, and certainly physiological investigation would furnish more exact results than the psychological investigations of the present time. Nevertheless, the above procedure can be justified. When an individual acts in a certain way because of some past experience which is perfectly well remembered by him but is so disagreeable that he does not like to speak of it, and when another individual has "repressed" an extremely painful experience so that it has become entirely "unconscious," both kinds of behavior can greatly resemble each other. The similarity becomes manifest in the fact that the observer can put himself psychologically in the place of both persons by means of empathy. The very possibility of empathy in both cases is the link by which conscious and unconscious processes are connected. This is the empirical reason why we are justified in using psychological terms and psychological methods in investigation of the unconscious. Physiological processes, on the other hand, work quite differently from unconscious ideas, wishes, and purposes and are not accessible to empathy.

We have made use of empathy as of an empirical symptom of the relationship between conscious and unconscious mental phenomena. It must be remarked, however, that it cannot be used as a definitive method of research. When we interpret the behavior of fellow-men by means of empathy, our interpretation mirrors experiences which we happen to be familiar with. Empathy, therefore, is subject to major errors and always needs farther examination by means of more reliable methods. We may strongly feel that a man is angry, and yet all inferences based on our feeling may later prove to be wrong. When, on the other hand, his further behavior conforms with our expectation, then and only then our opinions of him are verified. Since empathy works much faster than careful scientific examination, it supplies most of the psychological judgments in everyday life. Yet scientific predictions, based on empathy, may be relied on only in so far as they are confirmed by observable actions and reactions of the individuals concerned. The method of empathic interpretation may be used in scientific psychology as a preliminary heuristic tool. Certainly, it is fruitful if its results are tested later by observation of the perceivable behavior. But it is highly fallible, and the scientific content of all assertions obtained in this way consists solely in those components which can be confirmed by observation. The precariousness of empathy has led to the rise of behaviorism, which is discussed at another place in this *Encyclopedia*. At any rate, it is an empirical fact that man can experience empathy in certain cases and is unable or less able to experience it in other ones. The cases in which it can be experienced—that is, the conscious and the unconscious mental processes—obviously have certain empirical features in common.

To summarize: if and only if empathy is more or less feasible, may an unconscious process be named 'psychological' and be reckoned among mental phenomena. Nevertheless, empathy is a symptom only. With it a certain type of functioning, that is, of causal connection, becomes manifest that, apparently, is common to conscious and unconscious mental phenomena. Or, to put it in a different way: conscious and unconscious components of mind are subject to kindred laws; physiological processes, to entirely different ones. The unconscious elements of mind have been introduced into psychology in order to fill the gaps of its causal explanations and to complete the domain of validity of psychological laws. This method of completing the scientific domain is entirely legitimate and is used in the physical sciences as well. Astronomers, for example, do not hesitate to discuss multiple stars with partly bright and partly dark components. Psychology of unconscious mental phenomena is not less empirical than astronomy of invisible stars. Another point, which need not be discussed here, is that, in its present state, the method of exploring unconscious phenomena needs improvement. Psychology of the unconscious is young and fruitful; but it is as inexact as all young sciences have been initially. Greater exactness may be obtained in the future by experiments and by comparison with animal psychology. In this article, however, we are not concerned with empirical details and special questions. It has only been necessary to discuss the basic question of empirical testability.

8. The Social Sciences

Among empirical sciences, the social sciences are youngest. They have developed gradually from at least three different sources. Jurists, political writers, and philosophers of the seventeenth century dealt with public law and political philosophy. They were, however, more concerned with rationally establishing their political aims and theories than with careful and unbiased observation of empirical facts. Hobbes, for example, was more a rationalist than an empiricist in his theory of society. He was among the most consistent representatives of the doctrine of social contract, which dominated political philosophy until the end of the eighteenth century; and this doctrine's disregard of experience is obvious today. Primitive man subscribes to rational agreements just as little as he believes in the rational articles of "natural religion." With the advance in overseas trade, savage nations gradually became better known in eighteenth-century Europe, and an increasing number of authors began to criticize, with empirical arguments, the hypothesis of a social contract.

About the same period, French writers, among them Voltaire, Raynal, and Condorcet, started investigating the history of human civilization and the development of social institutions. In political historiography, also, the advance of empirical thinking and the gradual disappearance of theological ideas might be traced. Yet, the political historians still dealt with single events and individuals and hardly belong among the predecessors of the sociologists. On the other hand, the writers who cultivated "philosophical" history, as it was called in the period of the Enlightenment, helped to prepare the way for the social sciences. They discussed general processes in the development of civilization and the interdependence of these processes. The ideas of such writers were mixed with unexamined assumptions concerning human progress and contained a good deal of wishful thinking; but, in the last analysis, their philosophy was based on empirical observation and comparison: the "philosophical" components of their expositions were but vague—and often incorrect—formulations of sociological causes and laws. Reports on exotic peoples were also published more frequently in the eighteenth century; they greatly influenced the writings of the "philosophical" historians and, therefore, may be counted among them.

A third source of the social sciences springs from the practical needs of economy. In contrast to more primitive forms of economy, capitalism requires rational regulation of economic activities. This holds true of private as well as of political economy. Since the beginning of the modern era, the princes—or rather their secretaries and counselors—were compelled to form opinions on the question of how taxes and duties influence commerce and manufacture and are, in turn, influenced by the prosperity of the latter; their revenues depended on these questions. When, in the middle of the eighteenth century in France, problems of agriculture also became urgent, public officials and private writers began to study economics more systematically. And, finally, Adam Smith, the friend of Hume, published his comprehensive theory of economic processes (1776). He thus created the science of political economy and started a scientific development which has continued up to the present day. Political economy is the oldest among the social sciences. As it originated in the needs of an eminently rational and antitraditionalist form of economy, it was from its very beginnings based on reasoning and experience. It never went through an animistic stage and was spared controversy with Aristotelian entelechies.[11] Control and prediction of economic processes were among its aims from the beginning. Consequently, theoretical economists immediately began looking for causal explanations and very soon for economic laws. They obtained, and still obtain, the empirical material for their theories by comparative observation; in the nineteenth century statistical methods were added and proved to be highly successful. Experiments have not yet been performed in the field of economics. The hazardous steps, in politics and business life, that are often called economic experiments are done neither for scientific purposes nor with the necessary scientific precautions.

Rational deduction and mathematics do play a large part in certain economic theories. Political economy might be the only empirical science in which, for the reasons mentioned above, the empirical elements are likely to be impaired rather by excess of rational deduction than by prescientific tradition. It is still a rather imperfect science. It is about two centuries younger than physics, and its subject matter is more complex; economic experiments are not performed, and economic research is exposed to more selfish interests, political pressure, and wishful thinking than is the case in any other science. It is comprehensible, therefore, that in political economy scientific agreement could be reached only on comparatively unimportant questions; in fact, there are separate schools which do not even recognize each other. Some of them cling to experience; the results of their inquiries are collections of material rather than theories in which facts are causally explained. Others deal with nothing but laws of economy; they investigate them by means of rational analysis of a few basic concepts and construct large deductive systems based upon scanty observations. And yet, after all, political economy might be considered the most advanced among the social sciences.

Sociology originated in the beginning of the nineteenth century. Two rather different thinkers, Hegel (1770–1831) and Comte (1798–1857), may be called the first sociologists. Hegel comes from German Romanticism. His *Philosophy of History* employs the dialectical method, which is claimed to be rational and forms the backbone of his whole philosophical system. This method can be traced back to Plato. The metaphysical mill of thesis, antithesis, and synthesis, however, is as empty as inexact; it could yield almost any result. In Hegel's system it begins its work with the concept of being. It cannot be doubted that the method's rich output has, unconsciously, been supplied by and adapted to experience, for it would be sheer magic if reasoning alone were able to derive an abundance of concrete details from a concept which is devoid of content. However, while Hegel's exposition of the philosophy of nature is rather sterile and sometimes strangely contradicts later experiences, his philosophy of history, law, and religion abounds with fruitful results. Apparently, the artificial mechanism of dialectic fits, approximately, certain processes which often occur in history, society, and civilization. The exact and empirical description of those processes, and the demarcation of provinces in which they occur and in which they do not, has not yet been accomplished. Hegel discussed the general lines of development in law and ethics, in morals and political institutions, in art, religion, and philosophy. He did not describe isolated events but comprehended numerous single facts in one construct and saw relationships which no one before him had noticed. These relationships always refer to temporal succession of cultural phenomena. As they are repeated in an analogous way in various fields, they may be taken as preliminary intimations of laws of sociological succession. Regularity, however, was always interpreted as necessity by Hegel. Fruitful empirical knowledge often appears at first in vague and metaphysical expressions; Hegel's philosophy of history is possibly the most impressive case in point.

Comte, the disciple of Saint-Simon, stemmed from the philosophical historians of the French Enlightenment. His attitude was entirely empirical and antimetaphysical. He was well trained in the natural sciences and regarded prediction as the main goal of knowledge. For him, *to know* meant *to foresee*, and science, in his system, is identified with the investigation of laws. Like Hegel, he investigated only the main lines of historical development and comprehended numerous single facts in very general statements. Utilizing the ideas of predecessors, he affirmed the sequence of a theological, a metaphysical, and a positive stage in the history of civilization. This succession is supposed to appear in different fields in the same way. His theory of civilization, therefore, may be taken as a preliminary formulation of a sociological law of succession. Comte coined the names of both sociology and positivism. Compared with Hegel, his expositions contain less speculation, but their empirical results are probably poorer.

It is not necessary at this point to survey the development of sociology after Comte. And even achievements so important as the introduction of the concept of evolution by Spencer and the emphasis on economic facts and economic groups by Marx need not be discussed here. But a few methodological aspects should be pointed out. Among sociologists there are today various schools and many controversies; some schools even disregard the investigations of most of the others. It might be generally agreed that sociology is an empirical science. It is based on observation and comparison, if not yet on experiment. As it does not deal with individuals but investigates group and mass phenomena, the general sociological statements which appear in still rather uncritical forms in Hegel and Comte must be based on careful and complete collection of material if reliable results are to be achieved. It is here that statistical and sometimes even quantitative methods were successfully introduced in sociology. They were, however, largely applied to quite elementary phenomena and their use frequently resulted in mere collections of material. Causal and comprehensive sociological theories, based on statistics, are still lacking. Apart from the difficulty of putting them into practice, there is no reason why statistical methods might not be employed in the investigation of Hegel's and Comte's problems as well, that is, in investigation of cultural processes. Sociological processes are usually interpreted psychologically, that is, by empathy. It has already been pointed out that empathy proves to be fruitful as a heuristic method but that its results must always be tested by observable facts.

As to sociological laws, modern sociology does not restrict itself to the investigation of laws of succession but seeks laws of coexistence as well. There are, however, sociological schools which deny the possibility of any sociological laws. These maintain that in social research causes and laws have to be replaced by "types," by "understanding"—that is, empathy, by "wholeness," entelechies, and values. The preliminary methods of noncausal description and classification and, moreover, all prescientific and teleological concepts of the past are revived as ultimate goals of science in this rebellion against causality. No convincing reasons, however, have yet been given proving that causes and laws must be discarded. In everyday life we are wont to predict, more or less successfully, such social phenomena as overcrowdedness of railroad trains, the outcome of elections, and the trend of public opinion. Predictions like these presuppose that social phenomena are connected more or less regularly. It may be that in any society regularities practically always have exceptions and, consequently, hold only approximately. Social groups are seldom isolated and usually interact with one another; the number of their members is always comparatively small; the members are very different, and some of them exert disproportionate influence. These conditions do not favor group laws. With a gas inclosed in a vessel with permeable walls and consisting of only a million molecules, a few of them being extremely large, rather inexact gas laws could be ascertained. It is possible that in sociology, also, only very inexact regularities can be discovered. Yet, no physicist or astronomer would entirely disregard a regularity on the ground that it did not always hold.

One more point must be considered. The classical gas laws deal with the interdependence of temperature, pressure, and volume of the gas. Nevertheless, the single molecules of which the gas consists have neither temperature nor pressure; they whirl at random, have kinetic energy and impulse, and are, in classical mechanics, subject to laws entirely different from gas laws. In an analogous way sociological laws might connect quite different variables than psychological laws do, though social groups consist of human individuals. Thus we have come back once more to the question of empathy. If we look for social regularities by means of empathy, we may never find them, since ideas, wishes, and actions might not appear in them at all. That is, social regularities may belong to a type of connection entirely different from psychological ones. At any rate, if there are sociological laws, they can be discovered only by actually looking for them and not by discussing their possibility. And what methods are to be employed in this investigation? The empirical methods of causal research have, in all sciences, proved to be so fruitful that we shall not rashly give up hope of finding them successful in the field of sociology too.

We have surveyed the advance of empirical thinking in the domain of the special sciences. We have found that, though different means of inquiry were adapted to the various subjects of the various sciences, in the final analysis success was achieved by the same methods everywhere. These methods are: collection of the material, observation, and comparison; experiment wherever objects can be influenced by technological means; counting and measuring, if possible; causal investigation and investigation of laws. As to methods, there are no basic differences among empirical sciences. The empirical methods are best developed in physics, and physical patterns, consequently, have influenced the other sciences in an increasing degree, despite the remarkable countercurrents in sociology and biology of the last decades. As to methodological maturity and scientific success, biology is today next to physics. Sociology, on the other hand, is the least mature among the empirical sciences. Whether psychology or political economy is better developed is an open question. When such questions are to be decided, the deductive theories in the various sciences and the application of mathematics to their problems must be taken into account as well; the rational methods, however, have not been discussed here. As to historical sequence, the physical sciences are the oldest. It is noteworthy that causal psychology, which originated in the middle of the eighteenth century, and even causal political economy, which originated in the late eighteenth century, are considerably older than causal biology, which originated with Darwin and the physiology of the late nineteenth century. Yet biology has since then outstripped by far its older competitors. The historical sequence and historical development of the sciences do not exactly conform to the various degrees of complexity of their problems. They are greatly influenced by economic needs and the great struggle of ideas and can be explained only in connection with the general process of history.

IV. The Decline of the Mechanical Conception of Nature

9. The Elimination of Mechanical Models

In the second half of the nineteenth century important physical discoveries resulted in the breakdown of the mechanical theories of light, electricity, and magnetism. As philosophy, since the period of Galileo, had been influenced by physics to a higher degree than by any other empirical science, this physical revolution also reshaped philosophical thinking and the analysis of knowledge.

The process began with Maxwell's inquiries on electricity (1865). Using the experiments and ideas of Faraday (1791–1867), Maxwell succeeded in comprehending all laws of the spread of electromagnetic actions in two fundamental equations. Faraday, on the other hand, had interpreted electromagnetic processes mechanically and pictured electric and magnetic lines of forces as invisible ropes, acting in accordance with mechanical laws. Maxwell also obtained his equations by means of a mechanical (namely, a hydrodynamic) analogy, but the mechanical model turned out to be extremely complicated this time. Maxwell was forced to imagine a model consisting of rotating, invisible ether whirls and interspersed invisible particles pushed on by the rotations. In addition, his equations connected electrodynamics with optics, since they contained the velocity of light as an essential constant and covered all laws of propagation of light-waves as well. When H. Hertz twenty years later (1888), by his famous experiments, ascertained, for the first time, the existence of electromagnetic waves, the wave theory of light definitely became a part of Maxwell's theory of electromagnetism. Still, electromagnetic waves were supposed to spread as a kind of movement in the ether. The ether, however, had, on the one hand, to penetrate everything and, on the other, to behave mechanically as a solid body, if observable optical phenomena were to be represented correctly. Thus, mechanical models gradually came to be regarded as tools with which science might possibly dispense. Why should one derive both electromagnetic and optical from mechanical laws by means of highly complex models if all laws of optics can be derived directly from simply constructed electromagnetic equations? Possibly mechanisms do not have any preferred place over other physical phenomena. If equations, wherever they are derived from, present all the observable facts in the simplest way possible, they may very well fulfil the task of science better than theories attempting to reveal a "real" world behind the phenomena.

At first, ideas like these ventured forth hesitatingly, and their implications were either not realized or met with considerable resistance from physicists. Even before the experiments of Hertz, Kirchhoff, the founder of modern astrophysics, advocated, in his *Lectures on Mathematical Physics* (1874), the opinion that physics had to "describe" rather than explain phenomena. Similar ideas were supported by Helmholtz four years later (*Tatsachen der Wahrnehmung*, 1879). The physical and epistemological implications of such ideas were developed, with greater radicalism, by the philosophical physicist Ernst Mach (*Mechanik in ihrer Entwicklung*, 1883). Mach analyzed and consistently refuted the belief in the priority of mechanics to the other branches of physics. In his conception of science scientific explanation is equivalent to "economical description" of the observed facts; science has to represent, as he pointed out, as many facts as possible by as few concepts as possible, and it is irrelevant whether the concepts used are taken from mechanics or elsewhere. Moreover, Mach disclosed the philosophical pseudo-problems originating from the mechanical conception of nature. The opposition of an objective world of quantities and a subjective world of qualities, which had confused philosophy for three centuries, has been overcome in his analysis of scientific knowledge. Starting from different problems, the historian of physics, Pierre Duhem, and the chemist Wilhelm Ostwald helped to destroy the mechanical prejudice. Duhem (*La Théorie physique, son objet, sa structure,* 1906) pointed to the different styles of thinking favored by the English and the French, respectively, in their physical theories; he ascertained that the theories of the French, representing the facts solely by means of mathematical equations, were not less efficient than the English theories based on mechanical models. On the other hand, general energetics, advocated by Ostwald, was also suited to weaken the overestimation of mechanics; for energetics restricted itself to discussing transformations of energy in all fields equally and disregarded mechanical models.

Mechanism suffered its decisive defeat, however, in the theory of relativity and modern atomic physics. Influenced by Mach's ideas, Einstein published his fundamental papers on the special theory of relativity in 1905 and on the general theory in 1916. As is generally known, the theory of relativity pointed out the connection between certain paradoxes of light-propagation and of all spatial and temporal measurements and gravitation. Those highly general connections could no longer be represented by mechanical actions of one ether. A mechanical model of all relativistic laws has not yet been mathematically constructed in every detail. If, after the theory of relativity, a physicist still intended to be a mechanist, he would have to take refuge in a "great-" if not a "great-great-ether" behind the ether, in order to represent all relativistic facts. Evidently overcomplicated and patched up, a mechanism like this could not stand competition with the mathematical conciseness of the relativistic equations. And, finally, in 1913 Niels Bohr's paper appeared, which succeeded in representing the hydrogen atom by a planetary system of particles of electricity. From this day modern atomic physics has, with increasing success, derived all properties of matter itself—among them pushing and pulling—from actions of particles of electricity. These actions follow entirely nonmechanical laws. It can be explained only by history why modern microphysics has been named quantum *mechanics.* As is generally known, Bohr's original model of an electrical planetary system has since been given up and only equations are left. In Heisenberg's, Dirac's, and Schroedinger's quantum mechanics there is, for the most part, no question of even spatial movements of particles, let alone their pushing and pulling. Thus we have come to realize that it amounts to the same thing whether attraction of electric charges is explained by the pulling of invisible cords or the pulling of cords by the behavior of electrons and protons. We accept that theory which covers the observable phenomena and which, at the same time, is the more comprehensive, the more consistent, and the simpler. The mechanical prejudice has definitely been overcome. It need not be mentioned that the breakdown of the mechanistic conception does not mean at all a return to premechanical animistic or teleological ideas. The statistical laws of quantum mechanics are as rational and mathematical as the laws of classical mechanics and not a bit nearer to the behavior of living beings or spirits.[12]

The breakdown of mechanistic physics took place during a period of complete revolution in technology. The levers and pulleys of the seventeenth century had receded to the background long ago. The steam engine and, in the second half of the nineteenth century, the electromotor, the dynamo, and the internal-combustion engine had taken their place. The working of all of these is based on nonmechanical processes. Merely mechanical machines, as bicycles, typewriters, and foot-driven sewing machines, have been comparatively unimportant in the economy of the last sixty years. The textile industry, in which mechanisms played a comparatively greater part, had been the leading industry up to the early nineteenth century but was fifty years later surpassed by far in economic importance by the electrical, and then by the chemical, industry. Men, and even the children, of the twentieth century do not feel at all strange about the working of an electric bulb, a telephone, a photographic camera, a radio set; what is even more important, they do not feel any different toward the working of these implements than toward the working of a typewriter. The movements by which man influences the world around him have, with the technology of the twentieth century, shriveled down in many cases to turning on a switch; the electromagnetic, chemical, and caloric processes which are then initiated do the rest. If man is inclined to conceive natural processes after the pattern of how he himself influences nature, it is scarcely a surprise that the priority of mechanics has completely dwindled. Certainly, modern technology and economy have not only influenced modern physical theories but have been influenced by them as well. The electrical engineer, Marconi, did succeed the theorists Hertz, Maxwell, and Mach. On the other hand, however, Mach succeeded the steam-driven factories, the dynamo, and the electromotor, and, certainly, electro-engineering and radiobroadcasting helped immensely to make the nonmechanical conception of nature less paradoxical.

10. Final Remarks

The breakdown of mechanistic physics could not fail to give a new impetus to empirical thinking. With the failure of mechanistic physics, the assumption of a second world behind experience had lost its scientific support. Now the subject-object metaphysics, the pride of all philosophers, who looked down on the naïve laymen, was badly shaken; its problems began to appear as pseudo-problems. Since causes and laws were employed in the new physics as functional connections and mere regularities, the unempirical components of those concepts, already criticized philosophically by Hume, became suspect for scientists as well. All these implications were consistently developed by Mach. On the other hand, physical hypotheses and models had suddenly turned out to be unsuitable, though having proved fruitful for three centuries. Necessarily, general methodological questions arose as a result of that fact, and, for the first time in the history of modern physics, the whole internal construction of science became problematic. What part is played by simplicity in scientific theories? Which natural laws should one consider as fundamental, and which as derived, when constructing a theory? What service can be rendered to science by working hypotheses, by fictions, and by conventions? Most of these problems deal rather with the deductive side of theoretical knowledge than with its empirical components. They were raised by Mach, by fictionalism and conventionalism of the late nineteenth century, and were more or less suggested by the physical revolution. Poincaré's conventionalism, however, was influenced by modern mathematics as well as by the new physics. In the early twentieth century those mathematical and logical influences increased, united with the empiricist tradition, and resulted finally in logical empiricism—a subject which must be reserved for later treatment.

NOTES

1. The empirical achievements of Bacon and Albertus are often overestimated (cf. L. Thorndike, *A History of Magic and Experimental Science* [New York, 1923], Vol. II).

2. On the contemporary writings on inventors and discoverers cf. E. Zilsel, *Die Entstehung des Geniebegriffes* (Tübingen, 1926), pp. 130–43.

3. On the artist-engineers and their papers cf. Leonard Olschki, *Geschichte der neusprachlichen wissenschaftlichen Literatur*, I (Heidelberg, 1918), 45–447, and Zilsel, *op. cit.*, pp. 144–57.

4. On Galileo cf. Olschki, *op. cit.*, III (Halle, 1927), 117–467.

5. Galileo, *Opere* (ed. nazionale), XIX, 32 ff.

6. Before foundation of the Accademia del disegno, young artists had always received their education at the workshop, like all apprentices. The new school clearly shows how engineering gradually penetrated the province of academic instruction.

7. Galileo, *op. cit.*, XIX, 130 ff.

8. Cf. E. Zilsel, "The Origins of William Gilbert's Scientific Method," *Journal of the History of Ideas*, Vol. II (1941).

9. In fact, the metaphor is not older. In scholasticism the term 'natural law' had an entirely different and merely juristical and ethical meaning. The Middle Ages could not produce the modern concept of natural law for the simple reason that the feudal state was governed not by statute but by unwritten and loose traditional law. As far as medieval princes issued regulations at all, they were for the most part privileges given to single noblemen, monasteries, and towns. They compare, consequently, rather with exceptions than with regularities in nature. Also with the ancient authors the metaphor of natural law appears but seldom and in a rather vague form. Antiquity, however, was familiar with the mythological idea of "fate."

10. We need not distinguish here between unconscious and subconscious processes.

11. In the twentieth century, however, a few German economists attempted to introduce entelechies in political economy.

12. Cf. Philipp Frank, *Interpretations and Misinterpretations of Modern Physics.*

The Development of Logical Empiricism

Joergen Joergensen

The Development of Logical Empiricism

Contents:

The Development of Logical Empiricism

Joergen Joergensen

I. The Vienna Circle: Its Program and Presuppositions

1. Introductory Remarks

In Volume II, No. 8, of this *Encyclopedia* Edgar Zilsel outlined the evolution of empiricism in its broad sense up to the beginning of this century. The present article will continue this history, sketching the development until 1940[1] of the recent form of empiricism generally called "logical empiricism."

It is characteristic of this movement, which may presumably be said to have been the leading movement within the philosophy of the last two or three decades, that it is an expression of a need for clarification of the foundations and meaning of knowledge rather than of a need for justification of a preconceived view; that it attempts to make philosophy scientifically tenable through critical analysis of details rather than to make it universal by vague generalizations and dogmatic construction of systems; and that it is more interested in co-operation among philosophers and between philosophers and investigators in the special sciences than in the advancement of more or less striking individual opinions. What unites its members is, therefore, not so much definite views or dogmas as definite tendencies and endeavors. An evidence of this is the often considerable divergence and lively discussion between its members and the amendments in the fundamental views that have occurred several times in the course of its development. On the other hand, the constant exchange of opinion has led to an increasing convergence toward certain basic principles that have gradually taken shape and that now form the common basis for the further discussion of still unsettled questions. The nature of these fundamental principles will be clarified in the following exposition.

2. The Vienna Circle

The nucleus from which logical empiricism developed was the so-called "Vienna Circle," the origin of which is described by Herbert Feigl, one of its younger members, as follows:

"The Vienna Circle evolved in 1923 out of a seminar led by Professor Moritz Schlick and attended, among other students, by F. Waismann and H. Feigl. Schlick's teaching period in Vienna had begun in 1922, and by 1925 out of this nucleus a Thursday evening discussion group was formed. It is interesting to note that many of the participants were not professional philosophers. Even if some of them taught philosophy, their original fields of study lay in other disciplines. Schlick, for example, had specialized in physics, and his doctor's thesis, written under the guidance of Max Planck in Berlin, concerned a problem in theoretical optics. Among the other active members we may mention: Hans Hahn, mathematician; Otto Neurath, sociologist; Victor Kraft, historian; Felix Kaufmann, lawyer; Kurt Reidemeister, mathematician. An occasional but a most contributive visitor was the Prague physicist, Philipp Frank (now at Harvard). In 1927 and again in 1932 the brilliant Finnish psychologist and philosopher, E. Kaila, was present as an active and critical member of the group. Another visitor from Scandinavia was Å. Petzaell (Göteborg). Among the younger participants were K. Goedel (now at Princeton), T. Radakovic, G. Bergmann, M. Natkin, J. Schaechter, W. Hollitscher, and Rose Rand; and, among the visitors, C. G. Hempel, Berlin; A. E. Blumberg, Baltimore; and A. J. Ayer, Oxford. Among those more loosely affiliated with the group were K. Menger, E. Zilsel, K. Popper, H. Kelsen, L. v. Bertalanffy, Heinrich Gomperz, B. von Juhos.

"The most decisive and rapid development of ideas began in 1926 when Carnap was called to the University of Vienna. His contributions to axiomatics and particularly his theory of the constitution of empirical concepts (as published in *Der logische Aufbau der Welt*) proved a very stimulating source of discussions. In the same year also, Ludwig Wittgenstein's *Tractatus Logico-Philosophicus* was studied by the Circle. The philosophical position of Logical Positivism in its original form was the outcome of these profoundly incisive influences. Though many of the basic ideas had already been enunciated in a general manner by Schlick, they were formulated more precisely, stated more fully and radically, by Carnap and Wittgenstein, quite independently. These two men exerted an enormous influence upon Schlick, who was about ten years their senior.

"In contrast to Carnap, who became a regular and most influential participant in the group, Wittgenstein, then preoccupied with architecture, associated only occasionally with some of the members of the Circle. Even thus, more light was obtained on some of the rather obscurely written passages of his extremely condensed and profound *Tractatus*. A few years later Wittgenstein returned to his philosophical studies and was called to Cambridge, England, where he later became successor to G. E. Moore. Schlick, as a visiting Professor, went to California in 1929 and 1932. Carnap was called to Prague in 1930 (later, in 1936, to Chicago) and Feigl to the United States in 1930. Hans Hahn, who was an expert in *Principia Mathematica* and in general an enthusiastic follower of Russell, died prematurely in 1934. The Circle discussions, however, continued, with Schlick and Waismann leading, until Schlick's tragic death in 1936. A former student, for years under observation by psychiatrists and diagnosed as insane, assassinated Schlick. The passing of this kindly, truly great and noble man, was bitterly lamented by his many friends.

"The discussions of the Circle centered about the foundations of logic and mathematics, the logic of empirical knowledge, and only occasional excursions into the philosophy of the social sciences and ethics. Despite the many differences of opinion, there was a remarkable spirit of friendly cooperation in the Circle. The procedure was definitely that of a joint search for clarity."[2]

3. The Program

A more detailed exposition of the work of the Vienna Circle was given in 1929 in a publication entitled *Wissenschaftliche Weltauffassung: Der Wiener Kreis*, which marked the appearance of the Circle before the public as an organization with a scientific as well as an educational purpose. The publication was sent out by the Verein Ernst Mach, which was founded in November, 1928, with Schlick as a president, and had for its object to "further and propagate a scientific world view," as conceived by the members of the Circle, through public lectures and writings. According to this pamphlet, the main lines of their program may be described as follows:

The aim is to form an *Einheitswissenschaft*, i.e., a unified science comprising all knowedge of reality accessible to man without dividing it into separate, unconnected special disciplines, such as physics and psychology, natural science and letters, philosophy and the special sciences. The way to attain this is by the use of the *logical method of analysis*, worked out by Peano, Frege, Whitehead, and Russell, which serves to eliminate metaphysical problems and assertions as meaningless as well as to clarify the meaning of concepts and sentences of empirical science by showing their immediately observable content—"das Gegebene." In both respects the Vienna Circle continues the endeavors initiated by Ernst Mach; but, by the application of logical analysis, which is a distinctive feature of the new empiricism, and of positivism, as compared to the older forms of these movements, it obtains a hitherto unattained completeness and precision. The culmination so far has been reached in the "constitution theory" advanced by Rudolf Carnap in *Der logische Aufbau der Welt* (Berlin, 1928), according to which any tenable concept of real objects is constituted by being reduced to characteristics of that which is immediately given, and any meaningful statement is constituted by being reduced to a statement of the given. Thus a framework is created for the work of the Vienna Circle; its negative task is an expurgation of metaphysical-speculative statements as meaningless, while its positive task is to define ever more precisely and fully the meaning of scientifically tenable statements. "If anyone asserts: 'There is a God,' 'The first cause of the world is the Unconscious,' 'There is an entelechy which is the leading principle in living beings,' we do not say 'What you say is false'; rather, we ask him, 'What do you mean by your statements?' It then appears that there is a sharp division between two types of statements. One of the types includes statements as they are made in empirical science; their meaning can be determined by logical analysis, or, more precisely, by reduction to simple sentences about the empirically given. The other statements, including those mentioned above, show themselves to be completely meaningless, if we take them as the metaphysician intends them. Of course, we can frequently reinterpret them as empirical statements. They then, however, lose the emotional content which is the very thing which is essential to the metaphysician. The metaphysicians and theologians, misinterpreting their own sentences, believe that their sentences assert something, represent some state of affairs. Nevertheless, analysis shows that these sentences do not say anything, being instead only an expression of some emotional attitude. To express this may certainly be a significant task. However, the adequate means for its expression is art, for example, lyric poetry or music. If, instead of these, the linguistic dress of a theory is chosen, a danger arises: a theoretical content, which does not exist, is feigned. If a metaphysician or theologian wishes to retain the usual form in language, he should understand thoroughly and explain clearly that it is not representation but expression; not theory, information, or cognition, but rather poetry or myth. If a mystic asserts that he has experiences that transcend all concepts, he cannot be challenged. But he cannot speak about it, since speaking means grasping concepts and reducing to facts which can be incorporated into science."[3]

The view of the Vienna Circle as to how such incorporation should be undertaken is particularly evident in the above-mentioned theory of constitution, which will be more explicitly dealt with presently. First, however, it would be expedient to look at the development of the opinions and points of view forming the background of the Circle's conception of philosophy and knowledge generally.

4. Predecessors

The forerunners of logical empiricism are, in the opinion of the members of the movement themselves, all those philosophers and scientists who show a clear antimetaphysical or antispeculative, realistic or materialistic, critical or skeptical, tendency—as well as everyone who has contributed essentially to the development of their most important methodological instrument: symbolic logic. In antiquity the Sophists and the Epicureans are mentioned; in the Middle Ages the nominalists; and in modern times, Neurath[4] gives the following three lists of names, indicating the lines of development in England, France, and Germany that may be said to lead in the direction of logical empiricism: Bacon, Hobbes, Locke, Hume, Bentham, J. S. Mill, Spencer; Descartes, Bayle, D'Alembert, Saint-Simon, Comte, Poincaré; Leibniz, Bolzano, Mach. In their programs similar lines of development are noticed, only here the grouping has been made according to subject and not according to nationality:

1. Positivism and empiricism: Hume, the philosophers of the Enlightenment, Comte, Mill, Avenarius, Mach.
2. The basis, aims, and methods (hypotheses in physics, geometry, etc.) of the empirical sciences: Helmholtz, Riemann, Mach, Poincaré, Enriques, Duhem, Boltzmann, Einstein.
3. Logistics and its application to reality: Leibniz, Peano, Frege, Schröder, Russell, Whitehead, Wittgenstein.
4. Axiomatics: Pasch, Peano, Vailati, Pieri, Hilbert.
5. Eudaemonism and positivistic sociology: Epicurus, Hume, Bentham, Mill, Comte, Feuerbach, Marx, Spencer, Müller-Lyer, Popper-Lynkeus, Carl Menger (the economist).

The predecessors and teachers here mentioned were the ones especially studied and discussed by the Vienna Circle. Not until later did the Circle discover the American pragmatists, instrumentalists, and operationalists to whom they are closely related in several respects and with whom they have since co-operated, as well as with certain other affiliated groups. I shall return to this point in the following chapter, having mentioned here only the authors whose works have actually played an important part in the development of the views of the Circle. There are, however, three of these philosophers whose influence has been so significant that they must be more explicitly dealt with: Ernst Mach, Bertrand Russell, and Ludwig Wittgenstein.

5. The Positivism of Ernst Mach

Mach (1838–1916), who started his scientific career as a professor of mathematics and later of physics, in 1895 received a professorship in philosophy, especially the history and theory of the inductive sciences, at the University of Vienna. Owing to bad health, however, he had to retire in 1901; he was succeeded by the well-known physicist, L. Boltzmann. From his youth Mach had been vividly interested in philosophical and epistemological questions as well as in the historical development of the natural sciences. This appears clearly from his main works, *Die Mechanik in ihrer Entwicklung, historisch-kritisch dargestellt* (1883) (English trans., *The Science of Mechanics* [Chicago, 1893]), *Die Analyse der Empfindungen und das Verhältnis des Physischen zum Psychischen* (1886) (English trans., *The Analysis of Sensations* [Chicago, 1914]), *Die Prinzipien der Wärmelehre, historisch-kritisch entwickelt* (1896), *Populär-wissenschaftliche Vorlesungen* (1896) (English trans., *Popular Scientific Lectures* [Chicago, 1895]), *Erkenntnis und Irrtum: Skizzen zur Psychologie der Forschung* (1905), and *Die Prinzipien der physikalischen Optik, historisch und erkenntnispsychologisch entwickelt* (finished 1913, published 1921) (English trans., *Principles of Physical Optics* [New York, 1926]).

In these books Mach advanced his positivistic theory of knowledge, according to which human knowledge from its most primitive forms to the heights of science is a biological phenomenon, part of the history of man. Influenced by Darwin's theory of evolution, he conceived knowledge as a never ending process of adjustment of thoughts to reality and to each other. A priori and eternal truths do not exist, nor is there any difference in principle between axioms and deduced sentences. All statements concerning the world, particular as well as general rules, natural laws, theories, and principles, are subject to continuous control and modification by experience. Even geometrical sentences are, in so far as they are statements about reality, empirical sentences whose validity depends simply on immemorial observations of regularities in the spatial conditions and movements of things; so considered, geometry is a part of natural science of the same kind as mechanics or the theory of heat or the theory of electricity. Accordingly, we are not bound to follow any definite kind of geometry but may choose the one that appears to attain most expediently the most thought-saving description of experiences of the spatial relations of things. Space itself is merely the totality of the spatial relations of things, and not—as believed by Newton and Kant—an empty container in which things have been located in "absolute" places or in which they perform "absolute" movements.

On a closer examination, things, too, appear to be merely relatively constant complexes of so-called "qualities," which Mach identified with our sensations and called "elements." A "thing-in-itself" existing behind these elements is a metaphysical illusion, presumably due to the fact that the same names are used to designate things, even though these change, so that we are led to believe that the "same" thing persists throughout the changes. What we do observe is actually never any such hidden things but simply qualities of their mutual relations. Natural laws should therefore be formulated as functional relations between the elements, i.e., between sensations such as green, hot, hard, extended, continuous, etc. These sensations are not in themselves illusory or deceptive, but, on the contrary, they are all that we know of reality. What we call "deception of the senses" is merely certain unusual complexes of elements deceiving us because they resemble familiar complexes and so raise expectations that are not fulfilled. If they are correctly conceived, there is nothing deceptive about them. The point is to distinguish between the various contexts in which any given element may occur. These contexts are all equally real; but, if they are confused with one another, contradictions arise which we attempt to avoid by declaring one of their terms illusory. To a person measuring the site for a house, the surface of the earth is a plane, but to the person undertaking an exact measurement of the total surface of the globe, it is a spheroid. There is no contradiction in this, if only it is realized that the observations are made under different conditions and in different ways and that the words describing them take on different meanings when we pass from one standpoint to another. 'Up' and 'down', for instance, are everyday terms which have an easily understood sense in the world of our everyday life but which lose this sense when we proceed to describe the universe astronomically. And similarly the words 'red', 'yellow', and 'blue' are names designating sensations, and as such are well suited to describe the phenomena of our daily life; however, they must be replaced by words like 'wave movements' or 'corpuscle rays' if we want to describe the more subtle phenomena and contexts of phenomena observed by the physicist in his investigations of color. One kind of observation is not truer or more faithful to reality than is the other, but the contexts in which they occur differ and must be described by different words. Every scientific statement is a statement about complexes of sensations, and beyond or behind these there are no realities to be looked for, because the word 'reality' itself is merely a name for the sum total of the complexes of observable sensations.

What has been said here of physical things, according to Mach, also applies to the so-called "mental substances" of egos: they are merely special complexes of elements, even complexes with fluctuating boundaries, now expanding and now narrowing, but continuously changing in the course of life, disappearing in a dreamless sleep and altering completely in case of a mental disease or other abnormality. Although of great practical importance in our daily life, the word 'ego' does not signify any unchangeable or eternal object of a specific "mental" character; indeed, the difference between physical and psychical phenomena does not depend on the nature of the phenomena but solely on the context in which they occur: if a sensation is conceived as a link in a physical natural law, it is called a "physical phenomenon" or a quality of a physical thing, but if it is conceived as a link in a psychological law-directed regularity, i.e., as dependent according to natural law upon the condition of the observer, it is called a "psychical phenomenon." Physical and psychical phenomena are not essentially different, and all statements concerning them are of exactly equal rank, since they can all be reduced to statements about complexes of sensations which are all that is given or immediately observable.

In this the positivism of Mach differs from that of Comte. In Comte—who created the word 'positivism'—'positive', 'supposed', or 'given' signifies, in the first place, observable physical objects as opposed to fictive, speculatively constructed metaphysical entities; and in his system he found no room for psychology itself but classed it, without giving a detailed explanation, under biology. Further, the system of sciences assumed in Comte a hierarchic character, the six basic sciences—mathematics, astronomy, physics, chemistry, biology, and sociology, each presupposing the preceding ones without being capable of being deduced from them; this, expressed differently, means that the "higher" phenomena cannot be reduced to the lower; accordingly, the idea of a unity of science is incompatible with Comte's conception of the hierarchy of the sciences, whereas this idea was anticipated by Mach in his theory that all scientific statements should be reduced to statements of sensations. In this respect Comte's positivism is nearer to the dialectical materialism of Marx and the modern theory of emergent evolution than is the positivism of Mach—a fact which finds expression in Lenin's keen criticism of "Machism" in his *Materialismus und Empiriokriticismus* (1909) (English trans., *Materialism and Empirio-Criticism*, in *Collected Works of Lenin*, Vol. XIII [New York, 1927]). Although logical empiricists reject this criticism as being partially due to misunderstanding and consider themselves in accord with materialism[5] in all essentials, it cannot be denied that the positivism of the Vienna Circle is more closely related to the English empiricists than to the French materialists, with whom, from an epistemological point of view, it has, strictly speaking, in common only a strong aversion to speculative thinking. Among its great teachers we do not find the French encyclopedists or Comte, but Bertrand Russell;[6] Russell, the greatest living representative of English empiricism, may not unjustly be called the "father" of logical positivism, since in him is found for the first time the conscious and extensive application of logical analysis to the problems of epistemological empiricism,[7] a position which was reached by neither Comte nor Mach but which is characteristic of logical empiricism.[8]

6. The Logical Positivism of Bertrand Russell

Bertrand Russell (born 1872) is one of the great pioneers of modern logistics. In his *The Principles of Mathematics* (1903) and in *Principia mathematica*, Volumes I–III (1910–13), which he wrote in collaboration with Alfred North Whitehead (1861–1947), Russell made a more critical and comprehensive attempt than had yet been made to develop a symbolic logic and to show that all pure mathematics may be reduced to formal logic. This reduction he attempted to carry through by (*a*) trying to define all the main concepts of mathematics (such as the concepts of natural numbers and of the various kinds of numbers, the basic concepts of the theory of manifolds, and concepts like continuity and derivative) by means of half-a-dozen basic logical concepts and (*b*) by trying to prove all the axioms of mathematics by means of half-a-dozen logical axioms. In other words, the arithmetization of mathematical analysis, already carried through to a large extent by various mathematicians, Russell attempted to carry on by logicizing all mathematics, the concept of a natural number being, for instance, defined as "a class of similar classes." It is true that the attempt was not altogether successful; but its disadvantages as well as its advantages, and the wealth of ideas it contained, made it a unique source of inspiration to logicians, to investigators of the foundations of mathematics, and to philosophers. From among its many new features we shall mention here only three which came to play a special part in the formation of logical positivism.

The study of logical paradoxes and of paradoxes within the theory of sets led Russell to set forth the theory of logical types, according to which, for instance, every class is of a higher type than are its members and every statement about another statement is of a higher type than the one about which it is made. If the types are kept apart, paradoxes can be avoided, whereas there is a risk of such paradoxes if the types are confused. Russell maintained that statements containing confusions of types are meaningless, even if, according to the usual linguistic syntax, they are correctly constructed; and so he replaced the current logical division of statements into true and false by the tripartition: true, false, and meaningless.

Another important partition of statements introduced in connection with the theory of types was the division into (1) elementary statements, i.e., statements whose truth or falsity may be realized without any knowledge of individual objects or qualities or relations other than the ones whose names occur in the statement in question, and (2) generalized statements, i.e., statements presupposing classes of individuals, qualities, or relations (which must be divided into a hierarchy of ascending types). The elementary statements are subdivided into (*a*) atomic statements, i.e., statements containing no other statements as constituents, and (*b*) molecular statements, i.e., statements containing other statements as constituents. An especially important group of the latter are the so-called "truth-functions," i.e., molecular statements, the truth or falsity of which does not depend upon the meaning of the statements forming part of them but solely upon their truth-values, i.e., their truth or falsity, such as, for instance, negations, disjunctions, conjunctions, conditionals, and biconditionals of elementary statements.

Further, Russell and Whitehead introduced the so-called "principle of abstraction," that may equally well be called "the principle which dispenses with abstraction": "When a group of objects have that kind of similarity which we are inclined to attribute to possession of a common quality, the principle in question shows that membership of the group will serve all the purposes of the supposed common quality, and that therefore, unless some common quality is actually known, the group or class of similar objects may be used to replace the common quality which need not be assumed to exist."[9] Any statement of the common quality may be replaced by a statement saying that something is a member of the class. 'Red', for instance, may be defined by our pointing at a red object and saying that everything of the same color as that object is red, making it unnecessary to analyze this quality further; and, similarly, the cardinal number three may be defined as the class of all classes having the same number of members as, for instance, the class of Paris, London, and Berlin, making it unnecessary to assume that all these triads possess a common quality.

In *Principia mathematica* formal logic was generalized, systematized, and made precise to such an extent, and couched in such expedient symbolic language, that we understand the great expectations of Russell when he said: "The old logic put thought in fetters, while the new logic gives it wings. It has, in my opinion, introduced the same kind of advance into philosophy as Galileo introduced into physics, making it possible at last to see what kinds of problems may be capable of solution, and what kinds must be abandoned as beyond human powers. And where a solution appears possible, the new logic provides a method which enables us to obtain results that do not merely embody personal idiosyncrasies, but must command the assent of all who are competent to form an opinion."[10]

This statement is found in the lectures on our knowledge of the external world which Russell delivered in Boston in 1914 and in which for the first time he used his new logical-analytical method for the solution of epistemological problems. The leading principle was here a form of Occam's razor or law of parsimony: "Entia non sunt multiplicanda praeter necessitatem." Russell later stated this principle in this form: "Wherever possible, substitute constructions out of known entities for inferences to unknown entities";[11] in the lectures here referred to he formulates it in the following way: "In other words, in dealing with any subject-matter, find out what entities are undeniably involved, and state everything in terms of these entities. Very often the resulting statement is more complicated and difficult than one which, like common sense and most philosophy, assumes hypothetical entities whose existence there is no good reason to believe in. We find it easier to imagine a wall-paper with changing colours than to think merely of the series of colours, but it is a mistake to suppose that what is easy and natural in thought is what is most free from unwarrantable assumptions, as the case of 'things' very aptly illustrates."[12]

The example to which he here refers is the analysis of things given in his lectures. Briefly expressed, it says that what we call a "thing" is not a permanent substance different from its changing qualities or appearances but may be defined as "a certain series of appearances, connected with each other by continuity and by certain causal laws."[13] To this view he was led by the following reasoning: every philosophical investigation starts from certain data which we must assume as being, on the whole and in a certain sense, pragmatically true. The data resisting the influence of critical reflection Russell calls "hard data" and thinks they are of two sorts: "the particular facts of sense, and the general truths of logic,"[14] to which must be added certain facts of memory and some introspective facts. By "facts of sense" he means "facts of *our own* sense-data" and maintains that "*in so far* as physics or common sense is verifiable, it must be capable of interpretation in terms of actual sense-data alone. The reason for this is simple. Verification consists always in the occurrence of an expected sense-datum. . . . Now if an expected sense-datum constitutes a verification, what was asserted must have been about sense-data; or, at any rate, if part of what was asserted was not about sense-data, then only the other part has been verified."[15] In order to be verifiable, statements of the external world must accordingly be about our own sense-data. Or, in other words, verifiable statements of the external world must be capable of definition or construction in terms of our own sense-data.

"For instance, a thing may be defined as 'a certain series of appearances, connected with each other by continuity and by certain causal laws.' In the case of slowly changing things, this is easily seen. Consider, say, a wall-paper which fades in the course of years. It is an effort not to conceive of it as one thing whose colour is slightly different at one time from what it is at another. But what do we really *know* about it? We know that under suitable circumstances—i.e. when we are, as is said, 'in the room'—we perceive certain colours in a certain pattern: not always precisely the same colours, but sufficiently similar to feel familiar. If we can state the laws according to which the colour varies, we can state all that is empirically verifiable; the assumption that there is a constant entity, the wall-paper, which has these various colours at various times, is a piece of gratuitous metaphysics. We may, if we like, *define* the wall-paper as the series of its aspects. These are collected together by the same motives which led us to regard the wall-paper as one thing, namely a combination of sensible continuity and causal connection. More generally, a thing will be defined as a certain series of aspects, namely those which would commonly be said to be *of* the thing. To say that a certain aspect is an aspect *of* a certain thing will merely mean that it is one of those which, taken serially, *are* the thing. Everything will then proceed as before: whatever was verifiable is unchanged, but our language is so interpreted as to avoid an unnecessary metaphysical assumption of permanence."[16]

As will be seen, this theory is in accordance with that of Mach, only a little more cautiously and precisely expressed; just as natural numbers are analyzed or constructed or defined as classes of classes, so the things of the external world are analyzed, constructed, or defined as combinations of sense-data. Similar conditions apply to various other entities which Russell defines in terms of sense-data alone, so that statements about them are reduced to statements about sense-data; later on he extended and modified these analyses in his *Analysis of Mind* (1921), in *The Analysis of Matter* (1927), as well as in *An Outline of Philosophy* (1927), without, however, changing his basic views in principle.

In a similar way, Whitehead analyzed in *The Organization of*

Thought (1917), in *An Enquiry concerning the Principles of Natural Knowledge* (1919), and in *The Concept of Nature* (1920), various physical concepts, the only difference being that he did not reduce them to sense-data but to so-called "events." Altogether, philosophical analysis of objects and statements became a main preoccupation of the Cambridge analysts. As a leader of this group along with Russell, and perhaps before him, G. E. Moore (born 1873) should be mentioned. Moore is the author of the modern analytical method which he endeavored to formulate with increasing precision and subtlety without, however, drawing any positivistic conclusions from it; in the Preface to his *Principia ethica* (1903) he clearly indicated his method in the following words: "It appears to me that in ethics, as in all other philosophical studies, the difficulties and disagreements, of which its history is full, are mainly due to a very simple cause: namely, to the attempt to answer questions without first discovering precisely *what* question it is which you desire to answer. I do not know how far this source of error would be done away with if philosophers would *try* to discover what question they were asking before they set about to answer it; for the work of analysis and distinction is often very difficult: we may often fail to make the necessary discovery, even though we make a definite attempt to do so. But I am inclined to think that in many cases a resolute attempt would be sufficient to ensure success; so that, if only this attempt were made, many of the most glaring difficulties and disagreements in philosophy would disappear. At all events, philosophers seem, in general, not to make the attempt; and, whether in consequence of this omission or not, they are constantly endeavouring to prove that 'Yes' or 'No' will answer questions, to which *neither* answer is correct, owing to the fact that what they have before their minds is not one question, but several, to some of which the true answer is 'No,' to others 'Yes.' "[17] Through his penetrating endeavors to apply the method of analysis outlined above, Moore demonstrated convincingly the extreme difficulty of stating *exactly* the sense of many everyday assumptions and expressions, and his great carefulness and caution kept him from formulating far-reaching hypotheses or maintaining definite views. But his acute, painstaking endeavor itself exercised a profound influence on a large number of his pupils in England, among whom may be mentioned C. D. Broad, L. S. Stebbing, F. P. Ramsey, J. T. Wisdom, A. E. Duncan-Jones, A. J. Ayer, M. Black, H. B. Acton, R. B. Braithwaite, K. Britton, W. Kneale, H. Knight, M. MacDonald, C. A. Mace, A. M. MacIver, C. A. Paul, G. Ryle, J. W. Reeves, and, most important, Ludwig Wittgenstein, who attended Moore's lectures during the years 1912–14 and was later, from 1939 to 1948, his successor to the professorship at Cambridge.[18] As Wittgenstein has played a greater part than any other one philosopher in the development of the Vienna Circle, which seems not to have had any close knowledge of Moore, it will be necessary to deal a little more explicitly with his original philosophy.

7. Ludwig Wittgenstein's Logical-philosophical Treatise

The only published philosophical work by Wittgenstein, *Tractatus logico-philosophicus*, is in many respects a highly remarkable achievement. Appearing first in Oswald's *Annalen der Naturphilosophie* in 1921, it was published the following year in London as an independent book, containing the original German text together with an English translation, the latter being carefully revised by the author himself. Its contents are formulated as a series of aphorisms with comments. The comments contain specifications of the main propositions and more or less detailed justifications or illustrations of them. There is no proper deductive context, and often it may be difficult to see whether a proposition is put forward as a mere postulate or as the result of an unformulated argument. Consequently, the book is anything but easy to read, and to this day several parts of it are not completely understood. When, in spite of this, it has aroused such vital interest in expert circles, this is due to the fact that it contains, beyond doubt, remarkable discoveries and ideas of a logical and epistemological nature and that altogether it bears witness to a rare originality and acuteness. It is extremely regrettable that a commentary by Friedrich Waismann, announced for several years, has not been published. However, within the Vienna Circle, where Waismann and Schlick were periodically in touch with Wittgenstein, the book was thoroughly discussed; and through the publications of the Circle, as well as through Russell's Introduction to the English edition, certain of its fundamental new ideas gradually became more widely known. The most important of these, which may be considered largely as a critical development of Russell's thought, are contained in the main propositions of the book:

"1. The world is everything that is the case.

"2. What is the case, the fact, is the existence of atomic facts.

"3. The logical picture of the facts is the thought.

"4. The thought is the significant proposition.

"5. Propositions are truth-functions of elementary propositions. (An elementary proposition is a truth-function of itself.)

"6. The general form of truth-function is: $[\bar{p}, \xi, N(\xi)]$. This is the general form of proposition.

"7. Whereof one cannot speak, thereof one must be silent."

In order to understand these propositions, it is necessary to make clear what problem it is that Wittgenstein tries to solve in his treatise. To use the words of Russell, "What relation must one fact (such as a sentence) have to another in order to be *capable* of being a symbol for that other?"[19] To which Wittgenstein answers as follows:

The symbolizing fact must be a picture of what is symbolized, in the sense that it must be of the same form or structure as that which is symbolized. To every element in the one must correspond one and only one element in the other, and the elements of the two facts must be similarly arranged. They must be related to each other as a figure to its projection or as a gramophone record or the musical thought or the score or the waves of sound are related to one another, so that they can be deduced from each other mutually by means of a kind of "law of projection" (4.014; 4.0141). In the natural languages this relation of projection is highly imperfect, and that is just why our everyday language gives rise to many misunderstandings and senseless philosophical problems. "Most propositions and questions, that have been written about philosophical matters, are not false, but senseless. We cannot, therefore, answer questions of this kind at all, but only state their senselessness. Most questions and propositions of the philosophers result from the fact that we do not understand the logic of our language. (They are of the same kind as the question whether the Good is more or less identical than the Beautiful.) And so it is not to be wondered at that the deepest problems are really *no* problems" (4.003). "All philosophy is 'critique of language' (but not at all in Mauthner's sense). Russell's merit is to have shown that the apparent logical form of the proposition need not be its real form" (4.0031).

This basic thought and its consequences are further illustrated by the comments on the first six main propositions. As to No. 1, the comment simply says that the world consists of facts and that these are independent of one another. As to No. 2, it is observed that atomic facts are combinations of objects (entities, things) that are simple. The way in which objects hang together in an atomic fact is the form of the atomic fact. To the atomic facts correspond atomic propositions, the forms of which must be identical with those of the corresponding facts. "We make to ourselves pictures of facts" (2.1). "The picture is a model of reality" (2.12). "To the objects correspond in the picture the elements of the picture" (2.13). "The picture consists in the fact that its elements are combined with one another in a definite way" (2.14). "In order to be a picture a fact must have something in common with what it pictures" (2.16). "What the picture must have in common with reality in order to be able to represent it after its manner—rightly or falsely—is its form of representation" (2.17). "The picture, however, cannot represent its form of representation; it shows it forth" (2.172). "The picture has the logical form of representation in common with what it pictures" (2.2). "The picture agrees with reality or not; it is right or wrong, true or false" (2.21). "What the picture represents is its sense" (2.221). "In the agreement or disagreement of its sense with reality, its truth or falsity consists" (2.222). "In order to discover whether the picture is true or false we must compare it with reality" (2.223). "It cannot be discovered from the picture alone whether it is true or false" (2.224). "There is no picture which is apriori true" (2.225). A distinction must therefore be made between the truth of a picture and its sense: the sense is that which it represents, but whether it is true or false depends on whether it represents a fact or not, which cannot be decided a priori.

As to No. 3, Wittgenstein observes that "the totality of true thoughts is a picture of the world" (3.01) and that "in the proposition the thought is expressed perceptibly through the senses" (3.1). "In the proposition the name represents the object" (3.22), but "only the proposition has sense, only in the context of a proposition has a name a meaning" (3.3). "If we change a constituent part of a proposition into a variable, there is a class of propositions which are all the values of the resulting variable proposition. This class in general still depends on what, by arbitrary agreement, we mean by parts of that proposition. But if we change all those signs, whose meaning was arbitrarily determined, into variables, there always remains such a class. But this is now no longer dependent on any agreement; it depends only on the nature of the proposition. It corresponds to a logical form, to a logical prototype" (3.315). In order to avoid the errors due to the imperfection of our everyday language, we must employ a symbolism which excludes them, "a symbolism, that is to say, which obeys the rules of *logical* grammar—of logical syntax" (3.325). "In logical syntax the meaning of a sign ought never to play a rôle . . ." (3.33).

In the comments on No. 4 Wittgenstein develops his theory of the logical correspondence between propositions and reality. "The proposition is a picture of reality, for I know the state of affairs presented by it, if I understand the proposition. And I understand the proposition, without its sense having been explained to me" (4.021). "The proposition *shows* its sense. The proposition *shows* how things stand, *if* it is true. And it *says*, that they do so stand" (4.022). "To understand a proposition means to know what is the case, if it is true. (One can therefore understand it without knowing whether it is true or not.) One understands it if one understands its constituent parts" (4.024). "The meanings of the simple signs (the words) must be explained to us, if we are to understand them. By means of propositions we explain ourselves" (4.026). "One name stands for one thing, and another for another thing, and they are connected together. And so the whole, like a living picture, presents the atomic fact" (4.0311). "The possibility of propositions is based upon the principle of the representation of objects by signs. My fundamental thought is that the 'logical constants' do not represent. That the *logic* of the facts cannot be represented" (4.0312). "In the proposition there must be exactly as many things distinguishable as there are in the state of affairs, which it represents. They must both possess the same logical (mathematical) multiplicity (cf. Hertz's *Mechanics*, on Dynamic Models)" (4.04). "This mathematical multiplicity naturally cannot in its turn be represented. One cannot get outside it in the representation" (4.041). "Reality is compared with the propositions" (4.05). "Propositions can be true or false only by being pictures of the reality" (4.06). "Every proposition must *already* have a sense: assertion cannot give it a sense, for what it asserts is the sense itself. And the same holds of denial, etc." (4.064). "A proposition presents the existence and non-existence of atomic facts" (4.1). "The totality of true propositions is the total natural science (or the totality of the natural sciences)" (4.11). "Philosophy is not one of the natural sciences. (The word 'philosophy' must mean something which stands above or below, but not beside, the natural sciences.)" (4.111). "The object of philosophy is the logical clarification of thoughts. Philosophy is not a theory but an activity. A philosophical work consists essentially of elucidations. The result of philosophy is not a number of 'philosophical propositions,' but to make propositions clear. Philosophy should make clear and delimit sharply the thoughts which otherwise are, as it were, opaque and blurred" (4.112). "Propositions can represent the whole reality, but they cannot represent what they must have in common with reality in order to be able to represent it—the logical form. To be able to represent the logical form, we should have to be able to put ourselves with the propositions outside logic, that is, outside the world" (4.12). ". . . That which expresses *itself* in language, *we* cannot express by language. The propositions *show* the logical form of reality . . ." (4.121). "What *can* be shown *cannot* be said" (4. 1212).

These more or less obscure assertions were eagerly discussed within the Vienna Circle, and, as will appear from what follows, they have on essential points determined the view of the Circle on philosophy and its relation to the special sciences. However, for the time being, I will let this matter rest and pass to a short account of Wittgenstein's important theory of the truth-functions which is stated in the remaining part of the comments on No. 4 and in the comments on No. 5. He says here:

"The sense of a proposition is its agreement and disagreement with the possibilities of the existence and non-existence of the atomic facts" (4.2). "The simplest proposition, the elementary proposition, asserts the existence of an atomic fact" (4.21). "It is a sign of an elementary proposition, that no elementary proposition can contradict it" (4.211). "The elementary proposition consists of names. It is a connexion, a concatenation, of names" (4.22). "It is obvious that in the analysis of propositions we must come to elementary propositions, which consist of names in immediate combination . . ." (4.221). "If the elementary proposition is true, the atomic fact exists; if it is false the atomic fact does not exist" (4.25). "The specification of all true elementary propositions describes the world completely. The world is completely described by the specification of all elementary propositions plus the specification, which of them are true and which false" (4.26). "The truth-possibilities of the elementary propositions mean the possibilities of the existence and non-existence of the atomic facts" (4.3).

If the truth-values (truth or falsehood) of the elementary propositions are combined in every possible way, a survey is obtained of the total number of truth-possibilities, which corresponds to a survey of the total number of possible combinations of all atomic facts. To every combination of atomic facts corresponds a combined molecular proposition, expressing which combinations of atomic facts exist and which do not exist, i.e., which combinations of truth-values of the corresponding atomic propositions exist and which do not exist. This may be presented by a truth-table which corresponds to a molecular proposition and may appear as follows:

p	q	
T	T	T
F	T	T
T	F	F
F	F	T

or shorter: (TTFT)(p, q),

where 'p' and 'q' represent elementary propositions, and the 'T' and 'F' under each represent its possible truth-values, while the last column indicates whether the combinations of truth-values concerned exist or do not exist. Among the possible groups of molecular propositions, there are two extreme cases: the so-called "tautology," which is true for all truth-possibilities, and the so-called "contradiction," which is false for all truth-possibilities. They are without sense, but not senseless. "Tautology and contradiction are not pictures of the reality. They present no possible state of affairs. For the one allows *every* possible state of affairs, the other *none* . . ." (4.462).

According to No. 5, every proposition is now a truth-function of the elementary propositions, which are the truth-arguments of the proposition. In his comments Wittgenstein develops his theory of probability, his theory of the logical relation of inference, and his theory that all logical propositions are tautologies and therefore say nothing of the reality. Further, he criticizes Russell's theory of the relation of identity and develops his theory that it is possible to construct any truth-function from the elementary propositions by the successive application of single logical operations. "Every truth-function is a result of the successive application of the operation (- - - T) (ξ, . . .) to elementary propositions. This operation denies all the propositions in the right-hand bracket and I call it the negation of these propositions" (5.5). This development I shall now explain a little further.

As early as 1913 the American mathematician and logician, H. M. Sheffer,[20] showed that all the truth-functions used in *Principia mathematica* may be defined solely by means of the so-called "stroke-operation," which is written '$p|q$' and may be interpreted as 'not-p and not-q'. For instance, the negation of p and the disjunction of p and q may be defined as follows: $\sim p =_{Df} p|p$ and $p \vee q =_{Df} [(p|q)|(p|q)]$. Since the total number of other truth-functions of p and q may be defined by means of $\sim p$ and $p \vee q$, they may also be defined by repeated application of the stroke-operation. Wittgenstein's theory is a generalization of this from truth-functions with only one or two arguments to truth-functions with an arbitrary number of arguments. The principle is as follows:

If there is given only one elementary proposition, p, it is possible by means of the stroke-operation to construct: $p|p \; [=_{Df} \sim p]$, $(p|p)|p \; [=_{Df} p \cdot \sim p]$, $p|(p|p) \; [=_{Df} \sim p \cdot p]$, $(p|p)|(p|p) \; [=_{Df} p \cdot p]$, ...

If two and only two elementary propositions, p and q, are given, it is further possible to construct: $q|q \; [=_{Df} \sim q]$, $(q|q)|q \; [=_{Df} q \cdot \sim q]$, . . . $p|q \; [=_{Df} \sim p \cdot \sim q]$, $(p|q)|(p|q) \; [=_{Df} p \vee q]$, $(p|p)|(q|q) \; [=_{Df} p \cdot q]$, $((p|p)|q)|((p|p)|q) \; [=_{Df} p \supset q]$, ...

And it is possible to proceed in the same way no matter how many elementary propositions are given. Wittgenstein, writing instead of the above operation '(- - - T)(ξ, . . .)' its result '$N(\bar{\xi})$', i.e., the negation of the total number of variables of propositions ξ (5.502), expresses the method of construction in the general form of the truth-function $[\bar{p}, \bar{\xi}, N(\bar{\xi})]$, where

'$\bar{p}$' stands for the class of all elementary propositions,
'$\bar{\xi}$' stands for any class of propositions,
'$N(\bar{\xi})$' stands for the negation of all the propositions making up $\bar{\xi}$.

The whole symbol '$[\bar{p}, \bar{\xi}, N(\bar{\xi})]$' means whatever can be obtained by taking any selection of elementary propositions, negating them all, and then taking any selection of the class of propositions now obtained, together with any of the originals.[21] Or, in Wittgenstein's words: "This says nothing else than that every proposition is the result of successive applications of the operation $N'(\xi)$ to the elementary propositions" (6.001).

It is especially interesting that he also extends this method of construction to the so-called "generalized propositions," i.e., propositions of the form: "all x are f" (more exactly: "the propositional function fx is a true proposition for every value of x"), or of the form: "there are x's that are f" (more exactly: "the propositional function becomes a true proposition for at least one value of x"). In the current notation these propositions are written: '$(x)\ fx$' and '$(Ex)\ fx$', and Wittgenstein's theory is then expressed in the proposition: "If the values of ξ are the total values of a function fx for all the values of x, then $N(\xi) = \sim(Ex)\ fx$" (5.52). What this means, then, is that, by the application of the operation of negation $N'(\xi)$ to the class of propositions which are values of a given propositional function fx, one arrives at a proposition which says that it is false that there is at least one value of x for which fx is a true proposition. And by negation of this latter proposition one obtains the generalized existential proposition '$(Ex)\ fx$', that is, "there is at least one x, for which fx is a true proposition." Similarly, by starting from the negation of fx, not-fx, instead of from fx, it is possible to construct the generalized proposition '$(x)\ fx$', i.e., "fx is a true proposition for every value of x."

The remaining comments on No. 6 contain Wittgenstein's definition of natural numbers as exponents of operations and his theory that the propositions of mathematics are equations "and therefore pseudo-propositions" (6.2). It is impossible here to go into this question, but his view of logic must be briefly dealt with. He says:

"The propositions of logic are tautologies" (6.1). "The propositions of logic therefore say nothing. (They are the analytical propositions)" (6.11). "It is the characteristic mark of logical propositions that one can perceive in the symbol alone that they are true; and this fact contains in itself the whole philosophy of logic. And so also it is one of the most important facts that the truth or falsehood of non-logical propositions can *not* be recognized from the propositions alone" (6.113). "The fact that the propositions of logic are tautologies *shows* the formal—logical—properties of language, of the world . . ." (6.12). "The logical propositions describe the scaffolding of the world, or rather they present it. They 'treat' of nothing. They presuppose that names have meaning, and that elementary propositions have sense. And this is their connexion with the world. It is clear that it must show something about the world that certain combinations of symbols—which essentially have a definite character—are tautologies. Herein lies the decisive point. We said that in the symbols which we use much is arbitrary, much not. In logic only this expresses: but this means that in logic it is not *we* who express, by means of signs, what we want, but in logic the nature of the essentially necessary signs itself asserts. That is to say, if we know the logical syntax of any sign language, then all the propositions of logic are already given" (6.124). "Whether a proposition belongs to logic can be determined by determining the logical properties of the *symbol*. And this we do when we prove a logical proposition. For without troubling ourselves about a sense and a meaning, we form the logical propositions out of others by mere *symbolic rules*. We prove a logical proposition by creating it out of other logical propositions by applying in succession certain operations, which again generate tautologies out of the first. (And from a tautology only tautologies *follow*.) Naturally this way of showing that its propositions are tautologies is quite unessential to logic. Because the propositions, from which the proof starts, must show without proof that they are tautologies" (6.126). "Proof in logic is only a mechanical expedient to facilitate the recognition of tautology, where it is complicated" (6.1262). "All propositions of logic are of equal rank; there are not some which are essentially primitive and others deduced from these. Every tautology itself shows that it is a tautology" (6.127).

The remaining part of the comments on No. 6 are of a more sporadic character and are often merely short aphoristic remarks on various philosophical questions. I quote only the more important ones, which have exercised a certain influence on the philosophy of the Vienna Circle.

"Logical research means the investigation of *all regularity.* And outside logic all is accident" (6.3). "The law of causality is not a law but the form of a law" (6.32). "If there were a law of causality, it might run: 'There are natural laws . . .' " (6.36). "A necessity for one thing to happen because another has happened does not exist. There is only *logical* necessity" (6.37). "As there is only a *logical* necessity, so there is only a *logical* impossibility" (6.375). "It is clear that ethics cannot be expressed . . ." (6.421). "Death is not an event of life. Death is not lived through. . . . Our life is endless in the way that our visual field is without limit" (6.4311). "Not *how* the world is, is the mystical, but *that* it is" (6.44). "For an answer which cannot be expressed the question too cannot be expressed. *The riddle* does not exist. If a question can be put at all, then it *can* also be answered" (6.5). "The solution of the problem of life is seen in the vanishing of this problem . . ." (6.521). "There is indeed the inexpressible. This *shows* itself; it is the mystical" (6.522). "The right method of philosophy would be this: To say nothing except what can be said, i.e., the propositions of natural science, i.e., something that has nothing to do with philosophy: and then always, when someone else wished to say something metaphysical, to demonstrate to him that he had given no meaning to certain signs in his propositions. This method would be unsatisfying to the other—he would not have the feeling that we were teaching him philosophy—but it would be the only strictly correct method" (6.53). "My propositions are elucidatory in this way: he who understands me finally recognizes them as senseless, when he has climbed through them, on them, over them. (He must so to speak throw away the ladder, after he has climbed up on it.) He must surmount these propositions; then he sees the world rightly" (6.54).

And then the book concludes with No. 7 without comments: "Whereof one cannot speak, thereof one must be silent" (7).

The fascinating effect of this book on the members of the Vienna Circle will be understood, when it is kept in mind that it contained a series of important logical discoveries as well as a wealth of new philosophical views, the grounds for and consequences of which were often barely indicated and so left to be worked out in full by its readers. And, simultaneously, it contained an element of irritation because of its strange mixture of lucid clearness and obscure profundity. Logic and mysticism, elucidation and obscuration, were here found side by side and deeply impressed the members of the Circle and in particular Moritz Schlick, whose thoughts had already in several respects taken a similar trend. The book was eagerly discussed at the meetings of the Circle and contributed essentially to the formation of logical positivism and provoked both agreement and disagreement. In connection with the other influences formerly mentioned (pp. 848–49), it led, in the course of the twenties, to the crystallization of the philosophical view characteristic of the Vienna Circle, to which Wittgenstein himself did not belong.

8. Rudolf Carnap's Theory of The Constitution of Concepts

Another influence of considerable importance in the formation of the views of the Circle was Carnap's "theory of constitution," put forward in his *Der logische Aufbau der Welt* ("The Logical Construction of the World") (1928), the main lines of which I shall now state briefly. Carnap (born 1891), who in 1926 became lecturer in Vienna and, later, professor in Prague and, since 1936, has been a professor in Chicago, had earlier published several contributions to the philosophy of geometry and physics, among which may be mentioned *Der Raum: Ein Beitrag zur Wissenschaftslehre* (1922) and *Physikalische Begriffsbildung* (1926). In these he had shown himself to be an exceptionally stringent and lucid thinker, and he soon became one of the leading figures within the Vienna Circle. His great systematic gifts bore their first large fruit in the above-mentioned work, *Der logische Aufbau der Welt,* to which were added, as supplementary writings, some minor publications of the same year, namely, *Scheinprobleme in der Philosophie: Das Fremdpsychische und der Realismusstreit* (Berlin, 1928) and *Abriss der Logistik, mit besonderer Berücksichtigung der Relationstheorie und ihrer Anwendungen* (Vienna, 1928). In the first of these, certain applications of the theory of constitution propounded in the main work are explained and treated, and the latter contains a brief and clear account of the logistic method used in the formulation of theories.

The purpose of the theory of constitution will perhaps be best understood if viewed in the light of its philosophical applications and results. These results, which may be described as a continuation and clarification of certain ideas of Mach, Russell, and Wittgenstein, go to show, positively, that all the concepts of the natural and social sciences may be defined by means of so-called "elementary experiences," so that all statements of those sciences may be tested or checked by means of such elementary experiences and, negatively, that many traditional philosophical problems are merely pseudo-problems, since, being based on untestable assertions, they are, strictly speaking, senseless. More accurately, Carnap defines his criterion of meaning as follows:

"The meaning of a statement consists in its expressing a (thinkable, not necessarily also an actual) state of affairs. If an (alleged) statement expresses no (thinkable) state of affairs, it has no meaning and hence is only apparently an assertion. If a statement expresses a state of affairs, it is at all events meaningful, and it is true if this state of affairs exists, and false if it does not. One can know whether a statement is meaningful before one knows whether it is true or false.

"If a statement contains only concepts which are already known or recognized, it derives its meaning from these. On the other hand, if a statement contains a new concept or one whose legitimacy (scientific applicability) is in question, one must specify its meaning. In order to do this, it is necessary and sufficient to state the (only thinkable) experiential situations in which it would be called true (not: 'in which it is true'), and those in which it would be called false."[22]

In order to be factual, statements must be founded on an experience: nonfounded statements are empty or meaningless. This principle is acknowledged and practiced in all the natural and social sciences (natural science, psychology, cultural science). But this means that the objects of these sciences must be "constituted" in such a way that every statement about them can be written as, or "translated" into, a founded statement that is equivalent with, i.e., that has the same truth-value as, the original one. To show how this may be done is the aim of the theory of constitution.

To facilitate the understanding of Carnap's method, it may be expedient to recall a few of the logistic concepts used in the theory of constitution. There is, in the first place, the concept "propositional function," by which is understood an expression that contains one or more variables and which, by substitution of suitable arguments for these, becomes a true or a false proposition. Propositional functions with only one variable may be called properties, and their extension is then constituted by the objects satisfying them, that is to say, the names of which, when inserted instead of the variable, make the propositional function a true proposition. The totality of these objects is also called the class of objects defined by the propositional function. If two or more propositional functions are satisfied by exactly the same arguments, they are said to be of the same extension or to be coextensive; and, by the introduction of a special "sign of extension" for a group of coextensive propositional functions, it is possible to formulate propositions of the whole group without regarding the conceptual content ('red', for instance) contained in the propositional functions, but retaining the truth-values of the propositions resulting from the functions. Although not objects, the extensions are often spoken of as if they were, and Carnap therefore uses the designation "quasi-objects" as a convenient manner of speaking.

By means of these concepts Carnap is able to formulate his theory of constitution as follows: To constitute an object means to reduce it to other objects, i.e., to formulate a general rule (a "rule of constitution" or a "constitutional definition") indicating the way in which a statement containing the name of the first object may be replaced by an equivalent statement not containing it.[23] In simple cases this consists of a rule to the effect that, whenever the name of the first object appears, a certain expression containing the names of other objects but not that of the first object be substituted for it (explicit definition).[24] If an explicit definition in this sense is not possible, a contextual definition may be used, i.e., a rule of transformation stating generally how statements in which the expression which is not explicitly definable occurs may be replaced by other statements where it does not occur. Both explicit and contextual definitions may thus be used for the elimination of certain expressions, whether these are explicitly definable or not.

As, according to the definition just given, constitutional definitions concern only extensions, the formulation of constitutional definitions may be designated as an extensional method of definition. "It depends on the 'thesis of extensionality.' In every statement about a concept, the concept should be interpreted extensionally, i.e., it should be represented by its extension (class, relation); or, more precisely, in every statement about a propositional function, the propositional function may be replaced by its sign of extension."[25]

It is now the task of the theory of constitution to arrange the objects of every science according to their reducibility. The system forms, as it were, a genealogy of objects, the roots of which are the objects which cannot be reduced to others, and the trunk and branches show to which other objects any given object may be reduced. What Russell and Whitehead did in *Principia mathematica* with regard to mathematics (reduced all mathematical concepts to the logical fundamental concepts) Carnap in his theory of constitution attempts to do with regard to the natural and social sciences, although, as far as the greater part is concerned, only in outlines and with a limited application of symbolic logic.

As the various sciences do not generally avail themselves of the logistic language in which Carnap's definitions are formulated, but use a more everyday realistic language, he is obliged to replace his logistic criterion on the reducibility of objects by the following criterion of reducibility in realistic formulation: "We call an object *a* 'reducible to objects *b*, *c* . . .' if for the existence of every state of affairs with regard to *a*, *b*, *c* . . . a *necessary and sufficient condition* may be given which depends only on objects *b*, *c*. . . ."[26] By means of this criterion he is able to find out whether a given object is capable of being reduced to another object or not, and he can thus ascertain the order in which the objects must be constituted to form a connected and all-inclusive system of constitution. As this order is not uniquely determined on every point, however, by the said criterion of reducibility, Carnap also uses epistemological priority as a principle of arrangement; and this he defines as follows: "One object (or type of object) is called epistemologically prior with respect to another if the second is known by means of the first and, therefore, knowing the first object is a precondition to the knowing of the second object."[27] In consequence of the latter principle of arrangement, Carnap may also conceive of his whole system of constitution as a "rational reconstruction of the formation of reality, a formation which in the actual process of cognition is made intuitively."[28]

Being of the opinion that the four main kinds of objects are cultural (*geistige*) objects, other minds (*fremdpsychische*), physical objects, and data of our own minds (*eigenpsychische*), Carnap finds that between these objects reducibility is possible in the order mentioned,[29] so that cultural objects may be reduced to other minds, these may be reduced to physical objects, and these again to the data of one's own mind. Briefly, his course of reasoning is as follows: cultural objects (i.e., historical, sociological objects, such as religion, ways and customs, state, etc.) are known partly through their "mental manifestations" (human ideas, feelings and acts of volition), partly through their "documentations" (i.e., physical products, such as things, documents, and the like), and may therefore be constituted on the basis of these. And the objects of other minds are known partly—and mainly—through expressions of emotions and thought, partly—although, so far, very imperfectly—through the brain processes corresponding to the mental phenomena, and may therefore be constituted on the basis of physical objects. And the latter, finally, are known by observation, i.e., by data of our own minds, on the basis of which they may therefore be constituted.[30]

It should continuously be kept in mind, however, that the independence of the various kinds of objects is by no means eliminated by constitution. The higher objects are not composed of the lower but belong to quite different types of objects, which is shown by the fact that they cannot meaningfully be substituted for one another in given statements.[31] The constitution merely shows that statements of higher objects may be translated into statements of lower objects without their truth-value being altered and that statements of any kind of objects may accordingly be tested by means of statements of the lowest kind, the data of our own minds. Likewise, it should be emphasized that the order of arrangement here chosen is not the only possible one. It is, as already mentioned, based on epistemological priority. But if for other reasons it is found expedient, one may very well use the physical objects (i.e., the material things of our everyday life) as the basic element, since in principle the objects of our own mind may be constituted from the brain processes by means of the psychophysical relation. In view of the knowledge of the various kinds of objects and their mutual relations which actually exist in the sciences, Carnap chooses the data of our own minds as a basis for his system of constitution, and he describes them in the following way:

"The egocentric basis we shall also call the '*solipsistic*' *basis.* This does not mean that solipsism itself is here presupposed in the sense of regarding only one subject and his experiences as the sole reality, consequently denying the reality of other subjects. The distinction between real and unreal objects is not made at the beginning of the system of constitution. At the beginning there is no distinction made between those experiences which are, on the basis of later constitution, distinguished as perception, hallucination, dream, etc. This distinction as well as the consequent distinction between real and unreal objects occurs only at a rather high level of constitution. . . . The basic region can also be called 'the given'; it must be noted, however, that there is no intention to presuppose something or somebody to whom the given is given. . . .[32] The given is without a subject."[33]

Having chosen the data of the "own mind" as basis, Carnap determines the so-called "elementary" experiences as basic elements within this sphere and by elementary experiences he understands "experiences in their totality and closed unity."[34]

The function of elementary experiences within the system is similar to that of sensations within Mach's system, but they differ from the latter in not being the results of an analysis, but concrete, complicated units of experience. As such, they cannot be divided up but only be submitted to a so-called "quasi-analysis" on the basis of similarities and other relations holding among them.[35] As the basic relation Carnap uses the recollection of similarity, by which he understands the relation subsisting between two elementary experiences, when a comparison between a recollection of the first elementary experience and the second elementary experience shows that there is an approximate or complete agreement between a certain quality in the one and a certain quality in the other.[36] And from this fundamental relation between the elementary experiences the various objects and kinds of objects are constituted as outlined below.

It should be observed, to begin with, that in building up his system of constitution Carnap employs primarily the logistic symbolic language in the formulation of his constitutional definitions. To make them more easily understood, however, he writes them simultaneously in three other languages, viz., our everyday word-language, the realistic language which is the one current in the empirical sciences, and a fictitious constructive language containing the operational rules for the construction of the objects defined.[37] Although the logistic language is more exact, the following examples of Carnap's constitutional definitions will be given in the realistic language, which does not require so much specific knowledge in order to be understood and which seems sufficient to give an impression of the character of the system.

First, 'elementary experiences' are defined as the members of the relation "recollection of similarity," and next 'part-similarity' is defined as the relation subsisting between two elementary experiences, either of which contains a constituent part similar to a constituent part of the other one. Then 'a region of similarity' is defined as the greatest possible class of qualities between which a part-similarity exists, and 'a quality-class' as the quasi-object that represents something common to elementary experiences; further, he defines 'sense-classes' as classes of abstractions of chains of similarity of qualities, i.e., as what is common for a series of qualities passing evenly into one another. The sense of sight is thereupon defined as a sense-class having the dimensional number five (namely, color tone, saturation, lightness, height, and width). Then he defines 'sensation', 'place of field of vision', 'being in the same place', and 'neighborhood', as well as 'equicolored', 'color class', 'neighboring colors', and 'preceding in time'. And this concludes Carnap's treatment of the lowest stage, the data of the own mind in the system of constitution, from which he passes to the intermediate stages, beginning with physical objects.

By 'physical objects' he understands the material objects of our everyday life, which are characterized, in the first place, by filling a certain part of space at a certain point of time: "Place, shape, size, and position belong to the set of determinators of every physical thing. In addition, at least one sense-quality—e.g., color, weight, temperature—also belongs to this set of determinators."[38] Their constitution starts with the constitution of the space-time world, which is defined as the class of world-points to which are assigned colors (or other sense-qualities), i.e., the points in the *n*-dimensional space of real numbers in so far as they serve for the assignment of the qualities mentioned. For such assignments twelve rules are given, of which we shall mention here only Nos. 9 and 10, since they concern a controversial point in positivistic theories: "9. In so far as there is no reason to the contrary, it is presumed that a point of the external world that is seen once exists before and after it is seen; its positions form a continuous world line. 10. It is further presumed, in so far as there is no reason to the contrary, that such points of the external world have the same or similar color at other times as they had at the time they were seen."[39] The visual objects are defined as bundles of world-lines, the neighborhood relations of which remain much the same for a long period, and *my body* as a visual object with a series of special characteristics. With the help of these concepts, tactual-visual objects and the remaining senses and sense-qualities are defined, whereupon the whole domain of the own mind is determined as the sum total of elementary experiences thus arranged plus the unconscious objects constituted in analogy to the color points not seen at the present moment, so that these objects "consist of nothing but an appropriate rearrangement of immediately presented objects."[40]

The *world of perception* constituted in the whole space-time world by the attribution of sense-qualities to the individual world-points is completed by analogical attribution in a way that in a sense corresponds partly to a postulate of causality, partly to a postulate of substance. Between the world of perception and *the physical world* there is this difference: while the former is constituted by the attribution of sense-qualities to the world-points, the latter is constituted by the attribution of numbers, the physical quantities; this makes it possible to formulate laws mathematically and to achieve a unique noncontradictory intersubjectivation. Within the physical world *biological objects* may then be constituted, including especially *human beings* with their *expressive movements*, and they form the basis of the higher stage: other minds and cultural objects.

For the relation of expression is used the relation between certain observable physical processes in *my body* and a class of data frequently occurring at the same time in *my mind;* the mind of another person can then be constituted by the attribution of the latter class to similar processes in the body of another person. Accordingly, it is stated that "there are no other minds without bodies" and "the whole of the *experience of other people consists*, therefore, in nothing but *a reordering of my experiences and their constituent parts*."[41] Also the communications of other persons and the utterances of factual statements may be constituted on the basis of the signs generally applied to these processes. And thus the road is open for the constitution of *the world of the other person*, which, by comparison with my own world of observation, gives rise to the constitution of *the intersubjective world* that forms the proper domain of the objects of science—all of it, however, only as certain ramifications of "my" system of constitution, which does not mean, of course, that such ramifications exist only in my mind or in my body but merely that they may be constituted from objects in my own mind, i.e., that statements of them are capable of being transformed into statements of my own mind without any change in the truth-values of the statements. The same applies to the cultural objects constituted from their manifestations, i.e., from the mental processes in which they are actualized or make themselves known. Thus the object "state," for instance, may be constituted as follows: "A *state* is a system of relations among people characterized in such and such a way by its manifestations, viz., the mental behavior of these people and their dispositions for such behavior, especially the behavior dispositions of one individual as conditioned by the actions of other people."[42] As to *the values*, these may be constituted from an earlier point in the system of constitution, viz., from *experiences of value*, such as experiences of bad or good conscience, or duty, or responsibility, or aesthetic experiences, etc. "This does not mean that values are psychologized any more than the constitution of physical objects meant that these were psychologized. The system of constitution does not speak this realistic language, but is neutral with respect to the metaphysical components of realistic statements. However, it translates statements about the relation between value and value-feelings into the constitutional language in a way analogous to the way propositions concerning the relation between physical objects and perceptions are translated. . . .[43] With this we close our outline of the system of constitution."[44]

What, now, is achieved by this system? That it has been carried through in detail only as far as the fundamental part is concerned Carnap himself emphasizes repeatedly and that, accordingly, its universal application may involve several changes; but, apart from that, the question arises: What would be achieved by carrying it through completely?

The answer is that, if carried through, the system would show *that* and *how* the totality of statements about objects forming the subject matter of the various sciences are capable of being transformed into statements about immediate experiences having the same truth-values as the original statements. In other words, it would show that all scientific statements are capable of being verified or falsified by means of immediate experiences. This is the positive side of the matter. Negatively, it would show that it is superfluous to assume or apply other sources or means of knowledge than logic and immediate experience. Indeed, Carnap goes so far as to say that the allegation of such other sources of knowledge leads merely to metaphysical assumptions which are meaningless, i.e., incapable of being tested by experience. Only statements consisting solely of logical constants and terms capable of being constituted on the basis of experience have a meaning in the strict sense of this word. Therefore, the theory of constitution may be used to purge science and philosophy of meaningless statements and pseudo-problems. In principle all meaningful questions can be answered—in the affirmative or the negative. There are no insoluble riddles; the apparent insolubility of certain problems is due to the fact that they are based on meaningless assumptions.

"It is sometimes said that the answers to many questions cannot be put into concepts, that they cannot be expressed. But in that event even the question itself cannot be expressed. In order to see this, we will investigate more precisely *what constitutes the answer to a question.* In a strictly logical sense, the posing of a question consists in the presentation of a proposition and the setting of the task of establishing as true either this proposition itself or its negation. A proposition can be given only by the presentation of its sign, the sentence, which is composed of words or other symbols. It frequently occurs, especially in philosophy, that a series of words is given which is considered a sentence but which, in fact, is not. A series of words is not a sentence if it contains a word that is without meaning or (and this is more frequently the case) when all the individual words have a meaning but these meanings do not fit into the context of the sentence. . . . If a real question is presented, what is the situation with respect to the possibility of answering it? In that case a proposition is given, expressed in conceptual signs connected in a formally permissible manner. Every legitimate concept of science has, in principle, its definite place in the system of constitution ('in principle,' i.e., if not at present, then in a possible future stage of scientific knowledge); otherwise, it cannot be recognized as legitimate. Since we are concerned here only with answerability *in principle,* we disregard the momentary condition of science and consider the stage in which the concepts which appear in the given proposition are incorporated into the system of constitution. On the basis of its constitutional definition, we substitute for the sign of each of the concepts in the given sentence the defining expression and make, step by step, the further substitutions of constitutional definitions. . . . The sentence given in the posing of the question is thus so transformed that it expresses a definite (and, indeed, a formal and extensional) state of affairs in respect to the basic relation. We assume in the theory of constitution that it is in principle determinable whether or not a specified basic relation holds between two elementary experiences. However, the state of affairs mentioned is composed of such particular relational propositions; and, further, the number of elements among which the basic relations hold, viz., elementary experiences, is finite. From this it follows that the existence or nonexistence of the state of affairs in question is in principle determinable in a finite number of steps and thus the *question posed is in principle answerable.*"[45]

On these cardinal points the members of the Vienna Circle were in 1928 fairly well agreed. The discussions continued, however, and a further examination of the problems gave rise to difficulties which not only made Carnap modify his standpoint considerably but also resulted in certain divergencies of opinion within the Circle.[46] But before these divergent tendencies had made themselves felt, the Circle had become so firmly established and so convinced that its methods and fundamental views were basically correct that, after 1930, it decided to get into touch with similar-minded groups and persons in other countries, and its development during the decade which followed showed an increasing international activity and growing response from many different quarters. This led to an extension of the external frame, and the very fact that the circle of the participants in the discussions was so widened resulted, necessarily, in a broadening of the basis of discussion, since the persons attracted by the general attitude and program of the movement had, in many respects, very different standpoints and opinions regarding the details of *logical empiricism*, as the movement now came to be called; its adherents wanted to emphasize that they did not consider themselves tied to positivistic views in the more narrow and dogmatic sense. In the following chapter this energetic and comprehensive development, intensive as well as extensive, will be dealt with.

II. Logical Empiricism: Its Expansion and Elaboration

1. Publications, Congresses, and International Connections

Logical positivism was first introduced to an international forum of philosophical experts at the Seventh International Congress of Philosophy held at Oxford in 1930. Here Schlick read a paper on "The Future of Philosophy," in which, with as much enthusiasm for, as confidence in, the new method of philosophy, he heralded a new era in the history of philosophy. ". . . It appears that by establishing the natural boundaries of philosophy we unexpectedly acquire a profound insight into its problems; we see them under a new aspect which provides us with the means of settling all so-called philosophical disputes in an absolutely final and ultimate manner. This seems to be a bold statement, and I realize how difficult it is to prove its truth and, moreover, to make anyone believe that the discovery of the true nature of philosophy, which is to bear such wonderful fruit, has already been achieved. Yet it is my firm conviction that this is really the case and that we are witnessing the beginning of a new era of philosophy, that its future will be very different from its past, which has been so full of pitiful failures, vain struggles, and futile disputes."[1]

The new view of philosophy advocated was that of Wittgenstein, which Schlick expressed in two assertions, one negative and one positive: (1) philosophy is not a science and (2) it is the mental activity of clarification of ideas. Clarifying our thoughts means discovering or defining the real meaning of our propositions, which must be done before their truth can be established. This latter part is the task of the special sciences, with which philosophy cannot compete. All metaphysical attempts to do so have been vain and have only led to mutual conflicts between varying systems. And the reason for this is now understood: "Most of the so-called metaphysical propositions are no propositions at all, but meaningless combinations of words; and the rest are not 'metaphysical' at all, they are simply concealed scientific statements the truth or falsehood of which can be ascertained by the ordinary methods of experience and observation.[2]

"How will philosophy be studied and taught in the future?

"There will always be men who are especially fitted for analysing the ultimate meaning of scientific theories, but who may not be skillful in handling the methods by which their truth or falsehood is ascertained. These will be the men to study and to teach philosophizing, but of course they would have to *know* the theories just as well as the scientist who invents them. Otherwise they would not be able to take a single step, they would have no object on which to work. A philosopher, therefore, who knew nothing except philosophy would be like a knife without a blade and handle. Nowadays a professor of philosophy very often is a man who is not able to make anything clearer, that means he does not really philosophize at all, he just talks about philosophy or writes a book about it. This will be impossible in the future. The result of philosophizing will be that no more books will be written about philosophy, but that *all* books will be written in a philosophical manner."[3]

Schlick's prophecies, however, caused no great stir among the other members of the congress, and, if someone had foretold the immensely rapid development and the response which the movement was to evoke in the course of the coming decade, he would no doubt have been met with a skeptical shake of the head. Nevertheless, the movement gained speed very rapidly during the next few years. This was particularly due to the publication, begun in 1930, of the periodical *Erkenntnis* (edited by Hans Reichenbach and Rudolf Carnap) and to the various series of writings, among which must be especially mentioned the series "Schriften zur wissenschaftlichen Weltauffassung" (edited by Philipp Frank and Moritz Schlick) and "Einheitswissenschaft" (edited by Otto Neurath, Rudolf Carnap, Philipp Frank, and Hans Hahn until the death of the latter in 1934; thereafter by Neurath, Carnap, and Joergen Joergensen and, from 1938, Charles Morris). In the first series the following ten books were published:

R. von Mises, *Wahrscheinlichkeit, Statistik und Wahrheit* (1928); Eng. trans.: *Probability, Statistics, and Truth* (New York, 1939).
R. Carnap, *Abriss der Logistik* (1929).
M. Schlick, *Fragen der Ethik* (1930); Eng. trans.: *Problems of Ethics* (New York, 1939).
O. Neurath, *Empirische Soziologie* (1931).
P. Frank, *Das Kausalgesetz und seine Grenzen* (1932).
O. Kant, *Zur Biologie der Ethik: Psychopathologische Untersuchungen über Schuldgefühl und moralische Idealbildung, zugleich ein Beitrag zum Wesen des neurotischen Menschen* (1932).
R. Carnap, *Logische Syntax der Sprache* (1934); Eng. trans.: *Logical Syntax of Language* (London and New York, 1937).
K. Popper, *Logik der Forschung: Zur Erkenntnistheorie der modernen Naturwissenschaft* (1935).
J. Schächter, *Prolegomena zu einer kritischen Grammatik* (1935).
V. Kraft, *Die Grundlagen einer wissenschaftlichen Wertlehre* (1937).

In the second series the following seven monographs appeared:

H. Hahn, *Logik, Mathematik und Naturerkennen* (1933).
O. Neurath, *Einheitswissenschaft und Psychologie* (1933).
R. Carnap, *Die Aufgabe der Wissenschaftslogik* (1934).
P. Frank, *Das Ende der mechanistischen Physik* (1935).
O. Neurath, *Was bedeutet rationale Wirtschaftsbetrachtung* (1935).
Neurath, Brunswik, Hull, Mannoury, and Woodger, *Zur Enzyklopädie der Einheitswissenschaft. Vorträge* (1938).
R. von Mises, *Ernst Mach und die empiristische Wissenschaftsauffassung* (1939).

In 1938 this series was supplemented by the "Library of Unified Science Series," in which only two volumes have appeared:

R. von Mises, *Kleines Lehrbuch des Positivismus: Einführung in die empiristische Wissenschaftsauffassung* (1939).

H. Kelsen, *Vergeltung und Kausalität* (1941); Eng. trans.: *Society and Nature* (Chicago, 1943).

In 1938 was begun also the publication of the large *International Encyclopedia of Unified Science* (University of Chicago Press) long planned by Otto Neurath; a number of monographs in this work had appeared when World War II seriously slowed down the development of the enterprise. The monographs published are as follows:

Otto Neurath, Niels Bohr, John Dewey, Bertrand Russell, Rudolf Carnap, and Charles Morris, *Encyclopedia and Unified Science* (1938).

V. F. Lenzen, *Procedures of Empirical Science* (1938).

C. Morris, *Foundations of the Theory of Signs* (1938).

L. Bloomfield, *Linguistic Aspects of Science* (1939).

R. Carnap, *Foundations of Logic and Mathematics* (1939).

J. Dewey, *Theory of Valuation* (1939).

E. Nagel, *Principles of the Theory of Probability* (1939).

J. H. Woodger, *The Technique of Theory Construction* (1939).

G. de Santillana and E. Zilsel, *The Development of Rationalism and Empiricism* (1941).

O. Neurath, *Foundations of the Social Sciences* (1944).

P. Frank, *Foundations of Physics* (1946).

In addition, the followers of the movement, in various parts of the world, published in the course of the thirties a number of works of varying size, the most important of which will be mentioned below, and, at the initiative of Otto Neurath, the indefatigable organizer, co-operation was initiated between empiricist and logicist periodicals in various countries.

This extensive publishing activity contributed greatly to the development of the movement, as did also the arrangement of a number of international congresses that gave the members of the Vienna Circle an opportunity of stating and discussing their ideas with other philosophers and scientists feeling a need for international co-operation on the basis of empiricist-scientific fundamental views advocated by the movement. Detailed reports of all these congresses have been published in *Erkenntnis*, in the *Journal of Unified Science* (the periodical continuing *Erkenntnis*), and in a special report of the congress at Paris in 1935, *Actes du congrès international de philosophie scientifique, Sorbonne, Paris, 1935* (Paris, 1936).

The first two of these congresses (called "Tagungen") were kept within rather narrow limits and were attended by a relatively small number of participants from Austria, Czechoslovakia, and Germany. The first was held in Prague in 1929 and comprised, besides a number of papers (by Hahn, Neurath, and Frank) containing general information on the views of the Vienna Circle, a series of papers and discussions on causality and probability (Reichenbach, von Mises, Paul Hertz, Waismann, and Feigl) and on the foundations of mathematics and logic (Adolf Fraenkel and Carnap). The latter subject again formed the main subject at the second conference, held at Königsberg in 1930. Here Carnap lectured on the logicistic, Arend Heyting on the intuitionist, and Johann von Neumann on the formalist foundation of mathematics, while Otto Neugebauer read a paper on pre-Greek mathematics, Reichenbach one on the physical concept of truth, and Werner Heisenberg one on the principle of causality and quantum mechanics. Furthermore, there were at the two meetings vivid and stimulating discussions on the subjects and problems dealt with.

The next congress was held at Prague in 1934. It was called a preparatory meeting (the Paris congress was being planned for the following year under the name of the "Congrès international de philosophie scientifique"). Attended by people from various countries who later became more or less intimately connected with the movement and who met here for the first time, the preparatory meeting achieved a more international character than the preceding meetings had. Among the favorably interested participants were Lukasiewicz, Tarski, Ajdukiewicz, Janina Hosiasson, and Marja Kokoszynska, Poles; Ernest Nagel and Charles Morris, Americans; Louis Rougier, Frenchman; Eino Kaila, Finn; and Joergen Joergensen, Dane, all of whom read papers and took part in the discussions. The members of the group also took an active part in the Eighth International Congress of Philosophy held at Prague during the following days. Here Neurath read a paper on unified science, Schlick on the concept of wholeness, Carnap on the method of logical analysis, Reichenbach on the significance of the concept of probability for knowledge, and Joergensen on the logical foundations of science. These details are mentioned because they convey an impression of the kinds of problems with which the group was concerned.

This impression will become more complete if we consider the main subjects discussed the following year at the large congress at Paris: scientific philosophy and logical empiricism (Enriques, Reichenbach, Carnap, Morris, Neurath, Kotarbinski, Wiegner, Chwistek), unity of science (Frank, Du Nouy, Brunswik, Gibrat, Neurath, Hempel and Oppenheim, and Walther), language and pseudo-problems (Tarski, Kokoszynska, Massignon, Masson-Oursel, Richard, Chevalley, Padoa, Greenwood, Rougier, Matisse, Feigl, Vouillemin), induction and probability (Reichenbach, Schlick, Carnap, De Finetti, Zawirski, Hosiasson), logic and experience (Ajdukiewicz, Benjamin, Renaud, Petiau, Destouches, Métadier, Habermann, Chwistek, Braithwaite, Tranekjaer Rasmussen, Grelling), philosophy of mathematics (Gonseth, Lautman, Juvet, Bouligand, Destouches, Mania, Jaskowski, Raymond, Becker, Schrecker), logic (Tarski, Helmer, Sperantia, Lindenbaum, Bachmann, Padoa, Malfitano, Honnelaitre, Bollengier, Bergmann), and history of logic and scientific philosophy (Scholz, Jasinowski, Raymond, Bachmann, Padoa, Tegen, Hollitscher, Ayer, Zervos, Joergensen, Frank, Heinemann). At this congress also the above-mentioned *International Encyclopedia of Unified Science* was planned, at the initiative of Neurath, and a committee was set up for the drafting of a uniform international logical notation.

The following year, in 1936, the Second International Congress for the Unity of Science was held at Copenhagen and had the problem of causality as its main theme. This congress included a paper by Niels Bohr on causality and complementarity, while Frank spoke on philosophical interpretations and misinterpretations of the quantum theory, Lenzen on the interaction between subject and object in observation, J. B. S. Haldane on some principles of causal analysis in genetics, Rashevsky on physicomathematical methods in biological and social sciences, Rubin on our knowledge of other men, Neurath on sociological predictions, Somerville on logical empiricism and the problem of causality in social science, Hempel on a purely topological form of non-Aristotelian logic, and Popper on Carnap's logical syntax. During the congress, news was received of the assassination of Moritz Schlick, who had sent in a paper on the quantum theory and cognizability of nature.

The next year, in 1937, a Unity of Science Congress was held again in Paris in connection with the Ninth International Congress of Philosophy (Congrès Descartes). As, in the arrangement of its sections, this congress had shown itself particularly interested in the representatives of logical empiricism and had devoted a special section to the unity of science, the congress of the logical empiricists was confined to a conference on the problems of scientific co-operation, especially in connection with the *Encyclopedia* and the unification of logical symbolism. At the main congress (Congrès Descartes), Carnap spoke on the unity of science based on the unity of language; Neurath on prediction and terminology in physics, biology, and sociology; Reichenbach on the principal features of scientific philosophy; Frank on modern physics and the boundary between subject and object; Grelling on the influence of the antinomies in the development of logic in the twentieth century; Hempel on a system of generalized negations; Tarski on the deductive method; and Oppenheim on class concepts and order concepts. Several representatives of movements allied to logical empiricism took an active part. This was the last of the international congresses of philosophy before the outbreak of World War II.

But the movement of logical empiricism found time for two more congresses before the great catastrophe. The first was held in 1938 at Cambridge, England, and the second in 1939 at Harvard University, Cambridge, Massachusetts. The Cambridge congress of 1938 had for its main theme the language of science, and it included papers on language and misleading questions (L. S. Stebbing), relations between logical positivism and the Cambridge school of analysis (M. Black), the diverse definitions of probability (M. Fréchet), languages with expressions of infinite length (Helmer), mathematics as logical syntax and formalization of a physical theory (Strauss), the logical form of probability-statements (Hempel), the language of science (M. Fréchet), experience and convention in physical theory (Lenzen), autonomy of the language of physics (Rougier), the realistic interpretation of scientific sentences (Donald C. Williams), the formalization of a psychological theory (Woodger), the function of generalization (Arne Ness), the concept of Gestalt (Grelling and Oppenheim), the departmentalization of unified science (Neurath), propositional logic in the Middle Ages (K. Durr), the scope of empirical knowledge (Ayer), logic as a deductive theory (Waismann), two ways of definition by verification (Braithwaite), physics and logical empiricism (Frank), significant analysis of volitional language (Mannoury), and imperatives and logic (Joergensen).

The Harvard congress, the fifth and last congress before the war, was quite naturally attended predominantly by Americans, although some European philosophers and scientists were there. Further, in consequence of the anticultural and anti-Semitic politics of naziism, several of the leading figures of the movement had, in the course of the thirties, emigrated to the United States, which had thus become the new center of logical empiricism. The interest of the Americans had been stimulated by Morris and Nagel as well as by the men from Europe, among whom were Carnap, Reichenbach, Frank, von Mises, Feigl, Kaufmann, and Hempel. Some of the subjects of the papers read and discussed were: aims and methods for unifying science (Sarton, Bridgman, Kallen, Langer, Feigl, Nagel, Joergensen, von Mises, Gomperz), scientific method and the language of science (Swann, Carnap, Reichenbach, Hempel, Wundheiler, Williams, Senior, Felix Kaufmann, J. Kraft, Montague, Benjamin, Quine), methodology of the special sciences (Lindsay, Rougier, Pratt, Stevens, Leonard, Gerard, Henderson, Neurath, Morris,

Dennes, Somerville), problems in exact logic (Curry, Rosser-Kleene, Tarski, Church, Copeland, Margenau), science and society (Wirth, Zilsel, Brewster, Oboukhoff, Karpov, Byrne), and history of science (Jaeger, De Lacy, Santillana, Parsons, Davis, Kelsen, Frank). Although the outbreak of World War II, two days before the opening of this congress, made itself felt, the congress was carried through according to its program, and the lively and objective discussions of the subjects treated indicated a widespread interest in the views and methods of logical empiricism that promised well for the future of the movement in its new home. However, in consequence of the development of the war, and particularly when America joined in, conditions became so difficult that a slowing-down was inevitable. As I am yet uninformed as to details, I shall confine myself in the following exposition to the development of logical empiricism until the first of September, 1939, the date of the beginning of the war.

I shall, therefore, now go back to the twenties, or still further, and speak about the various circles with which the Vienna Circle gradually came to co-operate, owing to their common interest in one or several essential questions. They are principally the following: the Berlin group; the Lwow-Warsaw group; the Cambridge analysts, pragmatists, and operationalists; the Münster group, as well as various more isolated investigators in different countries.

2. The Berlin Group

Simultaneously with the gathering by Schlick of the Circle at Vienna, a similar group was formed in Berlin, which in 1928 was organized as the "Gesellschaft für empirische [later, following a proposal by David Hilbert, 'wissenschaftliche'] Philosophie." Among the leaders were Hans Reichenbach (born 1891), Alexander Herzberg, and Walter Dubislav; some other members of the society may be mentioned—Kurt Grelling (1886–1943?), Kurt Lewin (1890–1947), Wolfgang Köhler (born 1887), and Carl Gustav Hempel (born 1905). Its object was to promote scientific philosophy, by which was understood "a philosophical method which advances by analysis and criticism of the results of the special sciences to the stating of philosophical questions and their solutions."[4] The significance of such a scientific-analytical method had been emphasized by Reichenbach as early as 1920;[5] and, true to their program, he and those who agreed with him concerned themselves mainly with specific investigations of fundamental concepts, theories, and methods within the individual sciences, while they had some reservations about the tendency of the Vienna Circle to form systems and set up strict prescriptions and prohibitions.[6] Among their investigations may be mentioned Dubislav's detailed analysis of various methods of definition (*Über die Definition* [3d ed., 1926]); Grelling's inquiries into the paradoxes of the theory of sets and logic; Lewin's work on genidentity and scientific method; Köhler's on physical Gestalts; and Reichenbach's inquiries into the theory of relativity, the concepts of space and time, causality, probability, and the problem of induction. Being especially characteristic of the thought of the Berlin group, Reichenbach's work will be dealt with in more detail.

Through his inquiries into the general assumptions and the epistemological content of the theory of relativity, Reichenbach had reached the conclusion that the Kantian theory of the a priori character of space and time—as well as of other concepts of the natural and social sciences—was untenable. It is true that in his treatise on the theory of relativity and a priori knowledge he maintains that the world of experience is first constituted by means of a priori principles, but these are neither eternal nor deducible from an immanent scheme: "Our answer to the critical question is: there are, indeed, a priori principles which make the correlation of knowledge and observations univocal. But these may not be deduced from an immanent scheme. We must discover them only in the gradual labor of the analysis of science and cease questioning the duration of the validity of their special form."[7] In his remarkable work, *Philosophie der Raum-Zeit-Lehre* (1928), which contains a thorough epistemological analysis of the Euclidean and the non-Euclidean geometries and their relation to experience, he formulated the important principle of the relativity of geometry: "From this it follows that to say that a geometry is *true* is meaningless. We get only a proposition which characterizes something objective if, in addition to the geometry G of the space, we also specify the universal field of force K, which is connected with it. . . . Only the combination $G + K$ is an assertion of cognitive value."[8] This is partly due to the circumstance that all knowledge of reality presupposes so-called "correlative definitions" (*Zuordnungsdefinitionen*), i.e., statements as to what real things are designated by the concepts previously defined—a fact which had already been strongly stressed by Schlick in his *Allgemeine Erkenntnislehre* ([1918], pp. 55 f.) and by Reichenbach himself in his first book on the theory of relativity (pp. 32 f.). The first and most fundamental correlatives are a matter of definition and, in so far, arbitrary and of no epistemological value (which means that it is meaningless to regard them as true statements of the objects), but, on the basis of them, statements may be formulated, the truth and falsehood of which must be decided by experience. Formerly this fact was not fully realized, but the analysis of the theory of relativity has made it unambiguously clear "that correlative definitions are needed at many more places than the old theory of space-time believed; especially for the comparison of length at different places and in different inertial systems and for simultaneity. The core of the theory consists in the hypothesis that measuring bodies obey different correlative definitions from those which the classical theory of space-time assumed. This is, of course, an assertion of an empirical character and can be true or false; with it only the *physical theory of relativity* stands or falls. The *philosophical theory of relativity*, however, as the discovery of the definitional character of the system of measurement in all its particulars, is unaffected by any experience; to be sure, it was acquired through physical experiments; it is, however, a philosophical cognition, not subject to criticism from the special sciences."[9]

Unfortunately, it is impossible to enter into a detailed consideration of the many interesting and thorough analyses undertaken by Reichenbach in the work mentioned; and likewise space does not permit of a detailed account of his *Axiomatik der relativistischen Raum-Zeit-Lehre* (1924). However, a short exposition of his view on the relation between causality, probability, and induction is indispensable. His Doctor's thesis (1916) was *Der Begriff der Wahrscheinlichkeit für die mathematische Darstellung der Wirklichkeit,* and henceforth his interest has constantly centered on this subject, which he has treated in numerous papers and books, the most important of which are *Wahrscheinlichkeitslehre: Eine Untersuchung über die logischen und mathematischen Grundlagen der Wahrscheinlichkeitsrechnung* (1935) and *Experience and Prediction: An Analysis of the Foundations and the Structure of Knowledge* (1938). His fundamental thought is that natural science never confines itself merely to describing events of the past but also predicts coming events, which can never be done with absolute certainty but only with a smaller or greater degree of probability. The concept of probability, therefore, of necessity enters into the concept of knowledge of natural science. Even so-called "causal" statements are merely border cases of statements of probability: "For this reason, we must replace the strictly causal proposition by two propositions: (I) If an event is described by a certain number of parameters, a later event likewise characterized by a certain number of parameters can be predicted with probability p. (II) This probability p approaches 1 as more and more parameters are taken into consideration."[10] Accordingly, statements concerning the future are neither simply true nor simply false but more or less probable. In order to value their probability, a graduated scale for sentences must be constructed that, on the basis of previous facts, ascribes to every possible sentence about the future event a certain degree of truth. In the theory of statements concerning the future, two-valued logic, which operates only with the values truth and falsehood, should be replaced by a continuous scale of probability, and the theory of the Vienna Circle that the meaning of propositions consists in their method of verification should therefore be generalized; instead of maintaining that meaningful propositions must be either true or false, one should assert that they have a certain probability; and, instead of maintaining that propositions that are verified in the same way have the same meaning, one should assert that propositions to which any observable facts give the same value of probability have the same meaning. Only such a generalized probability-logic, containing two-valued logic as a boundary case, is capable of affording a satisfactory explanation of the statements of the natural and social sciences and their meaning. Reichenbach therefore developed a logic of probability, the basic concepts of which are definable by means of truth-tables that are generalizations of those of two-valued logic and contain these as special cases.[11] However, the basic elements of this probability-logic are not propositions but sequences of propositions, i.e., logical constructions obtained by co-ordinating with a propositional function a series of its arguments. In order to meet the difficulties arising when it is desired to fix the probability of a statement of a single future event, Reichenbach thinks it necessary to give a new interpretation of the meaning of statements of single events, and what he proposes is to assert such statements not as being true or false but as a "posit," the evaluation of which is fixed by the probability of the whole class of events of which the single event concerned is a member.[12] Having to choose among several relevant possibilities, we will choose the most probable and "posit" that one. The aim of the whole theory, which has been much debated and thoroughly discussed, is, then, in brief, to give an account of the meaning of a statement based not on its verifiability but on its probability, the latter being of such nature as to comprise verification and falsification as special cases. So far the theory has not, however, won the general assent of the adherents of logical empiricism.

The same applies to Reichenbach's theory of induction, which assumes that probability-logic can be applied to reality. But what right have we to assume that this is so? In answering this question Reichenbach attempts, first, to show that all the assumptions of probability-logic may be reduced to one, viz., the existence of a limit of relative frequency in a series of observable facts. If such limit exists, all the laws of the calculus of probability become tautological, and the question of the applicability of probability-logic is reduced to the question of whether the series of observable facts approach a limit or not. The assumption that they do so is decidedly no tautology, and already Hume has shown that the correctness of this assumption cannot be proved. In this Reichenbach agrees, but, arguing as follows, he still thinks there is a certain rational justification in maintaining the following assumption: Since we know neither whether the assumption is true nor whether it is false, we are justified in defending it in the same sense in which we make a "wager." We want to foresee the future, and we can do so if the assumption is justified—and so we wager on this assumption. Thus we have at least a *chance* of success, while, if we are skeptical and cautious and hold back, we are certain of obtaining nothing. "We are in the same situation as a man who wants to fish in an uncharted place in the sea. There is nobody to tell him whether or not there are fish in this place. Shall he cast his net? Well, if he wants to fish I would advise him to cast his net, at least to take the chance. It is preferable to try even in uncertainty than not to try and be certain of getting nothing."[13] This is his principal reply to the problem of induction, which he amplifies by considerations on the procedure of correction, according to which one does not merely consider a single series of inductions but connects it with the largest possible number of affiliated series, whereby it becomes possible to construct probabilities of higher order, which may increase the chance in every single case.[14] "The chances of our catching fish increase with the use of a more finely meshed net; we ought therefore to use such a net even if we do not know whether there are fish in the water or not. In these reflections, I submit, the problem of induction finds its solution. It does not presuppose a synthetic *a priori*, as Kant believed; for our characterization of induction as a necessary condition of prediction as well as the technique of refining inductive conclusions by the process of correction can be deduced from pure mathematics, that is, with tautological transformations only. This solution is due to a reinterpretation of the nature and meaning of scientific systems. A scientific system is not maintained as true, but only as our best wager on the future. To discover what is our *best* wager in any situation of inquiry is the aim of all scientific toil; never can we arrive at predictions which are certain. Science is the net we cast into the stream of events; whether fish will be caught with it, whether facts will correspond to it, does not depend on our work alone. We work and wait—if without success, well then, our work was in vain."[15]

The work of the Berlin group came to an end when the Nazis came into power in 1933. Its members were dispersed; some of them died, and others emigrated to the United States, where a number of them, including Reichenbach, R. von Mises, and Hempel, are continuing their work in the analysis of science.

3. The Lwow-Warsaw Group[16]

Under the influence of Kazimierz Twardowski (1866–1938), a pupil of Brentano, a vigorous opposition arose in Poland against the irrationalistic metaphysics of the Polish romanticists. Members of this opposition who should be mentioned are: Jan Lukasiewicz (born 1878), Tadeusz Kotarbinski (born 1886), Stanislaw Lesniewski (1886–1939), Zygmunt Zawirski (1882–1946), Kasimierz Ajdukiewicz (born 1890) and Leon Chwistek (1884–1944); and among the younger ones were: Alfred Tarski (born 1901), Janina Hosiasson-Lindenbaum (1899–1941), Mordechaj Wajsberg (died 1942?), Adolf Lindenbaum (died 1941), Marja Kokoszynska, Stanislaw Jaskowski, Izydora Dambska, Henryk Mehlberg, Edward Poznański, Alexander Wundheiler, M. Presburger, and Boleslaw Sobocinski. These investigators were almost all well trained in symbolic logic, and several of them have made valuable contributions to the development of this discipline. Although none of them were adherents of logical positivism, several of them began, in the thirties, to co-operate closely with the Vienna Circle in the scientific analytic work in which they were keenly interested and which they pursued under the name of "metatheory," i.e., the theory of scientific theories. There was an especially lively and fruitful exchange of thought between Carnap and Gödel, on the one side, and Tarski, on the other. In this connection attention should be drawn to

Tarski's important treatise, "Der Wahrheitsbegriff in den Sprachen der deduktiven Disciplinen" (in German in *Studia philosophica*, I [Lwow, 1935]; in Polish in *Travaux de la société de sciences*, Cl. III, No. 34 [Warsaw, 1933]). In this he showed that the truth-concept in formal languages may be defined purely morphologically, i.e., solely by means of the external forms of the expressions and their relations, but that this definition presupposes a metalanguage, containing expressions of a higher logical type than the language whose truth-concept is being defined. Further, Tarski demonstrated that the semantics (i.e., the theory of the relation between signs and their designata) of any formalized language can be built up as a deductive theory with its own axioms and its own fundamental concepts based on the morphology of the language alone. That such investigations were bound to be of the greatest importance to the further development of logical empiricism will appear from the subsequent exposition of Carnap's view of philosophy as the syntactical and, later also, the semantical analysis of the language of science.

4. Pragmatists and Operationalists[17]

In America the development of pragmatism had led to a philosophical view resembling the general viewpoints of European logical empiricism in many respects and well suited to form a natural supplement to them. In Charles Sanders Peirce (1839–1914) we find the combination of an interest in empiricist philosophy and symbolic logic that is characteristic of the movement. Even Wittgenstein's theory of the meaning of propositions consisting in their verifiability was in a way anticipated by Peirce: "It appears, then, that the rule for attaining the third grade of clearness of apprehension is as follows: consider what effects, which might conceivably have practical bearings, we conceive the object of our conception to have. Then, our conception of these effects is the whole of our conception of the object."[18] Although this rule was considered principally with reference to morals and religion by William James (1842–1910), other American investigators used it in a purely epistemological way. This is seen most clearly in the so-called "operationalists," whose most prominent representative is P. W. Bridgman, the physicist, who says in his *The Logic of Modern Physics* (1927): "In general, we mean by any concept nothing more than a set of operations; *the concept is synonymous with the corresponding set of operations.* If the concept is physical, as of length, the operations are actual physical operations, namely those by which length is measured; or if the concept is mental, as of mathematical continuity, the operations are mental operations, namely those by which we determine whether a given aggregate of magnitudes is continuous . . . the concepts can be defined only in the range of actual experiment, and are undefined and meaningless in regions as yet untouched by experiments. . . . Of course the true meaning of a term is to be found by observing what a man does with it, not by what he says about it."[19] This point of view, which puts the main stress on the practice and the acts of the investigator during his work of investigation, is characteristic of the whole pragmatic attitude. The way in which it was developed by John Dewey (born 1859) in, for instance, his *How We Think* (1910) and in *Experience and Nature* (1925) and *Logic, the Theory of Inquiry* (1939) gave this attitude a decidedly biosocial character that made the adoption of behavioristic viewpoints natural. In George Herbert Mead's (1863–1931) *Mind, Self, and Society* (1934) and *Philosophy of the Act* (1938), this tendency became dominant. By emphasizing the social nature of language and science, pragmatism led to a concept of meaning which, in Charles Morris' words, may be briefly stated as follows: "Seen in terms of the context of social behavior, meaning always involves a set of expectations aroused by the symbolic functioning of some object, while the object meant, whether past, present, or future, and whether confrontable by a particular person or not, is any object which satisfies the expectations. A self, as a social being, can for instance expect that other selves will verify its own expectations (a situation of constant occurrence in science), and in this sense at least meaning can outrun personal verification."[20] This view Morris himself later developed into a *semiotic* theory, according to which the meaning-situation is an organic whole with three closely interrelated dimensions: "the relation of sign to objects will be called M_E (to be read, 'the existential dimension of meaning,' or, in short, 'existential meaning'); the psychological, biological and sociological aspects of the significatory process will be designated M_P ('the pragmatic dimension of meaning,' or 'pragmatic meaning'); the syntactical relations to other symbols within the language will be symbolized by M_F ('the formal dimension of meaning,' or 'formal meaning'). The meaning of a sign is thus the sum of its meaning-dimensions: $M = M_E + M_P + M_F$."[21] Whereas the older form of empiricism concerned itself mainly with the first, pragmatism with the second, and logical positivism with the third of these dimensions of meaning, Morris thinks that, by considering all of them equally, a synthesis may be reached which signifies at the same time an expansion of the concept of meaning and an associated extended form of empiricism which he calls "scientific empiricism."

Of other American investigators working on related pragmatistic-operationalistic-scientific-analytic lines, among the best known are the following: Clarence Irving Lewis (born 1883), developed in his *Mind and the World Order* (1929) a "conceptualistic pragmatism" and undertook in his *Symbolic Logic* (1932), which he wrote together with C. H. Langford, significant investigations concerning the logic of modality; Morris R. Cohen (1880–1946), strongly influenced by Russell, expounded in his *Reason and Nature* (1931) a "realistic rationalism"; Victor F. Lenzen gave, in *The Nature of Physical Theory* (1931), an enlightening analysis of the concepts and theories of physics, emphasizing the importance of "successive definitions." Numerous other American philosophers and special scientists have worked along lines more or less related to the viewpoints of logical empiricism without having been in direct contact with this movement. Among its other adherents should be mentioned Ernest Nagel, A. E. Blumberg, Daniel J. Bronstein, and other members of the Harvard congress, and we must add to these the logical empiricists who have emigrated from Europe.

5. The Uppsala School[22]

Although the contact between logical empiricism and the members of the Uppsala school has so far been comparatively slight, this distinctive trend within Swedish philosophy should be mentioned because it is, in several essential respects, closely related to logical empiricism and because the connection between the two seems to be growing. The Uppsala school was founded about 1910 by Axel Hägerström (1868–1939) and his colleague Adolf Phalén (1864–1931) and gathered a number of adherents and pupils, among whom were: Karl Hedvall, Harry Meurling, Ejnar Tegen, Vilhelm Lundstedt, Karl Olivecrona, Gunnar Oxenstjerna, Konrad Marc-Wogau, Ingemar Hedenius, and Anders Wedberg. Although the reasoning and the views of the two movements are not identical, there is a far-reaching agreement between the Uppsala school and logical empiricism, in that they are both decidedly antimetaphysical and for both the main task of philosophy consists in the analysis of concepts. Further, both are opposed to epistemological idealism ("subjectivism," the nature and existence of that which is conceived depends on our conception thereof) and are adherents of the theory that statements of valuations are not true statements but merely expressions of certain feelings and, accordingly, have no factual meaning. Especially with regard to the two last points, the Uppsala school has performed a comprehensive and commendable piece of work which historically anticipates the work of logical empiricism without, however, having influenced it. As has already been mentioned, a certain contact was established during the course of the thirties, and there is every reason to expect a closer co-operation between the two movements to their mutual inspiration and a further development of the basic views. In this connection Marc-Wogau's recent study, *Die Theorie der Sinnesdaten: Probleme der neueren Erkenntnistheorie in England* (1945), should be mentioned.

6. The Münster Group

As the last of the groups working with logical empiricism, the logistic school in Münster, created by Heinrich Scholz (born

1884), may be mentioned. Scholz, who worked originally in theology and philosophy of religion, became interested in logistics; and in the thirties he inspired a number of young investigators—Bachmann, Hermes, and others—to undertake significant inquiries into the foundations of logic and mathematics, while he himself was eagerly engaged in the study of the historical development of logic, stressing in particular the work of Leibniz and Frege (*Geschichte der Logik* [1931]). He was not an adherent of the ideas of logical empiricism, and the co-operation was strictly limited to formalistic-logistic problems, with no mention of the positivistic applications and viewpoints condemned in Nazi Germany. Scholz and his school have resumed their work since the war.

7. Individuals

Besides the above-mentioned groups entering into a limited or extensive collaboration with logical empiricists, there were in the various countries a number of individuals who joined the movement and took an active part in its development. Among these were the following: Eino Kaila (born 1890 in Finland), developed logistic-empiricistic ideas in a series of remarkable writings (such as, for instance, *Über den physikalischen Realitätsbegriff* [1942]) and expounded his theory of knowledge based on those ideas in his *Den mänskliga Kunskapen* (1939); his pupil, Georg Henrik von Wright, besides making intensive studies in *The Logical Problems of Induction* (1941) and *Über Wahrscheinlichkeit* (1945), also wrote a simple and lucid exposition of the fundamental thought of logical empiricism (*Den logiska Empirismen* [1943]); the Frenchman, Louis Rougier (born 1889), has published writings on *Les Paralogismes du rationalisme* (1920), on *La Structure des théories deductives* (1921), on *La Scolastique et le thomisme* (1925), and various other subjects of scientific theory; the Englishman, J. H. Woodger, after having undertaken an epistemological analysis of biological concepts and theories in his *Biological Principles* (1929), has made valuable contributions to the development of formal theories with biological interpretations in his *Axiomatic Method in*

Biology (1937) and in his monograph in this *Encyclopedia* entitled *The Technique of Theory Construction;* Alfred Jules Ayer (also English), in his *Language, Truth, and Logic* (1936), gave a penetrating exposition of the fundamental principles and main results of logical empiricism and in numerous articles and papers defended them against various English critics. Later he expounded the theory in a revised form in his *The Foundations of Empirical Knowledge* (1940). In Germany, P. Oppenheim dealt with the systematization of science and (together with C. G. Hempel) wrote *Der Typusbegriff im Lichte der neuen Logik* (1936). The Swiss, F. Gonseth, formed an "idoneistic" theory of his own concerning the nature of mathematics; and Karl Dürr dealt with the logic of Leibniz and other historical predecessors of some of the fundamental thoughts of logical empiricism. The Norwegian, Arne Ness (born 1912), put forward a behavioristic theory of knowledge in terms referring to the verbal and nonverbal behavior of the scientist (*Erkenntnis und wissenschaftliches Verhalten* [1936]). In his work *'Truth' as Conceived by Those Who Are Not Professional Philosophers* (1938), he studied the actual use of the term 'truth' by psychological methods (questionnaires and interviews). He later developed a special form of philosophical analysis of language which he calls "precision analysis." It is elaborated in his forthcoming work, *Interpretation and Preciseness.* The Dane, Joergen Joergensen (born 1894), has written *A Treatise of Formal Logic,* Volumes I–III (1931), as well as a survey of the various sciences in their systematic contexts from encyclopedic points of view, and has used logical-empiristic viewpoints and methods in his *Psykologi paa biologisk Grundlag* ("Psychology Based on Biology") (1941–45).

Many other investigators in various countries have in various fields worked more or less along the lines of logical empiricism: the philosopher B. von Juhos in Austria, the logistician Uuno Saarnio in Finland, and, in Denmark, Alf Ross and Bent Schultzer, professors of law, and the sociologist, Svend Ranulf. It is impossible to make this an inclusive list because there is difficulty in deciding just where to draw the line.

8. The Question of the Nature of Philosophy

After this outline of the external development of logical empiricism, I shall now consider its internal development. As has already been indicated, important divergencies arose toward the end of the twenties within the Vienna Circle because of various doubts as to some of its own presuppositions. These divergencies concerned the question as to how their philosophical work should be rightly characterized on the basis of their own principles, on the question of the verifiability of statements, and, accordingly, on the theory of meaning as synonymous with verifiability.

As to the nature of philosophy, Wittgenstein maintained that the task of philosophy is a clarification of thought, not a theory but an activity, and that philosophical propositions are, strictly speaking, logically meaningless and "inexpressible," for which reason they should be discarded when their purpose has been attained, in the same way as a scaffolding is thrown away when a building is completed. While, to begin with, Schlick and most of the other members of the Vienna Circle immediately accepted this view, Neurath raised strong opposition against it toward the end of the twenties, as he feared that it would lead to a revival of metaphysics as "the philosophy of the inexpressible." In this he was strongly supported by Carnap, who gave the objections a precise form and put forward a new view of the nature of philosophy that was clearly expressed in his article "On the Character of Philosophical Problems."[23] According to Carnap, *philosophy is the theory of science* or "*the logic of science*, i.e., the logical analysis of the concepts, propositions, proofs, theories of science," and its propositions are not meaningless mediums for elucidation but constitute a legitimate field of study, which he called the "logical syntax of the language of science" and treated in detail in his great work *Logische Syntax der Sprache* (1934),[24] which we shall now consider.

9. Carnap's Logical Syntax of Language*

Carnap's concept of the logical syntax of a language is a generalization of Hilbert's metamathematics, in which, as is well known, the meaning of mathematical signs and formulas is completely disregarded and they are considered solely in a "formalistic" way, i.e., as figures written down and transformed according to certain definite rules. In other words, metamathematics is a theory the object of which is mathematical signs and formulas. Similarly, the logical syntax of language is a purely formal theory of the linguistic signs and their composition into sentences, proofs, and theories, particularly a theory of the signs or sign combinations occurring or acceptable in the sciences, including, of course, those occurring in mathematics; for this reason Carnap's syntax comprises Hilbert's metamathematics as a special part. While it was the object of Hilbert to establish through metamathematics the consistency of classical mathematics, Carnap's purpose is not so much to prove that all science is noncontradictory as to establish the following two theses: (*a*) an investigation of the logic of science need never pay regard to the meaning but only to the formal rules of linguistic expressions, and (*b*) the fixation of the formal rules of any language and the investigation of the consequences of such rules can be built up in exactly the same way as a scientific theory, namely, as a logical syntax of the language concerned, and can usually be formulated in that very language. Thus, considering philosophy as the syntax of the language of science, Carnap refutes Wittgenstein's view of philosophy as an activity that is able only to express itself in meaningless sentences; and at the same time he sharply delimits philosophy as something apart from the special sciences, since philosophy does not deal with the *objects* but only with the *sentences about the objects of such sciences*. The special sciences comprise all "object-questions," whereas philosophy is concerned only with the "logical questions" dealing with scientific concepts, propositions, the-

* [In fairness to Professor Joergensen it should be stated that considerations of space made it necessary to omit portions of the analysis of Carnap's earlier views.—The Editors.]

ories, etc., considered formally as complexes of signs constructed in accordance with certain rules for combinations of signs. These rules are partly rules of formation, partly rules of transformation. The former are the rules for the composition of the various kinds of signs of a language into sentences (i.e., corresponding roughly to usual grammar); the latter are rules for the deduction of a sentence from other sentences (i.e., rules of inference, so that the logical syntax will comprise what is generally called "logic"). For the different (from a logical point of view) languages, these two kinds of rules are different and may be fixed arbitrarily, because they, so to speak, define the language to which they are to apply. Here the "principle of tolerance" applies: "*It is not our business to set up prohibitions, but to arrive at conventions. . . . In logic, there are no morals.* Everyone is at liberty to build up his own logic, i.e. his own form of language, as he wishes. All that is required of him is that, if he wishes to discuss it, he must state his methods clearly, and give syntactical rules instead of philosophical arguments."[25] Thus we may, for instance, construct one language restricted to finitist mathematics and another language the mathematical part of which contains all of classical mathematics (and physics), and so reduce any dispute between intuitionists and formalists in mathematics to a mere disagreement as to which form of language they wish to use.

Carnap then proceeds to illustrate the syntax by means of two languages serving as instances of formal languages. Since it is a very complicated matter, although in principle possible, to set up the formal rules of our everyday language, Carnap chooses two artificial languages, viz., the language of finitist mathematics and the language of classical mathematics. He calls them "language I" and "language II," formulates the syntactical rules for each, and shows that the syntax of language I can be expressed in that language itself, so there is nothing to prevent a language in this sense describing its own form or structure (which had been declared impossible by Wittgenstein). Carnap demonstrates this by means of the method of arithmetization, introduced by Gödel into metamathematics, according to which a number is co-ordinated to every sign of the system in such a way that every expression can be translated into a corresponding arithmetical expression. And as language I contains arithmetic, this means that the syntax of this language can be expressed in the language itself.

In his investigation of the two model languages Carnap states a number of instances worked out in detail and confirming the above-mentioned thesis *b*. But, however thoroughly and subtly worked out, instances cannot, of course, prove the thesis generally, and Carnap accordingly undertakes a series of fundamental inquiries into general syntax, i.e., into syntactical rules concerning either all languages or comprehensive classes of languages. In these he defines and discusses a number of logical-syntactical concepts—the most important is the concept of "consequence," which for any language is defined by the rules of transformation of that language and which, therefore, varies from one language to another. Roughly speaking, it may be said that a sentence in a given language is a consequence of certain other sentences if, and only if, it can be constructed from these latter by application of the rules of transformation of the language concerned. As the rules of transformation are purely formal, the concept of "consequence" is so, too, and by means of the latter it is then possible to define various other important syntactical concepts, such as the concepts "analytic," "contradictory," and "synthetic," which give an exhaustive classification of all sentences occurring in the different branches of science, and also the concept "the content of a sentence," besides many other concepts that are worked out with extreme penetration and mathematical precision. For that very reason they are too complicated to be stated here, but an impression of Carnap's way of procedure may perhaps be obtained by considering his simple and lucid exposition in *Philosophy and Logical Syntax* (1935).

"Given any language-system, or set of formation rules and transformation rules, among the sentences of this language there will be true and false sentences. But we cannot define the terms 'true' and 'false' in syntax, because whether a given sentence is true or false will generally depend not only upon the syntactical form of the sentence but also upon experience; that is to say, upon something extra-linguistic. It may be, however, that in certain cases a sentence is true or false only by reason of the rules of the language. Such sentences we will call *valid* and *contravalid* respectively. Our definition of validity is as follows: a sentence is called *valid*, if it is a consequence of the null class of premises (i.e. if it presupposes no premise) . . . a sentence 'A' of a certain language-system is called *contravalid* if every sentence of this system is a consequence of 'A'. . . ."[26]

By means of the concepts here defined it is then possible to give a purely formal definition of the "sense" or "content" of a sentence. "If we wish to characterize the purport of a given sentence, its content, its assertive power, so to speak, we have to regard the class of those sentences which are consequences of the given sentence. Among these consequences we may leave aside the valid sentences, because they are consequences of every sentence. We define therefore as follows: the class of the non-valid consequences of a given sentence is called the *content* of this sentence."[27] The logical-philosophical significance of such definitions is that they show that all factual sentences may be submitted, by means of them, to a logical analysis without its being necessary at any point to consider anything but the purely formal properties of the sentences concerned. Their "sense" in the usual vague and undefined meaning of the word is logically irrelevant and may be replaced, according to Carnap's theory, in all logical investigations by the formal concept of "content" here defined.

There are, however, certain sentences with regard to which this is anything but evident, and as to these Carnap proceeds in a special way. They are the sentences which he calls "pseudo-object-sentences," i.e., sentences which *seem* to deal with the objects mentioned in them but which, upon a closer inspection, appear to be purely syntactical (e.g., 'The rose is a thing'). As these very sentences play an important part in current philosophy, Carnap attaches considerable weight to them, it being possible to maintain his above-mentioned thesis *a* only if it can be established that the pseudo-object-sentences are syntactical in character. He contrasts them with the *syntactical sentences* mentioned above and with *real object-sentences* (e.g., 'the rose is red'), which concern extra-linguistic objects and belong to the special sciences.

Although a pseudo-object-sentence may have the same grammatical subject as an object-sentence, it asserts no quality of the subject. We can, moreover, discover its truth without observing the object to which we refer but only by considering its syntactical status. Thus we see that a pseudo-object-sentence is really syntactical and can be translated into a syntactical sentence having the same content (e.g., 'The word 'rose' is a thing-word').

Syntactical sentences are also said to be sentences of *the formal mode of speech*, while pseudo-object-sentences are said to be sentences of *the material mode of speech.* According to Carnap, the material mode of speech often gives rise to pseudo-problems or disputes that may be avoided by translating the sentences concerned into the formal mode of speech.[28] Indeed, all traditional metaphysical problems seem to arise from the very circumstance that they have been discussed in the material mode of speech instead of having been analyzed syntactically. An example will illustrate this: "One frequent cause of dispute amongst philosophers is the question what *things* really are. The representative of a positivistic school asserts: 'A thing is a complex of sense-data.' His realistic adversary replies: 'No, a thing is a complex of physical matter'; and an endless and futile argument is thus begun. Yet both of them are right after all: the controversy has arisen only on account of the unfortunate use of the material mode. Let us translate the two theses into the formal mode. That of the positivist becomes: 'Every sentence containing a thing-designation is equipollent* with a class of sentences which contain no thing-designations, but sense-data-designations,' which is true; the transformation into sense-data-references has often been shown in epistemology. That of the realist takes the form: 'Every sentence containing a thing-designation

* Two sentences are *equipollent* if each is a consequence of the other.

is equipollent with a sentence containing no thing-designation, but space-time-coordinates and physical functions,' which is obviously also true. . . . There is no inconsistency. In the original formulation in the material mode the theses *seemed* to be incompatible, because they *seemed* to concern the essence of things, both of them having the form: 'A thing is such and such.' "[29]

In other cases the problem is solved by our showing that apparently contradictory theses do not belong to the same syntactically defined language. This applies, for instance, to the controversy between the logicists (Frege, Russell) and the axiomaticists (Peano, Hilbert) on the nature of numbers. The former assert: "Numbers are classes of classes of things," while the latter assert: "Numbers are a unique kind of entities." Translating these assertions into the formal mode, we get: "Numerical symbols are class-symbols of the second order," and "Numerical symbols are symbols of individuals (i.e., symbols of zero order that occur only as arguments)." These sentences may be conceived either as belonging each to its own separate arithmetical language or as proposals for separate languages, i.e., for different ways of talking about numbers. In either case the discussion is no longer a controversy about what is true and what is false but is reduced to a question of what language is best suited for talking about arithmetics.

On the basis of similar analyses of a long series of philosophical problems and assertions of various kinds, Carnap thinks himself justified in asserting that all the theorems of philosophy can be treated syntactically, that is to say, can be translated into the formal mode of speech whereby the problems attached to them will either be automatically revealed as being illusory pseudo-problems or be reduced to proposals for different languages, the expediency, but not the correctness, of which may be discussed. He finds no need to eliminate completely the material mode of speech: "This mode is usual and perhaps sometimes suitable. But it must be handled with special caution. In all decisive points of discussion it is advisable to replace the material by the formal mode; and in using the formal mode, reference to the language-system must not be neglected. It is not necessary that the thesis should refer to a language-system already put forward; it may sometimes be desired to formulate a thesis on the basis of a so far unknown language-system, which is to be characterized by just this thesis. In such a case the thesis is not an assertion, but a proposal or project, in other words a part of the definition of the designed language-system. If one partner in a philosophical discussion cannot or will not give a translation of his thesis into the formal mode, or if he will not state to which language-system his thesis refers, then the other will be well-advised to refuse the debate, because the thesis of his opponent is incomplete, and discussion would lead to nothing but empty wrangling."[30]

The main result is, therefore, that every indicative, meaningful sentence either is an object-sentence which, as such, belongs to one special science or another or is a syntactical sentence which belongs to logic or mathematics, and that, accordingly, philosophy may be defined as the sum total of the true syntactical sentences concerning the languages of the special sciences. This again leads to various new problems—or old problems in a new formulation—such as: Do the various special sciences speak the same language, and, if not, is it possible to construct a language common to all? And what is the criterion of the truth—or merely meaningfulness—of an object-sentence? The first of these questions leads to the discussion of one of the main points of logical empiricism: *the unity of science* and the associated thesis of *physicalism,* while the second is important with regard to the *theory of verifiability* and to the problem of the *basis of the system of constitution.* As these problems overlap to a large extent, a constant interaction took place in their treatment from the beginning of the thirties; but, for the purposes of clarity, they will here, as far as possible, be treated separately, and we shall start with the last problem: the problem of basis and verification.

10. Protocol-Sentences and Substantiations ("Konstatierungen")

In his *Der logische Aufbau der Welt* Carnap had, as we know, chosen as basis for his theory of constitution the immediately given experiences which he asserted were subjectless. This starting point was not, however, acceptable to Neurath. He was afraid that it might lead to a return to metaphysical absolutism, and, besides, he found the connection between the experiences and the sentences that were to describe them, and to be checked by them, metaphysical.

He consequently maintained that propositions were checked only by other propositions. A new proposition is compared with those already accepted and is called correct if it can be incorporated into the system. Sometimes one may, although this decision is not easily made, change the whole previously accepted system of propositions to allow for the new one. "Within unified science there are significant tasks of transformations,"[31] whereas *outside* unified science there is nothing with which a relation may be established. Not even direct statements of observation can be compared with the objects concerned but merely with other statements of observation or with statements of other kinds, and their truth does not depend on their agreement with the objects observed but solely on their agreement with the totality of all statements accepted at the given time. Neurath now proceeded to look for purely formal, syntactical characteristics of direct statements of observation, which he called (following a suggestion of Carnap) *protocol-sentences*, as he wanted to stress that these sentences were of the same kind as the ones which natural scientists use in making protocols (records of observations) and which form the starting point for their hypotheses and theories and the criterion for their validity.

As the debate proceeded, the prevailing view became that from a philosophical-epistemological standpoint there was no difference in principle between protocol-sentences and other legitimate scientific sentences; and the discussion concerning the syntactical form of protocol-sentences therefore subsided, the whole question then being considered a matter of convention.

The above "most radical form" of logical empiricism did not, however, win the general approval of all the followers of the movement. Schlick was in direct opposition. He maintained that, by introducing the concept "protocol-sentence," the aim had been from the beginning to sort out a group of sentences capable of serving as an absolutely firm basis for knowledge and as a means of testing all other sentences. And, since noncontradictoriness or incorporability in a system of sentences cannot enable us to distinguish between knowledge of reality and fairy tales that are consistently built up, the sentences characteristic of knowledge must be distinguished by special properties. But, as he admits that even protocol-sentences are hypothetical because for various reasons they *may* be doubted (the observer may, for instance, be guilty of a slip of the pen, or he may have lied) and are sometimes rejected (as, for instance, a single measurement that cannot be brought to agree with a certain series of measurements of the same magnitude), he thinks that knowledge—certain knowledge—cannot be based on protocol-sentences but must be based on certain sentences of observations or "Konstatierungen," by which he understands statements of what is being observed. These are always of the form "so-and-so here now." What these sentences have in common is that they fulfil the function of pointing. What 'here' and 'now', etc., mean cannot generally be given in verbal definitions but must be specified by gesture, pointing, etc. The meaning of a substantiation can be understood if, and only if, it is compared with the facts. However, as in the case of analytic judgments, establishing the meaning of a substantiation is not distinguishable from establishing its truth. "It makes as little sense to ask whether I can be mistaken about its truth as about that of a tautology. Both are absolutely valid. However, the analytic, tautological sentence is without content, while the substantiation provides us the satisfaction of genuine knowledge of fact."[32] Strictly speaking, however, substantiations cannot be written down at all, because when we do so the "pointing" words 'here' and 'now' lose their sense.[33]

Against this view Neurath asked how it would ever be possible to ascertain that Schlick had had an experience which he could not write down[34] and also called attention to the fact that the absolutely certain knowledge sought by Schlick is not an empirical fact but wishful thinking connected with certain metaphysical ideas of a difference between "the real world" and knowledge of it. Referring to the detailed and penetrating criticism of this metaphysical duplication which Frank had set forth in his *Das Kausalgesetz und seine Grenzen* (1932), chapter x, Neurath said: "Thus the attempt to achieve knowledge of fact is reduced to the attempt to bring the sentences of science into agreement with as many protocol-sentences as possible."[35] And as to consistent fairy tales versus knowledge of reality, he refers to the fact that "the practice of life" very quickly reduces the number of the systems of sentences having an equal right *from a logical point of view*, as most of them soon appear unsuited for predictions.[36] But this does not make the rest absolutely certain. "*There is no way to make absolutely certain protocol-sentences the point of departure for the sciences.*" There is no *tabula rasa*. We are like sailors who have to rebuild their ship on the open sea without ever being able to tear it down in a dry dock to rebuild it with new parts.[37]

As to the further course of this discussion, which was cut short by the death of Schlick in 1936, readers are referred to the instructive articles by C. G. Hempel, "On the Logical Positivists' Theory of Truth"[38] and "Some Remarks on 'Facts' and Propositions,"[39] in which he defends the standpoint of Neurath and Carnap and criticizes that of Schlick. It is impossible here to go into details, but a single aspect of the matter, namely, the development of the question of verifiability, will be dealt with below, as it signifies an essential generalization of the logical-empiricist view of the truth and meaning of sentences.

11. Verifiability and Testability

Keeping in mind that logical positivists identified the meaningfulness of reality-sentences with their verifiability, it will be understood that the question of how reality-sentences can be verified must be of the utmost interest to them. Reality-sentences that cannot be verified or falsified are, according to this view, meaningless. But now it became apparent that not even protocol-sentences, by means of which the truth of all other reality-sentences was to be tested, were capable of being verified in the strict sense of the word so that they become absolutely certain. And matters are still worse, of course, with regard to more complicated sentences, as, for instance, the general sentences of which the natural laws form a part, since from a logical point of view they are "general implications" in the simplest case of the form: for all x, if x has the property f, then x has the property g. As Popper says, "logical positivism destroys not only metaphysics but also natural science."[40] As a first way out of this difficulty Schlick had proposed a different conception of natural laws: "Natural laws are not (in the language of the logician) 'general implications,' because they cannot be verified for *all* cases; they are rather rules or directions for the investigator to find his way through the real world, to discover true sentences, to predict certain occurrences."[41] This way of escape, however, had not won the approval of the other logical empiricists, who felt bound to acknowledge natural laws as general implications. But how then could they be made to agree with the theory that the meaningfulness of sentences consists in their verifiability?

This could be done only by amending the theory. A first proposal in that direction was made by Popper in his *Logik der Forschung* (1935), where, as the criterion of the meaningfulness of a sentence, he uses not the verifiability but the falsifiability of the sentence. "Our formulation depends on an asymmetrical relation between verifiability and falsifiability, which is connected with the logical form of universal sentences; viz., these are never derivable from particular sentences but can be contradicted by particular sentences."[42] By this criterion of meaning, he proposed to sort out empirical-scientific sentences from a priori–analytical sentences (logic and mathematics) as well as from nonfalsifiable reality-sentences (metaphysics). But this suggestion was not approved either because universal sentences, logically viewed, seem to parallel existential sentences, which latter can never be falsified but in certain cases may be verified, and because some scientifically recognized sentences contain a combination of the peculiarities of universal and existential sentences (they contain both a universal and an existential quantifier) and, accordingly, can be neither verified nor falsified. This applies, for example, to statements concerning the limit of the relative frequency of a certain event in a series of events, that is, concerning probability-statements according to a commonly held view. Some logical empiricists were for a time inclined to regard these statements as "pseudo-sentences," or they sought another interpretation of sentences of probability. As this gave rise to great difficulties, the view was gradually accepted that scientific sentences may very well contain both universal and existential quantifiers, and consequently it became necessary to look for a new and more liberal criterion of meaning than that mentioned above. As he had done so often before, Carnap pulled the loose ends together, worked the matter through, and outlined a theory and a proposal that have since formed the starting point for the further discussion of this problem.

Carnap's exposition was published in a treatise called "Testability and Meaning."[43] Popper, in his *Logik der Forschung*, had strongly emphasized the necessity of distinguishing between various "degrees of testability (*Prüfbarkeit*)"; and Carnap, at the congress in Paris (1935), had stressed the importance of distinguishing between "truth" and "confirmation" (*Bewährung*): while truth is an absolute concept independent of time, confirmation is a relative concept, the degrees of which vary with the development of science at any given time. Further, he distinguished between two different kinds of reality-sentences, viz., those that are directly testable and those that are merely indirectly testable. "By a directly testable sentence we mean one for which, under imaginable conditions, on the basis of one or a few observations, we can with confidence regard either as so strongly confirmed that we accept it or as so strongly disconfirmed that we reject it. . . . Indirect testing of a sentence consists in directly testing other sentences which have a certain relation to it."[44] The most important testing operations are (*a*) confrontation of a sentence with observation and (*b*) confrontation of the sentence with sentences that have been previously recognized. Of these, the former operations are the more important, for in their absence there is no confirmation at all, while the latter are merely auxiliary operations, which mostly serve to eliminate unsuited sentences from the system of sentences of the science concerned.

These views are now further implemented and defined in "Testability and Meaning," which contains the following introductory remarks: "If by verification is meant a definitive and final establishment of truth, then no (synthetic) sentence is ever verifiable, as we shall see. We can only confirm a sentence more and more. Therefore we shall speak of the problem of *confirmation* rather than of the problem of verification. We distinguish the *testing* of a sentence from its confirmation, thereby understanding a procedure—e.g. the carrying out of certain experiments—which leads to a confirmation in some degree either of the sentence itself or of its negation. We shall call a sentence *testable* if we know such a method of testing for it; and we shall call it *confirmable* if we know under what conditions the sentence would be confirmed. As we shall see, a sentence may be confirmable without being testable; e.g. if we know that our observation of such and such a course of events would confirm the sentence, and such and such a different course would confirm its negation without knowing how to set up either this or that observation."[45]

Reichenbach thinks he has found a measure for the degree of confirmation in the limit of the relative frequency of the cases of confirmation, so that any sentence may be said to be a probability-sentence; this being a controversial point, however, Carnap prefers to distinguish between probability (in the frequency sense) and degree of confirmation;[46] he also advises us not to discuss matters in the material mode of speech, as this mode serves to veil the fact that the answering of certain pertinent questions depends on the choice of the structure of the language applied, which appears evident when the formal mode of speech is used. He therefore develops a logical-syntactical analysis of the pertinent concepts, which is too technical to be quoted here. It must suffice to state that the investigation results in various proposals or requirements, the fulfilment of which signifies more or less radical forms of empiricism and may be said to define different concepts of what empiricism is or should be. He makes four requirements in all, of which the first is the most rigid and the last the most liberal one. Being formulated in the formal mode of speech, the requirements concerned define four different empiricist languages. The four requirements are as follows:

"*Requirement of Complete Testability:* Every synthetic sentence must be completely testable." I.e., if any synthetic sentence *S* is given, we must know a method of testing for every descriptive predicate occurring in *S*.[47]

"*Requirement of Complete Confirmability:* Every synthetic sentence must be completely confirmable." I.e., if any synthetic sentence *S* is given, there must be for every descriptive predicate occurring in *S* the possibility of our finding out for suitable points whether or not they have the property designated by the predicate in question.[48]

"*Requirement of Testability:* Every synthetic sentence must be testable." This requirement admits incompletely testable sentences—these are chiefly universal sentences to be confirmed incompletely by their instances.[49]

"*Requirement of Confirmability:* Every synthetic sentence must be confirmable." Here both restrictions are dispensed with. Predicates which are confirmable but not testable are admitted; and generalized sentences are admitted. This is the most liberal of the four requirements. But it suffices to exclude all sentences of a nonempirical nature, e.g., those of transcendental metaphysics, inasmuch as they are not confirmable, even incompletely. Therefore, it seems to Carnap that it suffices as a formulation of the principle of empiricism; in other words, if a scientist chooses any language fulfilling this requirement, no objection can be raised against this choice from the point of view of empiricism.[50]

The main result is, then, that the discussion as to what sentences may be considered meaningful and what sentences meaningless, from the point of view of empiricism, has led, on the one hand, to a more precise definition of and distinction between various empirical languages and so to various concepts of meaning in an empirical sense and, on the other, to the acceptance of the most liberal of the alternatives compatible with empiricism. And this, Carnap thinks, has helped to smooth the way for the development of converging views and approaches to *scientific empiricism* as a movement comprising all allied groups—an ever more scientific philosophy. This aim is closely connected with the idea of a unity of science that has already been mentioned several times but which we shall now carry a little further.

12. Unity of Science and Physicalism

The expression 'unity of science' was introduced into logical empiricism by Neurath. He wanted thereby to mark his opposition to the view that there are different *kinds* of sciences (and, corresponding to them, different kinds of reality or being), such as natural sciences (*Naturwissenschaften*) versus the humanities (*Geisteswissenschaften*), or factual sciences (*Wirklichkeitswissenschaften*) versus normative sciences (*Normwissenschaften*). He also wanted, by the words 'unity of science', to sum up the objective aimed at by logical empiricists, viz., the formation of a science comprising all human knowledge as an epistemologically homogeneous ordered mass of sentences being of the same empiricist nature in principle, from protocol-sentences to the most comprehensive laws for the phenomena of nature and human life.[51] To use a traditional expression, unity of science might also be called "monism free from metaphysics." A first manifestation of this attitude in the Vienna Circle was Carnap's theory of constitution, which Neurath, however, for the above-mentioned (p.915) reasons found unacceptable. He thought it very important that the unification of the various special sciences into a unity of science should take place through the formation of a universal language of science, i.e., a language the logical syntax of which permitted sentences from the most different special sciences to be combined with one another so as to form a logical context. "The universal language of science becomes a self-evident demand, if it is asked, how can certain singular predictions be derived; e.g., 'the forest fire will soon subside'. In order to do this we need meteorological and botanical sentences and in addition sentences which contain the terms 'man' and 'human behavior'. We must speak of how people react to fire, which social institutions will come into play. Thus, we need sentences from psychology and sociology. They must be able to be placed together with the others in a deduction at whose end is the sentence: 'Therefore, the forest fire will soon subside'."[52] "*We must at times be able to connect all types of laws with one another.* All, whether they be chemical, climatological, or sociological laws, *must be conceived as parts* of one system, viz., unified science."[53] Without this, the practical application of science would be excluded in many domains, and the unity of science therefore forms the basis of the applications of science, depending on the combination of premises from different scientific disciplines into connected chains of inference.

In science as well as in our everyday life we do actually avail ourselves of a kind of "universal slang," whenever we want to reason or to think things over, and the aim of unified science is to make this universal slang homogeneous and universal, eliminating merely metaphysical absurdities. This, however, again raises the question: What language would be best suited for the performance of this task? In his theory of constitution Carnap used, as we know, an egocentric, phenomenological language, thinking that by constitution he would be able to reduce all other concepts to phenomenological basic concepts. But he and Neurath soon agreed that it would be more expedient to use a so-called "physical" or "physicalistic" language or, as it was later called, a "thing-language," by which they understood the language in which we, both in physics and in everyday life, speak of physical things (which again approximately means: material things, in the everyday understanding of that expression). The task then became to formulate the rules of formation and of transformation of such language so that all concepts and sentences can be expressed in it, if necessary, by suitable translations and so that all scientific theories can by means of it be reduced to as few deductive systems as possible, preferably to a single one. "In our discussions in the *Vienna Circle* we have arrived at the opinion that this physical language is the basic language of all science, that it is a universal language comprehending the contents of all other scientific languages. In other words, every sentence of any branch of scientific language is equipollent to some sentence of the physical language, and can therefore be translated into the physical language without changing its content. Dr. Neurath, who has greatly stimulated the considerations which led to this thesis, has proposed to call it the thesis of *physicalism*."[54]

The physical language, Carnap says further, is characteristic, in that it consists of sentences that give in their simplest form a quantitative description of a definite space-time-place (e.g., 'At such and such space-time-point the temperature is so and so many degrees'), or, expressed formally: Sentences that attribute to a certain series of values of space-time-co-ordinates a certain value of a definite physical function. In so far as rules are known for a unique translation of sentences of qualitative characteristics into sentences of quantitative characteristics, there is nothing to prevent the physical language from containing characteristics of the former kind.[55]

The reason for choosing such a language as that of unified science is that it is *intersensual, intersubjective, and universal.* What this means will now be outlined.

That the physical language is *intersensual* means that its sentences can be tested by means of various senses, because actually there is no physical function that can be co-ordinated solely with qualitative characteristics from a single sphere of sense. The characteristic "tone of such and such pitch, timbre, and loudness" may, for instance, be co-ordinated with the following characteristic in the physical language: "Material oscillation of such and such basic frequency with such and such harmonic frequencies with such and such amplitudes," which, by application of certain apparatuses, may be tested by means of the sense of sight or of touch. And, similarly, qualitative characteristics of color may be co-ordinated to physical characteristics of electromagnetic oscillations, which may, for instance, be tested by means of their place in the spectroscope, that may again, by suitable devices, be demonstrated by contact with an indicating needle or by ascertainment of a tone by listening, so that a blind physicist will very well be able to test the qualitative protocol-sentence, 'Here is now green of that and that shade'. It is therefore possible in principle to construct a physical language of such a kind that the qualitative characteristics of the protocol-language depend functionally and uniquely on the distribution of values of physical functions, so that *the physical characteristics may be said to apply intersensually.*[56]

That the physical language is *intersubjective* means that its sentences can be tested by various subjects and thus hold a meaning for all of them. This appears from the fact that a given subject (individual) can observe by means of experiments under what physical conditions various other subjects react by certain qualitative protocol-sentences, e.g., 'I now see green of that and that shade'. Thus a correspondence may be established between every single physical characteristic, on the one hand, and the qualitative characteristics contained in the protocol-sentences of the various subjects, on the other, so that *the physical characteristics may be asserted to apply intersubjectively.*[57]

That the above co-ordinations between physical and qualitative characteristics can be established is no logical necessity but depends on empirical "happy circumstances" connected with "a very general structural feature of experience."[58]

Finally, that the physical language is *universal* means that every scientifically acceptable sentence, whether originating from our everyday language or from a branch of science, can be translated into it. When investigating this view, it is necessary to distinguish between the question of the translatability of protocol-sentences and the question of the translatability of other sentences of the natural and social sciences.

As to the protocol-sentences, the assertion that they are in principle translatable depends on the so-called "logical behaviorism,"[59] which says that sentences concerning mental phenomena (experiences, observations, recollections, emotions, etc.) possess a meaning that can be intersubjectively tested only if they are conceived as sentences concerning the bodily condition and/or behavior of the individual concerned, such as, for example, the condition of his nervous system or his appearance and movements (including also movements of speech).

As regards the remaining sentences belonging to the natural and social sciences, it appears that the translation into the physical language cannot be accomplished solely by explicit definitions but also requires the application of the "reduction" defined by Carnap.[60] This operation has a function similar to that of "constitution" in *Der logische Aufbau der Welt,* but, contrary to constitution, reduction is defined in the formal mode of speech, and the definition is expressed exactly by means of logistic symbols. To go into details would take us too far, but, roughly speaking, it may be said that the content of the definition is that a term 'a' is "reducible" to other terms 'b', 'c', . . . , if it is possible by means of the latter to formulate the characteristics concerning the conditions under which we are going to use the term 'a'. "The simplest method of reducing, in this sense, one concept to another is by definition. If 'a' can be defined by 'b', 'c', . . . , then obviously 'a' is reducible to 'b', 'c'. . . . It can, however, be shown that the method of definition is not the only one but is the simplest special case of reduction. E.g., the concept 'electrical field' is reducible to the concepts 'body,' 'mass,' 'electric charge,' and spatiotemporal determinations. We can, by the use of these concepts, formulate rules for the application of the concept 'electrical field,' viz., describe an experimental test for this concept. On the other hand, we cannot give a definition of 'electrical field' which contains only those concepts. Therefore, we must distinguish the broader concept 'reducibility' from the narrower concept 'definability.' "[61]

While reduction by definition cannot be carried through even with regard to the concepts of physics, nothing seems to prevent the reduction of the total number of concepts of the natural and social sciences to the physical language, if reduction is taken in its wider and more liberal sense, and this procedure is therefore suggested for unified science. The crucial points, where special difficulties might be expected to arise, concern the translations of biological, psychological, and sociological sentences into the physical language. As is well known, the material mode of speech has often led us to conceive the objects of those sciences as different in principle and (in a not specified sense) mutually irreducible. But the translation into the formal mode of speech of the sentences about these sciences shows that a reduction of their concepts to the physical language is possible in principle, and that, accordingly, the advantages attached to the realization of the idea of the unity of science are within reach, although to carry it out in detail would, of course, require both much more special investigation and a greatly extended co-operation among the investigators of the special sciences, mutually, and between them and logicians. In order to convey an idea of the view of the logical empiricists, their attitude to the three crucial points mentioned should be dealt with briefly.

In regard to biological sentences, logical empiricists think that the possibility of their translation into the physical language is evident, considering that all biological concepts capable of being empirically tested concern conditions and processes in bodies that may be characterized by space-time-co-ordinates. The vitalistic concepts, such as "entelechy," etc., that are incapable of being tested must, of course, be dropped, while the analysis of concepts such as "whole" and "purposiveness" has not as yet been performed in detail but will hardly give rise to difficulties in principle.

As regards the translatability of psychological sentences, logical behaviorism is invoked. If this is tenable, the translatability in principle of psychological sentences into the physical language is evident. And, having once gone so far, there seems no possibility of insurmountable difficulties in connection with the translation of sociological sentences, which describe the relation of persons and other organisms to one another and to their surroundings.

In addition to these considerations concerning general principles, logical empiricists also, in support of the idea of a unity of science, refer to the many border sciences growing up in an increasing number and bearing witness to formerly unheeded cross-connections between the traditional branches of science, such as biophysics, biochemistry, psychophysiology, social psychology, etc.; and the result of their deliberations is that they think themselves entitled to assert that "the concepts of the thing-language provide a common basis to which all concepts of all the parts of science can be reduced"[62]—only, of course, in the above-mentioned sense of "reduction."

This result only justifies a continuation of the work of realizing the idea of the unity of science. A road has been opened for a detailed analysis of the concepts of the individual sciences for the purpose of showing how the sentences of every one of them may be reduced to the thing-language. But the analysis itself is, of course, a gigantic piece of work that can be performed only by the co-operation of logicians and the specialists within the various special sciences. Such co-operation has indeed also been eagerly sought by logical empiricists from the early days of the Vienna Circle, and the ever increasing number of special scientists participating in the congresses is impressive evidence of the fact that this need for co-operation is widely felt. Its most conspicuous expression has been, perhaps, in the *International Encyclopedia of Unified Science*, where scientists of the special sciences and philosophers work together in harmony, although they are completely free to express varying opinions on questions of doubt. Strictly speaking, the thesis of physicalism cannot be considered proved until the reduction to the thing-language of the total number of the concepts of the natural and social sciences is made, which means, of course, never. But this in no way makes the work on the creation of an ever greater connection within the sphere of sciences superfluous, let alone useless. And should it appear that concepts actually exist which cannot be reduced to the thing-language—well, then, that does not make the idea of unity less valuable but merely shows that it is necessary to choose another language than the thing-language as the language of unity, which may very well be possible, although care should be taken, of course, that a reasonable meaning can always be attached to the sentences of any such language.[63]

As regards the question of the reduction of scientific theories to a few or even a single deductive system, the prospects are, in the opinion of logical empiricists, much darker than where the question of the reduction of concepts to the physical thing-language is concerned. Not even all physical laws can at present be included in a single deductive theory, and the prospects for a derivation of biological from physical laws—let alone a derivation of psychological or sociological laws from the physical plus the biological laws—are distant, although not hopeless. Efforts are being made to create more comprehensive syntheses, and no limits can be set beforehand to these endeavors. Yet, in spite of the great advantage of a unity of laws, its importance is not so fundamental as that of the unity of language,[64] which is more easily achieved.

13. Present Tendencies and Tasks

In this exposition of the development of logical empiricism I have kept largely to the main lines of development, and I hope that I have succeeded in making them clear. I have had to leave completely out of consideration a great many penetrating individual analyses, some of which have been the conditions and others the fruits of this development. So far logical-mathematical and physical concepts, theories, and methods have been treated, whereas biological, psychological, and sociological subjects, as well as concepts and theories of value, have been only occasionally touched upon. But the very idea of a unity of science implies that all spheres of knowledge should be treated, and the predilection for the exact sciences is a defect, against which the gradually growing connection with the views developed within pragmatism should now serve as a useful remedy. As yet, we are in these domains at the beginner's stage, which is to some extent due to the relative backwardness of these sociohumanistic sciences. However, from the co-operation with pragmatists and operationalists especially interested in biology, psychology, and sociology results may be expected which are necessary in order to create a proper balance of things. At the Harvard congress of 1939 this tendency made itself plainly felt and also came to expression in Morris' paper, "Semiotic, the Sociohumanistic Sciences, and the Unity of Science," in which he showed that the so far neglected spheres of knowledge can, in principle, be incorporated into unified science via a comprehensive theory of signs, or semiotic, a point which he later developed in his *Signs, Language, and Behavior* (1946). The same tendency was also pronounced in Joergensen's paper at the Harvard congress, "Empiricism and Unity of Science," in which especially the incorporation of the formal sciences into unified science was sketched and an attempt made to establish this contact through psychology; this conception was later expounded in his *Psykologi paa biologisk Grundlag* ("Psychology Based on Biology") (1941–45). Since the precise apparatus of concepts and the refined methods which have resulted from the work with the analysis of the exact sciences have proved highly useful in the further realization of the program of the unity of science, it is possible that the development in this direction may proceed more quickly than might have been expected, in view of the relatively great complication of these spheres. Prospects seem bright for the further development of logical empiricism, which has had the good luck to be stimulated by the metaphysical verbosity re-emerging in the wake of the last war, in almost the same way as the Vienna school and its predecessors were in their day stimulated by the unverifiable nebulous speaking of speculative metaphysicians.

Another tendency came to expression at the Harvard congress, namely, an increasing interest in semantics, which had, in particular, been developed by the Lwow-Warsaw school but which, so far, in the works of the logical empiricists had been overshadowed by logical syntax. While syntax is exclusively concerned with purely formal relations between linguistic expressions *qua* mere figures, semantics treats of the relations between the expressions of a given language and their "designata," i.e., that which they designate. In his paper, "Science and Analysis of Language," Carnap stressed the great significance of semantics and characterized as semantical several main concepts which he had formerly regarded as syntactical. This applies, for example, to concepts like "consequence," "analytic," "contradictory," and others, all of which are based on the important semantical concept of truth. As to the connection between them, Carnap said: "I have explained the semantical analysis of a language as exhibiting the relation of designation. We might equally well regard it as exhibiting the truth-conditions of the sentences of the language in question. This is merely a different formulation of the aim of semantics. Suppose a sentence of the simplest form is given, consisting of a proper name combined with a one-place predicate (e.g. 'Switzerland is small'); if, now, we know which object is designated by the name and which property is designated by the predicate then we also know the truth-condition of the sentence: it is true if the object designated by the name has the property designated by the predicate. Thus the concept of *truth* turns out to be one of the fundamental concepts of semantics. We may say that the result of a semantical analysis of a sentence is the understanding of the sentence. To understand a sentence is to know what is designated by the terms occurring in the sentence and, hence, to know under what conditions it will be true. But the understanding of a sentence does not suffice, in general, for knowing whether those conditions are fulfilled, in other words, whether the sentence is true or not. But sometimes there is such a relation between two sentences that a semantical analysis of them, in other words, the understanding of the sentences, suffices to show that if the first sentence is true the second must also be true. In this case the second is called a *logical consequence* of the first. This concept, the basic concept of the theory of logical deduction and thereby of logic itself, is thus based upon a certain relation between truth-conditions, and hence is a semantical concept. The same holds for other logical concepts which are often applied in the logical analysis of science, e.g. logically true (analytic), logically false (contradictory), logically indeterminate or factual (synthetic, neither logically true nor logically false), logically compatible, etc."[65] In consequence of the above view, Carnap later, in his *Introduction to Semantics* (1943), replaced his earlier syntactical definitions of the concepts mentioned by semantical definitions and also found it necessary, in general, to supplement many of the former discussions and analyses by corresponding semantical ones. They were not incorrect, to be sure, but they were incomplete. Perhaps the awakening interest in the pragmatic views developed by Morris and others will show that semantical analyses are not exhaustive either; but this question belongs to the future.

To characterize in brief the value of the contribution of logical empiricism to the development of human knowledge can best be done, I believe, by emphasizing that it has led to the appearance of entirely new points of view as regards philosophical problems. These must today be posed in a way that differs in principle from the ones hitherto used, and their treatment requires much more exactness than has been exercised heretofore. The very fact that we have grown accustomed to ask for the *meaning* of words and sentences and have found useful criteria has intensified our criticism of statements made by ourselves and by others and has thus furthered the critical attitude which, combined with inventiveness and imagination, is the basic condition for a sensible approach to the practical problems of our day and to the promotion of scientific investigation. To return to past ways of thinking would be like ignoring the quantum theory in physics. Or, in other words, it is, as matters stand, impracticable. And to have made a contribution which it is impossible to ignore if scientific investigation is to proceed is presumably the utmost that may be expected from the pursuers of science. But this expectation the pioneers of logical empiricism have already fulfilled. And if today the still unsolved problems within the sphere of philosophy can be formulated and treated with a precision and clarity formerly unknown, the merit is theirs. They have not created a new philosophical system, which, indeed, would have been contrary to their highest intentions, but they have paved the way for a new and fruitful manner of philosophizing.

Notes and Bibliography

Chapter I

1. See Postscript, bringing this material up to date.

2. H. Feigl, "Logical Empiricism," in *Twentieth Century Philosophy*, ed. Dagobert D. Runes (New York, 1947), pp. 406–8. Compare Otto Neurath, *Le Développement du cercle de Vienne et l'avenir de l'empirisme logique* (Paris, 1935), chap. v, where he attempts to show the reason why the movement originated just in Vienna, where liberalistic and empiricist trends had made themselves felt for several decades, which had not been the case in Germany.

3. *Wissenschaftliche Weltauffassung*, pp. 16–17.

4. Neurath, *Le Développement du cercle de Vienne*, p. 58.

5. See, e.g., P. Frank, "Logisierender Empirismus in der Philosophie der U.S.S.R.," *Actes du congrès international de philosophie scientifiques, Sorbonne, Paris, 1935* (Paris, 1936), VIII, 68–76.

6. As a curiosity it may be noted that Comte is not mentioned at all in Russell's large *A History of Western Philosophy* (London, 1946). Even Ernst Mach is not mentioned.

7. Cf. Jean Nicod, "Les Tendences philosophiques de M. Bertrand Russell," *Revue de mét. et de mor.*, XXIX (1922), 77.

8. Concerning Mach's relation to logical empiricism see R. von Mises, "Ernst Mach und die empiristische Wissenschaftsauffassung," *Einheitswissenschaft* ("Library of Unified Science," No. 7 ['s Gravenhage, 1938]).

9. B. Russell, *Our Knowledge of the External World as a Field for Scientific Method in Philosophy* (Chicago, 1914), chap. ii.

10. *Ibid.*

11. B. Russell, "Logical Atomism," *Contemporary British Philosophy: Personal Statements, First Series* (London and New York, 1924), p. 363.

12. Russell, *Our Knowledge*, p. 112.

13. *Ibid.*, p. 111.

14. *Ibid.*, p. 78.

15. *Ibid.*, p. 89.

16. *Ibid.*, pp. 111–12.

17. G. E. Moore, *Principia ethica* (Cambridge, 1903), p. ix.

18. See, e.g., R. B. Braithwaite, "Philosophy", in *Cambridge University Studies*, VII (1933), pp. 1–32.

19. Russell's Introduction to Wittgenstein, *Tractatus* (London, 1922), p. 8.

20. H. M. Sheffer, "A Set of Five Independent Postulates for Boolean Algebras, with Application to Logical Constants," *Trans. Amer. Math. Soc.*, XIV (1913), 488–89.

21. Cf. Russell's Introduction to *Tractatus*, pp. 13–15.

22. Carnap, *Scheinprobleme*, pp. 27–29.

23. Carnap, *Der logische Aufbau der Welt*, p. 2.

24. *Ibid.*, p. 47.

25. *Ibid.*, pp. 57–58.

26. *Ibid.*, p. 65.

27. *Ibid.*, p. 74.

28. *Ibid.*, p. 139.
29. *Ibid.*, p. 79.
30. *Ibid.*, p. 80.
31. *Ibid.*, p. 86.
32. *Ibid.*, p. 87.
33. *Ibid.*, p. 92.
34. *Ibid.*, pp. 93–104.
35. *Ibid.*, p. 102.
36. *Ibid.*, pp. 109–10.
37. *Ibid.*, p. 133.
38. *Ibid.*, p. 23.
39. *Ibid.*, p. 169.
40. *Ibid.*, p. 176.
41. *Ibid.*, pp. 186–87.
42. *Ibid.*, p. 202.
43. *Ibid.*, p. 204.
44. *Ibid.*
45. *Ibid.*, pp. 245–55.

46. A brief survey of the first stages of this internal criticism will be found in A. Petzäll, *Logistischer Positivismus* ("Göteborgs Högskolas Årsskrift," Vol. XXXVII Göteborg, [1931]). Incidentally, it may be noted that the first thoroughgoing criticism of Carnap's *Der logische Aufbau der Welt* was advanced by another Scandinavian philosopher, Eino Kaila, who, like Petzäll, had studied in Vienna, in his penetrating essay, *Der logistische Positivismus: Eine kritische Studie* (Turku, 1930). Kaila later gave valuable contributions to certain parts of the theory of constitution in his *Das System der Wirklichkeitsbegriffe* (Helsingfors, 1936) and his *Den mänskliga Kunskapen* ("Human Knowledge") (Helsingfors, 1939). A comprehensive critical study which also considers the later development of the movement is J. R. Weinberg's *An Examination of Logical Positivism* (London, 1936).

Chapter II

1. M. Schlick, "The Future of Philosophy," *Seventh International Congress of Philosophy, Oxford, 1930* (Oxford, 1931), p. 112. A more detailed discussion under the same title, "The Future of Philosophy," was published in *Publications in Philosophy*, ed. P. A. Schilpp, Vol. I (College of the Pacific, 1932). This lecture was reprinted in *Gesammelte Aufsätze* (Vienna, 1938) and in *Basic Problems of Philosophy*, ed. D. J. Bronstein *et al.* (New York, 1947).

2. "The Future of Philosophy," p. 115.

3. *Ibid.*, p. 116.

4. See *Erkenntnis* I (1930), 72.

5. H. Reichenbach, *Relativitätstheorie und Erkenntnis apriori* (Berlin, 1920), p. 71.

6. H. Reichenbach, "Logistic Empiricism in Germany and the Present State of Its Problems," *Journal of Philosophy*, XXXIII (1936), 114.

7. Reichenbach, *Relativitätstheorie und Erkenntnis apriori*, p. 74.

8. Reichenbach, *Philosophie der Raum-Zeit-Lehre*, p. 45.

9. *Ibid.*, pp. 205–6.

10. Reichenbach, *Ziele und Wege der heutigen Naturphilosophie* (Leipzig, 1931), pp. 38–39.

11. Reichenbach, *Wahrscheinlichkeitslehre* (Leiden, 1935), p. 381.

12. *Ibid.*, p. 387.

13. Reichenbach, "Logistic Empiricism," p. 157; cf. also his *Experience and Prediction*, p. 363.

14. See Reichenbach, *Wahrscheinlichkeitslehre*, p. 305, and his *Experience and Prediction*, p. 363.

15. Reichenbach, "Logistic Empiricism," pp. 158–59.

16. Cf. Ajdukiewicz, "Der logistische Antiirrationalismus in Polen," *Erkenntnis,* V, 151; and Rose Rand, "Kotarbinski's Philosophie auf Grund seines Hauptwerkes: 'Elemente der Erkenntnistheorie, der Logik und der Methodologie der Wissenschaften,' " *Erkenntnis,* VII, 92. See also Z. Jordan's book (Postscript).

17. See, e.g., C. Morris, "Some Aspects of Recent American Scientific Philosophy," *Erkenntnis,* V, 142.

18. C. S. Peirce, "How To Make Our Ideas Clear," *Popular Science Monthly,* January, 1878, here quoted from Peirce, *Chance, Love, and Logic: Philosophical Essays* (London, 1923), p. 45.

19. P. W. Bridgman, *The Logic of Modern Physics* (New York, 1927), pp. 5–7; see also two articles by Herbert Feigl on operationism and explanation in *Psychological Review,* LII, 195; reprinted in *Readings in Philosophical Analysis* (see Postscript).

20. C. Morris, "The Concept of Meaning in Pragmatism and Logical Positivism," *Actes du huitième congrès international de philosophie à Prague, 2–7 septembre, 1934* (Prague, 1936), p. 133.

21. C. Morris, *Logical Positivism, Pragmatism, and Scientific Empiricism* (Paris, 1937), p. 65. Morris has recently given a comprehensive exposition of his semiotic in *Signs, Language, and Behavior* (New York, 1946).

22. See, e.g., K. Marc-Wogau's article, "Uppsala Filosofien och den logiska Empirismen," *Ord och Bild* (1944), p. 30, where similarities and differences between the two movements have been clearly stated.

23. In *Philosophy of Science,* I (1934), 5.

24. Published in a revised and enlarged English translation entitled *The Logical Syntax of Language* (1936). In his *Die Aufgabe der Wissenshaftslogik* (in the collection "Einheitswissenschaft," No. 3 [1934]) and in his London lectures, *Philosophy and Logical Syntax* (London, 1935), Carnap has given more popular expositions of his theory.

25. Carnap, *Logical Syntax of Language* (London, 1937), pp. 51–52.

26. Carnap, *Philosophy and Logical Syntax,* pp. 47–49.

27. *Ibid.,* p. 56.

28. This point of view Carnap had already advanced in his article, "Die physikalische Sprache als Universalsprache der Wissenschaft," *Erkenntnis,* II (1931), 432.

29. Carnap, *Philosophy and Logical Syntax,* pp. 81–82.

30. *Ibid.,* pp. 80–81.

31. O. Neurath, "Soziologie im Physikalismus," *Erkenntnis,* II (1931), 403; cf. also O. Neurath, "Physikalismus," *Scientia,* 1931, p. 299.

32. M. Schlick, "Über das Fundament der Erkenntnis," *Erkenntnis,* IV (1933), 96–97.

33. *Ibid.,* p. 98.

34. O. Neurath, "Radikaler Physikalismus und 'wirkliche Welt,' " *Erkenntnis,* IV (1933), 361.

35. *Ibid.,* p. 356.

36. *Ibid.,* p. 352.

37. O. Neurath, "Protokollsätze," *Erkenntnis,* III (1932), 206.

38. In *Analysis,* II (1935), 49.

39. *Ibid.,* p. 93.

40. K. Popper, *Logik der Forschung* (Vienna, 1935), p. 9.

41. M. Schlick, "Die Kausalität in der gegenwärtigen Physik," *Naturwissenschaften,* XIX (1931), 156.

42. Popper, *Logik der Forschung*, p. 13.

43. Published in *Philosophy of Science*, III (1936), 419, and IV (1937), 1.

44. R. Carnap, "Wahrheit und Bewährung," *Actes du congrès international de philosophie scientifique, Paris, 1935* (Paris, 1936), IV, 19.

45. *Philosophy of Science*, III, 20–21.

46. This question was discussed later in detail by C. G. Hempel in his "A Purely Syntactical Definition of Confirmation," *Journal of Symbolic Logic*, Vol. VIII (1943), and in "Studies in the Logic of Confirmation," *Mind*, LIV (1945), 1 and 97.

47. *Philosophy of Science*, IV, 33.

48. *Ibid.*, p. 34.

49. *Ibid.*

50. *Ibid.*, pp. 34–35. As to the view here mentioned, see also C. G. Hempel, "Le Problème de la vérité," *Theoria*, 1937, p. 206.

51. See, e.g., O. Neurath, *Empirische Soziologie: Der wissenschaftliche Gehalt der Geschichte und Nationalökonomie* (Vienna, 1931), p. 2.

52. O. Neurath, *Einheitswissenschaft und Psychologie* (Vienna, 1933), p. 7.

53. O. Neurath, "Soziologie im Physikalismus," *Erkenntnis*, II (1931), 395.

54. R. Carnap, *Philosophy and Logical Syntax*, p. 89.

55. R. Carnap, "Die physikalische Sprache als Universalsprache der Wissenschaft," *Erkenntnis*, II (1931), 441–42.

56. *Ibid.*, p. 445.

57. *Ibid.*, p. 447.

58. *Ibid.*

59. See, e.g., C. G. Hempel, "Analyse logique de la psychologie," *Revue de synthèse*, X (1935), 27; and R. Carnap, "Les Concepts psychologiques et les concepts physiques sont-ils foncièrement différentes?" *Revue de synthèse*, X (1935), 43.

60. R. Carnap, "Testability and Meaning," *Philosophy of Science*, III, 434.

61. Carnap, "Einheit der Wissenschaft durch Einheit der Sprache," *Travaux du IXe congrès international de philosophie* (Paris, 1937), IV, 54. Cf. Carnap, "Ueber die Einheitssprache der Wissenschaft: Logische Bemerkungen zum Projekt einer Enzyklopädie," *Actes du congrès international de philosophie scientifique* (Paris, 1936), II, 60.

62. Carnap, "Einheit der Wissenschaft," p. 57.

63. I could imagine, for instance, that one might go so far as to give up intersubjectivity and accordingly admit sentences as meaningful, if only they are introspectively testable (and so not mere sound-complexes with no designations) or, at any rate, introspective sentences that agree to such an extent as to fulfil the requirements generally made by psychologists for the admittance of their universal validity in psychology. In case a criterion of meaning as liberal as this is accepted, logical behaviorism will no longer be a necessary condition of such extended unity of science but merely a special means of testing, side by side with introspection. However, actual metaphysical sentences of untestable entities would be excluded as meaningless.

64. Cf. R. Carnap, *Logical Foundations of the Unity of Science*, in *Encyclopedia of Unified Science*, I, No. 1 (Chicago, 1938), 60–62.

65. Here quoted from a separate print distributed at the congress. A similar reaction to the prevalent formal-syntactical view came simultaneously to expression in J. Joergensen's "Reflexions on Logic and Language. I. Languages, Games, and Empiricism; II. Semantical Logic," *Journal of Unified Science* (*Erkenntnis*), VIII (1939), 218.

Postscript

By *Norman M. Martin*

World War II caused considerable disturbance in the movement of logical empiricism. Most of the European philosophers survived, and many of them left the Continent for Great Britain or the United States. Hosiasson and Lindenbaum died in Poland; Kurt Grelling and Karl Reach were deported by the Nazis and died or were killed. The *Journal of Unified Science* and the "Library of Unified Science" were discontinued because of the war. The work on the *International Encyclopedia of Unified Science* was hampered, although a number of monographs were issued (see chap. ii, part 1).

The war did not, however, put an end to the work of the logical empiricists. On the contrary, this work continued along several lines. The Sixth International Congress for the Unity of Science was held at the University of Chicago, September 2–6, 1941. Some of the main topics were the unification of science, the theory of signs, psychology, and valuation. Since then no congresses have been held.

Considerable progress has been made in semantics. Especially important here are the contributions of Carnap. His *Introduction to Semantics* (Cambridge: Harvard University Press, 1942) presents the problems involved in the construction of semantical systems, especially with the construction of L-concepts, i.e., concepts which are applicable on merely logical reasons as opposed to factual reasons, and with the relations between syntax and semantics. He there explains in detail how he would modify the views expressed in his *Logical Syntax of Language.* Further attention to the relation between semantics and syntax is paid in Carnap's *Formalization of Logic* (Cambridge: Harvard University Press, 1943). By "formalization" is meant the construction of a syntactical concept which applies whenever a given semantical concept applies in any semantical system which is a true interpretation of the constructed calculus. For example,

C-implication (derivability) in logical syntax is intended as a formalization of L-implication. The problem of the book is whether the calculi common today are full formalizations of logic and, if not, whether such a formalization can be made. Carnap shows that the usual propositional calculus is not a full formalization of propositional logic but that, with the introduction of a new type of syntactical concept called "junctives," a full formalization can be achieved. Similar results hold for functional logic.

In *Meaning and Necessity* (Chicago: University of Chicago Press, 1947) Carnap suggests the substitution of the method of extension and intension for the method of the name-relation, which had dominated earlier semantical discussion. By this method, instead of considering an expression as the name of an entity, it would be considered to have an intension and an extension; e.g., the predicate 'red' has the property of being red as its intension and the class of red things as its extension. Carnap proposes to use the concepts of intension and extension, which occurred now and again in the old logic, as key concepts in semantical analysis. He discusses the possibility of an adequate extensional metalanguage for semantics and finds it, in general, possible, although there are some doubtful features. He also outlines a system of modal logic which he constructs in greater detail in "Modalities and Quantification," *Journal of Symbolic Logic*, XI (1946), 33–64.

Several interesting contributions to semantic theory are contained in the "Symposium on Meaning and Truth" which appeared in Volumes IV (1944) and V (1945) of *Philosophy and Phenomenological Research*. C. I. Lewis, in "Modes of Meaning" (*ibid.*, Vol. IV), gives an analysis of language similar in many respects to the intension-extension distinction made by Carnap. He distinguishes (as terms): denotation—the class of all actual things to which a term applies; comprehension—the class of all consistently thinkable things to which a term applies; connotation—which is identified with a correct definition of the term; and signification—the comprehensive character such that everything that has that character is correctly namable by the term.

Analogous distinctions are made for sentences. G. Watts Cunningham, in "On the Linguistic Meaning-Situation" (*ibid.*, Vol. IV), attempts to specify the limits of conventionalism in semantics by asserting that, while the words and rules of syntax of a language are conventional, the syntactical structure is determined by the referent. Felix Kaufmann, in "Verification, Meaning, and Truth" (*ibid.*, Vol. IV), attempts to define truth in terms of agreement with the rules of scientific procedure—a position not unlike that of Neurath. C. J. Ducasse, in "Propositions, Truth, and the Ultimate Criterion of Truth" (*ibid.*, Vol. IV), defends the view that ultimate "undisbelievability" is the criterion of truth. Alfred Tarski, in "The Semantic Conception of Truth" (*ibid.*, Vol. IV), presents in English the main features of the definition of truth which he had presented earlier in Polish and German; he clarifies the nature of this concept and defends his views against his critics. Norman Dalkey, in "The Limits of Meaning" (*ibid.*, Vol. IV), gives an analysis of vagueness, pointing out three elements which he terms "confusion," "obscurity," and "incomplete determination." These views were discussed by the symposiasts and by Ernest Nagel, who came out strongly against Ducasse's formulations, in later issues of the journal.

Another significant line of work, which was pursued throughout the war years, was the analysis of science. A collection of articles by Philipp Frank, dating from 1908 to 1938, which dealt with the philosophy of physics was published under the name *Between Physics and Philosophy* (Cambridge: Harvard University Press, 1941). Felix Kaufmann presented his theory of scientific procedure in *Methodology of the Social Sciences* (New York: Oxford University Press, 1944). He emphasizes particularly the reliance of scientific procedure on rules (often implicit) of procedure. Thus, for him, the reversibility of the decision to accept a proposition into the body of knowledge (the principle of permanent control) and the necessity of having grounds for the acceptance of a proposition are extremely important. Rules of scientific procedure may in his opinion be changed, but only in connection with "rules of higher order." In line with this approach, Kaufmann discusses methodological issues, first of empirical science in general, and then, in particular, of problems in social science, such as value-statements, behaviorism, and the nature of social law.

Another important work is Hans Reichenbach's *Philosophical Foundations of Quantum Mechanics*, in which he discusses the mathematics of quantum mechanics and the problems of its interpretation, suggesting that a three-valued logic is more suitable for this purpose than the usual two-valued logic. Quantum mechanics can be formulated in one of three ways, in a wave language, a corpuscle language, or a neutral (three-valued) language. In the first two, sentences expressing causal anomalies appear. These do not appear in the neutral language; however, sentences about interphenomena do appear when this language is used.

In *Elements of Symbolic Logic* (New York: Macmillan Co., 1947) Reichenbach attempts to characterize the logic of scientific laws. He does this with the help of the concept of "original nomological statement," which he defines as "an all-statement that is demonstrably true, fully exhaustive, and universal." Then he is able to describe the common character of all laws. In the same book he also gives an analysis of grammar from the standpoint of modern logic.

An interesting contribution to the logical analysis of science is "Studies in the Logic of Explanation," by Carl G. Hempel and Paul Oppenheim, in *Philosophy of Science*, XV (1948), 135–75. Hempel and Oppenheim attempt to examine explanation by looking at the conditions that the explanans must fulfil. The principal requirements are: the explanandum must be a logical consequence of the explanans; the explanans must contain general laws, and these must actually be required in the derivation of the explanandum; the explanans must be capable, at least in principle, of test by experiment or observation, and the sentences constituting the explanans must be true. The authors hold that these criteria are general throughout science. In this connection they discuss the concept of "emergence," concluding that it must be purged of its connotations of absolute unpredictability. In line with these views they construct a precise logical theory of explanation.

A great deal of discussion on the philosophy of probability has occurred. In the "Symposium on Probability" which took place in *Philosophy and Phenomenological Research,* Volumes V (1945) and VI (1946), Donald Williams defends the Laplacean conception, which was attacked from a frequency point of view by Reichenbach, von Mises, and Margenau. Carnap, in "The Two Concepts of Probability" (*ibid.,* Vol. V [1945], 513–32), defends the view that there are actually two distinct concepts used under the name of "probability," both of which have a right to scientific treatment. The first of these concepts is sometimes also called "degree of confirmation," the second, "relative frequency in the long run." Carnap then illustrates at length the differences in logical nature between the two concepts. The first of these concepts is a semantical one dealing with relations between sentences; the second is an empirical concept. A basic sentence of the first is true by logic alone and one of the second by virtue of the facts. In "On Inductive Logic," *Philosophy of Science,* XII (1945), 72–97, Carnap elaborated his system of probability in the sense of degree of confirmation. The exposition of his system of inductive logic will constitute the bulk of his forthcoming two-volume work, *Probability and Induction.* Felix Kaufmann held, in "Scientific Procedure and Probability," in *Philosophy and Phenomenological Research,* VI (1945), 47–66, that degree of confirmation should be defined in terms of the process of accepting propositions into the body of accepted knowledge. For this reason he wants to distinguish sharply between the confirmation of a proposition not yet accepted and the corroboration of one already accepted. An alternative definition of degree of confirmation to that offered by Carnap was presented by Hempel and Oppenheim in "A Definition of 'Degree of Confirmation,'" *Philosophy of Science,* XII (1945), 98–115, and by Olaf Helmer and Oppenheim in "A Syntactical Definition of Probability and Degree of Confirmation," *Journal of Symbolic Logic,* X (1945), 25–60. This definition is of particular interest, since, when so defined, the degree of confirmation function is not a probability function, i.e., it does not have all the mathematical properties commonly associated with probability in mathematical theory. In the discussions centering around the symposium on probability, Nagel and Bergmann participated actively.

In addition to the above-mentioned works on more or less specific topics, several works of a more general nature on the theory of signs and theory of value were written by logical empiricists. Russell wrote *An Inquiry into Meaning and Truth* (New York: W. W. Norton & Co., 1940), in which he defends a causal theory of language. He holds the view that the proper individuals of an epistemologically correct language are universals. He maintains the necessity of basic propositions, i.e., propositions which are caused (and justified) by perception and which are known to be true. He defends the correspondence theory of truth against Dewey's "warranted assertibility" and similar opinions. Russell also published *A History of Western Philosophy and Its Connections with Political and Social Circumstances* (New York: Simon & Schuster, 1945), in which he attempts to analyze the major philosophers from the standpoint of the philosophy of logical analysis. In addition, Russell wrote *Human Knowledge: Its Scope and Limits* (New York: Simon & Schuster, 1948), in which he deals with problems of epistemology, semiotic, and the philosophy of science. For the first time he deals with the problem of probability; he holds that scientific inference needs some statement of the inductive principle which would be a synthetic statement but could not be established by any argument from experience. He concludes from his general study of knowledge that empiricism is not an adequate theory of knowledge, although less inadequate than previous theories. He also holds that these inadequacies can be discovered by adherence to the doctrine that "all human knowledge is uncertain, inexact and partial."

Charles Morris wrote *Paths of Life* (New York: Harper & Bros., 1942), in which he analyzes the principal patterns of value-preferences and suggests, as a possible means of uniting them in dynamic interaction, one which involves features of all of them. *The Open Self* (New York: Prentice-Hall, 1948) continues his empirical study of value-patterns. In *Signs, Language, and Behavior* (New York: Prentice-Hall, 1946) Morris analyzes meaning-phenomena at length. He distinguishes four modes of signifying: the designative (e.g., "the coin is round"), the appraisive (e.g., "the coin is good"), the prescriptive (e.g., "Come here"), and the formative (e.g., "the coin is a coin"). In addition, four uses of language (informative, valuative, incitive, and systemic) are distinguished. On the basis of these distinctions Morris classifies types of discourse. His point of departure is behavioral (more in the sense of Tolman and Mead than of Watson), and one of his principal results is the formulation of the theory of signs, or semiotic, in behavioral terms. He considers at length the importance and role of signs in individual and social life.

Another important contribution to the theory of meaning is C. I. Lewis' *Analysis of Knowledge and Valuation* (La Salle, Ill.: Open Court Publishing Co., 1946). Lewis begins with a general theory of meaning derived from Peirce, together with the distinction between empirical and analytic statements: the analytic ones are those which relate to meanings alone. Empirical sentences are of three types: first, expressive statements, which express an experience directly and which are, therefore, indubitable to the one who utters them, although they may be false in the case of a lying report—they are thus much like Schlick's substantiations; second, terminating judgments, which make predictions concerning experience specific as to time and place; and, finally, nonterminating judgments, which make predictions concerning experience general as to time and place. Lewis analyzes all knowledge into these types of empirical and analytic sentences. He also presents an analysis of value-sentences so that they are a species of empirical sentence.

Another study concerning the nature of valuation is Charles L. Stevenson's *Ethics and Language* (New Haven: Yale University Press, 1944). Stevenson attempts to distinguish between differences in belief and differences in attitude; valuational statements are analyzed into cognitive and prescriptive components (e.g., 'this is good' is interpreted as 'I approve of this, do so likewise'), and it is maintained that ethical differences are ultimately differences in attitude rather than in belief.

Within the "Library of Living Philosophers," edited by Paul A. Schilpp, two volumes were published under the titles of *The Philosophy of G. E. Moore* (Evanston: Northwestern University, 1942) and *The Philosophy of Bertrand Russell* (Evanston: Northwestern University, 1944). Each of these volumes contains a series of descriptive and critical essays by several authors, together with a reply by the philosopher concerned, and a complete bibliography. Among the contributors are C. D. Broad, C. L. Stevenson, Paul Marhenke, C. M. Langford, John Wisdom, Susan Stebbing, Hans Reichenbach, Kurt Gödel, Max Black, and Ernest Nagel.

An excellent summary of the work done by the Polish groups is presented in *The Development of Mathematical Logic and Logical Positivism in Poland between the Two Wars*, by Z. Jordan (New York: Oxford University Press, 1945). A good summary of the position of logical empiricism, including a selected bibliography of empiricist publications, can be found in Herbert Feigl's article on "Logical Empiricism," in *Twentieth Century Philosophy: Living Schools of Thought*, edited by Dagobert D. Runes (New York: Philosophical Library, 1943). This article and many others by writers of the logical-empiricist movement are reprinted in *Readings in Philosophical Analysis*, edited by Herbert Feigl and Wilfrid Sellars (New York: Appleton-Century-Crofts, 1949). This volume contains sections on semantics, confirmability, logic and mathematics, the a priori, induction and probability, logical analysis of philosophy, philosophy of science and ethics, from viewpoints allied to logical empiricism. Among the authors included, in addition to the editors, are Quine, Tarski, Frege, Russell, Carnap, Lewis, Schlick, Ajdukiewicz, Nagel, Waismann, Hempel, Broad, Ducasse, Reichenbach, and Stevenson, as well as a number of other empiricist philosophers.

The "Discussion on the Unity of Science," in which, among others, Neurath, Morris, and Horace Kallen participated, appeared in *Philosophy and Phenomenological Research,* Volume VI. In 1946, *Synthèse,* an international journal published in Holland, resumed the publication of its "Unity of Science Forum," which had been edited by Neurath until his death in 1945. *Analysis,* which represents, largely, the British analytic philosophers, has also resumed publication.

The movement of logical empiricism, having developed during the war even under unfavorable conditions, is now further expanding its activities.

EDITORS' NOTE: Since the Postscript written by Norman M. Martin early in 1949 a number of developments deserve notice. Among recent publications relevant to the field of this monograph, the following books may be mentioned: M. Black, *Language and Philosophy* (Ithaca: Cornell University Press, 1949); R. Carnap, *Logical Foundations of Probability* (Chicago: University of Chicago Press, 1950)—this is the first volume of the two-volume work, *Probability and Induction;* P. Frank, *Relativity—a Richer Truth* (Boston: Beacon Press, 1950); P. Frank, *Modern Science and Its Philosophy* (Cambridge: Harvard University Press, 1949)—this is an expanded version of his former work, *Between Physics and Philosophy;* A. Pap, *Elements of Analytic Philosophy* (New York: Macmillan Co., 1949); H. Reichenbach, *Theory of Probability* (Berkeley: University of California Press, 1949); P. Schilpp (ed.), *Albert Einstein: Philosopher-Scientist* (Evanston, Ill.: Library of Living Philosophers, 1949). A new journal, *Philosophical Studies,* edited by H. Feigl and W. Sellars, began publication in 1949 (Minneapolis: University of Minnesota).

Space does not permit reference to the recent developments of the philosophy of science in other countries, but the names of some new journals can at least be mentioned: *Methodos* (Italy); *Analisi* (Italy); *Sigma* (Italy); *Science of Thought* (Japan, and in Japanese); *British Journal for the Philosophy of Science.* The *Revue internationale de philosophie* devoted a special issue to logical empiricism (Vol. IV [1950]). There are articles by B. Russell, R. Carnap, C. G. Hempel, H. Feigl, and M. Barzin; and a selected bibliography of 216 items prepared by H. Feigl. There has just been received a book by V. Kraft, *Der Wiener Kreis: Der Ursprung des Neopositivismus* (Wien: Springer, 1950).

The Institute for the Unity of Science was incorporated in 1949 with Philipp Frank, of Harvard University, as president of the Board of Trustees. A grant from the Rockefeller Foundation made this incorporation possible. The *International Encyclopedia of Unified Science* will henceforth be owned and directed by the Institute for the Unity of Science. The Institute has been furnished quarters by the American Academy of Arts and Sciences, 28 Newbury Street, Boston 16, Massachusetts.

Bibliography and Index

Herbert Feigl and Charles Morris

Contents

Bibliography

The enormous and continually accelerated growth in the literature of the logico-methodological foundations and the history of the sciences prevents anything like completeness in the following list of books. A concerted effort has been made to include a variety of helpful references. Although there are countless valuable and pertinent articles and essays contained in periodicals, the following references are restricted exclusively to *books.* Fortunately, many an important paper or essay is included in one or another of the anthologies or collective volumes listed below.

It was often difficult to decide under which of the thirteen main headings to list a given item. Straight presentation of scientific problems and their attempted solutions often appear in conjunction with philosophical reflections. Books with predominantly scientific (rather than philosophical) contents are marked with an *asterisk.* Dates given are not always for first editions.

Many of the books listed contain valuable bibliographies of their own. The monographs in the *Encyclopedia* also provide lists of pertinent references (only a small subset of which are included in the present list). All of the monographs of the *Encyclopedia* are listed in this bibliography.

Professors Carnap, Hempel, and Nagel made suggestions for the improvement of an earlier version of the bibliography. For generous help in the compilation of the final version thanks are due to Dr. Jeffrey Bub, Dr. Roger Stuewer, Dr. Stephen Winokur, Dr. Michael Radner, Mr. Henry Lackner, and Mr. Robert Anderson. And for typing, to Mrs. Christopher Storer.

1. General Introductory Works

Benjamin, A. Cornelius. *Science, Technology, and Human Values.* Columbia, Mo.: University of Missouri Press, 1965.

Bibliography

Bergman, Gustav. *Philosophy of Science.* Madison: University of Wisconsin Press, 1957.

Braithwaite, R. B. *Scientific Explanation.* New York: Cambridge University Press, 1953.

Carnap, Rudolf. *Philosophical Foundations of Physics: An Introduction to Philosophy of Science.* Edited by M. Gardner. New York: Basic Books, 1966.

Caws, Peter. *The Philosophy of Science.* Princeton: Van Nostrand, 1965.

Frank, Philipp. *Philosophy of Science.* Englewood Cliffs, N.J.: Prentice-Hall, 1957.

Hanson, Norwood Russell. *Patterns of Discovery.* London: Cambridge University Press, 1958.

Hempel, Carl G. *Philosophy of Natural Science.* Englewood Cliffs, N.J.: Prentice-Hall, 1966.

Kemeny, John G. *A Philosopher Looks at Science.* Princeton: Van Nostrand, 1959.

Mises, Richard von. *Positivism, A Study in Human Understanding.* Cambridge, Mass.: Harvard University Press, 1951.

Nagel, Ernest. *The Structure of Science.* New York: Harcourt, Brace & World, 1961.

Neurath, Bohr, Dewey, Russell, Carnap, and Morris. Encyclopedia and unified science, *International Encyclopedia of Unified Science,* Vol. I, No. 1. Chicago: University of Chicago Press, 1938.

Rapoport, Anatol. *Operational Philosophy.* New York: Wiley, 1965.

Sachsse, Hans. *Naturerkenntnis und Wirklichkeit.* Brunswick: Vieweg, 1967.

———. *Erkenntnis des Lebendigen.* Brunswick: Vieweg, 1968.

Schrödinger, Erwin C. *Science, Theory and Man.* New York: Dover, 1957.

Smart, J. J. C. *Between Science and Philosophy.* New York:

Random House, 1968.

Toulmin, Stephen. *The Philosophy of Science.* London: Hutchinson's University Library, 1953.

Walker, Marshall. *The Nature of Scientific Thought.* Englewood Cliffs, N.J.: Prentice-Hall, 1963.

2. General Philosophy of Science (Including Logic and Methodology)

A. Books by Individual Authors

Brain, Sir Walter Russell. *Science, Philosophy and Religion.* Cambridge: At the University Press, 1959.

Broad, C. D. *Perception, Physics and Reality.* Cambridge: At the University Press, 1914.

Bunge, Mario. *Metascientific Queries.* Springfield, Ill.: Charles C. Thomas, 1958.

———. *Intuition and Science.* Englewood Cliffs, N.J.: Prentice-Hall, 1962.

———. *The Myth of Simplicity.* Englewood Cliffs, N.J.: Prentice-Hall, 1963.

———. *Scientific Research I: The Search for System.* New York: Springer-Verlag, 1967.

———. *Scientific Research II: The Search for Truth.* New York: Springer-Verlag, 1967.

Campbell, Norman R. *An Account of the Principles of Measurement and Calculation.* London: Longmans, Green & Co., 1928.

———. *What Is Science?* New York: Dover, 1952.

———. *Foundations of Science: The Philosophy of Theory and Experiment.* New York: Dover, 1957.

Carnap, Rudolf. *Die Aufgabe der Wissenschaftslogik,* in the series Einheitswissenschaft, No. 3. Vienna: Gerold, 1934.

———. *The Unity of Science.* London: Routledge & Kegan Paul, 1938.

Bibliography

Churchman, C. West. *Theory of Experimental Inference.* New York; Macmillan, 1948.

Churchman, C. West, and Ackoff, R. L. *Methods of Inquiry.* St. Louis: Educational Publishers, 1950.

Churchman, C. West, and Ratoosh, P. (eds.) *Measurement: Definitions and Theories.* New York: Wiley, 1959.

Clifford, William K. *The Common Sense of the Exact Sciences.* New York: Dover, 1955.

Cohen, M. R. *Reason and Nature.* New York: Harcourt, Brace & World, 1931.

———. *Studies in Philosophy and Science.* New York: Ungar, 1959.

Conant, J. B. *On Understanding Science.* New Haven: Yale University Press, 1947.

Dingle, H. *Science and Human Experience.* London: Williams & Norgate, 1931.

———. *Through Science to Philosophy.* Oxford: Clarendon Press, 1937.

Enriques, Federigo. *Problems of Science.* La Salle, Ill.: Open Court, 1914.

Frank, Philipp. *Modern Science and Its Philosophy.* Cambridge, Mass.: Harvard University Press, 1949.

———. *Relativity: A Richer Truth.* Boston: Beacon Press, 1950.

Good, I. J. (ed.) *The Scientist Speculates.** New York: Basic Books, 1962.

Goodman, Nelson. *Fact, Fiction and Forecast.* Indianapolis: Bobbs-Merrill, 1965.

Harré, R. *Matter and Method.* New York: St. Martin's Press, 1964.

Harris, Errol E. *The Foundations of Metaphysics in Science.* New York: Humanities, 1965.

Hawkins, David. *The Language of Nature.* San Francisco: W. H. Freeman & Co., 1964.

Helmholtz, Hermann von. *Schriften zur Erkenntnistheorie.* Edited by Moritz Schlick and Paul Hertz. Berlin: Springer, 1921.

———. *Counting and Measuring.* New York: Van Nostrand, 1930.

Hempel, Carl G. Fundamentals of concept formation in the empirical sciences, *International Encyclopedia of Unified Science,* Vol. II, No. 7. Chicago: University of Chicago Press, 1952.

———. *Aspects of Scientific Explanation.* New York: Free Press, 1965.

Hempel, Carl G., and Oppenheim, P. *Der Typusbegriff in Lichte der Neuen Logik.* Leiden: Sijthoff, 1936.

Humphreys, Willard C. *Anomalies and Scientific Theories.* San Francisco: Freeman, Cooper & Co., 1968.

Kaila, E. Beitrage zu einer Synthetischen Philosophie, *Annales Universitatis Aboensis,* 4, pp. 9-208, 1928.

Kraft, Victor. *Die Grundformen der Wissenschaftlichen Methoden.* S.-B. d. Österreich Akademie d. Wissenschaften, Vol. 203/3. Vienna: Hölder, 1925.

Kyburg, Jr., Henry E. *Philosophy of Science.* New York: Macmillan, 1968.

Lenzen, Victor F. Procedures of empirical science, *International Encyclopedia of Unified Science,* Vol. 1, No. 5. Chicago: University of Chicago Press, 1938.

Mach, Ernst. *Erkenntnis und Irrtum: Skizzen zur Psychologie der Forschung.* Leipzig: J. A. Barth, 1906.

Mandelbaum, Maurice. *Philosophy, Science, and Sense Perception.* Baltimore: Johns Hopkins Press, 1964.

Margenau, Henry. *Open Vistas: Philosophical Perspectives of Modern Science.* New Haven: Yale University Press, 1961.

Mehlberg, Henryk. *The Reach of Science.* Toronto: University of Toronto Press, 1968.

Nagel, Ernest. "On the Logic of Measurement." Ph. D. Dissertation, Columbia University, 1930.

———. *Freedom and Reason.* Glencoe, Ill.: Free Press, 1951.

———. *Logic Without Metaphysics.* Glencoe, Ill.: Free Press, 1953.

———. *Sovereign Reason and Other Studies in the Philosophy of Science.* Glencoe, Ill.: Free Press, 1954.

Nash, Leonard K. *The Nature of the Natural Sciences.* Boston: Little, Brown & Co., 1963.

Nielsen, H. A. *Methods of Natural Science: An Introduction.* Englewood Cliffs, N.J.: Prentice-Hall, 1967.

Northrop, F. S. C. *The Logic of the Sciences and the Humanities.* New York: Macmillan, 1947.

O'Neil, W. M. *Fact and Theory, An Aspect of the Philosophy of Science.* Sydney: University of Sydney Press, 1969.

Palter, Robert M. *Whitehead's Philosophy of Science.* Chicago: University of Chicago Press, 1960.

Pap, Arthur. *An Introduction to the Philosophy of Science.* New York: Free Press, 1962.

Pearson, Karl. *Grammar of Science.* New York: Meridian Books. 1957.

Peirce, C. S. *Collected Papers.* Cambridge, Mass.: Harvard University Press, 1931-58.

———. *Essays in the Philosophy of Science.* New York: Liberal Arts Press, 1957.

Poincaré, Henri. *Science and Hypothesis.* Translated by G. B. Halsted. New York: Science Press, 1905.

———. *The Foundations of Science.* New York: Science Press, 1929.

———. *Science and Method.* New York: Dover, 1958.

———. *The Value of Science.* New York: Dover, 1958.

Polanyi, Michael. *Personal Knowledge: Towards a Post-critical Philosophy.* Chicago: University of Chicago Press, 1958.

Popper, Karl R. *Logik der Forschung.* Vienna: Springer, 1935.

———. *The Logic of Scientific Discovery.* New York: Basic Books, 1959.

———. *Conjectures and Refutations.* New York: Basic Books, 1962.

Reichenbach, Hans. *La Philosophie scientifique, vues nouvelles sur ses buts et ses méthodes.* Actualités scientifiques et industrielles, Vol. 49. Paris: Hermann, 1932.

———. *The Rise of Scientific Philosophy.* Berkeley: University of California Press, 1956.

———. *Modern Philosophy of Science.* New York: Humanities, 1959.

Ruytinx, Jacques. *La Problématique philosophique de l'unité de la science.* Paris: Societé d'Edition, Les Belles Lettres, 1962.

Scheffler, Israel. *The Anatomy of Inquiry.* New York: Knopf, 1963.

———. *Science and Subjectivity.* New York: Bobbs-Merrill, 1967.

Schlick, Moritz. *Philosophy of Nature.* New York: Philosophical Library, 1949.

Schrödinger, Erwin. *Science and Humanism.* Cambridge: At the University Press, 1951.

———. *My View of the World.* New York: Cambridge University Press, 1964.

Sellars, Wilfrid. *Science, Perception and Reality.* New York: Humanities, 1963.

Smart, J. J. C. *Philosophy and Scientific Realism.* New York: Humanities, 1963.

———. *Between Science and Philosophy.* New York: Random House, 1968.

Topitsch, E. *Vom Ursprung und Ende der Metaphysik.* Vienna: Springer-Verlag, 1958.

Wartofsky, Marx W. *Conceptual Foundations of Scientific Thought.* New York: Macmillan, 1968.

Weyl, Hermann. *Philosophy of Mathematics and Natural Science.* New York: Atheneum, 1963.

Whitehead, A. N. *An Enquiry Concerning the Principles of Natural Knowledge.* Cambridge: At the University Press, 1925.

———. *The Concept of Nature.* Cambridge: At the University Press, 1926.

———. *Essays in Science and Philosophy.* New York: Philosophical Library, 1947.

———. *Science and the Modern World.* New York: Macmillan, 1950.

———. *The Principles of Natural Knowledge.* Cambridge: At the University Press, 1955.

Wiener, Norbert. *The Human Use of Human Beings.* Boston: Houghton-Mifflin, 1950.

Wilson, Edgar Bright. *Introduction to Scientific Research.* New York: McGraw-Hill, 1952.

B. Collections of Essays and Anthologies

Actes du Congrés International de Philosophie Scientifique. (Sorbonne, 1935.) Paris: Hermann, 1936.

Bar-Hillel, Yehoshua. (ed.) *Logic, Methodology and Philosophy of Science.** Amsterdam: North-Holland, 1965.

Baron, Salo W.; Nagel, Ernest; and Pinson, Koppel S. (eds.) *Freedom and Reason.* New York: Free Press, 1951.

Baumrin, Bernard. (ed.) *Philosophy of Science.* New York: Wiley. Delaware Seminar Vol. I, 1961-62; Vol. II, 1962-63.

Bunge, Mario. (ed.) *The Critical Approach to Science and Philosophy: Essays in Honor of Karl R. Popper.* New York: Free Press, 1964.

Cohen, Robert S., and Wartofsky, Marx W. (eds.) *Boston Studies in the Philosophy of Science.* New York: Humanities. Vol. I, 1962; Vol. II, 1965; Vol. III, 1968.

Colodny, Robert G. (ed.) *Frontiers of Science and Philosophy.* Pittsburgh: University of Pittsburgh Press, 1959.

———., (ed.) *Beyond the Edge of Certainty: Essays in Contemporary Science and Philosophy.* Englewood Cliffs, N.J.: Prentice-Hall, 1965.

———. (ed.) *Mind and Cosmos.* Pittsburgh: University of Pittsburgh Press, 1966.

Danto, Arthur, and Morgenbesser, Sidney. (eds.) *Philosophy of Science.* New York: Meridian Books, 1960.

Feigl, Herbert, and Brodbeck, May. (eds.) *Readings in the Philosophy of Science.* New York: Appelton-Century-Crofts, 1953.

Feigl, Herbert, and Maxwell, Grover. (eds.) *Current Issues in the Philosophy of Science.* New York: Holt, Rinehart & Winston, 1961.

Feigl, Herbert, and Scriven, Michael. (eds.) *Minnesota Studies in the Philosophy of Science.* Vol. I, The Foundations of science and the concepts of psychology and psychoanalysis. Minneapolis: University of Minnesota Press, 1956.

Feigl, Herbert; Scriven, Michael; and Maxwell, Grover. (eds.) *Minnesota Studies in the Philosophy of Science.* Vol. II, Concepts, theories and the mind-body problem. Minneapolis: University of Minnesota Press, 1958.

Feigl, Herbert, and Maxwell, Grover. (eds.) *Minnesota Studies in the Philosophy of Science.* Vol. III, Scientific explanation, space and time. Minneapolis: University of Minnesota Press, 1962.

Feyerabend, Paul K., and Maxwell, Grover. (eds.) *Mind, Matter and Method.* Minneapolis: University of Minnesota Press, 1966.

Frank, Philipp. (ed.) *The Validation of Scientific Theories.* Boston: Beacon, 1957.

Gregg, John R., and Harris, F. T. C. (eds.) *Form and Strategy in Science.* Dordrecht, Holland: D. Reidel, 1964.

Henkin, L.; Suppes, P.; and Tarski, A. *The Axiomatic Method, with Special Reference to Geometry and Physics.* Proceedings of an International Symposium at California, 1957-58. New York: Humanities, 1959.

Kockelmans, Joseph J. (ed.) *Philosophy of Science.* New York: Free Press, 1968.

Lakatos, Imre, and Musgrave, Alan. (eds.) *Problems in the Philosophy of Science.* Amsterdam: North-Holland, 1968.

Madden, Edward H. (ed.) *The Structure of Scientific Thought.* Boston: Houghton Mifflin, 1960.

Morgenbesser, Sidney. (ed.) *Philosophy of Science Today.* New York and London: Basic Books, 1967.

Nagel, E.; Suppes, P.; and Tarski, A. (eds.) *Logic, Methodology and Philosophy of Science.** Proceedings of the 1960 International Congress. Stanford: Stanford University Press, 1962.

Schilpp, P. A. (ed.) *Albert Einstein: Philosopher-Scientist.* Evanston, Ill.: Library of Living Philosophers, 1949.

———. (ed.) *The Philosophy of Bertrand Russell.* New York: Tudor, 1944.

———. (ed.) *The Philosophy of Rudolf Carnap.* LaSalle, Ill.: Open Court, 1963.

Wiener, Philip P. (ed.) *Readings in Philosophy of Science.* New York: Scribners, 1953.

3. History of Science

Agassi, Joseph. *Towards an Historiography of Science.* The Hague: Mouton, 1963.

Bernard, Claude. *An Introduction to the Study of Experimental Medicine.* Translated by Henry C. Greene. New York: Collier Books, 1961.

Berry, Arthur. *A Short History of Astronomy.* New York: Dover, 1898.

Butterfield, Herbert. *The Origins of Modern Science,* 1300-1800. New York: Free Press, 1965.

Carruccio, Ettore. *Mathematics and Logic in History and Contemporary Thought.* Chicago: Aldine, 1964.

Crombie, Alistair. (ed.) *Scientific Change.* New York: Basic Books, 1963.

Dampier-Whetham, William C. D. *A History of Science.* New York: Cambridge University Press, 1949.

Dijksterhuis, E. J. *Mechanization of the World Picture.* Translated by C. Dikshoorn. London: Oxford University Press, 1961.

Dingle, H. *The Sources of Eddington's Philosophy.* New York: Cambridge University Press, 1954.

Frank, Philipp. *Einstein, His Life and Times.* New York: Knopf, 1947.

Gillispie, Charles C. *The Edge of Objectivity.* Princeton: Princeton University Press, 1960.

Greene, John C. *The Death of Adam.* Ames, Iowa: Iowa State University Press, 1959.

Hall, A. R. *The Scientific Revolution, 1500-1800.* Boston: Beacon Press, 1966.

Ihdo, Aaron J. *Development of Modern Chemistry.* Evanston, Ill.: Harper & Row, 1964.

Jammer, Max. *Concepts of Space: The History of Theories of Space in Physics.* Cambridge: Harvard University Press, 1954.

———. *Concepts of Force.* Cambridge: Harvard University Press, 1957.

———. *Concepts of Mass in Classical and Modern Physics.* Cambridge: Harvard University Press, 1961.

———. *Conceptual Development of Quantum Mechanics.* New York: McGraw-Hill, 1966.

Jordan, Z. *On the Development of Mathematical Logic and of Logical Positivism in Poland.* London: Oxford University Press, 1946.

Joergensen, Joergen. The development of logical empiricism, *International Encyclopedia of Unified Science,* Vol. II, No. 9. Chicago: University of Chicago Press, 1951.

Koyré, Alexandre. *From the Closed World to the Infinite Universe.* Evanston, Ill.: Harper & Row, 1958.

Kraft, Victor. *Der Wiener Kreis: Der Ursprung des Neupositivismus.* Vienna: Springer, 1950.

———. *The Vienna Circle.* Translated by A. Pap. New York: Philosophical Library, 1953.

Bibliography

Kuhn, Thomas S. *The Copernican Revolution.* Cambridge: Harvard University Press, 1957.

———. The structure of scientific revolutions, *International Encyclopedia of Unified Science,* Vol. II, No. 2. Chicago: University of Chicago Press, 1962.

Lowinger, A. *The Methodology of Pierre Duhem.* New York: Columbia University Press, 1941.

McMullin, Ernan. (ed.) *Galileo.* New York: Basic Books, 1968.

Mason, Stephen F. *A History of the Sciences.* New York: Collier Books, 1962.

Merz, John T. *A History of European Thought in the Nineteenth Century.* 4 vols. New York: Dover, 1904.

Mises, Richard von. *Ernst Mach und die Empiristische Wissenschaftsauffassung,* in the series Einheitswissenschaft, No. 7. The Hague: van Stockum, 1938.

Morris, Charles. *Logical Positivism, Pragmatism, and Scientific Empiricism.* Paris: Hermann, 1937.

Neurath, Otto. *Le Développement du Cercle de Vienne et l'avenir de l'empirisme logique.* Paris: Hermann, 1935.

Neurath, Otto; Carnap, Rudolf; and Hahn, Hans. *Wissenschaftliche Weltauffassung: Der Wiener Kreis.* Vienna: Wolf, 1929.

Partington, James R. *A Short History of Chemistry.* Evanston, Ill.: Harper & Row, 1960.

Read, John. *Prelude to Chemistry.* New York: Macmillan, 1937.

Russell, Bertrand. *A Critical Exposition of the Philosophy of Leibniz.* London: Allen & Unwin, 1949.

Santillana, Giorgio de, and Zilsel, Edgar. The development of rationalism and empiricism, *International Encyclopedia of Unified Science,* Vol. 2, No. 8. Chicago: University of Chicago Press, 1941.

Sarton, George A. *Introduction to the History of Science.* 3 vols. Baltimore: Williams & Wilkins, 1927, 1931, 1947.

———. *A Guide to the History of Science.* New York: Ronald

Press Co., 1952.

Singer, Charles *A History of Biology.* New York: Abelard-Schuman, 1959.

Whittaker, Sir Edmund. *History of the Theories of Aether and Electricity.* 2 vols. London: Nelson & Son, 1951.

Wightman, William P. D. *The Growth of Scientific Ideas.* New Haven: Yale University Press, 1951.

Wolf, A. *A History of Science, Technology and Philosophy in the Sixteenth and Seventeenth Centuries.* 2 vols. New York: Harper, 1959.

4. Logic and Foundations of Mathematics

Ajdukiewicz, K. *Beitrage zur Methodologie der Deduktiven Wissenschaften.* Lvov: Verlag der Polnischen Philosophischen Gesellschaft in Lemberg, 1921.

Bachmann, Friedrich. *Untersuchungen zur Grundlegung der Arithmetik.* Leipzig: Meiner, 1934.

Bell, Eric T. *Development of Mathematics.** New York: McGraw-Hill, 1945.

Benacerraf, Paul, and Putnam, Hilary. *Philosophy of Mathematics.* Englewood Cliffs, N.J.: Prentice-Hall, 1964.

Bernays, Paul, and Fraenkel, A. A. *Axiomatic Set Theory.** Amsterdam: North-Holland, 1958.

Beth, Evert W. *The Foundations of Mathematics.* 2d rev. ed. Amsterdam: North-Holland. 1959.

———. *Mathematical Thought.* Dordrecht, Netherlands: D. Reidel, 1965.

Bonola, Roberto. *Non-Euclidean Geometry: A Critical and Historical Study of its Developments.** New York: Dover, 1954.

Cantor, Georg. *Contributions to the Founding of the Theory of Transfinite Numbers.** New York: Dover, 1915.

Carnap, Rudolf. *Philosophy and Logical Syntex.* London: Kegan Paul, Trench, Trubner, & Co., 1935.

———. *The Logical Syntax of Language.* New York: Humanities, 1937.

———. Foundations of logic and mathematics, *International Encyclopedia of Unified Science,* Vol. 1, No. 3. Chicago: University of Chicago Press, 1939.

———. *Einführung in die Symbolische Logik, mit besonderer Berücksichtigung ihrer Anwendungen.* Vienna: Springer, 1954.

———. *Meaning and Necessity.* 2d ed. Chicago: University of Chicago Press, 1956.

———. *Introduction to Symbolic Logic.* New York: Dover, 1958.

———. *Introduction to Semantics and the Formalization of Logic.* Cambridge, Mass.: Harvard University Press, 1959.

Church, Alonzo. *Introduction to Mathematical Logic.** Vol. I. Princeton, N.J.: Princeton University Press, 1956.

Chwistek, Leon. *The Limits of Science.* *New York: Humanities, 1949.

Cohen, M. R., and Nagel, E. *An Introduction to Logic and Scientific Method.* New York: Harcourt, Brace & World, 1934.

Cohen, Paul J. *Set Theory and the Continuum Hypothesis.** New York: W. A. Benjamin, 1966.

Curry, Haskell B. *Foundations of Mathematical Logic.* New York: McGraw-Hill, 1963.

Dantzig, T. *Number: The Language of Science.** 4th ed. New York: Macmillan, 1954.

Davis, Martin. *Computability and Unsolvability.** New York McGraw-Hill, 1958.

———. (ed.) *The Undecidable: Basic Papers on Undecidable Propositions, Unsolvable Problems, and Compatible Functions.** Hewlett, N. Y.: Raven Press, 1964.

Dedekind, Richard. *Essays on the Theory of Numbers.* *Translated by Wooster W. Beman. New York: Dover, 1901.

Eaton, Ralph M. *General Logic.* New York: Scribners, 1931.

Fitch, Frederick B. *Symbolic Logic.* New York: Ronald Press Co., 1952.

Fraenkel, Abraham A. *Abstract Set Theory.** 3d ed. New York: Humanities, 1965.

Fraenkel, Abraham A., and Bar-Hillel, Yehoshua; *Foundations of Set Theory.* Amsterdam: North-Holland, 1958.

Frege, Gottlob. *The Foundations of Arithmetic.* Translated by J. L. Austin. New York: Philosophical Library, 1950.

———. *The Basic Laws of Arithmetic.* Edited by M. Furth. Berkeley: University of California, 1965.

Gödel, Kurt. *The Consistency of the Continuum Hypothesis.** Princeton: Princeton University Press, 1940.

Gregg, John R., and Harris, F. T. C. (eds.) *Form and Strategy in Science.* New York: Humanities, 1964.

Hahn, Hans. *Logik, Mathematik und Naturerkennen,* in the series Einheitswissenschaft. Vienna: Gerold, 1933.

Hatcher, William S. *Foundations of Mathematics.* Philadelphia: W. B. Saunders Co., 1968.

Heyting, A. *Intuitionism: An Introduction.* New York: Humanities, 1966.

Hilbert, David. *The Foundations of Geometry.* Translated by E. J. Townsend. La Salle, Ill.: Open Court, 1959.

Hilbert, David, and Bernays, Paul. *Die Grundlagen der Mathematik.** Berlin: Springer. Vol. I, 1934, Vol. II, 1939.

Joergensen, Joergen. *A Treatise of Formal Logic.* London: Humphrey Milford, Oxford University Press, 1931.

Kaufmann, Felix. *Das Unendliche in der Mathematik und seine Ausschaltung.* Vienna: Deuticke, 1930.

Kleene, S. C. *Introduction to Metamathematics.* Princeton: Van Nostrand, 1952.

———. *Mathematical Logic.* New York: Wiley, 1967.

Kleene, S. C., and Vesley, R. E. *The Foundations of Intuitionistic Mathematics.**New York: Humanities, 1965.

Kneale, William C., and Martha, M. *The Development of Logic.* Oxford: Clarendon Press, 1962.

Körner, Stephan. *Philosophy of Mathematics.* New York: Hillary, 1960.

Ladrière, Jean. *Les Limitations internes des formalismes.* Paris: Gauthier-Villars, 1957.

Landau, Edmund. *Foundations of Analysis.** New York: Chelsea Publishing Co., 1957.

Lewis, C. I., and Langford, C. H. *Symbolic Logic.* New York: Dover, 1959.

Lorenzen, Paul. *Einführung in die Operative Logik und Mathematik.** Berlin: Springer, 1955.

———. *Formal Logic.* Translated by Frederick J. Crosson. New York: Humanities, 1965.

Martin, Richard M. *Truth and Denotation.* Chicago: University of Chicago Press, 1958.

———. *Intention and Decision.* Englewood Cliffs, N.J.: Prentice-Hall, 1963.

Menger, Karl. *Calculus: A Modern Approach.* Boston: Ginn, 1955.

Mostowski, Andrzej. *Thirty Years of Foundational Studies.* New York: Barnes and Noble, 1966.

Nagel, Ernest, and Newman, J. R. *Gödel's Proof.** New York: New York University Press, 1958.

Poincaré, Henri. *Mathematics and Science: Last Essays.* New York: Dover, 1963.

Polya, G. *Mathematics and Plausible Reasoning.* 2 vols. Princeton: Princeton University Press, 1954.

Quine, Willard Van O. *Mathematical Logic.* New York: W. W. Norton & Co., 1940; Cambridge, Mass.: Harvard University Press, 1947.

———. *From a Logical Point of View.* Cambridge, Mass.: Harvard University Press, 1953.

———. *Set Theory and its Logic.* Cambridge, Mass.: Harvard University Press, 1963.

———. *Selected Logic Papers.* New York: Random House, 1966.

———. *The Ways of Paradox and Other Essays.* New York: Random House, 1966.

Ramsey, F. P. *The Foundations of Mathematics and Other Logical Essays.* New York: Harcourt, Brace & World, 1931.

Reichenbach, Hans. *Nomological Statements and Admissible Operations.* New York: Humanities, 1954.

———. *Elements of Symbolic Logic.* New York: Free Press, 1966.

Rougier, Louis. *La Structure des théories déductives.* Paris: Alcan, 1921.

Russell, Bertrand. *Introduction to Mathematical Philosophy.* New York: Macmillan, 1930.

———. *The Principles of Mathematics.* 2d ed. New York: Norton, 1948.

———. *The Foundations of Geometry.* New York: Dover, 1956.

Suppes, Patrick. *Axiomatic Set Theory.* Princeton: Van Nostrand, 1960.

Suppes, Patrick, and Hill, Shirley. *First Course in Mathematical Logic.* New York: Blaisdell Publishing Co., 1964.

Tarski, Alfred. *Einführung in die Mathematische Logik und die Methodologie der Mathematik.** Vienna: Springer, 1937.

———. *Introduction to Logic and the Methodology of the Deductive Sciences.* Oxford: Oxford University Press, 1941.

———. *Logic, Semantics, Metamathematics.* Translated by J. H. Woodger. Oxford: Clarendon Press, 1956.

Van Heijencort, Jean. (ed.) *From Frege to Gödel: A Source Book in Mathematical Logic, 1879-1931.* Cambridge, Mass.: Harvard University Press, 1967.

Waismann, F. *Introduction to Mathematical Thinking.* Translated by J. Benac. New York: Ungar, 1951.

Wang, Hao. *A Survey of Mathematical Logic.** New York: (North Holland) Humanities, 1964.

Whitehead, A. N., and Russell, Bertrand. *Principia Mathematica.* 2d ed. 3 vols. Cambridge: At the University Press, 1925–27.

Wilder, Raymond L. *Introduction to the Foundations of Mathematics.* New York: Wiley, 1965.

Wittgenstein, Ludwig. *Remarks on the Foundations of Mathematics.* Oxford: Blackwell, 1956.

5. Philosophy of Language and Symbolism

Alston, William P. *Philosophy of Language.* Englewood Cliffs, N.J.: Prentice-Hall, 1964.

Austin, John. *How to Do Things with Words.* London: Oxford University Press, 1962.

Bar-Hillel, Yehoshua. *Language and Information.** Reading, Mass.: Addison-Wesley, 1964.

Berlyne, D. E. *Structure and Direction in Thinking. *New York: Wiley, 1965.*

Bloomfield, Leonard, Linguistic aspects of science, *International Encyclopedia of Unified Science,* Vol. I, No. 4. Chicago: University of Chicago Press, 1955.

———. *Language.** New York: Holt, Rinehart, & Winston, 1961.

Brown, Roger. *Words and Things.** Glencoe, Ill.: Free Press, 1958.

Cassirer, Ernst. *The Philosophy of Symbolic Forms.* Translated by Ralph Manheim. 3 vols. New Haven: Yale University Press, 1953–57.

Chomsky, Noam. *Syntactic Structures.* The Hague: Mouton, 1957.

———. *Aspects of the Theory of Syntax.* Cambridge, Mass.: M. I. T. Press, 1965.

———. *Cartesian Linguistics.* New York: Harper & Row, 1966.

Fodor, Jerry A., and Katz, Jerrold J. (eds.) *The Structure of Language.* Englewood Cliffs, N.J.: Prentice-Hall, 1964.

Gätschenberger, Richard. *Symbola.* Karlsruhe: G. Braun, 1920.

———. *Zeichen, die Fundamente des Wissens.* Stuttgart: Frommann, 1932.

Greenberg, Joseph H. *Essays in Linguistics.* Chicago: University of Chicago Press, 1957.

———. (ed.) *Universals of Language.** Cambridge, Mass.: M. I. T. Press, 1963.

Henle, Paul. (ed.) *Language, Thought and Culture.* Ann Arbor: University of Michigan Press, 1958.

Hockett, Charles F. *Course in Modern Linguistics.** New York: Macmillan, 1958.

Holloway, John. *Language and Intelligence.* New York: Macmillan, 1951.

Jesperson, Otto. *Philosophy of Grammar.* * New York: W. W. Norton & Co., 1965.

Johnson, Alexander B. *Treatise on Language.* Edited and with a critical essay by D. Rynin. Berkeley: University of California Press, 1947.

Katz, Jerrold J. *The Philosophy of Language.* New York: Harper & Row, 1966.

Katz, Jerrold J., and Postal, Paul M. *An Integrated Theory of Linguistic Descriptions.** Cambridge, Mass.: M. I. T. Press, 1964.

Lenneberg, Eric H. *Biological Foundations of Language.* * New York: Wiley, 1967.

———. (ed.) *New Directions in the Study of Language.** Cambridge, Mass.: M. I. T. Press, 1966.

Linsky, Leonard. (ed.) *Semantics and the Philosophy of Language.* Urbana: University of Illinois Press, 1957.

Logic and Language. (Studies dedicated to Professor Rudolf Carnap on the occasion of his seventieth birthday.) Dordrecht,

Netherlands: D. Reidel, 1962.

Morris, Charles. Foundations of the theory of signs, *International Encyclopedia of Unified Science,* Vol. 1, No. 2. Chicago: University of Chicago Press, 1938.

———. *Signs, Language and Behavior.* New York: Prentice-Hall, 1946. George Braziller, 1955.

———. *Signification and Significance.* Cambridge, Mass.: M. I. T. Press, 1964.

Mowrer, O. Hobart. *Learning Theory and the Symbolic Process.** New York: Wiley, 1960.

Naess, Arne. *Interpretation and Preciseness.* Totowa, N. J.: Bedminster Press, 1967.

Ogden, C. K., and Richards, I. A. *The Meaning of "Meaning".* 5th ed. New York: Harcourt, Brace & World, 1938.

Osgood, Charles E.; Diebold, A. Richard; and Miron, Murray S. *Psycholinguistics.* 2d ed. Bloomington: Indiana University Press, 1965.

Quine, Willard Van O. *Word and Object.* New York: Wiley, 1960.

Rorty, Richard. (ed.) *The Linguistic Turn.* Chicago: University of Chicago Press, 1967.

Ruesch, Jurgen. *Disturbed Communication.** New York: Norton, 1957.

———. *Therapeutic Communication.** New York: Norton, 1961.

Schaechter, J. *Prolegomena zu einer Kritischen Grammatik.* Vienna: Springer, 1938.

Sebeok, Thomas A. (ed.) *Style in Language.** Cambridge, Mass.: M. I. T. Press, 1960.

Sebeok, Thomas A.; Hayes, Alfred S.; and Bateson, Mary Catherine. (eds.) *Approaches to Semiotics.** The Hague: Mouton, 1964.

Stevenson, Charles L. *Ethics and Language.* New Haven: Yale University Press, 1944.

———. *Facts and Values.* New Haven: Yale University Press, 1963.

Waismann, F. *The Principles of Linguistic Philosophy.* Edited by R. Harré. New York: St. Martin's Press, 1965.

Ziff, Paul. *Semantic Analysis.* Ithaca, N.Y.: Cornell University Press, 1960.

6. Probability, Induction, and the Foundations of Statistics

Barker, S. F. *Induction and Hypothesis: A Study of the Logic of Confirmation.* Ithaca, N.Y.: Cornell University Press, 1957.

Boll, Marcel. *Elements de logique scientifique.* Paris: Danod, 1942.

Brown, G. Spencer. *Probability and Scientific Inference.** London: Longmans, Green & Co., 1957.

Carnap, Rudolf. *Logical Foundations of Probability.* Chicago: University of Chicago Press, 1950.

———. *The Continuum of Inductive Methods.* Chicago: University of Chicago Press, 1951.

———. *The Nature and Application of Inductive Logic,* consisting of six sections from *Logical Foundations of Probability.* Chicago: University of Chicago Press, 1951.

Carnap, Rudolf, and Stegmueller, W. *Inductive Logik und Wahrscheinlichkeit.* Vienna: Springer-Verlag, 1958.

Churchman, C. West. *Prediction and Optimal Decision.* Englewood Cliffs, N.J.: Prentice-Hall, 1961.

Cramér, Harald. *Mathematical Methods of Statistics.** Princeton: Princeton University Press, 1946.

———. *The Elements of Probability Theory and Some of Its Applications.** New York: Wiley, 1955.

Cox, Richard T. *The Algebra of Probable Inference.** Baltimore: Johns Hopkins Press, 1961.

Day, John P. *Inductive Probability.* New York: Humanities, 1961.

Feller, William. *An Introduction to Probability and Its Applications.** 2d ed. 2 vols. New York: Wiley, 1957, 1966.

Fisher, R. A. *The Design of Experiments.** Edinburgh and

London: Oliver and Boyd, 1951.

———. *Statistical Methods and Scientific Inference.** 2d ed. New York: Stechert-Hafner, 1956.

Foster, Marguerite H., and Martin, Michael L. (eds.) *Probability, Confirmation, and Simplicity.* New York: Odyssey Press, 1966.

Good, I. J. *Probability and Weighing of Evidence.** New York: Stechert-Hafner, 1950.

Hacking, Ian. *Logic of Statistical Inference.* London: Cambridge University Press, 1965.

Hintikka, Jaakko, and Suppes, Patrick. (eds.) *Aspects of Inductive Logic.* Amsterdam: North-Holland, 1966.

Hogben, Lancelot. *Statistical Theory: The Relationship of Probability, Credibility and Error.* New York: W. W. Norton & Co., 1957.

Jeffrey, Richard C. *The Logic of Decision.* New York: McGraw-Hill, 1965.

Jeffreys, Harold. *Theory of Probability.* 2d ed. Oxford: Clarendon Press, 1948.

———. *Scientific Inference.* Cambridge: At the University Press, 1957.

Katz, Jerrold J. *The Problem of Induction and Its Solution.* Chicago: University of Chicago Press, 1962.

Kendall, M. G., and Stuart, A. *The Advanced Theory of Statistics.* * 2 vols. New York: Stechert-Hafner, 1958, 1961.

Keynes, John Maynard. *A Treatise on Probability.* New York: Macmillan, 1929.

Kneale, William. *Probability and Induction.* Oxford: Clarendon Press, 1949, 1952.

Kolmogoroff, A. N. *Foundations of the Theory of Probability.** New York: Chelsea, 1951.

Kyburg, Jr., Henry E. *Probability and the Logic of Rational Belief.* Middletown, Conn.: Wesleyan University Press, 1961.

Kyburg, Jr., Henry E., and Nagel, Ernest. (eds.) *Induction: Some Current Issues.* Middletown, Conn.: Wesleyan University Press, 1963.

Kyburg, Jr., Henry E., and Smokler, Howard E. (eds.) *Studies in Subjective Probability.** New York: Wiley, 1964.

Lakatos, Imre. (ed.) *The Problem of Inductive Logic.* Amsterdam: North-Holland, 1968.

Le Blanc, Hugues. *Statistical and Inductive Probabilities.* Englewood Cliffs, N.J.: Prentice-Hall, 1962.

Lehmann, E. L. *Testing Statistical Hypotheses.** New York: Wiley, 1959.

Levi, Isaac. *Gambling with Truth.* New York: Knopf, 1967.

Loève, Michel. *Probability Theory.** New York: Van Nostrand, 1955.

Lukasiewicz, Julian. *Die Logischen Grundlagen der Wahrscheinlichkeitsrechnung.* Cracow: Krakauer Akademic d. Wissenschaften, 1913.

Mises, Richard von. *Probability, Statistics, and Truth.* New York: Macmillan, 1939. 2d revision. London: Allen & Unwin; New York: Macmillan, 1957.

Nagel, Ernest. Principles of the theory of probability, *International Encyclopedia of Unified Science,* Vol. I, No. 6. Chicago: University of Chicago Press, 1939.

Neyman, Jerzy. *First Course in Probability and Statistics.** New York: Holt, Rinehart & Winston, 1950.

———. *Lectures and Conferences on Mathematical Statistics and Probability.** Washington, D.C.: U. S. Department of Agriculture, 1952.

Nicod, Jean. *Le Problème logique de l'induction.* Paris: Alcan, 1924.

Reichenbach, Hans. *The Theory of Probability.* Los Angeles: University of California Press, 1949.

Rescher, Nicholas. *Hypothetical Reasoning.* New York: North-Holland, 1964.

Savage, Leonard J. *The Foundations of Statistics.** New York: Wiley, 1954.

Skyrms, Brian. *Choice and Chance.* Belmont, Calif.: Dickenson Publishing Co., 1968.

Von Wright, Georg Henrik. *The Logical Problem of Induction.* Helsinki: Acta Philosophica Fennica, fasc. 3, 1941. 2d rev. ed. Oxford: Blackwell, 1957.

———. *A Treatise on Induction and Probability.* London: Routledge and Kegan Paul, 1951.

7. Foundations of Physics

d'Abro, A. *The Evolution of Scientific Thought.* New York: Dover, 1950.

———. *The Rise of the New Physics.* 2 vols. New York: Dover, 1951.

Bergmann, Hugo. *Der Kampf um das Kausalgesetz in der Jungsten Physik.* Brunswick: F. Vieweg, 1929.

Bergmann, Peter G. *Basic Theories of Physics.* 2 vols. Englewood Cliffs, N.J.: Prentice-Hall, 1949.

Birkhoff, Garrett. *Hydrodynamics: A Study in Logic, Fact and Similitude.* Princeton, N.J.: Princeton University Press, 1961.

Bohm, David. *Quantum Theory.** Englewood Cliffs, N.J.: Prentice-Hall, 1951.

———. *Causality and Chance in Modern Physics.* Princeton: Van Nostrand, 1957.

Bohr, Niels. *Atomic Theory and the Description of Nature.* Cambridge: At the University Press, 1932.

———. *Atomic Physics and Human Knowledge.* New York: Wiley, 1958.

———. *Essays 1958-1962 on Atomic Physics and Human Knowledge.* New York: Wiley, 1963.

Boltzmann, Ludwig. *Lectures on Gas Theory.* Translated by Stephen G. Brush. Berkeley: University of California Press, 1964.

Bopp, Fritz. (ed.) *Werner Heisenberg und die Physik unserer Zeit.* Brunswick: F. Vieweg, 1961.

Born, Max. *Experiment and Theory in Physics.* Cambridge: At The University Press, 1944.

Bridgman, P. W. *The Logic of Modern Physics.* New York: Macmillan, 1927.

———. *The Nature of Physical Theory.* Princeton: Princeton University Press, 1936.

———. *The Nature of Thermodynamics.* Cambridge, Mass.: Harvard University Press, 1941.

———. *Dimensional Analysis.* rev. ed. New Haven: Yale University Press, 1943.

———. *Reflections of a Physicist.* New York: Philosophical Library, 1950, 1955.

———. *The Way Things Are.* Cambridge, Mass.: Harvard University Press, 1959.

Broglie, Louis de. *Matter and Light: The New Physics.* New York: W. W. Norton & Co. 1939. Translated by W. H. Johnston. New York: Dover, 1946.

———. *The Revolution in Physics.* New York: Noonday Press, 1953.

———. *Physics and Microphysics.* New York: Grossett & Dunlap, 1955.

———. *The Current Interpretation of Wave Mechanics.* Amsterdam: Elsevier Publishing Co., 1964.

Bunge, Mario. *Foundations of Physics.* New York: Springer, 1967.

———. (ed.) *Quantum Theory and Reality.* New York: Springer-Verlag, 1967.

Campbell, N. R. *Physics: The Elements.* Cambridge: At the University Press, 1921. Reprinted under new title, *Foundations of Science: The Philosophy of Theory and Experiment.* New York: Dover, 1957.

Capek, Milic. *Philosophical Impact of Contemporary Physics.* Princeton: Van Nostrand, 1961.

Carnap, Rudolf. *Physikalische Begriffsbildung.* Karlsruhe:

B. Braun, 1926.

———. *Philosophical Foundations of Physics.* Edited by M. Gardner. New York: Basic Books, 1966.

Cassirer, Ernst. *Determinism and Interdeterminism in Modern Physics.* Translated by O. Theodor Benfey. New Haven: Yale University Press, 1956.

Dirac, P. A. M. *The Principles of Quantum Mechanics.* 4th ed. New York: Oxford University Press, 1958.

Duhem, Pierre. *L'Evolution de la mécanique.* Paris: Joanin, 1903.

———. *The Aim and Structure of Physical Theory.* Princeton: Princeton University Press, 1954.

Eddington, Sir Arthur. *New Pathways in Science.* Cambridge: At the University Press, 1947; Ann Arbor: University of Michigan Press, 1959.

———. *Philosophy of Physical Science.* Ann Arbor: University of Michigan Press, 1958.

Ehrenfest, Paul, and Ehrenfest, Tatiana. *The Conceptual Foundations of the Statistical Approach in Mechanics.* Translated by Michael J. Moravcsik. Ithaca, N. Y.: Cornell University Press, 1959.

Einstein, Albert, and Infeld, Leopold. *The Evolution of Physics.* New York: Simon & Schuster, 1938.

Feigl, Herbert. *Theorie und Erfahrung in der Physik.* Karlsruhe: Braun, 1929.

Fermi, Enrico. *Elementary Particles.** London: Oxford University Press, 1951.

Feynman, Richard P. *Theory of Fundamental Processes.** New York: Benjamin, 1961.

Feynman, Richard P.; Leighton, Robert B.; and Sands, Matthew. *The Feynman Lectures on Physics.** Reading, Mass.: Addison-Wesley, 1963.

Frank, Philipp. *Das Ende der Mechanistischen Physik,* in the series Einheitswissenschaft, No. 5. Vienna: Gerold, 1935.

———. *Interpretations and Misinterpretations of Modern Physics.*

Paris: Hermann, 1938.

———. *Between Physics and Philosophy.* Cambridge, Mass.: Harvard University Press, 1941.

———. Foundations of physics, *International Encyclopedia of Unified Science,* Vol. 1, No. 7 Chicago: University of Chicago Press, 1946.

Gold, T., and Schumacher, D. L. (eds.) *The Nature of Time.* Ithaca, N. Y.: Cornell University Press, 1967.

Guillemin, Victor. *The Story of Quantum Mechanics.* New York: Charles Scribner's Sons, 1968.

Hanson, Norwood Russell. *The Concept of the Positron.* London: Cambridge University Press, 1963.

Harré, R. *Matter and Method.* New York: St. Martin's Press, 1964.

Heisenberg, Werner. *The Physical Principles of the Quantum Theory.* Chicago: University of Chicago Press, 1930.

———. *Physics and Philosophy.* Evanston, Ill.: Harper & Row, 1958.

———. *Introduction to the Unified Field Theory of Elementary Particles.** New York: Wiley, 1966.

Heitler, Walter. *Quantum Theory of Radiation.** London: Oxford University Press, 1944.

Hertz, Heinrich. *The Principles of Mechanics, 1894.* New York: Dover, 1956.

Hesse, Mary B. *Forces and Fields.* London: Nelson & Sons, 1961.

———. *Models and Analogies in Science.* London: Sheed & Ward, 1963.

Holton, Gerald J., and Roller, D. H. D. *Foundations of Modern Physical Science.** Reading, Mass.: Addison-Wesley, 1958.

Körner, Stephen L. (ed.) *Observation and Interpretation.* New York: Academic Press, 1957.

Landé, Alfred. *New Foundations of Quantum Mechanics.* London: Cambridge University Press, 1965.

Leech, John W. *Classical Mechanics.** London: Methuen, 1958.

Lindsay, Robert B., and Margenau, Henry. *Foundations of Physics.* * New York: Dover, 1957.

London, F., and Bauer, E. *La Théorie de l'observation en mécanique quantique.* Paris: Hermann, 1939.

McMullin, Ernan (ed.) *The Concept of Matter in Greek and Medieval Philosophy.* Notre Dame, Ind.: University of Notre Dame Press, 1963.

Mach, Ernst. *Die Prinzipien der Warmelehre, Historisch-Kritisch Entwickelt.* Barth: Leipzig, 1896.

———. *Popular Scientific Lectures.* La Salle, Ill.: Open Court, 1898.

———. *Principles of Physical Optics.* London: Methuen, 1926.

———. *The Science of Mechanics.* Translated by T. J. Mc Cormack. 6th ed. with a preface by K. Menger. La Salle, Ill.: Open Court, 1960.

Margenau, Henry. *The Nature of Physical Reality.* New York: McGraw-Hill, 1950.

Pap, Arthur. *The A Priori in Physical Theory.* New York: King's Crown Press, 1946.

Pauli, Wolfgang. (ed.) *Niels Bohr and the Development of Physics.* New York: Pergamon Press, 1955.

Pauling, Linus. *The Nature of the Chemical Bond.* * 3d ed. Ithaca, N. Y.: Cornell University Press, 1960.

Planck, Max. *The New Science.* (Three complete works: 1. Where is science going? 2. The universe in the light of modern physics. 3. The philosophy of physics.) New York: Meridian Books, 1959.

Reichenbach, Hans. *Atom and Cosmos: The World of Modern Physics.* New York: Macmillan, 1933.

———. *From Copernicus to Einstein.* New York: Philosophical Library, 1942.

———. *Philosophic Foundations of Quantum Mechanics.* Berkeley: University of California Press, 1944, German translation.

Basel: Birkhauser, 1949.

Roman, P. *Advanced Quantum Theory.* Reading, Mass.: Addison-Wesley, 1965.

Russell, Bertrand. *The Analysis of Matter.* New York: Harcourt, Brace & Co., 1927. With new introduction by L. E. Denonn. New York: Dover, 1954.

Schlesinger, George. *Method in the Physical Sciences.* New York: Humanities, 1963.

Sciama, Denis W. *The Unity of the Universe.* Garden City, N. Y.: Doubleday & Co., 1959.

Slater, Noel B. *The Development and Meaning of Eddington's Fundamental Theory.** Cambridge: At the University Press, 1957.

Stebbing, L. S. *Philosophy and the Physicists.* London: Methuen, 1937.

Tolman, Richard C. *The Principles of Statistical Mechanics.** London: Oxford University Press, 1938.

Truesdell, Clifford A. *Six Lectures on Modern Natural Philosophy.* Berlin: Springer, 1966.

Van der Waerden, B. C. *Sources of Quantum Mechanics.** New York: Dover, in prep.

Von Neumann, John. *The Mathematical Foundations of Quantum Mechanics.** Princeton: Princeton University Press, 1955.

Watson, William H. *On Understanding Physics.* New York: Harper & Row, 1959.

———. *Understanding Physics Today.* New York: Cambridge University Press, 1963.

Wax, Nelson. (ed.) *Selected Papers on Noise and Stochastic Processes.* New York: Dover, 1954.

Wheeler, J. A. *Geometrodynamics.** New York: Academic Press, 1962.

Yilmaz, Hüseyin. *Introduction to the Theory of Relativity and the Principles of Modern Physics.* New York: Blaisdell, 1965.

8. Space, Time, and Causality

Bergmann, Peter G. *Introduction to the Theory of Relativity.** Englewood Cliffs, N.J.: Prentice-Hall, 1942.

Bohm, David. *The Special Theory of Relativity.* New York: Benjamin, 1965.

Bondi, Hermann. *Cosmology.** New York: Cambridge University Press, 1960.

———. *Relativity and Common Sense.* New York: Doubleday & Co., 1964.

Bondi, H.; Bonner, W. B.; Lyttleton, R. A.; and Whitrow, G. J. *Rival Theories of Cosmology.* New York: Oxford University Press, 1960.

Born, Max. *Einstein's Theory of Relativity.* London: Methuen, 1924; New York: Dover, 1964.

———. *Natural Philosophy of Cause and Chance.* Oxford: Clarendon Press, 1949.

———. *Continuity, Determinism, and Reality.* Copenhagen: I Kommission hos Munksgaard, 1955.

Bridgman, P. W. *A Sophisticate's Primer on Relativity.* Middletown, Conn.: Wesleyan University Press, 1962.

Broglie, Louis de. *La Physique quantique restera-t-elle indéterministe?* Paris: Gauthier-Villard, 1953.

Bunge, Mario. *Causality.* Cleveland: World Publishing Co., 1963.

Carnap, Rudolf. *Der Raum: Ein Beitrag zur Wissenschaftslehre.* Berlin: Ergänzungshefte d. Kant-Studien, No. 56, 1922.

Cassirer, Ernst. *Substance and Function and Einstein's Theory of Relativity.* La Salle, Ill.: Open Court, 1923.

Eddington, Sir Arthur S. *The Mathematical Theory of Relativity.* * Cambridge: At the University Press, 1924.

Einstein, Albert. *Geometrie und Erfahrung.* Berlin: Springer, 1921.

———. *The Meaning of Relativity.* Princeton: Princeton University

Press, 1921.

———. *Sidelights of Relativity.* London: Methuen, 1922; New York: Dutton & Co., 1923.

Finlay-Freundlich, E. Cosmology, *International Encyclopedia of Unified Science,* Vol. I, No. 8. Chicago: University of Chicago Press, 1955.

Grünbaum, Adolf. *Philosophical Problems of Space and Time.* New York: Knopf, 1963.

———. *Modern Science and Zeno's Paradoxes.* Middletown, Conn.: Wesleyan University Press, 1967.

Kelsen, Hans. *Vergeltung und Kausalität.* The Hague: van Stockum, 1941.

———. *Society and Nature.* Chicago: University of Chicago Press, 1943; London: Kegan Paul, 1946.

McVittie, G. C. *General Relativity and Cosmology.* * Urbana: University of Illinois Press, 1965.

Mach, Ernst. *Space and Geometry.* La Salle, Ill.: Open Court, 1943.

Mehlberg, Henryk. Essais sur la théorie causale du temps, *Studia Philosophica I,* pp. 119-231, 1935. *Studia Philosophica II,* pp. 111-232, 1937.

Munitz, Milton. *Space, Time and Creation.* Glencoe, Ill.: Free Press, 1957.

———. (ed.) *Theories of the Universe.* Glencoe, Ill.: Free Press, 1957.

Nicod, Jean. *Foundations of Geometry and Induction.* New York: Harcourt, Brace & World, 1930; London: Kegan Paul, 1930.

Prior, Arthur N. *Time and Modality.* Oxford: Clarendon Press, 1957.

———. *Past, Present, and Future.* London: Oxford University Press, 1967.

Reichenbach, Hans. *Axiomatik der Relativistischen Raum-*

Zeit-Lehre. Brunswick: F. Vieweg, 1924.

———. *The Direction of Time.* Berkeley: University of California Press, 1956.

———. *The Philosophy of Space and Time.* New York: Dover, 1958.

———. *The Theory of Relativity and A Priori Knowledge.* Berkeley: University of California Press, 1965.

Robb. A. A. *Geometry of Time and Space.* Cambridge: At the University Press, 1936.

Ruyer, Raymond. *Esquisse d'une philosophie de la structure.* Paris: Alcan, 1930.

Schlegel, Richard. *Time and the Physical World.* East Lansing: Michigan State University Press, 1961.

———. *Completeness in Science.* New York: Appleton-Century-Crofts, 1967.

Schlick, Moritz. *Space and Time in Contemporary Physics.* 3d ed. Oxford: Clarendon Press, 1920.

Smart, J. J. C. (ed.) *Problems of Space and Time.* New York: Macmillan, 1964.

Synge, J. L. *Relativity: The Special Theory.** New York: Interscience, 1956.

———. *Relativity: The General Theory.** New York: Interscience. Amsterdam: North Holland, 1960.

Whitehead, A. N. *The Principle of Relativity.* Cambridge: At the University Press, 1922.

Whiteman, Michael. *Philosophy of Space and Time.* New York: Humanities, 1967.

Whitrow, G. J. *The Structure and Evolution of the Universe.** London: Hutchinson's University Press, 1959.

———. *The Natural Philosophy of Time.* London: Nelson & Sons, 1961.

9. Foundations of Biology

Allee, W. C.; Emerson, A. E.; Park T.; and Schmidt, K. P. *A Fundamental Treatise on Ecology.** Philadelphia: Saunders, 1949.

Beckner, Morton. *The Biological Way of Thought.* New York: Columbia University Press, 1959.

Bertalanffy, Ludwig von. *Problems of Life.* New York: Wiley, 1952.

———. *Modern Theories of Development.* Translated by J. H. Woodger. New York: Harper & Bros., 1962.

Blum, H. F. *Time's Arrow and Evolution.** Princeton: Princeton University Press, 1951.

Bonner, John Tyler. *The Ideas of Biology.** New York: Harper & Bros., 1962.

Dobzhansky, Theodosius. *Mankind Evolving.* * New Haven: Yale University Press, 1962.

Elsasser, Walter M. *The Physical Foundation of Biology.* New York: Pergamon Press, 1958.

Falconer, D. S. *Introduction to Quantitative Genetics.** New York: Ronald Press Co., 1964.

Fisher, Ronald A. *The Genetical Theory of Natural Selection.** London: Oxford University Press, 1930.

Goudge, Thomas A. *The Ascent of Life.* Toronto: University of Toronto Press, 1961.

Ingle, Dwight J. *Principles of Research in Biology and Medicine.** Philadelphia: J. B. Lippincott Co., 1958.

Mainx, Felix. Foundations of biology, *International Encyclopedia of Unified Science,* Vol. I, no. 9. Chicago: University of Chicago Press, 1955.

Mayr, Ernst. *Animal Species and Evolution.** Cambridge, Mass.: Harvard University Press, Belknap Press, 1963.

Medawar, P. B. *The Art of the Soluble.* London: Methuen, 1967.

Merrell, David J. *Evolution and Genetics.** New York: Holt, Rinehart, & Winston, 1962.

Bibliography

Morgan, Thomas H. *The Theory of the Gene.* * New Haven: Yale University Press, 1926.

Oparin, Alexander I. *Life: Its Nature, Origin, and Development.* Edinburgh and London: Oliver and Boyd, 1961.

Reiner, John. *The Organism as an Adaptive System.* * Englewood Cliffs, N.J.: Prentice-Hall, 1968.

Rensch, Bernhard. *Evolution above the Species Level.* * New York: Wiley, 1959.

Schrödinger, Erwin. *What Is Life?* Cambridge: At the University Press, 1944.

Simpson, George G. *The Meaning of Evolution.* New Haven: Yale University Press, 1949.

———. *This View of Life.* New York: Harcourt, Brace & World, 1966.

Sommerhoff, G. *Analytical Biology.* London: Oxford University Press, 1950.

Spemann, Hans. *Embryonic Development and Induction.* * New Haven: Yale University Press, 1938.

Srb, Adrian M.; Owen, Ray D.; and Edgar, Robert S. *General Genetics.* * 2d ed. San Francisco: W. H. Freeman & Co., 1965.

Stebbins, G. Ledyard. *Processes of Organic Evolution.* * Englewood Cliffs, N. J.: Prentice-Hall, 1966.

Thorpe, William H. *Science, Man and Morals.* Ithaca, N. Y.: Cornell University Press, 1966.

Waddington, C. H. *The Strategy of the Genes.* * New York: Macmillan, 1957.

———. *The Nature of Life.* * New York: Atheneum Publishers, 1962.

Watson, James D. *Molecular Biology of the Gene.* * New York: Benjamin, 1965.

Whyte, Lancelot Law. *Internal Factors in Evolution.* * New York: George Braziller, 1965.

Woodger, J. H. *Biological Principles.* London: Kegan Paul,

Trench & Co., 1929.

———. *Axiomatic Method in Biology.* London: Cambridge University Press, 1937.

———. The technique of theory construction, *International encyclopedia of Unified Science,* Vol. 2, No. 5. Chicago: University of Chicago Press, 1939.

———. *Biology and Language: An Introduction to the Methodology of the Biological Sciences, Including Medicine.* Cambridge: At the University Press, 1952.

———. *Physics, Psychology, and Medicine.* Cambridge: At the University Press, 1956.

10. Foundations of Psychology

Adrian, E. D. *The Physical Background of Perception.* * Oxford: Clarendon Press, 1947.

Adrian, E. D.; Bremer, F.; Jasper, H. H. (consulting eds.); and Delafresnaye, J. F. (ed. for the Council) *Brain Mechanisms and Consciousness: A Symposium.* Council for International Organization of Medicial Sciences. Springfield, Ill.: Charles C. Thomas, 1954.

Armstrong, D. M. *A Materialist Theory of Mind.* New York: Humanities Press, 1968.

Ashby, W. R. *Design for a Brain.* New York: Wiley, 1952.

Boring, E. G. *The Physical Dimensions of Consciousness.* * New York and London: Century Co., 1933.

———. *Sensation and Perception in the History of Experimental Psychology.* New York and London: Century Co., 1942.

———. *A History of Experimental Psychology.* 2d ed. New York: Appleton-Century-Crofts, 1957.

Brain, W. R. *The Contribution of Medicine to Our Idea of Mind.* Cambridge: At the University Press, 1952.

Broad, C. D. *The Mind and Its Place in Nature.* London: Routledge and Kegan Paul, 1925.

Bruner, J. S.; Goodnow, J. J.; and Austin, G. A. *A Study of Thinking.** New York: Wiley, 1956.

Brunswik, Egon. The conceptual framework of psychology, *International Encyclopedia of Unified Science,* Vol. 1, No. 10. Chicago: University of Chicago Press, 1952.

Calvin, Allen D. (ed.) *Psychology.* Boston: Allyn & Bacon, 1961.

Chappell, V. C. (ed.) *The Philosophy of Mind.* Englewood Cliffs, N.J.: Prentice-Hall, 1962.

Colby, Kenneth M. *Energy and Structure in Psychoanalysis.* New York: Ronald Press Co., 1955.

Craik, Kenneth J. W. *The Nature of Psychology.* Edited by Stephen L. Sherwood. New York: Cambridge University Press, 1966.

Culbertson, J. T. *Consciousness and Behavior.* Dubuque, Iowa: W. C. Brown Co., 1950.

De Vore, Irven. (ed.) *Primate Behavior: Field Studies of Monkeys and Apes.** New York: Holt, Rinehart, & Winston, 1965.

Estes, William K. et al. *Modern Learning Theory.** New York: Appleton-Century-Crofts, 1954.

Feigl, Herbert. *The "Mental" and the "Physical", The Essay and a Postscript.* Minneapolis: University of Minnesota Press, 1967.

Fenichel, Otto. *The Psychoanalytic Theory of Neurosis.** New York: Norton, 1945.

Frenkel-Brunswik, E. *Psychoanalysis and the unity of science, Proceedings of the American Academy of Arts and Sciences,* 80, pp. 271-350, 1954.

Guilford, J. P. *Psychometric Methods.** 2d ed. New York: McGraw-Hill, 1954.

Gustafson, Donald F. (ed.) *Essays in Philosophical Psychology.* Garden City, N. Y.: Doubleday & Co., 1964.

Hampshire, Stuart. (ed.) *Philosophy of Mind.* New York: Harper & Row, 1966.

Handy, Rollo. *Methodology of the Behavioral Sciences.*

Springfield, Ill.: Charles C. Thomas, 1964.

Hebb, D. O. *The Organization of Behavior: A Neuropsychological Theory.* * New York: Wiley, 1949.

Hilgard Ernest R., and Bower, Gordon H. *Theories of Learning.* * 3d ed. New York: Appleton-Century-Crofts, 1966.

Hinde, Robert A. *Animal Behavior.* * New York: McGraw-Hill, 1966.

Honig, Werner K. (ed.) *Operant Behavior: Areas of Research and Application.* * New York: Appleton-Century-Crofts, 1966.

Hook, Sidney. (ed.) *Psychoanalysis, Scientific Method, and Philosophy.* New York: New York University Press, 1959.

———. (ed.) *Dimensions of Mind.* New York: New York University Press, 1960.

Hull, Clark L. *Principles of Behavior.* * New York: Appleton-Century-Crofts, 1943.

———. *A Behavior System.* * New Haven: Yale University Press, 1952.

Hull, Clark L.; Hovland, C. I.; Ross, R. T.; Hall, M.; Perkins, D. T.; and Fitch, F. B. *Mathematico-Deductive Theory of Rote Learning.* * New Haven: Yale University Press, 1940.

Jespersen, Otto. *Language: Its Nature, Development, and Origin.* New York: Macmillan, 1922.

Koch, Sigmund. (ed.) *Psychology: A Study of a Science.* 7 vols. New York: McGraw-Hill, 1957-67.

Köhler, Wolfgang. *Dynamics in Psychology.* New York: Liveright, 1940.

———. *Gestalt Psychology.* New York: Liveright, 1947.

———. *The Place of Values in a World of Facts.* New York: Meridian Books, 1959.

Lewin, Kurt. *Principles of Topological Psychology.* * New York: McGraw-Hill, 1936.

Madden, Edward H. *Philosophical Problems of Psychology.*

New York: Odyssey Press, 1962.

Mandler, George, and Kessen, William. *The Language of Psychology.* New York: Wiley, 1959.

Marler, Peter R., and Hamilton, William J., III. *Mechanisms of Animal Behavior.** New York: Wiley, 1966.

Marx, Melvin H. (ed.) *Theories in Contemporary Psychology.* New York: Macmillan, 1963.

Meehl, Paul E. *Clinical Versus Statistical Prediction.** Minneapolis: University of Minnesota Press, 1954.

Miller, George A. *Language and Communication.** New York: McGraw-Hill, 1951.

———. *Mathematics and Psychology.** New York: Wiley, 1964.

Morris, Charles. *Six Theories of Mind.* Chicago: University of Chicago Press, 1932.

O'Neil, William M. *Introduction to Method in Psychology.* Melbourne, Australia: Melbourne University Press, 1957.

Osgood, Charles E.; Suci, George J.; and Tannenbaum, Percy H. *The Measurement of Meaning.** Urbana: University of Illinois Press, 1957.

Pratt, Carol C. *The Logic of Modern Psychology.* New York: Macmillan, 1939.

Presley, C. F. (ed.) *The Identity Theory of Mind.* St. Lucia, Australia: University of Queensland Press, 1967.

Pumpian-Mindlin, E. (ed.) *Psychoanalysis as a Science.* With essays by E. R. Hilgard, L. S. Kubie, and the editor. Stanford: Stanford University Press, 1952.

Russell, Bertrand. *The Analysis of Mind.* London: Allen & Unwin, 1921.

Ryle, Gilbert. *The Concept of Mind.* London: Hutchinson's University Library, 1949.

Scher, Jordan M. (ed.) *Theories of the Mind.* New York: Free Press, 1962.

Schrödinger, Erwin. *Mind and Matter.* New York: Cambridge University Press, 1958.

Shwayder, D. S. *The Stratification of Behavior.* New York: Humanities, 1965.

Sidman, Murray. *Tactics of Scientific Research.* New York: Basic Books, 1960.

Simon, Herbert A. *Models of Man.* * New York: Wiley, 1957.

Singer, E. A. *Mind As Behavior.* Columbus, Ohio: R. G. Adams & Co., 1924.

Skinner, B. F. *Science and Human Behavior.* New York: Macmillan, 1953.

———. *Verbal Behavior.* New York: Appleton-Century-Crofts, 1957.

———. *Cumulative Record.* * rev. ed. New York: Appleton-Century-Crofts, 1961.

Spence, K. W. *Behavior Theory and Learning.* * Englewood Cliffs, N.J.: Prentice-Hall, 1960.

Stevens, Stanley S. (ed.) *Handbook of Experimental Psychology.* * New York: Wiley, 1951.

Swartz, Robert J. (ed.) *Perceiving, Sensing, and Knowing.* Garden City, N. Y.: Doubleday & Co., 1965.

Taylor, Charles. *The Explanation of Behaviour.* New York: Humanities, 1964.

Tolman, E. C. *Purposive Behavior in Animals and Men.* * New York and London: Century Co., 1932.

———. *Collected Papers in Psychology.* * Berkeley: University of California Press, 1951.

Turner, Merle B. *Philosophy and the Science of Behavior.* New York: Appleton-Century-Crofts, 1965.

Vesey, G. N. A. *The Embodied Mind.* London: Allen & Unwin, 1965.

Wann, T. W. (ed.) *Behaviorism and Phenomenology.* Chicago: University of Chicago Press, 1964.

Whorf, Benjamin L. *Language, Thought, and Reality.* New York: Wiley, 1956.

Wisdom, John. *Other Minds.* Oxford: Blackwell, 1952.

Wolman, Benjamin B., and Nagel, Ernest. (eds.) *Scientific Psychology.* New York: Basic Books, 1965.

Wyburn, George M. (ed.) *Human Senses and Perception.* * Toronto: University of Toronto Press, 1964.

Zipf, George K. *Human Behavior and the Principle of Least Effort.* * Reading, Mass.: Addison-Wesley, 1949.

11. Bridge Sciences (Cybernetics, Information Theory, Etc.)

Apter, Michael. *Cybernetics and Development.* * New York: Pergamon Press, 1966.

Arbib, Michael A. *Brains, Machines, and Mathematics.* New York: McGraw-Hill, 1964.

Ashby, W. Ross. *An Introduction to Cybernetics.* * New York: Wiley, 1963.

Automatic Control. Scientific American Book. New York: Simon & Schuster, 1955.

Cherry, Colin. *On Human Communication.* * New York: Wiley, 1957.

Culbertson, J. T. *The Minds of Robots.* Urbana: University of Illinois Press, 1966.

Feigenbaum, Edward, and Feldman, Julian. *Computors and Thought.* New York: McGraw-Hill, 1963.

George, F. H. *The Brain as a Computer.* New York: Pergamon Press, 1962.

McCullough, Warren S. *Embodiments of Mind.* Cambridge, Mass.: M. I. T. Press, 1966.

Minsky, Marvin. *Computation, Finite and Infinite Machines.* Englewood Cliffs, N.J.: Prentice-Hall, 1967.

Pask, Allan. *An Approach to Cybernetics.* * London:

Hutchinson, 1961.

Reitman, Walter. *Cognition and Thought.** New York: Wiley, 1965.

Shannon, Claude E., and Weaver, Warren. *The Mathematical Theory of Communication.* * Urbana: University of Illinois Press, 1949.

Singh, Jaghit. *Great Ideas in Automation, Cybernetics, and Information Theory.** New York: Dover, 1966.

Von Foerster, H., and Zopf, G. W. (eds.) *Principles of Self Organization.* New York: Pergamon Press, 1962.

Von Neumann, John. *The Computor and the Brain.** New Haven: Yale University Press, 1958.

———. *Theory of Self-reproducing Automata.* Edited by Arthur W. Burks. Urbana: University of Illinois Press, 1966.

Wiener, Norbert. *Cybernetics.* * 2d ed. Cambridge, Mass.: M. I. T. Press, 1961.

12. Theory of Knowledge

Alexander, Peter. *Sensationalism and Scientific Explanation.* New York: Humanities, 1963.

Armstrong, D. M. *Perception and the Physical World.* New York: Humanities, 1961.

Aune, Bruce. *Knowledge, Mind, and Nature.* New York: Random House, 1967.

Avenarius, Richard. *Der Menschliche Weltbegriff.* Leipzig: Reisland, 1912.

Ayer, A. J. *The Foundations of Empirical Knowledge.* New York: Macmillan, 1940.

———. *Language, Truth, and Logic.* London: Gollancz, 1946.

———. *Thinking and Meaning.* London: K. K. Lewis & Co., Ltd., 1947.

———. *Philosophical Essays.* New York: Macmillan 1954.

———. *The Problem of Knowledge.* New York: St. Martin's Press, 1956.

———. *The Concept of a Person and Other Essays.* New York: St. Martin's Press, 1963.

———. (ed.) *Logical Positivism.* New York: Free Press, 1959.

Ayer, A. J., et al. *The Revolution in Philosophy.* New York: Macmillan, 1956.

Barnes, Winston H. F. *The Philosophical Predicament.* London: A. & C. Black, 1950.

Beloff, John. *The Existence of Mind.* New York: Citadel Press, 1964.

Bergmann, Gustav. *The Metaphysics of Logical Positivism.* New York: Longmans, Green & Co., 1954.

———. *Meaning and Existence.* Madison: University of Wisconsin Press, 1960.

———. *Logic and Reality.* Madison: University of Wisconsin Press, 1964.

———. *Realism: Critique of Brentano and Meinong.* Madison: University of Wisconsin Press, 1967.

Black, Max. *Language and Philosophy.* Ithaca, N. Y.: Cornell University Press, 1949.

———. *Problems of Analysis.* Ithaca, N. Y.: Cornell University Press, 1954.

———. (ed.) *Philosophical Analysis.* Ithaca, N. Y.: Cornell University Press, 1950.

Carnap, Rudolf. *Der Logische Aufbau der Welt.* Berlin: Weltkreis, 1928. English translation by R. A. George, including translation of *Scheinprobleme in der Philosophie* (1928). Berkeley: University of California Press, 1967.

———. *Scheinprobleme in der Philosophie.* Berlin: Weltkreis, 1928.

Collingwood, R. G. *An Essay on Metaphysics.* New York: Oxford University Press, 1940.

Dewey, John. Theory of valuation, *International Encyclopedia of Unified Science,* Vol. 2, No. 4. Chicago: University of Chicago Press, 1939.

Feigl, Herbert, and Sellars, Wilfrid. (eds.) *Readings in Philosophical Analysis.* New York: Appleton-Century-Crofts, 1949.

Frank, Philipp. *Das Kausalgesetz und seine Grenzen.* Vienna: Springer, 1932.

Goodman, Nelson. *The Structure of Appearance.* Cambridge, Mass.: Harvard University Press, 1951.

Hahn, Hans. *Überflüssige Wesenheiten.* Vienna: Wolf, 1929.

Hirst, R. J. *The Problems of Perception.* New York: Macmillan, 1959.

Hook, Sidney. (ed.) *Determinism and Freedom in the Age of Modern Science.* New York: New York University Press, 1958.

Körner, Stephan. *Experience and Theory.* London: Routledge & Kegan Paul, 1966.

Kraft, Victor. *Mathematik, Logik, und Erfahrung.* Vienna: Springer, 1947.

———. *Erkenntnislehre.* Vienna: Springer, 1960.

———. *Die Grundlagen der Erkenntnis und der Moral.* Berlin: Duncker & Humblot, 1968.

Lewis, C. I. *Mind and the World Order.* New York: Scribners, 1929.

———. *An Analysis of Knowledge and Valuation.* La Salle, Ill.: Open Court, 1946.

Mach, Ernst. *Contributions to the Analysis of Sensations.* La Salle, Ill.: Open Court, 1897.

———. *Analysis of Sensations, and the Relation of the Physical to the Psychical.* New York: Dover, 1959.

Pap, Arthur. *Elements of Analytic Philosophy.* New York: Macmillan, 1949.

———. *Analytische Erkenntnistheorie.* Vienna: Springer, 1955.

———. *Semantics and Necessary Truth.* New Haven: Yale University Press, 1958.

Passmore, John. *A Hundred Years of Philosophy.* New York: Basic Books, 1966.

Price, H. H. *Perception.* London: Methuen, 1932.

Reichenbach, Hans. *Experience and Prediction.* Chicago: University of Chicago Press, 1938.

Rougier, Louis. *Les Paralogismes du rationalisme.* Paris: Alcan, 1920.

———. *Traité de la connaissance.* Paris: Gauthier-Villars, 1955.

Russell, Bertrand. *Philosophical Essays.* New York: Longmans, Green & Co., 1910.

———. *Philosophy.* New York: W. W. Norton & Co., 1927.

———. *Mysticism and Logic.* New York: W. W. Norton & Co., 1929.

———. *Our Knowledge of the External World.* New York: W. W. Norton & Co., 1929.

———. *An Inquiry into Meaning and Truth.* New York: W. W. Norton & Co., 1940.

———. *Problems of Philosophy.* New York: Oxford University Press, 1946.

———. *Human Knowledge.* New York: Simon & Schuster, 1948.

———. *Logic and Knowledge.* London: Allen & Unwin, 1956.

———. *My Philosophical Development.* New York: Simon & Schuster, 1959.

Ruyer, Raymond. *La Conscience et le corps.* Paris: Presses Universitaires de France, 1950.

Schlick, Moritz. *Allgemeine Erkenntnislehre.* Berlin: Springer, 1925.

———. *Gesammelte Aufsaetze.* Vienna: Gerold, 1938.

Sellars, R. W. *The Philosophy of Physical Realism.* New York: Macmillan, 1932.

Sellars, Wilfrid. *Philosophical Perspectives.* Springfield, Ill.: Charles C. Thomas, 1967.

Skolimowski, Henryk. *Polish Analytical Philosophy.* New York:

Humanities Press, 1967.

Stebbing, L. S. *Logical Positivism and Analysis.* Proceedings of the British Academy. London, 1934.

Waismann, Friedrich. *Wittgenstein und der Wierner Kreis.* Edited by B. F. McGuinness. Oxford: Basil Blackwell, 1967.

———. *How I See Philosophy.* Edited by Harré. New York: St. Martin's Press, 1968.

Weinberg, Julius R. *An Examination of Logical Positivism.* New York: Harcourt, Brace & World, 1936.

Werkmeister, William H. *The Basis and Structure of Knowledge.* New York: Harper, 1948.

White, Morton. *Toward Reunion in Philosophy.* Cambridge, Mass.: Harvard University Press, 1956.

Wittgenstein, Ludwig. *Tractatus Logico-Philosophicus.* New York: Harcourt, Brace & World, 1922.

———. *Philosophical Investigations.* New York: Macmillan, 1953.

———. *The Blue and Brown Books.* Oxford: Blackwell, 1958.

Zilsel, E. *Das Anwendungsproblem: Ein Philosophischer Versuch über das Gesetz der Grossen Zahlen und die Induktion.* Leipzig: Barth, 1916.

13. Foundations of the Social Sciences

Albert, Hans. (ed.) *Theorie und Realität.* Tubingen, Germany: J. C. B. Mohr (Paul Siebeck), 1964.

Bolacchi, Giulio. *Teoria delle classi sociali.* Rome: Edizioni Ricerche, 1963.

Braybrooke, David. *Philosophical Problems of the Social Sciences.* New York: Macmillan, 1965.

Brodbeck, May. (ed.) *Readings in the Philosophy of the Social Sciences.* New York: Macmillan, 1968.

Brown, Robert. *Explanation in Social Science.* Chicago: Aldine, 1963.

Churchman, C. West; Ackoff, R. L.; and Arnoff, E. *Introduction* *to Operations Research.** New York: Wiley, 1957.

Dray, William H. *Law and Explanation in History.* Oxford: Clarendon Press, 1957.

———. (ed.) *Philosophical Analysis and History.* New York: Harper & Row, 1966.

Edel, Abraham. Science and the structure of ethics, *International Encyclopedia of Unified Science,* Vol. 2, no. 3. Chicago: University of Chicago Press, 1961.

Gardiner, Patrick. *The Nature of Historical Explanation.* London: Oxford University Press, 1952.

———. (ed.) *Theories of History.* Glencoe, Ill.: Free Press, 1959.

Gross, Llewellyn. (ed.) *Symposium on Sociological Theory.* New York: Harper & Row, 1959.

Hayek, F. A. *The Counter-revolution of Science.* Glencoe, Ill.: Free Press, 1952.

Hook, Sidney. *The Hero in History.* Boston: Beacon Press, 1943.

———. *Reason, Social Myths, and Democracy.* New York: Humanities, 1951.

———. (ed.) *Philosophy and History: A Symposium.* New York: New York University Press, 1963.

Kaplan, Abraham. *The Conduct of Inquiry.* San Francisco: Chandler Co., 1964.

Kaufmann, Felix. *Methodology of the Social Sciences.* London: Oxford University Press, 1944.

Kraft, Victor. *Die Grundlagen einer Wissenschaftlichen Wertlehre.* Vienna: Springer, 1937.

Krimerman, Leonard I. *The Nature and Scope of Social Science: A Critical Anthology.* New York: Appleton-Century-Crofts, 1969.

Lazarsfeld, Paul F. (ed.) *Mathematical Thinking in the Social Sciences.** New York: Free Press, 1954.

Lundberg, George A. *Foundations of Sociology.* New York: Macmillan, 1939.

Machlup, F. *Essays on Economic Semantics.* Englewood Cliffs,

N.J.: Prentice-Hall, 1963.

Merton, Robert K. *Social Theory and Social Structure.* * Glencoe, Ill.: Free Press, 1949.

Morris, Charles. *Varieties of Human Value.* Chicago: University of Chicago Press, 1956.

Nagel, Ernest, and Hempel, C. G. Symposium: problems of concept and theory formation in the social sciences, *Science, Language, and Human Rights.* American Philosophical Association, Vol. I. Philadelphia: University of Pennsylvania Press, 1952.

Natanson, Maurice. (ed.) *Philosophy of the Social Sciences: A Reader,* New York: Random House, 1963.

Neurath, Otto. *Antispengler.* Munich: Callwey, 1921.

———. *Empirische Soziologie: Der Wissenschaftliche Gehalt der Geschichte und Nationalökonomie.* Vienna: Schr. z. wiss. Weltauff., 1931.

———. *Einheitswissenschaft und Psychologie,* in the series Einheitswissenschaft, No. 1. Vienna: Springer, 1933.

———. *Was Bedeutet Rationale Wirtschaftsbetrachtung?* in the series Einheitswissenschaft, No. 4. Vienna: Gerold, 1935.

———. *Modern Man in the Making.* New York: Knopf, 1939.

———. Foundations of the social sciences, *International Encyclopedia of Unified Science,* Vol. 2, No. 1. Chicago: University of Chicago Press, 1944.

Papandreou, Andreas. *Economics as a Science.* Philadelphia: J. B. Lippincott, 1958.

Popper, Karl R. *The Open Society and Its Enemies.* London: George Routledge & Sons, 1945. (4th rev. ed. 1961.) Princeton, N.J.: Princeton University Press, 1950.

———. *Poverty of Historicism.* Boston: Beacon Press, 1957.

Robbins, Lionel C. *An Essay on the Nature and Significance of Economic Science.* New York: Macmillan, 1952.

Rose, Arnold M. *Theory and Method in the Social Sciences.* Minneapolis: University of Minnesota Press, 1954.

Rudner, Richard S. *Philosophy of Social Science.* Englewood Cliffs, N.J.: Prentice-Hall, 1966.

Schlick, Moritz. *Natur und Kultur.* Posthumous papers, edited by J. Rauscher. Vienna: Humboldt-Verlag, 1952.

Shackle, G. L. S. *The Years of High Theory.* New York: Cambridge University Press, 1967.

Thaon di Revel, Paolo. *Teorica del Bisogno: Saggio di Metaeconomia.* * Milan: Giuffre, 1967.

Tintner, Gerhard. Methodology of mathematical economics and econometrics, *International Encyclopedia of Unified Science,* Vol. 2, No. 6. Chicago: University of Chicago Press, 1968.

Von Neumann, John, and Morgenstern, O. *Theory of Games and Economic Behavior.* * 3d ed. Princeton: Princeton University Press, 1953.

Weber, Max. *The Methodology of the Social Sciences.* Glencoe, Ill.: Free Press, 1949.

Index
Volumes I and II

Index

Index

Index

Index